WELDING HANDBOOK

Eighth Edition

Volume 1

WELDING TECHNOLOGY

The Three Volumes of the Welding Handbook, Eighth Edition

1) WELDING TECHNOLOGY

2) WELDING PROCESSES

3) MATERIALS AND APPLICATIONS

WELDING HANDBOOK

Eighth Edition

Volume 1

WELDING TECHNOLOGY

Leonard P. Connor
Editor

AMERICAN WELDING SOCIETY
550 N.W. LeJeune Road
P.O. Box 351040
Miami, FL 33135

Library of Congress Number: 87-071919
International Standard Book Number: 0-87171-281-4

American Welding Society, 550 N.W. LeJeune Rd., P.O. Box 351040, Miami, FL
33135

THE WELDING HANDBOOK is a collective effort of many volunteer technical specialists to provide information to assist with the design and application of welding and allied processes.

Reasonable care is taken in the compilation and publication of the Welding Handbook to insure authenticity of the contents. No representation or warranty is made as to the accuracy or reliability of this information.

The information contained in the Welding Handbook shall not be construed as a grant of any right of manufacture, sale, use, or reproduction in connection with any method, process, apparatus, product, composition or system, which is covered by patent, copyright, or trademark. Also, it shall not be construed as a defense against any liability for such infringement. No effort has been made to determine whether any information in the Handbook is covered by any patent, copyright, or trademark, and no research has been conducted to determine whether an infringement would occur.

Printed in the United States of America

CONTENTS

WELDING HANDBOOK COMMITTEE

May 31, 1986

J.H. Hannahs, Chairman	Midmark Corporation
M.J. Tomsic, Vice Chairman	Plastronics Incorporated
L.P. Connor, Secretary	American Welding Society
D.R. Amos	Westinghouse Electric Corporation
C.W. Case	Inco Alloys International
J.R. Condra	E.I. duPont de Nemours and Company
E.H. Daggett	Babcock and Wilcox (Retired)
A.F. Manz	A.F. Manz and Associates
J.C. Papritan	Ohio State University
L.J. Privoznik	Westinghouse Electric Corporation
D.R. Spisiak	Gaymar Industries Incorporated

WELDING HANDBOOK COMMITTEE

June 1, 1987

M.J. Tomsic, Chairman	Plastronics Incorporated
D.R. Amos, 1st Vice Chairman	Westinghouse Electric Corporation
C.W. Case, 2nd Vice Chairman	Inco Alloys International
L.P. Connor, Secretary	American Welding Society
J.R. Condra	E.I. duPont de Nemours and Company
G.N. Fisher	Fisher Engineering Company
J.R. Hannahs	Midmark Corporation
A.F. Manz	A.F. Manz Associates
J.C. Papritan	Ohio State University
L.J. Privoznik	Westinghouse Electric Corporation
E.G. Shifrin	Detroit Edison Electric (Retired)
B.R. Somers	Lafayette College
P.I. Temple	Detroit Edison Electric

PREFACE

This is the first volume of the Eighth Edition of the *Welding Handbook*. The Eighth Edition will be published in three volumes. The Fourth through Seventh Editions of the *Welding Handbook* were published in five-volume sets. This volume, Welding Technology, addresses welding and engineering fundamentals. These subjects were last addressed in Volumes One and Five of the Seventh Edition.

Volume Two will address the welding processes and will update the content of Volumes Two and Three of the Seventh Edition. Volume Three will address materials and applications, and will update the content of Volume Four of the Seventh Edition and Volume Five of the Sixth Edition.

The chapters of this volume contain basic information useful to all individuals in the welding industry. Wherever appropriate, considerable practical application data is also included. This volume contains more illustrations, data, and specific examples than the volumes it replaces in the Seventh Edition. Volumes Two and Three will also contain additional application data. The intent is to help readers implement the new technologies.

A major subject index of the current volumes of this and all previous editions of the welding handbook precedes the subject index in this volume. It directs the reader to the appropriate chapter and volume that contains the needed information.

This volume, like the others, was a voluntary effort by the Welding Handbook Committee and the Chapter Committees. The Chapter Committee Members and the Handbook Committee Member responsible for each chapter are recognized on the title page of that chapter. Other individuals also contributed in a variety of ways, particularly in chapter reviews. All participants contributed generously of their time and talent, and the American Welding Society expresses herewith its appreciation to them and to their employers for supporting the work.

The Welding Handbook Committee expresses its appreciation to Richard French, Deborah Givens, former editor William Kearns and Hallock Campbell and other AWS Staff Members for their assistance in the production of this volume.

The Welding Handbook Committee welcomes your comments on the Handbook. Please address them to the Editor, Welding Handbook, American Welding Society, P.O. Box 351040, Miami, FL 33135.

J.R. Hannahs, Chairman
Welding Handbook Committee
1984-1987

L.P. Connor, Editor
Welding Handbook

SURVEY OF JOINING AND CUTTING PROCESSES

PREPARED BY A COMMITTEE CONSISTING OF:

C. E. Albright, Chairman
Ohio State University

J. E. Gould
Edison Welding Institute

R. L. Holdren
Welding Consultants, Inc.

J. Lowery
AGA Gas, Inc.

S. L. Ream
Battelle Columbus Labs.

N. L. Rundle
Union Carbide Coatings Service

WELDING HANDBOOK COMMITTEE MEMBER:
M. J. Tomsic
Plastronics Incorporated

SURVEY OF JOINING AND CUTTING PROCESSES

INTRODUCTION

THIS CHAPTER INTRODUCES the conventional, better known joining and cutting processes. The distinguishing features of the various processes, their attributes and limitations, and comparisons where applicable are identified here. Operational characteristics, power requirements, pertinent welding skills, and cost considerations, along with possible material and material thickness applications, are discussed. However, the information presented is a generalization of the subject and should not be used for selecting processes for specific applications. Where specific information and data are needed, the reader should consult additional sources including those listed at the end of this chapter and other volumes of the *Welding Handbook*.[1]

The arc welding group of joining processes is widely used in the industry. Other well known and often used joining processes are oxyfuel gas welding, resistance welding, flash welding, brazing, and soldering. Diffusion, friction, electron beam, and laser beam welding

along with adhesive bonding are being applied to an increasing number of applications in industry. Ultrasonic and explosive welding have only narrow fields of application, as does thermal spraying. Thermal cutting processes are also covered because of their applications in fabricating operations. Descriptions of several common methods of severing and removing metals are included.

Relatively few individuals develop an expertise in the use of all processes and probably no one has had the occasion to cut or join all of the currently available engineering materials. The purpose of this chapter, therefore, is to familiarize readers with all the major contemporary processes so that consideration may be given to those processes that might otherwise be overlooked.

Frequently, several processes can be used for any particular job. The major problem is to select the one that is the most suitable in terms of fitness for service and cost. These two factors, however, may not be totally compatible, thus forcing a compromise. Selection of a process can depend on a number of considerations, including the number of components being fabricated, capital equip-

---

1. In the 7th Edition of the *Welding Handbook*, welding processes are covered in detail in Volumes 2 and 3. In the 8th Edition, the processes will be included in Volume 2.

Table 1.1
Overview of Joining Processes*

Material	Thickness	SMAW	SAW	GMAW	FCAW	GTAW	PAW	ESW	EGW	RW	FW	OFW	DFW	FRW	EBW	LBW	TB	FB	IB	RB	DB	IRB	DFB	S
Carbon Steel	S	X	X	X		X				X	X	X			X	X	X	X	X	X	X	X	X	X
	I	X	X	X	X	X				X	X	X		X	X	X	X	X	X	X	X		X	X
	M	X	X	X	X					X	X	X		X	X	X	X	X	X				X	
	T	X	X	X	X			X	X		X	X		X	X			X					X	
Low Alloy Steel	S	X	X	X		X				X	X	X	X		X	X	X	X	X	X	X	X	X	X
	I	X	X	X	X	X				X	X		X	X	X	X	X	X	X	X	X		X	X
	M	X	X	X	X							X	X	X	X	X	X	X	X				X	
	T	X	X	X	X			X			X		X		X	X	X					X	X	
Stainless Steel	S	X	X	X		X	X			X	X	X			X	X	X	X	X	X	X	X	X	X
	I	X	X	X	X	X	X			X	X		X		X	X	X	X	X	X	X		X	X
	M	X	X	X	X	X					X		X		X	X	X	X	X	X			X	
	T	X	X	X	X			X			X		X		X	X	X		X				X	
Cast Iron	I	X										X					X	X	X				X	X
	M	X		X	X	X						X					X	X	X				X	X
	T	X	X	X	X							X						X					X	
Nickel and Alloys	S	X		X		X	X			X	X	X			X	X	X	X	X	X	X	X	X	X
	I	X	X	X		X	X			X	X			X	X	X	X	X	X	X	X		X	X
	M	X	X	X		X								X	X	X	X	X	X				X	
	T	X		X			X					X		X	X			X					X	
Aluminum and Alloys	S	X		X		X	X			X	X	X	X	X	X	X	X	X	X	X	X	X	X	X
	I	X		X	X					X	X	X	X	X	X	X	X	X	X			X	X	X
	M	X		X	X							X			X	X	X	X	X			X	X	
	T	X		X				X	X			X			X			X					X	
Titanium and Alloys	S			X		X	X			X	X		X		X	X		X	X				X	X
	I			X		X	X				X		X	X	X	X		X						X
	M			X		X	X				X		X	X	X	X		X						X
	T			X							X		X		X		X	X						X
Copper and Alloys	S			X		X	X				X				X	X	X	X	X	X	X		X	X
	I			X			X				X		X		X	X	X	X			X		X	X
	M			X							X		X		X	X	X	X					X	
	T			X							X				X			X					X	
Magnesium and Alloys	S			X		X				X					X	X	X	X			X		X	
	I			X		X				X	X			X	X	X	X	X			X		X	
	M			X							X			X	X	X		X					X	
	T			X							X				X								X	
Refractory Alloys	S			X		X	X			X	X				X		X	X	X	X		X	X	
	I			X			X				X				X		X	X					X	
	M									X	X													
	T																							

*This table presented as a general survey only. In selecting processes to be used with specific alloys, the reader should refer to other appropriate sources of information.

**See legend below:

LEGEND

Process Code

SMAW–Shielded Metal Arc Welding
SAW–Submerged Arc Welding
GMAW–Gas Metal Arc Welding
FCAW–Flux Cored Arc Welding
GTAW–Gas Tungsten Arc Welding
PAW–Plasma Arc Welding
ESW–Electroslag Welding
EGW–Electrogas Welding
RW–Resistance Welding
FW–Flash Welding
OFW–Oxyfuel Gas Welding
DFW–Diffusion Welding

FRW–Friction Welding
EBW–Electron Beam Welding
LBW–Laser Beam Welding
B–Brazing
  TB–Torch Brazing
  FB–Furnace Brazing
  IB–Induction Brazing
  RB–Resistance Brazing
  DB–Dip Brazing
  IRB–Infrared Brazing
  DFB–Diffusion Brazing
S–Soldering

Thickness

S–Sheet: up to 3 mm (1/8 in.)
I–Intermediate: 3 to 6 mm (1/8 to 3/4 in.)
M–Medium: 6 to 19 mm (1/4 to 3/4 in.)
T–Thick: 19 mm (3/4 in.) and up

X–Commercial Process

Table 1.2
Applicability of Cutting Processes to Materials*

Material	Cutting Processes			
	OC	**PAC**	**AAC**	**LBC**
Carbon steel	X	X	X	X
Stainless steel	X'	X	X	X
Cast iron	X'	X	X	X
Aluminum		X	X	X
Titanium	X'	X	X	X
Copper		X	X	X
Refractory metals		X	X	X

*This table should be regarded as only a very general guide to process applicability.
For processes to be used with specific alloys, appropriate sources should be consulted.

LEGEND

OC —Oxygen Cutting

PAC—Plasma Arc Cutting

AAC—Air Carbon
    Arc Cutting

LBC —Laser Beam Cutting

X —Commercially applied

X' —Process applicable with special
    techniques

ment costs, joint location, structural mass, and desired performance of the product. The adaptability of the process to the location of the operation or the type of shop, and the experience and abilities of the employees may also have an impact on the final selection. These considerations are touched upon as they relate to the different processes.

The information presented is complemented by Tables 1.1 for joining and 1.2 for cutting. Table 1.1 indicates those processes, materials, and material thickness combinations that are usually compatible. The left-hand column of the table lists a variety of engineering materials and four arbitrary thickness ranges. The major processes currently used by industry are listed across the top. Table 1.2 indicates the applicability of several major cutting processes to several materials. These tables should be regarded as only general indicators, and additional resources should be consulted for specific applications before final decisions are made or recommended. Nevertheless, the information presented can serve as a guide in screening processes for a joining or cutting requirement.

ARC WELDING

THE TERM *ARC WELDING* applies to a large and diversified group of welding processes that use an electric arc as the source of heat to melt and join metals. The formation of a weld between metals when arc welded may or may not require the use of pressure or filler metal.

The welding arc is struck between the workpiece and the tip of an electrode. The electrode will be either a consumable wire or rod or a nonconsumable carbon or tungsten rod which carries the welding current. The electrode is manually or mechanically moved along the joint, or it remains stationary while the workpiece is moved. When a nonconsumable electrode is used, filler metal can be supplied by a separate rod or wire if needed. A consumable electrode, however, will be designed not only to conduct the current that sustains the arc but also to melt and supply filler metal to the joint. It may also produce a slag covering to protect the hot weld metal from oxidation.

SHIELDED METAL ARC WELDING

SHIELDED METAL ARC welding (SMAW) is an early arc welding process. It is one of the simple and versatile

processes for welding ferrous and several nonferrous base metals. The process is illustated in Figure 1.1. It uses a covered electrode consisting of a core wire around which a concentric clay-like mixture of silicate binders and powdered materials (such as fluorides, carbonates, oxides, metal alloys, and cellulose) is extruded. The electrode is then baked to dry the covering. This covering is a source of arc stabilizers, gases to displace air, metal and slag to protect, support, and insulate the hot weld metal.[2]

The bare section of the electrode is clamped in an electrode holder, which in turn is connected to the power source by a welding lead (cable). The work is connected to the other power source terminal. The arc is initiated by touching the electrode tip against the work and then withdrawing it slightly. The heat of the arc melts the base metal in the immediate area, the electrode metal core, and the electrode covering. The molten base metal,

---

2. An excellent guide to the classification of shielded metal arc welding electrodes is provided in the appendices of the applicable AWS filler metal specifications. See for example ANSI/AWS A5.1, *Specification For Covered Carbon Steel Arc Welding Electrodes.*

core wire, and metal powders in the covering coalesce to form the weld.

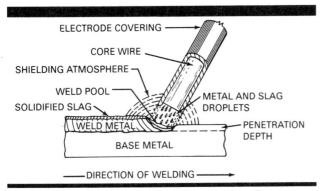

ELECTRODE COVERING
CORE WIRE
SHIELDING ATMOSPHERE
WELD POOL
SOLIDIFIED SLAG
METAL AND SLAG DROPLETS
WELD METAL
PENETRATION DEPTH
BASE METAL
DIRECTION OF WELDING

Figure 1.1—Shielded Metal Arc Welding Process

Covered electrodes are produced in a variety of diameters normally ranging from 1/16 to 5/16 in. (2 to 8 mm). The smaller diameters are used with low currents for joining thin sections and for welding in all positions. The larger diameters are designed for conducting high currents to achieve greater deposition rates in the flat and horizontal positions. Special alloy filler metal compositions can be formulated with relative ease by the use of metal powders in the electrode coating.

One group of heavily coated steel electrodes, often called "drag rods," can make welding in the flat and horizontal positions easier because they are allowed to touch the work. A relatively large amount of filler metal can be deposited by less experienced welders using this type of electrode. Electrodes designed for all-position welding require more experienced welders to obtain acceptable weld quality, particularly in the vertical and overhead positions.

The SMAW process has several advantages. Using the process, job shops can handle many welding applications with a relatively small variety of electrodes. Other advantages are the simplicity and lightness of the equipment, and its relatively low cost. Also, welds can be made in confined locations or remote from heavy power supplies.

An alternating current (ac) or direct current (dc) power source, welding leads (cables), an electrode holder, a welding helmet, and protective clothing are all that are required. The power source can be connected to a primary line providing about ten kilowatts (kW) or less. A gasoline engine driven power source can be used when portability is necessary. Solid-state power sources are available which are small and sufficiently light in weight to be manually carried to job locations. For these reasons, the SMAW process has wide application in the construction, pipeline, and maintenance industries. Uncomplicated, portable equipment is in common use for maintenance and field construction work.

The total cost of a typical production package ranges from about $500 to $1500, but portable, engine-driven equipment costs substantially more.

The SMAW process is suitable for joining metals in a wide range of thicknesses, but normally is best suited for sections of 1/8 to 3/4 in. (3 to 19 mm) in thickness. However, SMAW is frequently used for welding thick components which cannot be repositioned for flat or horizontal welding, but the deposition rates are lower in other positions. Sections thinner than 1/8 in. (3 mm) can be joined, but the required skill is much greater. Groove weld joints in plate thicknesses normally require edge preparation to allow proper access to the root of the joint. Fillet welds are very easy to make, particularly with heavily covered electrodes. Surfacing is another common application of the SMAW process.

The typical current range used for SMAW is between 50 and 300 A, although some special electrodes are designed to be used with currents as high as 600 A, and others as low as 30 A, allowing weld metal deposition rates of between 2 and 17 lb/h (1 and 8 kg/hr) in the flat position. However, a welder normally is only able to deposit up to about 10 lb (4.5 kg) per day in all-position welding because small diameter electrodes and low currents are used, and considerable electrode manipulation is necessary. Also, cleaning of the slag covering from the weld bead is required after each pass. As a result, labor costs are high. Material costs are also high since less than 60 percent of the weight of the purchased electrodes is deposited as filler metal. In spite of these deficiencies, SMAW maintains a position of dominance because of its simplicity and versatility. Many welders and engineers are comfortable with the process as a result of long experience with it. However, other more productive arc welding processes are replacing SMAW in many applications.

SUBMERGED ARC WELDING

IN SUBMERGED ARC welding (SAW), the arc and molten metal are shielded by an envelope of molten flux and a layer of unfused granular flux particles as shown in Figure 1.2. When the arc is struck, the tip of the continuously fed electrode is submerged in the flux and the arc is therefore not visible. The weld is made without the intense radiation that characterizes an open arc process and with little fumes.

The SAW process is used in both mechanized and semiautomatic operations, although the former is by far more common. High welding currents can be employed to produce high metal deposition rates at substantial cost savings. Welds can only be made in the flat and horizontal positions.

Mechanized equipment systems utilize welding power sources of 600 to 2000 A output, automatic wire feed devices, and tracking systems to move the welding head

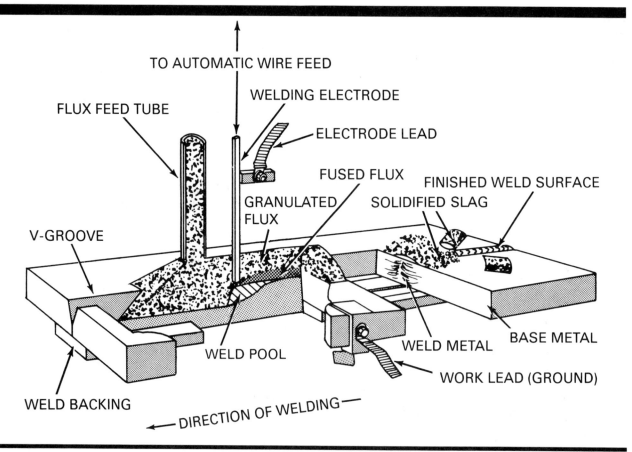

TO AUTOMATIC WIRE FEED

WELDING ELECTRODE

FLUX FEED TUBE

ELECTRODE LEAD

FUSED FLUX

FINISHED WELD SURFACE

GRANULATED FLUX

SOLIDIFIED SLAG

V-GROOVE

WELD POOL

WELD METAL

BASE METAL

WORK LEAD (GROUND)

WELD BACKING

DIRECTION OF WELDING

Figure 1.2—Submerged Arc Welding Process

or workpiece. The welding power sources are normally connected to standard three-phase primary power lines. When single phase power is required, it normally is derived from one phase of the three-phase line. Three-phase 220 or 440 V and single-phase 440 V systems are normally used.

Although a SAW system can be purchased for under $2000, a majority of single-electrode mechanized systems are priced in the $2000 to $10,000 range. Multiple electrode systems and systems installed on large fixtures could cost in excess of $50,000.

For mechanized equipment, welding operators can be trained to produce consistently high quality welds with a minimum of manual skill. Operation of semiautomatic systems is not difficult because the welder is not troubled by welding fumes or encumbered with a dark face shield and protective clothing.

The SAW process is useful for welding both sheet and plate. In welding relatively thin materials, speeds of up to 200 in./min (84 mm/s) can be achieved. In thick section applications, high metal deposition rates of 60 to 100 lb/h (27 to 45 kg/h) and reliability are the key advantages.

Many types of joints can be welded using the SAW process. Deep joint penetration can be achieved with direct current electrode positive (dcep). Therefore, edge preparation is not required in many applications. Joints may be backed with copper bars, flux, various types of tape, or integral steel members to support the molten weld metal.

The process is most widely employed for welding all grades of carbon, low alloy, and alloy steels. Stainless steel and some nickel alloys are also effectively welded or used as surfacing filler metals with the process. Various filler metal-flux combinations may be selected to provide specific weld metal properties for the intended service. The flux may contain ingredients that when melted react to contribute alloying additions to the weld metal. Approximately one pound of flux is consumed for every pound of electrode used.

If a particular job can be positioned for flat position welding, and if the job requires consistent weld quality and the deposit of large quantities of filler metal, then a mechanized welding system is justified. The SAW process should be considered for such applications. In general, if the job can be performed with SAW, this process will be one of the least expensive in terms of operating costs such as welding speed, materials, equipment, and reliability.

GAS METAL ARC AND FLUX CORED ARC WELDING

GAS METAL ARC welding (GMAW) and flux cored arc welding (FCAW) are two distinct processes, but they have many similarities in application and equipment. Both processes use a continuous solid wire or tubular electrode to provide filler metal, and both use gas to shield the arc and weld metal. In GMAW, the electrode is solid, and all of the shielding gas is supplied by an external source, as shown in Figure 1.3. With FCAW,

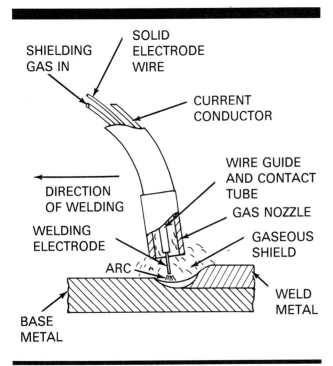

Figure 1.3—Gas Metal Arc Welding Process

the electrode is tubular and contains core ingredients that may supply some or all of the shielding gas needed. This process may also use auxiliary gas shielding, depending on the type of electrode employed, the material being welded, and the nature of the welding involved. Self-shielded and externally shielded FCAW are illustrated in Figures 1.4 and 1.5, respectively.

The shielding gases used by both processes have a dual purpose of protecting the arc and weld zones from air and providing desired arc characteristics. A variety of gases is used depending on the reactivity of the metal and the design of the joint to be welded. A variety of welding power sources is used with the two processes.

Gas Metal Arc Process Variations

IN GMAW, THE common variations of shielding gases, power sources, and electrodes have significant effects that can produce three different modes of metal transfer across the arc. These modes are known as spray, globular, and short-circuiting. Spray and globular transfer require relatively high welding currents while the short-circuiting transfer commonly uses low average currents. Conventional and pulsed GMAW operations with spray transfer require shielding gases rich in argon. Globular and short circuiting transfer are normally performed with carbon dioxide (CO_2) or gas mixtures rich in CO_2.

The GMAW process can be used to join virtually any metals using many joint configurations, and in all welding positions. However, each mode of metal transfer has specific advantages and limitations.

Spray Transfer. This mode of transfer describes an axial transfer of small discrete droplets of metal at rates of several hundred per second. Argon or argon-rich gas mixtures are necessary to shield the arc. Direct current electrode positive power is almost always used, and the amperage must be above a critical value related to the electrode diameter. The metal transfer is very stable, directional, and essentially spatter free.

Argon or argon-helium mixtures are used when joining reactive metals such as aluminum, titanium, and magnesium. When welding ferrous metals, small amounts of oxygen or carbon dioxide need to be added to these mixtures to stabilize the arc and prevent undercut and irregular welds. Unfortunately, the high arc energy associated with spray transfer limits the effectiveness of this mode of transfer for joining sheet metal and for welding steels in the vertical or overhead positions. It is used extensively for flat position welding of many metals and alloys including reactive metals such as aluminum, titanium, and magnesium.

Globular Transfer. In the globular-transfer process variation, carbon dioxide-rich gases are used to shield the arc and welding zone. Spray transfer cannot be obtained with these gases, regardless of the welding current level. The filler metal transfers across the arc in globules propelled by arc forces. Considerable spatter is produced, even when the arc is buried below the surface.

Carbon dioxide is reactive, and is used only for steels which can be properly alloyed to tolerate an oxidizing atmosphere and for those steels that do not require very low carbon content. The buried-arc method of operation using CO_2 shielding permits much higher welding speeds than possible with argon, and results in more effective and relatively inexpensive operations for many mechanized or repetitive manual applications. The arc is also more penetrating due to the high welding current used.

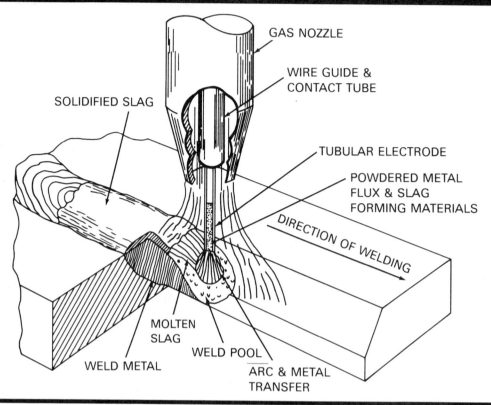

GAS NOZZLE

WIRE GUIDE & CONTACT TUBE

SOLIDIFIED SLAG

TUBULAR ELECTRODE

POWDERED METAL FLUX & SLAG FORMING MATERIALS

DIRECTION OF WELDING

MOLTEN SLAG

WELD POOL

WELD METAL

ARC & METAL TRANSFER

Figure 1.4—Gas Shielded Flux Cored Arc Welding

Short Circuiting Transfer. The need for relatively high currents with the spray and globular transfer process variations limits their applicability. These variations are not recommended for welding thin sections because the joint penetration and the deposition rates are too large resulting in melt through or overwelding. However, with short circuiting transfer, the average current and deposition rates can be limited by using power sources which allow metal to be transferred across the arc only during intervals of controlled short circuits occurring at rates in excess of 50 per second. Short-circuiting transfer can be used to weld both thin sections and out-of-position joints. This process variation has the advantage of being very easy to use. However, incomplete fusion can be a problem when welding sections thicker than 1/4 in. (6 mm) because the energy input is low. Good welding technique is very important when welding thick sections.

Pulsed Arc. If intermittent, high amplitude pulses of welding current are superimposed on a low level steady current that maintains the arc, the average current can be reduced appreciably while producing a metal spray transfer during the high amplitude pulses. Argon-rich gases are essential to achieve spray transfer. Pulsed arc operation is produced by the utilization of a program-med power source. Relatively large electrode diameters can be employed to weld thin as well as thick sections of many base metals in all positions.

Proper adjustment of the power source may require special training. Spray transfer welding in the flat position requires less skill to achieve sound welds than pulsed arc welding used in positions other than flat.

Flux Cored Arc Welding

FLUX CORED ARC welding (FCAW) uses cored electrodes instead of solid electrodes for joining ferrous metals. The flux core may contain minerals, ferroalloys, and materials that provide shielding gases, deoxidizers, and slag forming materials. The additions to the core promote arc stability, enhance weld metal mechanical properties, and improve weld contour. Many cored electrodes are designed to be used with additional external shielding. Carbon dioxide-rich gases are the most common. Weld metal can be deposited at higher rates, and the welds can be larger and better contoured than those made with solid electrodes, regardless of the shielding gas.

Another family of cored electrodes is the self-shielded variety. These electrodes are designed to generate protective shielding gases from additions in the core mate-

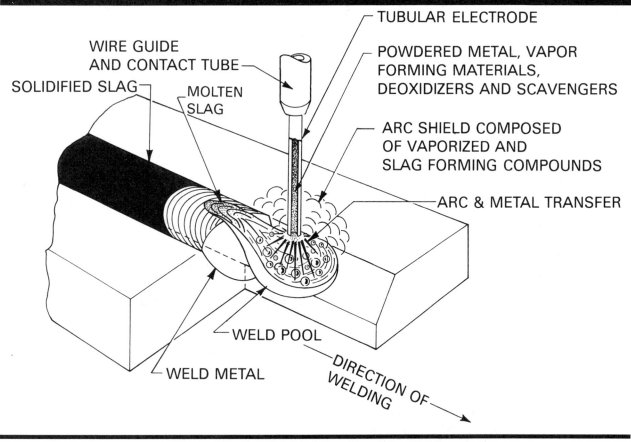

WIRE GUIDE
AND CONTACT TUBE

SOLIDIFIED SLAG

MOLTEN
SLAG

TUBULAR ELECTRODE

POWDERED METAL, VAPOR
FORMING MATERIALS,
DEOXIDIZERS AND SCAVENGERS

ARC SHIELD COMPOSED
OF VAPORIZED AND
SLAG FORMING COMPOUNDS

ARC & METAL TRANSFER

WELD POOL

WELD METAL

DIRECTION OF
WELDING

Figure 1.5—Self-Shielded Flux Cored Arc Welding

rial, similar to SMAW electrodes. These self-shielded electrodes do not require external shielding. Self-shielded electrodes are designed to use either direct current electrode positive or direct current electrode negative power. Self-shielded electrodes can be used in moderate cross ventilation with minimal disturbance of the gas shielding around the arc. Such electrodes are commercially available for welding carbon, low-alloy, and stainless steels. Proprietary hardfacing and surfacing electrodes are also available for maintenance applications.

Equipment and Personnel Requirements

ALL OF THE GMAW process variations and FCAW, with few exceptions, use similar equipment that includes the following:

(1) A variable speed motor and motor control to power feed rolls which drive the electrode at a preset and uniform rate.

(2) A welding gun which contains a switch to start and stop the electrode feed, the flow of shielding gas, the welding current, and cooling water if used. The gun also has a nozzle which directs the shielding gas to the arc and weld pool (except for self-shielded FCAW), and has a contact tube centered in the nozzle to transfer welding current to the electrode.

(3) A system of cables, hoses, electrical connections, and casings to direct the gas, electrode, power, and water.

(4) A mount for the spooled or coiled electrode.

(5) A control station containing the relays, solenoids, and timers needed to integrate the system.

(6) A source of shielding gas, if needed, and a device for metering gas flow rates.

(7) A power source to provide an appropriate amount and type of welding current.

(8) A water supply for cooling, if necessary.

The equipment is very flexible, and can also be used for SAW by substituting a gun with a flux hopper. It can be adapted for semiautomatic welding, or it may be mounted on fixtures for automatic or machine welding. The cost of the equipment is generally in the range of $1000 to $3000. The power sources for pulsed arc welding are more expensive than conventional ones, raising the cost of the equipment above this range. Power requirements range from 2 kW for short-circuiting arc welding to 20 kW for FCAW at high deposition rates. Preventive maintenance is required to assure that the shielding gas passages are clear and the contact tube is performing properly. The conduit or conduit liner between the wire feeder and the gun can easily collect dust and dirt causing rough wire feeding and an erratic arc condition. The conduit should be cleaned with compressed air periodically to avoid this problem. The wire feeders and control units are electromechanical devices and are quite reliable.

Less manipulative skill is needed to master these processes when compared with the SMAW process. If the welding is highly repetitive and the equipment controls are preset, only a relatively short time is needed to train a welder. Short-circuiting GMAW is quite easy to master for use in all positions. Flux-cored arc welding requires somewhat more skill, especially for making vertical and overhead welds. However, it can be used to produce vertical welds at deposition rates in excess of 5 lb/h (2 kg/h). The conventional and pulsed GMAW process variations are somewhat more difficult to use because the arc length needs to be more carefully controlled for best results.

Applications

SHORT-CIRCUITING GMAW IS most popular for welding ferrous metals in all positions if they are less than 1/4 in. (6 mm) in thickness and open root joint designs are used. This process variation is not generally used with nonferrous metals. Pulsed GMAW is used to weld sheet as thin as 3/64 in. (1 mm), as well as thick sections of all metals in all positions. Globular and spray transfer are restricted to welding steels in the flat and horizontal positions. Both the conventional and pulsed arc process variations can be used to weld aluminum alloys in all positions. Flux cored electrodes, 1/16 in. (1.6 mm) in diameter, are most commonly used to weld ferrous materials in all positions. Larger diameters, primarily 3/32 in. (2.4 mm) are used for flat and horizontal welding.

A variety of weld joints can be welded with these processes if the appropriate process and welding conditions are selected. Tight groove weld joints or lap joints in thicknesses of 3/16 in. (5.7 mm) or less can be welded. Thicker groove weld joints require edge preparation. Surfacing is common, particularly with mechanized equipment.

Deposition rates vary considerably depending upon the mode of metal transfer, and will range from 1 lb/h (0.5 kg/h) for short-circuiting arc welding to over 30 lb/h (13 kg/h) for flux cored arc welding.

In general, the GMAW and FCAW processes are among the most cost effective of all manual and semiautomatic welding processes using filler metals. The deposition efficiencies are particularly high, approaching 95 to 100 percent with solid electrodes (depending on the shielding gas), 85 to 90 percent with gas shielded cored electrodes, and 80 to 85 percent with the self shielded cored electrodes. Welders can work continuously with both GMAW and FCAW since the wire is fed continuously and only fatigue or changing position requires the arc to be interrupted.

GAS TUNGSTEN ARC WELDING

GAS TUNGSTEN ARC welding (GTAW) uses a nonconsumable tungsten electrode which must be shielded with an inert gas. The arc is initiated between the tip of the electrode and work to melt the metal being welded, as well as the filler metal, when used. A gas shield protects the electrode and the molten weld pool, and provides the required arc characteristics. This process is illustrated in Figure 1.6.

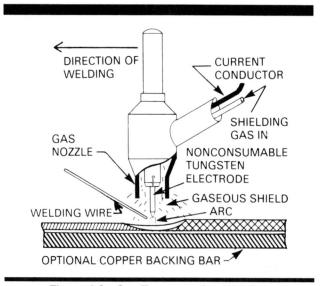

Figure 1.6—Gas Tungsten Arc Welding

The process may employ direct current with positive or negative electrode or alternating current. In general, ac is preferred for welding aluminum and magnesium. Direct current electrode negative is preferred for welding most other materials and for automatic welding of thick aluminum. Thin magnesium sometimes is welded with direct current electrode positive.

When ac is used with argon shielding, an arc cleaning action is produced at the joint surfaces on aluminum and magnesium. This cleaning action removes oxides and is particularly beneficial in reducing weld porosity when welding aluminum. When using dc, helium may be used as the shielding gas to produce deeper penetration. However, stringent precleaning of aluminum and magnesium parts is required with helium shielding. Argon and helium mixtures for gas shielding can provide some of the benefits of both gases.

Regardless of polarity, a constant current (essentially vertical volt-ampere characteristic) welding power source is required. In addition, a high-frequency oscillator is generally incorporated in power sources designed for GTAW. High-frequency can be employed with dc to initiate the arc instead of touch starting to minimize tungsten electrode contamination. Normally, the high frequency is turned off automatically after arc ignition. The high frequency power is normally operated continuously with ac to maintain ionization of the arc path as the arc voltage passes through zero.

Some special power sources provide pulsating direct current with variable frequency. This provision permits better control of the molten weld pool when welding thin sections, as well as when welding in positions other than flat.

Several types of tungsten electrodes are used with this process. Thoriated and zirconiated electrodes have better electron emission characteristics than pure tungsten, making them more suitable for dc operations. The electrode is normally ground to a point or truncated cone configuration to minimize arc wander. Pure tungsten has poorer electron emission characteristics but provides better current balance with ac welding. This is advantageous when welding aluminum and magnesium.

The equipment needed consists of a welding torch, a welding power source, a source of inert gas with suitable pressure regulators and flowmeters, a welding face shield, and protective clothing. Electric power requirements depend upon the type of material and the thicknesses to be welded. Power requirements range from 8 kW for a 200 A unit to 30 kW for a 500 A unit. Portable engine-driven power sources are also available.

A small 200 A welding equipment setup will cost about $1000, while a simple automatic unit of 500 A capacity may cost about $5000. The addition of arc voltage control, slope control, and other accessories will materially increase the cost.

Gas tungsten arc welding requires more training time, manual dexterity, and welder coordination than does SMAW or GMAW. The equipment is portable, and is applicable to most metals in a wide range of thickness and in all welding positions. Sound arc welds can be produced with the GTAW process when proper procedures are used.

The process can be used to weld all types of joint geometries and overlays in plate, sheet, pipe, tubing, and other structural shapes. It is particularly appropriate for welding sections less than 3/8-in. (10 mm) thick and also 1- to 6-in. (25.4- to 152.4-mm) diameter pipe. Thicker sections can be welded but economics generally indicate the choice of a consumable electrode process.

The combination of GTAW for root pass welding with either SMAW or GMAW is particularly advanta-geous for welding pipe. The gas tungsten arc provides a smooth, uniform root pass while the fill and cap passes are made with a more economical process.

Gas tungsten arc welding is generally more expensive than SMAW due to the cost of the inert gas, and is only 10 to 20 percent as fast as GMAW. However, GTAW will provide the highest quality root pass, while accommodating a wider range of thicknesses, positions, and geometries than either SMAW or GMAW.

PLASMA ARC WELDING

THE PLASMA ARC welding (PAW) process provides a very stable heat source for welding most metals from 0.001 to 0.25 in. (0.02 to 6 mm). This process has advantages over other open arc welding processes, such as SMAW, GMAW, and GTAW, because it has greater energy concentration, improved arc stability, higher heat content, and higher welding speeds. As a result, PAW has greater penetration capabilities than SMAW, GMAW, and GTAW.

The basic elements of the plasma arc torch, illustrated in Figure 1.7, are the tungsten electrode and the orifice.

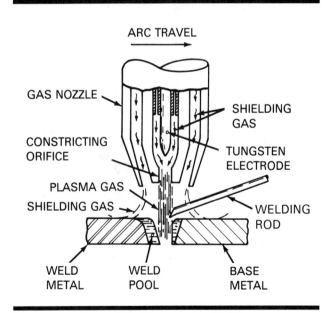

Figure 1.7—Plasma Arc Welding

A small flow of argon is supplied through the orifice to form the arc plasma. Shielding of the arc and weld zone is provided by gas flowing through an encircling outer nozzle assembly. The shielding gas can be argon, helium, or mixtures of argon with either hydrogen or helium. The plasma is initiated by an internal low current pilot arc between the electrode and the orifice. The pilot arc ionizes the orifice gas to ignite the primary arc between the electrode and the base metal. The arc plasma is constricted in size by the orifice around the electrode, and is called a *transferred arc*. If filler metal is used, it is fed into the arc as in the GTAW process.

Two welding techniques are possible with the PAW process: melt-in, and keyhole. The melt-in technique is normally used when welding thin sections in the range of 0.001 to 0.062 in. (0.025 to 1.5 mm). Narrower welds can be obtained with PAW rather than with GTAW because of the constricted arc. Manual welding of very fine sections of 0.01 in. (0.2 mm) and under is very difficult with the GTAW process but practical with low power PAW.

High current manual PAW can be done in the melt-in mode. Melt-in welds on 1/32 to 1/8 in. (0.8 to 3 mm) thick butt joints or lap joints can be used to join most materials. Titanium and other reactive metals can be joined using helium as a shielding gas. Generally, the operating current is 100 A or less.

The term *keyhole* has been applied to a penetrating hole at the leading edge of the molten weld pool. The weld metal flows around and solidifies behind the hole to form the weld bead. Keyhole welds are complete joint penetration welds with high depth-to-width ratios resulting in low distortion. Operation with a keyhole can be done on most metals in the thickness range of 3/32 to 3/8 in. (2.5 to 10 mm), and up to 3/4 in. (19 mm) in some titanium and aluminum alloys. This is one of the chief differences between the PAW and the GTAW processes with operating currents up to about 300 amperes.

The PAW process requires the following equipment: control console, power source, welding torch, cooling system, work clamp and lead assembly, shielding gas hose, orifice gas hose, inert gas regulators, and flowmeters. The operation of this equipment may be manual or mechanized. Typical costs for PAW equipment range from $3500 to $6000. Welder skill and training requirements are slightly greater than for oxyfuel gas welding. Maintenance skill and training would be greater than for GTAW because the equipment is more complex.

Out-of-position welding is possible and good uniform penetration can be obtained with PAW. Inert gas backing of the weld is normally used to minimize discontinuities at the root. Square-groove joints with little or no filler metal are commonly used up to 1/4 in. (6 mm) thick on most materials. Welding speeds range between 5 and 40 in./min. (2 to 17 mm/s) depending on the material and thickness. Welding speeds are generally 50 to 150 percent greater than those with GTAW on stainless steel.

ELECTROGAS WELDING

ELECTROGAS WELDING (EGW) is a mechanized welding process that uses either flux-cored or solid electrodes.

The process is normally performed in the vertical position. It is designed to weld a joint in a single pass by depositing weld metal into a cavity created by separated joint faces on two opposite sides and water-cooled molding dams or "shoes" on the other two sides. During welding, shielding of the molten weld pool is provided by a gas. The shielding gas may be from an external source or produced from a flux cored electrode, or both.

In addition to steels, the process has been successfully used on titanium and aluminum alloys. The process is illustrated in Figures 1.8 and 1.9.

Electrogas welding utilizes machine welding equipment. The vertical movement of the welding head is usually automatic and is controlled to maintain constant arc voltage, although other methods can also be used. For circumferential welds, the welding head is fixed while the vessel is rotated with the axis in the horizontal position.

Electrogas welding machines vary in size from light portable units weighing about 75 lb (35 kg) to the more common massive ones that are moved by cranes from one joint to another. The lightweight units are self-propelled while welding. Electrogas welding units are commonly self-aligning and pulled by chains connected to a drive mounted at the top of the joint. These units are generally used to weld sections ranging from 1/2 to 3 in. (13 to 76 mm) in thickness using a single electrode.

This process is generally used to make single-pass square groove welds, but may be modified to produce single-V-groove welds and T-joints. The ease of producing the desired edge preparation for square-groove welds minimizes joint preparation costs. The welding current depends on the electrode diameter and type. In general, electrogas welding machines operate at up to 400 A with solid electrodes and as high as 750 A with flux-cored electrodes. Approximately 20 kW of power is necessary for proper operation of the process over a range of operating settings. The cost of typical large systems ranges from $15,000 to $25,000, but can be greater if special features are required. The more portable units are significantly cheaper, costing about $5000, but they are unable to produce long uninterrupted welds.

Deposition rates for these machines are typically 15 to 30 lb/h (7 to 13 kg/h). The welding speed is related to section thickness and deposition rate. The weld soundness is generally excellent, and because transverse shrinkage is uniform, the joints are essentially free of angular distortion. When several welds are to be made, reduced labor costs, increased efficiency of depositing weld metal, and exceptional weld soundness can justify the relatively high equipment costs and setup time.

The equipment is reliable, and experienced welders and welding operators can be easily trained to understand and qualify to use this process.

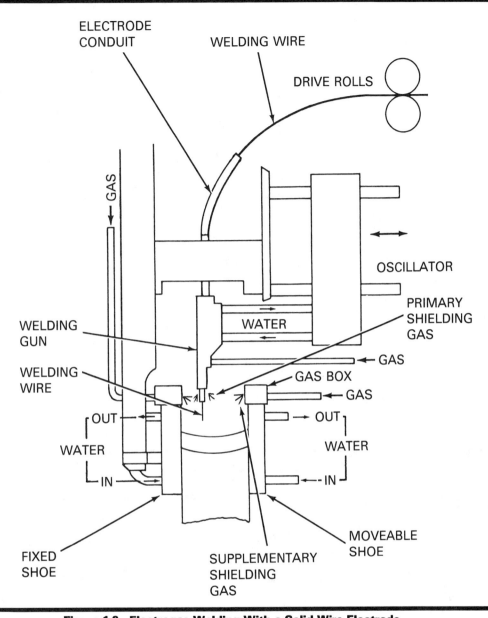

Figure 1.8—Electrogas Welding With a Solid Wire Electrode

ELECTROSLAG WELDING

IN MANY RESPECTS, electroslag welding (ESW) appears similar to electrogas welding. The major difference is that electroslag welding is not an arc welding process. The energy for melting and fusing of the base and filler metals is provided by a molten bath of slag that is resistance heated by the welding current. Like electrogas welding, electroslag welding is used to join the edges of two or more sections in a single pass, with the welding progression occurring vertically upward. At least two sides of the joint consist of the base metals to be joined.

The molten weld metal is contained in the joint by water-cooled dams or shoes.

The molten conductive slag floating on top of the molten weld pool protects the weld and melts the filler metal and the joint faces. The process is started with an arc between the electrode and the bottom of the joint. Welding flux is added and melted by the arc. When the molten slag reaches the tip of the electrode, the arc is extinguished. The slag is maintained in the molten state by the welding current passing from the electrode, through the slag, to the base metal. Unlike electrogas welding, multiple electrodes may be employed in electroslag welding.

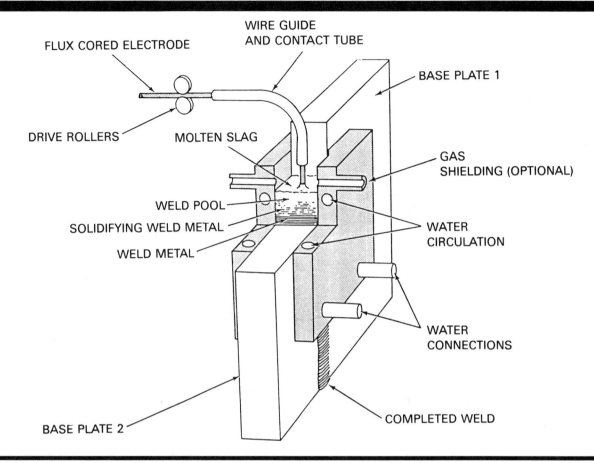

Figure 1.9—Electrogas Welding with a Flux Cored Electrode

Electroslag Welding Process Variations

THERE ARE TWO variations of the electroslag welding process: the nonconsumable guide or conventional method and the consumable guide method. These methods utilize different equipment and filler metal forms. The process variations are illustrated in Figures 1.10 and 1.11, respectively.

Noncomsumable Guide Method. The equipment for the nonconsumable guide method of ESW is quite similar to that for electrogas welding. The vertical rate of progression of the ESW equipment is either monitored by the welding operator or controlled automatically with a variety of devices. ESW equipment ranges from portable types weighing about 75 lb (35 kg) to more massive units which require cranes for moving from one location to another. The lightweight units may be self-propelled while welding. The heavier units are generally supported on a vertical column and move on tracks or screws while welding. ESW welding machines can incorporate single or multiple electrodes, and may incorporate electrode oscillation for making welds in sections

ranging in thickness from about 1/2 to 20 in. (13 to 500 mm).

Electroslag welding is used to weld the same types of joints as electrogas welding. The welding current

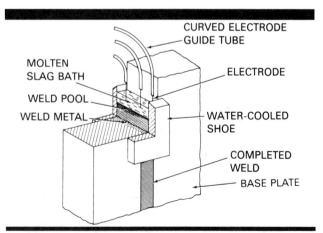

Figure 1.10—Nonconsumable Guide Method of Electroslag Welding (Three Electrodes)

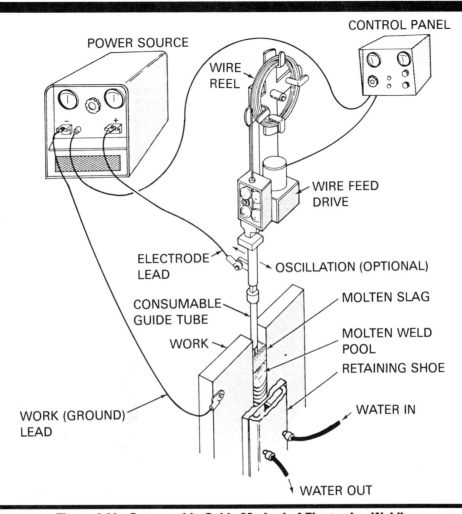

Figure 1.11—Consumable Guide Method of Electroslag Welding

varies appreciably depending on the number of electrodes used and the joint thickness. Electroslag welding machines commonly require several thousand amperes to weld thick sections with multiple electrodes. These units may require several hundred kilowatts of primary power.

The cost of typical large systems ranges from $15,000 to $25,000, and multiple electrode units can run in excess of $50,000. The smaller portable types cost about $5000, but they are not suitable for multi electrode welding.

Deposition rates for single electrode machines are typically 15 to 30 lb/h (7 to 13 kg/h). For multiple electrode machines the deposition rates are about 15 to 30 lb/h (7 to 13 kg/h) per electrode. The other advantages given for electrogas welding also apply to conventional electroslag welding.

Consumable Guide Method

THE CONSUMABLE GUIDE method uses a guide tube that is fixed at the top of the joint and extends to within about 1 in. (25 mm) of the bottom of the joint. Removable fixed dams or shoes are used to retain the molten weld metal and slag. The welding leads are connected to the electrode guide tube and a continuous wire electrode is fed through the tube. After the initial flux is melted and the molten slag pool is stabilized, the joint begins to fill. As the height of the slag pool rises, the guide tube also continuously melts and contributes to the weld metal. The system remains in equilibrium as long as the slag pool exists in sufficient depth. Loose flux can be added if the weld flashes, indicating arcing. Some guide tubes are manufactured with a flux coating so that flux in small quantities is continually added to the slag bath.

The process is flexible, particularly for relatively inaccessible joints where it is difficult to position moving equipment.

Advantages and Disadvantages

BOTH VARIATIONS OF ESW are used to fabricate thick weldments. By adding additional electrodes, virtually any thickness weldment can be fabricated. The welds are sound, and as the base metals increase in thickness, the ESW process becomes more cost effective over the multipass arc-welding process.

However, once started, the weld must continue to completion. Restarting an electroslag weld after a mid length stop would result in a large discontinuity. Because the weld travel speed is only about 1/2 to 1-1/2 in./min (0.2 to 0.6 mm/s) the weld metal and heat affected zone exhibit a coarse grain structure with relatively low notch toughness.

RESISTANCE WELDING

RESISTANCE WELDING (RW) incorporates a group of processes in which the heat for welding is generated by the resistance to the flow of electrical current through the parts being joined. It is most commonly used to weld two overlapping sheets or plates which may have different thicknesses. A pair of electrodes conduct electrical current to the joint. Resistance to the flow of current heats the faying surfaces, forming a weld. These electrodes clamp the sheets under pressure to provide good electrical contact and to contain the molten metal in the joint. The joint surfaces must be clean to obtain consistent electrical contact resistance to obtain uniform weld size and soundness.

The main process variables are welding current, welding time, electrode force, and electrode material and design. High welding currents are required to resistance heat and melt the base metal in a very short time. The time to make a single resistance weld is usually less than one second. For example, a typical practice for welding two pieces of 1/16 in. (1.6 mm) mild steel sheet requires a current of approximately 12 000 A and a time of 1/4 second while 1/8 in. (3.2 mm) sheet requires approximately 19 000 A and 1/2 second.

There are three major resistance welding processes: spot welding (RSW), projection welding (RPW), and seam welding (RSEW). Spot welding is illustrated in Figure 1.12. In RSW, the welding current is concentrated at the point of joining using cylindrical electrodes. Spot welds are usually made one at a time. In RPW, a projection or dimple is formed in one part prior to welding. The projection concentrates the current at the faying surfaces. Large, flat electrodes are used on both sides of the joint. Several projections may be formed in one of the components to produce several welds simultaneously. As an example, a stamped bracket may have three or four projections formed in it so that it can be welded to a sheet with one welding cycle.

In RSEW, leak-tight welds can be made by a series of overlapping spot welds. These are produced by introducing timed, coordinated pulses of current from rotat-

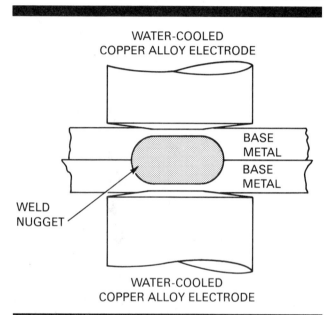

Figure 1.12—Resistance Spot Welding

ing wheel electrodes. Typical welding speeds are 60 in./min (25 mm/s) for 1/16 in. (1.6 mm) mild steel sheet and 40 in./min (17 mm/s) for 1/8 in. (3 mm) sheet. A variation of RSEW, known as roll spot welding, is identical in operation and equipment to RSEW except that the spot spacing is increased so that spots do not overlap and the weld is not leaktight. A roll spot weld is used for structural purposes only.

Resistance welds are made with either semiautomatic or automatic machines. With the semiautomatic machine, the welding operator either positions the work between the electrodes or positions a portable resistance welding gun around the work and pushes a switch to initiate the weld; the weld is completed by the execution of a preprogrammed sequence. In an automatic setup, the parts are automatically fed into a machine, then welded and ejected without welding operator assistance.

A typical stationary resistance welding system consists of a welding machine, an electronic contactor, and a control panel. The welding machine itself consists of a rigid frame, a stepdown transformer, an air cylinder or other means of applying a force to the upper electrode arm, a fixed but adjustable lower electrode arm, and heavy leads connecting the electrode arms to the secondary leads of the transformer. The electronic contactor may be a pair of ignitron tubes, thyratrons, or silicon controlled rectifiers. Older machines may be completely controlled by analog circuitry; however, newer machines allow digital input of welding data and are computer controlled. The simplest control sets a fixed current magnitude and duration. More sophisticated controls provide means of varying the current during a weld, and may also provide preheat and postheat after a cooling period. Preheat, postheat or both may be desirable for welding hardenable steels and other special alloys. Seam welding machine controls must provide an on-off sequencing of weld current and control of wheel electrode rotation.

Although the total power used is relatively small, the instantaneous power demand for RW is very high, and leads must be designed to carry the maximum kVA demand for the machine.

The cost of a complete RW machine including controls can vary over a wide range according to the type and thickness of material to be joined, the degree of sophistication required in the controls, and the amount of tooling needed. A 100 kVA single phase spot welding machine costs around $20,000 with a fully programmable controller. An RSEW machine costs approximately twice that amount.

A three-phase machine will generally cost about 2-1/2 times that of a single-phase machine of the same kVA rating, but it will have a better power factor. Large three-phase RW machines with complicated tooling and controls cost well over $45,000.

Other variations of resistance welding include dc secondary power and capacitor discharge power. The former uses single or three phase fully-rectified current. The latter uses a very high current, short duration, dc pulse provided by capacitors.

Resistance welding requires very little welding operator skill when the machine controls are preset. The welding operator only positions the work or the equipment (in the case of portable equipment) and initiates the start of the weld sequence. In RSEW, the welding operator guides the work into the rotating electrode wheels. The welding operator's maintenance duties would normally consist of replacing electrodes when the tips become worn or deformed and lubricating moving parts.

Robotics have been successfully adapted to resistance welding. Such units range from single welding gun configurations to systems of several welding machines. In the latter case, a single central computer can be used to control several welding machines. Many such systems are in use.

Resistance welding is used most frequently for joining equal or unequal metals up to 1/8 in. (3 mm) thick. Thicker sections can be joined, but larger machines are necessary, and arc welding may develop stronger, or more economical joints.

The cost of sophisticated automatic machines is high and amortization cost can be a significant part of the overhead burden for the process. However, if used in highly repetitive operations, RW is generally considered to be a low cost process. There are no consumables in the usual sense although electrode tip wear and power may be categorized as such. Machining of electrode tips and their periodic replacement may be classified as maintenance costs.

FLASH WELDING

FLASH WELDING (FW) is classified as a resistance welding process, but it is a unique process. Heat for welding is created at the faying surfaces of the joint by resistance to the flow of electric current, and by arcs across the interface. When the faying surfaces are heated to welding temperature, force is applied immediately to consummate a weld. Molten metal is expelled, the hot metal is upset, and a flash is formed. Filler metal is not added during welding.

The usual flash weld joins rods or bars end to end or edge to edge. Both components are clamped in electrodes which are connected to the secondary of a resistance welding transformer. One component is moved slowly towards the other, and when contact occurs at surface irregularities, the current flows and initiates the flashing action. This flashing action is continued until a molten layer forms on both surfaces. Then the components are forced together rapidly to squeeze out the molten metal and dross, and upset the adjacent hot base metal. This produces a hot worked joint free of weld metal. The mechanical properties of flash welds are often superior to other types of welds.

Flash welding is usually an automatic process. Parts are clamped in place by a welding operator who simply presses a button to start the welding sequence.

A flash welding machine is composed of a frame, a power transformer similar to a resistance welding transformer, a movable electrode clamp, a fixed electrode clamp, and a drive system for the movable electrode. A programmer establishes the flashing rate and time, and the upsetting time, current, and force.

Flash welding machines are available in a wide variety of sizes from about 10 kVA up to 1500 kVA. The power requirements and cost vary accordingly. Prices range from about $5000 for a 10 kVA welding machine to about $1,000,000 for a large 1500 kVA machine used to weld sheet in an automatic continuous mill. For large capacity continuous mills, much of the capital cost is for material handling and conveying equipment and for controls.

Minimum welding operator skill is needed to position the parts in the electrode clamps for correct alignment. After the parts are properly loaded in the machine, the flashing action and subsequent upset are initiated automatically by actuating a switch. Experience is needed, however, to establish the proper sequence.

Generally, welds can be made in sheet and bar thicknesses ranging from 0.01 to 1 in. (0.2 to 25 mm) and,

for rounds of diameters from 0.05 to over 3 in. (1 to over 76 mm). For small thicknesses or diameters, the main problems are to prevent the sheet or wire from bending during upset and to ensure that long joints are properly aligned. For large sizes, the main consideration is the capability of the machine to deliver the required power and provide the necessary upset force. Thick parts may require that one or both members be beveled to initiate flashing and to properly heat the center of the weld.

During flashing, metal from the ends being joined is expelled as a shower of sparks. Shortening takes place as a result of upsetting. These losses must be accounted for when designing parts which are to be flash welded.

The cost effectiveness of flash welding is similar to the other resistance welding processes. Other than the materials consumed in flashing, there are no consumables except for maintenance items such as the clamping surfaces of the electrodes. The major cost factors are labor, power, and amortization of equipment.

OXYFUEL GAS WELDING

OXYFUEL GAS WELDING (OFW) includes a group of welding processes that use the heat produced by a gas flame or flames for melting the base metal and, if used, the filler metal. Pressure may also be used. Oxyfuel gas welding is an inclusive term used to describe any welding process that uses a fuel gas combined with oxygen to produce a flame having sufficient energy to melt the base metal. The fuel gas and oxygen are mixed in the proper proportions in a chamber which is generally a part of the welding torch assembly. The torch is designed to give the welder complete control of the welding flame to melt the base metal and the filler metal in the joint.

Oxyfuel gas welding is normally done with acetylene fuel gas. Other fuel gases, such as methylacetylene propadiene and hydrogen, are sometimes used for oxyfuel gas welding of low melting metals. The welding flame must provide high localized energy to produce and sustain a molten weld pool. With proper adjustment, the flames also can supply a protective reducing atmosphere over the molten weld pool. Hydrocarbon fuel gases such as propane, butane, natural gas, and various mixtures employing these gases are not suitable for welding ferrous materials because the heat output of the flame is too low or the flame atmosphere is oxidizing.

Although the hydrogen flame temperature is about 4800°F (2660°C), hydrogen has limited use in OFW because the total heat content of the flame is low. The use of hydrogen is further complicated because the flame is essentially colorless. The lack of a visible cone makes the hydrogen-oxygen ratio very difficult to adjust. Oxy-

hydrogen welding (OHW) is used primarily for welding low melting metals, such as lead and, to a limited extent, thin sections and small parts.

Manual welding methods are most commonly used and require minimal equipment. A suitably sized torch, hoses, regulators, oxygen in a pressurized cylinder, fuel gas in a pressurized cylinder, welding rods, goggles, and protective clothing are needed.

In combination with pressure, oxyfuel gas flames can be used to make upset welds in butt joints without filler metals. This process is called pressure gas welding (PGW). In PGW, abutting surfaces are heated with oxyfuel gas flame(s) and forced together to obtain the forging action needed to produce a sound weld. The process is ideally adapted to a mechanized operation, and practically all commercial applications are either partly or fully mechanized.

The low cost of OFW equipment is one of the main reasons for its use. A complete manual welding unit including torch, hoses, regulators, tanks, and goggles costs about $200 to $500. However, larger units and special safety equipment are more expensive. Gas distribution systems vary widely in cost depending on the number of cylinders to be manifolded, the distance of the distribution lines, and the number of welding stations; thus the cost must be computed for each installation.

Since the OFW processes are primarily manual, it is essential that the welder be adequately trained and highly skilled for specific critical welding jobs such as pipe welding. The skill required by the welding operator for a fully mechanized PGW machine would be lower than that required by the manual welder since the machine control, when set, performs the complete operation.

Oxyfuel gas welding can be used for joining thick plate, but welding is slow and high heat input is required. Welding speed is adequate to produce economical welds in sheet metal and thin-wall and small diameter piping. Thus, OFW is best applied on material of about 1/4 in. (6 mm) maximum thickness.

Pressure gas welding is used to join sections up to 1 in. (25 mm) thick, and special machines have been made to join bar stock up to 3 in. (76 mm) in diameter. The PGW machines, while somewhat slower than flash welding machines, can provide satisfactory welds with less complex equipment. Slag entrapment can be a problem.

OFW equipment is very versatile and can be used with most construction materials. The equipment involved is easily portable. For these reasons, the cost effectiveness is good. However, when parts are to be made in quantity, other welding processes are usually more suitable. Weld quality is usually inferior to that obtained with arc welding because shielding is less effective. Pressure gas welding has limited applications because it is slower than comparable processes. The cost effectiveness is limited to small and medium size repetitive runs of limited production.

SOLID STATE WELDING

DIFFUSION WELDING

DIFFUSION WELDING (DFW) is a specialized process generally used only when the unique metallurgical characteristics of the process are required. Components to be diffusion welded must be specifically designed and carefully processed to produce successful joints. It is useful for applications concerned with (1) the avoidance of metallurgical problems associated with fusion welding processes, (2) the fabrication of shapes to net dimensions, (3) the maintenance of joint corrosion resistance with titanium and zirconium, and (4) the production of thick parts with uniform through-thickness properties, as with titanium laminates.

Diffusion welding occurs in the solid state when properly prepared surfaces are in contact under predetermined conditions of time, pressure, and elevated temperature. The applied pressure is set above the level needed to ensure essentially uniform surface contact but below the level that would cause macroscopic deformation. The temperature is generally well below the melting point. A filler metal, usually preplaced as an insert or plating, may be used. The function of the filler metal generally is to lower the required temperature, pressure, or time required for welding or to permit welding in a less expensive atmosphere.

Heating can be accomplished in a furnace, retort, or by resistance techniques. Pressure is applied by dead weight loading, a press, differential gas pressure, or by differential thermal expansion of the parts or of the tooling. The use of high pressure autoclaves or differential gas pressure techniques permits the welding of assemblies in which the joint surfaces are intersecting planar surfaces. Uniaxial methods of applying pressure are limited to welding parallel planar surfaces roughly perpendicular to the direction of load application. All techniques are essentially mechanized and require appropriate equipment. Encapsulating or canning of parts for welding is necessary when differential pressure techniques are practical and is useful when using other techniques.

The cost of equipment for DFW is roughly related to the joint area to be welded and ranges between $2000 and $4000 for each 1 in.^{2} (650 mm^2) of surface.

A high level of welding operator skill and training is required for most DFW operations. Only fully automated operations permit the use of semi-skilled personnel.

Diffusion welding is only economical when close dimensional tolerances, expensive materials, or special material properties are involved. Applications have been limited to the aerospace, nuclear, and electronics industries. Although the process has been used to fabricate complex, one-of-a-kind devices, it is more suitable for moderate volume production quantities. Consumable material costs are high if precious filler metals or inert gases are used.

FRICTION WELDING

FRICTION WELDING (FRW) machines are designed to convert mechanical energy into heat at the joint to be welded. As illustrated in Figure 1.13, the usual method of accomplishing this is to rotate one of the parts to be joined and force it against the other which is held in a stationary position. Normally, one of the two workpieces is circular or nearly circular in cross section, such as a hexagon. Frictional heat at the joint interface raises the metal to forging temperature. Axial pressure forces the hot metal out of the joint. Oxides and other surface impurities are removed with the soft, hot metal.

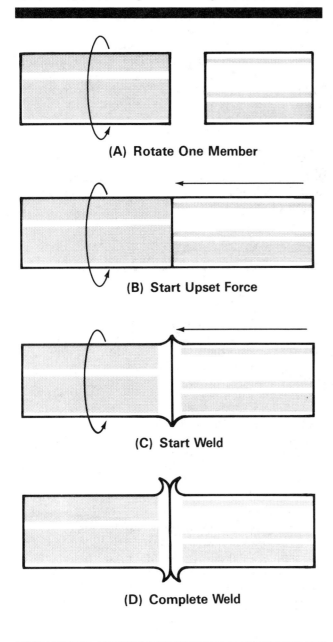

(A) Rotate One Member

(B) Start Upset Force

(C) Start Weld

(D) Complete Weld

Figure 1.13—Basic Steps in Friction Welding

Two major techniques for friction welding are available. With the conventional technique, the moving part is held in a motor- driven collet and rotated at a constant speed while axial force is applied to both parts. The fixed part must be held rigidly to resist the axial force and prevent it from rotating. Rotation is continued until the entire joint is heated sufficiently. Then simultaneously, the rotation is stopped and an upsetting force is applied to complete the weld. The process variables are rotational speed, axial force, welding time, and upset force. During the welding period, the drive motor must provide energy at the rate necessary to make the weld. Therefore, a relatively high powered motor is required.

The second FRW technique is called inertia welding. With it, energy is stored in a flywheel which has been accelerated to the required speed by a drive motor. The flywheel is coupled directly to the drive motor by a clutch and to the collet which grips the rotating member. A weld is made by applying an axial force through the rotating part while the flywheel decelerates, transforming its kinetic energy to heat at the joint. When properly programmed, the weld is completed when the flywheel stops. The welding variables are flywheel moment of inertia, flywheel rotational speed, axial force, and upset force if used.

The friction welding cycle is usually automatic. If automatic loading and unloading devices are installed, the machines are completely automatic.

Friction welding machines contain a driving head, a means for applying axial force, and a yoke or platform for mounting the tooling which holds the fixed part. They range in size from one which will weld a 0.5 in. (13 mm) maximum diameter steel bar to one which will weld a 5.0 in. (125 mm) maximum diameter steel bar. Power requirements for these machines vary from 25 kVA for smaller machines to 175 kVA for larger machines.

Inertia welding machines cost between $75,000 and $300,000 depending on their size and capacity. The cost of tooling depends on the specific parts to be welded, but it is generally about 35 percent of the basic machine cost.

Little skill in terms of manual dexterity is needed for friction welding since the process is fully automatic. However, a mechanical aptitude and an understanding of machine operation are required to set up jobs properly and to keep equipment in good operating condition.

The maximum diameter solid bar which can be welded depends upon the maximum axial force and driving energy available in the machine. The maximum size of hollow parts, such as tubes, depends upon the cross-sectional area to be joined. In inertia welding, the range can be extended downward by removing flywheel increments until only the inertia of the spindle and collet remain. This determines the minimum size which can be welded.

The time required to make friction welds is measured in seconds. Inertia welds require from less than 0.5 to approximately 15 seconds, exclusive of the time needed to accelerate the flywheel to the designated speed. In either type of machine, the time required to load and unload parts is longer than the actual weld time.

Factors contributing to the cost of friction welding are electric power, material loss as flash, labor, and the maintenance and amortization of equipment. The last two of these are the major cost items.

Other Solid State Welding Processes

THERE ARE NUMEROUS other solid state welding processes, including ultrasonic welding, explosive welding, and cold welding.

Ultrasonic welding employs a combination of a static normal force and a high-frequency oscillating shear force to cause coalescence at joint interfaces. Frequencies normally range from 10 000 to 60 000 Hz. The process has been very successfully applied to fine wire bonding in the electronic industry.

Explosive welding is accomplished by accelerating one of the components to be welded to an extremely high velocity with high energy explosives. The kinetic energy of motion is converted to heat at the joint interface. The high pressures generated at the interface cause coalescence and welding. The process is most commonly applied to the cladding of plates with a dissimilar metal. Slabs are also clad using explosive welding, and subsequently hot rolled to plates.

In cold welding, plastic deformation causes the generation of a new, clean surface at the joint interface, which in turn promotes solid-state welding. The process has been used to join soft, ductile metals in numerous applications including the welding of aluminum wire stock.

ELECTRON BEAM WELDING

ELECTRON BEAM WELDING (EBW) is accomplished with a stream of high-velocity electrons which is formed into a concentrated beam to provide a heating source. The process, illustrated in Figure 1.14, produces intense local heating through the combined action of the stream of electrons.[3] Each electron penetrates its own short distance and gives up its kinetic energy in the form of heat. With high beam energy, a hole can be melted through the material (a keyhole).

The hole is moved along the joint by moving either the electron gun or the workpiece. It is maintained as the metal at the front melts and flows around to the rear where it solidifies. Welds can be made without a keyhole, where melting takes place by conduction of heat from the surface, but welding speeds are lower.

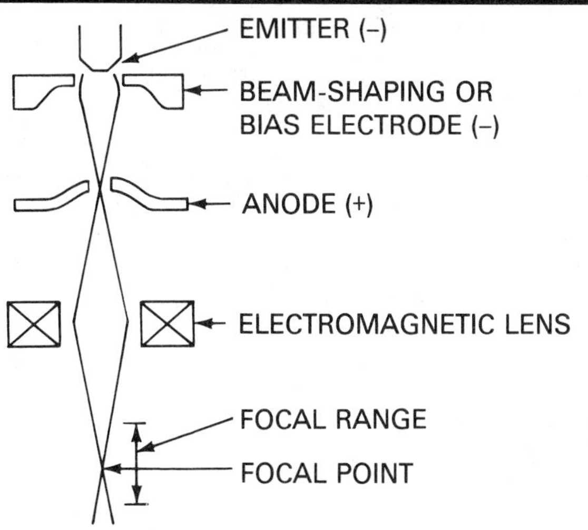

Electron beam welding has several major advantages. It can produce deep, narrow, and almost parallel-sided welds with low total heat input and comparatively narrow heat affected zones.

The depth to width ratio of electron beam welds can be greater than 10:1 and ratios of 30:1 are possible. The process involves only four basic variables: accelerating voltage, beam focus, welding speed, and beam current. Excellent control of penetration is possible using high-response closed-loop, servo controls on all welding variables. An evacuated work chamber creates a high purity environment for welding, resulting in freedom from impurities and contaminants. Finally, the process allows for high welding speeds of 300 to 500 in./min (125 to 200 mm/s) on sheet material, and generally requires no filler metal, flux, or shielding gas.

The resultant low total energy input minimizes distortion and shrinkage from welding and permits the welding of parts that have been finish machined. It also allows for welding in close proximity to heat sensitive components, and the welding of dissimilar metals which are metallurgically compatible. Also, by projecting the beam several inches to several feet, it is possible to make welds in otherwise inaccessible locations.

Electron beam welding generally is performed in a high vacuum at pressures of 10^{-6} to 10^{-3} torr (0.13 to 133 mPa), but the process can be adapted to weld in medium vacuum at pressure of 10^{-3} to 25 torr (0.13 to 3200 Pa), or at atmospheric pressure. However, penetration and depth-to-width ratio are reduced as the ambient pressure increases. The basic equipment required includes a vacuum chamber, controls, an electron beam gun (rated at from 30 kV to 175 kV and 50 mA to 1000 mA), a three-phase power supply, an optical viewing system or

EMITTER (–)

BEAM-SHAPING OR BIAS ELECTRODE (–)

ANODE (+)

ELECTROMAGNETIC LENS

FOCAL RANGE

FOCAL POINT

Figure 1.14—A Simplified Representation of an Electron Beam Gun Column

3. The actual flux of electrons in a 1 mA current stream is approximately 6.3 x 10^15 electrons per second.

tracking device, and work handling equipment. The operation of this equipment may be either semiautomatic or automatic. With high volume production equipment, the cycle time may be sufficiently reduced by the use of intermediate chambers or prepump stations.

The typical costs of an EBW facility range from $75,000 to $1,500,000. General-purpose high vacuum equipment falls in the $100,000 to $300,000 range, special-purpose high vacuum in the $150,000 to $1,500,000 range, general-purpose medium vacuum in the $80,000 to $150,000 range, and special-purpose medium vacuum in the $100,000 to $350,000 range. Indirect savings in welding costs can result from the reduced costs of joint preparation, ability to weld in a single pass, high welding speed, and low distortion of parts.

An automatic EBW facility requires only a skilled welding operator while general purpose equipment requires a skilled welding technician to control the four basic EBW variables. Mechanical and basic electrical backgrounds with solid state troubleshooting ability are necessary for equipment maintenance.

The EBW process can be used to weld almost any metal. Steel up to 4 in. (100 mm), aluminum up to 6 in. (150 mm), copper up to 1 in. (25 mm), and many other materials from foil to plate thickness are weldable in one pass.

Design is limited to either a tight butt or lap joint. As a general rule, fit-up is usually 0.005 in. (0.13 mm) or less. In most metals, the root opening should not exceed 0.010 in. (0.25 mm) for weld depths less than 1/2 in. (13 mm). Low voltage (low kV) equipment usually is capable of welding in any position.

LASER BEAM WELDING

SIMILAR TO THE electron beam, a focused high-power coherent monochromatic light beam used in laser beam welding (LBW) causes the metal at the point of focus to vaporize, producing a deep penetration column of vapor extending into the base metal. The elements of a typical laser beam welding or cutting station are shown in Figure 1.15. The vapor column is surrounded by a liquid

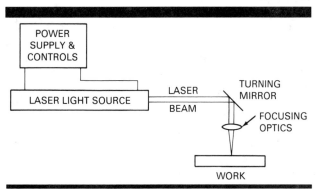

Figure 1.15—Elements of a Typical Laser Welding System

pool, which is moved along the joint producing welds with depth to width ratios greater than eight to one. Yttrium aluminum garnet (YAG) lasers may be used for spot welding thin materials, joining microelectronic components, and other tasks requiring precise control of energy input to the workpiece. Initial applications were limited by their low power, but later devices could produce 10 kW pulses having one millisecond duration. In seam welding, speeds are relatively low because the welds are formed by a series of overlapping spot welds.

Multi-kilowatt carbon dioxide gas laser systems are available with continuous power capability. Such systems produce an infrared beam which can be focused to provide power densities of 6.5 MW/in.^{2} (10 kW/mm^2) or greater. Continuous power provides a high power laser with deep penetration welding capability.

Although laser beam welding is most effective for autogenous welds, the process is amenable to filler metal addition. Because the laser allows precise control of energy delivery to highly localized regions, the process is ideally suited for square groove welds. Preplaced filler metal may be used in some applications.

Deep penetration welds produced with high power lasers are similar to electron beam welds, but LBW offers several advantages:

(1) A vacuum environment is not required for the workpiece because the laser beam is easily transmitted through air. However, reactive workpieces must be protected from the atmosphere by inert gas shields.

(2) X-rays are not generated by the beam.

(3) The laser beam may be readily shaped, directed, and focused with reflective optics, thereby allowing easy automation.

(4) The tendency for spiking, underbead spatter, incomplete fusion, and root bead porosity is less than with EBW because the power density of the focused laser beam is lower.

Laser beam welding has been used successfully to join a variety of metals and alloys including low alloy and stainless steels which do not exhibit high hardenability, aluminum alloys, lead, titanium, refractory metals, and high temperature alloys. Porosity-free, ductile welds can be attained with average tensile strengths equivalent to those of the base metal. Some change in hardness can be anticipated in alloys having microstructures related to cooling rates. Since laser beam welds typically involve low energy input per unit weld length, the cooling rates are high. In many cases, these high cooling rates offer substantial advantages. If undesirable, however, the cooling rates may be slowed by preheating, or the effects modified by postheating. The rapid cooling rate results in a fine dendritic weld metal microstructure that results in good notch toughness in low hardenability steels.

Typical CO_2 LBW performance at a power level of 5 kW is represented by welding speeds in excess of 150 in./min (65 mm/s) in 0.1 in. (2.5 mm) thick carbon steel or stainless steel. For 0.2 in. (5 mm) material, welding speed is about 100 in./min (42 mm/s) at 5 kW. For material of this thickness, welding speed is approximately proportional to power. At 10 kW, 3/16 in. (5 mm) thick aluminum can be welded at approximately 90 in./min (38 mm/s) and 3/16 in. (5 mm) thick titanium can be welded at approximately 135 in./min (57 mm/s). The maximum single pass penetration of about 0.7 in. (18 mm) has been achieved in Type 304 stainless steel at 15 kW and at a speed of 20 in./min

(8 mm/s). At 60 in./min (25 mm/s) penetration at 15 kW is still in excess of 0.6 in. (15 mm), but a substantially narrower fusion zone is obtained with the reduced energy input per unit length of weld. Furthermore, sound laser beam welds have been formed in 1 in. (25 mm) thick steel by a dual pass technique. Conversely, 0.008 in. (0.2 mm) material has been welded at 3000 in./min (1270 mm/s) at 6 kW. Industrial CO_2 lasers of 25 kW power have been built.

Laser beam welding is a high speed process ideally suited to automation, but it requires good joint fit-up. The high cost of equipment, which ranges from $40,000 to $850,000, currently relegates application to high volume production or to critical weldment applications requiring unique weld characteristics and reproducibility. The electrical efficiency of LBW equipment is between 8 and 15 percent. Welding equipment, with minor modifications to beam focusing provisions, can also be utilized for gas jet-assisted cutting and for surface heat treating and alloying applications.

Although the equipment is very sophisticated, it is designed to be used by machine operators that may not be skilled manual welders. Major maintenance must be handled by factory trained engineers.

BRAZING

BRAZING (B) IS a group of welding processes in which the joint is heated to a suitable temperature in the presence of a filler metal having a liquidus above 840°F (450°C) and below the solidus of the base metal. The filler metal is distributed between the closely fitted faying surfaces of the joint by capillary action. Braze welding is differentiated from brazing because the filler metal is deposited in a groove or fillet exactly at the point where it is to be used and capillary action is not a factor. Brazing is arbitrarily distinguished from soldering by the filler metal melting temperature. In soldering, filler metals melt below 840°F (450°C).

To produce acceptable brazed joints, considerations must be given to four basic elements: joint design, filler metal, uniform heating of the joint, and protective or reactive shielding. The various brazing processes are primarily designated according to the sources or methods of heating. Those which are currently of industrial significance include torch brazing, furnace brazing, induction brazing, resistance brazing, dip brazing, infrared brazing, and diffusion brazing.

TORCH BRAZING

TORCH BRAZING (TB) is accomplished by heating the parts to be brazed with one or more oxyfuel gas torches

using various fuels. While TB is normally done with hand-held torches, automated TB machines use preplaced fluxes and preplaced filler metal in paste, wire, or shim form.

Normally, torch and machine brazing are used to make lap joints in relatively thin sections ranging from 0.01 to 0.25 in. (0.25 to 6 mm). The joints can be brazed rapidly, but brazing speed decreases as the material thickness increases.

Manual torch brazing is most effective and economical when relatively few pieces need to be joined. Basic equipment costs are about $300. Automated torch brazing machines are very economical even though their costs range from about $15,000 to $50,000 depending on their complexity.

FURNACE BRAZING

FURNACE BRAZING (FB) uses a furnace to heat prefluxed or precleaned parts with the filler metal preplaced at the joints. Batch furnaces with protective atmospheres commonly are used, when relatively few brazements are involved or special heating and cooling cycles are required for distortion control or heat treatment. Continuous furnaces are most effective for very high production rates and simple joint designs. Both types of furnaces have a wide cost range from about $2,000 for an air atmosphere batch furnace to above $300,000 for an automated vacuum furnace. Furnace selection depends on the desired rate of production and the size of the assembly to be processed.

Furnace brazing processes are particularly suited for fabricating brazements of complex design. The brazing operator skill required for the batch and continuous furnaces is low and in many cases the operator has only to load, unload, or apply the filler metal.

INDUCTION BRAZING

INDUCTION BRAZING (IB) involves heat obtained from resistance to a high-frequency current induced in the part to be brazed. The parts are placed in or near a high-frequency coil, and are not connected to the primary electrical circuit.

The brazing filler metal is usually preplaced. Careful design of the joint and the coil setup are necessary to assure that the surfaces of all members of the joints reach brazing temperature at the same time. Flux is usually employed except when a protective or reducing atmosphere is used. Joints should be designed to be self-jigging rather than fixtured.

The cost of a single station unit is in the $5,000 to $15,000 range with smaller and larger units available. Equipment setup, power supply and controls, and coil design are critical to producing a good quality brazement.

The thickness of components at the joint normally does not exceed 1/8 in. (3 mm). Special attention must be given to the coil design and locations when various thicknesses are brazed. Production speeds are high: heating can be accomplished in four to ten seconds, and coils can be designed to allow a conveyor belt containing parts to pass through the coil.

IB is commonly used where many parts must be made continuously over a long period of time. However, small shops can handle short-run jobs very economically.

Resistance brazing (RB) uses heat produced by the resistance of the joint to electric current transferred to the work by electrodes. The brazing filler metal is preplaced or face-fed by some convenient method, and the flux must be partially conductive.

The parts to be brazed are held between two electrodes which transmit force and current to the joint. Pressure is maintained on the joint until the filler metal has solidified. The thickness of base metal commonly brazed with this process normally ranges from 0.005 to 0.50 in. (0.1 to 12 mm). The process is largely used to braze lap joints.

The cost of equipment depends upon the power required to heat the joint to the brazing temperature and the complexity of the electrical controls. Resistance brazing equipment varies from $1000 for very simple machines to $10,000 and above for the more sophisticated equipment with electronic controls and process monitoring equipment. Brazing operator skill for this type of equipment is minimal, although the more sophisticated equipment may require added experience for setup and control operations. The brazing speed is very fast, particularly when the power source and resistance tongs are adequate for the specific joint being made. Resistance brazing is used most economically on special application joints where the heat must be restricted to a very localized area without overheating surrounding parts.

DIP BRAZING

DIP BRAZING (DB) is performed in a molten salt or molten metal bath. Both types of baths are heated in suitable pots to furnish the heat necessary for brazing, provide protection from oxidation, and fluxing action if suitable salts are used.

The cost of DB equipment varies from $2,000 to $15,000 for small to average size molten metal or molten salt furnaces. Large installations with water and air pollution control equipment are in the $200,000 range.

Molten metal bath brazing is limited to small wires 0.005 to 0.200 in. (0.13 to 5 mm), sheet, and fittings that can be dipped into small heated pots. Molten salt baths can be used for thin sheet assemblies as well as heavier parts and very complex assemblies, such as aluminum radiators and other heat exchangers.

The production speed and efficiency of the dip baths are extremely good since the heating rates are very fast and many joints can be brazed at one time. Careful precleaning and postcleaning of parts are essential and may limit the production effectiveness in some cases. Brazing operator skill is minimal although extensive engineering may be required to set up large operations.

The cost effectiveness of this process is demonstrated by the fact that large parts may have as many as 100 to 1000 joints to be brazed at one time.

INFRARED BRAZING

INFRARED BRAZING (IRB) uses a high intensity quartz lamp as the source of heat. The process is particularly suited to the brazing of very thin materials such as honeycomb panels for aircraft, but is not normally employed on sheets thicker than 0.05 in. (1.3 mm). Assemblies to be brazed are supported in a position which enables the radiant energy to be focused on the joint. The assembly and the lamps can be placed in evacuated or controlled atmosphere retorts.

Small hand-held quartz lamps with focusing mirrors and without specific controls are available for approximately $500 while larger installations that would automatically heat both sides of large honeycomb panels are in the $30,000 range.

The cost effectiveness of the IRB process is best when the parts to be brazed are very thin, such as in the honeycomb applications. The cost of brazing relatively thick joints is comparable to FB in terms of cycle time, atmosphere, and equipment.

DIFFUSION BRAZING

DIFFUSION BRAZING (DFB), unlike the previous brazing processes, is not defined by its heat source, but by the mechanism involved. A joint is formed by holding the brazement at a suitable temperature for a sufficient time to allow mutual diffusion of the base metal and filler metal. The joint produced has a composition considerably different than either the filler metal or base metal, and no filler metal should be discernible in the finished microstructure. The DFB process produces stronger joints than the normal brazing process. Also, the DFB joint has a remelt temperature approaching that of the base metal.

The typical thicknesses of the base metal that are diffusion brazed range from thin foil to sections 1 to 2 in. (25 to 50 mm) in thickness. Relatively heavy parts can be brazed because the process is not sensitive to joint thickness.

Many brazements that are difficult to make by other processes can be diffusion brazed. Butt and lap joints having good mechanical properties can be produced. The parts are usually fixtured mechanically or tack welded together. Although DFB requires a relatively long brazing time (30 min. to 24 h) to complete, a number of assemblies can be brazed at the same time.

Furnaces are most frequently used for heating. They are similar to those used for the more sophisticated FB operations and range in cost from $50,000 for small furnaces to $300,000 for the large furnaces.

SOLDERING

SOLDERING (S) INVOLVES heating a joint to a suitable temperature and using a filler metal (solder) which melts below 840°F (450°C). The solder is distributed between the closely fitted surfaces of the joint by capillary action. Heat is required to raise the joint to a suitable temperature, melt the solder, and to promote the action of a flux on the metal surface so that the molten solder will wet and flow into the joint.

Successful soldering involves shaping the parts to fit closely together, cleaning the surfaces to be joined, applying a flux, assembling the parts, and applying the heat and the solder. Flux residues may be removed when the joint is cooled.

Various types of equipment, processes, and procedures, all having certain advantages and disadvantages, are employed for soldering operation. To make a satisfactory solder joint by these processes, it is essential to have a suitable joint design for the base metal and solder employed.

The soldering processes are primarily designated according to the source or method of heating. Those which are currently of industrial significance are dip soldering (DS), iron soldering (INS), resistance soldering (RS), torch soldering (TS), induction soldering (IS), and furnace soldering (FS). Infrared soldering (IRS) and ultrasonic soldering are also used.

The cost of equipment varies widely with the process and is similar to the costs given for similar brazing equipment. Industrial soldering irons and flame heating equipment can be obtained for less than $100. The DS, IS, RS, FS, and ultrasonic soldering equipment costs vary widely. Some small pieces of equipment cost under $1000 while high speed automated equipment, which can be very complex, cost in excess of $50,000.

A degree of skill is required to adequately perform the manual soldering operations, particularly when critical electronic equipment of large, complex, or critical components are to be soldered. The more automated soldering processes require little operator skill because the equipment is set up by skilled technicians prior to production.

A lap joint or some modification of the joint is commonly used. The electronics industry uses many special types of joints; some joints have no mechanical security prior to soldering while other types have partial or full mechanical security prior to soldering.

The soldering process can be used to join a wide range of metal thicknesses from very thin films to relatively thick components, such as bus bars and piping.

Soldering speeds for nonrepetitive manual operations are relatively slow, although a solderer can become quite adept at repetitive operations. Likewise, the electronics industry employs machines of various types for wave soldering of printed circuit boards. These machines produce many joints on a circuit board in a matter of seconds at high production rates. Manual dip soldering is much faster than manual iron soldering, and is useful for moderate production requirements.

Automated equipment for soldering produces the highest quality joints at a low cost per joint because many joints can be soldered simultaneously. The manual soldering iron process is much slower, and thus more costly. Manual soldering of components is economical when the production needs are low or the components are of complex design.

ADHESIVE BONDING

ADHESIVE BONDING IS a joining process which is gaining acceptance as an assembly method for joining metals. The method has several advantages and limitations. On one hand, it is capable of joining dissimilar materials, for example, metals to plastics; bonding very thin sections without distortion and very thin sections to thick sections; joining heat sensitive alloys; and producing bonds with unbroken surface contours. Furthermore, bonding can be accomplished at a low cost.

On the other hand, joints produced by this method may not support shear or impact loads. Also, such joints must have an adhesive layer less than 0.005 in. (0.13 mm) thick and must be designed to develop a uniform load distribution in pure shear or tension. The joints cannot sustain operational temperatures exceeding 500°F (260°C). Autoclaves, presses, and other tooling may be essential to achieve adequate bond strengths. The surfaces to be bonded must be clean, and curing time is needed to achieve full bond strength. Some adhesives must be used quickly after mixing. Finally, nondestructive testing of adhesive bonded joints is difficult.

A variety of adhesives can be used. Thermoplastic adhesives develop a bond through the evaporation of a solvent or the application of heat. The pressure-sensitive adhesives produce a relatively weak bond when pressure is applied to the joint. They retain flow characteristics allowing them to support only light loads. Some adhesives are stable in air but begin curing when the joint is assembled and air is excluded. Other adhesives, usually used for metals, react chemically with curing agents or catalysts. Generally, they require no pressure other than that needed to maintain intimate contact between the faying surfaces. Curing times range from hours at room temperature to seconds at temperatures up to 350°F (175°C). Some epoxy-based adhesives can produce joint strengths up to 10 000 psi (70 MPa) in shear or tension when cured at 350°F (175°C) for a few hours under pressures of about 150 psi (1030 kPa).

Minimal training is needed for production workers in adhesive bonding, but considerable skill is needed to design suitable joints and select the appropriate adhesive for the anticipated service requirements. Frequently, the joints are not properly cleaned and prepared to receive the adhesive. Metals in particular must be free of all contaminants. They must be treated to leave a surface which will be chemically receptive to the adhesive and will provide maximum wetting characteristics.

Equipment costs vary widely because some adhesives are cured by heat alone, some by pressure alone, and some by both. Heating oven size depends on the size of the assemblies being bonded. Presses to apply pressure to the joints may have to be specially designed to accommodate contoured parts. When properly used, adhesive bonding can do a very effective job and can be accomplished at low cost.

THERMAL SPRAYING

THERMAL SPRAYING (THSP) is a process in which a metallic or nonmetallic material is heated and then propelled in atomized form onto a substrate. The material may be initially in the form of wire, rod, or powder. It is heated to the plastic or molten state by an oxyfuel gas flame, an electric or plasma arc, or an explosive gas mixture. The hot material is propelled from the spray gun to the substrate by a gas jet. Most metals, cermets, oxides, and hard metallic compounds can be deposited by one or more of the process variations. The process can also be used to produce free-standing objects using a disposable substrate. It is sometimes called *metallizing* or *metal spraying*.

When molten particles strike a substrate, they flatten and form thin platelets that conform to the surface. These platelets rapidly solidify and cool. Successive layers are built up to the desired thickness. The bond between the spray deposit and substrate may be mechanical, metallurgical, chemical, or a combination of these. In some cases, a thermal treatment of the composite structure is used to increase the bond strength by diffusion or chemical reaction between the spray deposit and the substrate.

The density of the deposit will depend upon the material type, method of deposition, spraying procedures, and subsequent processing. The properties of the deposit may depend upon the density, the cohesion between the deposited particles, and the adhesion to the substrate.

Thermal spraying is widely used for surfacing applications to attain or restore desired dimensions; to improve resistance to abrasion, wear, corrosion, oxidation, or a combination of these; and to provide specific electrical or thermal properties. Frequently, thermal sprayed deposits are applied to new machine elements to provide surfaces with desired characteristics for the application.

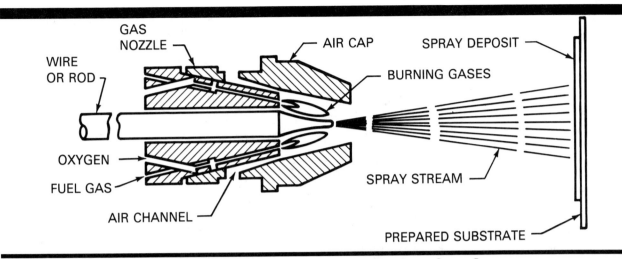

Figure 1.16—Cross Section of a Typical Wire Flame Spray Gun

PROCESS VARIATIONS

THERE ARE FOUR variations of thermal spraying:

(1) Flame spraying (FLSP)
(2) Plasma spraying (PSP)
(3) Arc Spraying (ASP)[4]
(4) Detonation flame spraying

These variations are based on the method of heating the spray material to the molten or plastic state and the technique for propelling the atomized material to the substrate.

In flame spraying, illustrated in Figure 1.16, the surfacing material is continuously fed into and melted by an oxyfuel gas flame. The material may be initially in the form of wire, rod, or powder. Molten particles are projected onto a substrate by either an air jet or the combustion gases.

In plasma spraying, the heat for melting the surfacing material is provided by a non-transferred plasma arc.[5] The arc is maintained between an electrode, usually tungsten, and a constricting nozzle. An inert or reducing

gas, under pressure, enters the annular space around the electrode where it is heated to a very high temperature (above about 15 000°F) by the arc. The hot plasma gas passes through and exits from the nozzle as a very high velocity jet. The surfacing material in powder form is injected into the hot gas jet where it is melted and projected onto the substrate.

The surfacing materials used with arc spraying are metals or alloys in wire form. Two continuously fed wires are melted by an arc operating between them. The molten metal is atomized and projected onto a substrate by a high velocity gas jet, usually air. This method is restricted to spraying of metals that can be produced in continuous wire form.

The detonation flame spraying method operates on principles significantly different from the other three methods. This method repeatedly heats charges of powder and projects the molten particles onto a substrate by rapid, successive detonations of an explosive mixture of oxygen and acetylene in a gun chamber. The particles leave the gun at much higher velocity than with the other methods.

THERMAL CUTTING PROCESSES

THERMAL CUTTING PROCESSES consist of oxygen, arc, and other cutting processes. Some of the cutting processes, such as oxygen arc, lance carbon arc, shielded metal arc, gas metal arc, and gas tungsten arc are diffi-

cult to regulate or are not capable of concentrating sufficient energy to make accurate, quality cuts. Nevertheless, some are used in special applications.

The most important cutting processes are oxyfuel, plasma arc, air carbon arc, and laser beam cutting. These processes provide low-cost, precision cuts of high quality in a variety of metals. The applicability of these cutting processes to industrial metals is summarized in

---

4. This method is commonly called *electric arc spraying.*

5. For information on plasma arc systems, refer to Chapter 9, Plasma Arc Welding, *Welding Handbook*, Vol. 2, 7th Ed. 1978.

Table 1.2. Each process has particular advantages and limitations.

OXYFUEL GAS CUTTING

OXYFUEL GAS CUTTING (OFC) is a commonly used method for severing or gouging metals which react chemically with oxygen. The metal is heated to its ignition temperature by a fuel-oxygen mixture. At this point a high velocity stream of oxygen is introduced to produce a chemical reaction and to blow the molten reaction products through the thickness. This severs the metal. Many steels and titanium alloys can be cut with this process. The torches used are designed to (1) provide a small diameter high velocity stream of oxygen to oxidize and remove metal from a narrow section and (2) a ring of flame to preheat the metal to its ignition temperature. The torch is moved at a speed that will maintain a continuous cutting action. Because the jet and flame are symmetrical, the torch can be moved in any direction, for cutting in more than one plane and in other than a straight line. The cut quality depends on the torch tip size, type, and distance from the plate; the oxygen and preheat gas flow rates; and the cutting speed. All of these factors are related to the material type and thickness. Supplies of oxygen and fuel gas are needed. Cutting torches are designed to use most fuel gases, including natural gas, propane, acetylene, and by-products of the chemical industries.

Oxyfuel gas cutting is almost always selected to cut mild steel plates up to 12 in. (300 mm) thick. Alloy steels are more difficult to sever, and the quality of cuts in such steels depends on the nature and amount of the alloying elements. Stainless steels cannot be cut easily, but severing cuts can be made by introducing flux or iron-rich powder into the oxygen cutting stream. Titanium can be cut faster than mild steel, but the oxidized cut surfaces must be removed before welding or the ductility of the welds will be lowered.

Modifications of OFC are used for scarfing metal in steel mills, for gouging defective welds, and for grooving plate edges in preparation for welding.

Manual OFC operations are common. A worker can be trained in a short time to make acceptable cuts with the process. However, considerable skill is necessary to produce cut surfaces suitable for welding. Light duty torches and regulators for cutting 1/2 in. (13 mm) steel can be purchased for less than $200. Heavy duty equipment will cost about $300 to $500.

Equipment for mechanized cutting is more expensive, and the skills required to use this equipment are dependent on design and application. Relatively little training is needed to produce good straight cuts since all of the operation conditions can be preset using tabulated data. Considerable skill is necessary to produce shape cuts with multiple torches because of the complexity of the

equipment. Equipment costs will vary depending on the nature of the drive and control systems, the sizes of plates to be cut, and the number of torches. Motorized OFC torches are available for about $500 which can produce consistently good cuts. A coordinated drive system with optical tracking for shape cutting will cost about $20,000. A computer controlled, multiple torch package of the type found in steel service centers could cost over $100,000.

PLASMA ARC CUTTING

PLASMA ARC CUTTING (PAC) is accomplished with an extremely hot, high-velocity plasma gas jet formed by an arc and inert gas flowing from a small diameter orifice. The arc energy is concentrated on a small area of the plate that melts and the jet of hot plasma gas forces the molten metal through the kerf (see Appendix A) and out of the back side. A supplementary gas shield of CO_2, air, nitrogen, or oxygen can be used to cool and further constrict the arc. This is called *dual flow plasma-arc cutting*, and it is illustrated in Figure 1.17. Water is also used for the same purpose. Water shrouds in combination with a water table will remove the fume and reduce the noise level. Torches are available that will cut under water. The cut quality of plate cut on a water table is not affected. However, the water tables are not used in manual cutting operations.

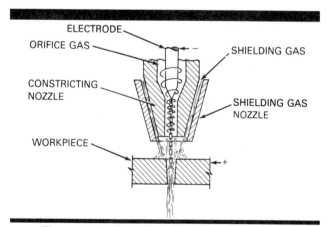

Figure 1.17—Dual Flow Plasma Arc Cutting

This process is versatile. All metals can be cut with it and mild steel can be severed faster than is possible with oxygen cutting. However, more metal is removed because the kerfs are wider and, in thick plates, the kerfs are smooth but slightly bevelled. Special nozzles are needed to produce parallel kerf walls, and special techniques are needed to produce shape cuts with controlled bevels.

The process requires high arc voltage provided by special power sources with high open-circuit voltage.

Depending on the material and thickness to be cut, between 25 kW and 200 kW of power are needed. Plasma gas mixtures of argon and hydrogen, or nitrogen and hydrogen, are generally used.

Plasma arc cutting torches are available for manual cutting. Skill requirements are similar to those needed for OFC, but more training is needed to operate the equipment. Complete PAC units, capable of cutting materials up to 3/8 inch thick, are available for as little as $2500.

Mechanized equipment is more common. The torch and other accessories are similar to those used for manual plasma arc welding. Additional costs may be necessary for the travel mechanisms which are similar to those used for oxygen cutting equipment, although they should be designed to travel at higher speeds. Multiple torch packages need several power sources and control panels, one for each torch. In addition, water shrouds or water tables may be used to absorb noise and fume. Therefore, mechanized equipment may cost between $20,000 and $100,000 depending on the tracking and drive systems and the plate thicknesses to be cut. Multiple torch numerical controlled equipment could exceed $100,000.

AIR CARBON ARC CUTTING

THE AIR CARBON arc cutting (AAC) process uses an arc to melt metal which is blown away by a high velocity jet of compressed air. The electrodes are rods made from a mixture of graphite and carbon, and most are coated with a layer of copper to increase their current-carrying capacity. Standard welding power sources are used to provide the current. Air is supplied by conventional shop compressors, and most applications require about 80 psi (550 kPa) at between 20 and 30 cubic feet per minute (cfm) (600 to 900 liters/min). Manual holders, Figure 1.18, are very similar in appearance to SMAW electrode holders, and supply both air and current.

Figure 1.18—Manual Air Carbon Arc Electrode Holder with the Electrode

In gouging operations, the depth and contour of the groove are controlled by the electrode angle, travel speed, and current. Grooves up to 5/8 in. (16 mm) deep

can be made in a single pass. In severing operations, the electrode is held at a steeper angle, and is directed at a point that will permit the tip of the electrode to pierce the metal being severed.

In manual work, the geometry of grooves is dependent on the cutting operator's skill. To provide uniform groove geometry, semiautomatic or fully automatic torches are used to cut "U" grooves in joints for welding. When removing weld defects or severing excess metal from castings, manual techniques are most suitable. Regardless of the equipment, an operator can be trained in one or two days.

The manual equipment, which is used in fabrication and maintenance shops, costs between $100 and $200, assuming that a power source and compressed air supply are available. Semiautomatic equipment, including the torch, the mount and drive unit, and accessories, cost about $1000. Voltage-controlled automatic torches and control units cost between $1000 and $2000. They are used for very precise gouging, with tolerances of less than 1/32 in. (0.8 mm) and, generally, are mounted on standard travel carriages.

The applicability of AAC to most metals is indicated in Table 1.2.

LASER BEAM CUTTING

AS IN LASER beam welding (LBW), the source of heat for laser beam cutting (LBC) is a concentrated coherent light beam which impinges on the workpiece to be cut. A combination of melting and evaporation provides the mechanism for removing material from the kerf. Beam cutting can also be used to cut certain nonmetallic materials, such as carbon and ceramics.

High power lasers exhibit unique advantages for cutting applications including:

(1) The ability to cut any metal and many nonmetals regardless of hardness
(2) Narrower kerf and heat affected zone than those produced by other thermal cutting processes
(3) High cutting speeds
(4) Ready adaptability to computer controlled contour cutting

These advantages are a result of the ease of beam transmission through the atmosphere and the high power density which can be obtained. Typically, a power density can be obtained which is sufficient to vaporize all materials.

Many cutting applications (particularly those involving metals) require the assistance of a gas jet to blow the metal from the kerf. The gas used may be inert to provide a smooth, clean kerf or a reactive gas, such as oxygen, to speed the cutting process. Smooth cuts can be achieved with appropriate selection of laser beam power, gas shielding condition, and cutting speed, and further edge finishing is often not required.

The absence of mechanical force is advantageous when cutting large and complex parts because minimal clamping and fixturing can be used. The ease with which the motion of the focused laser beam can be adapted to computer control renders the process suitable for cutting large, complex, multicontoured parts without requirements for template fabrication and inventory.

One of the principal current disadvantages of LBC is the relatively high capital cost of laser beam equipment. This factor, coupled with the decrease (compared to other thermal processes) in cutting speed as material thickness increases, presently limits the cost effectiveness of LBC to metals of approximately 1/2 in. (13 mm) or less in thickness. However, acceptable cuts have been made in 2 in. (50 mm) thick steel.

SUPPLEMENTARY READING LIST

American Society for Metals. *Metals handbook,* Vol. 6, 8th ed. Metals Park, Ohio: American Society for Metals, (1971).

American Welding Society. *Air arc gouging and cutting, recommended practices for,* C5.3-82. Miami, FL: American Welding Society, 1982.

———. *Brazing manual.* 3rd Ed. Miami, FL: American Welding Society, 1976.

———. *Design, manufacture, and inspection of critical brazed joints,recommended practices for,* C3.3-80. Miami, FL: American Welding Society, 1980.

———. *Electrogas welding, recommended practices for,* C5.7-81. Miami, FL: American Welding Society, 1981.

———. *Gas metal arc welding, recommended practices for,* C5.6-79. Miami, FL: American Welding Society, 1979.

———. *Gas tungsten arc welding, recommended practices for,* C5.5-80. Miami, FL: American Welding Society, 1980.

———. *Oxyfuel gas cutting, operators manual for,* C4.2-78. Miami, FL: American Welding Society, 1978.

———. *Plasma arc cutting, recommended practices for,* C5.2-83. Miami, FL: American Welding Society, 1983.

———. *Plasma arc welding, recommended practices for,* C5.1-73. Miami, FL: American Welding Society, 1973.

———. *Resistance welding, recommended practices for,* C1.1-66. Miami, FL: American Welding Society, 1966.

———. *Resistance welding coated low-carbon steel, recommendedpractices for,* C1.3-70. Miami, FL: American Welding Society, 1970.

———. *Resistance welding, theory and use.* Miami, FL: American Welding Society, 1956.

———. *Soldering manual.* Miami, FL: American Welding Society, 1978.

———. *Spot welding aluminum and aluminum alloys, recommendedpractices for,* C1.2-53. Miami, FL: American Welding Society, 1953.

———. *Standard welding terms and definitions,* A3.0-85. Miami, FL: American Welding Society, 1985.

———. *Thermal spraying: practice, theory, and application.* Miami, FL: American Welding Society, 1985.

———. *Welding Handbook,* Vol. 2, 7th ed. Miami, FL: American Welding Society, 1978.

———. *Welding Handbook,* Vol. 3, 7th ed. Miami, FL: American Welding Society, 1980.

H. B. Cary, *Modern welding technology.* Englewood Cliffs: Prentice-Hall, 1979.

Houldcroft, P. T. *Welding processes.* Cambridge: Cambridge University Press (1967).

The Lincoln Electric Company. *Principles of industrial welding.* Cleveland, OH: The Lincoln Electric Company.

The Lincoln Electric Company. *The procedure handbook of Arc welding,* 12 ed. Cleveland, OH: The Lincoln Electric Company (1973).

Linnert, G. E. *Welding metallurgy,* Vol. 1, 3rd ed. Miami, FL: American Welding Society, 1965.

Paton, B. E. *Electroslag welding,* 2nd ed. Translated from the Russian. Miami, FL: American Welding Society (1962).

———. *Electroslag welding and surfacing.* Translated from the Russian. Miami: American Welding Society (1983).

Vill, V. I. *Friction welding of metals.* Translated from the Russian. Miami, FL: American Welding Society (1962).

CHAPTER 2

PHYSICS OF WELDING

PREPARED BY A COMMITTEE CONSISTING OF:

K. A. Lyttle, Chairman
L-Tec Welding & Cutting Systems

D. G. Anderson
L-Tec Welding & Cutting Systems

H. D. Frick
Orgo-Thermit, Incorporated

J. McGrew
Babcock & Wilcox

D. W. Meyer
L-Tec Welding & Cutting Systems

R. A. Miller
Air Products

R. T. Telford
Linde Division, Union Carbide

WELDING HANDBOOK COMMITTEE MEMBER:
A. F. Manz
A. F. Manz and Associates

CHAPTER 2

PHYSICS OF WELDING

INTRODUCTION

MANY WELDING PROCESSES require the application of heat or pressure, or both, to produce a suitable bond between the parts being joined. The physics of welding deals with the complex physical phenomena associated with welding, including heat, electricity, magnetism, light, and sound.

Welding generally involves the application or development of localized heat near the intended joint. A common autogenous means of heating for welding is by the flow of current through electrical contact resistance at the faying surfaces of two workpieces. Friction heating and electrical discharges between members to be joined are also used successfully.

Friction or impact heating is generally regarded as incidental to such processes as ultrasonic and explosion welding, but in fact, it may contribute significantly to the joining action. Perhaps the only truly nonthermal welding process is cold welding, which accomplishes joining through the mechanism of controlled cold plastic deformation of the members being joined.

Almost every imaginable high energy density heat source has been used at one time or another in welding. Externally applied heat sources of importance include arcs, electron beams, light beams (lasers), exothermic reactions (oxyfuel gas and Thermit), and electrical resistance. Welding processes that acquire heat from external sources are usually identified with the type of heat source employed. The processes in this category are as follows:

(1) Arc welding
  (a) Shielded metal arc welding (SMAW)
  (b) Gas metal arc welding (GMAW)
  (c) Flux cored arc welding (FCAW)
  (d) Gas tungsten arc welding (GTAW)
  (e) Electrogas welding (EGW)
  (f) Plasma arc welding (PAW)
  (g) Submerged arc welding (SAW)
(2) Resistance welding (RSW, RSEW)
(3) Electroslag welding (ESW)
(4) Oxyacetylene welding (OAW)
(5) Thermit welding (TW)
(6) Laser beam welding (LBW)
(7) Electron beam welding (EBW)

As usually implemented, a high energy density heat source (used by one of the processes just listed) is applied to the prepared edges or surfaces of the members to be joined, and is moved along the path of the intended joint. The power and energy density of the heat source must be sufficient to accomplish local melting. Filler metal may be added, in which case the heat source has to melt the filler metal as it is delivered to the joint.

The transferred power is the rate at which energy is delivered per unit time from the heat source to the workpiece, typically expressed in watts. The energy density is the transferred power per unit area of effective contact between the heat source and the workpiece, generally expressed in watts per square meter or square millimeter (watts per square inch).[1] The energy density is a completely unambiguous measure of "hotness," applicable to all kinds of heat sources. (Heat sources are sometimes qualitatively compared in terms of temperature, a fairly satisfactory index of intensity for arcs and oxyfuel gas flames, but it is meaningless to speak of the "temperature" of an electron or laser beam.)

The significance of energy density as a heat source property cannot be overemphasized. In fact, the evolu-

---

1. The customary units of measure in the specific technology being discussed are the units in this volume. Secondary units are shown in parentheses.

tion of welding processes has, in large measure, been predicated on the development and adaptation of high energy density heat sources. Thus, the oxyacetylene flame, as used for welding, has been almost completely displaced by the higher energy density metal arc. Plasma arc, electron beam, and laser beam are unique welding processes that exhibit significantly higher energy density than conventional arc welding.

The transfer of energy to a workpiece from an arc, flame, or an electron or laser beam is a complex process, and the energy density of a welding heat source cannot generally be expressed as a precise number. It is difficult to define the area of contact between the heat source and the workpiece. In addition, the intensity of the heat is distributed nonuniformly over the contact area, typically exhibiting a maximum intensity near the center. Although the detailed nature of energy transfer is quite complicated, the concept of energy density contributes much to the understanding and comparison of welding heat sources.

One way of regarding a welding heat source, such as an arc, is to consider two distinct heat transfer processes. First, heat is transferred from the source to the surface of the workpiece. Then, heat is transferred by conduction, from the contact area to colder regions of the metal.[2] These two processes are somewhat competitive. With a very high energy density heat source, such as an electron beam, energy can be delivered to the contact area so rapidly that local melting occurs before there is any significant loss of heat by conduction. At the other extreme, a very low energy density source, such as an oxyfuel flame, can transfer a large quantity of energy to the workpiece with no melting at all if the heat is conducted away as rapidly as it is delivered. The effectiveness of a welding heat source depends fundamentally on the energy density of the source.

Fundamental to the study of heat flow in welding, primarily from the standpoint of how the heat source affects the material being welded, is the concept of *energy input*. In the case of arc welding, it is *arc energy input*. Arc energy input is the quantity of energy introduced per unit length of weld from a traveling heat source such as an arc, expressed in joules per meter or millimeter (joules per inch). The energy input is computed as the ratio of the total input power of the heat source in watts to its travel velocity.

$$H = P/v \tag{2.1}$$

where
H = energy input, joules per mm
P = total input power of heat source, watts
v = travel velocity of heat source, mm/sec.

2. Heat flow in welding is described in Chapter 3 of this volume.

If the heat source is an arc, to a first approximation:

$$H = EI/v \tag{2.2}$$

where
E = volts
I = amperes

If the objective is to make a precise determination of the heat effects of arcs on the materials being welded, the net energy input, H_{net} should be used:

$$H_{net} = f_1 H = \frac{f_1 P}{v} = \frac{f_1 EI}{v} \tag{2.3}$$

where f_1 is the heat transfer efficiency (heat actually transferred to the workpiece divided by the total heat generated by the heat source). With most consumable electrode arcs, the distinction between H and H_{net} is not important because the heat transfer efficiency, f_1, is generally greater than 0.8 and often close to 1.0.

The primary function of most welding heat sources is to melt metal. The quantity of metal which must be melted to produce a certain length of weld is dictated by (1) the size and configuration of the joint, (2) the number of weld passes used, and (3) the welding process. Almost without exception, it is preferable, for metallurgical reasons, to accomplish the necessary melting with minimum energy input. This objective is easier to accomplish using higher energy density heat sources.

Melting efficiency is the fraction of the net energy input, H_{net}, that is used for melting metal. The bead-on-plate weld cross section shown schematically in Figure 2.1 identifies three characteristic areas:

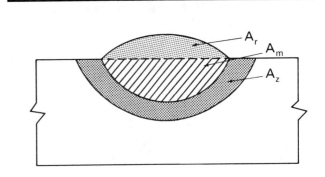

Figure 2.1—Bead on Plate Weld Cross Section

A_m = area of base metal that was melted.

A_r = area of the weld metal sometimes referred to as reinforcement. This area represents the volume of filler metal added (in the molten condition).

A_z = cross-sectional area of the heat affected zone.

The cross section of weld metal, A_w is the following:

$$A_w = A_m + A_r \qquad (2.4)$$

If no filler metal is added, then

$$A_w = A_m$$

There is a specific theoretical quantity of heat, Q, required to melt a given volume of metal (from a cold start). The quantity, Q, is a property of the metal or alloy, and is obtained by adding (1) the heat required to elevate the temperature of the solid metal to its melting point, to (2) the heat required to convert solid to liquid at the melting point (the heat of fusion).

A reasonable approximation of Q in J/mm³ is the following:

$$Q = \frac{(T_m + 273)^2}{300\,000} \qquad (2.5)$$

where T_m equals the melting temperature, °C of the metal. The melting efficiency, f_2, characterizing a weld pass, can be determined by measuring the weld metal cross section and the net energy input. Specifically, f_2 is the theoretical minimum arc heat required for melting, divided by the net energy input:

$$f_2 = \frac{QA_w}{H_{net}} = \frac{QA_w v}{f_1 P} = \frac{QA_w v}{f_1 EI} \qquad (2.6)$$

The melting efficiency depends on the welding process and the material being welded as well as other factors such as joint configuration and plate thickness. The higher the thermal conductivity of the metal being welded, the lower the melting efficiency because of the more rapid conduction of heat from the weld region.

The effect of thermal conductivity is more pronounced with a low energy density heat source. For example, when oxyfuel welding aluminum, only about two percent of the heat delivered to the workpiece is used for melting metal; the remaining heat is lost by conduction. For this reason, oxyfuel welding is seldom used to weld aluminum or other metals with high heat conductivity.

Very high energy density heat sources, such as electron and laser beams, accomplish melting with virtually 100 percent efficiency; heat is delivered so locally and so rapidly that melting takes place before any significant thermal conduction occurs. An energy density of 10 kW/mm² (76.5 MW/in.²) is close to the maximum which can be used in welding because, at higher intensities, boiling as well as melting occurs, and a cutting or erosive action results. In fact, electron and laser beams are used for drilling and cutting.

The submerged arc is a remarkably efficient welding heat source. Its energy density and its melting efficiency are higher than with any of the open arc processes.

By rearranging equation 2.6 and substituting for H_{net} from equation 2.3, a simple but important relationship can be found between the weld metal cross section, A_w and the energy input:

$$A_w = \frac{f_2 H_{net}}{Q} = \frac{f_1 f_2 H}{Q} \qquad (2.7)$$

For any particular welding process, the efficiencies of heat transfer, f_1, and melting, f_2, do not vary greatly with changes in specific welding variables, such as arc voltage, current, or travel speed. This means the cross section of a single weld pass is roughly proportional to the energy input. For example, when an arc weld pass is made on steel under the following conditions:

E = 20 V
I = 200 A
v = 5 mm/s
f_1 = 0.9
f_2 = 0.3
Q = 10 J/mm³

then, the cross-sectional area of the weld pass can be estimated based on equation 2.7:

$$A_w = \frac{0.9(0.3)20(200)}{5(10)} = 21.6 \text{ mm}^2 \ (0.03 \text{ in.}^2)$$

ENERGY SOURCES FOR WELDING

ALL WELDING PROCESSES require some form of energy. For many processes, the energy source provides the heat necessary for melting and joining. However, there are welding processes that do not utilize heat, but nevertheless require some form of energy to produce a bond. The welding processes discussed in this section are grouped under these five categories of energy sources: electrical sources, chemical sources, focused sources, mechanical sources, and solid state sources. While this section covers the most significant processes or process groups in each category, some important processes have been omitted. Flash welding, for instance, is not covered here, but it is an important process which uses electrical resistance as its source of energy. A general discussion of the process can be found in Chapter 1 of this volume.

Brazing and soldering, which are also omitted from the following discussion, cannot be categorized by their energy source (or heat source) because they can be performed using a variety of different energy sources. Their distinguishing feature is the temperature at which they are performed rather than their source of energy. For a general discussion of these processes, refer to the brazing and soldering sections of Chapters 1 and 4 of this volume.

ELECTRICAL SOURCES

Arc Welding

A LARGE NUMBER of welding processes use an electric arc as the source of heat for fusion because the heat of the arc may be effectively concentrated and controlled.[3] An electric arc consists of a relatively high current discharge sustained through a thermally ionized gaseous column called a plasma.

The power of an arc may be expressed in electrical units as the product of the current passing through the arc and the voltage drop across the arc. Given a typical value of 300 A and 25 V for current and voltage, the power dissipation is 7500 watts.[4] This power dissipation is equivalent to a heat generation of about 7500 J/s (7.11 Btu/s). However, not all of the heat generated in the arc can be effectively utilized in arc welding processes. The efficiency of heat utilization may vary from 20 to 85 percent. Convection, conduction, radiation, and spatter are responsible for heat losses. Efficiency is generally low for GTAW, intermediate for SMAW, and high for SAW.

Arc energy input has been used in weldability studies for travel speeds within a narrow range for covered electrodes. However, with higher travel speeds, the efficiency of heat transfer in the fusion zone is increased, and for a constant arc energy input, the volume of the fused metal increases as the travel speed is increased. As a result, with automatic welding processes where higher speeds and higher currents are utilized, the use of the arc energy input parameter may not be adequate in making comparisons in weldability studies.

In plasma arc heating sources, the arc is forced through a nozzle to constrict its diameter. Because a higher voltage is required to drive the arc through the nozzle and because the constriction results in a smaller diameter arc column, the temperature of the arc and the energy intensity (density) are significantly greater than for other nonconsumable electrode welding processes. The arc exits from the nozzle in the form of a high velocity, intensely hot, columnar, plasma jet. This plasma jet is referred to as a *plasma arc* and can be used as either a transferred arc or a nontransferred arc.[5]

Plasma arc cutting of 25 mm (1 in.) thick aluminum plate at a speed of 20 mm/s (50 in./min) provides an example of the use of a transferred arc. A typical operation might require an arc using 170 V and 400 A with a gas flow of 70 liters/min (150 ft³/h) through a 3 mm (1/8 in.) diameter nozzle. The 68 kW of power would produce a power density in the nozzle of approximately 8.5 kW/mm² (5.5 MW/in.²) and an average gas temperature of 9700 to 14 700°C (17 500 to 26 500°F). The resulting gas velocity approaches sonic velocity at these high temperatures. For plasma arc welding operations, a lower gas velocity is used to avoid spattering of the molten weld pool.

Resistance Welding

THE RESISTANCE WELDING processes employ a combination of force and heat to produce a weld between the workpieces.[6] Resistance heating occurs as electrical (welding) current flows through the workpieces. The workpieces are generally in the secondary circuit of a transformer which converts high voltage, low current commercial power into suitable high current, low voltage welding power.

The heat generated by current flow may be expressed by:

$$H = I^2 R t \qquad (2.8)$$

where

H = heat generated, joules
I = current, amperes
R = resistance, ohms
t = time of current flow, seconds

The welding current and time can be easily measured, but the resistance is a complex factor and difficult to

---

3. Refer to Chapter 1 of this volume for a general discussion of arc welding processes.

4. For dc, $W = EI$ but for ac, $W = EI \cos \theta$, where θ is the phase angle between sinusoidal current and voltage. In arc and resistance welding, $\cos \theta = 1$ for practical purposes since reactive losses may be assumed to be negligible. Cos θ is also called the "power factor."

5. For an explanation of transferred and nontransferred arcs, refer to the discussion of plasma arc welding in Chapter 1 of this volume.
6. Refer to Chapter 1 of this volume, for a general discussion of resistance welding.

measure. The resistance that is important in resistance spot welding is composed of several parts:

(1) The contact resistance between the electrodes and the work
(2) The contact resistance between the workpieces
(3) The body resistance of the workpieces
(4) The resistance of the electrodes

Contact resistance is greatly affected by surface conditions, such as cleanliness and freedom from oxides or other chemical compounds, and by the smoothness of the surface. In addition, contact resistance is directly related to the resistivities of the materials in contact and inversely related to the pressure on the contact area. With parts of uniform surface conditions, welding pressure becomes the major factor in the determination of contact resistance. Nonuniform surface oxides, such as mill scale on steel, make the uniform control of energy in resistance welding difficult. It is therefore preferable to remove these oxides chemically or mechanically prior to welding.

Base metal resistance is proportional to the resistivity of the material and the length of the current path, and is inversely proportional to the area of the current path. For materials of high resistivity and heavier gages, the base metal resistance becomes more important, and the contact resistance becomes less important. For high conductivity materials, contact resistance is the most important. Differences in resistivity are reflected in the rather widely differing currents that are required to make the same size weld in various materials.

In general, the magnitudes of the resistances involved are about 100 microhms. As a result, the currents are large, running into the thousands and tens of thousands of amperes. In the case of capacitor-discharge power sources, the current may be as high as several hundred thousand amperes, but the weld time is low.

The contact resistance between the steel workpieces largely disappears during the first half cycle of alternating current flow. However, heat generated at the faying surfaces during the first cycle raises the temperature of the base metal and causes a significant increase in base metal resistance which rises with temperature. The increase in base metal resistance makes the welding current more effective in producing a weld. The importance of contact resistance in materials of medium resistivity, such as steel, is borne out by the fact that higher welding forces, which reduce contact resistance, require higher welding currents.

The quantity of energy that is required to produce a given resistance weld is determined by several factors. The desired weld area (heated volume), peak temperature, specific heat of the workpieces, and heat losses into the surrounding metal and electrodes are key factors. An increase in magnitude of one or more of these factors requires a corresponding increase in energy to produce the weld.

The particular resistance welding process and weld schedule selected also significantly affect the above energy factors. For example, heat losses become significantly greater as the duration of current flow increases. Hence, long weld times require a corresponding increase in input energy to the weld to compensate for the heat losses.

It is possible to estimate the heat generated in the spot welding of two sheets of 1.0 mm (0.04 in.) thick steel that required a current of 10 000 A for 0.1 second by using equation 2.8. An effective resistance of 100 microohm will be assumed:

$$H = (10\ 000)^2 (0.0001) (0.1) = 1000\ \text{J}$$

Approximately 1381 joules are required to heat and melt one g of steel (600 Btu/lb). Assuming that the fusion zone for the above weld is a cylinder of 5 mm (0.2 in.) diameter and 1.5 mm (0.06 in.) high, the fused metal would have a volume of approximately 31 mm³ $(1.89 \times 10^{-3}\ \text{in.}^3)$, and a mass of 0.246 g (0.0005 lb). The heat required to melt this mass is only 340 joules. The remaining heat $(1000 - 340 = 660\text{J})$ would be absorbed by surrounding metal.

By comparison, consider the use of a capacitor-discharge power supply in making a projection weld between two 1 mm thick (0.04 in.) sheets of steel. A weld current pulse of 30 000 A and weld time of 0.005 seconds would be typical. In this case,

$$H = (30\ 000)^2 (0.0001) (0.005) = 450\ \text{J}$$

The lesser quantity of heat, in this instance, is a result of lower heat losses and the localizing of heat at the weld interface.

Electroslag Welding

IN ELECTROSLAG WELDING (ESW), an electrode such as a wire is fed through an electrically conductive bath of molten slag.[7] The resistance of the slag bath to the flow of current produces heat, the bulk of which is concentrated primarily in the slag area immediately surrounding the electrode tip. The quantity of heat, H, produced in the slag pool, can be expressed as

$$H = EIt \qquad (2.9)$$

where

t is time in seconds.

---

7. Refer to Chapter 1 of this volume, for a general discussion of electroslag welding.

This heat melts the wire and base metal to form the weld. As the electrode wire or consumable nozzle and wire melt, weld metal is deposited through the molten slag which dissolves some impurities and protects the weld metal from the atmosphere. The weld metal then solidifies upward as heat is extracted by the surrounding base metal and the containing shoes or dams.

The manner in which the heat is transferred is complex. In a practical sense, however, the properties of the weldment can be determined, to a large extent, by controlling the pool shape. The pool shape is influenced by the nature and depth of the slag, the electrical variables, the electrode and base metal dimensions, and the geometry of the weldment.

While the slag is used primarily for converting electrical energy to thermal energy, its properties are important. The slag should be conductive and stable at reasonable operating temperatures; it should possess low volatility and provide the proper chemical reactivity required for producing suitable weld metal. The slag conductivity should vary as little as possible over the normal operating range.

For ESW to be stable, an essentially constant slag temperature should be maintained to provide a proper balance between heat generated and heat lost. The slag should have the proper viscosity so as to perform adequate refining without being so fluid that it leaks through shoe or backing bar openings too readily.

Generally, the best weld properties are obtained if the weld pool is shallow with a large radius of curvature. A shallow pool promotes vertical freezing with an acute angle between weld metal grains. Shallow pools are produced by high voltages and low current values. Deep, and thus less desirable, weld pools are produced by high currents, high wire feed levels, and low voltages. The slag bath should also be relatively shallow since deep slag baths result in incomplete fusion.

ESW can be performed using either alternating or direct current (ac or dc). The choice of current type is more significant for slag-metal reactions than for operating characteristics. Most ESW done in the United States uses direct current electrode positive (dcep). Constant potential power sources and wire feeders are commonly used. To produce the same surface appearance with ac welding, the welding voltage must be set slightly higher than with dc welding. This results in higher welding heat according to equation 2.9.

CHEMICAL SOURCES

CHEMICAL ENERGY STORED in a wide variety of forms can be converted to useful heat. The temperature and rate of reaction are two major characteristics which determine the application of the various energy sources.

Oxyfuel Gas Welding

THE WELDING OF steel with an oxyfuel gas requires two flame characteristics: a high flame temperature capable of melting and controlling the weld metal and a neutral or reducing atmosphere surrounding the molten metal to minimize contamination before solidification.[8] Most commonly used fuel gases achieve maximum flame temperatures over 2760°C (5000°F) when mixed with oxygen and burned in an open flame. At maximum temperature, the flames are oxidizing in nature and are therefore not suitable for welding due to the formation of oxides in the weld metal. Adjusting the flames to neutral normally lowers the flame temperatures, as shown in Table 2.1.

Table 2.1
Flame Temperatures of Oxyfuel Gases

	Max Temp.		Neutral Flame Temp.	
	°C	°F	°C	°F
Acetylene	3100	5615	3100	5600
MAPP†	2900	5255	2600	5100
Propylene	2860	5174	2500	4900
Hydrogen	2870	5260	2390	4800
Propane	2780	5030	2450	4500
Natural gas/methane	2740	4967	2350	4260

†Methylacetylene-propadiene (stabilized)

With the exception of the oxyacetylene flame, the flame temperatures are reduced to the point where it is difficult to control the weld pool for other than thin sheet metal. Methylacetylene-propadiene gas (stabilized) can be used for welding, providing special procedures are followed. The fuel-to-oxygen ratio must be adjusted to be slightly oxidizing so that the flame temperature is moderately increased, and a highly deoxidized filler metal must be used to produce sound weld metal.

The combustion of acetylene in oxygen at the orifice of a torch takes place in two steps. The first step is the burning of carbon to carbon monoxide, the hydrogen remaining unconsumed. This burning takes place in a small bluish white cone close to the torch in which the gases are mixed. This reaction provides the heat which is most effective for welding. In the second step, the conversion of carbon monoxide to carbon dioxide, and of the hydrogen to water vapor takes place in a large blue flame that surrounds the welding operation, but contributes only a preheating effect. The equations representing these two steps of combustion are as follows:

---

8. Refer to Chapter 1 of this volume for a general discussion of oxyfuel gas welding.

$$C_2H_2 + O_2 \rightarrow 2CO + H_2$$
$$2CO + H_2 + 1.5\,O_2 \rightarrow 2CO_2 + H_2O(g)$$

The first reaction generates heat by the breaking up of acetylene as well as by the formation of carbon monoxide. The dissociation of acetylene liberates 227 kJ/mol at 15°C (50°F). The combustion of carbon to form carbon monoxide liberates 221 kJ/mol. The total heat supplied by the first reaction is therefore 448 kJ/mol (501 Btu/ft³) of acetylene.

The second reaction liberates 242 kJ/mol of water vapor by the burning of hydrogen. The combustion of carbon monoxide provides 285 kJ/mol or an additional 570 kJ/mol for the reaction. The total heat supplied by the second reaction is therefore 812 kJ/mol (907 Btu/ft³).

The total heat supplied by the two reactions is 1260 kJ/mol (1408 Btu/ft³) of acetylene. The concentrated heat liberated by the first reaction in the small inner cone of the flame is 35.6 percent of the total heat. The remaining heat is developed in the large brush-like outer envelope of the flame, and is effective for preheating. It is this heat that reduces the thermal gradient and cooling rate for oxyacetylene welding (OAW).

One volume of oxygen is used in the first step of the combustion to burn one volume of acetylene. This oxygen must be supplied in the pure state through the torch. The one and one-half volumes of oxygen required in the second step are supplied from the atmosphere. When just enough oxygen is supplied to burn the carbon to carbon monoxide, as indicated in the first step, the resulting flame is said to be neutral. If less than enough oxygen is supplied to complete the combustion of the carbon, the flame is said to be a reducing or an excess acetylene flame. With more than enough oxygen for the first reaction, the flame is oxidizing.

The combustion of hydrogen is a simple reaction with oxygen to form water vapor. If sufficient pure oxygen is supplied to burn all of the hydrogen, the flame temperature would be approximately 2870°C (5200°F). However, this would not provide a protective outer envelope of reducing atmosphere to shield the weld pool. If only enough pure oxygen to burn half of the hydrogen were provided, the concentrated heat would be 121 kJ/mol (135 Btu/ft³). In the latter case, the flame temperature would drop to a little over 2480°C (4500°F).

The temperature of an oxyhydrogen flame exceeds the melting temperature of steel by approximately 930°C (1614°F), and the temperature of the oxyacetylene flame exceeds it by about 1700°C (3060°F). Consequently, oxyhydrogen welding of steel (OHW) is much slower than oxyacetylene. It may be noted that by adjusting the amount of oxygen supplied, the relative proportions of concentrated heat and outer flame preheat may be varied. A similar control of the acetylene flame is not suitable because reducing the relative amount of the outer envelope of heat results in an oxidizing flame.

Oxyacetylene torches of medium size are provided with a variety of orifices since acetylene flows from 0.9 to 142 liters/min (2 to 300 ft³/h). At a flow rate of 28 liters/min (60 ft³/h), the total heat available would be 25 kJ/s (1408 Btu/min). However, remembering that approximately one-third of the heat is liberated in the small inner core of the flame, only 8.8 kJ/s (500 Btu/min) would be available at the torch tip. The efficiency of heat utilization is very low in most cases. Typically, heat transfer rates for oxyacetylene torches are on the order of 1.6 to 16 J/mm² · s (1 to 10 Btu/in.² · s)

Thermit Welding

THERMIT WELDING (TW) is a process that uses heat from exothermic reactions to produce coalescence between metals.[9] The process name is derived from "thermite," the generic name given to reactions between metal oxides and reducing agents. Thermit mixtures consist of oxides with low heats of formation and metallic reducing agents which when oxidized have high heats of formation. The excess heats of formation of the reaction products provide the energy source required to form the weld.

If finely divided aluminum and metal oxides are blended and ignited by means of an external heat source, the aluminothermic reaction will proceed according to the following general equation:

Metal oxide + aluminum → aluminum oxide + metal + heat

The chemical reaction of the most common Thermit welds and the respective heat of reactions are as follows:

$$3/4 Fe_3O_4 + 2Al \rightarrow 9/4 Fe + Al_2O_3 \quad (\Delta H = 838\ kJ) \quad (1)$$
$$3FeO + 2Al \rightarrow 3Fe + Al_2O_3 \quad (\Delta H = 880\ kJ) \quad (2)$$
$$Fe_2O_3 + 2Al \rightarrow 2Fe + Al_2O_3 \quad (\Delta H = 860\ kJ) \quad (3)$$
$$3CuO + 2Al \rightarrow 3Cu + Al_2O_3 \quad (\Delta H = 1210\ kJ) \quad (4)$$
$$3Cu_2O + 2Al \rightarrow 6Cu + Al_2O_3 \quad (\Delta H = 1060\ kJ) \quad (5)$$

The first reaction represents the most common method for Thermit welding steel and cast iron parts. Reactions (4) and (5) are commonly used for joining copper, brasses, bronzes, and copper alloys to steel.

In Thermit welding, the parts to be welded are aligned with a gap between them, and a mold, either built on the parts or formed on a pattern of the parts, is placed in position. The next step varies according to the size of the parts. If they are large, preheating within the mold cavity is necessary to bring the parts to welding temperature

---

9. Thermit is the term commonly used to identify this welding process even though it is a registered trademark. Thermit welding is discussed in more detail in the *Welding Handbook*, Vol. 3, 3rd Ed.

and to dry out the mold. If the parts are small, however, preheating is often eliminated. The superheated products of a thermit reaction are next allowed to flow into the gap between the parts with sufficient heat to melt both faces of the base metal. When the filler metal has cooled, all unwanted excess metal may be removed by oxygen cutting, machining, or grinding.

The maximum temperature to be obtained from such a reaction can only be estimated. The theoretical maximum temperature of the reaction is aproximately 3200°C (5800°F). Practically, the temperature varies between 2200°C (4000°F) and 2400°C (4350°F).

Thermit reactions are nonexplosive. The ignition temperature for thermit granules suitable for welding is in the area of 1200°C (2200°F). Therefore, a Thermit mixture does not constitute a fire hazard, nor does it require special storage conditions, as long as it is not in the proximity of an open heat source (flame or fire). To start the reaction, a special starting powder or rod incorporates peroxides, chlorates, or chromates as the oxidizing agent is required.

FOCUSED SOURCES

Two sources of welding energy, laser and electron beam, are focused beams that operate according to the laws of optics. The laser beam is focused by various lens arrangements. Electrostatic and electromagnetic lenses focus the electron beam. Consequently, high power densities can be achieved with these methods.

Lasers

A high degree of spectral purity and low divergence of the laser radiation permit focusing a laser beam on extremely precise areas, resulting in power densities often greater than 10 kW/mm² (6.45 MW/in.²). The unfocused beam exiting from a laser source may be typically 1 to 10 mm (0.04 to 0.4 in.) in diameter and must be focused to be useful for laser beam welding (LBW) applications.[10] The focused spot size, d, of a laser beam is given by

$$d = f\theta \qquad (2.10)$$

where f is the focal length of the lens and θ is the full angle beam divergence.

The power density, PD, at the focal plane of the lens is given by

$$PD = \frac{4 P_1}{\pi d^2} = \frac{4 P_1}{\pi (f\theta)^2} \qquad (2.11)$$

where P_1 is the input power.

10. Refer to the *Welding Handbook*, Vol. 3, 7th Ed., 1980.

Therefore, power density is determined by the laser power, P, and beam divergence, θ. For a laser beam operating in the fundamental mode where the energy distribution across the beam is gaussian, the beam divergence is

$$\theta \cong \frac{\lambda}{\alpha} \qquad (2.12)$$

where α is a characteristic dimension of the laser, and λ is the wavelength of laser radiation. Combining equations 2.11 and 2.12 shows that the power density is inversely proportional to the square of the wavelength of the laser radiation.

$$PD \cong \frac{4 P_1 \alpha^2}{f^2 \lambda^2} \qquad (2.13)$$

Two types of lasers are used for welding: solid state lasers and gas lasers. Solid state lasers are single crystals or glass doped with small concentrations of transition elements (chromium in ruby) or rare earths (neodymium in YAG or glass).[11] Industrial gas lasers are carbon dioxide lasers. Ruby and Nd-glass lasers are capable of high energy pulses but are limited in maximum pulse rate. Nd-YAG and CO_2 lasers can be operated continuously or pulsed at very high rates.

Most metallic surfaces reflect appreciable amounts of incidental laser radiation. In practice, however, sufficient energy is usually absorbed to initiate and sustain a continuous molten puddle. Ruby and Nd-glass lasers, because of their high energy outputs per pulse, can overcome most metal welding reflectivity problems. However, due to their inherently low pulse rates, typically 1 to 50 pulses per second, welding speeds in thin gage metals are extremely slow. In contrast, Nd-YAG and CO_2 lasers are capable of high continuous wave outputs, or they can be pulsed at several thousand pulses per second, giving rise to high speed continuous welding.

Pulsed Laser Beam Welding

During pulsed laser beam welding, a high energy focused light beam strikes the plate surface. Part of the beam energy is reflected, and the balance is absorbed at the surface of the base metal. The weld penetration is dependent upon the total energy absorbed and the thermal properties of the base metal. A pulsed laser beam makes a continuous seam weld by overlapping spot welds from each pulse. Each pulse results in a complete melt-and-freeze cycle.

The competing base metal thermal properties that control penetration are the boiling temperature and the

11. Lasers in which single crystals of Yttrium-Aluminum-Garnet or glass are doped with neodymium ions are abbreviated as Nd-YAG and Nd-glass, respectively.

thermal diffusivity. Thus, if the thermal diffusivity is poor, the energy is absorbed by the small volume of metal on the surface, the temperature increases rapidly, and some metal may vaporize. If the thermal diffusivity is high, the heat energy is conducted away and a larger volume of metal reaches the melting temperature prior to surface vaporization.

For a single pulse of a laser, the time constant τ is the pulse duration required to penetrate through a sheet of thickness x.

$$\tau = \frac{x^2}{4k} \qquad (2.14)$$

where k is the thermal diffusivity.

The time constants for several thicknesses and metals are shown in Table 2.2.

Continuous Wave Laser Beam Welding

ND-YAG AND CO_2 lasers are capable of making high speed continuous metal welds. Lasers in excess of 500 W output are capable of welding steel sheet 0.25 mm (0.010 in.) thick at several millimeters per second, while CO_2 lasers of 10 kW continuous wave output power can produce deep penetration welds in 13 mm (1/2 in.) thick steel at 25 mm/s (60 in./min).

Electron Beam Welding

IN ELECTRON BEAM welding (EBW), energy is developed in the workpiece by bombarding it with a focused beam of high velocity electrons.[12] The power density PD in watts per unit area is given by

$$PD = \frac{ne\text{E}}{A} = \frac{\text{E}I}{A} \qquad (2.15)$$

where
$n =$ total number of electrons per second in the beam
$e =$ The charge on an individual electron (1.6 × 10^{-19} coulombs)
$\text{E} =$ the accelerating voltage on the electrons, V, in volts
$I =$ the beam current, A
$A =$ the area of the focused beam at the workpiece surface

The depth of penetration of a focused beam is determined by the beam current, accelerating voltage, and speed of welding.

Power concentrations of 1 to 100 kW/mm² (0.65 to 65 MW/in.²) are routinely achieved, and 10 MW/mm² (65 000 MW/in.²) can be obtained. The energy concentration of the beam is dependent on the accelerating voltage. Electron beam welding is generally performed at voltages between 20 kV and 150 kV, with the higher voltages having higher power densities.

The advantages associated with the EBW process include a high depth-to-width ratio, high strength, and the ability to weld thick sections in a single pass with relatively low heat input, low distortion, and narrow heat-affected zones. These advantages are the result of a large power density.

The high power densities, achievable with electron beams, cause practically instantaneous volatilization of metal. This produces a needlelike vapor-filled cavity or keyhole in the workpiece through which the beam can penetrate. This cavity is held open by the vapor pressure of the workpiece material. The molten metal flows from the front to the rear of the keyhole where it solidifies to form the weld.

There are three commercial variants of the EBW process, distinguished by the degree of vacuum used.

(1) High vacuum EBW, the pioneering process that operates at a pressure of 13 MPa (10^{-4} torr) or lower
(2) Medium vacuum or soft vacuum EBW, that operates at pressures of 13.0 Pa (10^{-1} torr)
(3) Nonvacuum EBW which operates at 100 kPa (1 atm)

Each of these has different performance capabilities and has excelled in different application areas. The penetration capability is inversely related to the pressure.

---

12. Refer to *Welding Handbook*, Vol. 3, 7th Ed. for a complete discussion of electron beam welding.

Table 2.2
Thermal Time Constants—Laser Beam Welding

Material	Time, in seconds		
	Thickness 0.13 mm (0.005 in.)	Thickness 0.64 mm (0.025 in.)	Thickness 2.5 mm (0.100 in.)
Copper	0.035	0.884	14.1
Brass	0.119	2.970	47.5
Aluminum	0.047	1.170	18.8
Carbon steel	0.333	8.330	133.3
Stainless steel	1.004	25.100	401.7
Nickel	0.260	6.500	104.1
Inconel	0.948	23.700	379.3
Titanium	0.593	14.800	237.3
Tungsten	0.060	1.509	34.1

MECHANICAL SOURCES

THE THREE WELDING processes discussed in this category—friction, ultrasonic, and explosion welding—are solid state processes because there is no melting of the base metal. These processes all involve some type of mechanical movement which produces the energy for welding. This feature distinguishes them from other solid state processes, such as diffusion welding, that are characterized by atomic motion.

Friction Welding

IN FRICTION WELDING (FRW), a bond is created between a stationary and a rotating member by generating frictional heat between them while they are subjected to high, normal forces on the faying surfaces.[13] Application of this process requires that the rotating member be essentially symmetrical about the axis of rotation while the other member can be of any suitable geometry that is within the clamping limitations of the welding machine.

Practical application of this welding heat source utilizes one of three approaches: (1) relatively slow rotational speeds and high normal force, (2) high speed and relatively low normal force, or (3) a flywheel that is disengaged from the rotating drive before the start of welding. In the latter case, the rotational speed is continuously decreasing during the welding cycle. The first two approaches are referred to as conventional FRW while the latter, because the energy is provided by a flywheel, is called inertia welding. Both make use of the frictional heating of the faying surfaces to accomplish welding.

Since frictional heat is related to speed and normal force, the time required to produce a bond is a function of both of these variables. Also, the radial temperature distribution is nonuniform, being highest near the outer surface where the surface speed is highest.

The average interface temperature is always below the melting point of either member being joined. Thus, the bond is metallurgically achieved by diffusion rather than by fusion. Because of this, the process is admirably suited for joining dissimilar metals, particularly those that form undesirable phases when joined by melting processes. The width of the diffusion zone may vary from a line that cannot be accurately defined in width by present techniques to one readily detected by low power magnification.

Ultrasonic Welding

ULTRASONIC WELDS ARE produced by the introduction of high frequency vibratory energy into the weld zone of

13. Refer to Chapter 1 of this volume, for a general discussion of FRW.

metals to be joined. Ultrasonic welding (USW) is generally used to produce spot, straight, and circular seam (ring) welds between members of which at least one is of sheet or foil thickness. The workpieces are clamped together between two tips or jaws, and the vibratory energy is transmitted through one or both sonotrode tips which oscillate in a plane essentially parallel to the weld interface. This oscillating shear stress results in elastic hysteresis, localized slip, and plastic deformation at the contacting surface which disrupt surface films and permit metal-to-metal contact. The process thus produces a metallurgical bond between similar or dissimilar metals without melting. The elastic and plastic deformations induce a very localized, transient temperature rise at the weld interface. Under proper conditions of clamping force and vibratory power, the temperatures reached are usually in the range of 35 to 50 percent of the absolute melting point of the metals joined.

By means of a frequency converter, 60 Hz electrical power is transformed into high frequency power generally within the range of 15 000 to 75 000 Hz, although higher or lower frequencies may be used. The high frequency electrical power is converted into acoustical power at the same frequency by one or more transducers of either the magnetostrictive or piezoelectric ceramic types. Appropriate acoustical coupling members transmit the acoustical power from the transducer to the sonotrode tip and into the metal workpieces.

With recently developed solid state frequency converters, more than 90 percent of the line power is delivered electrically as high frequency power to the transducer. The electromechanical conversion efficiency of the transducer, that is, its efficiency in converting electrical power into acoustical power, is in the range of 25 to 35 percent for magnetostrictive nickel-stack transducers and is frequently in excess of 75 percent for piezoelectric transducers such as lead-zirconate-titanate ceramics. Thus, in the case of ceramic transducers, as much as 65 to 70 percent of the input electrical line power may be delivered to the weld metal as acoustical power.

The amount of acoustical energy required to weld a given material, increases with material hardness and thickness. This relationship for ultrasonic spot welding can be expressed, as a first approximation, by the equation:

$$E_a = K_1 (Ht)^{3/2} \qquad (2.16)$$

where
$E_a =$ acoustical energy, J
$H =$ Vickers microindentation hardness number
$t =$ thickness of the material adjacent to the ultrasonically active tip, in millimeters or inches
$K_1 =$ 8000 for t in millimeters and 63 for t in inches

This empirical equation is reasonably valid for common metals such as aluminum, steel, nickel, and copper in thicknesses up to at least 0.8 mm (0.03 in.). Some of the more exotic materials are less responsive to this relationship. Experimentation over a range of welding machine settings is usually recommended to establish precise settings that will produce satisfactory welds in a given material and material thickness.

Explosion Welding

IN EXPLOSION WELDING (EXW), the detonation of an explosive is utilized to accelerate a component (called the flyer) to a high velocity as it collides with a stationary component. At the moment of impact, the kinetic energy of the flyer plate is released as a compressive stress wave on the surfaces of the two components. During explosive welding, the collision progresses across the surface of the plates being welded forming an angle between the two colliding components. The surface films are liquefied, scarfed off the colliding surfaces, and jetted out of the interface leaving perfectly clean, oxide-free surfaces. Under these conditions, the normal interatomic and intermolecular forces create a weld. The result of this process is a weld without a heat affected zone.

The energy for the EXW process is provided by an explosive. The detonation velocity of the explosive must fall within limits to produce the necessary impact velocity and angle between the two components. The maximum velocity of the explosive detonation should not exceed the highest sonic velocity within the materials being joined. The physical forms of the explosives utilized include plastic flexible sheet, cord, and pressed, cast, granulated, and liquid shapes. These explosives also vary in detonation velocity. They are usually detonated with a standard commercial blasting cap.

Most explosive welding applications use a low detonation velocity explosive, which is usually placed in direct contact with the flyer plate. Low velocity explosives develop relatively lower pressures than high velocity explosives, and can be used without causing shock damage. These explosives make it easier and more practical to achieve explosive welding. They should have a detonation velocity of approximately 2400 to 3600 m/s (8000 to 12 000 ft/s). The detonation velocity depends on the thickness of the explosive layer and the packing density.

When the detonation velocity is less than the sonic velocity of the metal, the pressure generated in the metal by the expanding gases moves faster than the detonation and is spread out ahead of the detonation front. A shock wave is not produced. If the detonation velocity of the explosive is slightly greater than the sonic velocity in the metal, a detached shock wave may be created.

In this case, the detonation velocity slightly exceeds the metal sonic velocity and a shock wave is formed which moves slightly ahead of the detonation. High velocity explosives are difficult to use for high quality welding because they can cause considerable shock wave damage resulting in spalling along the edges and fissuring at the weld interface. When high velocity explosives are used, thick buffers are required between the explosive and the cladding plate.

The velocity of the flyer plate can be changed by changing the explosive charge per unit of area. If it is increased, several things can happen:

(1) The angle of incidence at which weld waviness begins increases.

(2) Larger waves are produced with the same angle of incidence.

(3) The range of angles within which waves are produced also increases.

(4) Tendency for the formation of intermetallic compounds in the weld interface increases.

Two components will not weld if the explosive charge per unit area is too low. Thus, there is a minimum flyer plate velocity.

The detonation velocity tends to be constant throughout the entire explosion. Since the energy release of most explosives depends on the thickness of the explosive and the degree of confinement, the detonation velocity may vary as these quantities are changed. The velocity may also be varied by selection of the explosive ingredients and by changing the packing density of the explosive.

OTHER SOURCES

Diffusion

THE SOURCE OF energy for diffusion welding is provided by furnaces. The furnaces may be electrical or chemical, but the diffusion welding process is unlike arc, resistance, oxyfuel or Thermit welding; therefore, it has not been included in the sections for electrical or chemical sources for welding energy.

Because diffusion plays an essential role in solid state welding, a short description of the theory and the mechanism of diffusion is in order.

The diffusion phenomenon is of prime importance in metallurgy because many phase changes take place in metal alloys involving a redistribution of the atoms. These changes occur at rates that are dependent on the speed of the migrating atoms.

Diffusion in metal systems is usually categorized into three different processes, depending on the path of the diffusing element. Each of these processes—volume diffusion, grain boundary diffusion, and surface diffusion—has different diffusivity constants. The specific rates for grain boundary and surface diffusion are higher than the rate for volume diffusion.

Fick's first law gives the basic equation for diffusion in metals as:

$$\frac{1}{A} \frac{dm}{dt} = -D \frac{\partial c}{\partial x} \qquad (2.17)$$

where

$\frac{dm}{dt}$ = rate of flow (g/s) of metal across a plane perpendicular to the direction of diffusion

D = diffusion coefficient whose values depend on the metallic system being considered (mm^2/s); minus sign expresses a negative concentration gradient

A = the area (mm^2) of the plane across which diffusion occurs

$\frac{\partial c}{\partial x}$ = the concentration gradient that exists at the plane in question (c is expressed in g/mm^3)

x = distance, mm

The diffusion coefficient, D, is not generally constant. It is a function of such dynamic variables as temperature, concentration, and crystal structure.

Several mechanisms can account for the diffusion of atoms in metals. Two of these are the interstitial mechanism and the vacancy mechanism. The former is concerned with the movement of atoms having small atomic radii compared to the matrix atoms. These elements move from one location to another along the interstices of the crystal lattice, hence the name interstitial elements. These moves occur within the crystal without distorting or permanently displacing the matrix atoms.

The matrix or substitutional atoms use the vacancy mechanism for their mode of transportation. Because of their size, it is literally impossible for these atoms to migrate along the interstices. The only path open to them is the vacancy sites. Although the energy required to move a matrix atom is equal to that for an interstitial element, the rate is considerably slower due to the fewer vacant locations available to the atoms.

The pronounced effect that temperature has on diffusion may be evaluated by the rule of thumb that an 11°C (20°F) rise in temperature will double the diffusion constant. It has been found that the diffusion constant changes with variation in concentration. For example, the diffusion constant of carbon in iron at 930°C (1700°F) will show a three fold increase over a range of carbon from 0 to 1.4 percent. Crystal structure has also been found to influence the diffusion constant at a given temperature. Self-diffusion of iron occurs 100 times more rapidly in ferrite than in austenite. Directionality of a crystal also influences the diffusion constant. It has also been shown that the rate of diffusion is greater in cold worked structures than in annealed structures.

Diffusion Welding

DIFFUSION WELDING (DFW) is a solid state welding process that produces coalescence at the faying surfaces by the application of pressure at elevated temperatures.[14] Some nonstandard terms for this process are pressure bonding, diffusion bonding, self welding, and cold welding.

The essential ingredients of the process are time, temperature, and pressure. Atomic vibration and mobility increase with increasing temperature. As the faying surfaces are brought into close contact by applied pressure, diffusion across the faying surfaces is enhanced in both directions resulting in the coalescence required for welding. Because solid state diffusion is an atomic process, the welding time is long compared to electric arc, resistance, oxyfuel, or Thermit welding. Because temperature of the assemblies is constant during DFW, and because no melting takes place, very close tolerance assemblies with low residual stresses can be fabricated.

ARC CHARACTERISTICS

DEFINITIONS

FOR ALL PRACTICAL purposes, a welding arc can be considered a gaseous conductor which changes electrical energy into heat. The arc is the heat source for many welding processes because it produces a high intensity of heat and is easy to control through electrical means.

Arcs are sources of radiation as well as heat sources. When used in welding processes, an arc may help remove surface oxides in addition to supplying heat. The arc also influences the mode of transfer of metal from the electrode to the work.

A welding arc is a particular group of electrical discharges that are formed and sustained by the development of a gaseous conduction medium. The current carriers for the gaseous medium are produced by thermal means and field emission. Many kinds of welding arcs have been conceived, each with a unique application in the field of metal joining. In some cases, the welding arc is a steady state device. More frequently, it is intermittent, subject to interruptions by electrical short circuiting, or continuously nonsteady, being influenced by an alternating directional flow of current or by a turbulent flow of the conducting gas medium.

---

14. Diffusion Welding is discussed in Chapter 1 of this volume and in the *Welding Handbook*, Vol. 3, 7th Ed. 1980.

THE PLASMA

THE ARC CURRENT is carried by a plasma, the ionized state of a gas composed of nearly equal numbers of electrons and ions. The electrons, which support most of the current conduction, flow out of a negative terminal (cathode) and move toward a positive terminal (anode). Mixed with the plasma are other states of matter, including molten metals, slags, vapors, neutral and excited gaseous atoms, and molecules. The establishment of the neutral plasma state by thermal means, that is, by collision processes, requires the attainment of equilibrium temperatures according to the ionization potential of the material from which the plasma is produced. The formation of plasma is governed by an extended concept of the Ideal Gas Law and the Law of Mass Action. A basic equation is the following:

$$\frac{n_e n_i}{n_o} = \frac{2 Z_i (2\pi m_e k T)^{3/2}}{Z_o h^3} \, exp \left\{ \frac{-V_i}{kT} \right\} \qquad (2.18)$$

where

n_e, n_i, and n_o = particle densities (number per unit volume for electrons, ions, and neutral atoms, respectively)

V_i = the ionization potential of the neutral atom

T = temperature in degrees absolute

Z_i and Z_o = partition functions for ions and neutral particles

h = Planck's constant

m_e = the electron mass

k = Boltzmann's constant

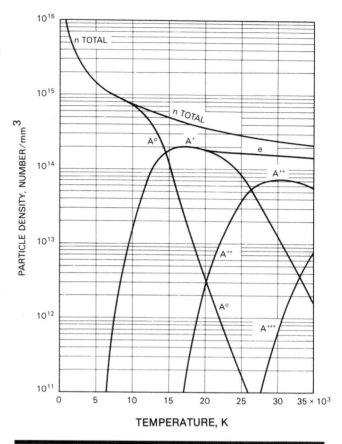

Figure 2.2—Argon Shielded Arc Plasma Composition—100 kPa (1 atm) pressure

The particle densities of three kinds of particles can be determined by assuming the plasma is electrically neutral and that the ions have a single positive charge. Then the number of electrons is equal to the number of ions.

$$n_e = n_i$$

The density and distribution of argon gas particles and electrons in an argon plasma between 0 and 35 000 K is shown in Figure 2.2.

The expression of thermal equilibrium of the heated gas in an arc means that all kinetics and reactions of the particles in a microvolume may be represented by the same temperature. Thermal equilibrium in welding arcs is closely approached, but may be considered only approximate because of the influence of dominant processes of energy transport, including radiation, heat conduction, convection, and diffusion. The heated gas of the arc attains a maximum temperature of between 5000 and 50 000 K, depending on the kind of gas and the intensity of the current carried by the plasma. The degree of ionization is between 1 and 100 percent; complete ionization is based on all particles being at a temperature corresponding to the first ionization potential.

The attainment of a very close approximation to thermal equilibrium is more questionable in the region near to the arc terminals, where current-conducting electrons are accelerated so suddenly by a high electrric field that the required number of collisions does not occur. It is in the arc terminal regions that an explanation of current conduction based wholly on thermal ionization is insufficient and must be augmented by the theory of field emission or some other concept.

TEMPERATURE

MEASURED VALUES OF welding arc temperatures normally fall between 5000 and 30 000 K, depending on the nature of the plasma and the current conducted by it. As a result of a high concentration of easily ionized materials such as sodium and potassium that are incorporated in the coatings of covered welding electrodes, the maximum temperature of a shielded metal arc is about 6000 K. In pure inert gas arcs, the axial temperature may approach 30 000 K. Some special arcs of extreme power loading may attain an axial temperature of 50 000 K. In most cases, the temperature of the arc is determined by measuring the spectral radiation emitted. An isothermal map of a 200 A arc in argon between a tungsten electrode and a watercooled copper anode is shown in Figure 2.3.

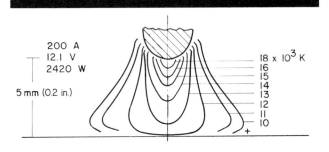

Figure 2.3—Isothermal Diagram of an Argon Shielded Tungsten Arc

The temperature attainable in arcs is limited by heat leakage rather than by a theoretical limit. The energy losses (due to heat conduction, diffusion, convection, and radiation) characteristic of an arc plasma of specific composition and mass flow are in balance with the electrical power input. The energy losses from arcs vary in a complex way according to the magnitude of the temperatures and the influences of thermal conduction, convection, and radiation. In the classical Elenbaas-Heller equation of energy balance, the radial loss due to thermal conduction in the cylindrical geometry of the arc plasma is expressed as follows:

$$\sigma E^2 = -\frac{1}{r}\frac{d}{dr}\left(rk\frac{dT}{dr}\right) \tag{2.19}$$

where

σ = the electrical conductivity
E = the electric field strength
k = the thermal conductivity coefficient
r = the column radius
T = the temperature

To include all loss mechanisms in the energy balance requires a more involved differential equation. The thermal conductivities, of several gases for one atmosphere pressure are shown in Figure 2.4.

The data shown for the gases, hydrogen and nitrogen, show peaks due to the effect of thermal dissociation and association of the molecular and atomic forms respectively.

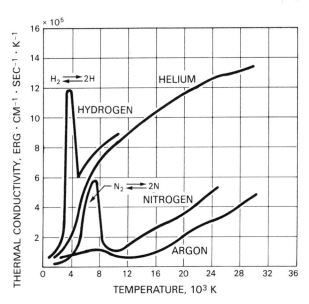

Figure 2.4—Thermal Conductivity of Some Representative Gases as a Function of Temperature

RADIATION

THE AMOUNT AND character of radiation emitted by arcs depend upon the atomic mass and chemical structure of the gas, the temperature, and the pressure. Spectral analysis of arc radiation may show banks, lines, and continua. The analysis of radiation from organic type covered electrodes shows molecular bands revealing the existence of vibrational and rotational states, as well as line and continuum emissions from excited and ionized states. The inert gas arcs radiate predominantly by atomic excitation and ionization. As the energy input to arcs increases, higher states of ionization occur, giving radiation of higher energy levels.

Radiation loss of energy may be over 20 percent of the total input in the case of argon welding arcs, while in other welding gases the radation loss is not more than about 10 percent. Intense radiation in the ultraviolet, visible, and infrared wavelengths is emitted by all exposed welding arcs. Ultraviolet radiation from argon shielded arcs is particularly strong because of mass effects and because little or no self-absorption occurs within the plasma volume. The visible spectrum and a portion of the infrared spectrum emanating from an argon shielded gas tungsten arc are shown in Figure 2.5.

ELECTRICAL FEATURES

A WELDING ARC is an impedance to the flow of current, as are all normal conductors of electricity. The specific

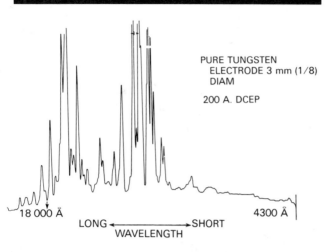

PURE TUNGSTEN
ELECTRODE 3 mm (1/8)
DIAM

200 A. DCEP

18 000 Å   4300 Å

LONG ◄──────────► SHORT
WAVELENGTH

**Figure 2.5—Spectrum of an Argon Shielded Gas
Tungsten Arc**

impedance is inversely proportional to the density of the charge carriers and their mobility, with the total impedance depending on the radial and axial distribution of the carrier density. The plasma column impedance is a function of temperature, but generally not in the regions of the arc near its terminals. The electrical power dissipated in each of the three spaces or regions of the arc is the product of the current flowing and the potential across the region. The current and potential across each region are expressed according to:

$$P = I(E_i + E_c + E_p) \qquad (2.20)$$

where

P = Power, W
E_c = anode voltage, V
E_c = cathode voltage, V
E_p = plasma voltage, V
I = Current, A

The regions are referred to as the cathode fall space, the plasma column fall space, and the anode fall space. The potential distribution across the arc is shown in Figure 2.6. However, there are intermediate regions taken up in expanding or contracting the cross section of the gaseous conductor to accommodate each main region. As a consequence, welding arcs assume bell or cone shapes, elliptical contours or some other noncylindrical configuration. Many factors may contribute to the various shapes, including the configuration of the arc terminals, gravitational and magnetic forces, and interactions between the plasma and ambient pressures. The area over which the current flows into the arc terminals (anode and cathode spots) has a strong effect on the arc configuration and on the flow of heat energy into these

terminals. The current density at the workpiece terminal is of utmost importance to the size and shape of the fusion zone, and to the depth of fusion in a welded joint.

The total potential of an arc falls with increasing current and rises again with a further increase in current. Typical curves are shown in Figure 2.7. The decrease in total arc potential with increasing current can be attributed to a growth of thermal ionization and thermally induced electron emission at the arc cathode. The total potential of arcs generally increases as the spacing between the arc terminals increases. Because the arc column is continually losing charge carriers by radial migration to the cool boundary of the arc, lengthening the arc exposes more of the arc column to the cool boundary, imposing a greater requirement on the charge carrier maintenance. To accommodate this loss of energy and maintain stability, the applied voltage must be increased.

Much of the foregoing concerned the plasma column which is best understood. Although the mechanisms effective at the arc terminals have even more importance in welding arcs, they are less understood. The arc terminal materials must, in most cases, provide the means for achieving a continuity of conduction across the plasma column.

It is essential that the cathode material provide electrons by emission of sufficient density to carry the current. In the GTAW process, the tungsten electrode is chosen because it readily emits electrons when only a portion of the electrode tip is molten. Other cathode materials that are melted and transferred through the arc must also provide sufficient density of electrons to carry the arc current. In the case of consumable electrodes,

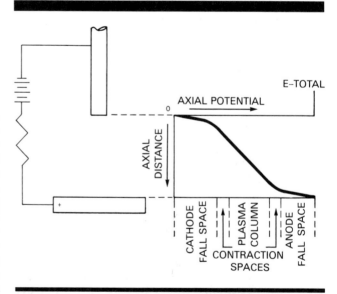

AXIAL POTENTIAL

E-TOTAL

AXIAL DISTANCE

CATHODE FALL SPACE

PLASMA COLUMN

ANODE FALL SPACE

CONTRACTION SPACES

**Figure 2.6—Arc Potential (Volts) Distribution Between
Electrode and Work**

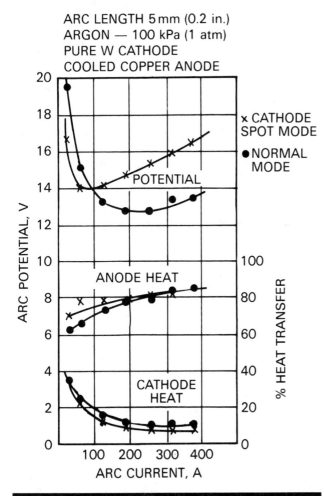

ARC LENGTH 5 mm (0.2 in.)
ARGON — 100 kPa (1 atm)
PURE W CATHODE
COOLED COPPER ANODE

Figure 2.7—Typical Volt—Ampere and Percent Heat Transfer Characteristics of an Argon Shielded Tungsten Arc

additives in the form of coatings may be used to insure stable or spatter-free transfer.

INFLUENCE OF MAGNETIC FIELDS ON ARCS

MAGNETISM HAS INTERESTING effects on welding arcs; some are detrimental, and others are beneficial. Magnetic fields, whether induced or permanent, interact with the arc current to produce force fields that cause arc deflection commonly called arc blow. Arc blow, plasma streaming, and metal transfer are some of the welding arc characteristics strongly influenced by the presence of magnetic fields. Magnetic flux may be self-induced and associated with the arc current, or it may be produced by either residual magnetism in the material being welded or an external source.

Like gravity and electricity, magnetism is a field phenomenon, defined by vectorial measures of flux density. Since a welding arc always has its own associated magnetic field, any effects of external magnetic fields come as a consequence of interaction with the self-field.

The effects of external magnetic fields on welding arcs are determined by the Lorentz force which is proportional to the vector cross product of the external field strength and the arc current. The usual effect of external magnetic fields on arcs is to cause arc deflection. In a macroscopic sense and within the limits of stable deflection, an arc behaves as a flexible conductor having an elastic stiffness that resists the overall Lorentz force. The arc deflects in a smooth curve from a fixed point at the electrode to the base metal. The magnitude of arc deflection is proportional to the applied field strength.

The direction of the Lorentz force, and consequently arc deflection, is determined by Fleming's left-hand rule for arc deflection, illustrated in Figure 2.8. Arc deflection may be understood intuitively if one thinks of the flux lines encircling a conductor, adding vectorially to the applied field lines on one side and canceling the applied field lines on the other side. The arc will be deflected towards the weak flux side.

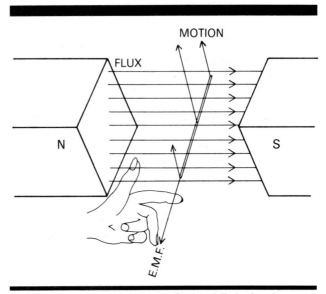

Figure 2.8—Fleming's Left-Hand Rule for Arc Deflection

Arc deflection in the direction of travel (forward deflection) results in a more uniform weld that may be wider with less penetration. Controlled shallow penetration has application for welding thin sections. Forward arc deflection results in improved bead appearance and reduced undercut at higher welding speeds. The use of forward arc deflection to improve bead appearance is illustrated in Figure 2.9.

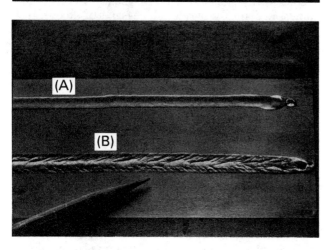

Figure 2.9—Appearance of Gas Tungsten Arc Welding Beads on Aluminum Alloy. Weld (A) Subjected to Transverse Magnetic Field to Produce Forward Arc Deflection. Weld (B) Deposited Without an External Magnetic Field

Arc deflection can also be affected by the proximity of multiple arcs. A two- or three-wire submerged arc welding operation utilizes the magnetic fields of neighboring arcs to obtain higher travel speeds without undercut.

Heavy undercutting and extensive reinforcement associated with backward arc deflections have little use in practical welding. Alternating magnetic fields, however, cause the arc to oscillate back and forth across the weld axis with a frequency equal to that of the applied field. This phenomenon is used to advantage in the GTAW hot wire process where the arc is deflected in a controlled manner by the magnetic field around the hot filler wire.

Arc Blow

UNDER CERTAIN CONDITIONS, the arc has a tendency to be forcibly directed away from the point of welding, thereby making it difficult to produce a satisfactory weld. This phenomenon, called *arc blow*, is the result of magnetic disturbances surrounding the welding arc. In general, arc blow is the result of two basic conditions:

(1) The change in direction of current flow as it enters the work and is conducted toward the work lead

(2) The asymmetric arrangement of magnetic material around the arc, a conditon that normally exists when welding is done near the end of ferromagnetic materials

Although arc blow cannot always be eliminated, it can be controlled or reduced to an acceptable level through knowledge of these two conditions.

The first condition is illustrated in Figure 2.10. The heavy dotted line traces the path of the current through the electrode, the arc, and the work. Magnetic lines of force surround the current path. The lines of force are shown here schematically as circles concentric with the current path. They are concentrated on the inside of the bend in the current path and are sparse on the outside curve. Consequently, the magnetic field is much stronger on the side of the arc toward the work connection than on the other side and, according to Fleming's left-hand rule, this force is always in a direction away from the work connection. Thus, welding away from the work connetion produces more favorable forward arc blow.

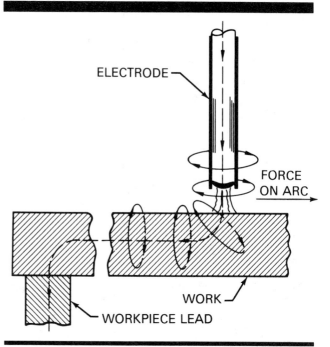

Figure 2.10—Force on the Arc by the Induced Magnetic Field Resulting From the Ground Location

The second condition is illustrated in Figure 2.11. It is much easier for magnetic flux to pass through a magnetic material than through air. Therefore, it will concentrate in the steel base metal and take the shortest air distance, which is between the beveled edges of the seam. When the arc is near one end of the seam, the magnetic flux lines become more concentrated between the arc and the ends of the plates. The resulting effect on the arc is shown in Figure 2.12.

As a rule, the force on the arc due to the magnetic material around the arc acts toward the best magnetic path. When welding, the total force tending to cause the arc to blow is nearly always a combination of the two forces illustrated in Figures 2.10 and 2.11. When welding with ac, the magnetic effect on the arc is lessened by eddy currents induced in the work, as shown in Figure 2.13.

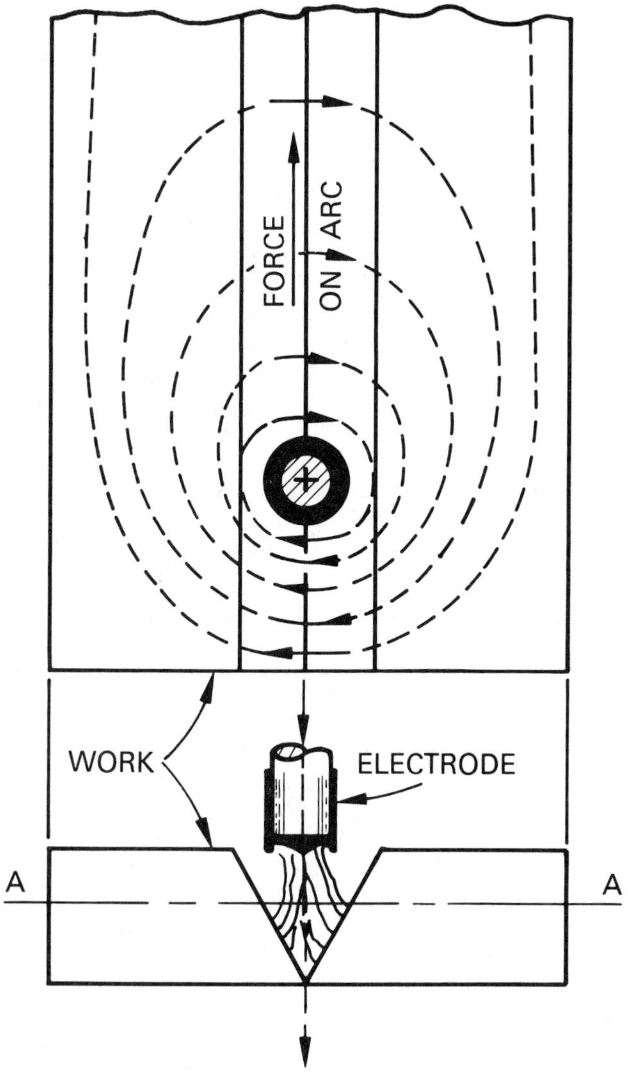

Low voltage results in a short, stiffer arc that resists arc blow better than a long higher voltage arc.

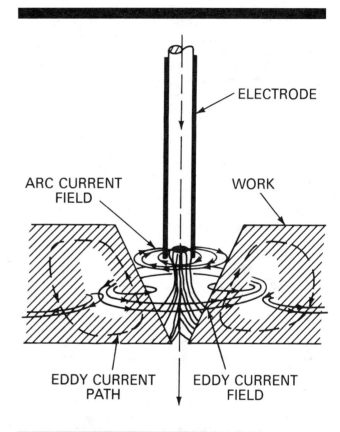

Figure 2.13—Effect of Eddy Currents in Neutralizing Magnetic Field Induced by AC Current

Figure 2.11—Distortion of Induced Magnetic Field at the Edge of a Steel Plate

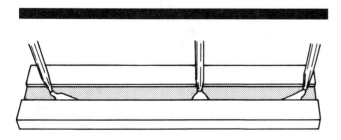

Figure 2.12—Arc Blow at Ends of a Ferromagnetic Workpiece

METAL TRANSFER

CONSUMABLE ELECTRODE ARC welding processes are used extensively because filler metal is deposited more efficiently and at higher rates than is possible with other welding processes. To be most effective, the filler metal needs to be transferred from the electrode with small losses due to spatter. Furthermore, uncontrolled short circuits between the electrode and the work should be avoided; otherwise, the welder or welding operator will have difficulty controlling the process. In the case of the GMAW process, arc instability caused by erratic transfer can generate pressure fluctuations that draw air into the vicinity of the arc.

The different types of transfer have been studied with motion pictures and by analysis of the short circuit oscillograms. Transfer through the arc stream of covered electrodes can be characterized as globular (massive drops) or as a showery spray (large number of small drops). These modes are rarely found alone. More generally, material is transferred in some combination of both. Transfer with the GMAW process varies greatly when used with argon shielding. When the current is above the transition level, the transfer mechanism can be best described as an axial spray, and short circuits are nonexistent. However, when helium or an active gas such as carbon dioxide is used for shielding, the transfer is globular, and some short circuiting is unavoidable. The GMAW short circuiting arc process has been adapted to use only short circuits for the transfer of metal to the pool.

The physics of metal transfer in arc welding is not well understood. The arcs are too small, and their temperatures are too high for easy study, and metal transfers at high rates. Because of the difficulty involved in establishing the mechanisms that regulate the process, a great number of mechanisms have been suggested. These forces have been considered:

(1) Pressure generated by the evolution of gas at the electrode tip

(2) The electrostatic attraction between the electrodes

(3) Gravity

(4) The "pinch effect" caused by electromagnetic forces on the tip of the electrode

(5) Explosive evaporation of the necked filament between the drop and electrode due to the very high density of the conducting current

(6) Electromagnetic action produced by a divergence of current in the plasma around the drop

(7) Friction effects of the plasma jet

In all probability, a combination of these forces functions to detach the liquid drop from the end of the electrode.

EFFECT OF POLARITY ON METAL TRANSFER IN ARGON

Electrode Positive

AT LOW WELDING currents in argon, liquid metal from the electrode is transferred in the form of drops having a diameter greater than that of the electrode. With electrode positive, the drop size is roughly inversely proportional to the current, and the drops are released at the rate of a few per second. With a sufficiently long arc to minimize short circuits, drop transfer is reasonably stable and associated with a relative absence of spatter.

Above a critical current level, however, the characteristics of this transfer change to the axial spray mode. In axial spray transfer, the tip of the electrode becomes pointed, and minute drops are transferred at the rate of hundreds per second. The current at which this occurs is called the transition current. Often, as in the case of steel, this change is very abrupt.

The axial spray transfer is unique not only because of its good stability but also because of the absence of spatter. Furthermore, the drops are transferred in line with the electrode rather than along the shortest path between the electrode and workpiece. The metal, therefore, can be directed where needed when making vertical or overhead welds.

The key to spray transfer is the pinch effect which automatically squeezes the drops off the electrode. This occurs as a result of the electromagnetic effects of the current, as illustrated in Figure 2.14.

The transition current is dependent on a number of variables, including the electrode composition, diameter, electrode extension, and the shielding gas composi-

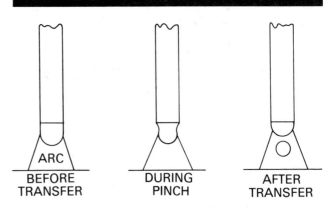

Figure 2.14—Individual Drop Formation Sequence in Spray Transfer

Table 2.3
Approximate Arc Currents for Transition from Drop to Spray Metal Transfer

Electrode Diameter		Transition Current*,A	
		Steel	Aluminum
mm	in.	Ar + 2%O_2	Argon
0.75	0.030	155	90
0.90	0.035	170	95
1.15	0.045	220	120
1.6	0.062	275	170

*The transition current varies with electrode extension, alloy content, and shielding gas composition.

tion. A great difference in transition current is found with various metal systems. Transition currents for various sizes of steel and aluminum electrodes are shown in Table 2.3.

The transition current is approximately proportional to the diameter of the electrode, as shown in Figure 2.15. Transition current is not dependent on current density, but it is mildly dependent on the electrode extension. An increase in the extension allows a slight decrease in the current at which spray transfer develops. [In practical welding operations, electrode extension is usually 13 to 25 mm (1/2 to 1 in.)].

In GMAW of steels, the spray arc mode is most often used with argon-based shielding gas. Small additions of oxygen to the shielding gas lower the transition current slightly while CO_2 additions raise it. The transition current defines the lower limit of useful current for spray transfer.

At high welding current densities, a rotary arc mode takes place. With appropriate mixtures of shielding gases, wire feeding controls, and welding guns that perform well at high wire feed speeds, rotating mode GMAW can be used to deposit 7 kg/h (16 lbs/h) or more steel weld metal. The useful upper current limit is the value where the rotational arc becomes unstable with loss of puddle control and high amounts of spatter.

The welding current at which axial spray disappears and rotational spray begins is proportional to the electrode diameter and varies inversely with electrode extension. The significance of these variables is illustrated in Figure 2.16.

Spray transfer can also be achieved at average current levels below the transition current using pulsed welding current. One or more drops of filler metal are transferred at the frequency of the current pulses. This technique increases the useful operating range of a given electrode size.

Solid state power sources that simplify the set-up for pulsed power welding have increased the applications of pulsed spray welding. The relatively low average current levels permit out-of-position welding of steel at rela-

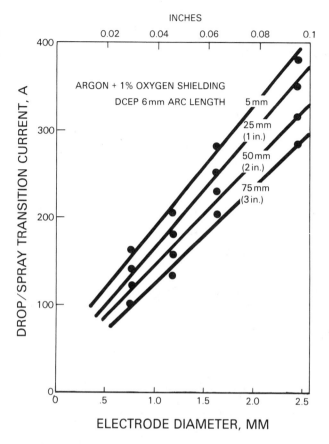

Figure 2.15—Influence of Electrode Diameter and Extension on Drop-to-Spray Transition Current for Mild Steel

tively high deposition rates. It is also utilized for aluminum welding with larger diameter electrodes and lower average currents and wire feed rates.

Electrode Negative

THE GMAW PROCESS is normally used with direct current electrode positive power. When the electrode is negative, the arc becomes unstable, and spatter is excessive. The drop size is large, and arc forces propel the drops away from the workpiece. This action appears to result from a low rate of electron emission from the negative electrode. If the thermionic properties of the electrode are enhanced by light coatings of alkali metal compounds, metal transfer is significantly improved. Although the use of emissive coatings allows spray transfer in GMAW with dcen, commercial filler metals are not generally available with such coatings.

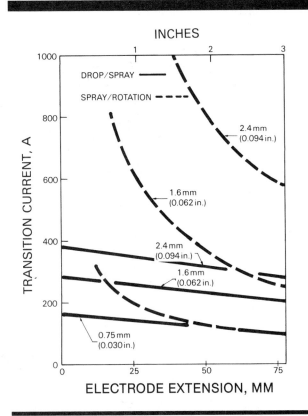

Figure 2.16—Effect of Electrode Extension and Diameter on Transition Current of Steel Filler Metals

EFFECT OF OTHER GASES ON METAL TRANSFER

ALTHOUGH HELIUM IS inert, it is unlike argon for shielding a welding arc because it does not usually produce an axial spray arc. Instead, the transfer is globular at all current levels and with both polarities. Helium shielded arcs are useful, nevertheless, because they provide deep penetration. Spray transfer is produced in helium by mixing relatively small quantities of argon with it. Using dilute mixtures, the deep penetration is not adversely changed. Although 20 percent argon in helium is sufficient to achieve these results, the normal commercial mixtures contain 25 percent argon as a safety factor. Argon-helium mixtures are used for welding nonferrous materials such as aluminum and copper. Generally, the thicker the material to be joined, the higher the percentage of helium in the shielding gas used.

Active gases such as carbon dioxide and nitrogen are much like helium in their effects on the arc. Spray transfer cannot be achieved without treatment of the wire surface. In addition, greater instabilities in the arc and chemical reactions between the gas and superheated metal drops cause considerable spatter. The difficulty

with spatter can be minimized by welding with the buried arc technique. This technique is common when carbon dioxide is used to shield copper, and when nitrogen is mixed with argon to shield aluminum alloys.

To offset the harsh globular transfer and spatter associated with CO_2 shielding, argon may be added to stabilize the arc and improve metal transfer characteristics. Short circuiting transfer is optimized using mixtures of 20 to 25 percent CO_2 in argon. Higher percentages of CO_2 are used for joining thick steel plate.

Small amounts of oxygen (2 to 5 percent) or carbon dioxide (5 to 10 percent) are added to argon to stabilize the arc, alter the spray transition current, and improve wetting and bead shape. These mixtures are commonly used for welding steel.

SHORT CIRCUITING TRANSFER

SHORT CIRCUITING TRANSFER of metal from the electrode tip to the molten weld pool has several advantages. Metal deposited in this way is less fluid and less penetrating than that formed with the spray transfer. It is easily handled by the welder in all positions, and it is particularly useful for joining thin materials. The spatter normally associated with short circuits is minimized by using electrical inductance to control the rate of current rise when the wire and pool are in contact. As a result, the peak value of current at short circuit is relatively low. The average current is kept low by using small diameter electrodes.

With the proper adjustment of equipment, the rate of short circuiting is high (on the order of hundreds of shorts per second). There is little time available to melt the electrode, so the drops formed on the tip are very small. The drops are transferred to the weld by surface tension when the electrode tip and weld pool come in contact

The changes in current and voltage that characterize short circuiting transfer are shown in Figure 2.17. When the wire contacts the weld pool, the current surges to a level high enough to cause the interface between the solid wire and liquid pool to neck down and finally vaporize, transferring a metal drop to the molten pool. An arc is formed with relatively high current and voltage. The high current and voltage cause the electrode tip to melt. However, immediately after the arc is established, the current decreases from its short circuit peak. The electrode advances toward the weld pool, eventually causing another short circuit, and the cycle repeats itself.

PULSED CURRENT CONSUMABLE ELECTRODE TRANSFER

PULSED CURRENT TRANSFER is achieved by pulsing the welding current back and forth between the globular- and spray-transfer current ranges. To suppress globular transfer, the time period between consecutive pulses

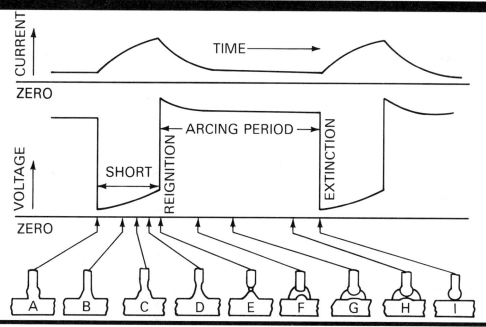

Figure 2.17—Schematic Representation of Short Circuiting Metal Transfer

must be less than that required for transfer by the globular mode. The time period between pulses produced by positive half cycles from a 60 Hz power source is short enough to suppress globular transfer at all current levels. Conversely, the pulse duration is long enough to ensure that transfer by the spray mode will occur with an appropriate current in the spray transfer range. The pulsed current mode of transfer differs from normal spray transfer in that metal transfer is interrupted between the current pulses.

Pulse shape and frequency may be varied over a wide range. Using microcomputer technology and solid state power sources, the pulse-power can be coordinated with the electrode feed rate such that a single drop of molten metal is transferred with each pulse. Both synergic and adaptive systems have been developed to employ pulsed power GMAW with relative ease on both ferrous and nonferrous materials. A one knob system can be used when the proper relationship between the pulse rate and the feed rate is maintained.

There are many ways of generating a modulated dc current for pulsed current transfer. The current is comprised of a background and a pulse current. Pulsating currents illustrated in Figure 2.18, which also shows the metal transfer sequence. With pulsed power, currents and deposition rates can be decreased to permit welding

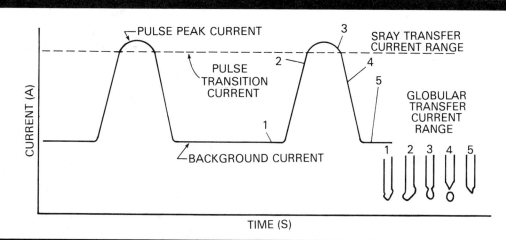

Figure 2.18—Output Current Wave Form of a Pulsed Current Power Source and Metal Transfer Sequence

of sheet as thin as 1.0 mm (0.04 in.) or even thinner with mechanized welding.

SUBMERGED ARC TRANSFER

DIRECT OBSERVATION OF metal transfer in the SAW process is impossible because the arc is completely obscured by a flux blanket. The submerged arc plasma is essentially a slightly ionized vapor column with a core temperature of about 6000 K. This central core is surrounded by thin concentric zones at lower temperatures with a steep radial temperature gradient. This gradient terminates at an indefinite vapor-liquid phase boundary at the boiling temperature of the flux components. Oscillographic studies indicate that the current may be carried simultaneously through the ionized vapor and liquid phases. The major portion of the current in commercial fluxes is through the vapor phase. The transfer of the metal from the electrode to the molten weld pool undoubtedly is in the form of globules and fine droplets, depending upon the current.

SHIELDED METAL ARC

THE MECHANISM OF metal transfer with covered electrodes is difficult to establish because the arc is partially obscured by fume and particles of slag. In many cases, a deep cavity formed by the covering hides the tip of the electrode from view. In general, the metal during transfer consists of either globules that short circuit the arc or a fine, showery spray of metal and slag particles that do not create a short circuit. The showery spray transfer is desirable. In some cases, however, spray transfer cannot be used because it is associated with great quantities of spatter. Covered electrodes are fed manually and used with considerable manipulation. Therefore, arc stability and metal transfer mode depend on the skill of the welder.

Most electrodes contain cellulose or metal carbonates that dissociate in the arc, forming a gas shield to protect the weld metal from atmospheric contamination. This shield consists primarily of the active gases carbon dioxide, carbon monoxide, hydrogen, and oxygen. These gases do not develop a highly conductive arc plasma. The current distribution is such that the liquid metal is forced away from the arc and weld pool in massive drops and spatter. Because these reactions are more intense when the electrode is negative, reverse polarity is generally used with the covered electrodes that do not contain cathode stabilizers (E6010, E7015).

Coverings are intended to make the electrode thermionic. Rutile, lime, and iron oxide are generally used in combination for this purpose. Such electrodes produce a more stable arc, less spatter, and form smaller drops with direct current electrode negative. Included in this type are the E6012, E6020, and the high iron powder varieites.

The stability of arcs with ac is dependent upon reignition of the arc during the interval when polarity is changed and the current has been reduced to zero. Stability frequently is achieved by substituting potassium silicate for sodium silicate. The potassium forms a lower ionization path between the electrode and work and increases the cathode emissivity to permit an easy reignition of the arc. Electrodes containing large quantities of rutile or lime are also thermionic and do not require a potassium silicate binder for ac welding.

FLUX CORED ARC TRANSFER

METAL TRANSFER WITH flux cored wire is a combination of the basic GMAW transfer types and the shielded metal arc (covered electrode) transfer. Metal can be transferred in the globular, short-circuiting, spray, and pulse modes. The type of transfer depends on the formulation of the flux as well as the arc voltage and current.

MELTING RATES

GENERAL CONTROLLING VARIABLES

AS DISCUSSED EARLIER, heat in an arc is generated by electrical reactions at the anode and cathode regions and within the plasma. Portions of this energy will melt the electrode which supports the arc unless it is adequately cooled. The greatest portion of the arc energy used for melting is obtained from cathode or anode reactions, but substantially more heating can be released from the cathode. Furthermore, when the electrode is the cathode (negative) terminal, good control of the energy release is possible. Little can be done to modify the release of

energy at the anode (positive) terminal. It is related mostly to the current magnitude and less to composition and other factors.

Most commercial metals and their alloys form what is called a cold cathode; its area is rather small, but great quantities of energy are generated to release the electrons needed to support an arc. However, metals having very high melting points easily supply electrons to sustain the arc at high temperature. These metals are called thermionic. Included in this category are molybdenum and tungsten. The cold cathode, low melting point metals can be made to supply electrons more easily by coating them with compounds that reduce the surface work function. They then become thermionic at lower temperatures. In some cases, the mechanism of electron release depends on surface oxide emission. In others, it is due to the formation of an atomic film of an alkali metal on the surface. The degree of change is regulated by the selection of the oxides or metals and their quantities. In general, the change from cold cathode to thermionic emission is accompanied by a lowering of the heating energy, and therefore, a reduction in the melting rate. Any improvement in the mode of metal transfer with dcen or in arc stability with ac is associated with a reduction in melting rate.

The arc plasma may or may not supply radiant heat to melt the electrode. When it does, the effect is relatively small by comparison with other sources. Therefore, those variables affecting the plasma, such as the shielding gas, flux, or the arc length, do not directly affect melting rate. If a change in melting rate is demonstrated, it is more likely caused by another factor, such as a change in current or cathode heating.

In addition to the energy supplied by the welding arc, electrical resistance heating of the electrode by welding current affects the melting rate of the electrode. This effect is particularly significant in welding processes that use small diameter electrodes. This heating is caused by the resistance of the electrode to the flow of current. Electrical resistance is greater with small electrode diameters, long electrode extensions, and low-conductivity metals and alloys.

The relationship of all the aforementioned variables on electrode melting rate can be written simply as follows:

$$MR = aI + bLI^2 \qquad (2.21)$$

where

$MR =$ the electrode melting rate, kg/h (lb/h)

$a =$ constant of proportionality for anode or cathode heating. Its magnitude is dependent upon polarity, composition, and, with direct current electrode negative, the emissivity of the cathode, kg/h · A (lb/h · A)

$b =$ constant of proportionality for electrical resistance heating and includes the electrode resistivity, kg/h·A^2·mm (lb.h·A^2·in.)

$L =$ the electrode extension or stickout, mm (in.)

$I =$ the welding current, A

Arc power is not a term in this equation, nor is it essential, since the plasma drop and voltage drop at the workpiece are not part of the relationship affecting melting rate.

GAS METAL ARC WELDING

THE MELTING RATE with the GMAW process is controlled by (1) the electrode diameter and extension and (2) the cathode or anode heating. Heating is dependent on the electrode polarity as well as the magnitude of the welding current. The shielding gas, the arc length, and the arc voltage have no significant effect on melting rate.

Equation 2.21 for melting rate can be used to calculate melting rates with electrode positive. Problems develop with dcen because the cathode heating value is so sensitive to the presence of oxides and alkali and alkaline-earth compounds.

The first term of the equation is more significant at low currents and with short extensions of the electrode. The influence of the second term becomes progressively greater with smaller electrodes, increased electrode extension, and higher welding current. The relative magnitude of the heating coefficients with 1.6 mm (1/16 in.) diameter wires is shown in Table 2.4.

The values of a and b of equation 2.21 are dependent on the electrode composition. For example, the first term is of greater significance with aluminum electrodes because the electrical resistance is low. It is also more important when the electrode is negative since the use of any additive that affects the cathode emissivity will also reduce the magnitude of term a in equation 2.21

An example of the effect of an additive on melting rate is shown in Figure 2.19. The electrode can be made suf-

Table 2.4
Relative Magnitude of Heating Coeffiicients in the Melting Rate of 1.6 mm (1/16 in.) Diam Wire Electrode

	a^*		b^*	
	kg/h · A	lb/h · A	kg/h · A^2·mm	lb/h · A^2·in.
Aluminum (dcep)	5.4×10^{-3}	1.2×10^{-2}	4.4×10^{-6}	9.7×10^{-6}
Mild steel (dcep)	8.6×10^{-3}	1.9×10^{-2}	2.5×10^{-5}	5.5×10^{-5}
Mild steel (dcen)	1.8×10^{-2}	4.0×10^{-2}	2.5×10^{-5}	5.5×10^{-5}

*Values given are for a and b in equation 2.21 when the electrode diam is 1.6 mm (1/16 in.).

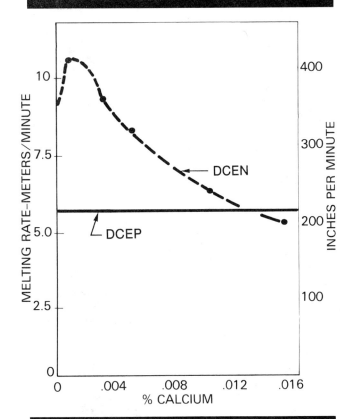

Figure 2.19—Effect of the Percent of Calcium in the Electrode Coating Compound on the Melting Rate of 1.6 mm (1/16 in.) Diameter Mild Steel Electrodes

rotation and, since the onset of unstable arc rotation increases with the electrode diameter, high current arcs generally are sustained with the large diameter electrodes. The extent of these ranges is shown in Figure 2.20 for steel. This factor is particularly important with steel but is of little significance with aluminum. A rough weld surface condition prevents the use of very high currents in welding aluminum.

The current limits of arcs in active gas shields are not determined on the basis of metal transfer since the transfer is always globular. The lower level of current is established by random short circuiting, absence of wetting, and other conditions that result in poor weld quality. The upper limit of current is determined in substantially the same way—spatter, poor bead appearance, and porosity.

SUBMERGED ARC WELDING

THE PRECEDING COMMENTS for GMAW apply generally to the SAW process. The melting rate of the electrode increases as the current increases. Changes in the anode or cathode voltages produced by changes in flux composition or changes in voltage level, travel speed, or in electrode preheat will influence the exact melting rates. The melting rate for mild steel electrodes in the SAW process

ficiently thermionic to reduce the heating effect represented by the term a for dcen power below that of dcep power. Direct current electrode negative arcs have great appeal because their melting rates can be so high. Unfortunately, when melting rates are high, the transfer of metal can be globular and spattery. Improvements in transfer gained by making the cathode more thermionic are accompanied by a reduction in melting rate. When ac is used, the magnitude of the term a in equation 2.21 is an average of the values obtained at direct current with electrode negative and electrode positive.

The range of melting rate is limited by a number of undesirable effects. When argon shielding is used, the lower limit of melting rates is defined by the current at which drop transfer begins. This varies with the electrode diameter since lower currents can be used with smaller wire diameters. The melting rate is not always reduced significantly by this expedient because the resistance of the electrode increases as the diameter decreases, and the second term of equation 2.21 contributes significantly to the melting rate. The upper limit of melting rate is defined by the formation of unstable arc

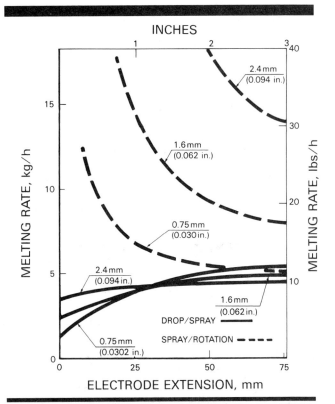

Figure 2.20—Useful Ranges of Melting Rates of Mild Steel Electrodes

is often reported as 13.5 kg/h · kA (30 lb/h · kA) and will range from approximately 11 to over 16 kg/h · kA (24 to over 36 lb/h · kA) depending upon electrode extension, polarity, and composition.

SHIELDED METAL ARC WELDING

THE SMAW PROCESS is the least efficient of the three in converting electrical energy to useful weld heat. This inefficiency is due in part to the need for melting a flux along with the core wire. Most important, however, is the absence of extensive electrical resistance heating. The diameters of the electrodes are so large that anode or cathode heating is the primary source of energy. Control of melting rate, therefore, is achieved largely by adjusting current. The range of current that can be used with covered electrodes is more limited than with the GMAW or SAW processes. As the current requirements increase, the electrode diameter must increase.

The lower limit of current is defined by incomplete fusion, high viscosity of the flux, or an unstable globular transfer. The upper limit is caused by a cumulative rise in temperature of the core wire of the electrode due to electrical resistance heating. The heated core wire dam-

ages the electrode covering. This heating of some materials in the covering, such as cellulose materials and carbonates, will cause these constituents to break down before reaching the arc where the products of dissociation are needed for shielding. Overheating may also cause the covering to spall. Some coverings are more susceptible to damage by overheating than others. For example, the cellulose-containing E6010 electrode of 6 mm (1/4 in.) diameter is useful in the range between 200 and 300 A, while for the same diameter, the rutile-base E6012 that does not rely on gas formers has a useful range between 200 and 400 amperes.

Iron powder electrodes were introduced to provide even higher melting rates and better material transfer than was possible with the conventional mineral coverings. Because iron is contained in the covering, the melting rate for a given current is higher. However, the optimum current needed to obtain an acceptable weld appearance is also higher with iron powder electrodes. The increase in current is proportional to the amount of iron contained in the covering. Therefore, higher melting rates for a given electrode diameter are achieved by a combination of increased efficiency and higher current. The results of these effects are shown in Figure 2.21.

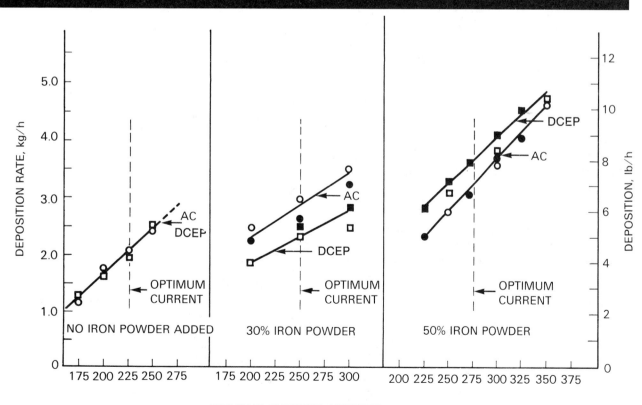

Figure 2.21—Effect of Welding Current on Deposition Rates for Three Levels of Iron Powder Content in Coating

PHYSICAL PROPERTIES OF METALS AND SHIELDING GASES

THE PHYSICAL PROPERTIES of the metals or alloys being joined influence the efficiency and applicability of the various joining processes. The nature and properties of the shielding gases and of the contaminants from the atmosphere may have a pronounced effect on the resulting weld. The shielding gases may be generated either by the decomposition of fluxing materials or by direct introduction into the arc stream and the area surrounding the arc plasma.

Both thermal conductivity and thermal expansion have a direct effect on distortion of the weldment. Base metal electrical resistivity and thermal conductivity have a pronounced effect on the application of both resistance and arc welding to the various metals. In the case of resistance welding, base metal resistivity, thermal conductivity, and specific heat influence the power requirements.

In the case of arc welding, arc starting and arc stability are greatly influenced by the ionization potentials of the metal and flux vapors as well as by the various electronic transitions that occur in the shielding gases under the extreme temperature conditions that exist in the arc. The thermionic work function of the electrode material, and to a lesser extent, that of the materials being welded, have a direct bearing upon the efficiency of the energy transferred by a welding arc. Electrical resistivity also plays an important role in these processes as a result of resistance heating of the electrode between the contact tube or electrode holder and the work. Resistance heating of the electrode may be an important contribution to the total energy input to the weld zone.

Weld bead shape is dependent to varying degrees upon the interfacial energy between the surrounding atmosphere and the molten metal. The surrounding atmosphere may consist either of a gas or a liquid flux. Elements in the surrounding medium may control the shape of the bead.

Another important material property that should be considered when determining the relative weldability of alloys is the rate of oxidation of the base metal. This rate is important in determining the degree of shielding required. A corollary to this is the relative stability of oxides that may be present.

The specific heat and density of the shielding gases affect the heat of the arc and the shielding coverage.

ELECTRICAL RESISTIVITY

THE IMPORTANCE OF electrical resistivity to most welding processes cannot be overemphasized. Its role in resistance welding is obvious because the resistance of the materials being welded is directly related to the heat generated for a given welding current. Electrical resistivity also contributes to preheating the electrode in consumable electrode processes. The resistivities of a number of the metals and alloy classes of interest in welding are listed in Table 2.5.

The values shown in the table were determined at room temperature, but in all cases, the resistivity to a first approximation increases almost linearly as the temperature increases. Although the rate of increase is different for different materials, the room temperature resistivity is a useful guide in determining the relative values for various materials at elevated temperatures.

THERMAL CONDUCTIVITY

THE ROOM TEMPERATURE value of thermal conductivity of a number of metals and alloys are shown in Table 2.5. Thermal conductivity decreases as the working temperature is increased in a manner analogous to the electrical conductivity values. Similarly, pure metals have the highest conductivity, and the addition of alloying elements tends to decrease the values of this property.

COEFFICIENT OF EXPANSION

THE THERMAL COEFFICIENT of expansion of the materials being welded is important in analyzing distortion problems involved in welded assemblies. Coefficient of expansion values at room temperature are shown in Table 2.5. The same general considerations previously described for thermal conductivity and electrical resistivity are applicable to the coefficient of expansion.

IONIZATION POTENTIALS

IN THE CASE of arc welding, both the ease of arc initiation and the stability of the arc are related to the minimum ionization potential of the elements in the arc atmosphere. This atmosphere consists of flux materials and metal vapors, as well as gases introduced externally for shielding purposes. It is believed that, in the cases of helium and argon, the stability is achieved through the transitions from the metastable excited stages to the ionized state. The ionization potentials for metal vapors and gases are shown in Tables 2.6 and 2.7, respectively.

THERMIONIC WORK FUNCTION

THE THERMIONIC WORK function is the energy required for an electron to escape a solid surface. Since the ease of starting and maintaining an arc is exponentially related to the thermionic work function, representative values

Table 2.5
Physical Properties of Selected Metals and Alloys [at 20 ° C (68 ° F)]

Metal	Melting Temperature			Resistivity	Thermal Conductivity	Coefficient of Expansion	Density
	°F	°C	K	$10^{-8}\Omega m$	W/cm·K	$10^{-6}K^{-1}$	g/cm^{-3}
Aluminum							
99.9 + pure	1220.67	660.37	933.52	2.65	2.37	23.1	2.70
Wrought alloys	1170-1200	630-650	900-920	2.8-6.7	1.2-2.1	22-23	2.6-2.8
Casting alloys	930-1200	500-650	770-920	3.0-7.0	1.2-1.8	22-23	2.6-3.0
Cobalt							
99.9 + pure	2721	1494	1767	4.3	1.01	13.0	8.86
Co-Cr-W alloys	2400-2500	1320-1370	1590-1640	8.5-17.6	0.12-0.25	11.2	8.0-9.0
Copper							
99.9 + pure	1983.9	1084.4	1357.6	1.68	4.01	16.5	8.93
90 Cu, 10 Al (alum. bronze)	1920-1960	1050-1070	1320-1340	12-14	1.60	17	8.34
90 Cu, 10 Zn (commercial bronze)	1920-1960	1050-1070	1320-1340	3.92	1.92	18	8.92
Iron							
99.9 + pure	2800	1538	1811	9.69	0.808	11.8	7.87
Low alloy steel	2610-2730	1430-1500	1700-1770	10-20	0.32-0.66	11.4	7.8-8.0
Martensitic stainless steel	2700-2800	1480-1540	1750-1810	54-69	0.25	9-10	7.6-7.8
Austenitic stainless steel	2500-2650	1370-1450	1640-1730	71-79	0.15	14-15	7.8-8.0
Magnesium							
99.9 + pure	1200	649	922	4.30	1.56	24.8	1.74
Wrought alloys	930-1200	500-650	770-920	4.5-12.5	0.60-1.4	25-28	1.8
Cast alloys	840-1200	450-650	720-920	4.5-16	0.46-1.4	25-28	1.8-1.9
Molybdenum	4750	2621	2894	5.09	1.38	4.8	10.24
Nickel							
99.95 Ni + Co	2651	1455	1728	7.47	0.915	13.4	8.9
60 Ni, 33 Cu, 6.5 Fe	2370-2460	1300-1350	1570-1620	48	0.26	13	8.4
Ni-Cr-Mo-Fe alloys (Hastelloy alloy series)	2050-2460	1120-1350	1390-1620	110-140	0.10-0.20	11-12	7.8-9.1
Tantalum	5468	3020	3293	12.8	0.575	6.3	16.6
Titanium alloys	2780-3040	1530-1670	1800-1940	43-230	0.024-0.24	7-9	4.4-4.8
Tungsten	6129	3387	3360	5.42	1.75	4.5	19.3

Based on the International Practical Temperature Scale of 1968; melting temperature may indicate a single temperature or a range of temperatures.
This Table was prepared by the Thermophysical Properties Research Center, Purdue University, for the Office of Standard Reference Data of the National Bureau of Standards, Washington, DC 20234.

for a number of the elements are listed in Table 2.8. Generally, the values do not differ significantly. However, lowering this value by even one electron volt significantly affects the arc characteristics.

METAL OXIDES

THE DIFFICULTY OF transferring some alloying elements across the arc, as well as the susceptibility of oxide inclu-

sions in the weld metal, is a direct function of the oxidation potential of the metal. The relative stability of several metal oxides and the reactivity of several metals is shown in Table 2.9.

SPECIFIC HEAT

PHYSICAL PROPERTIES OF the common gases are shown in Table 2.10, and specific heats of several metals are

Table 2.6
Ionization Potentials of Metal Vapors

Element	eV	Element	eV
Aluminum	5.986	Potassium	4.341
Barium	5.212	Lithium	5.392
Boron	8.298	Magnesium	7.646
Carbon	11.260	Molybdenum	7.099
Calcium	6.113	Nickel	7.635
Cobalt	7.86	Silicon	8.151
Chromium	6.766	Sodium	5.139
Cesium	3.894	Titanium	6.82
Copper	7.726	Tungsten	7.98
Iron	7.870		

Note: Table 2.6 was prepared by the Atomic Energy Levels Data Center, National Bureau of Standards, Washington, DC 20234.

Table 2.7
Ionization Potentials of Gases

Element or Compound	Symbol	eV
Argon	Ar	15.760 (11.548)
Hydrogen	H_2	15.43
	H	13.598
Helium	He	24.5876 (20.96430) (19.8198)
Nitrogen	N_2	15.58
	N	14.534
Oxygen	O_2	12.07
	O	13.618
Carbon Dioxide	CO_2	13.77
Carbon Monoxide	CO	14.1

Note: Table 2.7 was prepared by the Atomic Energy Levels Data Center, National Bureau of Standards, Washington, DC 20234.

Table 2.8
Electron Thermionic Work Functions

Element	Range, eV	Element	Range, eV
Aluminum (Al)	3.8 to 4.3	Magnesium (Mg)	3.1 to 3.7
Barium (Ba)	4.1 to 4.4	Manganese (Mn) $(\alpha, \beta, \text{ or } \gamma)$	3.8 to 4.4
Barium oxide (BaO)	4.9 to 5.2	Molybdenum (Mo)	4.0 to 4.8
Cerium (Ce)	1.7 to 2.6	Neodymium (Nd)	4.1 to 4.5
Cesium (Cs)	1.0 to 1.6	Nickel (Ni)	2.9 to 3.5
Cesium film or W	2.7 to 3.1	Palladium (Pd)	4.5 to 5.3
Chromium (Cr)	4.4 to 5.1	Platinum (Pt)	4.9 to 5.7
Cobalt (Co)	3.9 to 4.7	Samarium (Sm)	5.2 to 5.9
Columbium (Cb)*	1.8 to 2.1	Scandium (Sc)	3.3 to 3.7
Copper (Cu)	1.1 to 1.7	Silver (Ag)	2.4 to 3.0
Europium (Eu)	4.4 to 4.7	Strontium (Sr)	2.1 to 2.7
Gadolinium (Gd)	2.2 to 2.8	Titanium (Ti)	3.8 to 4.5
Gold (Au)	4.2 to 4.7	Tungsten (W)	4.1 to 4.4
Hafnium (Hf)	2.9 to 3.3	Vanadium (V)	4.3 to 5.3
Iron (Fe) $(\alpha \text{ or } \gamma)$	3.5 to 4.0	Yttrium (Y)	2.9 to 3.3
Lanthanum (La)	3.3 to 3.7	Zirconium (Zr)	3.9 to 4.2

*Also called niobium.
Note: Table 2.8 was prepared by the Alloy Data Center, National Bureau of Standards, Washington, DC 20234.

shown in Table 2.11. These values are a measure of the amount of heat required to raise the temperature of a unit weight of material one degree. They are expressed in units of joules per kilogram-kelvin. It may be regarded as a measure of the ability of a metal or gas to absorb or store heat. For dilute alloys such as plain carbon or low alloy steels, the values of the major element given in Table 2.11 may be used with reasonable accuracy.

Table 2.9
Oxidation Potential of Metals and Metal Oxides

Relative Stability[1]	Relative Reactivity[2]
1. CaO	1. aluminum
2. MgO	2. magnesium
3. Al_2O_3	3. cobalt and titanium
4. TiO_2	
5. SiO_2	4. tungsten
6. V_2O_3	5. manganese
7. MnO	6. vanadium
8. Cr_2O_3	7. molybdenum
9. WO_2 and MoO_2	8. iron
10. Fe_2O_3	9. chromium

1. Listed in decreasing order of stability.
2. Listed in decreasing order or reactivity.

Table 2.10
Physical Properties of Shielding Gases

Name of Gas	N_2	Ar	He	H_2	CO_2
Molecular weight	28.0134	39.948	4.0026	2.01594	44.011
Normal boiling point					
K	77.347	87.280	4.224	20.268	194.65
°C	− 195.81	− 185.88	− 268.94	− 252.89	− 78.51
°F	− 320.44	− 302.57	− 452.07	− 423.19	− 109.3
Density at 21.1°C (70°F), 1 atm:					
kg/m^3	1.161	1.656	0.1667	0.0841	1.833
lb/ft^3	0.07249	0.1034	0.01041	0.00525	0.1144
Specific volume at 21.1°C (70°F), 1 atm:					
m^3/kg	0.8613	0.6039	5.999	11.89	0.5455
ft^3/lb	13.79	9.671	96.06	190.5	8.741
Specific gravity at 21.1°C (70°F), 1 atm: (air = 1)	0.9676	1.380	0.1389	0.0700	1.527
Specific heat-constant pressure at 21.1°C (70°F), 1 atm:					
J/kg · K	1041	521.3	5192	1490	846.9
Btu/lb · °F	0.2487	0.1246	1.241	3.561	0.2024
Specific heat-constant volume at 21.1°C (70°F), 1 atm:					
J/kg · K	742.2	312.1	3861	1077	653.4
Btu/lb · °F	0.1774	0.0746	0.7448	2.575	0.1562

This Table was prepared by the Cryogenic Data Center National Bureau of Standards, Institute for Basic Standards, Boulder, Colorado 80302.

Table 2.11
Specific Heat of Selected Elements

Element	Symbol	Atomic Weight	Temperature*			Specific Heat kJ/kg · K
			°F	°C	K	
Aluminum	Al	26.98154	68	20	293	0.898
			212	100	373	0.937
			1220.67**	660.37**	933.52**	1.250
Carbon (gas)	C	12.011	68	20	293	1.735
			932	500	773	1.731
			1832	1000	1273	1.731
			3632	2000	2273	1.756
			5432	3000	3273	1.822
Cobalt	Co	58.9332	68	20	293	0.419
			783	417	690	0.524 (α)
			783	417	690	0.517 (β)
			2052†	1122†	1395†	0.936
			2721**	1494**	1767**	0.640
Chromium	Cr	51.996	68	20	293	0.483
			1832	1000	1273	0.705
			2732	1500	1773	0.870
			3380**	1860**	2433**	0.974
Columbium (Niobium)	Cb (Nb)	92.9064	68	20	293	0.273
			932	500	773	0.290
			1832	1000	1273	0.313
			3632	2000	2273	0.358
			4480**	2471**	2744**	0.381
Iron	Fe	55.847	68	20	293	0.444
			1416†	769†	1043†	1.498
			1672	911	1185	0.742 (α)
			1672	911	1185	0.608 (γ)
			2541	1394	1667	0.679 (γ)
			2541	1394	1667	0.737 (δ)
			2800**	1538**	1811**	0.763
Magnesium	Mg	24.305	68	20	293	1.019
			212	100	373	1.102
			932	500	773	1.253
			1200**	649**	922**	1.327
Molybdenum	Mo	95.94	68	20	293	0.259
			1832	1000	1273	0.313
			3632	2000	2273	0.419
			4750**	2621**	2894**	0.546
Nickel	Ni	58.71	68	20	293	0.440
			1168†	358†	631†	0.663
			1832	1000	1273	0.604
			2651**	1455**	1728**	0.616
Tantalum	Ta	180.9479	68	20	293	0.140
			1832	1000	1273	0.156
			3632	2000	2273	0.181
			5468**	3020**	3293**	0.243
Titanium	Ti	47.90	68	20	293	0.519
			1621	883	1156	0.707 (α)
			1621	883	1156	0.611 (β)
			3043**	1673**	1946**	0.782
Tungsten	W	183.85	68	20	293	0.426
			1832	1000	1273	0.479
			3632	2000	2273	0.537
			6129	3387	3660	0.617

*Based on the International Practical Temperature Scale of 1968.
**Melting point.
†Curie temperature.
This Table was prepared by the Thermophysical Properties Research Center, Purdue University, for the Office of Standard Reference Data of the National Bureau of Standards, Washington, DC 20234.

SUPPLEMENTARY READING LIST

Albom, M. J. Solid state bonding. *Welding Journal*. 43 (6): June 1964; 491–504.

Anderson, J. E. and Jackson, C. E. Theory and application of pulsed laser welding. *Welding Journal*. 44 (12): Dec. 1965; 1018–1026.

Bakish, R. and White, S. S. *Handbook of electron beam welding*. New York, NY: J. Wiley & Sons, (1964).

Bunshaw, F. F. High power electron beams. *International science and technology*, April (1962).

Campbell, H. C. Electroslag, electrogas, and related welding processes. *Welding Research Council Bulletin 154*. 1970.

Chase, T. F. and Savage, W. F. Effect of anode composition on tungsten arc characteristics. *Welding Journal*. 50 (11): Nov. 1971; 467s–473s.

D'Annessa, A. T. The solid-state bonding of refractory metals. *Welding Journal*. 43 (5): May 1964; 232s–240s.

Duvall, D. S., Owczarski, W. A., Paulonis, D. F., and King, W. H. Methods for diffusion welding superalloy Udimet 700. *Welding Journal*. 51 (2): Feb. 1972; 41s–49s.

Gottlieb, M. B. The fourth state of matter. *International science and technology*. (August) 1965: 42–47.

Hicken, G. K. and Jackson, C. E. The effects of applied magnetic fields on welding arcs. *Welding Journal*. 45 (11): Nov. 1966; 515s–524s.

Holtzman, A. H. and Cowan, G. R. Bonding of metals with explosives. *Welding Research Council Bulletin 104*. 1965.

Jackson, C. E. The science of arc welding. *Welding Journal*. 39 (4–6): Apr. 1960; 129s–140s: June 1960; 177–190s.

Jones, J. B., Maropis, N., Thomas, J. G., and Bancroft, D. Phenomenological considerations in ultrasonic welding. *Welding Journal*. 40 (7): July 1961; 289s–305s.

Jones, W. and March, N. H. *Theoretical solid state physics*, Vol. 2. London: Wiley-Interscience, 1973.

Locke, E., Hoag, E., and Hella, R. Deep penetration welding with high power CO_2 lasers. *Welding Journal*. 51 (5): May 1972; 245s–249s.

Manz, A. F. *The welding power handbook*. Tarrytown, NY: Union Carbide Corporation, (1973).

Needham, J. C. and Carter, A. W. Material transfer characteristics with pulsed current. *British Welding Journal*. 12: Dec. 1965; 229–241.

Nestor, O. H. High intensity and current distributions at the anode of high current, inert gas arcs. *Journal of Applied Physics*. 33: 1962; 1638–1648.

Nishiguchi, Kimiuki and Takahashi, Yasuo. *Fundamental study of solid state bonding*, 1986; IA-336-86-OE, Annual Assembly of IIW in Tokyo.

O'Brien, R. L. *Plasma arc metalworking process*. New York, NY: American Welding Society, 1967.

Quigley, M. B. C. and Webster, J. M. Observations of exploding droplets in pulsed-arc GMA welding. *Welding Journal*. 50 (11): Nov. 1971; 461s–466s.

Rager, D. D. Direct current, straight polarity gas tungsten-arc welding of aluminum. *Welding Journal*. 50 (5): May 1971; 332–341.

Reed, T. B. Plasma torches. *International science and technology*. (June 1962): 42–48.

Shifrin, E., G. and Rich, M. I. Effect of heat source width in local heat treatment of piping. *Welding Journal*. 52 (12): Dec. 1973; 792–799.

Stoeckinger, G. R. Pulsed dc high frequency GTA welding of aluminum plate. *Welding Journal*. 52 (12): Dec. 1973; 558s–567s.

Tseng, Chao-Fang and Savage, W. F. Effect of arc oscillation. *Welding Journal*. 50 (11): 1971; 777–786.

Union Carbide Corporation. *The oxyacetylene handbook*, 3rd ed. New York, NY: Union Carbide Corporation, (1976).

Vill, V. I. *Friction welding of metals*. Translated from the Russian. New York: American Welding Society, 1962.

Wood, F. W. and Beall, R. A. Studies of high-current metallic ores. *Bureau of mines bulletin 625*. 1965.

Woods, R. A. and Milner, D. R. Motion in the weld pool in arc welding. *Welding Journal*. 50 (4): Apr. 1971; 163s–173s.

Woollard, K. F. and Gugdel, L. L. Welding and brazing developments utilizing stabilized methylacetylene propadiene. *Welding Journal*. 45 (2): Feb. 1966; 123–129.

HEAT FLOW IN WELDING

PREPARED BY A COMMITTEE CONSISTING OF:

S. S. Glickstein, Chairman
Westinghouse Electric Corporation

H. S. Ferguson
Duffers Scientific Inc.

E. Friedman
Westinghouse Electric Corporation

C. E. Jackson*
Ohio State University

S. Kou
University of Wisconsin

WELDING HANDBOOK COMMITTEE MEMBER:
J. C. Papritan
Ohio State University

*Deceased

CHAPTER 3

HEAT FLOW IN WELDING

INTRODUCTION

THE THERMAL CONDITIONS in and near the weld metal must be maintained within specific limits to control metallurgical structure, mechanical properties, residual stresses, and distortions that result from a welding operation. Of particular interest in this connection are (1) the solidification rate of the weld metal, (2) the distribution of maximum or peak temperatures in the weld heat-affected zone, (3) the cooling rates in the weld metal and in the heat-affected zone, and (4) the distribution of heat between weld metal and heat-affected zone.

In the past, experimental results of weld heat transfer were fit to simplified equations suggested by heat transfer theory, and techniques were developed to calculate cooling rates, heat-affected-zone peak temperatures and dimensions, and solidification rates. With these equations, methods to control the heat flow in welding to achieve specific desirable properties could be predicted. However, with the advent of modern computer technology, much of the curve fitting to experimental data is not necessary. The geometric weldment configuration can be modeled, and a general solution to the heat flow equations can be solved numerically.

The fundamentals of heat flow and the basic features of welding heat transfer are discussed here. The information applies mainly to arc welding where the input energy is distributed over a specific area on the surface of the weldment. The determination of the effective energy transfer from the welding arc to the weldment, often referred to as arc efficiency, is an important variable for determining heat transfer. While arc efficiency has been assumed to be constant for a given welding process, several investigators have shown that the arc efficiency is a function of welding technique. The subjects of arc efficiency and energy distribution are also addressed in this chapter.

Computers can be used to solve the complex differential equations that model heat flow in welding. However, simplified equations supported by engineering data still play an important role in predicting cooling rates and other weld characteristics. Equations for determining cooling rates and peak temperatures and methods for calculating the width of the heat-affected zone are provided. Several illustrative examples are also included.

The effects of several welding variables on the welding process are also discussed in this chapter. Until the advent of computer simulation of a welding process, most of these effects had been determined experimentally. With a computer model, the effects of small changes in welding variables on the final weldment can be predicted qualitatively and quantitively.

The application of heat flow analysis to weld metal solidification is reviewed. This includes a discussion of pool shape, grain structure, nucleation mechanisms, and weld heat-afffected-zone structure.

The mathematical symbols and units of measure used in this Chapter are shown in Table 3.1.

Table 3.1
Guide to Mathematical Symbols

Symbol	Definition	Unit of Measurement	Equation(s) where used
T	General temperature variable	°C	3.1
T_p	Peak temperature	°C	3.9
T_o	Uniform initial temperature	°C	3.6, 3.7, 3.8, 3.9
T_m	Melting temperature	°C	3.9
T_c	Temperature at which cooling rate is measured	°C	3.6, 3.7
$Q\,(x,y,z)$	Internal power generation	W/mm³	3.1, 3.4
$k\,(T)$*	Thermal conductivity of the metal	J/mm·s·°C	3.1, 3.1, 3.7
ρ	Density of material	g/mm³	3.1, 3.7, 3.8, 3.9
$C\,(T)$**	Specific heat of solid metal	J/g·°C	3.1, 3.7, 3.8, 3.9
$\rho C\,(T)$**	Volumetric specific heat	J/mm³ °C	3.1, 3.7, 3.8, 3.9
y	Distance from fusion line to peak temperature	mm	3.9
Y_z	Width of heat-affected zone	mm	3.9
R	Cooling rate	°C/s	3.6, 3.7
h	Plate or sheet thickness	mm	3.7
τ	Relative plate thickness	Dimensionless	3.8
S_t	Solidification time, time interval between the onset and completion of freezing	s	3.10
$q\,(x,y)$	Rate of surface heat generation	W/mm²	3.2, 3.4, 3.5
H_{net}***	Net energy input	J/mm	3.6, 3.7, 3.8
f_1	Heat transfer efficiency	Dimensionless	3.6, 3.7, 3.8
E	Arc voltage	V	3.6, 3.7, 3.8
I	Current	A	3.6, 3.7, 3.8
v	Weld travel speed	mm/S	3.6, 3.7, 3.8
L	Heat of fusion	J/mm³	3.10
$\bar{r}$	Heat distribution parameter that defines region of 95% energy deposition	mm	3.4, 3.5
x,y,r,z	Distance variables	mm	
t	time	s	3.3

*$k\,(T)$ is a function of temperature, but for the purposes of computation in this Chapter, average values are used. $k(T)$ or k for steel is 0.028 J/mm·s·°C.
**In this Chapter, $C\,(T)$ always appears as ρC. The product of the density and specific heat is called the volumetric specific heat. In this Chapter, the average value of 0.0044J/mm³·°C will be used for the volumetric specific heat of steel.
*** From Chapter 2, $H_{net} = f_1 \dfrac{EI}{v}$

FUNDAMENTALS OF HEAT FLOW

THE BASIC FEATURES of the transfer of thermal energy from the welding arc to the workpiece and within the workpiece itself are reviewed in this section. These features define the applicable heat transfer mechanisms that interact to affect the distribution of peak temperatures in the weldment, the size and shape of the weld pool and the heat-affected zone, and the cooling rates of the weld metal and the heat-affected zone.

BASIC FEATURES OF WELDING HEAT TRANSFER

THE TRANSFER OF heat in the weldment is governed primarily by the time-dependent conduction of heat, which is expressed by the following equation:

$$\frac{\partial}{\partial x}\left[k(T)\frac{\partial T}{\partial x}\right] + \frac{\partial}{\partial y}\left[k(T)\frac{\partial T}{\partial y}\right] + \frac{\partial}{\partial z}\left[k(T)\frac{\partial T}{\partial z}\right]$$
$$= \rho C(T)\frac{\partial T}{\partial t} - Q \tag{3.1}$$

where
x = coordinate in welding direction, mm
y = coordinate transverse to weld, mm
z = coordinate normal to weldment surface, mm
T = the temperature in the weldment, °C
$k(T)$ = thermal conductivity of the metal, J/mm·s·°C
ρ = density of the metal, g/mm³
C = specific heat of the metal, J/g·°C
Q = rate of internal heat generation, W/mm³

The internal heat generation Q is important when the welding process involves input of energy below the surface of the workpiece. Examples of such processes are submerged arc and electron beam welding using high power densities. With low power densities, the heat is directed onto the weldment surface. Surface heating is most often characterized by a heat-flux distribution, $q(x,y)$ applied over a relatively small area of the weldment surface, rather than the internal heat generation Q. This distribution is governed by the following equation:

$$k(T)\frac{\partial T}{\partial z} = q\,(x,y) \qquad (3.2)$$

where $q(x,y)$ is expressed in watts per square millimeter (W/mm^2), and is directed onto the weldment heat surface $(z = O)$. Heat is lost to the surrounding environment by a combination of radiation and convection mechanisms, or it is transferred to the weld fixture or other attached components. Therefore, the heat input to form the weld bead may be either deposited on a small portion of the weldment surface as a heat flux (q) or may be applied internal to the weldment (Q). The thermal energy is then distributed within the weldment primarily by conduction and is finally lost to the surrounding environment, the weld fixturing, and the attached structural components.

Heat Input Deposition

THE AREA OF heat input is small, relative to the overall dimensions of the workpiece. Three variables govern the input of heat to the workpiece whether the heat is applied on the weldment surface or internal to the weldment. These variables are (1) the magnitude of the rate of input energy (the product of the efficiency and the energy per unit time produced by the power source is usually expressed in watts), (2) the distribution of the heat input, and (3) the weld travel speed. The expression heat input is used because not all of the welding energy enters the workpiece. The heat input is often characterized by the single variable H_{net}, where H_{net} is the ratio of the arc power entering the workpiece to the weld travel speed. However, certain conditions require that the heat input rate and the weld speed should be treated separately when describing the weld thermal cycle in the vicinity of the weld metal and heat-affected zones.

Efficiency. The arc efficiency, f_1, is defined as the ratio of the energy actually transferred to the workpiece to the energy produced by the power source. (The arc efficiencies for four welding processes are summarized in Figure 3.1A.)[1] It is lowest in gas tungsten arc welding (GTAW),

intermediate in shielded and gas metal arc welding (SMAW and GMAW), and highest in submerged arc welding (SAW).

In GTAW, the electrode is nonconsumable and the arc is not thermally insulated from its surroundings. As a result, the total heat loss from the electrode to the surroundings can be high. In consumable-electrode welding processes (such as GMAW, SMAW, and SAW), almost all of the energy consumed in melting the electrode is transferred to the workpiece with the liquid metal droplets. As a result, the arc efficiency is higher for consumable electrode welding processes than for a nonconsumable electrode process. In the SAW process, the arc efficiency is further increased, because the arc is covered by an insulating blanket of molten and granular flux, and the heat loss to the surrounding area is thus minimized.

More recently, other investigators have reported GTAW arc efficiencies between 60 and 90 percent, as illustrated in Figure 3.1(B).[2] Arc efficiency was measured during seam welding of aluminum pipes by using a calorimeter made of the pipe itself, as shown in Figure 3.2(A). The arc efficiency, f_1, can be determined using the following equation:

$$\int_0^\infty WC\,(T_{out} - T_{in})dt = f_1 EI\,t_{weld} \qquad (3.3)$$

where

$$
\begin{aligned}
W &= \text{mass flow rate of water, g/s} \\
C &= \text{specific heat of water, J/g} \cdot {}^\circ\text{C} \\
T_{out} &= \text{outlet water temperature, } {}^\circ\text{C} \\
T_{in} &= \text{inlet water temperature, } {}^\circ\text{C} \\
t &= \text{time, s} \\
f_1 &= \text{arc efficiency, \%} \\
E &= \text{welding arc voltage, V} \\
I &= \text{welding current, A} \\
t_{weld} &= \text{welding time, s}
\end{aligned}
$$

In the water temperature range of the studies conducted, C and W can be considered a constant and equation 3.3 becomes

$$WC \int_0^\infty (T_{out} - T_{in})dt = f_1 EI\,t_{weld} \qquad (3.3a)$$

The arc efficiency in GTAW has been reported to be much lower with alternating current than with direct

1. Christensen, N., Davis V. del., and Germundsen, K. (article title). Brit. Weld. J. 12(54): 1965.

2. Ghent, H. W., Roberts, D. W., Hermance, C. E., Kerr, H. W., and Strong, A. B. Arc physics and weld pool behavior. London: The Welding Institute, 1980; Smartt, H. B., Stewart, J. A., and Einerson, C. J. "Heat transfer in gas tungsten arc welding." Metals Park, OH: Am. Soc. Met.; ASM Metals/Materials Technology Series Paper No. 8511-011, 1985; Kou, S., Le, Y. Heat flow during the autogenous GTA TIG welding of aluminum alloy pipes. Metallurgical Transactions A. 15A(6): June 1984, 1165-1171; Glickstein, S. S. and Friedman, E. Temperature transients in gas tungsten-arc weldments. Welding Review. 2(2): May 1983, 72-76; Glickstein, S. S. and Friedman, E. Weld modeling applications. Welding Journal 63(9): September 1984; 38-42.

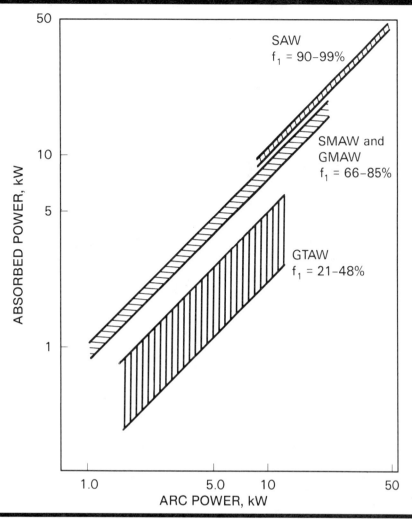

SAW
f_1 = 90–99%

SMAW and
GMAW
f_1 = 66–85%

GTAW
f_1 = 21–48%

ABSORBED POWER, kW

ARC POWER, kW

Figure 3.1A—Measured Arc Efficiencies in SAW, GMAW, SMAW and GTAW

current electrode negative (DCEN).[3] As shown in Figure 3.3 the arc efficiency increases at constant welding current and voltage as the arc length decreases from 5 to 2.2 mm (0.2 to 0.09 in.).[4] If the arc length remains constant and the voltage increases from 12 to 14 volts at constant current, then the efficiency remains unchanged.

Because weld keyholes tend to trap most of the energy from the heat source, the efficiency of the electron beam welding process can be 80 to 95 percent. The efficiency of laser beam welding can range from very low on pol-ished aluminum or copper to very high in deep penetration keyhole welding or in materials with low reflectivity. The efficiency can be improved in some cases by coating the workpiece surface with thin layers of graphite and zinc phosphate to enhance absorption of the beam energy. The effect of the surface layers on the properties of the weld should be investigated before this technique is used in production.

Energy-Density Distribution

THE DISTRIBUTION OF the heat applied on the surface of the workpiece or in a volume internal to the weldment has an important effect on the heat distribution pattern in the vicinity of the weld. It is in this region that the weld bead and the heat-affected zone are formed, and it is important to identify how the heat input distribution influences their size and shape. If the weld thermal cycle

3. Ghent, H. W., Roberts, D. W., Hermance, C. E., Kerr, H. W., and Strong, A. B. *Arc physics and weld pool behavior.* London: The Welding Institute, 1980; Smartt, H. B., Stewart, J. A., and Einerson, C. J. "Heat transfer in gas tungsten arc welding." Metals Park, OH: Am. Soc. Met.; ASM Metals/Materials Technology Series Paper # 8511-011, 1985.

4. Tsai, N. S. and Eager, T. W. Distribution of the heat and current fluxes in gas tungsten arcs. *Metallurgical Transactions B.* 16B(12): 1985.

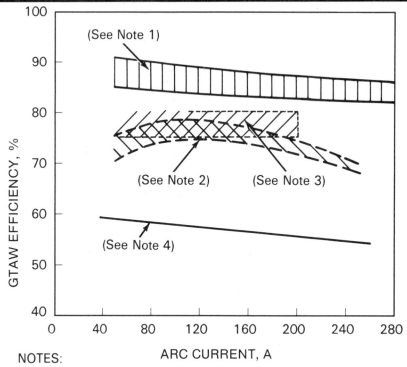

NOTES:

1. Refer to Tsai, N. S. and Eager, T. W. Distribution of the heat and current fluxes in gas tungsten arcs. *Metallurgical Transactions B*. 16B(12): 1985.

2. Refer to Smartt, H. B., Stewart, J. A., and Einerson, C. J. "Heat transfer in gas tungsten arc welding". Metals Park, OH: American Society for Metals. ASM Metals/Materials Technology Series Paper No. 8511-011.

3. Refer to Glickstein, S. S. and Friedman, E. Weld modeling applications. *Welding Journal*. 63(9): Sept. 1984; 38–42.

4. Refer to Eager, T. W. and Tsai, N. S. Temperature fields produced by traveling distributed heat sources. *Welding Journal*. 62(12): 1983; 346s–355s.

Figure 3.1B—Comparison of GTAW Efficiency from Several Investigators

in the vicinity of the weld metal and heat-affected zone is not of interest, the heat input distribution is not important, and the heat input can be treated as a concentrated heat source. The heat input and the weld speed are then sufficient to determine the thermal cycle.

For arc welding processes, the deposition of heat may be characterized as a distributed heat flux on the weldment surface. Assuming that the heat from the welding arc is applied at any given instant of time as a normally distributed heat flux then the rate of the heat generation is given by

$$q(r) = \frac{3IE\,f_1}{\pi\,\bar{r}^2} \exp\left\{-3(r/\bar{r})^2\right\} \tag{3.4}$$

where r is the distance from the center of the heat source on the surface and $\bar{r}$ is a characteristic radial dimensional distribution parameter that defines the region in which 95 percent of the heat-flux is deposited.[5] This type of thermal energy-density distribution is often used as an approximation as illustrated in Figure 3.4.[6] For an arc

5. Friedman, E. Thermomechanical analysis of the welding process using the finite element method. *Trans. ASME, J. Pressure Vessel Techn.* 97 Series J (3): August 1975, 206-213; Pavelic, R., Tanbuachi, R., Uyehara, O. A., and Myers, P. S. Experimental and computed temperature histories in gas tungsten-arc welding of thin plates. *Welding Journal, Research Supplement*. 48(7): July 1969; 295s-305s.

6. Tsai, N. S., and Eager, T. W. Distribution of the heat and current fluxes in gas tungsten arcs. *Metallurgical Transactions B*. 16B(12): 1985.

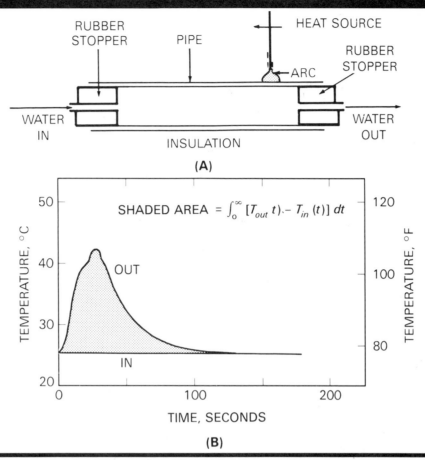

Figure 3.2—Measurement of Arc Efficiency in GTAW Welding: (A) Calorimeter; (B) Temperature Variations

moving at speed v, the time-dependent power density at a cross-section of the weldment normal to the welding direction is given by[7]

$$q(x,t) = \frac{3IE\,f_1}{\pi\,\bar{r}^2} \exp\left\{-3\,(x/\bar{r})^2\right\} \exp\left\{-3\,(vt/\bar{r})^2\right\} \quad (3.5)$$

where x is the distance from the center of the heat source on the weldment surface in this cross-section and t is the time after the center of the heat source passes over the cross-section of interest.

With GTAW, the arc becomes more constricted and the weld depth-to-width ratio becomes greater as the effect of vertex angle of the conical tip of the tungsten electrode increases as shown in Figure 3.5.[8] As the arc becomes more constricted, the heat source is more con-centrated. This increases the joint penetration. The electromagnetic force in the weld pool also increases as the arc becomes more constricted, and produces a change in weld pool motion. This favors convection heat transfer from the arc to the bottom of the weld pool, thus producing deeper joint penetration.[9] The shielding gas composition and minor alloying elements can also affect the energy-density distribution of the arc (and weld pool convection), and hence the shape of the resultant weld.[10]

7. Friedman, E. Thermomechanical analysis of the welding process using the finite element method. *Trans. ASME J. Pressure Vessel Techn.* 97 Series J (3): August 1975; 206-213.
8. Key, J. F. Anode/cathode geometry and shielding gas interrelationships in GTAW. *Welding Journal.* 59(12): December 1980; 364s-370s.

9. Kou, S. and Sun, K. K. Fluid flow and weld penetration in stationary arc welds. *Metallurgical Transactions A.* 16A(2): February 1985; 203-213.
10. Key, J. F. Anode/cathode geometry and shielding gas interrelationships in GTAW. *Welding Journal.* 59(12): December 1980, 364s-370s; Glickstein, S. S. and Yeniscavich, W. "A review of minor element effects on the welding arc and weld penetration." New York: Welding Research Council Bulletin No. 226, May 1977; Glickstein, S. S. Arc modeling for welding analysis. *International Conference on Arc Physics and Weld Pool Behavior.* London: The Welding Institute, May 1979; Heiple, C. R., Burgardt, P. and Roper, J. R. Control of GTA weld pool Shape. *Advances in Welding Science and Technology.* (1986).

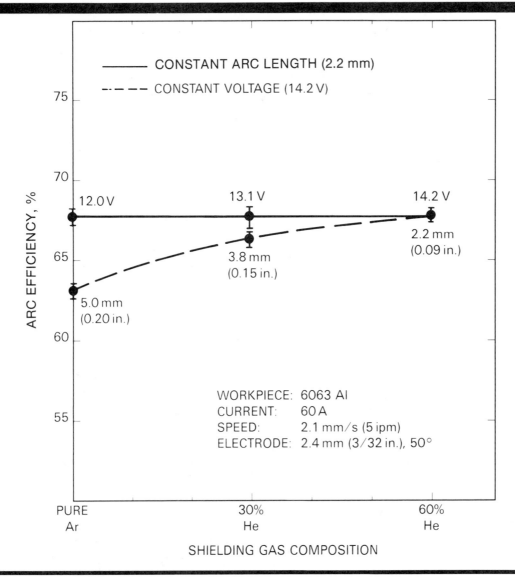

Figure 3.3—Results of Arc Efficiency Measurement in GTAW Welding

HEAT TRANSFER IN A WELDMENT

THE HEAT FLOW pattern resulting from a moving heat source is three-dimensional; temperature gradients exist through the thickness of the weld as well as in directions parallel and transverse to the welding direction. However, this heat flow pattern is somewhat simplified when

(1) The energy from the welding heat source is applied at a uniform rate.

(2) The heat source is moving at constant speed on a straight-line path relative to the workpiece.

(3) The cross-section of the weld joint is constant.

(4) The end effects resulting from initiation and termination of the weld are neglected.

Under these conditions, the time-dependent temperature distribution at any cross-section of the weld normal to the welding direction is the same as that at any other cross-section except for a time lag between the responses at the two cross-sections. This reduces the complexity of determining the weld thermal cycle to that of a pseudo-two-dimensional treatment. This type of weld thermal cycle is termed *quasistationary* and most often has been the basis for analytical models to calculate welding temperatures. If any of the above four conditions are not met, the simplified two dimensional treatment is not valid, and a three-dimensional treatment is necessary.

A typical weld thermal cycle is such that the temperature at a point in the workpiece increases rapidly as heat is transferred from the heat source. The rate of increase

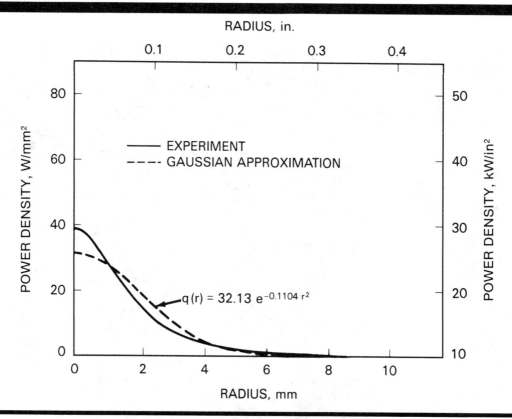

Figure 3.4—Power Density Distribution in a Gas Tungsten Arc

varies inversely with the distance from the weld center-line. When the temperature reaches a peak value, cooling commences at a rate that also varies inversely with the distance from the weld centerline. The cooling rate then decreases with time. Results of calculations of weld thermal cycles are illustrated in Figure 3.6.[11]

The thermal cycles vary from location to location in the weldment, with the variations depending on a host of variables related to heat input, weldment geometry, and material properties. The variation of the peak temperature with both the magnitude and the distribution at the heat input for a nickel base alloy bead-on-plate weld is shown in Figure 3.7. This figure shows that the distribution variable (r) had a small effect on the peak temperature, while the heat input had a significant effect on the peak temperature.

The effect of weld speed on peak temperature at constant heat input (H_{net}) is illustrated in Figure 3.8. The distribution of the peak temperature in the vicinity of the centerline of the weld determines both the size and the shape of the weld pool and the heat-affected zone.

When the peak temperature does not exceed the melting temperature, the material remains solid during the entire welding cycle, and the heat is transferred by conduction. The molten weld pool is stirred by a number of mechanisms, and within the pool, heat is transferred by convection. Heat is lost from the pool to the atmosphere by evaporation or boiling, and by radiation. Heat is lost to the base metal by conduction. These are complex phenomena and they can be extremely important in determining weld penetration.[12] After the weld metal freezes, heat is transferred exclusively by conduction.

As the metal melts, it absorbs heat called the *heat of fusion*. The heat of fusion is liberated as the metal freezes, and during these reactions, the heating and cooling rates are affected. Curves passing through the phase change temperature range of Figure 3.6 illustrate the effect.

DISSIPATION OF WELDING HEAT

THERMAL ENERGY APPLIED to the weld zone is distributed by conduction in the weldment causing a tempera-

11. Friedman, E. Thermomechanical analysis of the welding process using the finite element method. *Trans. ASME, J. Pressure Vessel Techn.* 97 Series J (3): August 1975; 206-213.

12. Heiple, C. R., Burgardt, P. and Roper, J. R. Control of GTA Weld pool shape. *Advances in Welding Science and Technology* (1986).

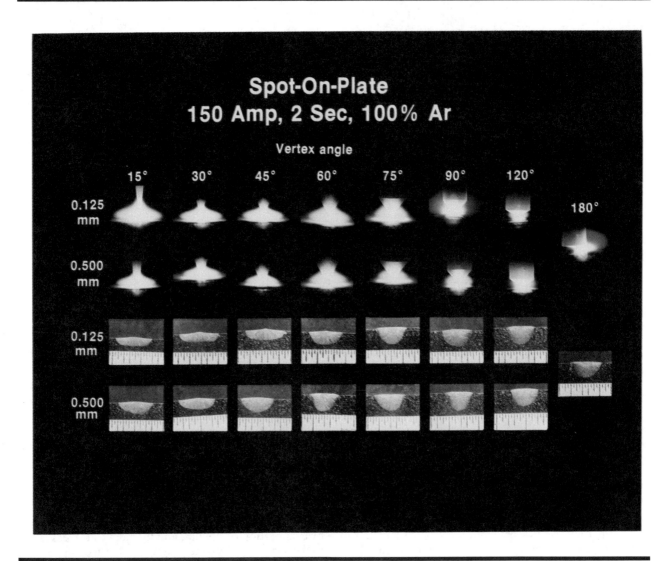

Figure 3.5—Arc Shape and Weld Bead Geometry as a Function of Electrode Tip Angle in a Pure Argon Shield

ture rise followed by a combination of heat loss to the environment and heat transfer to the base metal and the welding fixtures. The cooling rate experienced in the weldment is a function of the rate of energy dissipation.

The welding heat is lost to the surrounding atmosphere by radiation and convection. Radiation losses are dominant close to the weld pool because the losses are proportional to the fourth power of the absolute temperature (T^4) of the metal. At some distance from the weld pool, convection is the primary mechanism for heat loss to the atmosphere. The remaining heat is conducted throughout the workpiece until thermal equilibrium is established. Eventually, all of the heat is lost to the atmosphere and fixtures. When fixtures are in close proximity to the weld zone, the heat flows into the fixture and increases the rate of heat

dissipation in the workpiece. A backing bar is an example of such a fixture.

The rate of energy loss from the weld zone is a direct function of the heat conductivity and temperature gradient and an inverse function of the heat capacity. The difference in thermal properties of argon and helium which are sometimes used as shielding gases can also have an important affect on the cooling rate.

WORKING SOLUTIONS TO HEAT FLOW EQUATIONS

SOLUTIONS TO HEAT flow equations resulting from models of energy distributions require the use of computers and programming expertise. This is particularly true for

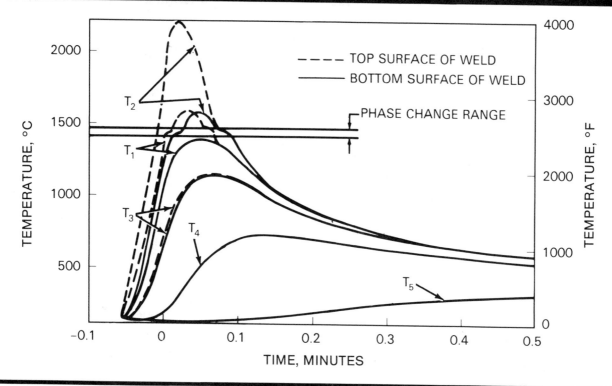

Figure 3.6—Temperature Histories on Top and Bottom Surfaces at Various Distances from the Weld Centerline

the welding processes such as gas tungsten arc, plasma arc (keyhole), laser and electron beam welding where the weld nugget shape is significantly different from that of conventional consumable electrode arc welding processes. However, for conditions that result in approximately semicircular or oval weld bead cross sections, the equations can be simplified by assuming a point heat source. Under this assumption, the ratio of the welding energy entering the base metal per unit time to the weld travel speed (H_{net}) becomes a principal variable in the heat flow equations.

The following sections provide practical working equations for consumable electrode welding applications and other weld processes. The equations can be used for estimating some of the important quantities such as:

(1) Cooling rates
(2) Peak temperatures
(3) Width of the heat-affected zone
(4) Solidification rates

COOLING RATES

THE FINAL METALLURGICAL structure of the weld zone is primarily determined by the cooling rate from the maxi-

mum temperature achieved during the weld cycle (peak temperature). Cooling rate is of particular importance with the heat-treatable steels. The critical cooling rate for the formation of martensite in these steels is often commensurate with those likely to be encountered in welding.[13] In general, this condition is not true of other metals.

For example, in welding aluminum, the cooling rates developed are almost always much higher than any which might be regarded as metallurgically "critical"; therefore, calculation of cooling rates actually contributes very little to the understanding of aluminum welding metallurgy.

With carbon and low alloy steels, the temperature at which the cooling rate is calculated is not critical, but should be the same for all calculations and comparisons. Cooling rate calculations in this chapter will be made at 550°C (1020°F). This temperature (T_c) is quite satisfactory for most steels.

The major practical use of cooling rate equations is in the calculation of preheat requirements. For instance, consider a single weld pass in making a butt joint between two plates of equal thickness. If the plates are relatively thick, requiring several passes (more than six)

---

13. Martensite and other steel microstructures are discussed in Chapter 4, Welding Metallurgy.

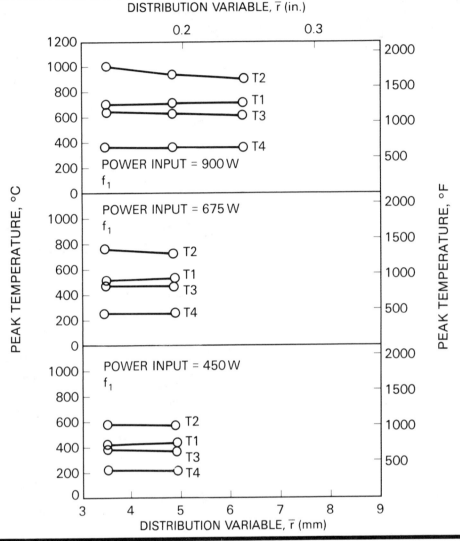

Figure 3.7—Calculated Peak Temperatures at Different Positions T₁ in the Weldment. Plate Thickness = 6.35 mm (1/4 in.) Weld Speed = 4.133 mm/s (9.8 in./s).

to complete the joint, the cooling rate, R, can be approximated by:[14]

$$R = \frac{2\pi k \, (T_c - T_o)^2}{H_{net}} \tag{3.6}$$

where
  R = the cooling rate at the weld center line, °C/s
  k = thermal conductivity of the metal, J/mm·s·°C
  T_o = the initial plate temperature

14. Adams, C. M. Jr. Cooling rates and peak temperatures in fusion welding. *Welding Journal Research Supplement.* 37(5): 1958; 210s-215s.

  T_c = the temperature at which the cooling rate is calculated, °C

For the calculations of this section, k is assumed to be 0.028 J/mm·s·°C

Strictly speaking, the cooling rate is a maximum on the weld center line, and it is this maximum which is given by the equation. However, at the temperature of interest (i.e., $T_c = 550$°C) the cooling rate near the weld fusion boundary is only a few percent lower than on the center line. Accordingly, the cooling rate equation applies to the entire weld and the immediate heat-affected zone.

At temperatures near the melting point [T =1600°C (2900°F)], the cooling rates at the centerline and the

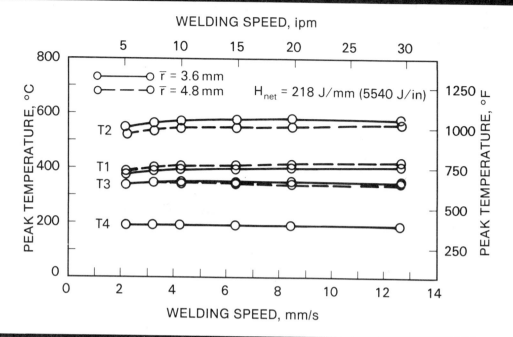

Figure 3.8—Variation of Peak Temperatures with Welding Speed

fusion boundary may differ significantly, and equation 3.6 is only valid at the centerline.

If the plates are relatively thin, requiring fewer than four passes:[15]

$$R = 2\pi k\rho C\left(\frac{h}{H_{net}}\right)^2\left(T_c - T_o\right)^3 \qquad (3.7)$$

where
 h = thickness of the base metal, mm
 ρ = density of base metal, g/mm³
 C = specific heat of base metal, J/g °C
 ρC = volumetric specific heat, 0.0044 J/mm³·°C
   for steels

The distinction between thick and thin requires some explanation. The thick plate equation is used when heat flow is three-dimensional—downward as well as lateral from the weld. The thick plate equation would apply, for example, to a small bead-on-plate weld pass deposited on thick material.

The thin plate equation would apply to any weld in which the heat flow is essentially lateral; that is, a base metal thin enough that the difference in temperature between the bottom and top surfaces is small in comparison to the melting temperature. Cooling rates for

any single pass, complete penetration welding (or thermal cutting) application should be calculated using the thin plate equation, 3.7.

Sometimes, it is not obvious whether the plate is thick or thin because the terms have no absolute meaning. For this reason, it is helpful to define a dimensionless quantity called "the relative plate thickness, τ":

$$\tau = h\sqrt{\frac{\rho C(T_c - T_o)}{H_{net}}} \qquad (3.8)$$

The thick plate equation applies when τ is greater than 0.9, and the thin plate equation when τ is less than 0.6. When τ falls between 0.6 and 0.9, the upper bound of the cooling rate is given by the thick plate equation, and the lower bound by the thin plate equation. Relative plate thicknesses are illustrated in Figure 3.9.

If an arbitrary division is set when τ is equal to 0.75, with larger values regarded as thick and smaller as thin, the maximum error may not exceed 15 percent in a cooling rate calculation. The error in applying the equations to the calculation of preheat requirements is minor.

PREHEAT TEMPERATURE AND CRITICAL COOLING RATE

FROM THE COOLING rate equations, increasing the initial uniform temperature, T_o, of the base metal being welded has the effect of reducing the cooling rate. Preheat is

15. Jhaveri, P., Moffatt, W. G., and Adams, C. M. Jr. The effect of plate thickness and radiation on heat flow in welding and cutting. *Welding Journal.* 41(1): 1962; 12s-16s.

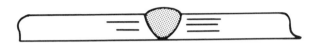

(A) Three Dimensional Heat Flow
$\tau \geq 0.9$

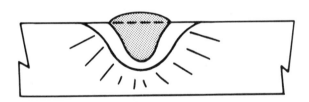

(B) Two Dimensional Heat Flow
$\tau \leq 0.6$

(C) Intermediate Condition — Neither Thick Nor Thin. $0.6 < \tau < 0.9$

Figure 3.9—Effect of Weld Geometry and Relative Plate Thickness on Heat Flow Characteristics

often used for this purpose in welding hardenable steels. For each steel composition, there is a critical cooling rate; if the actual cooling rate in the weld metal exceeds this critical value, hard martensitic structures may develop in the heat-affected zone, and the risk of cracking under the influence of thermal stresses in the presence of hydrogen is enhanced. The cooling rate equation can be used (1) to determine the critical cooling rate (under welding conditions), and (2) to calculate preheat temperatures that avoid the formation of hard heat-affected zones.

When welding hardenable steel, the first problem is to determine the critical cooling rate. The simplest and most direct way of doing this is to make a series of bead-on-plate weld passes in which all variables, except the arc travel speed, are held constant. Suppose, for example:

$E = 25$ V
$I = 300$ A
$h = 6$ mm (1/4 in.)
$f_1 = 0.9^{**}$
$T_o = 25$ °C (76 °F)
$T_c = 550$ °C (1022 °F)

The bead-on-plate weld passes are deposited at 6, 7, 8, 9, and 10 mm/s (14, 17, 19, 21, and 24 in./min). Hardness tests are performed on the weld cross sections, and it is found that structures having high hardnesses have developed in the heat-affected zones of the welds deposited at 9 and 10 mm/s (21 and 24 in./min), but not in the others. Thus, the critical cooling rate was encountered at some travel speed above 8 mm/s (19 in./min). More to the point, the pass deposited at 8 mm/s experienced a cooling rate which is a maximum "safe" value. In this circumstance, according to Table 3.1, the energy input is

$$H_{net} = f_1 \frac{EI}{v} = 844 \text{ J/mm}$$

The relative plate thickness can be calculated from equation 3.8 as:

$$\tau = 0.31$$

Therefore, the thin plate equation applies, and the cooling rate can be calculated using equation 3.7:

$$R = 5.7 \text{ °C/s (10.3 °F/s)}$$

This result means that approximately 6 °C/s (11 °F/s) is the maximum safe cooling rate for this steel. Preheat can be used in actual welding operations to reduce the cooling rate to 6 °C/s or less.

For the following welding conditions, the minimum preheat can be determined to result in a maximum cooling rate of 6 °C/s (11 °F/s):

$E = 25$ V
$I = 250$ A
$v = 7$ mm/s (16.5 in./min)
$h = 9$ mm (3/8 in.)
$f_1 = 0.9$

Assuming the thin plate equation still applies, 6 °C/s (11 °F/s) is substituted for R and equation 3.7 is solved for T_o. Thus

$$T_o = 162 \text{ °C (320 °F)}$$

** Approximate value assumed as 0.9 for purposes of calculation.

The thin plate assumption is verified to be correct by calculating using equation 3.8,

$$\tau = 0.41$$

If it had been incorrectly assumed that the thick plate condition was applied, equation 3.6 would have been used, and a minimum required preheat of 384 °C (723 °F) would have been calculated. However, on substituting the appropriate values into equation 3.8, the relative plate thickness would be

$$\tau = 0.27$$

Thus, the assumption of a thick plate condition would be shown to be incorrect.

Therefore, a minimum preheat temperature of 162 °C (324 °F) will result in a maximum cooling rate of 6 °C/s (11 °F/s), and high-hardness martensite will not form in the weld heat-affected zone.

If the same energy input were to be used on a 25 mm (1 in.) thick plate, the preheat temperature that would avoid a high hardness heat-affected zone could be calculated using the thin plate formula equation (3.7) and

$$T_o = 354 \text{ °C (670 °F)}$$

The relative plate thickness can then be determined as 0.82, which is neither thick nor thin.

Using the equation 3.6 shows that a thick plate assumption is not valid either.

$$T_o = 389 \text{ °C (733 °F)}$$
and $\tau = 0.74$

Neither of the two calculated preheat values is exactly correct. However, the difference is not of practical importance. In this case, selecting the higher preheat temperature is prudent.

Repeating the exercise for a 50 mm (2 in.) plate, but with the same energy input, the thick plate equation (3.6) applies, and again:

$$T_o = 389 \text{ °C (733 °F)}$$

Using this preheat temperature to calculate the relative thickness in equation 3.8 results in:

$$\tau = 1.48$$

Because τ is greater than 0.9 the assumption that equation 3.6 applies is valid.

Selecting preheat temperatures to avoid high hardness heat-affected zones should be guided by experience as well as by calculation whenever possible. The optimum preheat temperature is the lowest one that maintains the cooling rate somewhat below the critical temperature (allowing for a margin of safety).

The purpose of these calculations is to determine the minimum preheat necessary to achieve a maximum cooling rate of 6 °C/s (11 °F/s). Thus, if the calculation shows that the minimum preheat is −20 °C (−5 °F) then a 25 °C (76 °F) preheat would also result in a cooling rate less than 6 °C/s. In actual practice, even when cooling rate considerations do not so dictate, a modest preheat is frequently used with the hardenable steels to remove condensed moisture on the work surfaces.

The cooling of a weld depends on the available paths for conducting heat into the surrounding cold base metal. It may therefore be necessary to modify the heat input for equation 3.7 and 3.8 to account for an additional path. Thus, if a weld is deposited as a fillet weld in a 9 mm (3/8 in.) T-joint (Figure 3.10A) with a heat input of 804 J/mm (20 400 J/in.), the web plate would absorb some of the heat, and the apparent thickness would be greater than indicated. To compensate for the extra heat sink, the heat input should be reduced by 1/3 to 536 J/mm, (13 600 J/in.), and the minimum preheat to avoid martensite formation determined from equation 3.7 as:

$$T_o = 254 \text{ °C (490 °F)}$$

That a thin plate condition exists can be verified by using equation 3.8. Thus τ is 0.44, and the thin plate assumption is correct.

However, if the thick plate condition exists, the extra plate cannot provide additional apparent thickness, as shown in Figure 3.10B, and the heat input should not be reduced.

The base metal of multipass welds can be preheated by the initial weld passes. The effect of multiple passes on the calculated and measured weld thermal cycles is illustrated in Figure 3.11, and the effect on cooling is shown in Figure 3.12.[16] The effects are greatest on the second pass. As the preheat condition stabilizes after the second pass, the change in thermal cycle and the resultant change in cooling rate are less significant.

CHARACTERISTIC WELD METAL COOLING CURVES

TO INVESTIGATE THE effect of welding technique on weld metal cooling rates, cooling curves have been mea-

16. Glickstein, S. S. and Friedman, E. Weld modeling applications. *Welding Journal.* 63(9): September 1984; 38-42.

**(A) Thin Plates — Extra Path Increases
Apparent Base Metal Thickness**

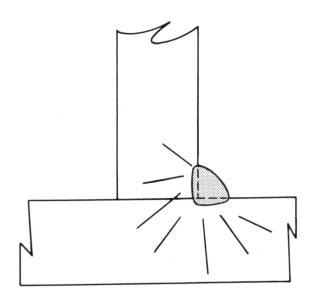

**(B) Thick Plates — Extra Path Does Not
Increase Apparent Base Metal Thickness**

**Figure 3.10—The Effect of the Number of Heat-Flow
Paths on Apparent Plate Thickness**

sured for a wide range of submerged arc welding techniques.[17] Temperature-time curves can be obtained by

17. Dorschu, K. E. and Lesnewich, A. Development of a filler metal for high toughness alloy plate steel with a minimum yield strength of 140 ksi. *Welding Journal.* 43(12): 1964, 564s-576s; Krause, Gregory T. Heat flow and cooling rates in submerged arc welding. Master's Thesis. Ohio State University, Department of Welding Engineering, 1978.

plunging a platinum-platinum 13% rhodium thermocouple into the melted weld pool behind the arc. A typical temperature time curve is shown in Figure 3.13. This curve is typical of all arc welding studies, and results from the common situation of a high local temperature being reduced by conduction of heat to the surrounding relatively large, cold base metal.

Therefore, in the temperature range from 500 to 1000 °C (930 to 1830 °F) the choice of critical temperature T_c is not significant. As shown in Figure 3.14, a linear relationship exists between the cooling rate at 538 °C (1000 °F) and the average cooling rate between 800 and 500 °C (1470 and 930 °F). The relationship between the weld cross-sectional area and cooling rate, Figure 3.15, is consistent whether the cooling rate is measured at 538 °C (1000 °F) or 704 °C (1300 °F). Heating and cooling rates in metals undergoing phase changes are affected by the heat of transformation. This is shown in Figure 3.6 by the change in slope of the cooling curve in the phase change temperature range.

PEAK TEMPERATURES

Peak Temperature Equation

PREDICTING OR INTERPRETING metallurgical transformations at a point in the solid metal near a weld requires some knowledge of the maximum or peak temperature reached at a specific location. For a single pass complete joint penetration butt weld in sheet or plate, for example, the distribution of peak temperatures in the base metal adjacent to the weld is given by[18]

$$\frac{1}{T_p - T_o} = \frac{\sqrt{2\pi e}\rho C\,hy}{H_{net}} + \frac{1}{T_m - T_o} \qquad (3.9)$$

where
$T_p =$ the peak or maximum temperature, °C, at a distance, y (mm), from the weld fusion boundary (The peak temperature equation does not apply at points within the weld metal, but only in the adjacent heat-affected zone.)

$T_m =$ melting temperature, °C (specifically, liquidus temperature of the metal being welded)

$e = 2.718$, base of the natural logarithms

At the fusion boundary ($y = 0$), the peak temperature (T_p) is equal to the melting temperature (T_m).

18. Adams, C. M. Jr., Cooling rates and peak temperatures in fusion welding *Welding Journal Research Supplement.* 37(5): 1958; 210s-215s.

The peak temperature equation can be used for several purposes including: (1) the determination of peak temperatures at specific locations in the heat-affected zone, (2) estimating the width of the heat-affected zone, and (3) showing the effect of preheat on the width of the heat-affected zone.

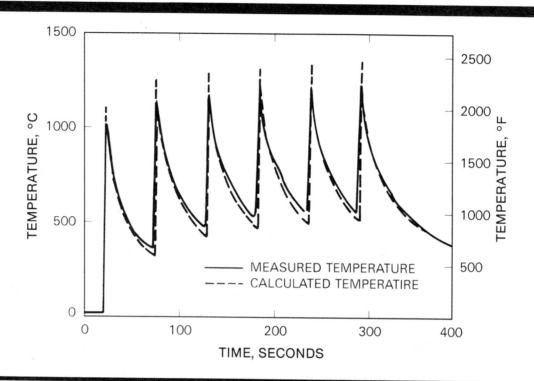

Figure 3.11—The Effect of Multiple Passes on Weld Thermal Cycles

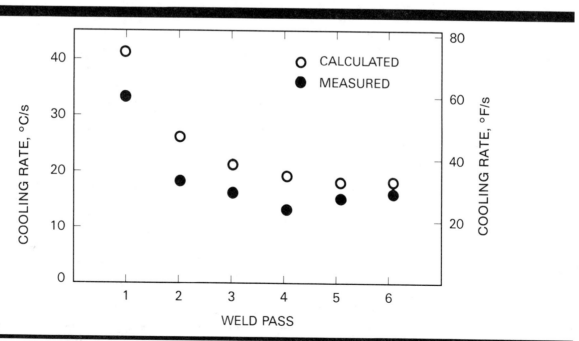

Figure 3.12—Average Cooling Rates for Each Pass in a Multipass Weld

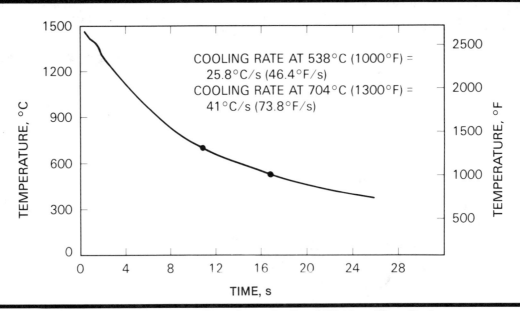

Figure 3.13—Temperature-Time Curve for Bead Weld on Plate Surface

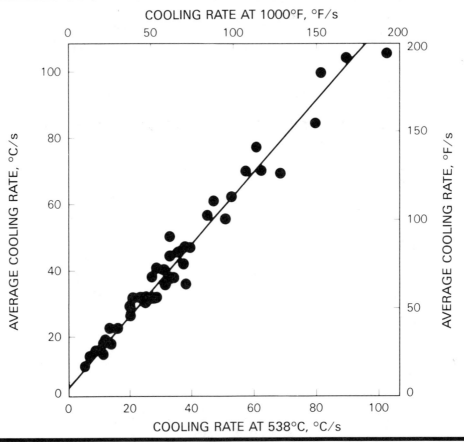

Figure 3.14—Comparison of Cooling Rate as Measured by the Average Cooling Rate from 800 to 500°C and the Tangent at 540°C

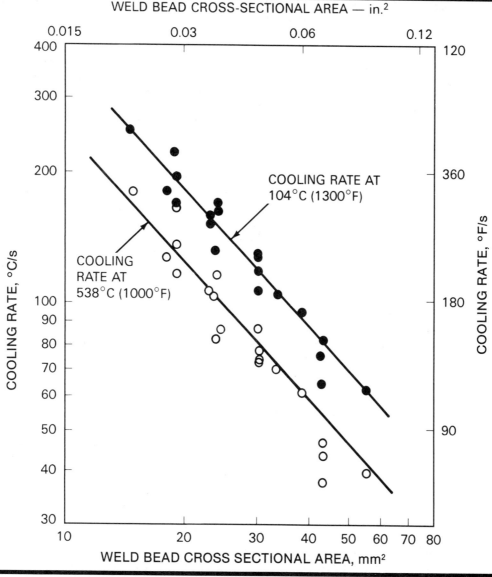

Figure 3.15—Correlation Between Weld Cross Section Area and Cooling Rate at 538 °C and 704 °C

For example, a single, complete joint penetration weld pass is made on steel using the following conditions:

$E = 20$ V

$I = 200$ A

$v = 5$ mm/s (12 in./min)

$T_o = 25$ °C (76 °F)

$T_m = 1510$ °C (2750 °F)

$C = 0.0044$ J/mm³ · °C (40J/in.³ · °F)

$h = 5$ mm (3/16 in.)

$f_1 = 0.9$

$H_{net} = 720$ J/mm (18 300 J/in.)

The peak temperature at 1.5 and 3.0 mm (1/16 and 1/8 in.) from the fusion boundary can be calculated from equation 3.9. Thus:

at $y = 1.5$ mm (1/16 in.), $T_p = 1184$ °C (2160 °F)

and

at $y = 3.0$ mm (1/8 in.), $T_p = 976$ °C (1790 °F)

Calculating Width of Heat-Affected Zone

IF ONE DEFINES a peak temperature below which the welding heat does not affect the properties of the base metal, then equation 3.9 can be used to calculate the width of the heat-affected zone. For most plain carbon and low alloy steels, the mechanical properties are not affected if the peak temperature is below approximately 730 °C (1350 °F). Using 730 °C for T_p in equation 3.9, the width of the heat-affected zone Y_z is 5.9 mm (0.23 in.).

If a heat-treated steel is tempered at 430 °C (806 °F), then any region heated above 430 °C will, in theory, be "over tempered", and would exhibit modified properties. Substituting 430 °C into equation 3.9 will calculate the width of the heat-affected zone including all of the over-tempered region.

Steels which respond to a quench-and-temper heat treatment are frequently preheated prior to welding. From equation 3.9, this treatment has the side effect of widening the heat-affected zone.

One of the simplest and most important conclusions that may be drawn from the peak temperature equation is that the width of the heat-affected zone is proportional to the heat input. However, the peak temperature equation is an approximation that is reasonably accurate within the normal ranges of heat input encountered in production welding operations. Within those limits, the heat-affected-zone width increases proportionally with the net heat input.

Although the peak temperature equation can be very instructive and useful, it is important to recognize certain other restrictions in its application. The most important of these is that the equation is derived for the so-called "thin-plate" condition in which heat conduction takes place along paths which are parallel to the plane of the plate. The equation thus applies to any single pass, complete penetration (welding or thermal cutting) process, regardless of plate thickness. Only small errors result in applying the equation to any complete joint penetration arc weld that can be fabricated with fewer than four passes. The equation must be applied on a per pass basis; however, the interpass temperature (that temperature to which the weld region cools between passes) should be inserted as an accurate value for T_o in the peak temperature equation.

SOLIDIFICATION RATES

THE RATE AT which weld metal solidifies can have a profound effect on its metallurgical structure, properties, and response to heat treatment. The solidification time, S_t (in seconds), of weld metal depends on the net energy input:

$$S_t = \frac{L \, H_{net}}{2\pi \, k\rho C \, (T_m - T_o)^2} \qquad (3.10)$$

where

S_t = solidification time; the time lapse from beginning to end of solidification at a fixed point in the weld metal, s

L = heat of fusion, J/mm³

For steels the heat of fusion is approximately 2 J/mm³.

Solidification time is a function of the energy input and the initial temperature of the metal. If, for example, a weld pass of 800 J/mm (20 300 J/in.) net energy input is deposited on a steel plate, with initial temperature 25 °C (76 °F), then from equation 3.10, the solidification time, S_t, is 0.94 seconds.

By comparison with any casting process, weld metal solidification is extremely rapid. The final microstructure of a weld and a casting are both dendritic, but the differences in structure are much greater than the similarities. The mass of liquid weld metal is extremely small relative to the mass of solid metal in virtually all weldments. The thermal contact between the liquid and solid metal is excellent, and as a result of this combination of factors, the heat is rapidly removed from the weld pool.

In contrast, the liquid metal of most commercial castings is large in comparison to the surrounding mold, and the thermal contact with the mold is not as efficient as in the case of weld metals. This results in long solidification times for castings relative to weld metal. For this reason, it is incorrect to describe weld metal as having a cast structure.

SOLIDIFICATION RATES AND DENDRITE SPACING

SOLIDIFICATION TIME DIRECTLY affects the structure of weld metal. Most alloys of commercial importance freeze dendritically, and an important structural feature of weld metal is the dendrite spacing (more exactly, dendrite arm spacing). The dendrite spacing increases with increasing solidification time. Because weld metal solidification times are rapid, the spacing between dendrites is small, and a very fine microstructure results. Longer solidification times result in larger dendrite spacing and coarser microstructures. The mechanical properties of weld metals with fine microstructures are generally better than those with coarse microstructures. Thus, the final weld metal properties are a result of the solidification time and the weld metal cooling rate.

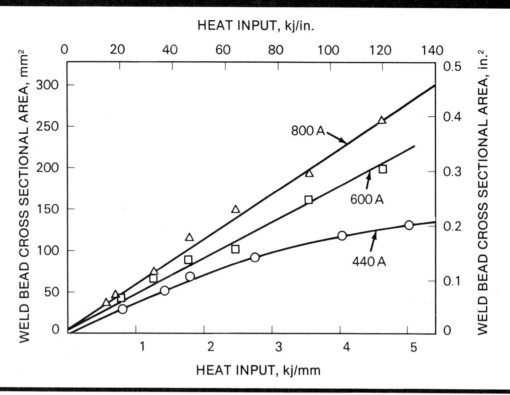

Figure 3.16—Effect of Heat Input on Area

Effects of Heat Input

ACCORDING TO EQUATION 3.10, the solidification time, S_t, is proportional to the heat input. For the normal ranges of arc power and weld travel speeds encountered in commercial welding, the net heat input, H_{net}, can be treated as a single variable. However, if the heat input is increased by reducing the travel speed and increasing the arc voltage at a constant current, the normal relationship between dependent variables such as weld cross-sectional area and cooling times can be distorted.

The weld bead cross sectional area is linearly related to the heat input in Figure 3.16 for increasing heat input at 600 and 800 A. For 440 A, the linear relationship is not valid above about 2000 J/mm (50 800 J/in.). For most commercial work the heat input can be considered as a single variable, but for continuing research studies it is essential that the heat input distribution $Q(x,y)$ and the weld travel speed be considered independently.

Considering the heat distribution, a weld thermal cycle for GTAW in aluminum was calculated for the weld centerline and the fusion boundary. The results are shown in Figure 3.17.[19] This calculation shows that the solidification time at the fusion boundary is greater than

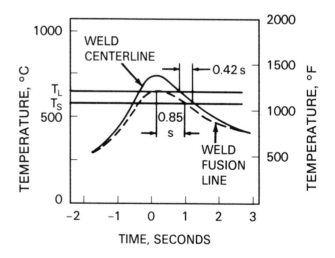

Figure 3.17—Calculated Thermal Cycles and Local Solidification Times at the Centerline and the Fusion Line of a GTA Weld in 3.2 mm (1/8 in.) thick 6061 Aluminum Plate

19. Kou, S. and Le, Y.: Metallurgical Transactions, *14A*: 2245 (1983).

at the centerline. The grain structure at the fusion boundary and centerline for an aluminum plate welded under the same conditions, is shown in Figure 3.18 A and B respectively. The microstructure is finer at the centerline than at the weld fusion line, which is in agreement with the calculated solidification times (S_t). If the heat input is increased, the solidification times at both the fusion boundary and the weld center line increase, and the dendrite spacing increases accordingly.

The mechanical properties of the weld metal such as yield and tensile strength are the average properties of all of the zones of the weld metal tensile specimens, and therefore, when considering overall weldment performance it is useful to consider the heat input as a single variable. However, if the details of specific locations within a weld are of concern, then the heat distribution as well as the heat input should be addressed.

The Effect of Initial Plate Temperature

THE OTHER WELDING variable that affects solidification time and dendrite spacing is the initial plate temperature. However, when the melting temperature, T_m, of a metal is large compared to the initial plate temperature T_o, the effect of changes in T_o in equation 3.10 is small. For materials such as steels, changes in preheat within the usual ranges encountered in production work do not affect the solidification time and therefore, the dendrite spacing. For low melting temperature metals such as aluminum and magnesium, the preheat temperature may have a significant effect on dendrite spacing.

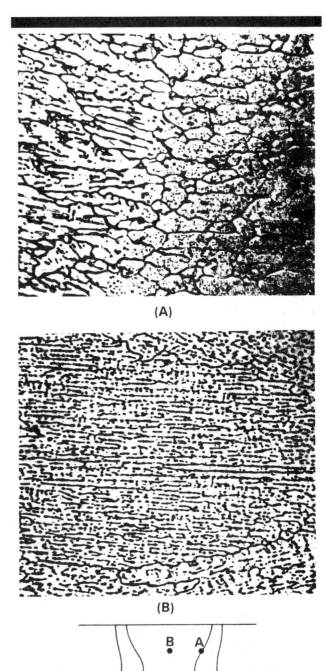

(A)

(B)

Figure 3.18—Subgrain Structure of the Weld Mentioned in Figure 3.17: (A) Fusion Boundary; (B) Weld Centerline

SUPPLEMENTARY READING LIST

Davis, G. J. and Garland, J. G. International Metallurgical Reviews, 20: 83 (1975).

Friedman, E. On the calculation of temperatures due to arc welding. *Nuclear Metallurgy,,* Edited by R.J. Arsenault, J. R. Beeler, Jr, and J. A. Simmons (Apr. 1976): 20, Part II; 1160-70.

Friedman, E. Numerical simulation of the gas tungsten-arc welding process. *Numerical Modeling of Manufacturing Process.* ASME Publication PVP-PB-025. Nov. 1977: 35-47.

Glickstein, S. S. and Friedman, E. *Temperature and distortion transients in gas tungsten-arc weldments.* WAPD-TM-1428. Bettis Atomic Power Laboratory, (Oct. 1979).

Kohns, R., and Jones, S. B. *An initial study of arc energy and thermal cycles in the submerged arc welding of steel.—Research Report.* London: The Welding Institute (1978).

Kou, S. *Welding Metallurgy.* New York: John Wiley and Sons, in press.

Kou, S. and Le, Y. *Metallurgical Transactions. 14A:* 2245 (1983).

Masubuchi, K., *Analysis of welded structures.* Oxford: Pergamon Press (1980).

Myers, P. S., Uyehara, O. A., and Borman, G. L., Fundamentals of heat flow in welding. *Welding Research Council Bulletin.* No. 123: July 1967.

Roest, C. A. and Rager, D. Resistance welding parameter profile for spot welding aluminum. *Welding Journal.* 53 (12): Dec. 1974; 529s-536s.

Rosenthal D. and R. Schmerber, Thermal study of arc welding. *Welding Journal.* 17 (4): April 1938; 2s-8s.

Rosenthal, D. Mathematical theory of heat distribution during welding and cutting. *Welding Journal.* 20 (5): May 1941; 220s-234s.

Rosenthal, D. The theory of moving sources of heat and its applications to metal treatment. *Transactions.* ASME 68 (1946): 819-866.

Rykalin, N. N. *Calculation of heat flow in welding.* Translated by Zvi Paley and C. M. Adams, Jr. Document 212-350-74. Contract No. UC-19-066-001-C-3817. London: International Institute of Welding, (1974).

Wang, K. K. and Lin, W. Flywheel friction welding research. *Welding Journal.* 53 (6): June 1974; 233s-241s.

Wells, A. A. Heat flow in welding. *Welding Journal.* 31 (5): May 1952; 263s-267s.

WELDING METALLURGY

PREPARED BY A COMMITTEE CONSISTING OF:

L. M. Friedman, Chairman
Westinghouse Electric Corporation

W. W. Canary
Teledyne CAE

P. J. Konkol
USX Corporation

S. D. Reynolds Jr.
Westinghouse Electric Corporation

WELDING HANDBOOK COMMIITTEE MEMBER:
L. J. Privoznik
Westinghouse Electric Corporation

CHAPTER 4

WELDING METALLURGY

INTRODUCTION

WELDING INVOLVES MANY metallurgical phenomena. These phenomena such as melting, freezing, solid state transformations, thermal strains and shrinkage stresses can cause many practical problems. The problems can be avoided or solved by applying appropriate metallurgical principles to the welding process.

Welding metallurgy differs from conventional metallurgy in certain important respects, but an understanding of welding metallurgy requires a broad knowledge of general metallurgy. For this reason, general metallurgy is addressed first, and then the specific aspects of welding metallurgy are discussed. The survey of general metallurgy is by no means exhaustive, and those who wish to increase their knowledge of the discipline are directed to specific references in the Supplementary Reading List.

GENERAL METALLURGY

STRUCTURE OF METALS

SOLID METALS HAVE a crystalline structure in which the atoms of each crystal are arranged in a specific geometric pattern. This orderly arrangement of the atoms, called a lattice, is responsible for many of the properties of metals. The most common lattice structures found in metals are listed in Table 4.1, and their atomic arrangements are illustrated in Figure 4.1.

In the liquid state, the atoms composing metals have no orderly arrangement. As the liquid metal approaches the solidification temperature, solid particles called nuclei begin to form at preferred sites, as shown in Figure 4.2(A). Solidification proceeds, Figure 4.2(B), as the individual nuclei grow into larger solid particles called grains. As the amount of solid metal increases, of course, the amount of liquid metal decreases proportionately, and the grains grow larger until there is no liquid between them. The grains meet at irregular boundaries called grain boundaries, Figure 4.2(C).

Each grain in a pure metal at any particular temperature has the same crystalline structure and the same atomic spacing as all other grains. However, each grain grows independently of every other grain, and the orientation of the grain lattice differs from one grain to another. The periodic and orderly arrangement of the atoms is disrupted where the grains meet, and the grain boundaries form a continuous network throughout the metal. Because of this grain boundary disorder, there often are differences in the behavior of the metal at those locations.

Up to this point, only pure metals have been considered. However, most common engineering metals contain residual or intentionally added metallic and nonmetallic elements dissolved in the matrix. These ingredients, called alloying elements, affect the properties of the base metal. The atomic arrangement (crystal structure), the chemical composition, and the thermal and mechanical history have an influence on the properties of an alloy.

Table 4.1
Crystal Structures of Common Metals

A. Face Centered Cubic [Figure 4.1(A)]

Aluminum	Iron[b]
Cobalt[a]	Lead
Copper	Nickel
Gold	Silver

B. Body Centered Cubic [Figure 4.1(B)]

Chromium	Titanium[c]
Iron[b]	Tungsten
Molybdenum	Vanadium
Columbium	Zirconium[c]

C. Hexagonal Close Packed [Figure 4.1(C)]

Cobalt[a]	Titanium[c]
Magnesium	Zinc
Tin	Zirconium[c]

a. Cobalt is face-centered cubic at high temperature and transforms to hexagonal close packed at lower temperatures.
b. Iron is body-centered cubic near the melting temperature and again at low temperatures, but at intermediate temperatures iron is face-centered cubic.
c. Titanium and Zirconium are body-centered cubic at high temperature and hexagonal close packed at lower temperatures.

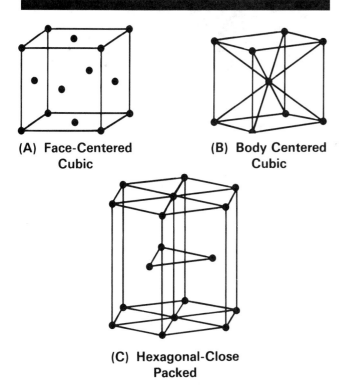

(A) Face-Centered Cubic

(B) Body Centered Cubic

(C) Hexagonal-Close Packed

Figure 4.1—The Three Most Common Crystal Structures in Metals

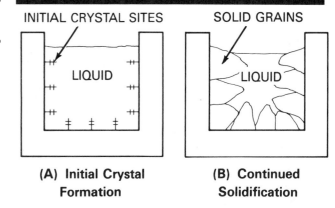

(A) Initial Crystal Formation

(B) Continued Solidification

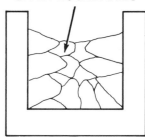

(C) Complete Solidification

Figure 4.2—Solidification of a Metal

Alloying elements, called *solutes*, are located in the parent metal matrix in one of two ways. The solute atoms may occupy lattice sites replacing some atoms of the parent metal atoms, called the *solvent*. Alternatively, if the solute atoms are small enough, they may fit into spaces between the solvent atoms.

Substitutional Alloying. If the solute atoms occupy sites at the lattice locations as shown in Figure 4.3(A), then the type of alloy is called a substitutional solid solution. Examples of substitutional solid solutions are gold dissolved in silver, and copper dissolved in nickel.

Interstitial Alloying. When the alloying atoms are small in relation to the parent atoms, they can locate (or dissolve) in the spaces between the parent metal atoms without occupying lattice sites. This type of solid solution is called *interstitial*, and is illustrated in Figure 4.3(B). Small amounts of carbon, nitrogen, and hydrogen can alloy interstitially in iron and other metals.

Multiphase Alloys

FREQUENTLY, THE ALLOYING atoms cannot dissolve completely, either interstitially or substitutionally. The

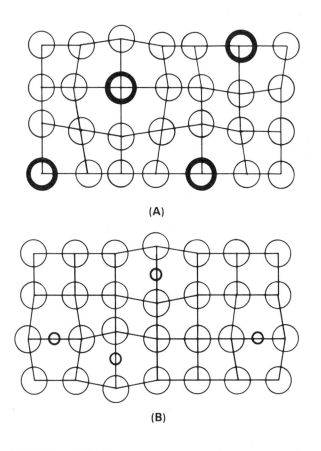

(A)

(B)

Figure 4.3—Schematic Illustration of Substitutional and Interstitial Solid Solutions

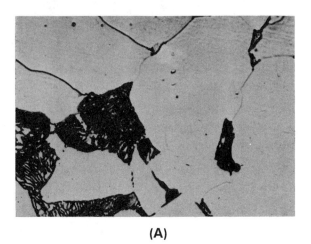

(A)

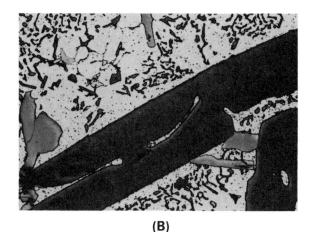

(B)

Figure 4.4—Typical Microstructure of Two Phase Pearlitic Low Carbon Steel. (A) Light Areas are Ferrite, and Dark Areas are Pearlite. (B) Fine Grain Sample with Small Pearlite Patches

result, in such cases, is the formation of mixed atomic groupings (different crystalline structures) within a single alloy. Each different crystalline structure is referred to as a phase, and the alloy is called a multiphase alloy. The individual phases may be distinguished one from another, under a microscope at magnifications of 50 to 2000 times, when the alloy is suitably polished and etched. The process of polishing, etching, and examining metals at some magnification is called metallography. Metallographic examination is one way of studying the many characteristics of metals and alloys.

Two examples of multiphase alloys are shown in Figure 4.4. The typical microstructure of low-carbon pearlitic steel is shown in Figure 4.4(A). The light areas are ferrite and the dark areas are pearlite. The latter structure is composed of two phases, ferrite and iron carbide. Figure 4.4(B) shows multiple phases within the grains of an aluminum-silicon alloy.

Commercial metals consist of a primary or basic element and smaller amounts of one or more alloying elements. The alloying elements may be intentionally

added, or they may be residual (tramp) elements. Commercial metals may be single or multiphase alloys. Each phase will have its own characteristic crystalline structure.

The overall arrangement of the grains, grain boundaries, and phases present in a metal alloy is called the *microstructure* of the alloy. The microstructure is largely responsible for the physical and mechanical properties of the metal. It is affected by chemical composition, thermal treatment, and mechanical history of the metal. Microstructure is affected by welding because of the thermal or mechanical effects, or both, but the changes are confined to the local region of the weld. The metallurgical changes in the local region of the base metal (called the heat-affected zone) can have a profound effect on the service performance of a weldment.

Many unique phenomena that affect the mechanical properties of an alloy at both low and high temperatures take place at grain boundaries where the arrangement of atoms is irregular. There are many vacancies or missing atoms at grain boundaries. The spaces between atoms may be larger than normal, permitting individual atoms to move about with relative ease. Because of this, diffusion of elements (the movement of individual atoms) through the solvent structure generally occurs more rapidly at grain boundaries than it does within the grains. The resulting disarray makes it easier for odd-sized atoms to segregate at the boundaries. Such segregation frequently leads to the formation of undesirable phases that adversely affect the properties of a metal, such as reducing its ductility or increasing its susceptibility to cracking during welding or heat treatment.

Fine-grained metals generally have better mechanical properties for service at room and low temperatures. Conversely, coarse-grained metals generally perform better at high temperatures.

PHASE TRANSFORMATIONS

Critical Temperatures

AT SPECIFIC TEMPERATURES, many metals change their crystallographic structure. For example, the crystalline structure of pure iron at temperatures up to 1670°F (910°C) is body-centered cubic, Figure 4.1(B). From 1670°F to 2535°F (910°C to 1390°C), the structure is face-centered cubic, Figure 4.1(A), and from 2535°F (1390°C) to 2795°F (1535°C), the melting temperature, it is again body-centered cubic. A phase change in crystal structure in the solid state is known as an *allotropic transformation.*

Other metals that undergo allotropic transformations include titanium, zirconium, and cobalt. Chemical composition, cooling rate, and the presence of stress influence the temperature at which transformation takes place.

A metal also undergoes a phase change when it melts or solidifies. Pure metals melt and solidify at a single temperature. Alloys, on the other hand, usually melt and solidify over a range of temperatures. The exception to this rule is the eutectic composition of certain alloys, which will be discussed later.

Phase Diagrams

METALLURGICAL EVENTS SUCH as phase changes and solidification are best shown by means of a drawing called a *phase diagram* (sometimes referred to as an *equilibrium diagram* or a *constitution diagram*).

From a phase diagram of a given alloy system, one can find the phases and percentages of each phase that are present for various alloy compositions at specified temperatures. One can also determine what phase changes tend to take place with a change in either composition or temperature, or both. Phase diagrams only approximately describe commercial alloys because most published phase diagrams are based on two component systems at equilibrium. Most commercial alloys have more than two components, and equilibrium conditions are approached only at high temperatures. Phase diagrams for systems with more than two components are complex and are more difficult to interpret. Nevertheless, they still are the best way to study most alloy systems.

A very simple phase diagram for the copper-nickel alloy system is shown in Figure 4.5. This is a binary system in which both elements are completely soluble in each other in all proportions at all temperatures, in both the liquid and the solid states. Phase diagrams are drawn with the alloy content plotted on the horizontal axis and temperature on the vertical axis. The extreme left hand edge of Figure 4.5 represents 100 percent (pure) nickel, while the extreme right hand edge represents 100 percent (pure) copper.

At temperatures above Curve A in Figure 4.5, called the *liquidus*, the only phase present is liquid metal. At temperatures below Curve B, called the *solidus*, the only phase present is solid metal. All solid alloys in this diagram are homogeneous single-phase solid solutions because copper and nickel are completely soluble in each other in all proportions. One important commercial alloy (30% copper - 70% nickel) remains solid up to 2425°F (1330°C) where it begins to melt. Melting is complete at 2490°F (1365°C). In the region between Curves A and B, solid and liquid phases coexist. Unlike pure metals, most alloys melt and freeze over a range of temperatures. In this system, only pure copper and pure nickel melt and freeze at a constant temperature.

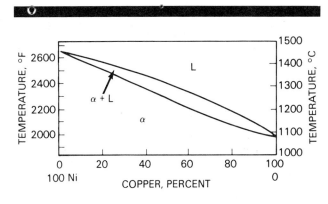

Figure 4.5—Nickel-Copper Phase Diagram

The silver-copper system exhibits a more complex phase diagram, Figure 4.6. This diagram is used extensively in designing brazing alloys. All compositions in this system are entirely liquid at temperatures above the liquidus. Similarly, all compositions are solid at temperatures below the solidus. However, the solid exists as a single phase in two areas of the diagram and as two phases in another area. The silver-rich phase is called alpha (α), and the copper-rich phase is called beta (β). Both phases are face-centered cubic, but the chemical compositions and the crystal dimensions are different. In the region between the solidus and liquidus lines, the liquid solution is in equilibrium with either α or β phase. Finally, the area labeled $\alpha+\beta$ contains grains of both alpha and beta.

This phase diagram illustrates another feature—the *eutectic point*. Alloys of eutectic composition solidify at a constant temperature. The eutectic composition solidifies differently than pure metals in that small quantities of alpha and beta phases freeze alternately exhibiting intermingled grains of α and β in the microstructure. For this reason, eutectic composition microstructures have a distinctive appearance.

The boundary between the β and the $\alpha+\beta$ regions in Figure 4.6 represents the solubility limit of silver in copper. The β solubility increases with increasing temperature, which is typical for most liquid and solid solutions.

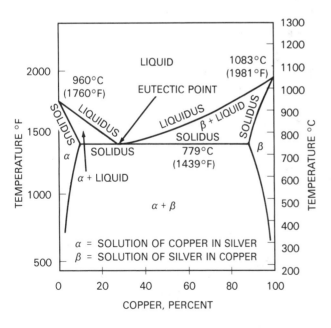

Figure 4.6—Silver-Copper Phase diagram

EFFECTS OF DEFORMATION AND HEAT TREATMENT

Deformation and Annealing of Metals

WHEN METALS ARE plastically deformed at room temperature, a number of changes take place in the microstructures. Each individual grain must change shape to achieve the overall deformation. As deformation proceeds, each grain is deformed, and as a result becomes stronger, making it more difficult to deform it further. This behavior is called *work hardening*. The effect of cold working on the strength and ductility of a metal is illustrated in Figure 4.7(A). The original properties are partially or completely restored by heat treatment as shown in Figure 4.7(B). The microstructures of mildly deformed, heavily deformed, and stress-relieved metals are shown in Figures 4.8(A), (B), and (C) respectively. When the metal is deformed below a critical temperature, there is a gradual increase in the hardness and strength of the metal and a decrease in ductility. This phenomenon is known as *cold working*.

If the same metal is worked moderately [Figure 4.8(A)] or severely [Figure 4.8(B)] and then heated to progressively higher temperatures, several things happen. At temperatures up to about 400°F (205°C), there is a steady decline in the residual stress level but virtually no change in microstructure or properties. At about 400 to 450°F (205 to 230°C), the residual stress has decreased to a relatively low level, but the microstructure has not changed [Figure 4.8(A) and (B)]. The strength of the metal is still relatively high and the ductility, while improved, is still rather low. The reduction in stress level and the improvement in ductility are attributed to the metallurgical phenomenon called *recovery*, a term indicating a reduction in crystalline stresses without accompanying microstructural changes.

When the cold-worked metal is heated to a temperature above 450°F (230°C), mechanical property changes become apparent, as do changes in microstructure. In place of the deformed grains found in Figure 4.8(A) or (B), a group of new grains form and grow [Figure 4.8(C)]. These grains consume the old grains, and eventually all signs of the deformed grains disappear. The new microstructure resembles the microstructure prior to cold-working, and the metal is now softer and more ductile than it was in the cold-worked condition. This process is called *recrystallization*, a necessary part of annealing procedures. (Annealing refers to a heating and cooling process usually applied to induce softening.) When heated to higher temperatures, the grains begin to grow and the hardness and strength of the metal are significantly reduced. Metals are often annealed prior to further cold working or machining.

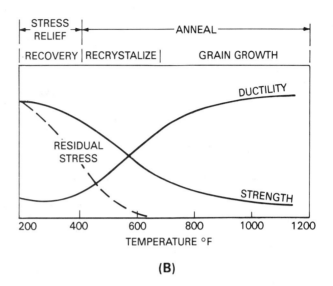

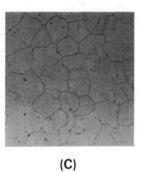

(A)

(B)

Figure 4.7—(A) The Effect of Cold Work on Strength and Ductility of Metals. (B) The Effect of Post Cold Work Heat Treatment on the Strength and Ductility of Metals

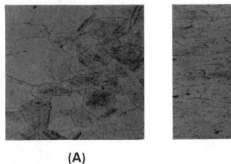

(A)

(B)

(C)

Figure 4.8—Grain Structure of (A) Lightly Cold Worked, (B) Serverely Cold Worked, and (C) Cold Worked and Recrystallized

Phase Transformations in Iron and Steel

STEEL AND OTHER iron alloys are the most common commercial alloys in use. The properties of iron and steel are governed by the phase transformations they undergo during processing. Understanding these transformations is essential to the successful welding of these metals.

Pure iron, as mentioned earlier, solidifies as a body-centered cubic structure called delta iron or delta *ferrite*. On further cooling, it transforms to a face-centered cubic structure called gamma iron or *austenite*. The austenite subsequently transforms back to a body-centered cubic structure known as alpha iron or alpha *ferrite*.

Steel is an iron alloy containing less than two percent carbon. The presence of carbon alters the temperatures at which freezing and phase transformations take place. The addition of other alloying elements also affects the transformation temperatures. Variations in carbon content have a profound affect on both the transformation temperatures and the proportions and distributions of the various phases (austenite, ferrite, and cementite). The iron-carbon phase diagram is shown in Figure 4.9.

Delta Ferrite to Austenite (on cooling). This transformation occurs at 2535°F (1390°C) in essentially pure iron, but in steel, the transformation temperature increases with increasing carbon content to a maximum

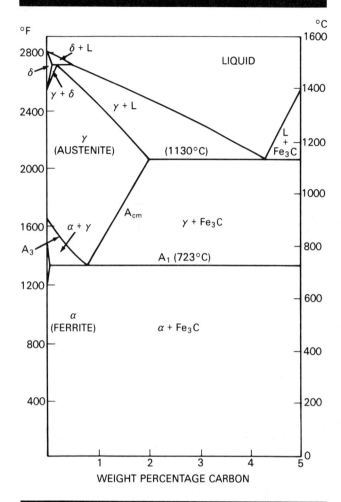

Figure 4.9—The Iron-Carbon Phase Diagram

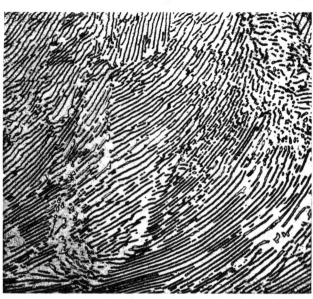

Figure 4.10—Typical Lamellar Appearance of Pearlite

of 2718°F (1492°C). Steels with more than 0.5 percent carbon freeze directly to austenite at a temperature below 2718°F (1492°C), and therefore, delta ferrite does not exist in these steels.

Austenite to Ferrite Plus Iron Carbide (on cooling). This is one of the most important transformations in steel. Control of it is the basis for most of the heat treatments used for hardening steel.

This transformation occurs in essentially pure iron at 1670°F (910°C). In steel with increasing carbon content, however, it takes place over a range of temperatures between boundaries A_3 and A_1, Figure 4.9. The upper limit of this temperature range (A_3) varies from 1670°F (910°C) down to 1333°F (723°C). For example, the A_3 of a 0.10 percent carbon steel is 1600°F (870°C), while for a 0.50 percent carbon steel it is 1430°F(775°C). Thus, both at high and low temperature the presence of carbon promotes the stability of aus-

tenite at the expense of delta and alpha ferrite. The lower temperature of the range (A_1) remains at 1333°F (723°C) for all plain carbon steels, regardless of the carbon level.

Austenite can dissolve up to 2.0 percent of carbon in solid solution, but ferrite can dissolve only 0.025 percent. At the A_1 temperature, austenite transforms to ferrite and an intermetallic compound of iron and carbon (Fe_3C), called *cementite*. Ferrite and cementite in adjacent platelets form a lamellar structure. The characteristic lamellar structure, known as pearlite, is shown in Figure 4.10.

Most of the common alloying elements added to steel further alter the transformation temperatures. Room temperature microstructures of iron-carbon alloys at the equilibrium conditions covered by this diagram include one or more of the following constituents:

(1) Ferrite—A solid solution of carbon in alpha iron
(2) Pearlite—A mixture of cementite and ferrite that forms in plates or lamellae
(3) Cementite—Iron carbide, Fe_3C, present in pearlite or as massive carbides in high carbon steels

When carbon steels are slowly cooled from the austenitic temperature range, the relative amounts of these three constituents at room temperature depend on the chemical composition. However, austenite decomposition is suppressed when the cooling rate is accelerated. When transformation does begin, it progresses more rapidly, and larger volumes of pearlite are formed. As

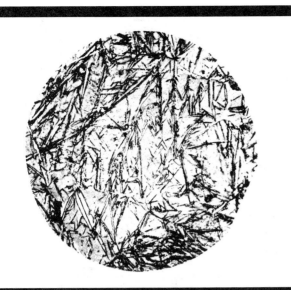

Figure 4.11—As-Quenched Martensite

the cooling rate is further increased, the pearlite lamellae become finer (closely spaced platelets).

At fast cooling rates, still lower transformation temperatures are encountered, and a feathery distribution of carbides in ferrite is formed instead of pearlite. This feathery arrangement of shear needles with fine carbides in a ferrite matrix is called *bainite*. It has significantly higher strength and hardness and lower ductility than fine pearlitic structures.

With very fast cooling rates (severe quenching), *martensite* forms. Martensite is the hardest austenite decomposition product. When the cooling rate is fast enough to form 100 percent martensite, no further increases in hardness can be achieved by faster quenching. A typical martensitic microstructure is shown in Figure 4.11.

The decomposition of austenite is an important consideration in the welding of steel alloys because the weld metal and parts of the heat-affected zone undergo this transformation.

Certain commonly used heat treating terms refer in part to the rate at which austenite is cooled to room temperature. The following treatments are listed in the order of increased cooling rates from above the A_3 temperature:

(1) Furnace annealing—slow furnace cooling
(2) Normalizing—cooling in still air
(3) Oil quenching—quenching in an oil bath
(4) Water quenching—quenching in a water bath
(5) Brine quenching—quenching in a salt brine bath

If treated in the order of (1) to (5), the hardness and tensile strength of a specific steel would become increasingly greater until the microstructure is nearly 100 percent martensite.

The Isothermal Transformation or TTT Diagrams

THE IRON-CARBON PHASE diagram is very useful. However, it does not (1) provide information about the transformation of austenite to any structure other than equilibrium structures, (2) furnish details on the suppression of the austenite transformation, nor (3) show the relationship between the transformation products and the transformation temperature. A more practical diagram is the *isothermal transformation* or *time-temperature-transformation diagram*. It is also known as a *TTT diagram*. This diagram graphically describes the time delay and the reaction rate of the austenite transformation to pearlite, bainite, or martensite. It also shows the temperature at which these transformations take place. A TTT diagram for 0.80 percent plain carbon steel is shown in Figure 4.12.

To produce this diagram, samples of 0.80 percent carbon steel were austenitized at 1550°F (845°C). The samples were then quenched to a variety of temperatures below 1300°F in molten salt baths. Each specimen was held at its chosen reaction temperature for a specified length of time, removed from the salt bath, quenched to room temperature, polished, etched, and examined under a microscope. Metallographic examination revealed the amount of austenite that transformed in the salt bath. The reaction start times and completion times were plotted as shown in Figure 4.13.

As shown in Figure 4.12, austenite at 1300°F (700°C) begins to transform after about 480 seconds (8 minutes), and the reaction is complete after about 7200 seconds (2 hours). At 1000°F (540°C) (the *nose* of the curve), the reaction begins after an elapsed time of only one second and proceeds rapidly to completion in about seven seconds. The temperature at the nose of TTT diagrams of many carbon and low alloy steels is about 1000°F (540°C).

At temperatures below the nose, the transformation products change from pearlite to bainite and martensite with their characteristic feathery and acicular structures.

As carbon and alloy content increases, the TTT curves shift to the right. When the curves move to the right, the steels can transform to martensite at slower cooling rates. These steels are said to have higher hardenability.

Continuous Cooling Transformation Diagrams

TTT DIAGRAMS HELP to understand the isothermal transformation of austenite. In most heat treating processes, including welding, austenite transforms during the cooling process. A diagram, similar to TTT curves, called a *continuous cooling transformation diagram (CCT)* can be constructed. CCT diagrams apply directly to the transformation of austenite during cooling.

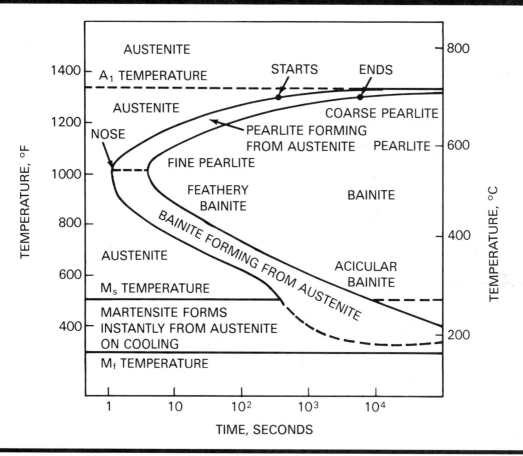

Figure 4.12—TTT Diagram of a Eutectoid Plain Carbon Steel

The most important difference is that during continuous cooling, the start of transformation takes place after a longer delay time and at a lower temperature than would be predicted using a TTT diagram. In other words, the curves of a CCT diagram are displaced downward and toward the right. A CCT and a TTT diagram for AISI 8630 steel are shown together in Figure 4.14.

TRANSFORMATION OF STEEL AUSTENITIZED BY THE WELDING PROCESS

THE TTT DIAGRAM provides a basis for estimating the microstructures that would be produced in a steel heat-affected zone during welding. Therefore, it provides the basis for welding plans based on the metallurgical characteristics of the steel. However, TTT diagrams do not provide an exact basis for predicting microstructures because there are significant differences in austenite transformation characteristics developed during a welding thermal cycle and those resulting from a heat treating cycle.

In the case of heat treatment, a steel is held at temperature for a sufficient time to dissolve the carbides and develop a homogeneous austenitic phase having relatively uniform grain size. In a weld thermal cycle, the austenitizing temperature (peak temperature) varies from near the melting point to the lower critical temperature, and the duration of the cycle is very short relative to normal soaking times in heat treating cycles.

At high peak temperatures near the fusion line, diffusion is more rapid, and the solute atoms (especially carbon) disperse uniformly in the austenite.[1] In addition, austenite grain growth occurs. At lower peak temperatures slightly above the austenite transformation temperature, carbides may not be completely dissolved in the austenite. Furthermore, the solute atoms that do dissolve because of the relatively low temperature may not diffuse far from the original site of the carbide. Thus, the austenite at these lower peak temperatures contain areas of high alloy content (enriched) and low alloy content (impoverished). In addition, the austenite micro-

---

1. The relationship between peak temperature and the distance from the fusion line is discussed in Chapter 3 of this volume.

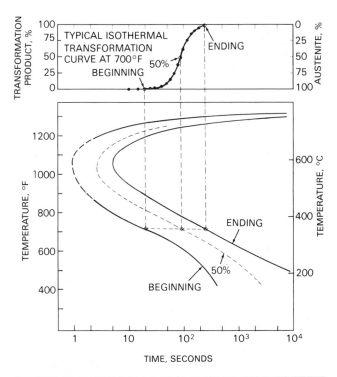

Figure 4.13—Method of Constructing TTT Diagram

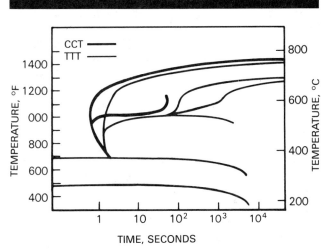

Figure 4.14—CCT and TTT Diagrams for a 8620 Type Steel

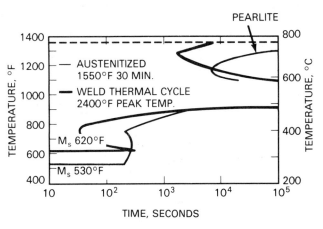

Figure 4.15—CCT Diagrams for Heat-Treated Bar and a Simulated Weld Thermal Cycle

structure is fine grained. At intermediate peak temperatures, the homogeneity and the grain size of austenite are between these extremes.

On cooling, the austenitic decomposition temperature and decomposition products depend on the local chemical composition and the grain size as well as the cooling rate. Nonhomogeneous austenite has a different transformation behavior than homogenous austenite.

Two transformation diagrams of the same heat of Ni-Cr-Mo steel are shown in Figure 4.15. One of the diagrams characterizes isothermal transformation of austenite after a 30 minute soak at 1550°F (840°C). The other diagram characterizes austenite transformation during the cooling portion of a weld thermal cycle to a peak temperature near the fusion line [2400°F (1310°C)]. There are marked differences in the transformation characteristics for the two thermal conditions. In the heat-affected-zone microstructure, the start of pearlite and bainite transformations was delayed by a factor of six, and the M_s temperature was depressed by 90°F (50°C).

If the peak temperature of the weld thermal cycle was reduced to below 2400°F (1300°C), the austenite would be less homogeneous, have a finer grain size, and exhibit a different transformation behavior on cooling.

It is obvious that austenitizing conditions have a significant effect on transformation characteristics.

FACTORS AFFECTING AUSTENITE TRANSFORMATION

THE IMPORTANT FACTORS in determining the austenite transformation of steels are the chemical composition, austenitic grain size, and the degree of homogeneity of the austenite. These factors will be described in relation to weld thermal cycles.

All of the normal transformations in solid steel (above the temperature where martensite forms) take place by the process of nucleation and grain growth. New phases

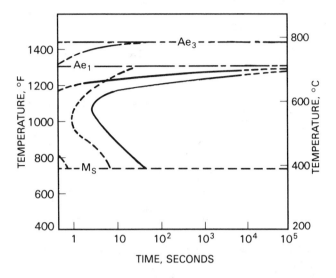

Figure 4.16—TTT Diagram for a 0.35% Carbon-0.37% Manganese Steel

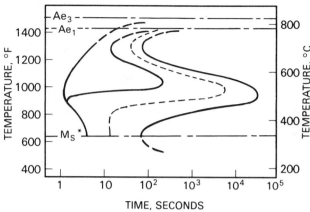

*CALCULATED TEMPERATURE

Figure 4.18—TTT Diagram for a High Carbon 2% Chromium Steel

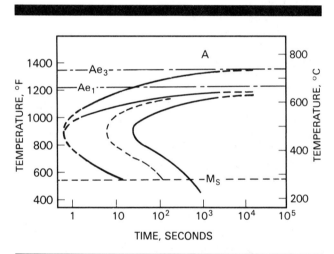

Figure 4.17—TTT Diagram for a 0.37% Carbon-0.68% Manganese Steel

first form at certain favorable locations, usually at grain boundaries, and grow in volume until transformation is complete.

The composition of the steel is most important in determining transformation behavior. Carbon, nickel, manganese below one percent, silicon below one-and-one-half percent, and copper move the transformation curve to the right but do not change its shape. Thus, the TTT diagram of a plain carbon steel shown in Figure 4.16 is similar to that of a nickel alloy steel except that the austenite decomposition begins and ends later for the

nickel steel as shown in Figure 4.17. Chromium, molybdenum, vanadium and other strong carbide forming elements move the curve to the right too, but they also change the shape of the curve as shown in Figure 4.18. Only a few alloying elements, such as cobalt and tellurium, that are not normally used in carbon and low-alloy steels, move the lines of the transformation diagram to the left.

Pearlite nucleates at austenitic grain boundaries. Therefore, fine grained austenites provide many more nucleation sites than course grain austenite. As a result, more nuclei form and the austenite transforms more rapidly. Bainite formation seems to be little affected by austenite grain size.

As previously mentioned, the transformation behavior is related to the local austenite chemical composition. If the austenite is not homogeneous, then the alloy enriched areas behave as higher hardenability steels than do the impoverished areas.

Undissolved carbides promote the austenite transformation in two ways: (1) undissolved carbides do not contribute to the alloy content of the austenite, and (2) undissolved carbides serve as nucleation sites for austenite decomposition.

The martensite start temperature, M_s, and the martensite finish temperature, M_f, are generally lowered by the addition of alloying elements to steel. In some cases, the M_f point may be below room temperature, and some austenite will be retained in the steel after cooling to room temperature. The M_s and M_f temperatures are also influenced by austenite homogeneity, but grain size seems to have little effect.

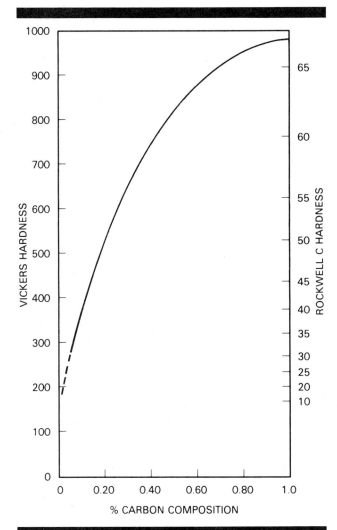

Figure 4.19—Effect of Carbon on the Maximum Hardness of Martensite

HARDENABILITY

ALTHOUGH TTT AND CCT diagrams portray the transformation characteristics of steels, the concept of hardenability presents another method of describing the transformation of austenite in various steels. Hardenability should not be confused with hardness. The maximum hardness of a steel is a function of its carbon content, as shown in Figure 4.19. Hardenability, on the other hand, is a measure of the amount of martensite that forms in the microstructure on cooling. Certain steels with high hardenability will form martensite when they are cooled in air. Other steels with low hardenability require fast cooling rates to transform to martensite. Hardenability characteristics are important to the welding engineer because they will determine the extent to which a steel will harden during welding.

TEMPERING OF MARTENSITE

MARTENSITE, IN THE as-quenched condition, is generally unsuitable for engineering applications because it can be quite brittle. However, a tempering heat treatment can effectively increase its ductility and toughness while only moderately reducing its strength. Tempering consists of reheating the steel to an appropriate temperature (always below the A_1) and holding at that temperature for a short time. This heat treatment allows the carbon to precipitate in the form of tiny carbide particles. The resulting microstructure is called tempered martensite. The desired compromise between hardness and toughness can be obtained by choosing the proper tempering temperature and time. The higher the tempering temperature, the softer and tougher is the steel.

The effect of tempering temperature on the mechanical properties of quenched AISI 4140 Cr-Mo steel is shown in Figure 4.20. Quenching and tempering are frequently used to enhance the properties of machinery,

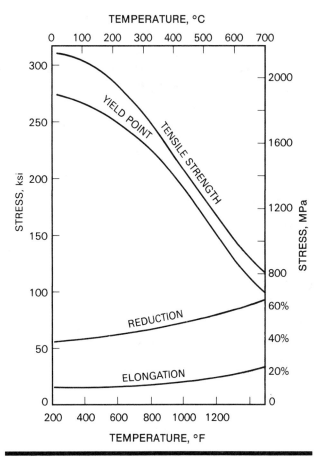

Figure 4.20—Effect of Tempering Temperature on the Strength and Ductilligy of a Chromium-Molybdenum Steel (AISI 4140)

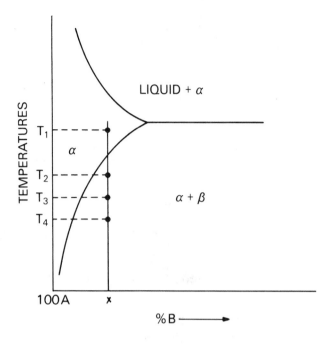

Figure 4.21A—Phase Diagram of a Precipitation Hardening System

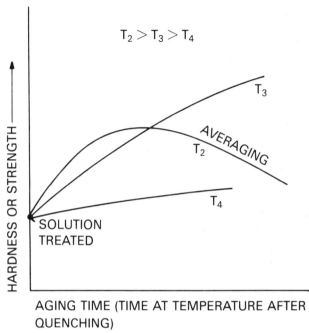

Figure 4.21B—Effect of Aging Time and Temperature on the Hardness of Precipitation Hardening System

pressure vessel, and structural steels. Quenched and tempered low alloy steels exhibit high yield and tensile strengths, high yield-to-tensile strength ratios, and improved notch toughness compared to the same steel in the hot rolled, annealed, or normalized condition.

OTHER GRAIN STRUCTURE CHANGES

THE AUSTENITIC GRAIN size of a steel depends on the austenitizing temperature. Grain refinement occurs when a steel that will transform is heated to a temperature slightly above the A_3 temperature and is then cooled to room temperature. A fine grain size is desirable for improved toughness and ductility. Steel forgings and castings are frequently normalized specifically to produce grain refinement.

At higher austenitizing temperatures [over 1800°F (1000°C)], steels usually develop a coarse austentic grain structure. Coarse-grained steels usually are inferior to fine-grained steels in strength, ductility, and toughness.

PRECIPITATION HARDENING

PRECIPITATION HARDENING (AGE hardening) is another method of developing high strength and hardness in some steels. It is also important in the hardening of nonferrous alloys.

The principles of precipitation hardening are illustrated in Figure 4.21A and B. In Figure 4.21A, if a two-phase alloy of composition X is heated to temperature T_1, the β phase will dissolve into the α phase. If the alloy is then quenched rapidly to room temperature, the α phase will not have time to transform to β phase. The resulting alloy will remain single phase, homogeneous and relatively soft.

When the alloy is reheated to temperatures between T_2 and T_4 for a time, the β phase forms as fine precipitates within the α grains. The fine β precipitates in the grains can significantly strengthen the alloy as shown in Figure 4.21B. The mechanical properties of the alloy depend on the aging temperature and time. Excessive temperature or time at temperature will not develop maximum strength and hardness.

THE METALLURGY OF WELDING

WELDING IS A very effective method of joining metals. It is a complex, metallurgical process involving melting, solidification, gas-metal reactions, slag metal reactions, surface phenomena, and solid state reactions. These reactions occur very rapidly during welding in contrast to most metallurgical reactions in metal manufacturing, casting, and heat treatment. A welded joint consists of weld metal (which has been melted), heat-affected zones, and unaffected base metals.

The metallurgy of each weld area is related to the base and weld metal compositions, the welding process, and the procedures used. Most typical weld metals are rapidly solidified, and usually have a fine grain dendritic microstructure. The weld metal is an admixture of melted base metal and deposited (filler) metal, if used. Some welds are composed of only remelted base metal (autogenous). Examples of autogenous welds are gas tungsten arc and electron beam welds made without filler metal, and resistance welds. In most arc welding processes, filler metal is added.

To achieve mechanical and physical properties that nearly match those of the base metal, the weld metal is often similar in chemical composition to the base metal. This is not a universal rule, and sometimes, the weld metal composition is deliberately made significantly different from that of the base metal. The intent is to produce a weld metal having properties compatible with the base metal. Therefore, variations from the base metal composition are not uncommon in filler metals.

When a weld is deposited, the first grains to solidify are nucleated by the unmelted base metal, and these grains maintain the same crystal orientation. Depending upon composition and solidification rates, the weld solidifies in a cellular or a dendritic growth mode. (Dendrite growth is discussed later.) Both modes cause segregation of alloying elements. Consequently, the weld metal may be less homogeneous than the base metal. Cellular and dendrite growth modes, and the resulting segregation patterns, are illustrated in Figure 4.22.

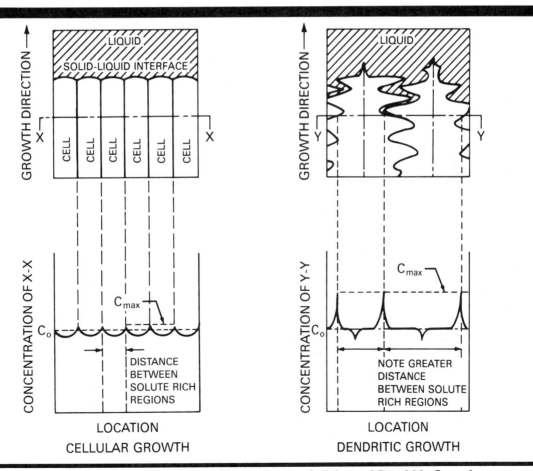

Figure 4.22—Schematic of Solute Distribution for Cellular and Dendritic Growth

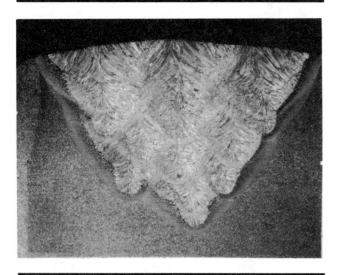

Figure 4.23—Macrosection of Low Alloy Steel Weld

The weld heat-affected zone is adjacent to the weld metal. The heat-affected zone is that portion of the base metal that has not been melted, but whose mechanical properties or microstructure have been altered by the heat of welding. The width of the heat-affected zone is a function of the heat input.[2]

The heat-affected zone may in theory include all regions heated to any temperature above the ambient. From a practical viewpoint, however, it includes those regions which are actually influenced by the heat of the welding process.

For a plain carbon as-rolled steel, the heat of welding has little influence on those regions heated to less than about 1350°F (700°C). For a heat-treated steel that was quenched to martensite and tempered at 600°F (315°C), heating above 600°F would change the mechanical properties of the metal. For a heat-treated aluminum alloy age hardened at 250°F (120°C), any portion of a welded joint heated above this temperature is the heat-affected zone.

Heat-affected zones are often defined by the response of the welded joint to hardness variation or microstructural changes. Thus, changes in microstructure produced by the welding heat which are seen in etching or in hardness profiles may be used to establish the heat-affected zone. In many cases, these are arbitrary measures of the heat-affected zone, although they may be of practical value in testing and evaluating welded joints.

A weld cross section of a multipass weld in low alloy steel is shown in Figure 4.23. Each weld pass has a heat-

affected zone; the overlapping heat-affected zones along the fusion faces extend through the full thickness of the plate. (Weld metal affected by the heat of welding is not considered as heat-affected zone.)

Adjacent to the heat-affected zone is the unaffected base metal. The base metal is selected by the designer for the specific application based on a specific property or combination of properties, such as yield or tensile strength, notch toughness, corrosion resistance, or density. One job of the welding engineer is to select the welding consumables and process and to develop welding procedures that allow the design properties to be fully utilized in service. The characteristic of a metal to be welded without losing desirable properties is called *weldability*, which is discussed later.

WELD METAL

THE MICROSTRUCTURE OF a weld metal is markedly different from that of base metal of similar composition, as shown in Figure 4.23. The difference in microstructure is not related to chemical compositions, but to different thermal and mechanical histories of the base metal and the weld metal. The structure of the base metal is a result of a hot rolling operation and multiple recrystallization of the hot-worked metal. In contrast, the weld metal has not been mechanically deformed, and therefore has an as-solidified structure. This structure and its attendant mechanical properties are a direct result of the sequence of events that occur as the weld metal solidifies. These events include reactions of the weld metal with gases in the vicinity of the weld and with nonmetallic liquid phases (slag or flux) during welding, and also reactions that took place in the weld after solidification.

Solidification

THE UNMELTED PORTIONS of grains in the heat-affected zone at the solid-liquid interface serve as nucleation sites for weld metal solidification. Metals grow more rapidly in certain crystallographic directions. Therefore, favorably oriented grains grow for substantial distances, while the growth of others that are less favorably oriented is blocked by faster growing grains.

As a result, weld metal often exhibits a macrostructure, described as columnar, in which the grains are relatively long and parallel to the direction of heat flow. This structure is a natural result of the influence of favorable crystal orientation on the competitive nature of solidification grain growth.

Weld metal solidification of most commercial metals involves microsegregation of alloying and residual elements. This action is associated with and in large measure responsible for the formation of *dendrites*. A dendrite is a structural feature which reflects the com-

2. The heat input as described in Chapter 2 and 3 can be $H_{net} = 60\,f_1 EI/v$ J/in. ($= f_1\,EI/v$ J/mm) where f_1 is the arc efficiency, E and I are the voltage and current respectively, and v is the travel speed in in./m (mm/s).

plex shape taken by the liquid-solid interface during solidification (see Figure 4.22).

As the primary dendrites solidify, solutes that are more soluble in the liquid are rejected by the solid material and diffuse into the remaining liquid, lowering the freezing point. As the solute alloys concentrate near the solid-liquid interface, crystal growth is arrested in that direction. The grains then grow laterally, producing the dendrite arms characteristic of as-solidified metals. Many dendrites may grow simultaneously into the liquid from a single grain during solidification. Therefore, each of these dendrites has the same crystal orientation, and they will all be part of the same grain. However, a solute rich network will exist among the dendrites in the final structure, as indicated in Figure 4.22. Thus, the weld metal structure appears coarse at low magnification, Figure 4.23, because only the grain structure is visible. At high magnification, a fine dendritic structure is exhibited, as shown in Figure 4.24.

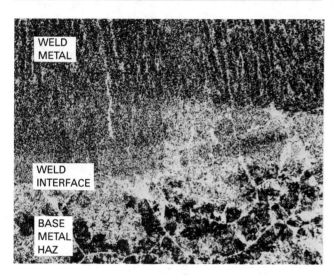

Figure 4.24—Weld-Metal and Heat-Affected Zone Microstructure

The spacing between dendrite arms is a measure of alloy segregation, and is determined by the rate of solidification. The more rapid the solidification, the more closely spaced are the dendrites.

The general tendency is for weld-metal grain size to increase with heat input, but there is no fixed relationship. The grain size may be influenced by nucleating agents, vibration, or other process variables, but the dendrite arm spacing is exclusively a function of solidification rate which is controlled by heat input.[3]

---

3. See the discussion of solidification rates in Chapter 3 of this volume.

Gas-Metal Reactions

GAS-METAL REACTIONS DEPEND on the presence of oxygen, hydrogen, or nitrogen, individually or combined, in the shielding atmosphere. There are many sources for these elements. Oxygen is intentionally added to argon in gas metal arc welding (GMAW) of steel to stabilize the arc. It can also be drawn in from the atmosphere or result from the dissociation of water vapor, carbon dioxide, or a metal oxide. Air is the most common source of nitrogen, but there are many souces of hydrogen. In shielded metal arc (SMAW) and submerged arc welding (SAW), hydrogen may be present as water in the electrode coating and the loose flux, respectively. Hydrogen may be present in solid solution in nonferrous metals or in surface oxides and lubricating compounds from wire drawing operations. Welding grade argon and helium are high purity gases, and therefore are rarely the source of reactive gases in gas tungsten arc (GTAW) and gas metal arc welding. Other processes, such as submerged arc or electron beam welding (EBW) use a liquid slag or a vacuum to prevent reactions with atmospheric gases.

Welding of Ferrous Metals. Gas-metal reactions in welding steels occur in several steps. First, the diatomic gas molecules are broken down in the high temperature of the welding atmosphere, and then the gas atoms dissolve in the liquid metal. Reaction rates are exceedingly fast at temperatures of 3000°F (1650°C) and higher. Once dissolved in the steel, oxygen and nitrogen will generally react with intentionally added deoxidizers such as manganese, silicon, and aluminum. These oxides will form a slag and float to the surface of the weld or precipitate in the metal as discrete oxides. Oxygen contents of 250 to 600 parts per million (ppm) are not uncommon in steels joined by consumable electrode arc welding processes. Oxides and nitrides are present as small discrete particles. Although they reduce the ductility and notch toughness of steel weld metal, the resulting mechanical properties are satisfactory for most commercial applications.

In consumable electrode welding, the oxide content of steel weld metal is significantly greater than the nitrogen content because oxygen is intentionally present in arc atmospheres, whereas nitrogen is not. If the weld metal does not contain sufficient deoxidizers, the soluble oxygen will react with soluble carbon to produce CO or CO_2 during solidification. The gas molecules will be rejected during solidification and produce porosity in the weld metal.

Hydrogen is always present in the arc atmosphere if only in small quantities. Hydrogen atoms are soluble in liquid steel and less soluble in solid steel. Excess hydrogen that is rejected during solidification will cause porosity. A more significant problem is created by the hydrogen that remains dissolved in the solid steel.

Welding of Nonferrous Metals. The primary gas-metal reactions of concern are the solution, reaction, and evolution of hydrogen or water vapor. These gases, therefore, should be excluded from the molten weld pool. With aluminum and magnesium, hydrogen is often introduced into the weld pool from hydrated oxides on the surfaces of the filler wire or workpieces, or both. It is rejected from the metal during solidification to produce porosity. For this reason, aluminum and magnesium filler metals should be stored after cleaning in sealed, desiccated containers. Mechanical cleaning or vacuum heating at 300°F (150°C) is recommended for workpieces or filler metals which have been exposed to moist air. The hydrogen solubility difference between the liquid and solid states for magnesium is less than that for aluminum. Consequently, the tendency for hydrogen-produced porosity is lower in magnesium.

In the case of copper and copper alloys, hydrogen will react with any oxygen in the molten weld pool to produce water vapor, and thus, porosity during solidification. The filler metals for copper alloys contain deoxidizers to prevent this reaction. Porosity caused by water vapor will not form in alloys of zinc, aluminum, or beryllium because these elements form stable oxides. Porosity from water vapor can form in nickel-copper and nickel alloy weld metal, and filler metals for these alloys generally contain strong deoxidizers.

Titanium alloys are embrittled by reaction with a number of gases including nitrogen, hydrogen, and oxygen. Consequently, these elements should be excluded from the arc atmosphere. Welding should be done using carefully designed inert gas shielding or in vacuum. Titanium heat-affected zones are also significantly embrittled by reaction with oxygen and nitrogen. Titanium weldments should be shielded so that any surface heated to over 500°F (260°C) is completely protected by an inert gas. The surface appearance of the titanium weld zone can indicate the effectiveness of the shielding. A light bronze color indicates a small amount of contamination; a shiny blue color indicates more contamination, and a white flaky oxide layer indicates excessive contamination.

Hydrogen is the major cause of porosity in titanium welds. The hydrogen source, as in other nonferrous and ferrous metals, can be the filler metal surface. In addition, soluble hydrogen in the filler metal and the base metal can contribute significantly to the total hydrogen in the molten weld pool.

Liquid-Metal Reactions

DURING THE WELDING process, nonmetallic liquid phases are frequently produced that interact with the molten weld metal. These liquid phases are usually slags formed by the melting of an intentionally added flux.

The slags produced in the shielded metal arc, submerged arc and the electroslag welding processes are designed to absorb deoxidation products and other contaminants produced in the arc and molten weld metal. The quantity and type of nonmetallic deoxidation products generated when arc welding steel are directly related to the specific shielding and deoxidants used. These products, primarily silicates of aluminum, manganese, and iron, may float to the surface of the molten weld pool and become incorporated in the slag, but some can be trapped in the weld metal as inclusions. The cleanliness of the weld metal is influenced by the quantity of nonmetallic products produced and the extent to which they can be removed with the slag. Clearly, the more strongly oxidizing the arc environment is, the more deoxidation is required, and the greater the quantity of nonmetallic products produced.

Another important effect that results from the interaction of the liquid and solid state is the weld defect referred to as hot cracking. This phenomenon arises during the solidification of the weld metal whenever the interdendritic liquid (the last region to freeze) has a substantially lower freezing temperature than the previously solidified metal. Under these conditions, shrinkage stresses produced during solidification become concentrated in this small liquid region and produce microcracks between the dendrites. These cracks are called *hot cracks* because they occur at temperatures close to the solidification temperature. They are promoted by any compositional variations in the weld metal that produce a low melting interdendritic liquid.

The most common cause of hot cracking in welds is the presence of low melting phases that wet the dendrite surfaces. This may be caused by the presence of sulphur, lead, phosphorus, and other elements in the metal. In some ferrous alloys, such as stainless steels, silicates have been found to produce cracking. The control of cracking in ferrous alloys is usually accomplished by controlling both the amount and type of sulfides that form and the minor alloy constituents that may promote cracking. In carbon and low alloy steel welds, manganese-to-sulfur composition ratios of 30 or more are used to prevent hot cracking in weld metal. In austenitic stainless steel weld metal, a duplex microstructure such as austenite and delta ferrite, will effectively prevent cracking. For this reason, austenitic stainless steel filler metals are formulated so that at room temperature, the weld metal will contain two to eight percent ferrite. Ferrite in greater amounts may adversely affect weld metal properties.

Solid State Reactions

IN TERMS OF the behavior of weld metals, there are a number of solid state reactions that are important as strengthening mechanisms in the weld metal itself.

These will be discussed in detail later in this chapter. In terms of interactions occurring during the welding process, however, there are some important phenomena involving solid state transformations and subsequent reactions with dissolved gases in the metal. The most significant of these phenomena is the formation of cold cracks in steel weld metal or heat-affected zones, often referred to as *delayed cracking*. The steels most susceptible to this type of cracking are those that transform to martensite during the cooling portion of the weld thermal cycle.

Cracking occurs after the weld has cooled to ambient temperature, sometimes hours or even days after welding. It is always associated with dissolved hydrogen in the weld metal which remains there during solidification and subsequent transformation to martensite. Because this type of cracking is always associated with dissolved hydrogen, the following two precautions are universally used to minimize the risk of delayed cracking:

(1) Preheating the base metal to slow the cooling rate
(2) Use of low-hydrogen welding processes

The use of preheat prevents the formation of a crack susceptible microstructure and also promotes the escape of hydrogen from the steel by diffusion.

Hydrogen is relatively soluble in austenite, and virtually insoluble in ferrite. Upon rapid cooling, the austenite transforms either to an aggregate of ferrite and carbide or to martensite, and hydrogen is trapped in solution. In a plain carbon steel, this transformation takes place at a relatively high temperature (near 1300°F [700°C]), even if cooling is rapid. Consequently, the hydrogen atoms have sufficient mobility to diffuse out of the metal. Moreover, the high temperature transformation product (ferrite plus carbide) which forms in the weld metal and heat-affected zone is relatively ductile and crack resistant.

A rapidly cooled hardenable steel transforms at a much lower temperature where the hydrogen atoms have lower mobility. Moreover, the microstructure is martensitic and more crack sensitive. It is this combination which is most likely to cause cracking. The association of hydrogen with delayed cracking led to the development of low-hydrogen covered electrodes.

Low-hydrogen electrode coverings must be kept as nearly moisture free as possible because water is a potent source of hydrogen. For this reason, the electrodes frequently are supplied in hermetically sealed containers. Exposed electrodes should be stored in a desiccated environment or at a temperature of about 200° to 500°F (100 to 250°C). Electrodes that have become hydrated by exposure to the atmosphere must be rebaked (dried) using procedures recommended by the manufacturer.

Another solid state reaction that affects weld joint mechanical properties in ferrous and nonferrous alloys is the precipitation of second phases during cooling. Precipitation of a second phase in grain boundaries is particularly deleterious because the grain boundaries are continuous throughout the metal. A concentration of a second phase at grain boundaries may significantly reduce ductility and toughness.

Strengthening Mechanisms in Weld Metals

THE PRACTICAL METHODS for strengthening weld metals are fewer than for base metals. Weld metals are not usually cold worked for example. However, there are four mechanisms for strengthening weld metals, and where applicable the mechanisms are additive: (1) solidification grain structure, (2) solid solution strengthening, (3) transformation hardening, and (4) precipitation hardening. The first mechanism is common to all welds, and the second is applicable to any alloy type, but the third and fourth apply only to specific groups of alloys. These processes will be considered separately.

Solidification Grain Structure. As mentioned previously, weld metals freeze rapidly creating a segregation pattern within each grain. The resulting microstructure consists of fine dendrite arms in a solute rich network, as shown in Figure 4.22. This type of microstructure impedes plastic flow during tensile testing. As a result, weld metals typically have higher yield-to-tensile strength ratios than base metals.

Solid Solution Strengthening. Weld metals are strengthened by alloy additions. Both substitutional and interstitial alloying elements will strengthen ferrous and nonferrous weld metals.

Transformation Hardening. Transformation hardening can take place in ferrous weld metals even if the austenite decomposition product is not martensite. The rapid cooling rates achieved during the cooling portion of weld thermal cycles decreases the austenite transformation temperature. The ferrite-carbide aggregate formed at low transformation temperatures is fine and stronger than that formed at higher transformation temperatures. The effect of transformation temperature on the ultimate tensile strength of steel weld metal is shown in Figure 4.25.

Precipitation Hardening. Weld metals of precipitation hardening alloy systems can often be strengthened by an aging process. In most commercial applications, the precipitation hardened weldments are aged after welding without the benefit of a solution heat treatment. In multipass welds, some zones of weld metal will be aged or overaged from the welding heat. The heat-affected zone will also contain overaged metal. In spite of the presence of some overaged metal, an aging heat treatment will strengthen the weld metal and the heat-affected zone.

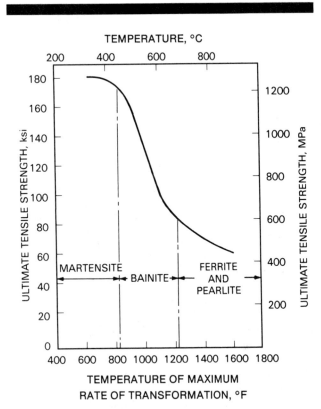

Figure 4.25—Effect of Transformation Temperature on Strength

However, they may not strengthen to the same level as the base metal because of the presence of overaged metal. Some precipitation hardening weld metals of aluminum will age naturally at room temperature.

THE HEAT-AFFECTED ZONE

THE STRENGTH AND toughness of the heat-affected zone in a welded joint are dependent on the type of base metal, the welding process, and the welding procedure. Base metals most influenced by welding will be those strengthened or annealed by heat treatments because a weld thermal cycle involves high temperatures. The temperatures in the weld heat-affected zone vary from ambient to near the liquidus temperature. Metallurgical processes that proceed slowly at lower temperatures can proceed rapidly to completion at temperatures close to the liquidus.

Thus, to understand the various effects of welding heat on the heat-affected zone, it is simplest to discuss these effects in terms of four different types of alloys which may be welded: (1) alloys that are strengthened by solid solution, (2) alloys that are strengthened by cold work, (3) alloys that are strengthened by precipitation hardening, and (4) alloys that are strengthened by transformation (martensite). Some alloys can be strengthened by more than one of these processes, but for simplicity, the processes will be considered separately.

Solid Solution Strengthened Alloys

SOLID-SOLUTION alloys normally exhibit the fewest weld heat-affected zone problems. If they do not undergo a solid state transformation, the effect of the thermal cycle is small, and the properties of the heat-affected zone will be largely unaffected by welding. Grain growth will occur next to the fusion line as a result of the high peak temperature. This will not significantly affect mechanical properties if the grain coarsened zone is only a few grains wide.

Commonly used alloys strengthened by solid solution are aluminum alloys, copper alloys, and hot rolled low carbon steels. Ferritic and austenitic stainless steels come under essentially the same category.

Strain Hardened Base Metals

STRAIN HARDENED BASE metals will recrystallize when heated above the recrystallization temperature. The heat of welding will recrystallize the heat-affected zones in cold worked metals and soften the metal considerably. The weld cross sections shown in Figure 4.26 illustrate the effect of weld thermal cycles on cold worked microstructures. The unaffected base metal (A) shows the typical elongated grains resulting from mechanical deformation. Fine equiaxed grains were formed where the heat-affected-zone temperature exceeded the recrystallization temperature (B), and grain growth took place at higher temperatures near the fusion zone. The recrystallized heat-affected zone is softer and weaker than the cold worked base metal, and the strength cannot be recovered by heat treatment.

If the cold worked materials undergo an allotropic transformation when heated, the effects of welding are even more complex. Steel, titanium, and other alloys that exhibit allotropic transformations may have two recrystallized zones, as illustrated in Figure 4.26. The first fine-grained zone results from recrystallization of the cold worked alpha phase. The second fine-grained zone results from the allotropic transformation to the high temperature phase.

Precipitation Hardened Alloys

ALLOYS THAT ARE strengthened by precipitation hardening respond to the heat of welding in the same manner as work hardened alloys; that is, the heat-affected zone undergoes an annealing cycle. The response of the heat-affected zone is more complex because the welding ther-

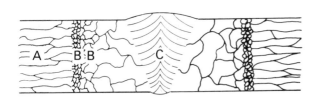

(A) No phase change when heated

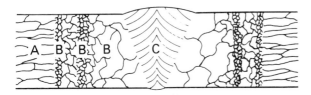

(B) Allotropic transformation when heated

Figure 4.26—Recrystallization of Cold Worked Grains in Heat-Affected Zone

mal cycle produces different effects in different regions. The heat treating sequence for precipitation hardening is: solution treat, quench, and age. The welding heat will resolution treat the heat-affected-zone regions closest to the weld and produce a relatively soft single phase

solid solution with some coarse grains. This region can be hardened by a post weld aging treatment.

Those regions of the heat-affected zone that are heated to temperatures below the solution treatment temperature will be overaged by the welding heat. A postweld aging treatment will not reharden this region. If the welding heat does not raise the heat-affected-zone temperature above the original aging temperature, the mechanical properties are not significantly affected. The heat-affected-zone structure of a precipitation hardened alloy is illustrated in Figure 4.27.

It is difficult to weld high strength precipitation hardenable alloys without some loss of strength, but three techniques may be used to minimize the loss. The most effective of these techniques is to resolution treat, quench, and age the weldment. This technique is expensive and, in many cases, may not be practical. A second approach would be to weld precipitation hardened base metal and then re-age the weldment.

This raises the strength of the solution treated region of the heat-affected zone, but does not improve the strength of the overaged zone. Another alternative is to weld the base metal in the solution-treated condition and age the completed weldment. The overaged zone is still the weakest link, but the overall effect may be an improvement over the previous approaches.

Since it is the weld thermal cycle that lowers the strength of the heat-treated base metal, high heat input welding processes are not recommended for these alloys. Low heat input will minimize the width of the heat-affected zone and the amount of softened base metal.

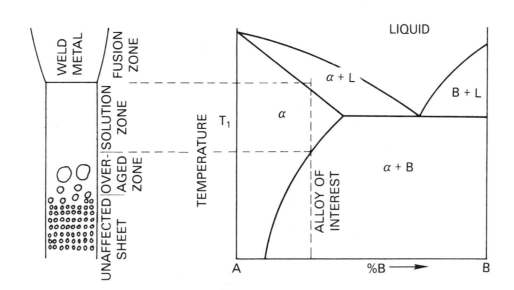

Figure 4.27—Growth of Precipitates in Heat-Affected Zone of Precipitation-Hardened Alloy

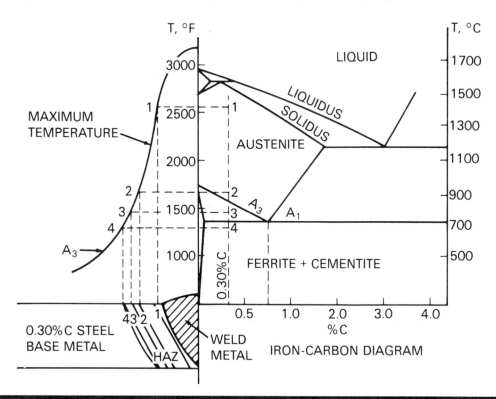

Figure 4.28—Approximate Relationships Among Peak Temperature, Distance from Weld Interface, and the Iron-Carbon Phase Diagram

Transformation Hardening Alloys

THE TRANSFORMATION HARDENING alloys of interest are the steels with sufficient carbon and alloy content to transform to martensite upon cooling from welding. These may be steels that are already heat treated to tempered martensite prior to welding or that have adequate hardenability to transform to martensite during a weld thermal cycle, even though they may not have been heat treated. In either case, the heat-affected zone is affected by the weld thermal cycle in approximately the same manner. The heat-affected zones together with the iron-carbon phase diagram are illustrated in Figure 4.28.

The grain coarsened region is near the weld interface, (Region 1). Rapid austenitic grain growth takes place in this region when exposed to the near melting point temperatures. The large grain size increases hardenability, and this region can readily transform to martensite on cooling. Region 2 is austenitized, but the temperature is too low to promote grain growth. The hardenability of Region 2 will not be significantly increased by grain growth but may still transform to martensite if the cooling rate is fast enough or if the alloy content is great enough. In Region 3, some grains transform to austenite and some do not. The austenite grains are very fine. No

austenitic transformation takes place in Region 4, but the ferrite grains may be tempered by the heat of welding.

The width of the heat-affected zone and the widths of each region in the heat-affected zone are controlled by the welding heat input. High heat inputs result in slow cooling rates, and therefore, the heat input may determine the final transformation products.

High carbon martensite is hard and strong, and it can create problems in the heat-affected zone. The hardness of a weld heat-affected zone is a function of the base metal carbon content. The hardness and crack susceptibility increase and the toughness decreases with increasing carbon content. Martensite alone will not cause cracking; dissolved hydrogen and residual stresses are also necessary.

The same precautions used to prevent delayed cracking in weld metal will also prevent cracking in the heat-affected zone. These precautions were given previously. The hardness of a weld-heat-affected zone is usually a good indication of the amount of martensite present and the potential for cracking. Cracking rarely occurs when the Brinell hardness is below 250 HB, but is common when the hardness approaches 450 HB and no precautions are taken.

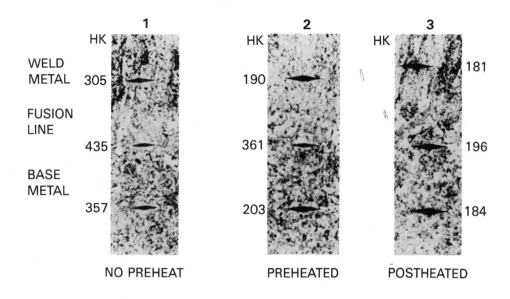

WELD METAL | **FUSION LINE** | **BASE METAL**

	1	2	3
	NO PREHEAT	PREHEATED	POSTHEATED
HK (WELD METAL)	305	190	181
HK (FUSION LINE)	435	361	196
HK (BASE METAL)	357	203	184

Figure 4.29—Base-Metal, Heat-Affected Zone, and Weld-Metal Hardness of 0.25% Carbon Steel

The microstructure and microhardness of three weld beads on 0.25% carbon steel made under different conditions of preheat and postheat are shown in Figure 4.29. The heat-affected-zone Knoop hardness of the weld made without preheat was 435 HK. The second weld bead was deposited on a preheated plate, and the heat-affected-zone Knoop hardness was 361 HK. The third plate was tempered after welding at 1100°F (600°C), and the heat-affected-zone hardness was 196 HK. The heat-affected-zone hardness of the second and third weld beads were markedly lower then that of the first weld. A postweld heat treatment may be required to reduce the hardness of the weld metal and the heat-affected zone in hardenable steels.

Special precautions may be necessary when welding hardenable steels that have been intentionally heat treated to produce a tempered martensitic microstructure. It is usually desirable to use a low welding heat input to control the size of the heat-affected zone, and a high preheat temperature to control the cooling rate of the weld. The welding recommendations of the steel manufacturer should be followed in preparing welding procedures for low-alloy high-strength steels.

BASE METAL

THE THIRD COMPONENT in a welded joint is the base metal. Many of the common engineering alloys available today are readily weldable. However, some alloys are more difficult to weld, and require special precautions.

Weldability is the capacity of a material to be welded under the imposed fabrication conditions into a specific suitably designed structure and to perform satisfactorily in the intended service. According to this definition, the weldability of some systems may be poor under some conditions but satisfactory under other conditions. For example, all grades of ASTM A514 steel (a heat treated 100 ksi [689 MPa] yield strength constructional alloy steel) have satisfactory weldability, provided the base metal is sufficiently preheated, low-hydrogen welding procedures are followed, and the allowable heat input limitations are not exceeded.

The primary factor affecting the weldability of a base metal is its chemical composition. Each type of metal has welding procedural limits within which sound weldments with satisfactory properties can be fabricated. If these limits are wide, the metal is said to have good weldability. If the limits are narrow, the metal is said to have poor weldability. If extraordinary precautions are necessary, then the material is often called "unweldable". Yet, in some cases and industries, "unweldable" materials are routinely welded under tight controls with vigorous inspection procedures and acceptance criteria. These methods are followed because welding may be the only (or at least the best) method to achieve the desired function within the design criteria for the whole assembly.

WELDABILITY OF COMMERCIAL ALLOYS

MANY alloys are used commercially in welded products. Several of these are described briefly here.[4]

STEELS

MOST COMMERCIAL STEELS can be classified according to three groups: plain carbon, low alloy, and high alloy steels. Typical mechanical properties of some commercial steel grades are shown in Table 4.2.

Plain Carbon Steels

PLAIN CARBON STEEL consists of iron, usually with less than 1.0 percent carbon and minor amounts of manganese, phosphorous, sulfur, and silicon. The properties and weldability of these steels depend mainly on carbon content, although other alloying and residual elements do influence these properties to a limited extent. Plain carbon steels frequently are categorized as low, medium, and high carbon steels.[5]

The weldability of carbon steels is excellent for low carbon steels, good to fair for medium carbon steels, and poor for high carbon steels.

Low Alloy Steels

THE CARBON CONTENT of low alloy steels intended for welded applications is usually less than 0.25 percent and frequently is below 0.15 percent. Other alloying elements (nickel, chromium, molybdenum, manganese, silicon) are added to increase the strength of these steels at room temperatures, as well as to impart better notch toughness at low temperatures. These elements also alter the response of the steels to heat treatment, and can improve their corrosion resistance. Alloy additions adversely affect the crack susceptibility of low alloy steels, therefore low hydrogen welding processes should be used on these steels. Preheat may also be required.

Modern design often uses steels that have higher strength and toughness than the plain carbon and structural steels. A yield strength of 50 000 psi (345 MPa) and a tensile strength of 70 000 psi (480 MPa) are achieved in the as-rolled condition by adding two or more alloying elements; adequate weldability is maintained by restricting the carbon to 0.20 percent maximum. Some of these steels are heat treated up to 100 000 psi (690 MPa) yield strength, and they have better notch toughness than ordinary carbon steel. Proper choice of filler metal and welding procedures will develop comparable properties in welded joints in these steels.[6]

A wide variety of steels has been developed for use in machinery parts. These steels have been classified according to chemical composition by the AISI-SAE classification system. The weldability of these steels depends on their chemical composition. Other steels have been developed to meet the needs of the cryogenic industry. Those steels have good notch toughness at temperatures well below zero. Fine-grained aluminum-killed steels with up to 10 percent nickel are frequently used for cryogenic service.

---

4. For a complete discussion of metals and their weldability, see the *Welding Handbook, Vol. 4, 7th Ed.* (Metals and their weldability will be discussed in Volume 3 of the 8th Ed.)
5. The American Welding Society has prepared specifications for carbon steel filler metals for use with all of the arc welding processes.

6. The American Welding Society publishes low alloy steel filler metal specifications for all of the common arc welding processes.

Table 4.2
Typical Mechanical Properties of Some Commercial Steels at Room Temperature

Material	Type	Conditions	Ultimate Tensile Strength ksi	Yield Strength 0.2% Offset ksi	Elongation In 2 in. %	Hardness Brinell	Hardness Rockwell
Mild steel	Plain carbon	Hot rolled	55	30	30	110	62B
A36	Plain carbon	Hot rolled	65	38	28	135	74B
A285	Plain Carbon	Hot rolled	70	39	27	140	77B
Medium Carbon Steel	Plain carbon	Oil quenched and tempered at 400°F	120	93	18	242	100B 23C
		tempered at 1200°F	80	62	30	202	90B 98B
A514	Low alloy	Water quenched and tempered	115	95	18	230	21C
A240 Type 304	Stainless Steel	Annealed	90	40	60	160	83B

Low alloy steels have also been developed for high temperature service in welded structures such as steam boilers, oil refining towers, and chemical processing retorts. Additions of chromium and molybdenum give these steels structural stability and provide high creep and stress rupture values at temperatures up to 1100°F (595°C).

High Alloy Steels

Stainless Steels. Stainless steels are the most important commercial high alloy steels. Stainless steels resist attack by many corrosive media at atmospheric or elevated temperatures. They contain at least 12 percent chromium and many of the grades have substantial amounts of nickel. Other alloying elements are added for special purposes.

There are three basic types of stainless steel: austenitic, ferritic, and martensitic. Some are precipitation hardenable. The martensitic stainless steels contain the smallest amount of chromium and exhibit high hardenability. The cutlery grades are martensitic stainless steels.

Care must be taken when welding the martensitic steels because the martensitic heat-affected zone is susceptible to cracking. Preheating and postheating treatments are necessary to prevent cracking.

The ferritic stainless steels contain 12 to 27 percent chromium with small amounts of austenite-forming elements. The ferrite phase is stable up to the melting temperature.

Austenitic stainless steels are produced by adding alloying elements that stabilize austenite at room temperature. Nickel is the most important austenite stabilizing element. Manganese, carbon, and nitrogen also stabilize austenite.

Chromium, nickel, molybdenum, nitrogen, titanium, and columbium provide the austenitic stainless steels with special properties of corrosion resistance, oxidation resistance, and elevated temperature strength. Carbon can contribute to elevated temperature strength, but it may reduce corrosion resistance by forming a chemical compound with chromium. The austenitic alloys cannot be hardened by heat treatment, and therefore they do not harden in the weld-heat-affected zone. The austenitic stainless steels have excellent weldability.

The stainless steels are readily joined by arc, electron beam, laser beam, resistance, and friction welding processes. Gas metal arc, gas tungsten arc, flux-cored arc, and shielded metal arc commonly are used. Plasma arc and submerged arc welding are suitable methods.[7]

Oxyacetylene welding is seldom recommended, and then not for sections thicker than 0.13 in.

Other High Alloy Steels. High alloy steels have been specially designed for applications requiring outstanding mechanical properties or elevated temperature strength and ductility. These steels range from chromium-molybdenum steels (ASTM A387) and nickel steels (ASTM A353 and A553) to nickel-cobalt maraging steels and tool steels. Important compositions are discussed in Chapters 1, 3, and 4 of the *Welding Handbook*, Vol. 4, 7th Ed.

The chromium-molybdenum and nickel steels are generally welded under controlled conditions by arc processes (GTAW, GMAW, SMAW, SAW). Dies, punches, shears, etc., are made from high carbon tool steels that also contain moderate amounts of other alloying elements. These tool steels may be fabricated and repaired using any of the common arc welding processes. Premium quality high strength alloys may depend on gas tungsten arc or plasma arc welding to achieve clean weld deposits with suitable mechanical properties.

ALUMINUM ALLOYS

ALUMINUM AND ALUMINUM alloys are face-centered cubic metals at all temperatures up to their melting temperatures. The alloys feature low density, about one-third that of steel or copper, with excellent corrosion resistance. Aluminum resists corrosion by air, water, oils, and many chemicals because it rapidly forms a tenacious, refractory oxide film on a clean surface in air. The oxide is virtually insoluble in the molten aluminum, and therefore inhibits wetting by molten filler metals.

Aluminum conducts thermal and electrical energy approximately four times faster than steel. As a result, (1) fusion welding requires high heat input, (2) thick sections may require preheating, and (3) resistance spot welds require higher current and shorter weld time than steel welds of equivalent thickness.

The metal is non-magnetic, so arc blow is not a problem. It is highly reflective of radiant energy, and does not change color prior to melting at about 1200°F (650°C).

Aluminum is strengthened by alloying, cold working, heat treatment, and combinations of these methods. Heating during welding, brazing, or soldering may soften those alloys strengthened by heat treatment or cold working. This must be considered when designing the component and selecting the joining process and manufacturing procedures.

Aluminum is alloyed principally with copper, magnesium, manganese, silicon, and zinc. Small additions of

---

7. Welding precautions and recommendations will be found in the *Welding Handbook*, Vol, 4, 7th Ed. The American Welding Society publishes three stainless steel filler metal specifications covering electrodes for the common arc welding processes, and also AWS D10.4,

Recommended Practices for Welding Austenitic Chromium-Nickel Steel Piping and Tubing.

chromium, iron, nickel, titanium, and lithium are added to specific alloy systems to obtain desired properties and to refine the grain. Magnesium, manganese, silicon, and iron, singly or in various combinations, are used to strengthen aluminum by solid solution or by dispersing intermetallic compounds within the matrix. Silicon addition also lowers the melting point and increases fluidity.

Copper, magnesium, silicon, zinc, and lithium additions can produce alloys that are heat treatable. These elements become more soluble in aluminum with increasing temperature. Such alloys can be strengthened by appropriate thermal treatments, which may be supplemented by cold working. However, the heat treatment and the cold work may be negated by the thermal cycle of a joining operation. Heat treatment in conjunction with or following the welding or brazing step should be provided to restore optimum mechanical properties.[8]

MAGNESIUM ALLOYS

MAGNESIUM AND ITS alloys have a close-packed hexagonal crystal lattice structure. At room temperature they can sustain limited deformation compared to aluminum. Fortunately, their workability increases rapidly with temperature, and they can be severely worked between 400 and 600°F (200 and 310°C). Forming and straightening operations are generally performed at an elevated temperature.

Magnesium will oxidize rapidly at welding and brazing temperatures. The oxide inhibits wetting and flow during welding, brazing, or soldering. A protective shield of inert gas or flux must be used to prevent oxidation during exposure to elevated temperatures.

Magnesium is well known for its extreme lightness, machinability, weldability, and the high strength-to-weight ratio of its alloys. On an equal volume basis, it weighs about one-fourth as much as steel and two-thirds as much as aluminum.

Magnesium requires relatively little heat to melt it because its melting point, latent heat of fusion, and specific heat per unit volume are all comparatively low. On an equal volume basis, the total heat of fusion is approximately two-thirds that for aluminum and one-fifth that for steel. The high coefficients of thermal expansion and

thermal conductivity tend to cause considerable distortion during welding. Fixtures for welding magnesium must be more substantial than those for steel. Fixtures similar to those used for aluminum welding are adequate.[9]

COPPER ALLOYS

COPPER AND MOST of its alloys have a face-centered cubic crystal lattice. Most commercial copper alloys are solid solutions (single phase alloys). Fourteen common alloying elements are added to copper. Copper alloys show no allotropic or crystallographic changes on heating and cooling, but several have limited solubility with two phases stable at room temperature. Some of these multiple phase alloys can be hardened by precipitation of intermetallic compounds.

Two-phase copper alloys harden rapidly during cold working, but they usually have better hot working and welding characteristics than do solid solutions of the same alloy system.

The corrosion resistance of copper alloys is often improved by small additions of iron, silicon, tin, arsenic, and antimony.

The most important age-hardening reactions in copper alloys are obtained with additions of beryllium, boron, chromium, nickel, silicon, and zirconium. Care must be taken in welding and heat treating such age-hardenable copper alloys. The proper welding process and filler metal must be used. Whenever possible, complete resolution; aging should be provided after welding.

Note that copper alloy weld metals do not flow well without high preheat. The high heat conductivity of the base metal removes welding heat rapidly from the deposited bead making it flow sluggishly. The beads will not blend smoothly with the base metal if preheating is inadequate.[10]

NICKEL-BASE ALLOYS

NICKEL AND ITS alloys have a face-centered cubic crystal structure, at all temperatures up to their melting temperatures. This makes them readily weldable. Specific alloys are noted for their resistance to corrosion and for their high temperature strength and toughness.

Alloying elements added to commercial nickel alloys are principally copper, chromium, iron, molybdenum, and cobalt. These major additions, with the exception of

---

8. For more detailed information on aluminum welding, the *Welding Handbook, Vol. 4, 7th Ed.* should be consulted. *AWS A5.10, Specification for Aluminum and Aluminum Alloy Bare Welding Rods and Electrodes* and *AWS A5.3, Specification for Aluminum and Aluminum Alloy Covered Arc Welding Electrodes* are the aluminum filler metal specifications published by the American Welding Society. The appendices of the two filler metal specifications (A5.3 and A5.10) also provide useful information on classifications and applications. The American Welding Society also publishes *D10.7, Recommended Practices for Gas Shielded Arc Welding of Aluminum and Aluminum Alloy Pipe.*

9. Magnesium filler metals are covered in *AWS A5.19, Specification for Magnesium Alloy Welding Rods and Bare Electrodes*. The *Welding Handbook*, Vol. 4, 7th Ed., discusses magnesium and magnesium alloys at length.

10. Copper alloy welding is discussed extensively in the *Welding Handbook, Vol. 4, 7th Ed*. Copper alloy filler metals are specified in *AWS A5.6, Covered Copper and Copper Alloy Arc Welding Electrodes* and *AWS A5.7, Specification for Copper and Copper Alloy Bare Welding Rods and Electrodes.*

molybdenum, form binary solid solutions with nickel in the commercial alloys and have relatively little effect on weldability. Molybdenum present above 20 percent forms a second phase with nickel; the two-phase commerical alloys are weldable.

Alloying elements added in smaller amounts include aluminum, carbon, magnesium, manganese, columbium, silicon, titanium, tungsten, and vanadium.

High nickel alloys are strengthened by solid-solution alloying, by dispersion hardening with a metal oxide, and by precipitation-hardening heat treatments. Precipitation-hardening is achieved by controlled precipitation in the microstructure of a second phase, essentially the compound Ni_3Al. Commerical alloy systems have been developed with nickel-copper, nickel-chromium, nickel-molybdenum, nickel-chromium-molybdenum, and nickel-iron-chromium compositions.

The welding of nickel alloys is similar to the welding of austenitic stainless steels. Arc welding is broadly applicable.[11] Oxyacetylene welding is generally not recomended. Resistance welding is readily performed. Electron beam welding may achieve greater joint efficiency than gas tungsten arc welding, but heat-affected-zone cracking may occur in thick section welds. Nickel alloys can be brazed with proper base metal preparation, brazing environment and filler metal selection.

COBALT ALLOYS

COBALT HAS A close-packed hexagonal crystal structure up to 750°F (400°C) and has a face-centered cubic crystal lattice above 750°F. Nickel additions to cobalt will stabilize the face-centered cubic structure to below room temperature. Most cobalt alloys contain nickel to retain the inherent ductility of the face-centered cubic structure. Commercial cobalt based alloys are widely used in high temperature applications.

The cobalt alloys commonly fabricated by welding generally contain two or more of the elements nickel, chromium, tungsten, and molybdenum. The latter three form a second phase with cobalt in the commercial alloys, but that does not adversely affect weldability.

Other alloying elements such as manganese, columbium, tantalum, silicon, and titanium are not detrimental to welding if kept within specified limits. Sulfur, lead, phosphorus, and bismuth, which are insoluble in cobalt or undergo eutectic reactions with it, may initiate weld

hot cracking. Therefore these impurities are maintained at low levels in commercial cobalt alloys.[12]

REACTIVE AND REFRACTORY METALS

THE GROUPS OF metals known as the reactive and refractory metals include titanium, zirconium, hafnium, columbium (niobium), molybdenum, tantalum and tungsten. These metals oxidize rapidly when heated in air making it necessary to shield the weld region, including large portions of the heated base metal, from the atmosphere. A high degree of mutual solid solubility exists between all pairs of these metals. Two-component alloys are solid solution strengthened. The reactive and refractory metals and their alloys can be welded to each other, but ductility may be reduced in some combinations. They cannot be fusion welded directly to alloys of iron, nickel, cobalt, copper, and aluminum because extremely brittle intermetallic compounds are formed in the fusion zone. These welds crack as thermal stresses develop on cooling.

Titanium Alloys

PURE TITANIUM IS a silver-colored metal that has a close-packed hexagonal crystal structure known as alpha (α) phase, up to 1625°F (885°C). Above 1625° the crystal structure of titanium is body-centered cubic, called beta (β) phase. The transformation temperature, sometimes called the *beta transus*, is a function of the chemical composition of the beta phase. Aluminum, oxygen, nitrogen and carbon stabilize the alpha phase. Tin and zirconium are neutral, and the other metallic elements in titanium are beta stabilizers. Hydrogen also stabilizes beta titanium. The room temperature stability of the α and β phases is a function of the chemical composition. The relative amounts and distribution of α and β phases control the properties of titanium and titanium weld metals.

The decomposition of the beta phase is similar to that of austenite in steel. The tranformation of beta is normally a diffusion-controlled nucleation and growth process, but can become a martensitic type shearing transformation on rapid cooling.

Unlike steel the beta and alpha-beta titanium alloys are relatively soft and strengthen during an aging treatment. During aging a precipitation hardening reaction

---

11. AWS A5.11, *Specification for Nickel and Nickel Alloy Covered Welding Electrodes*, defines the filler metal requirements for shielded metal arc welding of nickel alloys. AWS A5.14, *Specification for Nickel and Nickel Alloy Bare Welding Rods and Electrodes*, defines filler metal requirements for gas metal arc, gas tungsten arc, plasma arc, and submerged arc welding of nickel alloys. The appendices of these filler metal specifications provide a wealth of information on the classification and application of the electrodes. For more complete discussions of the welding of nickel alloys see the *Welding Handbook*, Vol. 4, 7th Ed.

12. Cobalt alloy filler metals are not covered by any American Welding Society specifications. The Society of Automotive Engineers publishes material specifications for use by the aerospace industry. Several Aerospace Material Specifications (AMS) cover cobalt welding materials. Two are AMS 5797, *Covered Welding Electrodes— 52Co20Cr10Ni15W*, and AMS 5801, *Welding Wire— 39Co22Cr22Ni14.5W0.07La*. For helpful information about cobalt alloys see the *Welding Handbook*, Vol. 4, 7th Ed.

occurs in which fine alpha particles form within the beta grains.

Titanium quickly forms a stable, tenacious oxide layer on a clean surface exposed to air, even at room temperature. This makes the metal naturally passive and provides a high degree of corrosion resistance to salt or oxidizing acids, and an acceptable resistance to mineral acids.

The strong affinity of titanium for oxygen increases with temperature, and the surface oxide layer increases in thickness. Above 1200°F (650°C) oxidation increases rapidly, and the metal must be well shielded from air to avoid contamination and embrittlement by oxygen and nitrogen.

Pure titanium has low tensile strength and is extremely ductile. Dissolved oxygen and nitrogen markedly strengthen the metal, so do iron and carbon to a lesser degree. Hydrogen embrittles titanium. These elements may unintentionally contaminate the metal during processing or joining.[13]

Zirconium

ZIRCONIUM HAS CHARACTERISTICS similar to titanium except that it is 50 percent more dense. It has the same hexagonal close-packed crystal structure (alpha phase) up to about 1600°F (870°C) and there similarly transforms to body-centered cubic (beta phase). The beta phase is stable to the melting point.

Zirconium is highly resistant to corrosive attack because it forms a stable, dense, adherent, and self-healing zirconium oxide film on its surface. A visible oxide forms in air at about 400°F (200°C), becoming a loose white scale on long time exposure over 800°F (425°C). Nitrogen reacts with zirconium slowly at about 700°F (370°C) and more rapidly above 1500°F (810°C). Zirconium is strongly resistant to corrosive attack by most organic and mineral acids, strong alkalies, and some molten salts. It resists corrosion in water, steam, and sea water. At elevated temperatures and pressures, zirconium also resists corrosion by liquid metals.

Zirconium weldments are used in the petrochemical, chemical process, and food processing industries. Because zirconium has a low neutron absorption, it is used in nuclear reactors. In nuclear applications, boron and hafnium contents are maintained at low levels because they have high neutron absorption. Nuclear welding operations must be performed in a vacuum-purged welding chamber which can be evacuated and back filled with high purity inert gas. An alternative to this procedure is to continuously purge the chamber with high purity inert gas.[14]

Hafnium

HAFNIUM IS A sister element to zirconium, with hexagonal close-packed crystal structure up to 3200°F (1760°C) and body-centered cubic structure above 3200°F (1760°C). It is three times as dense as titanium. It is superior to zirconium in resisting corrosion in water, steam, and molten alkali metals. It resists dilute hydrochloric and sulfuric acids, various concentrations of nitric acid, and boiling or concentrated sodium hydroxide.

Hafnium is primarily used to resist virulent corrosion, to absorb neutrons (in nuclear reactor control rods), and to contain spent nuclear fuel in reprocessing plants.

Hafnium can be annealed at about 1450°F (790°C) for 10 minutes. Recrystallization takes place at 1475 to 1650°F (800 to 900°C), depending on the amount of cold work. Weldments can be stress relieved at about 1000°F (540°C).

Hafnium is readily welded by the processes and procedures used for titanium. Its low coefficient of expansion causes little distortion. Its low modulus of elasticity assures low residual welding stresses. Welded joints in metal 3/4 in. (19 mm) thick normally do not crack unless grossly contaminated. Hafnium is severely embrittled by relatively small amounts of nitrogen, oxygen, carbon, or hydrogen. Joint faces to be welded should be abraded using stainless steel wool or a draw file. Then the parts should be cleaned with a suitable solvent and immediately placed in a vacuum-purge chamber. The chamber should at once be evacuated to 19th torr (13.3 mPa) and back purged with inert gas. High frequency arc starting will avoid tungsten contamination from the GTAW electrode.

Sample welds made prior to production welding should be capable of being bent 90 degrees around a radius three times the thickness of the weld sample, to verify the welding procedure and the purity of the welding atmosphere.

Tantalum

TANTALUM HAS A body-centered cubic crystal structure up to the melting point. It is an inherently soft, fabricable metal that is categorized as a refractory metal because of its high melting temperature. Unlike many body-centered cubic metals, tantalum retains good ductility to very low temperatures, and does not exhibit a ductile-to-brittle transition temperature.

---

13. The titanium filler metal specification is *AWS A5.16, Specification for Titanium and Titanium Alloy Bare Welding Rods and Electrodes.* Procedures for welding titanium and its alloys are discussed in the *Welding Handbook, Vol. 4, 7th Ed.* The American Welding Society also publishes *AWS D10.6, Recommended Practices for Gas Tungsten Arc Welding of Titanium Piping and Tubing.*

14. Zirconium welding filler metals are covered by *AWS A5.24, Specification for Zirconium and Zirconium Alloy Bare Welding Rods and Electrodes.* Zirconium welding and brazing is discussed in more detail in the *Welding Handbook, Vol. 4, 7th Ed.*

Tantalum has excellent corrosion resistance to a wide variety of acids, alcohols, chlorides, sulfates, and other chemicals. It also is used in electrical capacitors and for high temperature furnace components.

Tantalum oxidizes in air above about 570°F (300°C) and is attacked by hydrofluoric, phosphoric, and sulfuric acids, and by chlorine and fluorine gases above 300°F (150°C). Tantalum also reacts with carbon, hydrogen, and nitrogen at elevated temperatures. When dissolved interstitially, these elements and oxygen increase the strength properties and reduce the ductility of tantalum.

Tantalum is available as powder metallurgy, vacuum arc melted, and electron beam melted products. Welding of powder metallurgy material is not recommended because the weld would be very porous. Electron beam or vacuum arc melted material is recommended for welding applications.

Tantalum alloys are strengthened by solid solution, by dispersion or precipitation, and by combinations of these methods. Some tantalum alloys have intentional carbon additions that respond to thermal treatments during processing. Their strength comes in part from carbide dispersion and also from solution strengthening.

Tantalum and its alloys should be thoroughly cleaned prior to welding or brazing. Rough edges to be joined should be machined or filed smooth prior to cleaning. Components should be degreased with a detergent or suitable solvent and chemically etched in mixed acids. A solution of 40 percent nitric acid, 10 to 20 percent hydrofluoric acid, up to 25 percent sulfuric acid, and remainder water is suitable for pickling, followed by hot and cold rinsing in deionized water and spot-free drying.

The cleaned components should be stored in a low-humidity clean room until ready for use.

Tantalum and tantalum alloys are readily welded using the processes and procedures described for titanium. Contamination by oxygen, nitrogen, hydrogen, and carbon should be avoided to prevent embrittlement of the weld. Preheating is not necessary.

The high melting temperature of tantalum can result in metallic contamination if fixturing contacts the tantalum too close to the weld joint. Copper, nickel alloys, or steel fixturing could melt and alloy with the tantalum. If fixturing is required close to the joint, a molybdenum insert should be used in contact with the tantalum. Graphite should not be used for fixturing because it will react with hot tantalum to form carbides.

Resistance spot welding of tantalum is feasible, but adherence and alloying between copper alloy electrodes and tantalum sheet is a problem. Welding under water might be helpful because of the improved cooling of the copper electrodes. Weld time should not exceed 10 cycles (60 Hz).

Tantalum can be explosion welded to steel for cladding applications. Special techniques are required for welding the tantalum clad steel to avoid contamination of the tantalum.

Successful brazing of tantalum and its alloys depends upon the application. For corrosion applications, the brazed joint must also be corrosion resistant. Commercially available filler metals are not so corrosion resistant as tantalum. For high temperature applications, the brazed joint must have a high remelt temperature and possess adequate mechanical properties at the service temperature.

Tantalum must be brazed in a high purity inert atmosphere or in high vacuum. Special equipment is usually required for high temperature brazing. For low temperature brazing, the tantalum can be plated with another metal, such as copper or nickel, which are readily wet by the brazing filler metal. The brazing filler metal should alloy with and dissolve the plating during the brazing cycle.

Tantalum forms brittle intermetallic compounds with most commercial brazing filler metals. The composition of the filler metal, the brazing temperature, and the heating cycle affect the degree of interaction between the two metals. In general, the brazing time should always be minimum unless diffusion brazing techniques are used.

Nickel-base filler metals, such as the nickel-chromium-silicon alloys, form brittle intermetallic compounds with tantalum. Such filler metals are satisfactory for service temperatures only up to about 1800°F (980°C).[15]

Columbium

COLUMBIUM, SOMETIMES CALLED niobium, is both a reactive metal and a refractory metal with characteristics similar to those of tantalum. It has a body-centered cubic crystal structure, and does not undergo allotropic transformation. Its density is only half that of tantalum and its melting temperature is lower.

Columbium oxidizes rapidly at temperatures above about 750°F (400°C) and absorbs oxygen interstitially at elevated temperatures, even in atmospheres containing small concentrations. It absorbs hydrogen between 500° and 1750°F (260 and 950°C). The metal also reacts with carbon, sulfur, and the halogen gases at elevated temperatures. It forms an oxide coating in most acids that inhibits further chemical attack. Exceptions are diluted strong alkalis and hydrofluoric acid.

Heat treatment should be performed in a high-purity inert gas or in high vacuum to avoid contamination by the atmosphere. Vacuum is generally the more practical.

Columbium alloys containing zirconium respond to aging after a solution treatment. Fusion welds in such

---

15. For an extensive discussion of brazing techniques for tantalum see the *Welding Handbook*, Vol. 4, 7th Ed.

alloys are sensitive to aging during service in the range of 1500° to 2000°F (810 to 1100°C).

Columbium and its alloys can be cleaned and pickled using the solvents and etchants described for tantalum. Cleaned components should be stored in a clean room under low humidity conditions.

Most columbium alloys have good weldability, provided the tungsten content is less than 11 percent. With higher tungsten content in combination with other alloying elements, weld ductility at room temperature can be low.

The processes, equipment, procedures, and precautions generally used to weld titanium are also suitable for columbium. One exception is welding in the open with a trailing shield. Although columbium can be welded with this technique, contamination is more likely because of the high temperature of the weld zone. Therefore, it is not recommended.

Gas tungsten arc welding of columbium should be done in a vacuum-purged welding chamber backfilled with helium or argon. Helium is generally preferred because of the higher arc energy. Contamination by the tungsten electrodes should be avoided by either using a welding machine equipped with a suitable arc initiating circuit or striking the arc on a run-on tab.

Those alloys that are prone to aging should be welded at relatively high travel speeds with the minimum energy input needed to obtain desired penetration. Copper backing bars and hold downs can be used to extract heat from the completed weld. The purpose of these procedures is to minimize aging.

Resistance spot welding of columbium presents the same problems encountered with tantalum. Electrode sticking can be minimized by making solid-state pressure welds between two sheets rather than actual nuggets. Projection welding might be an alternative to spot welding, but shielding might be required to avoid contamination by the atmosphere.

Cooling the spot welding electrodes with liquid nitrogen substantially decreases electrode sticking and deterioration.

Columbium alloy brazements are used in high temperature service. Brazing temperatures range from 1900 to 3450°F (1040 to 1900°C), and the filler metals readily wet and flow on the columbium in a vacuum. In some systems the brazement strength is improved by a diffusion heat treatment.

Molybdenum and Tungsten

MOLYBDENUM AND TUNGSTEN are two metals that have very similar properties and weldability characteristics. Both metals have a body-centered cubic crystal structure and will show a transition from ductile to brittle behavior with decreasing temperature. The transition temper-ature is affected by strain hardening, grain size, chemical composition, and other metallurgical factors.

The ductile-to-brittle transition temperature of recrystallized molybdenum alloys may vary from below to well above room temperature. The transition temperature of tungsten will be above room temperature. Consequently, fusion welds in these metals and their alloys will have little or no ductility at room temperature. In addition, preheating to near or above the transition temperature may be necessary to avoid cracking from thermal stresses.

These metals and their alloys are consolidated by powder metallurgy and sometimes by melting in vacuum. Wrought forms produced from vacuum melted billets generally have lower oxygen and nitrogen contents than those produced from billets manufactured by powder metallurgy methods.

Molybdenum and tungsten have low solubilities for oxygen, nitrogen, and carbon at room temperature. Upon cooling from the molten state or from temperatures near the melting point, these interstitial elements are rejected as oxides, nitrides, and carbides. If the impurity content is sufficiently high, a continuous, brittle grain boundary film is formed that severely limits plastic flow at moderate temperatures. Warm working below the recrystallization temperature breaks up the grain boundary films and produces a fibrous grain structure. This structure will have good ductility and strength parallel to the direction of working but may have poor ductility transverse to the working direction.

The interstitial compounds may dissolve in the grain coarsened heat-affected zone and the weld metal of warm-worked molybdenum and tungsten alloys. Upon cooling, the compounds may precipitate at the grain boundaries. At the same time, grain growth and an accompanying reduction in grain boundary surface area will take place. Then, the weld metal and heat-affected zones will be weaker and less ductile than the warm-worked base metal. The ductility of a welded joint is intimately related to the amount of interstitial impurities present and to the recrystallized grain size. Tungsten is inherently more sensitive to interstitial impurities than is molybdenum. Welds that are ductile at room temperature have been produced in molybdenum. However, tungsten welds are brittle at room temperature.

Oxygen and nitrogen may be present in the metal, or they could be absorbed from the atmosphere during welding. Welding should be done in a high-purity inert atmosphere or in high vacuum. Because grain size influences the distribution of the grain boundary films and the associated brittleness, welding should be controlled to produce fusion and heat-affected zones of minimum widths.

Molybdenum and tungsten are sensitive to the rate of loading and to stress concentration. The ductile-to-brittle transition temperature of molybdenum and tungsten increases with increasing strain rates. Welded joints are notch sensitive and, therefore, the weld surface should be finished smooth and faired gradually into the base metal wherever possible. Notches at the root of the weld should be avoided.

The mechanical properties of molybdenum and tungsten welds are not improved by heat treatment, but the alloys can be stress relieved at temperatures just below the recrystallization temperature. This should reduce the likelihood of cracking during subsequent handling.

Many brazing filler metals may be used to join molybdenum or tungsten. Their brazing temperatures range from 1200 to 4500°F (650 to 2480°C).

The chosen brazing filler metal must be evaluated by suitable tests for a specific application. In many cases, the service temperature limits the choice. Effects of brazing temperature, diffusion, and alloying on the base metal properties must also be determined. The brazing time should be as short as possible to minimize recrystallization and grain growth in the base metal.

BERYLLIUM

BERYLLIUM HAS A hexagonal close-packed crystal structure, which partly accounts for its limited ductility at room temperature. Its melting point and specific heat are about twice those of aluminum or magnesium. Its density is about 70 percent that of aluminum, but its modulus of elasticity is about 4 times greater. Therefore, beryllium is potentially useful for lightweight applications where good stiffness is needed. It is used in many nuclear energy applications because of its low neutron cross section.

Beryllium mill products are normally made by powder metallurgy. Wrought products are produced from billets manufactured by casting or powder metallurgy techniques. Cold-worked material may have good ductility in only one direction, and poor ductility perpendicular to that direction. Tensile properties may vary greatly, depending on the manner of processing.

An adherent refractory oxide film rapidly forms on beryllium, as with aluminum and magnesium. This oxide film inhibits wetting, flow, and fusion during welding and brazing. Therefore, parts must be adequately cleaned prior to joining. The joining process must be shielded by inert gas or vacuum to prevent oxidation.

Intergranular microcracking in beryllium welds is caused by grain broundry precipitation of binary and ternary compounds containing residual elements such as aluminum, iron and silicon. If aluminum and iron are present as $AlFeBe_4$ then they are less detrimental. Therefore by controlling the ratio of aluminum and iron atoms and maintaining low levels of these residuals microcracking can be reduced.

Those fusion welding processes that produce the smallest weld zones usually provide the best results in welding beryllium. Electron beam welding with its characteristic low heat input produces a narrow heat-affected zone, minimal grain growth, and low distortion.

Beryllium can be joined by brazing and by braze welding. However because of the oxide film on the surface of the metal good capillary flow is difficult to achieve. Low melting temperature filler metals such as aluminum-silicon and silver base alloys are normally recommended. Filler metals should be preplaced in the joint for best results, and brazing times should be short to minimize alloying and grain broundry penetration.

Beryllium may also be joined by diffusion welding. Silver coated samples can be successfully joined at low pressures and low temperatures in as little as 10 minutes.

WELDABILITY TESTING

THE TERM WELDABILITY is a qualitative term, and it is affected by all the significant variables encountered in fabrication and service. There are too many variables in the design, fabrication, and erection of real structures. No single test or combination of tests can duplicate the conditions of a real structure. Some of the variables are joint restraint, fit-up, surface condition, erection and service stresses. Furthermore, these variables are different on each assembly in each structure, and therefore, laboratory weldability tests can only provide an index to compare different metals, procedures, and processes.

Within these limitations, weldability testing can provide valuable data on new alloys, welding procedures, and welding processes. The data should be used in comparison to the weldability of existing alloys, procedures, or processes, for which the fabrication and service performance is known. However, the results are only qualitative, and laboratory testing regardless of the extent cannot quantitatively predict fabrication experience.

Numerous weldability tests have been devised, all of which can be classified as either simulated, or actual welding.

SIMULATED TESTS

TO SIMULATE THE heat effect of welding on base metals, and thus create a synthetic weldability test, two general types of apparatus are available: (1) a unit which heats and cools a metal specimen over a small area according to a predetermined cycle, and (2) a unit which not only heats and cools the specimen, but which also can apply a controlled tension load to the specimen at any time during the cycle.

While these tests provide very useful information regarding the mechanical properties of various areas within a heat-affected zone, during as well as after a welding cycle, they cannot account for residual and reaction stresses, contamination, and other conditions which may be imposed on production welds.

ACTUAL WELDING TESTS

AN EXTENSIVE VARIETY of actual welding tests has been devised to investigate the weldability characteristics of base metals. In general, these tests serve two purposes. First, they may be used to evaluate the weldability of particular grades or individual heats of base metals. For this purpose, the specimen dimensions and welding conditions remain constant to make the base metal sample the only variable. Second, they may be used to establish compatible combinations of base metal, filler metal, and welding conditions that will produce acceptable results in the test. In effect, these tests fall into two groups: those used to assess fabrication weldability and those used to measure service weldability. Fabrication tests predict if a particular material and process can be used to produce a joint acceptably free of defects. The service tests measure the properties of the weldments.[16]

Fabrication Weldability Tests

THE TESTS IN this category determine the susceptibility of the welded joint to cracking and can be grouped according to the type of cracking that they produce.

Hot-Cracking Tests. Hot cracks are formed at high temperatures and are often the result of solidification segregation and shrinkage strains. Several tests have been devised to study hot cracking.

One of the more common tests in use is the Varestraint test. This test, shown in Figure 4.30, utilizes external loading to impose plastic deformation in a plate while an autogenous weld bead is being made on the long axis of the plate. The severity of deformation is varied by changing the bend radius.

Delayed Cracking Tests. A large number of tests have been devised to investigate delayed cracking in steel weldments. All of the tests use large specimens relative to other laboratory investigations. The larger test specimens are required to simulate service conditions. The following are three restraint cracking tests:

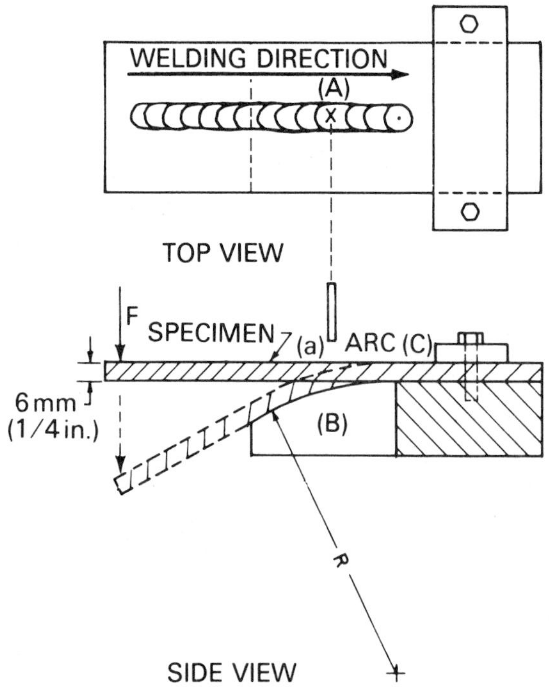

Figure 4.30—The Varestraint Hot Cracking Test

(1) The Lehigh restraint test is shown in Figure 4.31. The normal size of the Lehigh specimen is 8 in. by 12 in. (200 mm by 300 mm). The restraint can be reduced by cutting slits along the long edge of the plate, and it can be increased by reducing the weld groove length, or by increasing the specimen size.

(2) The controlled-thermal-severity test, Figure 4.32, consists of a plate bolted and anchor-welded to a second plate in a position to provide two fillet (lap) welds. The fillet located at the plate edges has two paths of heat flow (bithermal weld). The lap weld located near the middle of the bottom plate has three paths of heat flow (trithermal weld), thus inducing faster cooling. Further control of the cooling rate is possible by varying the plate thicknesses, or by using preheat.

(3) The cruciform cracking test, Figure 4.33, comprises three plates with ground surfaces tack welded at the ends to form a double T-joint. Four test fillet welds are deposited in succession in the order shown, with complete cooling between deposits. Cracking, detected by cross sectioning, is most likely to occur in the third bead. Careful fit-up of the plates is necessary to obtain reproducibility.

---

16. Service tests are discussed in Chapter 12 of this volume.

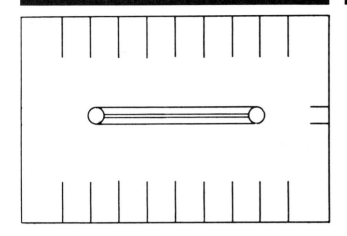

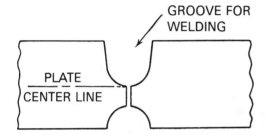

GROOVE FOR
WELDING

PLATE
CENTER LINE

Figure 4.31—The Lehigh Restraint Cracking Test

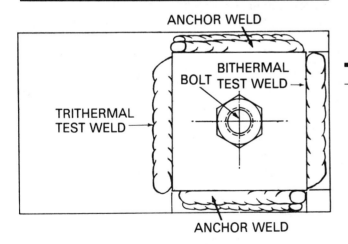

ANCHOR WELD

TRITHERMAL
TEST WELD

BOLT

BITHERMAL
TEST WELD

ANCHOR WELD

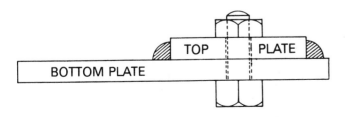

TOP PLATE

BOTTOM PLATE

Figure 4.32—The Controlled Thermal Severity Test

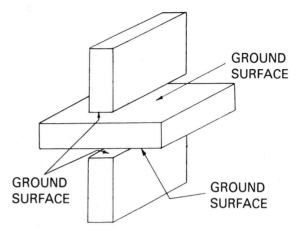

GROUND
SURFACE

GROUND
SURFACE

GROUND
SURFACE

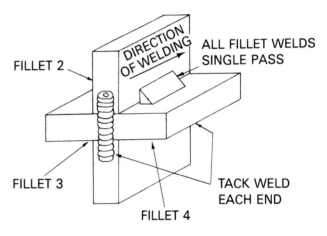

FILLET 2

DIRECTION
OF WELDING

ALL FILLET WELDS
SINGLE PASS

FILLET 3

FILLET 4

TACK WELD
EACH END

Figure 4.33—The Cruciform Cracking Test

BRAZING AND SOLDERING METALLURGY

BRAZING AND SOLDERING are performed at temperatures below the solidus of the base materials.[17] Metallurgical considerations that affect these processes range from the properties and solidification of liquid metal to base metal surface interactions and the physical and environmental conditions under which the joints are made.

Brazing and soldering have several advantages. Joining temperatures are below the base-metal melting temperature and therefore several beneficial effects result: (1) base metal properties are less affected by the process, (2) residual stresses and distortion are lower, and (3) whole assemblies can be exposed to the process temperature creating economic benefits of large production lots.

Several disadvantages can be anticipated from brazing and soldering. Brazed and soldered joints cannot easily be tested nondestructively. Depending upon the complexity of the components, jigs and fixtures may have to be used to control the fit of parts to the close tolerances required for brazing and soldering.

Brazing filler metals have a liquidus above and soldering filler metals a liquidus below 840°F (450°C). In both processes, the filler metal is distributed between the closely fitted surfaces of the joint by capillary action.

The capillary flow of the liquid metal into the joint generally depends upon its surface tension, wetting characteristics, and physical and metallurgical reactions with the base material and oxides involved. The flow is also related to the generation of hydrostatic pressure within the joint. Figure 4.34 is an idealized presentation of the wetting concept.

A contact angle of less than 90 degrees measured between the solid and liquid usually identifies a positive wetting characteristic. Contact angles greater than 90 degrees are usually an indication of no wetting.

In some brazing and soldering processes, wetting and spreading are assisted by the addition of flux. In vacuum brazing, however, flow and wetting are dependent upon the surface interactions between the liquid metal and base metals themselves. Fuel gases, hydrogen, and vacuum systems provide the most common types of controlled atmospheres.

Fluxes perform functions similar to controlled atmospheres in providing surfaces receptive to wetting and spreading. Most oxides are readily displaced or removed by flux. Oxides of chromium, aluminum, titanium, and manganese are more difficult and may require special treatments.

A brazing or soldering cycle generally consists of heating to, residing at, and cooling from a peak temperature.

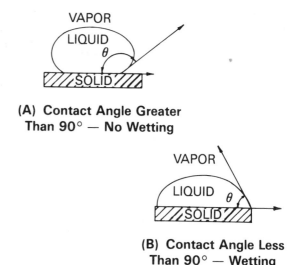

(A) Contact Angle Greater Than 90° — No Wetting

(B) Contact Angle Less Than 90° — Wetting

Figure 4.34—Wetting Angles of Brazing and Soldering Filler Metals

During the time at peak temperature when liquid filler metal is present in the joint, metallurgical reactions can occur with the base metal. These reactions are generally referred to as erosion or dissolution. The rate of dissolution of the base metal by the filler metal depends on the mutual solubility limits, the quantity of brazing filler metal available to the joint, the brazing cycle, and the potential formation of lower temperature eutectics.

In some important metallurgical systems, an interlayer of intermetallic compound forms between the filler metal and the base metal during the joining operation. The degree of intermetallic growth and the type of phases present can substantially alter the joint properties. Phase diagrams may be used to predict intermetallic compound formation.

Once the filler metal has solidified to form the joint, subsequent effects are controlled by diffusion phenomena within the joint. For example, it is possible through high temperature aging of a soldered joint to convert the filler metal to an intermetallic compound. When titanium, for example, is joined with a pure tin filler metal, it is possible by subsequent aging or heat treatment to completely diffuse the tin into the base metal so that a joint effectively no longer exists. This method of metallurgical joining has been defined as liquid-activated diffusion welding, but actually, it is an extension of the joining mechanism in brazing and soldering.

Impurities and contaminants are important factors to be considered with these processes. Contaminants

---

17. Brazing and soldering are discussed in Chapter 1 of this volume.

caused by mishandling materials or as a result of preparation for joining can affect the formation of the joint and the joint properties. For example, residual sulfur compounds can cause poor flow or result in hot cracking in certain brazed or soldered joints.

The properties of the resulting joints are dependent upon successful metallurgical bonding at the interfaces and on the final composition of the brazed metal in the joint area. Most joints of this type are designed with a large factor of safety to insure satisfactory performance in service. However, the metallurgical properties of the joint can be important when high temperature service or exposure to corrosive media is to be expected.

Strength measurements are made in shear, in tension using a peel method, or by hardness traverses across the joint area. Dynamic properties of joints are measured in creep, fatigue, and under stress corrosion conditions. Consideration must be given to all these factors when selecting a suitable brazing or soldering filler metal. Extreme care is necessary in the interpretation of destructive tests of brazed and soldered joints. Results may only reflect the properties of the actual joint such as on the joint design, the joint soundness, and the testing procedure.

Liquid filler metal penetration between the grain boundries of the base metal has previously been recognized within joints. If excessive, this penetration can lead to embrittlement problems.

Base materials in a stressed state are particularly susceptible to liquid metal penetration. For example, where copper-based filler metals are used on high iron-nickel alloys under stress, rapid failure can result. The diffusion rate of alloying elements is greater in grain boundries than in the crystal lattice. Therefore, intergranular penetration of brazing filler metal atoms will be greater than transgranular penetration.

If a eutectic is formed it may fill any grain boundary crack as it separates, then little damage may be done. This is known as an intrusion. Where high solubility exists, intergranular attack generally is not so intense. When rapid failure of a joint or adjacent material occurs during the manufacturing operation, boundary penetration phenomena should be suspected.

The dynamic characteritics of the brazing and soldering processes are receiving increasing recognition, and careful consideration is being given to the subsequent diffusion and metallurgical changes that can occur in service. At elevated temperatures, intermetallic compounds still grow in the solid state as a direct result of diffusion. This means that the metallurgical and mechanical properties of these joints can change in service.

SUPPLEMENTARY READING LIST

Australian Welding Research Association. *Welding stainless steels*. Australia: Australian Welding Research Association. AWRA Technical Note No. 16, (Dec. 1985).

David, S. A. and Slaughter, G. M. *Welding technology for energy applications*. (Proceedings International Conference, Gatlinburg, Tennessee, May 16-19, 1982) Oak Ridge, TN: Oak Ridge National Laboratory (Sept. 1982): 673–685.

Dolby, R. E. Advances in Welding metallurgy of steel. *Metals Technology*. 10(9): Sept. 1983; 349–362.

Easterling, K. E. *Introduction to the physical metallurgy of welding*. Seven Oaks Kent, UK: Butterworths and Company Limited (1983).

Hart, P. H. M. Effects of steel inclusions and residual elements on weldability. *Metal Construction*. 18(10): Oct. 1986; 610–616.

Hulka, K. and Heisterkamp, F. *HSLA steels technology and applications* (Proceedings International Conference, Philadelphia, October 3-6, 1983). Metals Park, OH: American Society for Metals (1984): 915–924.

Kirkwood, P. R. *Welding of niobium containing micro-alloyed steels* (Proceedings International Symposium, San Francisco, November 8-11, 1981). Warrendale, PA: Metallurgical Society of AIME (1984): 761–802.

Kou, S. Welding metallurgy and weldability of high strength aluminum alloys. *Welding Research Council Bulletin*. No. 320: Dec. 1986; 1–20.

Larsson, B. and Lundquist, B. Fabrication of ferritic-austenitic stainless steels, Part B. *Materials and Design*. 7(2): March-April 1986, 81–88.

Linnert, G. E. *Welding Metallurgy*. 3d Ed, Miami: American Welding Society Vol. 1. (1965): Vol. 2 (1967).

Masubuchi, K. *Underwater Welding of offshore platforms and pipelines* (Proceedings Conference, New Orleans, November 5-6, 1980). Miami: American Welding Society (1981) 81–98.

Paxton, H. W. *Alloys for the eighties* (Proceedings Conference, Ann Arbor, MI, June 17-18, 1980). Greenwich, CT: Climax Molybdenum Company, (1981).

Pickens, J. R. The weldability of lithium-containing aluminum alloys. *Journal of Materials Science.* 20(12): Dec. 1985, 4247–4258.

Robinson, S. Welding aluminum—basics & theory. *FWP Journal.* 26(10): Oct. 1986; 7, 10, 12–14.

Stout, R. D. *Weldability of Steels* 4th Ed. Miami: American Welding Society, 1987.

DESIGN FOR WELDING

PREPARED BY A COMMITTEE CONSISTING OF:

W. A. Milek, Chairman
Consultant

P. B. Dickerson
Aluminum Company of America

D. D. Rager
Reynolds Metals Company

W. W. Sanders, Jr.
Iowa State University

WELDING HANDBOOK COMMITTEE MEMBER:
D. R. Amos
Westinghouse Electric Corporation

CHAPTER **5**

DESIGN FOR WELDING

GENERAL CONSIDERATIONS

A WELDMENT IS an assembly that has component parts joined by welding. It may be a bridge, a building frame, an automobile, a truck body, a trailer hitch, a piece of machinery, or an offshore tubular structure.

The base objectives of weldment design[1] are ideally to provide an assembly that

(1) Will perform its intended functions
(2) Will have the required reliability and safety
(3) Is capable of being fabricated, inspected, transported, and placed in service at minimum total cost

Total cost includes the costs of:

(1) Design
(2) Materials
(3) Fabrication
(4) Erection
(5) Inspection
(6) Operation
(7) Repair
(8) Product maintenance

THE DESIGNER

DESIGNERS OF WELDMENTS should have some knowledge and experience in the following areas in addition to the basic design concepts for the product or structure:[2]

(1) Cutting and shaping of metals
(2) Assembly of components
(3) Preparation and fabrication of welded joints
(4) Weld acceptance criteria, inspection, mechanical testing, and evaluation

Designers also need general knowledge of the following subjects and their effects on the design of weldments:[3]

(1) Mechanical and physical properties of metals and weldments
(2) Weldability of metals
(3) Welding processes, costs, and variations in welding procedures
(4) Filler metals and properties of weld metals
(5) Thermal effects of welding
(6) Effects of restraint and stress concentrations
(7) Control of distortion
(8) Communication of weldment design to the shop, including the use of welding symbols
(9) Applicable welding and safety standards

Several of these topics involve special areas of knowledge and experience. Therefore, designers should not rely entirely upon their own knowledge and experience, which may be limited, but should consult with welding experts whenever appropriate.

Engineers who are responsible for designing welded structures and machine parts must have and routinely apply knowledge of each of the following areas:

---

1. Of necessity, the topics discussed in this chapter have not been exhaustively developed. For more complete information on any individual topic, the reader should refer to available textbooks, manuals, and handbooks, several of which are given in footnotes or listed in the Supplementary Reading List at the end of the chapter.
2. There is similarity between the design of weldments and brazements, except for joint designs and joining processes. Much informa-

tion presented here can be applied to brazement design. Also, refer to the *Welding Handbook*, Volume 2, 7th Edition 1978.
3. The listed subjects are covered in this volume and in Vols. 2, 3, and 4, *Welding Handbook*, 7th Ed.

(1) Efficient use of steel, aluminum, and other metals in weldments

(2) Design for appropriate stiffness or flexibility in welded beams and other structural members

(3) Design for torsional resistance

(4) Effects of thermal strains induced by welding in the presence of restraints

(5) Effects of stress induced by welding in combination with design stresses

(6) Practical considerations of welding and selection of proper joint designs for the application

One common mistake in the design of weldments is copying the over-all shape and appearance of a casting or other form that is to be replaced by a weldment. A weldment should (1) have a shape appropriate for the transfer of applied loads and (2) be capable of fabrication by means of welding. When a change is made from a casting to a weldment, both appearance and function may be improved because weldment design generally involves a more economical and strategic use of materials.

PROPERTIES OF METALS

STRUCTURE SENSITIVE PROPERTIES

THE PROPERTIES OF metals can be divided into five general groups, (1) mechanical, (2) physical, (3) corrosion, (4) optical, and (5) nuclear properties. Typical metal properties of each group are shown in Table 5.1. Not all of these properties are discussed herein. The specific properties in each group are divided into structure-insensitive properties and structure-sensitive properties. This distinction in properties is commonly made in most textbooks on metals to emphasize the considerations that should be given to reported property values.

Structure-insensitive properties are well defined properties of a metal. They do not vary from one piece of a metal to another of the same kind. This is true for most engineering purposes, and is verified by the data obtained from standard engineering tests. These properties often can be calculated or rationalized by consideration of the chemical composition and the crystallographic structure of the metal.[4] They commonly are listed in handbooks as constants for the particular metals.

Structure-sensitive properties are dependent upon not only chemical composition and crystallographic structure but also on microstructural details that may be affected in subtle ways by the manufacturing and processing history of the metal. Even the size of the sample can influence the test results obtained for a structure-sensitive property, and they are likely to vary to some degree if there are differences in the treatment and preparation of the samples.

The most important mechanical properties of metals in the design of weldments, with the exception of elastic moduli, are structure-sensitive. Consequently, single values of these properties published in a handbook must

Table 5.1
Properties of Metals

General Groups	Structure-Insensitive Properties	Structure-Sensitive Properties
Mechanical	Elastic Moduli	Ultimate strengty Yield strength Fatigue strength Impact strength Hardness Ductility Elastic limit Damping capacity Creep strength Rupture strength
Physical	Thermal expansion Thermal conductivity Melting point Specific heat Emissivity Thermal evaporation rate Density Vapor pressure Electrical conductivity Thermoelectric properties Magnetic properties Thermionic emission	Ferromagnetic properties
Corrosion	Electrochemical potential Oxidation resistance	
Optical	Color Reflectivity	
Nuclear	Radiation absorbtivity Nuclear cross section Wavelength of characteristic x-rays	

---

4. General and welding metallurgy of metals is covered in Chapter 4 of this volume.

be accepted with reservation. It is not uncommon for plates or bars of a metal, which represent unusual sizes or conditions of treatment, to have significant deviations in mechanical properties from those published for the metal. Also, the mechanical properties, as determined by standard quality acceptance tests in an American Society of Testing and Materials (ASTM) specification, do not guarantee identical properties throughout the material represented by the test sample. For example, the direction in which wrought metal is tested (longitudinal, transverse, or through-thickness) may give significantly different values for strength and ductility. Although the physical and corrosion properties of metals are considered to be structure-insensitive for the most part, some of the values established for these properties apply only to common polycrystalline metals.

MECHANICAL PROPERTIES

METALS ARE THE most useful material for construction because they are generally strong, tough, and ductile. This combination of properties is not often found in nonmetallic materials. Most nonmetallic materials of construction depend upon composite action with metals for their usefulness. Furthermore, the strength, toughness, and dutility of metals can be varied individually by alloy selection or by heat treatment. Joining of metals by welding or brazing also affects their mechanical properties. The applied heat, cooling rates, filler metal addition, and metallurgical structure of the joint are some of the factors that affect mechanical properties.[5]

Metals not only offer many useful mechanical properties and characteristics, but they can also develop a large number of combinations of these properties. The versatility of metals has allowed designers to select the best combination of properties to insure good service performance. In addition to service performance, the base metals and welding consumables should be selected to facilitate fabrication. To solve these selection problems, or to effect a compromise, it is necessary to examine first the governing properties and then consider their combined effect upon the design and service behavior of the weldment.

Modulus of Elasticity

A CONVENIENT WAY of appraising the ability of a metal to resist stretching (strain) under stress in the elastic range is by the ratio E between the stress and the corresponding strain. This ratio is known as *Young's modulus* or the *modulus of elasticity*. It is commonly expressed by the following formula:

$$E = \frac{\sigma}{\epsilon} \qquad (5.1)$$

where

σ = the stress, psi
ϵ = the strain, in./in.

Young's modulus is a constant characteristic of a metal as measured in polycrystalline metals during standard tensile, compression, bending, and other engineering tests in which stress and strain are correlated. The elastic modulus is a structure-insensitive property; it is virtually unaffected by grain size, cleanliness, or condition of heat treatment. In fact, for a given metal, the modulus of elasticity often remains unchanged even after substantial alloying additions have been made. Table 5.2 lists the modulus of elasticity for a number of metals.

Table 5.2
Modulus of Elasticity of Metals

Metal	Modulus of Elasticity, psi*	
Aluminum	9.0	
Beryllium	42.0	
Columbium	15.0	
Copper	16.0	
Iron	28.5	
Lead	2.0	$\Big\} \times 10^6$
Molybdenum	46.0	
Nickel	30.0	
Steel, carbon & alloy	29.0	
Tantalum	27.0	
Titanium	16.8	
Tungsten	59.0	

*To convert to Pa, multiply the psi value by 6,895.

The elastic modulus can be used practically to determine the level of stress created in a piece of metal when it is forced to stretch elastically a specified amount. The stress can be determined by multiplying the strain by the elastic modulus. It is important to point out that the modulus of elasticity decreases with increasing temperature, and that the change in elastic modulus with temperature varies with different metals.

Elastic Limit

THE ELASTIC BEHAVIOR of metal reaches a limit at a level of stress called the *elastic limit*. The elastic limit of a metal is structure-sensitive and dependent on strain rate. It is the upper bound of stress where the member will return to its original dimensions when the load is

---

5. The testing and evaluation of welded joints are discussed in Chapter 12 of this volume.

released. When the elastic limit is exceeded, the member permanently deforms.

An engineer usually needs to know the capability of a metal to carry loads without plastic deformation, except self-limiting deformation in the region of connections. Therefore, several properties closely related to the elastic limit have been defined for guidance. These properties can be determined easily from a stress-strain diagram that is commonly plotted for a tensile test (see Figure 5.1).

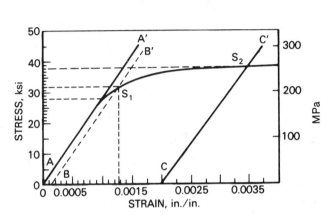

Figure 5.1—Typical Tensile Stress-Strain Diagram for a Metal Stressed Beyond the Limit of Elastic Behavior

The stress-strain curve is initially a straight line, A-A'. The slope of this line is the modulus of elasticity (Young's Modulus) for the metal under test. As the line proceeds upward, a point is reached where the strain exceeds the amount predicted by the earlier straight line relationship. It is difficult to state exactly where the proportionality ends between stress and strain because the clarity and interpretation of the curve may vary. The elastic limit (also called the proportional limit) on the stress-strain curve in Figure 5.1 is about 28 ksi (190 MPa). This is the maximum stress at which the strain remained directly proportional to stress.

Strain in a metal below the elastic limit is recoverable upon removal of the load. When metal is stressed beyond the elastic limit, the additional strain is plastic in nature and results in permanent deformation. As an example, if the tensile specimen depicted in Figure 5.1 were loaded to 32 ksi (220 MPa) (S1), the specimen would elongate 0.00125 in./in. (mm/mm) of length. Upon removal of the load, the specimen would not return to its original length, but would display a permanent stretch of about 0.00015 in./in., line B-B'.

Yield Strength

THE YIELD STRENGTH of a metal is the stress level at which the metal exhibits a specified deviation from proportionality of stress and strain. A practicable method of determining the yield strength of a metal is illustrated in Figure 5.1. The line C-C' is drawn parallel to the elastic line A-A' from a point on the abscissa representing 0.2 percent (0.0020 in./in.) (mm/mm) elongation. The line C-C' will intersect the stress-strain curve at S_2 where the stress level is about 38 ksi (260 MPa). This stress is the yield strength of the tested metal.

While 0.2 percent offset yield strength is commonly used in engineering design, offsets of 0.1 and 0.5 percent are sometimes employed in the same manner for some metals.

Most commercial low-carbon steel grades exhibit a stress strain curve feature called the *yield point*. The yield point is the first stress in a material, less than the maximum obtainable stress, at which an increase in strain occurs without an increase in stress. For these grades, a minimum yield point is used as an acceptance criterion.

Tensile Strength

THE RATIO OF the maximum load sustained by a tensile test specimen to the original cross-sectional area is called the *ultimate tensile strength* (UTS), which is the value regularly listed for the strength of a metal. The UTS represents a convenient value calculated from a standard engineering test. The "true" tensile strength of the metal which is the ratio of the breaking load to the final cross-sectional area, is substantially higher than the reported tensile strength.

Tensile strength values obtained for metals are influenced by many factors. Tensile strength is a structure-sensitive property, and is dependent upon chemical composition, microstructure, orientation, grain size, strain history, and other factors. The size and shape of the specimen and the rate of loading can also affect the result. For these reasons, the ultimate tensile strength of the heat-affected zone may be different from that of the unaffected base metal.

Fatigue Strength

BEHAVIOR UNDER CYCLIC loading is an important aspect of the strength of metals and welded joints. Fatigue fractures develop because each application of the applied tensile stress, even at nominal tensile stresses lower than yield point stress, causes the tip of a crack to advance a minute amount (stable crack growth). The rate of advance increases as the area of section ahead of the crack decreases with each application of applied load until the crack reaches a critical size. Then unstable crack growth initiates, and sudden, complete failure occurs.

Crack growth does not occur when the net stress at the crack tip is compressive. A crack may initiate due to high residual tensile stresses, but the formation of the crack will relieve the local stress condition. Thus, if the applied stresses are compressive, the crack will not grow to a critical length.

The stress that a metal can endure without fracture decreases as the number of repeated stress applications increases. The *fatigue strength* is generally defined as the maximum stress that can be sustained for a stated number of cycles without failure. As the desired number of stress repetitions is increased, the corresponding fatigue strength becomes smaller.

For steel, the fatigue strength is usually almost constant beyond about two million cycles. Many million additional cycles are required to cause a significant reduction in fatigue strength. The *fatigue limit*, therefore, practically becomes the maximum stress or stress range which the metal will bear for an infinite number of cycles without fracture. Such a limiting stress level is often called the *endurance limit*. The *fatigue life*, accordingly, is the number of cycles of stress that can be sustained by a metal under stipulated conditions.

The endurance limits reported for metals in engineering handbooks usually have been determined with polished round specimens tested in air. These data are valid and useful for design in applications such as shafts in rotating machinery and other uniform members. However, the data generally may have little relevance in the design of weldments, because weldments are characterized by abrupt changes in cross section, geometrical and metallurgical discontinuities, and residual stresses. All of these characteristics adversely affect fatigue life.

The life (cycle life) of a welded structural member subject to repeated variation of tensile or alternately tensile and compressive stresses within the elastic range of the material is primarily dependent on the stress range and the configuration of the welded assembly. *Life* is defined as the number of times a member can be subjected to a specific load prior to the initiation and growth of a fatigue crack to size that causes failure of the component.

The *stress range* is the algebraic difference between the maximum and minimum stresses in a cycle. Structural weldments often include details that result in significant differences in section thickness, and these differences create stress concentrations at the location of the section change. Thus, fatigue failures usually occur at these locations.

When designing welded built-up members and welded connections for structures subject to fatigue loading, the applicable standard governing the subject structure must be followed. In the absence of a specific standard, designs of existing welded components should be used as a guide.

Localized stresses within a structure may result entirely from external loading, or there may be a combination of applied and residual stresses. Residual stresses are not cyclic, but they may augment or detract from applied stresses, depending upon their respective signs. For this reason, it may be advantageous to induce, if possible, compressive residual stress in critical areas of a weldment where cyclic applied tensile stresses are expected. This may be accomplished by a welding sequence that controls the residual stresses from welding, or by a localized treatment that acts to place the surface in compression.

Thermal stresses also must be considered in the same light as an applied stress because thermal cycling can lead to fatigue failure if the thermal gradients are steep or if the thermal stresses are concentrated by a stress raiser.

The rate of repetition of loading is important because it determines the time required for the number of cycles that will cause a crack to initiate and propagate to a critical size. Weldments in rotating equipment are particularly prone to fatigue failure. Pressure vessels can fail by fatigue also when pressurization is cyclic and stress above the fatigue strength is concentrated at some point.

Designers of weldments need to thoroughly understand the fatigue characteristics of metals as used in weldments. The most common cause of fracture in weldments is fatigue. One of the reasons for this is the frequent presence of stress raisers (changes in cross section, discontinuities, etc.) that concentrate imposed cyclic stresses to levels above the fatigue limit of the metal for the existing conditions.

Ductility

THE AMOUNT OF plastic deformation that an unwelded or welded specimen undergoes in a mechanical test carried to fracture is considered a measure of the ductility of the metal or the weld. Values expressing ductility in various mechanical tests are meaningful only for the relative geometry and size of the test specimen. Thus, they do not measure any fundamental characteristic, but merely provide relative values for comparing the ductilities of metals subjected to identical test conditions. The plasticity exhibited by a specimen is simply the deformation accomplished during the yielding process.

Ductility, regardless of the method of measurement, is a structure-sensitive property, and is affected by many of the conditions of testing. Size and shape of the specimen, ambient temperature, the strain rate, microstructure, and surface conditions all influence the amount and location of plastic deformation prior to fracture.

Ductility values obtained from precise or elaborate tests are not used directly in design. Most structures are designed to operate at stresses below the yield strength, and any significant deformation usually makes the unit or article unfit for service. Ductility values give an indication of the ability of a metal to yield and relieve high secondary stresses. These values also may give some idea of the reserve of plasticity available to insure against sudden fracture under unexpected overloading. However, ductility values do not necessarily indicate the amount of plastic deformation that will take place under all conditions of loading. Most structures are sensitive to both loading rate and ambient temperature.

Fracture Toughness

A METAL THAT is judged ductile by a standard tensile test or slow bend test may perform in a brittle manner in another type of test or when exposed to service conditions. The only forecast that can be made with reasonable certainty from tensile or bend test results is that a metal with very little ductility is not likely to behave in a ductile manner in any other type of mechanical test carried to fracture.

A metal that displays good ductility in a tensile or bend test may or may not behave in a ductile manner in other types of mechanical tests. In fact, there have been numerous cases where ductile metals (as judged by tensile and bend tests) have fractured in service with little or no plastic deformation. The lack of deformation and other aspects of such failures usually show that little energy was required to produce the fracture. This general experience prompts the metallurgist to speak of the toughness of metal as a property distinct from ductility.[6]

Toughness is the ability of a metal to resist fracture in the presence of a notch, and to accomodate the loads by plastic deformation. Three conditions markedly influence the behavior of a metal; namely, (1) the rate of straining, (2) the nature of the load, that is, whether the imposed stresses are uniaxial or multiaxial, and (3) the temperature of the metal.

Many metals can absorb energy and deform plastically under the simple circumstances represented in tensile or bend tests (and therefore would be judged ductile); a lesser number of these metals are considered to have good toughness when tested under conditions of high stress concentration. The toughness displayed by a metal tends to decrease as (1) the rate of straining increases, (2) the stresses become multiaxial, and (3) the temper-ature of the metal is lowered. Weld zones in service easily may be exposed to one or more of these conditions. Consequently, there is good reason to be concerned about the toughness of weld metal and weld heat-affected zones.

In designing with ductile metals, the fail-safe load carrying capability of an engineering structure, including the welds, is normally based on a stress analysis to assure that the nominal stresses are below the yield strength. Failures that occur at stresses below the yield strength are broadly classified as brittle fractures. These failures can result from the effects of discontinuities or crack-like defects of critical size in the weld or the base metal, that do not greatly alter the nominal stress distribution and are customarily neglected in the stress analysis.

When structural grade steel is tested in uniaxial tension, it deforms in a ductile manner prior to rupture at the ultimate load. Because the volume of metal must remain constant, any elongation in one direction must be accompanied by contraction in one or both of the other directions. The uniaxial tensile test is free to contract in the other direction, and ductile behavior results. If the necessary lateral contraction is severely restrained or prevented and the longitudinal stresses are sufficiently large, the same material that exhibited ductile behavior in a tensile test may fail in a brittle manner.

In real structures where separate elements are joined by welding, conditions of restraint and stress concentrations usually are very different from those of simple uniaxial tension. Large material thickness by itself may provide the sufficient restraint to prevent lateral contractions. Thus, structural details that proved satisfactory in long usage and service may not necessarily have satisfactory ductile characteristics if dimensions are proportionately increased to a large degree.

A complete fracture-safe analysis requires proper attention to the role of discontinuities. For many classes of structures, such as ships, bridges, and pressure vessels, experience with specific designs, materials, and fabrication procedures has established a satisfactory correlation between notch test standards for base and weld metals and acceptable service. The problem is to insure the soundness and integrity of a new design.

One of the motivations for the application of fracture mechanics concepts and tests to welded joints is the possibility of designing safely against the effects of common weld discontinuities. It is widely recognized that welded joints almost always contain some discontinuities, and this places the designer using welded joints in a dilemma.

The designer plans joints that are entirely free from discontinuities, but this is not realistic. The practical approach is to recognize that discontinuities are present, and to place a reasonable limit on their existence. The problem is how to determine the types and extent of discontinuities that are acceptable.

While conventional toughness testing procedures cannot deal directly with this problem, fracture mechanics tests, where applicable, specifically define a relationship between flaw size and fracture stress for a given base metal or weld joint. Thus, the tests permit a direct estimate of allowable flaw sizes for different geometrical configurations and operating conditions.

For the allowable stress, fracture mechanics can establish the minimum or critical crack-like flaw size that

---

6. For information on fracture toughness, see Chapter 12 of this volume, and also Rolfe, S. T. and Barsom, J. M. *Fracture and fatigue control in structures: applications of fracture mechanics.* Englewood Cliffs, NJ: Prentice-Hall, 1977.

would initiate unstable crack propagation. However, in members subject to cyclic loading or corrosion, or both, cracks may initiate at stress raisers that are considered acceptable discontinuities. These small cracks could grow by stable crack extension with each application of tensile stress until the crack reaches the critical size. For such conditions, information relative to crack growth rate is essential to establish continuing inspection frequencies and acceptance criteria.

The specification of allowable flaw sizes and inspection procedures for flaw detection can be referred to a rational and logical procedure rather than one based solely on experience or opinion. As weldments of more complicated design and higher strength requirements are introduced, the designer will need totake an analytical approach to the problem of weld discontinuities.

Low Temperature Properties

LOWERING THE TEMPERATURE of a metal profoundly affects fracture characteristics, particularly if the metal possesses a body-centered cubic crystalline structure (carbon steel is an example). Strength, ductility, and other properties are changed in all metals and alloys as the temperature decreases.

The designer must consider the properties of metals at very low temperatures. Pressure vessels and other welded products sometimes are expected to operate at low temperatures. Very low temperatures are involved in cryogenic service that entails the storage and use of liquefied industrial gases, such as oxygen and nitrogen.

As the temperature is lowered, a number of changes in properties take place in metals. The elastic modulus, for example, increases. In general, the tensile and yield strengths of all metals and alloys increase as the temperature is lowered.

The ductility of most metals and alloys tends to decrease as temperature is lowered. However, some metals and alloys retain considerable ductility at very low temperatures. Because of the notch-toughness transition behavior of carbon and low alloy steels and certain other metals, the suitability of metals for low temperature service is judged by tests that evaluate propensity to brittle fracture rather than simple tensile or bend ductility. The most common specimen for low-temperature testing is the notched-bar impact specimen.

The principal factors that determine the low temperature behavior of a metal during mechanical testing are (1) crystal structure, (2) chemical composition, (3) size and shape of the test specimen, (4) conditions of manufacture and heat treatment, and (5) rate of loading. The notched-bar impact strengths for five common metals are listed in Table 5.3. Iron and steel suffer a marked reduction in impact strength at low temperatures. The addition of alloying elements to steel, especially nickel and manganese, can markedly improve notch toughness at low temperatures, while increased carbon and phosphorus can greatly decrease low-temperature notch toughness.

ELEVATED TEMPERATURE PROPERTIES

PERFORMANCE OF A metal in service at an elevated temperature is governed by other factors in addition to strength and ductility. Time becomes a factor because metals will creep at high temperatures, that is, the section under stress will continue to deform even if the load is maintained constant. The rate at which a metal creeps under load increases rapidly with increasing temperature and increasing load. Consequently, the time over which a metal under load will deform too much to be usable can vary from many years at a slightly elevated temperature to a few minutes at a temperature near the melting point.

The creep rates of metals and alloys differ considerably. If the temperature and stress are sufficiently high, the metal will creep until rupture occurs. The term

Table 5.3
Notch-Bar Impact Strengths of Metals at Low Temperatures

Temperature, °F	Charpy V-Notch Impact Strength, ft · lb									
	Aluminum[a]		Copper[a]		Nickel[a]		Iron[b]		Titanium[c]	
	Ft lb	J	Ft lb	J	Ft lb	J	Ft lb	J	Ft lb	J
75 (room temperature)	20	27	40	54	90	122	75	102	15	20
0	20	27	42	57	92	125	30	41	13	18
− 100	22	30	44	60	93	126	2	3	11	15
− 200	24	33	46	63	94	128	1	1	9	12
− 320	27	37	50	68	95	129	1	1	7	10

a. Face-centered cubic crystal structure.
b. Body-centered cubic crystal structure.
c. Hexagonal close-packed crystal structure.
Source: Linnert, G. E. *Welding metallurgy,* Vol. 1, 3rd Ed. Miami, Florida: American Welding Society, 1965.

creep-rupture is used to identify the mechanics of deformation and failure of metals under stress at elevated temperatures.

PHYSICAL PROPERTIES

THE PHYSICAL PROPERTIES of metals seldom receive the attention regularly given to mechanical properties in a general treatise on welding. Nevertheless, the physical properties are an important aspect of metal characteristics. The welding engineer may be unaware that the success of the joining operation depends on a particular physical property. The physical properties of regular polycrystalline metals are not so structure-sensitive as the mechanical properties. Constant values usually are provided for metals and alloys, and they serve satisfactorily for most engineering purposes. Only the physical properties that may require some consideration in designing or fabricating a weldment are discussed here.[7]

Thermal Conductivity

THE RATE AT which heat is transmitted through a material by conduction is called *thermal conductivity* or *thermal transmittance*. Metals are better heat conductors than nonmetals, and metals with high electrical conductivity generally have high thermal conductivity.

Metals differ considerably in their thermal conductivities. Copper and aluminum are excellent conductors, which makes it difficult to weld these metals using a relatively low-temperature heat source, like an oxyacetylene flame. Conversely, the good conductivity of copper makes it a good heat sink when employed as a hold down or backing bar.

Melting Temperature

THE HIGHER THE melting point or range, the larger is the amount of heat needed to melt a given volume of metal. Hence, the temperature of the heat source in welding must be well above the melting range of the metal. However, two pieces of a metal, such as iron, may be joined with a metal of lower melting point, such as bronze. The brazing filler metal must wet and adhere to the steel faces to which it is applied. Welding of two metals of dissimilar composition becomes increasingly difficult as the difference in melting ranges widens.

Thermal Expansion and Contraction

MOST METALS INCREASE in volume when they are heated, and decrease in volume when they are cooled.

The *coefficient of thermal expansion* is the unit change in linear dimensions of a body when its temperature is changed by one degree. The coefficient also serves to indicate contraction when the temperature is decreased. Engineers usually are concerned with changes in length in metal components, and most handbooks provide a linear coefficient of thermal expansion rather than a coefficient for volume change.

Metals change in volume when they are heated and cooled during welding. The greater the increase in volume and localized upsetting during heating, the more pronounced will be the distortion and shrinkage from welding. Distortions resulting from welding must be considered during weldment design so that the final weldment dimensions are correct.

Electrical Conductivity

METALS ARE RELATIVELY good conductors of electricity. Increasing the temperature of a metal interferes with electron flow; consequently, electrical conductivity decreases. Adding alloying elements to a metal or cold working also decreases conductivity. These characteristics are important variables affecting resistance welding processes.

CORROSION PROPERTIES

THE CORROSION PROPERTIES of a metal determine its mode and rate of deterioration by chemical or electrochemical reaction with the surrounding environment. Metals and alloys differ greatly in their corrosion resistance. Corrosion resistance often is an important consideration in planning and fabricating a weldment for a particular service. Therefore, the designer should know something about the behavior of weld joints under corrosive conditions.[8]

Weld joints often display corrosion properties that differ from the remainder of the weldment. These differences may be observed between the weld metal and the base metal, and sometimes between the heat-affected zone and the unaffected base metal. Even the surface effects produced by welding, like heat tint formation or oxidation, fluxing action of slag, and moisture absorption by slag particles can be important factors in the corrosion behavior of the weld metal.

Welds made between dissimilar metals or with a dissimilar filler metal may be subject to electrochemical corrosion. Brazed joints can be particularly vulnerable. Appropriate protective coatings are required to avoid corrosion in sensitive environments.

---

7. Physical properties of metals are discussed in Chapter 2 of this volume.

8. Types of corrosion and corrosion testing of welds are discussed in Chapter 12 of this volume.

DESIGN PROGRAM

A WELDMENT DESIGN program starts with a recognition of a need. The need may be for improving an existing machine or for building an entirely new product or structure using advanced design and fabrication techniques. In any event, many factors must be taken into account before a design is finalized. These considerations involve numerous questions and considerable research into the various areas of marketing, engineering, and production.

ANALYSIS OF EXISTING DESIGN

WHEN DESIGNING AN entirely new machine or structure, information should be obtained about similar units, including those of other manufacturers or builders. If a new design is to replace an existing design, the strengths and weaknesses of the existing design should be determined first. The following questions can help:

(1) What are the opinions of customers and the sales force about the existing products?
(2) What has been the performance history of the existing products?
(3) What features should be retained, discarded, or added?
(4) What suggestions for improvements have been made?

DETERMINATION OF LOAD CONDITIONS

THE SERVICE REQUIREMENTS of a weldment and the conditions of service that might result in overloading should be ascertained. From such information, the load on individual members can be determined. As a starting point for calculating loads a designer may find the following methods useful:

(1) Determine the torque on a shaft or revolving part from the motor horsepower and speed.
(2) Calculate the forces on members caused by the dead weight of parts.
(3) Determine the maximum load on members of a crane hoist, shovel, lift truck, or similar material handling equipment from the load required to tilt the machine.
(4) Use the maximum strength of critical cables on such equipment to determine the maximum loads on machine members.
(5) Consider the force required to shear a critical pin as an indication of maximum loading on a member.
(6) Determine the frequency of application of design applied loads and desired service life. (Designing for fatigue at lesser loads may be more critical than design-

ing for maximum strength to resist infrequently applied maximum loads.)

If a satisfactory starting point cannot be found, design for an assumed load, and adjust from experience and tests.

MAJOR DESIGN FACTORS

IN DEVELOPING A design, the designer should consider how decisions will affect production operations, manufacturing costs, product performance, appearance, and customer acceptance. Many factors far removed from engineering considerations per se become major design factors. Some of these, along with other relevant rules, are as follows:

(1) The design should satisfy strength and stiffness requirements. Overdesigning wastes materials and increases production and shipping costs.
(2) The safety factor should be realistic.
(3) Good appearance may be necessary, but only in areas that are exposed to view. The drawing or specifications should specify those welds that must be ground or otherwise conditioned to enhance appearance.
(4) Deep, symmetrical sections should be used to efficiently resist bending.
(5) Welding the ends of beams rigidly to supports increases strength and stiffness.
(6) Rigidity may be provided with welded stiffeners to minimize the weight of material.
(7) Tubular sections or diagonal bracing should be used for torsion loading. A closed tubular section is significantly more effective in resisting torsion than an open section of similar weight and proportions.
(8) Standard rolled sections, plate, and bar should be used for economy and availability.
(9) Accessibility for maintenance must be considered during the design phase.
(10) Standard, commercially available components should be specified when they will serve the purpose. Examples are index tables, way units, heads, and columns.

DESIGNING THE WELDMENT

TO A DESIGNER familiar only with castings or forgings, the design of a weldment may seem complex because of the many possible choices. However, flexibility is one of the advantages of welded design; opportunities for saving are presented. The following are general pointers for effective design:

(1) Design for easy handling of materials, inexpensive tooling, and accessibility of the joints for reliable welding.

(2) Check with the shop for ideas that can contribute to cost savings.

(3) Establish realistic tolerances based on end use and suitability for service. Excessively close tolerances serve no useful purpose, and increase costs.

(4) Minimize the number of pieces. This will reduce assembly time and the amount of welding.

Part Preparation

THERMAL CUTTING, SHEARING, sawing, blanking, nibbling, and machining are methods for cutting blanks from stock material. Selection of the appropriate method depends on the available material and equipment and the relative costs. The quality of edges needed for good fit-up and the type of edge preparation for groove welds must also be kept in mind. The following points should be considered:

(1) Dimensioning of a blank may require stock allowance for subsequent edge preparation.

(2) The extent of welding must be considered when proposing to cut the blank and prepare the edge for welding simultaneously.

(3) Weld metal costs can be reduced for thick plate by specifying J- or U-groove preparations.

(4) Consider air carbon arc gouging, oxygen gouging, or chipping for back weld preparation.

Forming

FORMING OF PARTS can sometimes reduce the cost of a weldment by avoiding joints and machining operations. The base metal composition, part thickness, over-all dimensions, production volume, tolerances, and cost all influence the choice of forming methods. The following suggestions may be helpful in making decisions in this area:

(1) Create a corner by bending or forming rather than by welding two pieces together.

(2) Bend flanges on the plate rather than weld flanges to it.

(3) Use a casting or forging in place of a complex weldment to simplify design and reduce manufacturing costs.

(4) Use a surfacing weld on an inexpensive component to provide wear resistance or other properties in place of an expensive alloy component.

Cold forming reduces the ductility and increases the yield strength of metals. Heat treating the metal to restore ductility may increase or decrease the strength, depending on the alloy type. Heat from arc welding on cold formed material can also affect the mechanical properties of the base metal. The mechanical properties of the heat affected zone of cold formed materials may be reduced by the heat of welding. Generally, the relevant standard will provide maximum cold forming allowances, and minimum strength properties of cold formed weldments. For example, Section VIII of the ASME Boiler and Pressure Vessel Code requires that under certain circumstances cold forming that results in extreme fiber elongation over 5 percent in carbon-, low alloy-, and heat-treated-steel plates must be stress relieved.

Weld Joint Design

THE JOINT DESIGN should be selected primarily on the basis of load requirements. However, variables in design and layout can substantially affect costs. Generally, the following rules apply:

(1) Select the joint design that requires the least amount of weld metal.

(2) Use, where possible, square-groove and partial joint penetration welds.

(3) Use lap and fillet welds instead of groove welds if fatigue is not a design consideration.

(4) Use double-V- or U-groove instead of single-V- or U-groove welds on thick plates to reduce the amount of weld metal and to control distortion.

(5) For corner joints in thick plates where fillet welds are not adequate, beveling both members should be considered to reduce the tendency for lamellar tearing.

(6) Design the assembly and the joints for good accessibility for welding.

Size and Amount of Weld

OVERDESIGN IS A common error, as is over-welding in production. Control of weld size begins with design, but it must be maintained during the assembly and welding operations. The following are basic guides:

(1) Adequate but minimum size and length should be specified for the forces to be transferred. Oversize welds may cause excessive distortion and higher residual stress without improving suitability for service. (They also contribute to increased costs.) The size of a fillet weld is especially important because the amount of weld required increases as the square of the weld size increases.

(2) For equivalent strength, a continuous fillet weld of a given size is usually less costly than a larger sized intermittent fillet weld. Also, there are fewer weld terminations that are potential sites of discontinuities.

(3) An intermittent fillet weld can be used in place of a continuous fillet weld of minimum size when static load

conditions do not require a continuous weld. An intermittent fillet weld should not be used under cyclic loading conditions.

(4) To derive maximum advantage of automtic welding, it may be better to use one continuous weld rather than several short welds.

(5) The weld should be placed in the section of least thickness, and the weld size should be based on the load or other requirements of that section.

(6) Welding of stiffeners or diaphragms should be limited to that required to carry the load, and should be based on expected out-of-plane distortion of the supported components under service loads as well as during shipment and handling.

(7) The amount of welding should be kept to a minimum to limit distortion and internal stresses and, thus, the need and cost for stress-relieving and straightening.

Subassemblies

IN VISUALIZING ASSEMBLY procedures, the designer should break the weldment into subassemblies in several ways to determine the arrangement that offers the greatest cost savings. The following are advantages of subassemblies:

(1) Two or more subassemblies can be worked on simultaneously.

(2) Subassemblies usually provide better access for welding, and may permit automatic welding.

(3) Distortion in the finished weldment may be easier to control.

(4) Large size welds may be deposited under lesser restraint in subassemblies which, in turn, helps to minimize residual stresses in the completed weldment.

(5) Machining of subassemblies to close tolerances can be done before final assembly. If necessary, stress relief of certain sections can be performed before final assembly.

(6) Chamber compartments can be leak tested and painted before final assembly.

(7) In-process inspection and repair is facilitated.

(8) Handling costs may be much lower.

When possible, it is desirable to construct the weldment from standard sections, so that the welding of each can be balanced about the neutral axis.

WELDING PROCEDURES

ALTHOUGH DESIGNERS MAY have little control of welding procedures, they can influence which procedures are used in production. The following guidelines can help to effect the ultimate success of weldment design:

(1) Backing strips increase the speed of welding when making the first pass in groove welds.

(2) The use of low-hydrogen electrodes or welding processes eliminates or reduces preheat requirements for steel.

(3) If plates are not too thick, a joint design requiring welding only from one side should be considered to avoid manipulation or overhead welding.

(4) Reinforcement of a weld is generally unnecessary to obtain a full-strength joint.

(5) With T-joints in thick plate subject to tensile loading in the through-thickness direction, the surface of the plate should be buttered with weld metal for some distance beyond the intended weld terminations. Research has shown that buttering is the most effective technique available to the fabricator for minimizing the possibility of lamellar tearing in T-joints.

(6) Joints in thick sections should be welded under conditions of least restraint; for example, prior to installation of stiffeners.

(7) Sequencing of fit up, fixturing, and welding is particularly important for box members made of plates because correction of distortion after completion of welding is virtually impossible.

RESIDUAL STRESSES AND DISTORTION

RESIDUAL STRESSES ARE the consequence of contraction of solidified weld metal in the presence of the surrounding cool base metal; hence, they are unavoidable with the welding process. Residual stresses may be minimized by joint design and welding technique, but they cannot be eliminated. In applications where the effects of residual stresses are significant, the effects are normally accounted for in the design rules for the application. For example, residual stresses do not affect shear strength.

In tension members, peak residual tension stresses, which may be as large as yield point, have no effect upon ultimate strength. This is because the tensile residual stress must be internally balanced by compressive residuals to maintain equilibrium. Thus, when external tensile force is applied, the net tensile stresses in the region of the tensile residual stress will increase while the compressive residual stresses will be reduced. As the applied force is further increased, the net tensile stresses will reach yield point of the material, but no higher. Even higher applied loads will cause spread of yielding controlled by the remaining net compressive stresses. If the load is reduced or removed at this stage, the stresses will reduce linearly until, at rest, the peak tensile residual stress will be blunted and reduced. That is, applied tensile loads tend to wipe out residual stresses. This fact has been recognized in some specifications which require that certain structures be subjected to an initial, one time, application of an overload to eliminate residual stresses. The ultimate strength of a member is unaffected by initial residual stresses.

In the case of compression members of a length such that buckling is a potential failure mode, the converse effect takes place, but in this case, the component performance is affected. As compression load is applied, the net tensile stress in the region of the original tensile residual will be reduced while the net compressive stresses will be increased. When the net compression stress reaches yield point, the affected area is no longer able to participate in resisting the tendency to buckle. Fortunately, this effect has been fully accounted for in the column strength curves recommended by the Structural Stability Research Council which are the basis for the compression stress provisions of design specifications. Hence, compression stress need not be considered further by the user. The same is true to a lesser degree for lateral buckling of laterally unsupported beams.

In the case of members subject to cyclic loading, residual stresses may have an effect on the fatigue life. Total fatigue life may be divided into a crack initiation phase and a stable crack growth phase leading to unstable crack growth at which time sudden rupture occurs.

The maximum cyclic tensile stress affects the time to crack initiation. The difference between the maximum and the minimum stress (stress range) controls the rate of crack growth. Hence, the presence of tensile residual stresses affects the time required to initiate a crack at a stress raiser, but once initiated, the residual stresses do not affect the crack growth rate.

The crack growth rate is a function of the stress range only. With applied cyclic compressive stress, a crack may initiate as a result of the net tensile residual stress, but once the crack is initiated, the residual tensile stresses are eliminated. The crack will not propagate in the remaining compressive stress field. Again, because a very significant portion of the total fatigue life is involved in the crack initiation phase, some project specifications have stipulated a controlled one-time initial overload to wipe out residual stresses to provide for increased service life.

It is well for the designer to keep residual stress in mind because it does have a bearing upon the total fatigue life. It is equally important that the design criteria in specifications be based upon extensive testing of full size welded members and details containing typical residual stresses, or that a postweld heat treatment be incorporated to reduce residual stresses.

LAMINATIONS AND LAMELLAR TEARING

THE DESIGNER MUST understand the true significance of residual stresses resulting from contraction of the cooling weld metal. These contractions result in yield level stresses with a significant amount of strain. The shrinkage strains provide the potential for lamellar tearing and underbead cracking. When lamellar tears or cracks occur, they most often occur during fabrication.

Because the strains associated with allowable design stresses are so small in comparison to the weld shrinkage strains which are responsible for lamellar tearing, it is unlikely that either cyclic or static service loads would initiate a lamellar tear. The detail geometry, material thickness, and weld size selected may lead to difficult or virtually impossible fabrication conditions.

In connections where a member is welded to the outside surface of a main member, the capacity to transmit through-thickness tensile stresses is essential to the proper functioning of the joint. Laminations (preexisting planes of weakness) or lamellar tears (cracks parallel to the surface caused by high localized through-thickness thermal strains induced at restrained corner and T-joint welds) may impair this capacity.

Consideration of the problem of lamellar tearing must include design aspects and welding procedures that are consistent with the properties of the base material. In connections where lamellar tearing might be a problem, consideration should be given in design to provide for maximum component flexibility and minimum weld shrinkage strain.

The following precautions should help to minimize the problems of lamellar tearing in highly restrained welded connections. It is assumed that procedures producing low-hydrogen weld metal would be used in any case.

(1) On corner joints, where feasible, the edge preparation should be on the through-thickness member.

(2) The size of the weld groove should be kept to a minimum consistent with the design, and overwelding should be avoided.

(3) If possible, welds in corner and T-joints should be completed early in the fabrication process to minimize restraint in such joints.

(4) A predetermined welding sequence should be selected to minimize overall shrinkage of the most highly restrained elements.

(5) The lowest strength weld metal available, consistent with design requirements, should be used to promote straining in the weld metal rather than in the more sensitive through-thickness direction of the base metal.

(6) Buttering with weld metal, approximately 3/16 in. (5 mm) thick and extending beyond the limits of the joint, should be considered prior to fit-up to provide an area of material less susceptible to lamellar tearing at locations of the most severe strains.

(7) Specification of material with improved through-thickness ductility should be considered for critical connections.[9]

In critical joint areas subject to through-thickness direction loading, material with preexisting laminations

---

9. Improved quality steel does not eliminate weld shrinkage and, by itself, will not necessarily avoid lamellar tearing in highly restrained joints. Thus, it should not be specified in the absence of comprehensive design and fabrication considerations.

and large metallic inclusions should be avoided. In addition, the following precautions should be taken:

(1) The designer should specify ultrasonic inspection, after fabrication or erection, or both, of those specific highly-restrained connections that could be subject to lamellar tearing and which are critical to the structural integrity.

(2) The designer must consider whether minor weld flaws or base metal imperfections can be left unrepaired without jeopardizing structural integrity. Gouging and repair welding will add additional cycles of weld shrinkage to the connection, and may result in the extension of existing flaws or the generation of new flaws by lamellar tearing.

CLEANING AND INSPECTION

DESIGN SPECIFICATIONS CAN have some effect on cleaning and inspection costs. The safety requirements of the weldment determine the type and amount of inspection required. The following shop practices also affect these costs:

(1) As-welded joints that have uniform appearance are acceptable for many applications. Therefore, the surface of a weld need not be machined smooth or flush unless that is required for another reason. Smoothing a weld is an expensive operation.

(2) Undesirable overwelding should be noted during inspection because it can be costly and also contributes to distortion. Corrective action should be directed at work in progress rather than at completed weldments.

(3) The type of nondestructive inspection to be used on weldments must be capable of detecting the types and sizes of weld discontinuities that are to be evaluated for acceptability.

WELDED DESIGN CONSIDERATIONS

THE PERFORMANCE OF any member of a structure depends on the properties of the material and the characteristics of section. If a design is based on the efficient use of these properties, the weldment should be functionally good and conservative of materials.

Engineers assigned to design welded members need to know (1) how to select the most efficient structural section and determine required dimensions, and (2) when to use stiffeners and how to size and place them.

The mathematical formulas for calculating forces and their effects on sections, and for determining the sections needed to resist such forces appear quite forbidding to the novice. With the proper approach, however, it is possible to simplify the design analysis and the use of those formulas. In fact, as will be explained later, it is often possible to make correct design decisions merely by examining one or two factors in an equation, without using tedious calculations. On the whole, the mathematics of weld design is no more complex than in other engineering fields.

THE WELDED DESIGN APPROACH

CONSIDERATIONS OTHER THAN the engineer's wishes may prevail; for example, when a machine is to be converted from a cast to a welded design. Management may favor the redesign of one or more components to weldments, and conversion of the design over a period of years to an all-welded product. Gradual conversion avoids the obsolescence of facilities and skills, and limits the requirement for new equipment. Available capital and personnel considerations often limit the ability of a company when changing to welded design. Supplementing these considerations is the need to maintain a smooth production flow, and to test the production and market value of the conversion as it is made step by step.

From the standpoint of performance and ultimate production economics, redesign of the machine or structure as a whole is preferable. The designer then is unrestricted by the previous design, and in many cases is able to reduce the number of pieces, the amount of material used, and the labor for assembly. A better, lower-cost product is realized immediately. When the adjustment to changes in production procedures is complete, the company is in a position to benefit more fully from welded design technology.

STRUCTURAL SAFETY CONSIDERATIONS

A SAFE WELDED structure depends upon a combination of good design practices, good workmanship during fabrication, and good construction methods. In design, the selection of a safety factor and the appropriate analytical procedures require experience and sound engineering judgement. Deterioration from corrosion or other service conditions during the life of the structure, variations in material properties, potential imperfections in materials and welded joints, and many other factors also need consideration.

A rational approach to structural safety is a statistical evaluation of the random nature of all the variables that determine the strength of a structure and also the variables that may cause it to fail. From these data, the risk of failure may be evaluated and the probability of occurrence kept at a safe level for the application considering the risk of injury, death, property damage, or unsatisfactory service performance.

The choice of materials and safe stress levels in members may not produce the most economical structure. However, safety must take precedence over cost savings when there is a question of which should govern. Great skill, care, and detailed stress analyses are needed when the designer attempts a new design or structural concept. Laboratory tests of models or sections of prototype structures should be used to verify the design.

SELECTING A BASIS FOR WELDED DESIGN

A REDESIGN OF a product may be based on the previous design or on loading considerations solely. Following a previous design has the advantages of offering a "safe" starting point, if the old design is known to perform satisfactorily. This approach, however, has disadvantages in that it stifles creative thinking toward developing an entirely new concept to solve the basic problem. Little demand is made on the ingenuity of the designer when the welded design is modeled on the previous product. Tables of equivalent sections or nomographs can be used to determine required dimensions for strength and rigidity.

A design based on the loading requires designers to analyze what is needed and come up with configurations and materials that best satisfy the need. They must know or determine the type and amount of load, and the values for stress allowables in a strength design, or deflection allowables in a rigidity design.

DESIGNING FOR STRENGTH AND RIGIDITY

A DESIGN MAY require strength only, or strength and rigidity to support the load. All designs must have sufficient strength so that the members will not fail by breaking or yielding when subjected to normal operating loads or reasonable overloads. Strength designs are common in road machinery, farm implements, motor brackets, and various types of structures. If a weldment design is based on calculated loading, design formulas for strength are used to dimension the members.

In certain weldments, such as machine tools, rigidity as well as strength is important because excessive deflection under load would result in lack of precision in the product. A design based on rigidity also requires the use of design formulas for sizing members.

Some parts of a weldment serve their design function without being subjected to loadings much greater than their own weight (dead load). Typical members are fenders and dust shields, safety guards, cover plates for access holes, and enclosures for esthetic purposes. Only casual attention to strength and rigidity is required in sizing such members.

DESIGN FORMULAS

THE DESIGN FORMULAS for strength and rigidity always contain terms representing the load, the member, the stress, and the strain or deformation. If any two of the first three terms are known, the others can be calculated. All problems of design thus resolve into one of the following:

(1) Finding the internal stress or the deformation caused by an external load on a given member
(2) Finding the external load that may be placed on a given member for any allowable stress, or deformation
(3) Selecting a member to carry a given load without exceeding a specified stress or deformation

A load or force causes a reaction in the member such as tension, compression, bending, torsion, or shear stresses. The result is a strain measured by relative displacements in the member as elongation, contraction, deflection, or angular twist. A useful member must be designed to carry a certain type of load within an allowable stress or deformation. In designing within the allowable limits, the designer should generally select the most efficient material section size and section shape. The properties of the material and those of the section determine the ability of a member to carry a given load.

Common design formulas, developed for various conditions and member types, are much too numerous for inclusion here. However, some are used to illustrate specific design problems.[10]

The use of design formulas may be illustrated by the problem of obtaining adequate stiffness in a cantilever beam. The amount of vertical deflection at the end of the beam under a concentrated load at the end, Figure 5.2, can be determined using the following deflection formula:

$$\Delta = \frac{FL^3}{3EI} \tag{5.2}$$

10. The properties of sections for standard structural shapes are given in the *Manual of steel construction*, Chicago: American Institute of Steel Construction (AISC); latest edition and *Engineering data for aluminum structures (ED-33)*, Washington, DC: The Aluminum Association; latest edition.

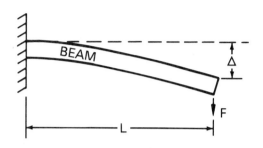

Figure 5.2—Deflection (Δ) of a Cantilever Beam Under Concentrated Load F

where

    Δ = the deflection

    F = the applied force or load, lb (N)

    L = the length of the beam, in. (mm)

    E = the modulus of elasticity of the metal, psi (Pa)

    I = the area moment of inertia of the beam section, in.^{4} (mm^4)

It is normally desirable to have the least amount of deflection. Therefore, E and I values should be as large as possible. The common structural metal having the highest modulus of elasticity (E) is steel, with a value of about 30×10^6 psi (200 000 MPa). The other factor requiring a decision relative to deflection is I, the moment of inertia of the cross section. The moment of inertia of a member is a function of the geometry of the member. The beam must have a cross section with a moment of inertia about the horizontal axis large enough to limit the deflection to a permissible value. A section with adequate in-plane moment of inertia will satisfy the vertical deflection requirement, whatever the shape of the section. However, the out-of-plane stability of the beam may also need consideration especially if forces transverse to the principle axis or torsion are involved. The designer must then decide the shape to use for the best design at lowest cost.

TYPES OF LOADING

THERE ARE FIVE basic types of loading: tension, compression, bending, shear, and torsion. When one or more of these types of loading are applied to a member, they induce stress in addition to any residual stresses.

The stresses result in strains or movements within the member, the magnitudes of which are governed by the modulus of elasticity, E, of the metal. Some deformation always takes place in a member when the load is applied because the associated stress inevitably causes strain.

Tension

PURE TENSION LOADING is generally the simplest type of loading from the standpoint of design and analysis. The axial tensile load causes axial strains and elongation but no secondary effects tending to cause instability and presence of residual stresses does not reduce the ultimate strength of the bar. The principal requirement is adequate net cross-sectional area to carry the load.

Compression

A COMPRESSIVE LOAD may require designing to prevent buckling. Few compression members fail by crushing or by exceeding the ultimate compressive strength of the material. If a straight compression member, such as the column in Figure 5.3(A), is loaded through its center of gravity, the resulting stresses are simple axial compressive stresses. In the case of a slender column, it will start to bow laterally at a stress lower than the yield strength. This movement is shown in Figure 5.3(B). As a result of bowing, the central portion of the column becomes increasingly eccentric to the axis of the force and causes a bending moment on the column, as shown in Figure 5.3(C). Under a steady load, the column will remain stable under the combined effect of the axial stress and the bending moment. However, with increasing load and associated curvature, Figure 5.3(D), a critical point will be reached where the column will buckle and fail. The presence of residual stresses developed during manufacturing may also reduce the failure load. As a column deflects under load, a bending moment can develop in restrained end connections.

Two properties of a column are important to compressive load: cross-sectional area (A) and radius of gyration (r). The area is multiplied by the allowable compressive stress (σ_c) to arrive at the compressive load that the column can support in the absence of buckling. The radius of gyration indicates, to a certain extent, the ability of the column to resist buckling. It is the distance from the neutral axis of the section to an imaginary line in the cross section about which the entire area of the section could be concentrated and still have the same moment of inertia about the neutral axis of the section. The formula for the radius of gyration is as follows:

$$r = \frac{(I)^{1/2}}{A} \qquad (5.3)$$

where

    r = radius of gyration

    I = moment of inertia about the neutral axis

    A = cross-sectional area of the member

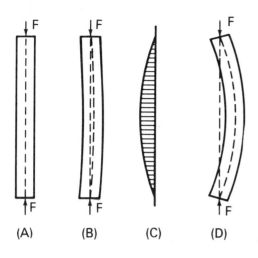

Figure 5.3—(A) Straight Column Under Compressive Load; (B) Column Deflects Laterally with Increasing Load; (C) Bending Moment Diagram; (D) Increased Deflection with Higher Loading

Table 5.4
Allowable Axial Compressive Stresses (AISC)

(1) When the effective slenderness ratio, KL/r, is less than C_c:

$$\sigma_c = \frac{\left(1 - \frac{(KL/r)^2}{2C_c{}^2}\right)\sigma_y}{\frac{5}{3} + \frac{3(KL/r)}{8C_c} - \frac{(KL/r)^3}{8C_c{}^3}}$$

(2) When the effective slenderness ratio, KL/r, exceeds C_c:

$$\sigma_c = \frac{12\,\pi^2\,E}{23\,(KL/r)^2}$$

where

σ_c = allowable compressive stress, psi (Pa)
K = effective length factor (ratio of the length of an equivalent pinned-end member to the length of the actual member)
L = unbraced length of the member, in. (mm)
r = radius of gyration, in. (mm)
C_c = $(2\pi^2 E/\sigma_y)^{1/2}$
E = modulus of elasticity, psi (Pa)
σ_y = minimum yield strength, psi (Pa)

The worst condition is of concern in design work and, therefore, it is necessary to use the least radius of gyration relative to the unbraced length. Thus an overloaded, unbraced, wide flange column section will always buckle as shown in Figure 5.3(D). However, if the column is braced at the midlength to prevent buckling, the column could buckle toward a flange. The critical slenderness ratio of a column is the larger ratio L_x/r_x or L_y/r_y where L_x and L_y are the distances between braced points, and r_x and r_y are the radii of gyration about the x and y axis respectively.

The design of a long compression member is by trial and error. A trial section is selected, and the cross-sectional area and the least radius of gyration are determined. A suitable column table (AISC *Manual of Steel Construction*, Appendix A) gives the allowable compressive stress for the column length and the radius of gyration (L/r ratio). This allowable stress is then multiplied by the cross-sectional area, A, to give the allowable total compressive load that may be placed on the column. If this value is less than the load to be applied, the design must be changed to a larger section and tried again. Table 5.4 gives the AISC column formulas, and Figure 5.4 gives allowable compressive stresses with various slenderness ratios.

Bending

FIGURE 5.5 ILLUSTRATES bending of a member under uniform loading. Loads may also be nonuniform or con-

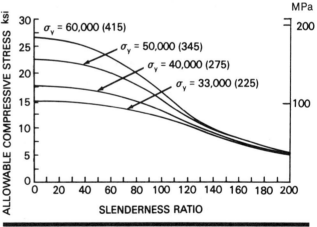

Figure 5.4—Allowable Compressive Stress for Steel Columns of Various Yield Strengths and Slenderness Ratios

centrated at specific locations on the beam. When a member is loaded in bending within the elastic range, the bending stresses are zero along the neutral axis and increase linearly to a maximum value at the outer fibers. The bending stress at any distance, d, from the neutral axis in the cross section of a straight beam, Figure 5.6, may be found by the formula:

$$\sigma = \frac{Md}{I} \tag{5.4}$$

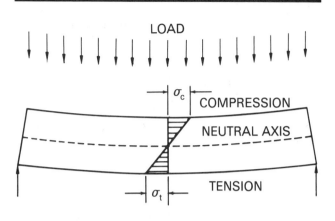

Figure 5.5—Bending of a Beam with Uniform Loading

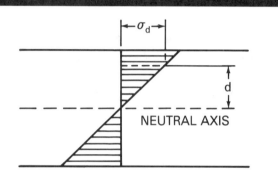

Figure 5.6—Bending Stress at Any Point (d) in the Cross Section of a Straight Beam Elastically Loaded

where

σ = bending stress (tension or compression) psi
M = bending moment at the point of interest, lb·in. (N·mm)
d = distance from neutral axis of bending to point d, in. (mm)
I = moment of inertia about the neutral axis of bending, in.⁴ (mm⁴)

In most cases, the maximum bending stress is of greatest interest, in which case the formula becomes:

$$\sigma = \frac{Mc}{I} = \frac{M}{S} \qquad (5.5)$$

where

σ = maximum bending stress, psi (kPa)
M = bending moment, lb·in. (N·mm)
c = distance from the neutral axis of bending to the extreme fibers, in. (mm)
I = moment of inertia, in.⁴ (mm⁴)
S = section modulus (I/c), in.³ (mm³)

As the bending moment decreases along the length of a simply supported beam toward the ends, the bending stresses (tension and compression) in the beam also decrease. If a beam has the shape of an I-section, the bending stress in the flange decreases as the end of the beam is approached. If a short length of the tension flange within the beam is considered, a difference exists in the tensile forces, F_1 and F_2, at the two locations in the flange, as shown enlarged above the beam in Figure 5.7 (The tensile force, F, is the product of the tensile stress, σ, and the flange cross-sectional area, A.)

The decrease in the tensile force in the flange results in a corresponding shearing force between the flange and the web. This shearing force must be transmitted by the fillet welds joining the two together. The same reaction takes place in the upper flange, which is in compression. The change in tensile force in the lower flange transfers as shear through the web to the upper flange, and is equal to the change in compression in that flange.

A common bending problem in machinery design involves the deflection of beams. Beam formulas found in many engineering handbooks are useful for quick approximations of deflections of common types of beams where the span is large compared to the beam depth. An example of a typical beam with applicable formulas is shown in Figure 5.8. Beams that are supported or loaded differently have other applicable design formulas. To meet stiffness requirements, the beam should have as large a moment of inertia as practical.

Information concerning the design of a compression member or column also may apply to the compression flange of a beam. The lateral buckling resistance of the compression flange must also be considered. It should have adequate width and thickness to resist local buckling, be properly supported to prevent twisting or lateral movement, and be subjected to compressive stresses within allowable limits.

Shear

FIGURE 5.9 ILLUSTRATES the shear forces in the web of a beam under load. They are both horizontal and vertical, and create diagonal tensile and diagonal compressive stresses.

The shear capacity of a member of either an I or a box beam cross section is dependent upon the slenderness proportions of the web or webs. For virtually all hot rolled beams and welded beams of similar proportions where the web slenderness ratio (h/t) is less than 260, vertical shear load is resisted by pure beam shear without lateral buckling. This is true for a level of loading well above that where unacceptable deflections will develop. Thus, design is based upon keeping the shear stress on the gross area of the web below the allowable value of 0.4 σ_y to prevent yielding in shear.

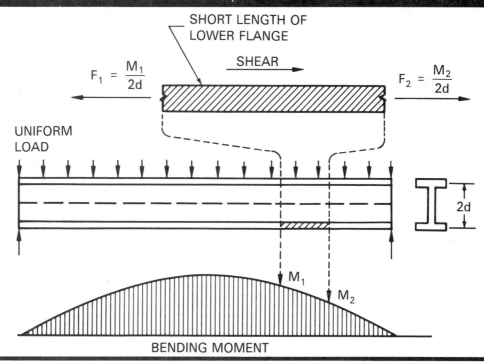

Figure 5.7—Approximate Tensile Force on a Section of the Lower Flange of a Loaded Beam

In plate girders with slender webs, shear is resisted by plane beam shear up to a level of stress which will cause shear buckling. However, webs stiffened by transverse stiffeners subject to shear stress have considerable post-buckling strength. Current design specifications take this strength into account. After buckling occurs, the web resists larger shear loads by a combination of beam shear and diagonal tension in the panels between stiffeners. If the length-to-depth ratio of the panel is approximately three or more, the direction of the diagonal tension becomes too near to the horizontal for it to be effective in providing significant post buckling strength, and the shear strength is limited to beam shear strength.

The onset of diagonal compression buckling in a web panel has negligible structural significance. With properly proportioned transverse stiffeners, the web continues to resist higher levels of shear loading up to the point where diagonal tension yielding occurs.[11] However, there are a few practical considerations independent of maximum strength that may govern a design.

For architectural reasons, especially in exposed fascia girders, the waviness of the web caused by controlled compression buckles may be deemed unsightly. In plate girders subject to cyclic loading to a level which would

initiate web shear buckling, each application of the critical load will cause an "oil canning" or breathing action of the web panels. This action causes out-of-plane bending stresses at the toes of web-to-flange fillet welds and the stiffener welds. These cyclic stresses will eventually initiate fatigue cracking. For these cases, web stresses should be limited to the beam shear strength values which preclude shear buckling.

In the case of a beam fabricated by welding, the shear load per unit of length on the welds joining the flanges of the beam to the web can be calculated by the formula:

$$W_s = \frac{Vay}{In} \tag{5.6}$$

where

W_s = shear load per unit length of weld lb/in. (N/mm)

V = external shear force on the member at this location lb (N)

a = cross-sectional area of the flange in.^{2} (mm^2)

y = distance between the center of gravity of the flange and the neutral axis of bending of the whole section in. (mm)

I = moment of inertia of the whole section about the neutral axis of bending in.^{4} (mm^4)

n = number of welds used to attach the web to the flange

---

11. Reliable criteria for the design of plate and box girder webs to achieve full maximum strength from beam shear and tension field action may be found in the *Manual of steel construction*, Section 5. Chicago: American Institute of Steel Construction (AISC), latest edition.

The required weld size can then be determined from the value of W_s.

Torsion

TORSION CREATES GREATER design problems for bases and frames than for other machine members. A machine with a rotating unit may subject the base to torsional loading. This becomes apparent by the lifting of one corner of the base, if the base is not anchored to the floor.

If torsion is a problem, closed tubular sections or diagonal bracing should be used, as shown in Figure 5.10.

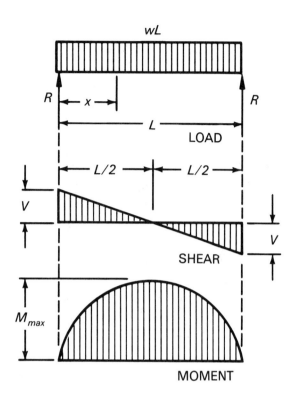

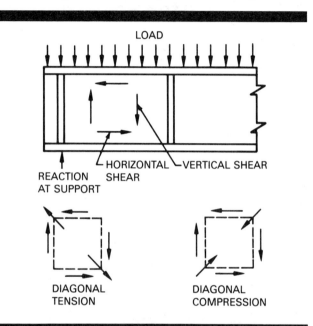

Figure 5.9—Shear Forces in the Web of a Beam Under Uniform Load

$$\text{Reaction Force } R = V_{max} = \frac{wL}{2}$$

$$\text{Shear Force } V_x = w\left(\frac{L}{2} - x\right)$$

$$\text{Maximum Moment at Center, } M_{max} = \frac{wL^2}{8}$$

$$\text{Moment at Any Point } (x) \, M_x = \frac{wx}{2}(L - x)$$

$$\text{Deflection at Center } \Delta_{max} = \frac{5wL^4}{384EI}$$

$$\text{Deflection at Any Point } (x) \, \Delta_x = \frac{wx}{24EI}(L^3 - 2Lx^2 + x^3)$$

$$\text{Slope of Beam at Ends, } \theta = \frac{wL^3}{24EI}$$

Figure 5.8—Formula for a Simply Supported Beam with a Uniformly Distributed Load on the Beam Span

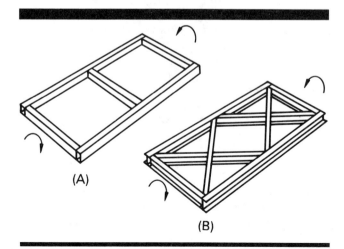

Figure 5.10—Application of (A) Closed Tublar Sections or (B) Open Structural Sections with Diagonal Bracing to Resist Torsion

Closed tubular sections are as much as 1000 times better for resisting torsion than comparable open sections. Closed members can easily be made from channel or I-sections by intermittently welding flat plate to the toes of the rolled sections. The torsion effect on the perimeter of an existing frame may be eliminated or the frame stiffened for torsion by adding cross bracing. Torsion problems in structures can be avoided by judicious arrangement of members to transmit loads by direct stresses or bending moments.

Torsional Resistance. The torsional resistance of a flat strip or open section (I-beam or channel) is very low. The torsional resistance of a solid rectangular section having a width of several times the thickness may be approximated by the following formula:

$$R = \frac{bt^3}{3} \tag{5.7}$$

where

R = torsional resistance, in.^{4} (mm^4)
b = width of the section, in. (mm)
t = thickness of the section, in. (mm)

The total angular twist (rotation) of a member can be estimated by the following formula:

$$\theta = \frac{TL}{GR} \tag{5.8}$$

where

θ = angle of twist, radians
T = torque, lb·in. (N·mm)
L = length of member, in. (mm)
G = modulus of elasticity in shear, psi (kPa)
R = torsional resistance, in.^{4} (mm^4)

The unit angular twist, θ, is equal to the total angular twist divided by the length, L, of the member.

The torsional resistance of an open structural member, such as an I-beam or a channel, is approximately equal to the sum of the torsional resistances of the individual flat sections into which the member can be divided. This is illustrated in Table 5.5, listing the actual and calculated angle of twist of a flat strip and an I-shape made up of three of the flat strips. The applied torque was the same for both sections.

Torsional resistance increases markedly with closed cross sections, such as circular or rectangular tubing. As a result, the angular twist is greatly reduced because it varies inversely with torsional resistance. The torsional resistance, R, of any closed box shape enclosing only one cell can be estimated by the following procedure. Draw a dotted line through the mid-thickness around the section, as shown in Figure 5.11. The area enclosed by the dot-dash lines is A. Divide the cross section of the member into convenient lengths, L_N having thicknesses t_N. Determine the ratios of these individual lengths to their corresponding thicknesses. Torsional resistance is then obtained from the following formula:

$$R = \frac{4A^2}{\Sigma\left(\frac{L_N}{t_N}\right)} \tag{5.9}$$

Table 5.5
Calculated and Actual Angle of Twist

	Angle of Twist, Degrees	
	Strip[a]	I-Section[b]
Calculated using torsional resistance	21.8	7.3
Actual twist	22	9.5

a. 0.055 in. by 2 in.
b. Made of three of the strips.

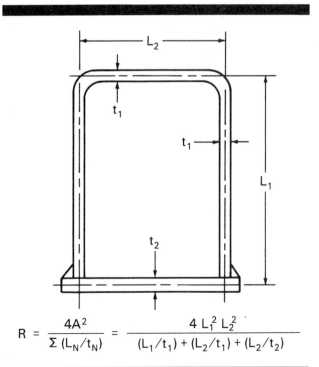

$$R = \frac{4A^2}{\Sigma(L_N/t_N)} = \frac{4\,L_1^2\,L_2^2}{(L_1/t_1) + (L_2/t_1) + (L_2/t_2)}$$

Figure 5.11—Torsional Resistance of a Closed Box Section

The maximum shear stress in a rectangular section in torsion is on the surface at the center of the long side. When the unit angular twist is known, the following formula will give the maximum shear stress at the surface of a rectangular part.

$$\tau = \phi tG = \frac{Tt}{R} \qquad (5.10)$$

where

τ = maximum shear stress, psi (kPa)
ϕ = unit angular twist, radians/in. (rad/mm)
G = modulus of elasticity in shear, psi (kPa)
T = applied torque, lb·in. (N·mm)
t = thickness of the section, in. (mm)
R = torsional resistance, in.^{4} (mm^4)

This formula can be applied to a flat plate or a rectangular area of an open structural shape (channel, angle, I-beam). In the latter case, R is the torsional resistance of the whole structural shape.

Diagonal Bracing. Diagonal bracing is very effective in preventing the twisting of frames. A simple explanation of the effectiveness of diagonal bracing involves an understanding of the directions of the forces involved.

A flat bar of steel has little resistance to twisting but has exceptional resistance to bending (stiffness) about its major axis. Transverse bars or open sections at 90 degrees to the main members are not effective for increasing the torsional resistance of a frame because, as shown in Figure 5.12(A), they contribute only relatively low torsional resistance. However, if the bars are oriented diagonally across the frame, as in Figure 5.12(B), twisting of the frame is resisted by the stiffness of the bars. To be effective, the diagonal braces must have good stiffness perpendicular to the plane of the frame.

TRANSFER OF FORCES

LOADS CREATE FORCES that must be transmitted through the structure to suitable places for counteraction. The designer needs to know how to provide efficient pathways.

One of the basic rules is that a force applied transversely to a member will ultimately enter that portion of the section that lies parallel to the applied force. Figure 5.13, for example, shows a lug welded parallel to the length of a beam. The portion of the beam that is parallel to the applied force is the web. The force in the lug is easily transferred through the connecting welds into the web. No additional stiffeners or attaching plates are required.

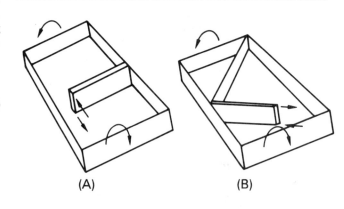

Figure 5.12—Frames Subjected to Torsion with (A) Transverse Rib Bracing and (B) Diagonal Bracing

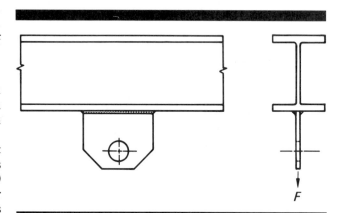

Figure 5.13—Lug Welded Parellel to the Length of a Beam. No Additional Stiffeners Required

Suppose, however, that the lug was welded to the beam flange at right angles to the length of the beam, Figure 5.14(A). The outer edges of the flange tend to deflect rather than support much load. This forces a small portion of the weld in line with the web to carry a disproportionate share of the load, Figure 5.14(B). To uniformly distribute the load on the attachment weld, two stiffeners can be aligned with the lug and then welded to the web and to the adjacent flange of the beam, as shown in Figure 5.15. The stiffeners reinforce the bottom flange and also transmit part of the load to the web of the beam. The welds labeled A, B, and C in Figure 5.15 must all be designed to carry the applied force, F.

If a force is to be applied to a beam parallel to the flanges by a plate welded to the web, as indicated in Figure 5.16, unacceptable distortion may result. The required weld (or web) strength must be estimated by a yield line analysis of the web.

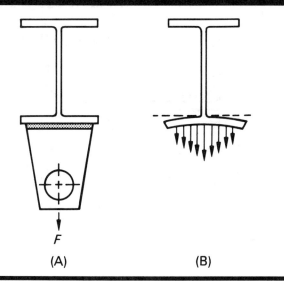

Figure 5.14—(A) Lug Welded to Flange Transverse to Beam Length, (B) Resulting Loading and Deflection of the Flange

(A)   (B)

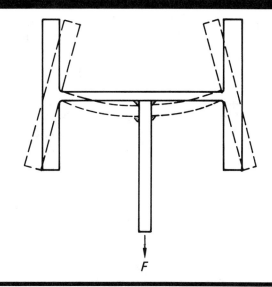

Figure 5.16—Distortion of a Beam from Loaded Attachment to the Web

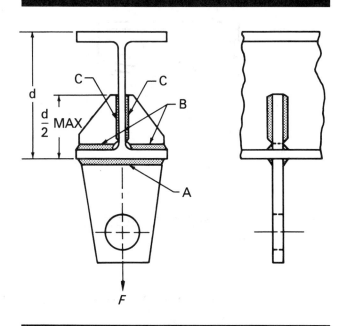

Figure 5.15—Additional Stiffener Required to Transmit Load to Web of Beam

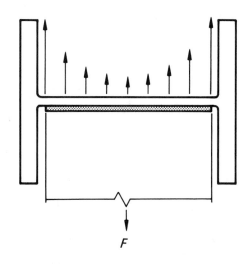

Figure 5.17—Stress Distribution with Attachment Welded to Web Only

On the other hand, orienting the attachment plate transverse to the web, with welding only to the web, results in a low strength connection because of high concentrations of stress at the ends of the weld, as shown in Figure 5.17. Similiarly, if the attachment is welded to both flanges and the web without a stiffener on the opposite side of the web, the weld must not be sized on the assumption of uniform stress along the total length of weld. Only negligible loads are transferred across most of the web width, as shown in Figure 5.18. Most of the load is transmitted to the two flanges through the attachment welds by shear.

If attachment is made by welding to the flanges only, the problem is not one of distribution of stress in the weld, but one of the effect of shear lag in the attachment

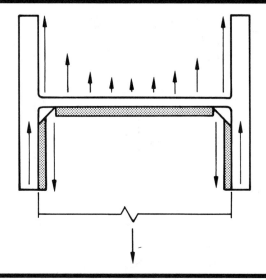

Figure 5.18—Stress Distribution with Attachment Welded to Web and Flanges

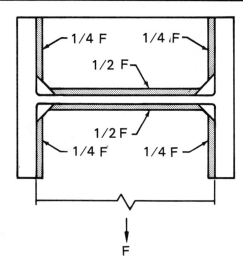

Figure 5.20—Stiffener on Opposite Side of Attachment, Welded as Shown, for Large Loads

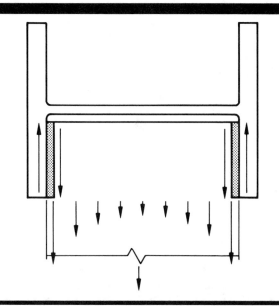

Figure 5.19—Stress Distribution with Attachment Welded to Flanges Only

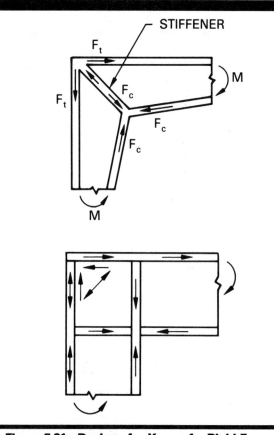

Figure 5.21—Designs for Knee of a Rigid Frame

causing high stress concentrations in the plate near the flange edges as illustrated in Figure 5.19. The tension plate should be designed on the basis of reduced allowable stresses or reduced effective area. An exception would be if the length of each weld along the edges of the plate is more than the width of the plate as required by the AISC specifications.

For large forces, it might be necessary to place a stiffener on the opposite side of the web (see Figure 5.20). In this case, both the plate and the stiffener must be welded to the web as well as the flanges. With this detail, the length of welds parallel to the direction of applied stress is effectively doubled, and the shear lag effect in the attachment is greatly reduced.

In all cases, the fillet welds joining the plate to the beam flanges must not extend around the plate along the edges of the flanges. The abrupt changes in weld direction on two planes can intensify stress concentrations.

When a force in a structure changes direction, a force component is involved. This is illustrated in Figure 5.21A and B, a knee of a rigid frame subjected to a bend-ing moment. The compressive force in the interior flanges must change direction at the knee. To prevent buckling of the web and failure of the structure, diagonal stiffeners are placed on both sides of the web at the intersections of the two flanges as shown in Figure 5.21A. An alternate detail using vertical and horizontal stiffeners to accomplish the same effect is shown in Figure 5.21B. The compressive force component in the web and stiffeners balances the change in direction of the tensile force in the outer flanges.

DESIGN OF WELDED JOINTS

TYPES OF JOINTS

THE LOADS IN a welded structure are transferred from one member to another through welds placed in the joints. The types of joints used in welded construction and the applicable welds are shown in Figure 5.22. The configurations of the various welds are illustrated in Figures 5.23, 5.24, and 5.25. Combinations of welds may be used to make a joint, depending on the strength requirements and load conditions. For example, fillet and groove welds are frequently combined in corner and T-joints.

Welded joints are designed primarily to meet the strength and safety requirements for the service conditions under which they must perform. The manner in which the stress will be applied in service, whether tension, shear, bending, or torsion, must be considered. Different joint designs may be required, depending on whether the loading is static or dynamic where fatigue must be considered. Joints may be designed to avoid stress-raisers and to obtain an optimum pattern of residual stresses. Conditions of corrosion or erosion require joints that are free of irregularities, crevices, and other areas that make them susceptible to such forms of attack. The design must reflect consideration of joint efficiency, which is the ratio of the joint strength to the base metal strength, generally expressed as a percentage.

Certain welding processes in conjunction with certain related types of joints have a long record of proven satisfactory performance. Therefore, these processes, providing the weld procedures meet other specific requirements of *ANSI/AWS D1.1, Structural Welding Code—Steel*, enjoy prequalification status. However, prequalification rests on a history of satisfactory performance, and therefore, the use of a prequalified welding procedure and joint geometry on a unique design or application may not fulfill the designer's obligations. The designer must consider the following:

(1) Is the joint accessible to welders and inspectors?

(2) Does the design consider the economic factors of welding?

(3) If the joint cannot be postweld heat treated, will it lead to excessive residual stresses?

GROOVE WELDS

GROOVE WELDS OF different types are used in many combinations: the selection of which is influenced by accessibility, economy, adaptation to the particular design of the structure being fabricated, expected distortion, and the type of welding process to be used.

A square-groove weld, Figures 5.23(A) and 5.24(A), is economical to use, provided satisfactory soundness and strength can be obtained. However, its use is limited by the thickness of the joint.

For thick joints, the edge of each member must be prepared to a particular geometry to provide accessibility for welding and ensure desired soundness and strength. In the interest of economy, those joint designs should be selected with root openings and groove angles that require the smallest amount of weld metal and still give sufficient accessibility for sound welds. The selection of root opening and groove angle also is greatly influenced by the metals to be joined, the location of the joint in the weldment, and the required performance.

Welds in J- and U-groove joints may be used to minimize weld metal requirements when the savings are sufficient to justify the more costly preparation of the edges. These joints are particularly useful in the welding of thick sections. Single-bevel- and J-groove welds are more difficult to weld than V- and U-groove welds because one edge of the groove is vertical.

The amount of joint penetration and the strength of the weld metal control the strength of the welded joint. Welded joints must be designed to provide strength adequate to transfer the design forces. Frequently, this requires that the welded joint provide strength equal to the base metal. To accomplish this, designs that permit

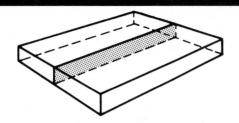

Applicable Welds

Square-groove	J-groove
V-groove	Flare-V-groove
Bevel-groove	Flare-bevel-groove
U-groove	Edge-flange
	Braze

(A) Butt Joint

Applicable Welds

Fillet	Flare-bevel-groove
Square-groove	Edge-flange
V-groove	Corner-flange
Bevel-groove	Spot
U-groove	Projection
J-groove	Seam
Flare-V-groove	Braze

(B) Corner Joint

Applicable Welds

Fillet	J-groove
Plug	Flare-bevel-groove
Slot	Spot
Square-groove	Projection
Bevel-groove	Seam
	Braze

(C) T-Joint

Applicable Welds

Fillet	J-groove
Plug	Flare-bevel-groove
Slot	Spot
Bevel-groove	Projection
Seam	Braze

(D) Lap Joint

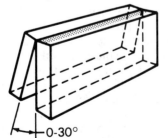

0-30°

Applicable Welds

Square-groove	Edge-flange
Bevel-groove	Corner-flange
V-groove	Seam
U-groove	Edge
J-groove	

(E) Edge Joint

Figure 5.22—Types of Joints

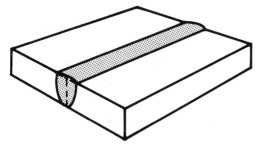

(A) Single-Square-Groove Weld

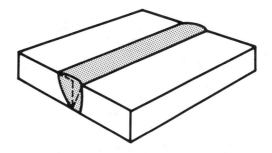

(B) Single-Bevel-Groove Weld

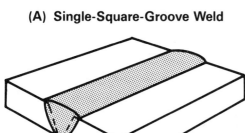

(C) Single-V-Groove Weld

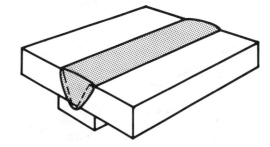

(D) Single-V-Groove Weld (with Backing)

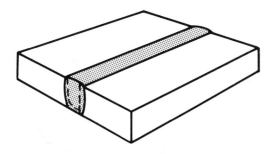

(E) Single-J-Groove Weld

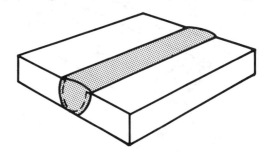

(F) Single-U-Groove Weld

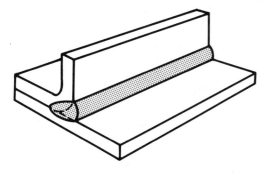

(G) Single-Flare-Bevel-Groove Weld

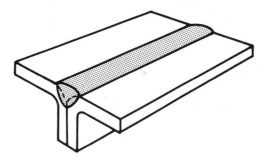

(H) Single-Flare-V-Groove Weld

Figure 5.23—Single Groove Welds

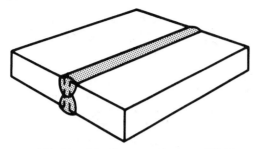

(A) Double-Square-Groove Weld

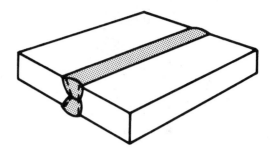

(B) Double-Bevel-Groove Weld

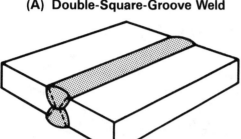

(C) Double-V-Groove Weld

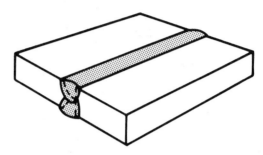

(D) Double-J-Groove Weld

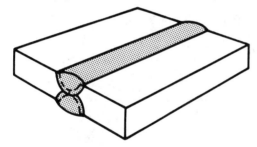

(E) Double-U-Groove Weld

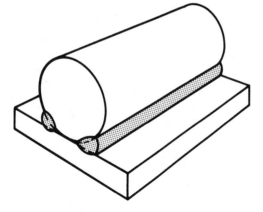

(F) Double-Flare-Bevel-Groove Weld

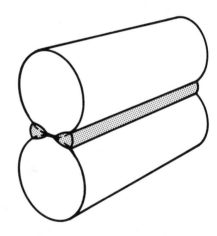

(G) Double-Flare-V-Groove Weld

Figure 5.24—Double Groove Welds

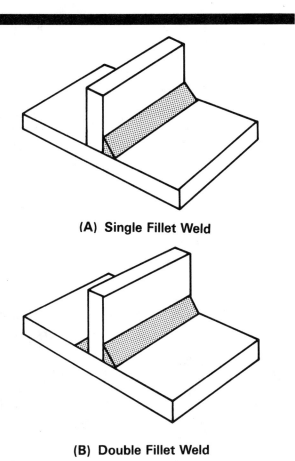

(A) Single Fillet Weld

(B) Double Fillet Weld

Figure 5.25—Fillet Welds

penetration completely through the members being joined are commonly used.

The selection of details of welding grooves (groove angle, root face, root opening, etc.) depends upon the welding process and type of power to be used, and the physical properties of the base metal(s) being joined.[12] Some welding processes characteristically provide deeper joint penetration than others. Some metals, such as copper and aluminum, have relatively high thermal conductivities or specific heats, or both. Those metals require greater heat input for welding than other metals with lower thermal properties.

The various types of groove welds have certain advantages and limitations with respect to their applications. In the following discussion, comments on joint penetration or effective throat apply to joining of carbon steel by shielded metal arc (SMAW), gas metal arc (GMAW), flux cored arc (FCAW), and submerged arc welding (SAW). Joint penetration with other processes and base

metals may be different when their physical properties vary significantly from those of carbon steel.

Complete Joint Penetration

GROOVE WELDS WITH complete joint penetration are suitable for all types of loading, provided they meet the acceptance criteria for the application. In most cases, to ensure complete joint penetration with double-groove welds and single-groove welds without a backing strip, the root of the first weld must be back gouged to sound metal before making a weld pass on the other side.

Partial Joint Penetration

A PARTIAL JOINT penetration groove weld has an unwelded portion at the root of the weld.

The unwelded portions of groove welded joints constitute a stress raiser having significance when cyclic loads are applied transversely to the joint. This fact is reflected in their low allowable fatigue stress range. However, when the load is applied parallel to the weld axis, a higher stress range is permitted.

Regardless of the rules governing the service application of partial joint penetration groove welds, the eccentricity of shrinkage forces in relation to the center of gravity of the section can result in angular distortion on cooling after welding. This same eccentricity also tends to cause rotation of a transverse axial load across the joint. Therefore, means must be applied to restrain or preclude such rotation both during fabrication and in service.

For static loading, the allowable stresses in partial joint penetration groove welds depend upon the type and direction of loading and the applicable code requirements. Under *ANSI/AWS D1.1-87, Structural Welding Code—Steel*, the allowable tensile stress transverse to the weld axis is 0.3 times the nominal tensile strength of the weld metal. Under the *AISC Manual of Steel Construction*, the tensile stress transverse to the weld may not exceed the allowable stress in the base metal.[13] In both standards, for tension or compression parallel to the weld axis, the allowable stress can be the same as that for the base metal.

The allowable weld metal shear stress in both standards may not exceed 0.30 times the nominal tensile strength of the weld metal. The actual shear stress at a point may exceed the normally allowable shear stress due to weld joint geometry. However, such peak stresses of yield strength levels are of little consequence, because they are highly localized. Thus, very small strains can transfer the loads to the adjacent material.

12. Groove-weld joint preparations used with various welding processes and base metals are discussed in the *Welding Handbook*, Vols. 2, 3, and 4, 7th Ed., 1978, 1980, and 1982.

13. "Specification for the design, fabrication, and erection of structural steel for buildings." *Manual of Steel Construction*, Eighth Ed. Chicago: American Institute of Steel Construction; 1980.

Joints welded from one side only should not be used in bending with the root in tension nor should they be subjected to transverse fatigue nor impact loading. Partially penetrated joints should not be exposed to corrosive conditions.

For design purposes, the effective throat is never greater than the depth of joint penetration.

Double-V-Groove Welds

THE STRENGTH OF a double-V-groove weld depends upon the complete joint penetration. For all types of loading, full strength joints can be obtained with complete joint penetration. This type of weld is economical when the depth of the groove does not exceed 3/4 in. (19 mm). Double-V-groove welds with complete joint penetration may be costly when the joint thickness exceeds 1-1/2 in. (38 mm); double-U-groove welds may be more economical. Partial joint penetration groove welds are recommended only for static loads transverse to the weld axis and for other types of loads parallel to the weld axis.

Single-Bevel-Groove Welds

SINGLE-BEVEL-GROOVE welds have characteristics similar to single-V-groove welds with respect to properties and applications. The bevel type requires less joint preparation and weld metal; therefore, it is more economical.

One disadvantage of this type of weld is that proper welding procedures are required to obtain complete fusion with the perpendicular face of the joint. In the horizontal position, that face should be placed on the lower side of the joint to obtain good fusion.

Double-Bevel-Groove Welds

DOUBLE-BEVEL-GROOVE welds have the same characteristics as double-V-groove welds. The perpendicular joint faces make complete fusion more difficult. Also, satisfactory back gouging of the root of the first weld may be harder to accomplish.

A double-bevel joint design is economical when the depth of the groove does not exceed about 3/4 in. (19 mm), and the joint thickness is 1-1/2 in. (38 mm) or less.

Single- and Double-J-Groove Welds

J-GROOVE WELDS HAVE the same characteristics as similar bevel-groove welds. However, they may be more economical for thicker sections, provided the savings in deposited weld metal exceed the cost of machining or gouging the edge preparation. Their use may be best suited to the horizontal position in some applications.

Single- and Double-U-Groove Welds

U-GROOVE WELDS and J-groove welds are used for similar applications. However, complete fusion is easier to obtain, and root gouging is more readily accomplished with U-grooves.

FILLET WELDS

WHERE THE DESIGN permits, fillet welds are used in preference to groove welds for economy. Fillet welded joints are very simple to prepare from the standpoint of edge preparation and fit-up, although groove welded joints sometimes require less welding.

When the smallest continuous fillet weld that is practicable to make results in a joint strength greater than that required, intermittent fillet welding may be used to avoid overwelding, unless continuous welding is required by the service conditions.

Fillet weld size is measured by the length of the legs of the largest right triangle that may be inscribed within the fillet weld cross section as shown in Figure 5.26. The effective throat, a better indication of weld shear strength, is the shortest distance between the root of the weld and the weld face. The strength of a fillet weld is based on the effective throat and the length of the weld (effective area of the weld).

The actual throat may be larger than the theoretical throat by virtue of joint penetration beyond the root of the weld. Submerged arc and flux cored arc welding are deep penetrating processes. Under certain conditions, some standards allow the extra penetration that these processes provide to be used as part of the effective throat in fillet welds.

Applications

FILLET WELDS ARE used to join corner, T, and lap joints because they are economical. Edge preparation is not required, but surface cleaning may be needed. Fillet welds are generally applicable for transfer of shear forces parallel to the axis of the weld and for the transfer of static forces transverse to the axis of the weld when the required weld size is less than about 5/8 in. (16 mm). If the load would require a fillet weld of 5/8 in. (16 mm) or larger, a groove weld should be used possibly in combination with a fillet weld to provide the required effective throat. Fillet welds may be used in skewed T- or corner joints having a dihedral angle between 60 and 135 degrees. Beyond these limits, these welds are considered partial joint penetration groove welds.

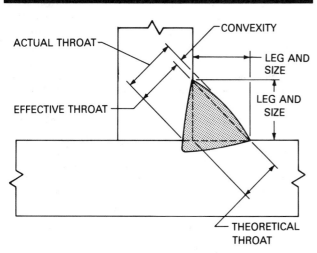

(A) Convex Fillet Weld

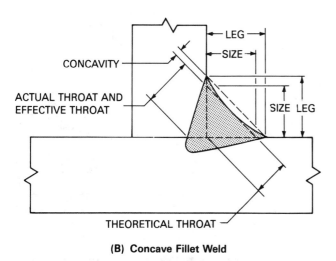

(B) Concave Fillet Weld

Figure 5.26—Fillet Weld Sizes

Fillet welds are always designed on the basis of shear stress on the throat regardless of the direction of applied force relative to the axis of the weld. The maximum shear stress is calculated on the basis of the effective area of the weld. In the case of steel, the maximum shear stress is normally limited to about 30 percent of the nominal (classification) tensile strength of the weld metal.

Weld Size

FILLET WELDS MUST be large enough to carry the applied load and to accommodate shrinkage of the weld metal during cooling if cracking is to be avoided, particularly with highly restrained, thick sections. To minimize dis-

tortion and welding costs, however, the specified fillet weld size should not be excessive. Welds in lap joints cannot exceed in size the thickness of the exposed edge, which should be visible after welding.

Fillet welds may be designed with unequal leg sizes to provide the required effective throat or the needed heat balance for complete fusion with unequal base metal thicknesses.

Single Fillet Welds

SINGLE FILLET WELDS are limited to low loads. Bending moments that result in tension stresses in the root of a fillet weld should not be permitted because of the notch condition. For this reason, single fillet welds should not be used with lap joints that can rotate under load. The welds should not be subjected to impact loads. When used with fatigue loading, the allowable stress range must be subject to stringent limitations.

Double Fillet Welds

SMALLER DOUBLE FILLET welds are preferred to a large single fillet weld. Full plate strength can be obtained with double fillet welds under static loading. Double fillet welding of corner and T-joints limits rotation of the members about the longitudinal axis of the joint, and minimizes tension stresses at the root of the welds. These types of joints can be cyclically loaded parallel to the weld axes.

Double-welded lap joints should have a minimum overlap of about five times the base metal thicknesses to limit joint rotation under load.

Fillet welds can be combined with complete and partial joint penetration groove welds, as shown in Figure 5.27.

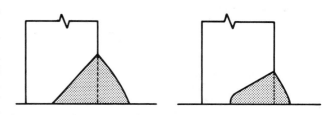

Figure 5.27—Combination of Groove Weld with Reinforcing Fillet

SELECTION OF WELD TYPE

THE DESIGNER IS frequently faced with the question of whether to use fillet or groove welds. Cost is a major consideration. Double fillet welds, Figure 5.28(A), are

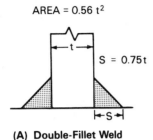

AREA = 0.56 t²

t

S = 0.75 t

S

(A) Double-Fillet Weld

AREA = 0.25 t²

t

**(B) Double-Bevel-
Groove Weld**

AREA = 0.50 t²

t

(C) Single-Bevel-Groove Weld

**Figure 5.28—Comparison of Weld Quantities for
Three Conditions of T-Welds**

easy to apply and require no special edge preparation. They can be made using large diameter electrodes with high welding currents for high deposition rates.

In comparison, the double-bevel groove weld, Figure 5.28(B), has about one half the cross-sectional area of the fillet welds. However, it requires edge preparation and the use of small diameter electrodes to make the root pass.

Referring to Figure 5.28(C), a single-bevel-groove weld requires about the same amount of weld metal as the double fillet weld, Fig. 5.28(A); thus, it has no apparent economic advantage. There are also some disadvantages, one of them being that the single-bevel weld

requires edge preparation and a low-deposition root pass. From a design standpoint, however, it does offer direct transfer of force through the joint, which means that it is probably better than fillet welds under cyclic loading.

Double fillet welds having a leg size equal to 75 percent of the plate thickness would be sufficient for full strength. However, some standards have lower allowable stress limits than other standards for fillet welds, and may require a leg size equal to the plate thickness.

The cost of a double fillet welded joint may exceed the cost of a single-bevel-groove weld in thick plates. Also, if the joint can be positioned so that the weld can be made in the flat position, a double fillet weld would be more expensive than a single-bevel-groove weld.

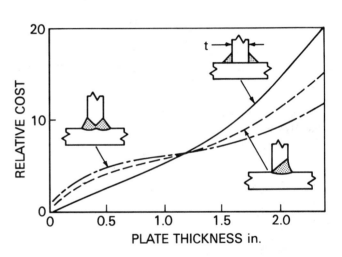

**Figure 5.29—Estimated Relative Costs of Fillet and
Groove Welds for Various Thicknesses**

The construction of curves based on the best determination of the actual cost of joint preparation, positioning, and welding, is a technique for determining the plate thicknesses where a double-bevel-groove weld becomes less costly. A sample curve is illustrated in Figure 5.29. The intersection of the fillet weld curve with the groove weld curve is the point of interest. The validity of the information is dependent on the accuracy of the cost data at a particular fabricating plant.

The combined double-bevel-groove and fillet weld joint, shown in Figure 5.30, is theoretically a full strength weld. The plate edge is beveled to 60 degrees on both sides to a depth of 30 percent of the thickness of the plate. After the groove on each side is welded, it is reinforced with a fillet weld of equal area and shape. The

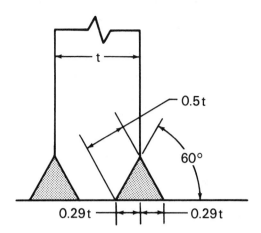

Figure 5.30—Combined Groove and Fillet Welds with Partial Joint Penetration, but Capable of Full Strength

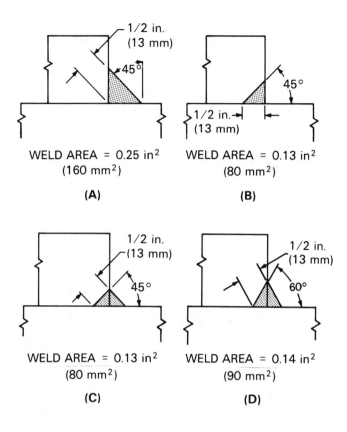

Figure 5.31—Equal Throat T-Welds with Several Joint Configurations and Resulting Weld Deposits

total effective throat of weld is equal to the plate thickness. This partial joint penetration groove weld has only about 60 percent of the weld metal in a full-strength, double fillet weld. It does require joint preparation, but the wide root face permits the use of large electrodes and high welding currents. It is recommended for submerged arc welding to achieve deep joint penetration.

Full strength welds are not always required in a weldment, and economies can often be achieved by using smaller welds where applicable and permissible. With equal effective throats, (shear area), a fillet weld, Figure 5.31(A), requires twice the weld metal needed for a 45 degree partial joint penetration, single-bevel-groove weld, Figure 5.31(B). The latter weld may not be as economical as a fillet weld, however, because of the cost of edge preparation. Also, some welding standards limit the effective throat of this type of weld to less than the depth of the bevel with certain welding processes because of possible incomplete root penetration.

If a single-bevel-groove weld is combined with a 45 degree fillet weld, Figure 5.31(C), the cross-sectional area for the same effective throat is also about 50 percent of the area of the fillet weld in Figure 5.31(A). Here, the bevel depth is smaller than it is with the single-bevel-groove weld in Figure 5.31(B). A similar weld with a 60 degree groove angle and an unequal leg fillet, but with the same effective throat, Figure 5.31(D), also requires less weld metal than a fillet weld alone. This joint allows the use of higher welding currents and larger electrodes to obtain deep root penetration.

The desired effective throat of combined groove and fillet welds can be obtained by adjustment of the groove dimensions and the fillet weld leg lengths. However, consideration must be given to the accessibility of the root of the joint for welding, and to stress concentra-

tions at the toes of the fillet weld. When a partial joint penetration groove weld is reinforced with a fillet weld, the minimum effective throat is used for design purposes. The effective throat of the combined welds is not the sum of the effective throats of each weld. The combination is treated as a single weld when determining the effective throat.

CORNER JOINTS

CORNER JOINTS ARE widely used in machine design. Typical corner joint designs are illustrated in Figure 5.32. The corner-to-corner joint, Figure 5.32(A), is difficult to position and usually requires fixturing. Small electrodes with low welding currents must be used for the first weld pass to avoid excessive melt-through. Also, the joint requires a large amount of weld metal.

The corner joint shown in Figure 5.32(B) is easy to assemble, does not require backing, and needs only about half of the weld metal required to make the joint shown in Figure 5.32(A). However, the joint has lower strength because the effective throat of the weld is smaller. Two fillet welds, one outside and the other inside, as in Figure 5.32(C), can provide the same total effective throat as with the first design but with half the weld metal.

With thick sections, a partial joint penetration, single-V-groove weld, Figure 5.32(D), is often used. It requires joint preparation. For deeper joint penetration, a J-groove, Figure 5.32(E), or a U-groove may be used in preference to a bevel groove. A fillet on the inside corner, Figure 5.32(F), makes a neat and economical corner. The inside fillet weld can be used alone or in combination with any of the outside corner joint configurations of Figure 5.32.

The size of the weld should always be designed with reference to the thickness of the thinner member. The joint cannot be stronger than the thinner member, and the weld metal requirements are minimized for low cost.

Lamellar tearing at the exposed edges of corner joints in thick steel plates must always be considered during the design phase. Weld joint designs that significantly reduce the through-thickness shrinkage stresses are shown in Figure 5.32(D) and (E). Such joint designs exhibit a lower tendency to lamellar tearing than that of Figure 5.32(F) in which edge preparation is limited to base metal not stressed in through-thickness direction.

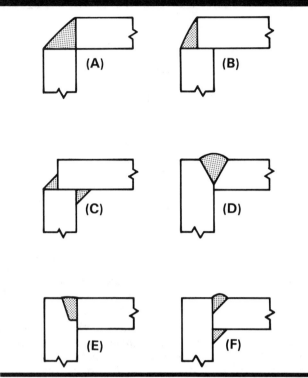

Figure 5.32—Typical Corner Joint Designs

SIZING OF STEEL WELDS

WELDS ARE SIZED for their ability to withstand static or cyclic loading.[14] Allowable stresses in welds for various types of loading are normally specified by the applicable standard for the job. They are usually based on a percentage of the tensile or the yield strength of the metal to make sure that a soundly welded joint can support the applied load for the expected service life. Allowable stresses or stress ranges are specified for various types of welds under static and cyclic loads. For the most part, the allowable stress ranges for welded joints subjected to cyclic loading specified in current standards are based on testing of representative full-size welded joints in actual or mockup structures.

STATIC LOADING

EXAMPLES OF ALLOWABLE stresses for static loading con-

ditions for steel weld metals are given in Table 5.6. The various types of loading for the welds in Table 5.6 are illustrated in Figure 5.33.

Complete joint penetration groove welds, illustrated in Figure 5.33(A), (B), (C), and (D), are considered full-strength welds because they are capable of transferring the full strength of the connected members.

The allowable stresses in such welds are the same as those in the base metal, provided weld metal of the proper strength is used. In complete joint penetration groove welds, the mechanical properties of the weld metal must at least match that of the base metal. If two base metals of different strengths are welded together, the weld metal strength must match or exceed the strength of the weaker base metal.

Partial joint penetration groove welds, illustrated in Figure 5.33(B), (C), (E), and (F), are widely used for economical welding of thick sections. Such welds can provide required joint strength, and in addition, accomplish savings in weld metal and welding time. The minimum weld sizes for prequalified partial joint penetration

---

14. Unless otherwise noted, the rules for sizing of steel welds in this chapter are from *ANSI/AWS D1.1-86, Structural Welding Code—Steel.*

Table 5.6
Allowable Stress and Strength Levels for Steel Welds in Building Construction per AWS D1.1.

Type of Weld	Stress in Weld[a]		Allowable Stress	Required Weld Strength Level
Complete joint penetration groove welds	Tension normal to the effective area		Same as base metal	Matching weld metal must be used
	Compression normal to the effective area		Same as base metal	Weld metal with a strength level equal to or one classification (10 ksi) less than matching weld metal may be used
	Tension or compression parallel to the axis of the weld		Same as base metal	Weld metal with a strength level equal to or less than matching weld metal may be used
	Shear on the effective area		0.30 nominal tensile strength of weld metal (ksi), except shear stress on base metal shall not exceed 0.40 yield strength of base metal	
Partial joint penetration groove welds	Compression normal to the effective area	Joint not designed[b] to bear compression	0.50 nominal tensile strength[b] of weld metal (ksi), except stress on base metal shall not exceed 0.60 yield strength of base metal	Weld metal with a strength level equal to or less than matching weld metal may be used
		Joint designed to[b] bear compression	Same as base metal	
	Tension or compression parallel to the axis of the weld		Same as base metal	
	Shear parallel to the axis of the weld		0.30 nominal tensile strength of weld metal (ksi), except shear stress on base metal shall not exceed 0.40 yield strength of base metal	
	Tension normal to the effective area		0.30 nominal tensile strength of weld metal (ksi), except tensile stress on base metal shall not exceed 0.60 yield strength of base metal	
Fillet welds	Shear on the effective area		0.30 nominal tensile strength of weld metal (ksi), except shear stress on base metal shall not exceed 0.40 yield strength of base metal	Weld metal with a strength level equal to or less than matching weld metal may be used
	Tension or compression parallel to the axis of weld		Same as base metal	
Plug and slot welds	Shear parallel to the faying surfaces (on effective area)		0.30 nominal tensile strength of weld metal (ksi), except shear stress on base metal shall not exceed 0.40 yield strength of base metal	Weld metal with a strength level equal to or less than matching

a. The effective weld area is the effective weld length multiplied by the effective throat.
b. AISC Specification stipulates allowable stress to be same as base metal without distinction as to whether joint is milled to bear or not based upon results of full size column tests.

groove welds are shown in Table 5.7. The minimum weld size should provide adequate process heat input to counteract the quenching effect of the base metal to avoid cracking in the weld or the heat-affected zone.

Various factors should be considered in determining the allowable stresses on the throat of partial joint penetration groove welds. Joint configuration is one factor.

The effective throat of a prequalified partial joint penetration groove weld is the depth of the groove when the groove angle is 60 degrees or greater at the root of the weld. For groove angles less than 60 degrees, the effective throat depends on the welding process, the welding position, and the groove angle at the root. *ANSI/AWS D1.1 Structural Welding Code—Steel*, latest edition,

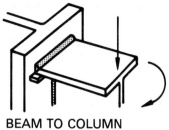

BEAM TO COLUMN

BUTT JOINT

(A) Complete Joint Penetration Groove Weld in Tension

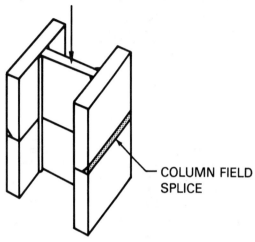

COLUMN FIELD
SPLICE

PARTIAL JOINT PENETRATION

COLUMN
FIELD
SPLICE

COMPLETE JOINT PENETRATION

(B) Compression Normal to Axis of Weld

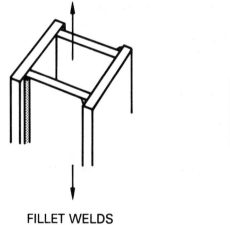

FILLET WELDS

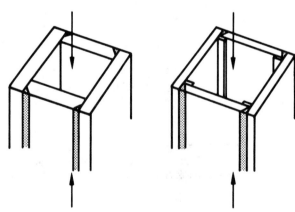

GROOVE WELDS

(C) Tension or Compression Parallel to Weld Axis

Figure 5.33—Examples of Welds with Various Types of Loading

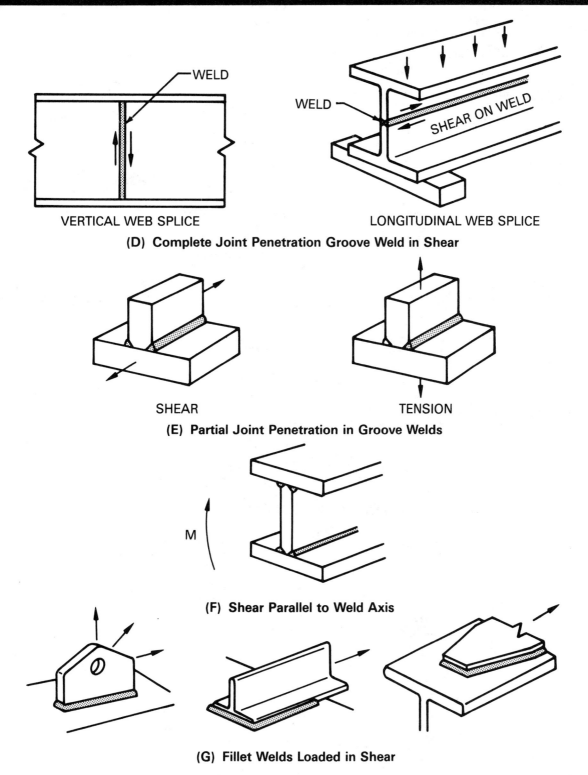

VERTICAL WEB SPLICE

LONGITUDINAL WEB SPLICE

(D) Complete Joint Penetration Groove Weld in Shear

SHEAR

TENSION

(E) Partial Joint Penetration in Groove Welds

(F) Shear Parallel to Weld Axis

(G) Fillet Welds Loaded in Shear

Figure 5.33 (Continued)—Examples of Welds with Various Types of Loading

Table 5.7
Minimum Effective Throat for Partial Joint Penetration Groove Welds in Steel

Thickness of Base Metal, [a] in.	Minimum Effective Throat[b], in.
1/8 to 3/16	1/16
Over 3/16 to 1/4	1/8
Over 1/4 to 1/2	3/16
Over 1/2 to 3/4	1/4
Over 3/4 to 1-1/2	5/16
Over 1-1/2 to 2-1/4	3/8
Over 2-1/4 to 6	1/2
Over 6	5/8

a. Thickness of thicker section with unequal thicknesses.
b. The effective throat need not exceed the thickness of the thinner part.

should be consulted to determine if an allowance for poor penetration is required for the conditions of a particular weld.

The allowable shear stress in steel weld metal in groove and fillet welds is about 30 percent of the nominal tensile strength of the weld metal.[15] Table 5.8 gives the allowable unit loads on various sizes of steel fillet welds of several strength levels. These values are for equal-leg fillet welds where the effective throat is 70.7 percent of the weld size. For example, the allowable unit force on a 1/2-in. (13 mm) fillet weld made with an electrode that deposits weld metal of 70,000 psi minimum tensile strength is determined as follows:

Allowable shear stress, $\tau = 0.30 \ (70\ 000)$
$= 21\ 000$ psi (145 MPa)
Allowable unit load, $f = 0.707 \ (1/2) \ (21\ 000)$
$= 7420$ lb/in.
(1300 N/mm) of weld

The minimum fillet weld sizes for structural welds are shown in Table 5.9. Where sections of different thickness are being joined, the minimum fillet weld size is governed by the thicker section. However, the weld size does not need to exceed the thickness of the thinner section unless a larger size is required by the loading conditions.

The minimum fillet weld size is intended to ensure sufficient heat input to reduce the possibility of cracking in either the heat heat-affected zone or weld metal, especially in a restrained joint. The minimum size applies if it is greater than the size required to carry design stresses.

CYCLIC LOADING

WHEN METALS ARE subjected to cyclic tensile or alternating tensile-compressive stress, they may fail by fatigue. The performance of a weld under cyclic stress is an important consideration in structures and machinery. The specifications relating fatigue in steel structures have been developed by the American Institute of Steel Construction (AISC), the American Association of State Highway and Transportation Officials (AASHTO), and

15. The validity of this relationship was established by a series of fillet weld tests conducted jointly by the American Institute of Steel Construction and the American Welding Society.

Table 5.8
Allowable Load Per Inch of Length of Steel Fillet Weld

Weld Size in.	mm	Classification Strength Level of Weld Metal, ksi						
		60	70	80	90	100	110	120
		Allowable unit loads, 1000 lb/in.*						
1/16	2	0.795	0.930	1.06	1.19	1.33	1.46	1.59
1/8	3	1.59	1.86	2.12	2.39	2.65	2.92	3.18
3/16	5	2.39	2.78	3.18	3.58	3.98	4.38	4.77
1/4	6	3.18	3.71	4.24	4.77	5.30	5.83	6.36
5/16	8	3.98	4.64	5.30	5.97	6.63	7.29	7.95
3/8	10	4.77	5.57	6.36	7.16	7.95	8.75	9.54
7/16	11	5.57	6.50	7.42	8.35	9.28	10.21	11.14
1/2	13	6.37	7.42	8.48	9.54	10.61	11.67	12.73
5/8	16	7.96	9.28	10.61	11.93	13.27	14.58	15.91
3/4	19	9.55	11.14	12.73	14.32	15.92	17.50	19.09
7/8	22	11.14	12.99	14.85	16.70	18.57	20.41	22.27
1	25	12.73	14.85	16.97	19.09	21.21	23.33	25.45

*To convert 1000 lb/inch to MN/m multiply the quantity in the table by 0.175.

Table 5.9
Recommended Minimum Fillet Weld Sizes for Steel

Section Thickness, in.ᵃ (mm)	Minimum Fillet Weld Size, in.ᵇ	
	in	mm
Over 1/8 to 1/4	1/8	3
Over 1/4 to 1/2	3/16	5
Over 1/2 to 3/4	1/4	6
Over 3/4	5/16	8

a. Thickness of thicker section whensections of unequal thicknesses are joined.
b. Single-pass welds or first pass ofmultiple pass welds. The size ofthe weld need not exceed the thickness of the thinner section being joined, provided sufficient pre-heat is used to ensure weld soundness.

the American Railway Engineering Association (AREA). The applicable standards of these organizations are essentially the same as the following material; however, reference should be made to the latest edition of the appropriate standard for specific information.

Although sound weld metal may have about the same fatigue strength as the base metal, any change in cross section at a weld lowers the fatigue strength of the member. In the case of a complete joint penetration groove weld, any reinforcement, undercut, incomplete joint penetration, or cracking acts as a notch or stress raiser. Each of these conditions is detrimental to fatigue life. The very nature of a fillet weld transverse to the stress field provides an abrupt change in section that limits fatigue life.

The fatigue stress provisions for the design of new bridges in *ANSI/AWS D1.1 Structural Welding Code—Steel* are presented in Table 5.10. Examples of each situation in Table 5.10 are illustrated in Figure 5.34, and curves for allowable design stress ranges for each stress category are plotted in Figure 5.35A and B for redundant and nonredundant structures respectively. A *stress range* is the magnitude of the change in stress that occurs with the application or removal of the cyclic load that causes tensile stress or a reversal of stress. Loads that cause only changes in the magnitude of compressive stress do not cause fatigue.

The allowable stress ranges of Figure 5.35A and B are independent of yield strength, and therefore apply equally to all structural steels. The allowable stress ranges for nonredundant structures, Figure 5.35B, are more conservative than those of redundant structures, because fracture resulting from fatigue crack growth in a nonredundant member would cause catastrophic collapse of the structure.

When fatigue conditions exist, the anticipated cyclically applied loads, the number of cycles, and the desired life must be given. The designer then selects materials and details to accommodate the design conditions for each member and situation (see Table 5.10 and Figure

5.34). The designer then calculates the maximum stress in each member to verify that the stresses do not exceed those allowable for static conditions. If the calculated stresses under cyclic conditions exceed the allowable stresses under static conditions, then the member sections must be increased to bring the calculated stress within that allowed.

Comparison of Figures 5.35A and B will indicate that for stress categories A, B, and C (for girder web and flange stiffeners only), the long term (high cycle) allowable stress range is the same for the redundant and nonredundant structures. However, below about 2.5 million cycles the allowable stress range of redundant structures is significantly higher than for nonredundant structures. For stress categories C (except for girder web and flange stiffeners), D, E, and F, the allowable stress range is greater for redundant structures than for nonredundant structures at all cycle lives.

Partial joint penetration groove welds are not normally used in fatigue applications, but their response to fatigue stresses is similar to that of fillet welds.

RIGID STRUCTURES

IN MACHINE DESIGN work, the primary design requirement for some members is rigidity. Such members are often thick sections so that the movement under load will be within close tolerances. The resulting stresses in the members are very low. Often, the allowable stress in tension for mild steel is 20 000 psi (138 MPa), but a welded machine base or frame may have a working stress of only 2000 to 4000 psi (14 to 28 MPa). In this case, the weld sizes need to be designed for rigidity rather than load conditions.

A very practical method is to design the weld size to carry one-third to one-half of the load capacity of the thinner member being joined. This means that if the base metal is stressed to one-third to one-half of the normal allowable stress, the weld would be strong enough to carry the load. Most rigid designs are stressed below these values. However, any reduction in weld size below one-third of the normal full-strength size would give a weld that is too small in appearance for general acceptance.

PRIMARY AND SECONDARY WELDS

WELDS MAY BE classified as primary and secondary types. A primary weld transfers the entire load at the point where it is located. Primary welds must have the same strength properties as the members. If the weld fails, the member fails. Secondary welds are those that simply hold the parts together to form a built-up member. In most cases, the stresses on these welds are low.

When a full-strength, primary weld is required, a weld metal with mechnical properties that match or exceed

Table 5.10
Fatigue Stress Provisions—Tension or Reversal Stresses*

General Condition	Situation	Stress Category (See Fig. 5.34)	Example (See Fig. 5.34)
Plain material	Base metal with rolled or cleaned surfaces. Oxygen-cut edges with ANSI smoothness of 1000 μin. (25 μm)or less	A	1.2
Built-up members	Base metal and weld metal in members without attachments, built up of plates or shapes connected by continuous complete or partial joint penetration groove welds or by continuous fillet welds parallel to the direction of applied stress	B	3, 4, 5, 7
	Calculated flexural stress at toe of transverse stiffener welds on girder webs or flanges	C	6
	Base metal at end of partial length welded cover plates having square or tapered ends, with or without welds across the ends	E	7
Groove welds	Base metal and weld metal at complete joint penetration groove welded splices of rolled and welded sections having similar profiles when welds are ground[1] and weld soundness established by RT or UT Inspection	B	8, 9
	Base metal and weld metal in or adjacent to complete joint penetration groove welded splices at transitions in width or thickness, with welds ground[1] to provide slopes no steeper than 1:2-1/2 and weld soundness established by RT or UT Inspection	B	10.11

General Condition	Situation	Longitudinal loading	Transverse loading[2]			Example (See Fig. 5.33A)
			Materials having equal or unequal thickness, sloped,[1] welds ground,[1] (web connections excluded)	Materials having equal thickness, not ground: (web connections excluded)	Materials having unequal thickness, not sloped or ground, including web connections	
Groove welded connections	Base metal at details of any length attached by groove welds subjected to transverse or longitudinal loading, or both, when weld soundness transverse to the direction of stress is established by RT or UT Inspection and the detail embodies a transition radius, R, with the weld termination ground[1] when					
	(a) R $\geq$ 24 in.	B	B	C	E	13
	(b) 24 in. > R $\geq$ 6 in.	C	C	C	E	13
	(c) 6 in. > R $\geq$ 2 in.	D	D	D	E	13
	(d) 2 in. > R $\geq$ 0[3]	E	E	E	E	12, 13

Table 5.10 (Continued)

General Condition	Situation	Stress Category (See Fig. 5.34)	Example (See Fig. 5.34)
Groove welds	Base metal and weld metal or in adjacent to complete joint penetration groove welded splices either not requiring transition or when required with transsitions having slopes no greater than 1:2.5 and when in either case reinforcement is not removed and weld soundness is established by RT or UT inspection	C	8, 9, 10, 11
Groove or fillet welded connections[3]	Base metal at details attached by groove or fillet welds subject to longitudinal loading where the details embodies a transition radius, R, less than 2 in., and when the detail length, L, parallel to the line of stress is		
	(a) L < 2 in.	C	12, 14, 15, 16
	(b) 2 in. ≤ L < 4 in.	D	12
	(c) L ≥ 4 in.	E	12
Fillet welded connections	Base metal at details attached by fillet welds parallel to the direction of stress regardless of length when the detail embodies a transition radius. R, 2 in. or greater and with the weld termination ground[1]		
	(a) When R ≥ 24 in.	B[4]	13
	(b) When 24 in. > R ≥ 6 in.	C[4]	13
	(c) When 6 in. > R ≥ 2 in.	D[4]	13
Fillet welds	Shear stress on throat of fillet welds	F	8a
	Base metal at intermittent welds attaching transverse stiffeners and stud-type shear connectors	C	7, 14
	Base metal at intermittent fillet welds attaching longitudinal stiffeners	E	–
Stud welds	Shear stress on nominal shear area of Type B shear connectors	F	14
Plug and slot welds	Base metal adjacent to or connected by plug or slot welds	E	–

*Except as noted for fillet and stud welds.
1. See AWS Dl.l, Structural Welding Code–Steel.
2. Applicable only to complete joint penetration groove welds.
3. Radii less than 2 in. need not be ground.
4. Shear stress on throat of weld (loading through the weld in any direction) is governed by Category F.

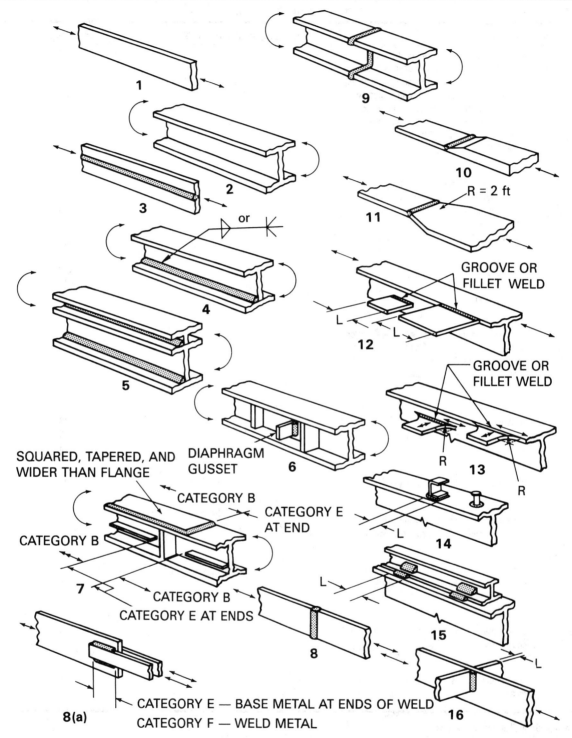

Note: The numbers below each example are referenced in Table 5.11

Reprinted with the permission of the American Association of State Highway and Transportation Officials

Figure 5.34—Examples of Various Fatigue Categories

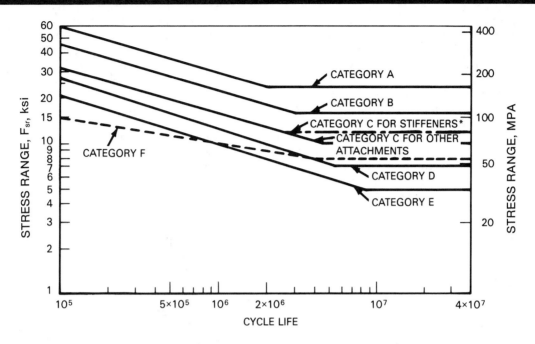

*Transverse Stiffener Welds on Girder Webs or Flanges

Figure 5.35A—Design Stress Range Curves for Categories A to F—Nonredundant Structures

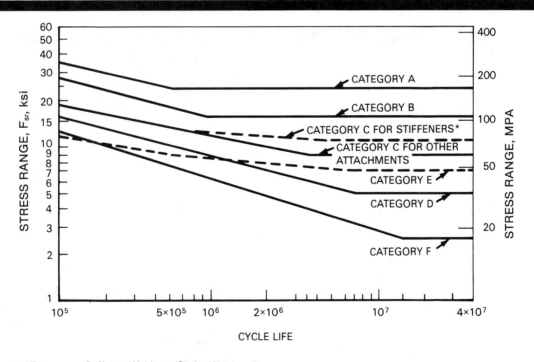

*Transverse Stiffener Welds on Girder Webs or Flanges

Figure 5.35B—Design Stress Range Curves for Categories A to F—Nonredundant Structures

those of the base metal is required. Generally, it is unnecessary for the weld metal and the base metal compositions to be exactly alike. For low alloy chromium-molybenum and stainless steels and for most nonferrous alloys, the weld metal compositions are similar to those of the base metals. However, for carbon and most low alloy steels, the weld metal compositions are generally not similar to the base metal compositions. For materials strengthened by heat treatment, the manufacturer's recommendations should be followed to avoid degradation of mechanical properties by the heat of welding.

In welding high strength steels, full-strength welds should not be used unless they are required. High strength steel may require preheat and special welding procedures because of its tendency for weld cracking, especially if the joint is restrained.

Secondary welds in high strength steels can be made with weld metal of lower strength than the base metal. Low-hydrogen weld metal with 70 000 to 90 000 psi (480 to 620 MPa) minimum tensile strength is preferred because the likelihood of cracking is lower than with matching weld metal. In any case, the weld must be sized to provide a joint of sufficient strength.

A comparison of behaviors of full-strength and partial-strength welds in quenched-and-tempered ASTM A514 steel is shown in Figure 5.36. The full strength weld is transverse to and the partial-strength weld is parallel to the tensile load. The plate has a tensile strength of 110 000 psi, and it is welded with an E11018 covered electrode to provide a full strength weld [see Figure 5.36(A)]. When the stress is parallel to the weld axis, Figure 5.36(B), a weld made with an E7018 covered electrode (70,000 psi minimum tensile strength) is adequate, so long as there is sufficient weld to transmit any shear load from one member to the other.

In the full-strength welded joint, both the plate and the weld metal have equivalent strengths and their behavior under load is shown by the stress-strain curve shown in Figure 5.36(A). If a transversely loaded test weld were pulled in tension, it is likely that the plate would neck down and fail first.

In the partial-strength weld loaded axially, Figure 5.36(B), both the plate and the weld would be strained together. As the member is loaded, the strain increases from 1 to 2 on the stress-strain plot with a corresponding increase in the stress in both the plate and weld from 1 to 3. At this point, the E7018 weld metal has reached its yield strength. On further loading, the strain is increased to 4. The weld metal is stressed beyond its yield strength at 5, and flows plastically; however, the plastic deformation is controlled and limited by the still elastic base material. The stress in the plate, however, is still below its yield strength at 6. With still further loading, the strain will reach 7 where the ductility of the plate will be exhausted. The plate will fail first because the weld metal has greater ductility. The weld will not fail until its unit strain reaches 8.

It is obvious in the example that the 70 000 psi (480 MPa) has sufficient strength to carry an axial load because it carries only a small portion of the total axial

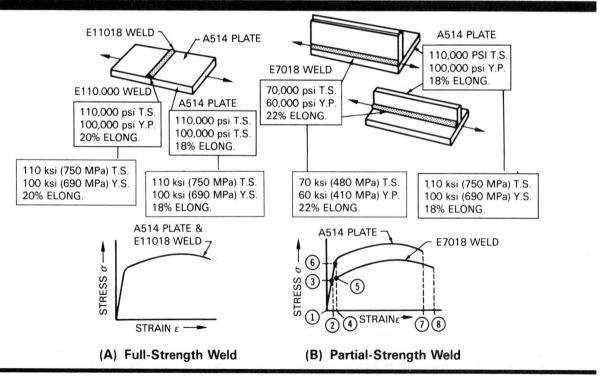

(A) Full-Strength Weld   **(B) Partial-Strength Weld**

Figure 5.36—Stress-Strain Characteristics of Full and Partial Strength Welds

load on the weldment. When a weld must transmit the total load, it has to be as strong as the base metal.

SKEWED FILLET WELDS

A SPECIAL CONDITION exists when members come together at an angle other than 90 degrees, and fillet welds are to be used to make the connection. Ordinary specifications for the weld leg at some joint angles could result in excessive waste of weld metal, along with difficulty in depositing the weld on the acute side of the joint.

Figure 5.37 shows skewed fillet welds and the relationships between the dihedral angle, Ψ, the leg size, b, and the effective throat, t, of each weld. Formulas are given to determine the proper effective throat for each weld to deposit a minimum area, A_t, of weld metal in the joint. The leg sizes, b_1 and b_2 can be determined for the respective effective throats.

TREATING A WELD AS A LINE

WHEN THE TOTAL length of weld in a connection is large compared to its effective throat, the weld can be assumed to be a line having a definite length and configuration rather than an area. The proper size of weld required for adequate strength can be determined using this concept. Referring to Figure 5.38, the welded connection is considered as a single line having the same outline as the connection area. The welded connection now has length, not effective area. Instead of determining the stress on a weld, which cannot be done until the weld size is known, the problem becomes simply one of determining the force per unit length on the weld.

The property of a welded connection, when the weld is treated as a line, can be substituted in the standard design formula used for the particular type of load, as shown in Table force per unit length on the weld may be calculated with the appropriate modified formula.

Problems involving bending or twisting loads may be satisfactorily and conservatively handled by treating the loads as vectors and adding the vector. The actual strength of welded connections in which external load does not pass through the shear center of the weld requires use of a more complex approach. The more complex method recognizes that when an eccentric load is applied, there will be relative rotation as well as translation between the welded parts. The actual center of rotation will not be about the center of gravity of the weld group but about a center dependent upon the relative magnitude of shear and moment reactions, weld geometry, and deformations of obliquely loaded incremental lengths of weld.[16]

The geometrical properties of common joint configurations can be determined using the formulas shown in

Table 5.12. Moment of inertia, I, section modulus, S, polar moment of inertia, J_w, and distance from the neutral axis or center of gravity to the extreme fibers, c, are included.

For a given connection, two dimensions are needed, width b, and depth d. Section modulus, S_w, is used for welds subjected to bending loads; polar moment of inertia, J_w, and distance, c, for twisting loads. Section moduli are given for maximum force at the top and bottom or right and left portions of the welded connections. For the unsymmetrical connections shown in Table 5.12, maximum bending force is at the bottom.

If there is more than one force applied to the weld, they are combined vectorially. All forces that are combined must be vectored at a common location on the welded joint.

Weld size is found by dividing the resulting unit force on the weld by the allowable strength of the type of weld used.

The steps in applying this method to any welded construction are as follows:

(1) Find the position on the welded connection where the combined forces are maximum. There may be more than one combination that should be considered.

(2) Find the value of each of the forces on the welded connection at this position.

(3) Select the appropriate formula from Table 5.12 to find the unit force on the weld.

(4) Use Table 5.12 to find the appropriate properties of the welded connection treated as a line.

(5) Combine vectorially all of the unit forces acting on the weld.

(6) Determine the required effective throat size by dividing the total unit force by the allowable stress in the weld.

The following example illustrates the steps in calculating the size of a weld considered as a line.

Assume that a bracket supporting an eccentric load of 18 000 lb (80 000 N) is to be fillet welded to the flange of a vertical column, as shown in Figure 5.39.

Step 1. The point of maximum combined unit forces is at the right ends of the top and bottom horizontal welds.

Step 2. The twisting force caused by the eccentric loading is divided into horizontal (f_h) and vertical (f_v) components. The distance from the center of gravity to the point of combined stress, c_{yr}, is calculated from the formula in Table 5.12 for this shape of connection (the fourth configuration).

$$c_{yr} = \frac{b(b+d)}{2b+d} = \frac{5(5+10)}{2(5)+10} = \frac{75}{20} = 3.75 \text{ in. (95 mm)}$$

16. Butler, Dal and Kulak, "Eccentrically loaded welded connections". *Journal of Structural Division*. ASCE, Vol. 98, No.ST 5; May 1972.

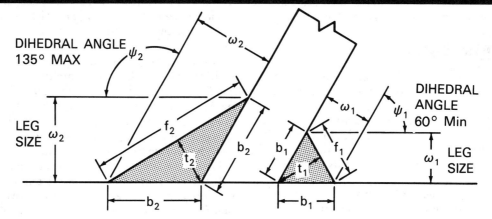

For Each Weld:

$$t = \frac{\omega}{2 \sin\left(\dfrac{\psi}{2}\right)} \text{ or } \omega = 2\,t \sin\left(\frac{\psi}{2}\right)$$

$$f = \frac{\omega}{\cos\left(\dfrac{\psi}{2}\right)} = 2\,t \tan\left(\frac{\psi}{2}\right)$$

$$b = \frac{t}{\cos\left(\dfrac{\psi}{2}\right)}$$

$$A = \frac{\omega^2}{4 \sin\left(\dfrac{\psi}{2}\right) \cos\left(\dfrac{\psi}{2}\right)} = t^2 \tan\left(\frac{\psi}{2}\right)$$

If $b_1 = b_2$, Then for $t = t_1 + t_2$

$$t_1 = t\;\frac{\cos\left(\dfrac{\psi_2}{2}\right)}{\cos\left(\dfrac{\psi_1}{2}\right) + \cos\left(\dfrac{\psi_2}{2}\right)} \qquad t_2 = t\;\frac{\cos\left(\dfrac{\psi_1}{2}\right)}{\cos\left(\dfrac{\psi_1}{2}\right) + \cos\left(\dfrac{\psi_2}{2}\right)}$$

For Minimum Total Weld Metal:

$$t_1 = \frac{t}{1 + \tan^2\left(\dfrac{\psi_1}{2}\right)} \qquad t_2 = \frac{t}{1 + \tan^2\left(\dfrac{\psi_2}{2}\right)}$$

$$A_t = \frac{t^2 \tan\left(\dfrac{\psi_1}{2}\right)}{1 + \tan^2\left(\dfrac{\psi_1}{2}\right)}$$

Figure 5.37—Formulae for Analyzing Skewed T-Joints

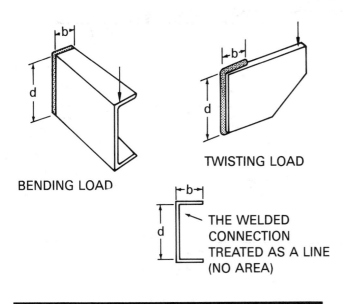

BENDING LOAD

TWISTING LOAD

THE WELDED
CONNECTION
TREATED AS A LINE
(NO AREA)

Figure 5.38—Treating a Weld as a Line

Table 5.11
Formulas for Calculating Force per Unit Length on Welds

Type of Loading	Standard Formula for Unit Stress	Formula for Force per Unit Length
Tension or compression	$\sigma = \dfrac{P}{A}$	$f = \dfrac{P}{L_w}$
Vertical shear	$\tau = \dfrac{V}{A}$	$f = \dfrac{V}{L_w}$
Bending	$\sigma = \dfrac{M}{S} = \dfrac{Mc}{I}$	$f = \dfrac{M}{S_w} = \dfrac{Mc}{I_w}$
Torsion	$\tau = \dfrac{Tc}{J}$	$f = \dfrac{Tc}{J_w}$

σ = normal stress
τ = shear stress
f = force per unit length
P = concentrated load
V = vertical shear load
A = total area of cross section
L_w = total length of a line weld
c = distance from neutral axis to the extreme fibers of a line weld
T = torque on the weld joint
S = sections modulus of an area =
S_w = section modulus of a line weld
J = polar mement of inertia of an area
J_w = polar moment of inertia of a line weld
I = moment of inertia
I_w = moment of inertia of a line weld

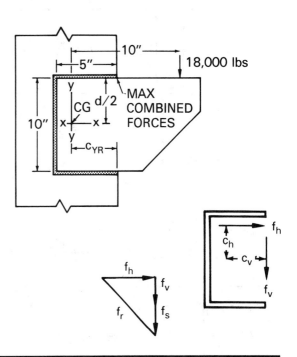

Figure 5.39—Bracket Joined to a Column Face with Fillet Weld

The polar moment of inertia:

$$J_w = \frac{b^3}{3}\frac{(b+2d)}{2b+d} + \frac{d^2}{12}\ (6b+d)$$
$$J_w = \frac{5^3}{3}\frac{(5+20)}{10+10} + \frac{(10)^2}{12}(30+10)$$
$$J_w = 385.4 \text{ in.}^3\ (6.3 \times 10^6 \text{ mm}^3)$$

Horizontal component of twisting from the torsion formula, Table 5.11:
Torque, $T = 18\ 000 \times 10 = 180\ 000$ in.·lb

$$f_v = \frac{(T)(d/2)}{J_w} = \frac{(180\ 000)(10/2)}{385.4} = \begin{matrix}2\ 340 \text{ lb/in.}\\(410 \text{ N/mm})\end{matrix}$$

Vertical component of twisting from the torsion formula Table 5.11:

$$f_v = \frac{Tc_{yr}}{J_w} = \frac{(180\ 000)(3.75)}{385.4} = \begin{matrix}1\ 750 \text{ lb/in.}\\(306 \text{ N/mm})\end{matrix}$$

Vertical shear force from Table 5.11:

$$f_s = \frac{P}{L_w} = \frac{18\ 000}{20} = 900 \text{ lb/in.}(158 \text{ N/mm})$$

Table 5.12
Properties of Welded Connections Treated as a Line

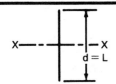

$$I_x = \frac{d^3}{12} \qquad S_x = \frac{d^2}{6}$$

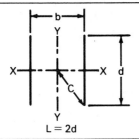

$$I_x = \frac{d^3}{6} \qquad S_x = \frac{d^2}{3} \qquad J_w = \frac{d}{6}(3b^2 + d^2)$$

$$I_y = \frac{b^2 d}{2} \qquad S_y = bd \qquad C = \frac{(b^2 + d^2)^{1/2}}{2}$$

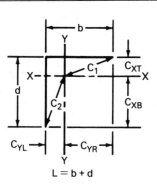

$$I_x = \frac{d^3}{12}\left(\frac{4b + d}{b + d}\right) \qquad S_{XT} = \frac{d}{6}(4b + d) \qquad S_{XB} = \frac{d^2}{6}\left(\frac{4b + d}{2b + d}\right)$$

$$I_y = \frac{b^3}{12}\left(\frac{b + 4d}{b + d}\right) \qquad S_{YL} = \frac{b}{6}(b + 4d) \qquad S_{YR} = \frac{b^2}{6}\left(\frac{b + 4d}{2d + b}\right)$$

$$J_w = \frac{b^3 + d^3}{12} + \frac{bd(b^2 + d^2)}{4(b + d)}$$

$$C_{XT} = \frac{d^2}{2(b + d)} \qquad C_{XB} = \frac{d}{2}\left(\frac{2b + d}{(b + d)}\right) \qquad C_1 = (C_{XT}^2 + C_{YR}^2)^{1/2}$$

$$C_{YL} = \frac{b^2}{2(b + d)} \qquad C_{YR} = \frac{b}{2}\left(\frac{b + 2d}{b + d}\right) \qquad C_2 = (C_{XB}^2 + C_{YL}^2)^{1/2}$$

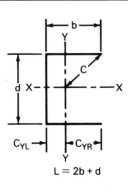

$$I_x = \frac{d^2}{12}(6b + d) \qquad S_x = \frac{d}{6}(6b + d)$$

$$I_y = \frac{b^3}{3}\left(\frac{b + 2d}{2b + d}\right) \qquad S_{YL} = \frac{b}{3}(b + 2d)$$

$$C_{YL} = \frac{b^2}{2b + d} \qquad C_{YR} = \frac{b(b + d)}{2b + d} \qquad S_{YR} = \frac{b^2}{3}\left(\frac{b + 2d}{b + d}\right)$$

$$C = \left[C_{YR}^2 + \left(\frac{d}{2}\right)^2\right]^{1/2} \qquad J_w = \frac{b^3}{3}\left(\frac{b + 2d}{2b + d}\right) + \frac{d^2}{12}(6b + d)$$

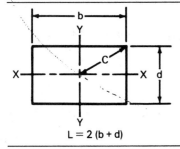

$$I_x = \frac{d^2}{6}(3b + d) \qquad S_x = \frac{d}{3}(3b + d)$$

$$I_x = \frac{d^2}{6}(b + 3d) \qquad S_y = \frac{b}{3}(b + 3d)$$

$$J_w = \frac{(b + d)^3}{6} \qquad C = \frac{(b^2 + d^2)^{1/2}}{2}$$

Table 5.12 (Continued)

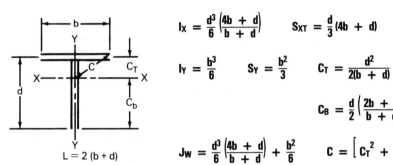

$$I_X = \frac{d^3}{3}\left(\frac{2b + d}{b + 2d}\right) \qquad S_{XT} = \frac{d}{3}(2b + d) \qquad S_{XB} = \frac{d^2}{3}\left(\frac{2b + d}{b + d}\right)$$

$$I_Y = \frac{b^3}{12} \qquad S_Y = \frac{b^2}{6} \qquad C_T = \frac{d^2}{b + 2d}$$

$$J_W = \frac{d^3}{3}\left(\frac{2b + d}{b + 2d}\right) + \frac{b^3}{12} \qquad C_b = d\left(\frac{b + d}{b + 2d}\right)$$

$$C = \left[C_T^{\,2} + \left(\frac{b}{2}\right)^2\right]^{1/2}$$

L = b + 2d

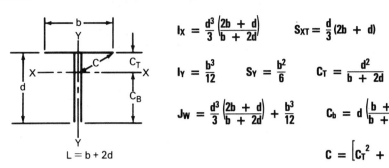

$$I_X = \frac{d^3}{6}\left(\frac{4b + d}{b + d}\right) \qquad S_{XT} = \frac{d}{3}(4b + d) \qquad S_{XB} = \frac{d^2}{3}\left(\frac{4b + d}{b + d}\right)$$

$$I_Y = \frac{b^3}{6} \qquad S_Y = \frac{b^2}{3} \qquad C_T = \frac{d^2}{2(b + d)}$$

$$C_B = \frac{d}{2}\left(\frac{2b + d}{b + d}\right)$$

$$J_W = \frac{d^3}{6}\left(\frac{4b + d}{b + d}\right) + \frac{b^2}{6} \qquad C = \left[C_T^{\,2} + \left(\frac{b}{2}\right)^2\right]^{1/2}$$

L = 2 (b + d)

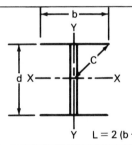

$$I_X = \frac{d^2}{6}(3b + d) \qquad S_X = \frac{d}{3}(3b + d)$$

$$I_Y = \frac{b^3}{6} \qquad S_Y = \frac{b^2}{3}$$

$$J_W = \frac{d^2}{6}(3b + d) + \frac{b^3}{6} \qquad C = \frac{(b^2 + d^2)^{1/2}}{2}$$

L = 2 (b + d)

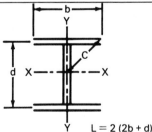

$$I_X = \frac{d^2}{6}(6b + d) \qquad S_X = \frac{d}{3}(6b + d)$$

$$I_Y = \frac{b^3}{3} \qquad S_Y = \frac{2}{3}\, b^2$$

$$J = \frac{d^2}{6}(6b + d) + \frac{b^3}{3} \qquad C = \frac{(b^2 + d^2)^{1/2}}{2}$$

L = 2 (2b + d)

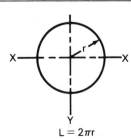

$$I = \pi r^3 \qquad S_W = \pi r^2 \qquad J_W = 2\pi r^3$$

L = 2πr

Step 3.   Determine the resultant force:

$$f_v = [f_h{}^2 + (f_v + f_s)^2]^{1/2}$$
$$= [(2\,340)^2 + (1\,750 + 900)^2]^{1/2}$$
$$= 3\,540 \text{ lb/in. (620 N/mm)}$$

Step 4.   The allowable shearing stress on the effective area of weld metal having an ultimate tensile strength of 60 000 psi (413 MPa) is as follows (Table 5.7):

$$\tau = 0.30(60\,000) = 18\,000 \text{ psi (124 MPa)}$$

The effective throat:

$$(E) = f_v = \frac{3\,540}{18\,000} = 0.197 \text{ in. (5 mm)}$$

Assuming an equal leg fillet weld, the fillet weld size:

$$S = \frac{(E)}{0.707} = \frac{0.197}{0.707} = 0.279 \text{ in. (7mm)}$$

A 5/16-in. (8 mm) fillet weld should be specified on the welding symbol.

Procedures for determining the allowable eccentric loads for various welded connections used in structural steel construction are given in the *Manual of Steel Construction*, published by the American Institute of Steel Construction.

STRUCTURAL TUBULAR CONNECTIONS

TUBULAR MEMBERS ARE being used in structures such as drill rigs, space frames, trusses, booms, and earthmoving and mining equipment.[17] They have the advantage of minimizing deflection under load because of their greater rigidity when compared to standard structural shapes. Various types of welded tubular connections, the component designations, and nomenclature are shown in Figure 5.40.

With structural tubing, there is no need for cutting holes at intersections, and as a result, the connections can have high strength and stiffness. However, if a connection is to be made with a complete joint penetration groove weld, the weld usually must be made from one side only and without backing because the small tube size or configuration, or both, will prevent access to the root side of the weld. Special skill is required for making tubular connections using complete joint penetration groove welds from one side.

With relatively small, thin-wall tubes, the end of the brace tube may be partially or fully flattened. The end of the flattened section is trimmed at the appropriate angle to abut against the main member where it is to be welded. This design should only be used with relatively low load conditions because the load is concentrated on a narrow area of the main tube member. The flattened section of the brace member must be free of cracks.

WELD JOINT DESIGN

WHEN TUBULAR MEMBERS are fit together for welding, the end of the branch member or brace is normally contoured to the shape of the main member. The members may be joined with their axes at 80 to 100 degrees in the case of T-connections [Figure 5.40(C)], or at some angle less than 80 degrees in Y- and K-connections [Figure 5.40(D) and (E)]. The tubes may have a circular or box shape, and the branch member may be equal in size or smaller than the main member. Consequently, the angle (Ψ) between the adjacent outside tube surfaces or their tangents, in a plane perpendicular to the joint (local dihedral angle), can vary around the joint from about 150 to 30 degrees. To accommodate this, the weld joint design and welding procedures must be varied around the joint to obtain a weld with adequate penetration (throat).

Tubular joints are normally accessible for welding only from outside the tubes. Therefore, the joints are generally made with single groove or fillet welds. Groove welds may be designed for complete or partial joint penetration, depending upon the load conditions. To obtain adequate joint penetration, shielded metal arc, gas metal arc, and flux cored arc welding are generally used to make tubular joints in structures.

Suggested groove designs[18] for complete joint penetration with four dihedral angle ranges are shown in Figure 5.41. The areas of the circular and box connections to which the groove designs of Figure 5.41 apply are shown in Figure 5.42(A) and (B), respectively. The specified root opening, R, or the width of a backing weld, W, in Figure 5.41 depends upon the welding process and groove angle. The purpose of the backing welds, W in Figure 5.41, is to provide a sound root condition for the deposition of the production weld. The backing weld is not counted as part of the throat of the joint design.

---

17. The welding of steel tubular structures is covered by ANSI/AWS D1.1, *Structural Welding Code—Steel*. Miami: American Welding Society;latest edition.

18. These groove designs meet the joint geometry requirements for prequalified welding procedures in ANSI/AWS D1.1, *Structural Welding Code—Steel*.

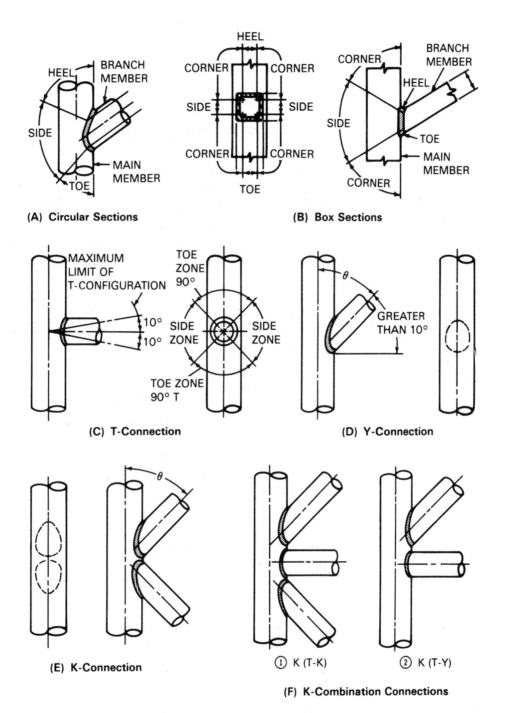

(A) Circular Sections

(B) Box Sections

(C) T-Connection

(D) Y-Connection

(E) K-Connection

① K (T-K)   ② K (T-Y)

(F) K-Combination Connections

Figure 5.40—Welded Tubular Connections, Components, and Nomenclature

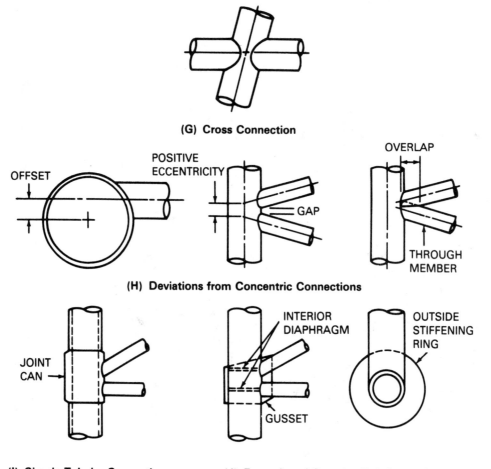

(G) Cross Connection

(H) Deviations from Concentric Connections

(I) Simple Tubular Connections   **(J) Examples of Complex Reinforced Connections**

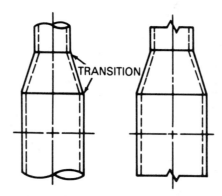

(K) Flared Connections and Transitions

Figure 5.40 (Continued)—Welded Tubular Connections, Components, and Nomenclature

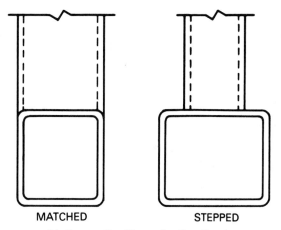

(L) Connection Types for Box Sections

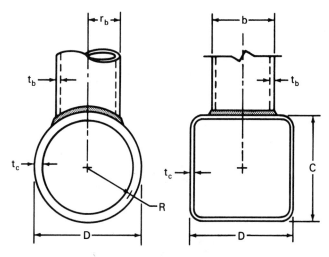

(M) Geometric Parameters

Parameter	Circular sections	Box sections
β	r_b/R	b/D
η	—	a_x/D [See (B)]
γ	R/t_c	$D/2t_c$
τ	t_b/t_c	t_b/t_c
θ	Angle between member center lines	
ψ	Local dihedral angle at given point on welded joint	

Figure 5.40 (Continued)—Welded Tubular Connections, Components, and Nomenclature

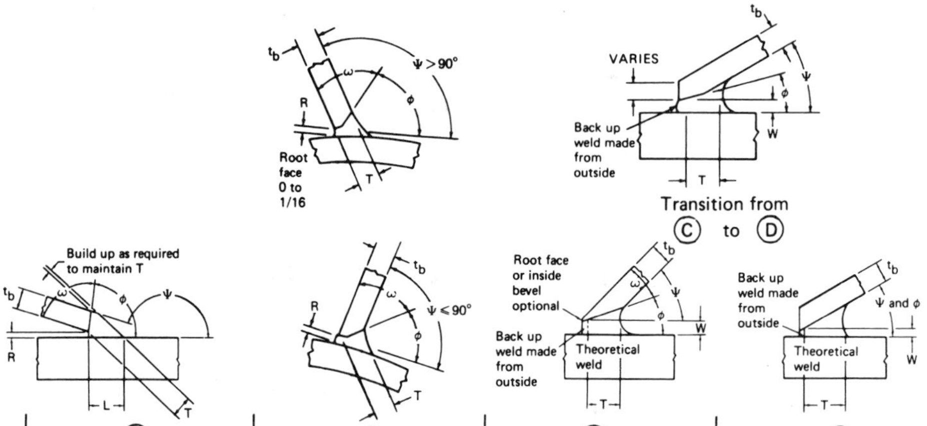

		A (3) $\Psi=180°-135°$		B (3) $\Psi = 150°-50°$		C (3) $\Psi = 75°-30°$	D (3) $\Psi = 37\text{-}1/2°-15°$
End preparation (ω)	max	90°		90°		(a)	
	min	45°		10° or 45° for$\Psi>105°$		10°	
		FCAW SMAW (1)	GMAW FCAW (2)	FCAW SMAW (1)	GMAW FCAW (2)	FCAW SMAW (1) $\quad$ W max. (b) $\quad$ ϕ $\left\{\begin{array}{l}\text{1/8 in.}\\\text{3/16 in.}\end{array}\right.$ $\begin{array}{l}22\text{-}1/2° - 37\text{-}1/2''\\15° - 20\text{-}1/2''\end{array}$	
Fitup or root opening (R)					1/4 in. for $\phi>45°$ 5/16 in. for $\phi\leqslant45°$		
	max	3/16 in.	3/16 in.	1/4 in.		GMAW FCAW (2) $\left\{\begin{array}{l}\text{1/8 in.}\\\text{1/4 in.}\\\text{3/8 in.}\\\text{1/2 in.}\end{array}\right.$ $\begin{array}{l}30° - 37\text{-}1/2''\\25° - 30°\\20° - 25°\\15° - 20°\end{array}$	
	min	1/16 in. No min for $\phi>90°$	1/16 in. No min for $\phi>120°$	1/16 in.	1/16 in.		
Joint included angle ϕ	max			60° for $\Psi\leqslant105°$		37-1/2° if more use B	
	min			37-1/2° if less use C		1/2 Ψ	
Completed weld	T	$\geqslant t_b$		$\geqslant t$ for $\Psi>90°$		$\geqslant t/\sin \Psi$ but need not exceed 1.75t	$\geqslant2t_b$
	L	$\geqslant t/\sin \Psi$ but need not exceed 1.75t		$\geqslant t/\sin \Psi$ for $\Psi\leqslant90°$		Weld may be built up to meet this	

(a) Otherwise as needed to obtain required ϕ.

(b) Initial passes of back up weld discounted until width of groove (W) is sufficient to assure sound welding; the necessary width of weld groove (W) provided by back up weld.

Notes:

1. These root details apply to SMAW and FCAW (self-shielded).
2. These root details apply to GMAW (short circuiting transfer and FCAW (gas sheilded).
3. See Figure 5.42 for locations on the tubular connection.

Figure 5.41—Joint Designs for Complete Joint Penetration in Simple T-, K-, and Y-Tubular Connections

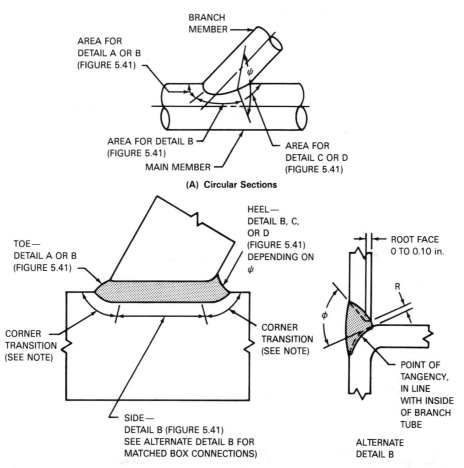

BRANCH
MEMBER

AREA FOR
DETAIL A OR B
(FIGURE 5.41)

ψ

AREA FOR DETAIL B
(FIGURE 5.41)

MAIN MEMBER

AREA FOR
DETAIL C OR D
(FIGURE 5.41)

(A) Circular Sections

TOE—
DETAIL A OR B
(FIGURE 5.41)

HEEL—
DETAIL B, C,
OR D
(FIGURE 5.41)
DEPENDING ON
ψ

CORNER
TRANSITION
(SEE NOTE)

CORNER
TRANSITION
(SEE NOTE)

SIDE—
DETAIL B (FIGURE 5.41)
SEE ALTERNATE DETAIL B FOR
MATCHED BOX CONNECTIONS)

ROOT FACE
0 TO 0.10 in.

R

φ

POINT OF
TANGENCY,
IN LINE
WITH INSIDE
OF BRANCH
TUBE

ALTERNATE
DETAIL B

Note:
Joint preparation for welds at corner shall provide a smooth transition from one detail to another. Welding
shall be carried continuously around corners, with corners fully built up and all starts and stops within flat
faces.

(B) Box Sections

Figure 5.42—Locations of Complete Joint Penetration Groove Weld Designs on Tubular Connections

Suggested groove designs for partial joint penetration groove welds for circular and box connections are shown in Figure 5.43. The sections of circular and box connections to which they apply are shown in Figure 5.44.

Because of the variation of dihedral angles around the joint, the inaccessibility for welding from inside and the differences in penetration of different processes, a separate consideration of the questionable root area of the weld is required as contrasted to direct tabulation of effective throat with more conventional prequalified partial joint penetration welds. An allowance, Z, called a loss factor, should be made for incomplete fusion at the throat of partial joint penetration groove welds. This allowance assures that the actual throat of the weld is not less than the design requirement. The loss factor, Z, is shown in Table 5.13 for various local dihedral angles (Ψ) and welding processes.

Suggested fillet weld details for T-, K-, and Y-connections in circular tubes are shown in Figure 5.45. The recommended allowable stress on the effective throat of partial joint penetration groove welds and fillet welds in steel T-, K-, and Y-connections is 30 percent of the specified minimum tensile strength of the weld metal. The stress on the adjoining base metal should not exceed that permitted by the applicable code.

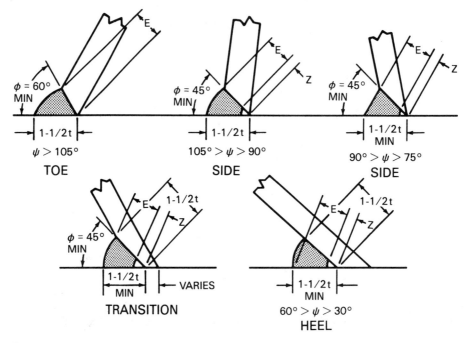

Notes:
1. t = thickness of thinner section.
2. Bevel to feather edge except in transition and heel zones.
3. Root opening: 0 to 3/16 in. (5 mm).
4. Effective throat = E, where ≥ t; see Table 5.13 for loss factor (z).
5. See Figure 5.44 for orientations of tubes and grooves.

Figure 5.43—Joint Designs for Partial Joint Penetration Groove Welds in Simple T-, K-, and Y-Tubular Connections

A welded tubular connection is limited in strength by four factors:

(1) Local or punching shear failure
(2) Uneven distribution of load on the welded connection
(3) General collapse
(4) Lamellar tearing

LOCAL FAILURE

WHERE A CIRCULAR or stepped-box T-, K-, or Y-connection (Figure 5.45) is made by simply welding the branch member to the main member, local stresses at a potential failure surface through the main member wall may limit the useable strength of the welded joint. The shear stress at which failure can occur depends upon both the geometry of the section and the strength of the main member. The actual localized stress situa-tion is more complex than simple shear; it includes shell bending and membrane stress as well. Whatever the mode of main member failure, the allowable punching shear stress is a conservative representation of the average shear stress at failure in static tests of simple welded tubular connections. The method for determining the punching shear stress in the main member is given in *ANSI/AWS D1.1, Structural Welding Code—Steel*, latest edition.

The term *punching shear* describes a local failure con-dition in which the main member fails adjacent to the weld by shear. As a result, a plug welded to the branch is sheared from the main member. When the failure occurs in compression, it more closely resembles a punching operation than when it occurs in tension.

The actual punching shear stress in the main member caused by the axial force and any bending moment in the branch member must be determined and compared with the allowable punching shear stress. The effective area and length of the weld, as well as its section modulus,

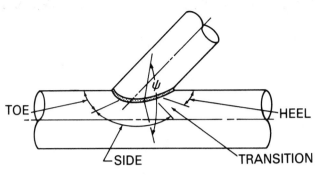

(A) Circular Connection

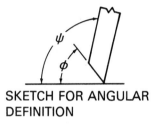

SKETCH FOR ANGULAR
DEFINITION

$$150° \geq \psi \geq 30°$$
$$90° \geq \phi \geq 30°$$

(B)

**Figure 5.44—Locations of Partial Joint Penetration
Groove Weld Designs on Tubular Connections**

must be determined to treat the axial force and bending moment on the joint. These joint properties are factored into the stress and force calculations as described in *ANSI/AWS D1.1, Structural Welding Code—Steel.*

UNEVEN DISTRIBUTION OF LOAD

ANOTHER CONDITION THAT can limit the strength of a welded connection is uneven distribution of load on the weld. Under load, some bending of the main member could take place, which might cause uneven distribution of the force applied to the weld. As a result, some yielding and redistribution of stresses may have to take place for the connection to reach its design load. To provide for this, welds in T-, K-, and Y-connections [Figure 5.40 (C), (D), and (E)] must be capable, at their ultimate breaking strength, of developing the lesser of (1) the yield strength of the branch member or (2) the ultimate punching shear strength of the shear area of the main member. These conditions are illustrated in Figure 5.46. This particular part of the design is best handled by working with terms of unit force (pounds per linear inch).

The ultimate breaking strength, Figure 5.46(A), of fillet welds and partial joint penetration groove welds is computed at 2.67 times the basic allowable stress for 60 ksi (413 MPa) and 70 ksi (480 MPa) tensile strength weld metal, and at 2.2 times for higher strength weld metals.

The unit force on the weld from the branch member at its yield strength, Figure 5.46(B), is as follows:

**Table 5.13
Loss Factors for Incomplete Fusion at the Root of Partial Joint Penetration Groove Welds**

Groove Angle, ϕ	Welding Process[2] (V or OH)[1]	Loss Factor[3]		Welding Process[2] (H of F)[1]	Loss Factor	
		in.	mm		in.	mm
$\phi \geq 60°$	SMAW	0	0	SMAW	0	0
	FCAW	0	0	FCAW	0	0
	FCAW-G	0	0	FCAW-G	0	0
	GMAW	NA[4]	NA[4]	GMAW	0	0
	GMAW-S	0	0	GMAW-S	0	0
$60° \geq \phi \geq 45°$	SMAW	1/8	3	SMAW	1/8	3
	FCAW	1/8	3	FCAW	0	0
	FCAW-G	1/8	3	FCAW-G	0	0
	GMAW	NA[4]	NA[4]	GMAW	0	0
	GMAW-S	1/8	3	GMAW-S	1/8	3
$45° > \geq 30°$	SMAW	1/4	6	SMAW	1/4	6
	FCAW	1/4	6	FCAW	1/8	3
	FCAW-G	3/8	10	FCAW-G	1/4	6
	GMAW	NA[4]	NA[4]	GMAW	1/4	6
	GMAW-S	3/8	10	GMAW-S	1/4	6

1. Position of welding F = Flat; H = Horizontal; V = Vertical; OH = Overhead.
2. Processes: FCAW = Self shielded flux cored arc welding GMAW = Spray transfer or globular transfer
FCAW-G = Gas shielded flux cored arc welding   GMAW-S = Short circuiting transfer
3. Refer to Figure 5.45
4. NA = Not applicable.

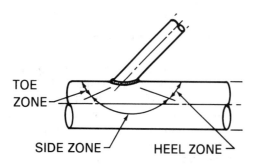

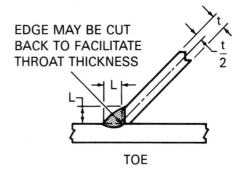

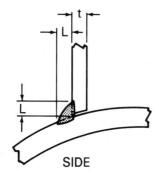

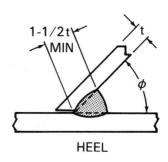

Notes:
1. t = thickness of thinner part
2. L = minimum size = t
3. Root opening 0 to 3/16 in.
4. ϕ = 15° min

Figure 5.45—Fillet Weld Details for T-, K-, and Y-Connections

$$f_1 = \sigma_y t_b \qquad (5.11)$$

where
> f_1 = unit force, lb/in.
> σ_y = yield strength of branch member
> t_b = thickness of branch member, in.

The ultimate shear on the main member shear area, Figure 5.46(C), at failure is as follows:

$$f_2 = 1.8\,\tau_a t \qquad (5.12)$$

where
> f_2 = ultimate unit shear normal to the weld, lb/in.
> τ_a = allowable shear stress, psi
> t = thickness of the main member, in.

The unit shear force per inch on the weld, f_3, is

$$f_3 = \frac{f_2}{\sin\theta} = \frac{1.8\tau_a t}{\sin\theta} \qquad (5.13)$$

where θ is the angle between the axes of the two members.

GENERAL COLLAPSE

AS NOTED PREVIOUSLY, the strength of the connection also depends on what might be termed *general collapse*. The strength and stability of the main member in a tubular connection should be investigated using the proper technology and in accordance with the applicable design code. If the main member has sufficient thickness required to resist punching shear and this thickness extends beyond the branch members for a distance of at least one quarter of the main member diameter, general collapse should not be a limiting factor.

THROUGH-THICKNESS FAILURES

IN TUBULAR CONNECTIONS, such as those shown in Figures 5.40 through 5.46, the force must be transmitted through the thickness of the main member when the axial force on the branch member is tension. The ductility and notch toughness of rolled metals is significantly lower in the through-thickness (short transverse) direction than in the longitudinal or (long) transverse directions. Thus, a tubular member could delaminate from tensile stresses transmitted through the thickness. To avoid this condition, interior diaphragm or continuity plates in combination with gusset plates or stiffening rings, as shown in Figure 5.40(J), can be employed at highly stressed connections. To further reduce the through-thickness tensile stresses, the diaphragm plate can penetrate the shell of the main member as shown in Figure 5.40(J) (left). The resulting single-bevel-groove weld, in which the main member is grooved, transfers the delaminating forces from the primary structural member to the secondary structural member. This follows the principles suggested earlier in the chapter of beveling the through-thickness member to avoid lamellar tears from weld shrinkage.

FATIGUE

THE DESIGN OF welded tubular structures subject to cyclic loading is handled in the same manner as discussed previously. The specific treatment may vary with the applicable code for the structure.[19] Stress categories are assigned to various types of pipe, attachments to pipe, joint designs, and loading conditions. The total cyclic fatigue stress range for the desired service life of a particular situation can be determined.

Fatigue behavior can be improved by one or more of the following actions:

(1) A capping layer can be added to provide a smooth contour with the base metal.
(2) The weld face may be ground transverse to the weld axis.
(3) The toe of the weld may be peened with a blunt instrument to cause local plastic deformation and to smooth the transition between the weld and base metals.

19. Such codes include ANSI/AWS D1.1, *Structural welding code—steel*. Miami: American Welding Society; latest edition and API RP 2A, *Recommended practice for planning, designing, and constructing fixed offshore platforms*, 11th Ed. Dallas: American Petroleum Institute; 1980.

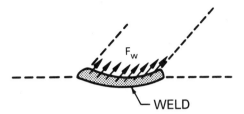

(A) Ultimate Strength of Welded Connection

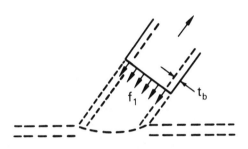

(B) Brace Member at Yield Strength

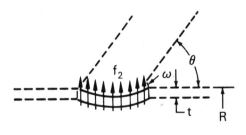

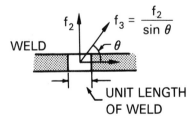

(C) Ultimate Shear on the Main Member Shear Area at Failure

Figure 5.46—Loads on Welded Tubular Connections

ALUMINUM STRUCTURES

DESIGNING FOR WELDING

THE METHODS EMPLOYED to design structures in aluminum are generally the same as those used with steel or other metals.[20] The methods and stress values recommended for structural aluminum design are set forth in the Aluminum Association's *Specification for Aluminum Structures*.[21]

Aluminum is available in many product forms and shapes, both cast and wrought. The designer can take advantage of the light weight of aluminum by utilizing available aluminum structural forms. Proper engineering design minimizes the number of joints and amount of welding without affecting product requirements. This, in turn, provides for good appearance and proper functioning of the product by limiting distortion caused by heating. To eliminate joints, the designer may use castings, extrusions, forgings, or bent or roll-formed shapes

to replace complex assemblies. Special extrusions that incorporate edge preparations for welding may provide savings in manufacturing costs. Typical designs are shown in Figure 5.47. An integral lip can be provided on the extrusion to facilitate alignment and serve as a weld backing.

For economical fabrication, the designer should employ the least expensive metal-forming and metal-working processes, minimize the amount of welding required, and place welds at locations of low stress. A simple example is the fabrication of an aluminum tray, Figure 5.48. Instead of using five pieces of sheet and eight welds located at the corners, Figure 5.48(A), such a unit could be fabricated from three pieces of sheet, one of which is formed into the bottom and two sides, Figure 5.48(B), thereby reducing the amount of welding.

Further reduction in welding could be achieved by additional forming, as in Figure 5.48(C). This forming also improves performance as butt or lap joints can be welded instead of corners. However, some distortion would likely take place in the two welded sides because all of the welds are in those two planes. The refinement of a design to limit only the amount of welding

20. Welding requirements applicable to welded aluminum structures are given in *ANSI/AWS D1.2, Structural welding code—aluminum.* Miami, FL: American Welding Society; latest edition.

21. *Specification for aluminum structures, Construction manual series.* Washington, DC: Aluminum Association; latest edition.

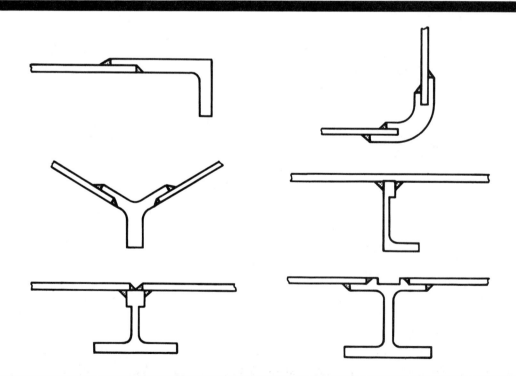

Figure 5.47—Typical Extrusion Designs Incorporating Desired Joint Geometry, Alignment, and Reinforcement Between Different Thicknesses

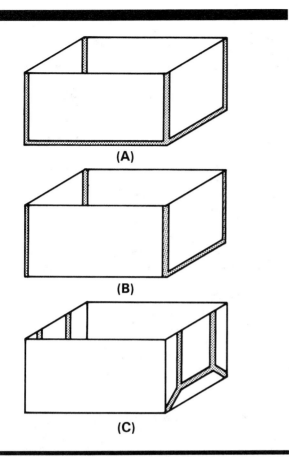

Figure 5.48—Designs for an Aluminum Tray

could lead to problems in fabrication, end use, or appearance. Therefore, the extent of welding should not be the single consideration in weldment design.

WELD JOINTS

BUTT, LAP, EDGE, corner, and T-joints may be used in aluminum design. For structural applications, edge and corner joints should be avoided because they are harder to fit, are weaker, and are more prone to fatigue failure than the other joints. However, they are commonly used in sheet metal fabrications.

Butt Joints

BUTT JOINTS ARE generally easy to design, present good appearance, and perform better under cyclic loading than other types of joints. However, they require accurate alignment and usually require joint edge preparation on thicknesses above 1/4 in. (6 mm) to permit satisfactory root penetration. In addition, back chipping and a backing weld are recommended to ensure complete fusion on thicker sections.

Sections of different thicknesses may be butted together and welded. However, it is better to bevel the thicker section before welding to reduce stress concentration, particularly when the joint will be exposed to cyclic loading in service.

When thin aluminum sheets are to be welded to thicker sections, it is difficult to obtain adequate depth of fusion in the thicker section without melting away the thin section. This difficulty can be avoided by extruding or machining a lip on the thicker section equal in thickness to that of the thin part as shown in Figure 5.49(B). This design will also provide a better heat balance and further reduce the heat related distortion. If the thicker section is an extrusion, a welding lip can be incorporated in the design as described previously.

Lap Joints

LAP JOINTS ARE used more frequently on aluminum alloys than is customary with most other metals. In thicknesses up to 1/2 in.(13 mm), it may be more economical to use single-lap joints with fillet welds on both sides rather than butt joints welded with complete joint penetration. Lap joints require no edge preparation, are easy to fit, and require less jigging than butt joints. The efficiency of lap joints ranges from 70 to 100 percent, depending on the base metal composition and temper. Preferred types of lap joints are shown in Figure 5.50.

Lap joints do create an offset in the plane of the structure unless the members are in the same plane and strips

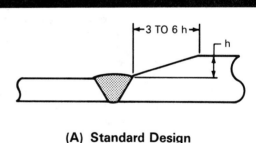

(A) Standard Design

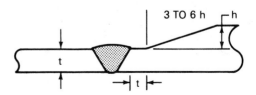

(B) Alternate Design

Figure 5.49—Recommended Transition Joints

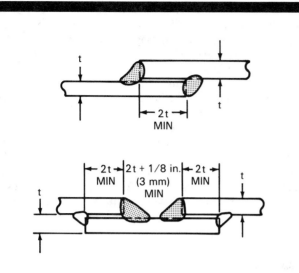

Figure 5.50—Preferred Types of Lap Joints

are used on both sides of the joint. Those with an offset tend to rotate under load. Lap joints may be impractical if the joint is not accessible on both sides.

T-Joints

T-JOINTS SELDOM REQUIRE edge preparation because they are usually connected by fillet welds. The welds should have complete fusion to or beyond the root (corner) of the joint. A single- or double-bevel-groove weld in combination with fillet welds may be used with thicknesses above 3/4 in. (19 mm) to reduce the amount of weld metal. T-joints are easily fitted and normally require no back chipping. Necessary fixturing is usually quite simple.

A T-joint with a single fillet weld is not recommended. Although the joint may have adequate shear and tensile strength, it is very weak when loaded with the root of the fillet weld in tension. Small continuous fillet welds should be used on both sides of the joint, rather than large intermittent fillet welds on both sides, or a large continuous fillet weld on one side. Continuous fillet welding is recommended for better fatigue life and for avoiding crevice corrosion and crater cracks. Suggested allowable shear stresses in fillet welds for building and bridge structures are given in the *Specification for Aluminum Structures* published by The Aluminum Association.

JOINT DESIGN

IN GENERAL, THE designs of welded joints for aluminum are similar to those recommendations for steel

joints.[22] However, aluminum joints normally have smaller root openings and larger groove angles. To provide adequate shielding of the molten aluminum weld metal, larger gas nozzles are usually employed on welding guns and torches. The excellent machinability of aluminum makes J- and U-groove preparations economical to reduce weld metal volume, especially on thick sections.

EFFECTS OF WELDING ON STRENGTH

ALUMINUM ALLOYS ARE normally used in the strain-hardened or heat-treated condition, or a combination of both, to take advantage of their high strength-to-weight ratios. The effects of strain hardening or heat treatment are wholly or partially negated when aluminum is exposed to the elevated temperatures encountered in welding. The heat of welding softens the heat-affected zone in the base metal. The extent of softening is related to the section thickness, original temper, heat input, and rate of cooling. The lower strength heat-affected zone must be considered in design. The orientation of the heat-affected zone with respect to the direction of stress and its proportion of the total cross section determines the allowable load on the joint.

The variation in tensile or yield strength across a welded joint in aluminum structures is illustrated in Figure 5.51. With plate, the extent of decreased properties

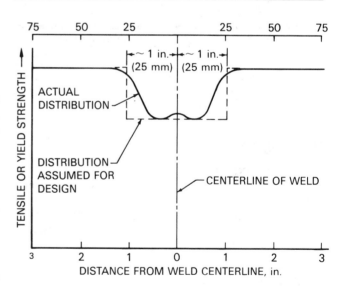

Figure 5.51—Distribution of Tensile or Yield Strength Across a Weld in Aluminum Member to Reduce Stress Concentration at End of Connection

22. Suggested groove weld joint designs are given in the Welding Handbook, Vol. 4, 7th Ed., 1982: and in ANSI/AWS D1.2, Structural Welding Code—Aluminum, latest edition, published by the American Welding Society.

is considered to be a 2 in. (50 mm) wide band with the weld in the center. When joining sheet gages with an automatic welding process, the band will be narrower. The minimum mechanical properties for most commonly used welded aluminum alloys are given in the *Specification for Aluminum Structures* published by The Aluminum Association. The minimum tensile properties for those alloys approved for work covered by *ANSI/ AWS D1.2, Structural Welding Code—Aluminum* are shown in Table 5.14.

Transverse welds in columns and beams should be located at points of lateral support to minimize the effects of welding on buckling strength. The effects of longitudinal welds in structural members can be neglected if the softened zone is less than 15 percent of the total cross-sectional area. Circumferential welds in piping or tubing may reduce bending strength; longitudinal welds usually have little effect on buckling strength when the heat affected zone is a small percentage of the total area of cross section.

With the proper choice of filler metal, a weldment fabricated from a heat-treatable aluminum alloy can be solution heat treated and aged after welding. The welded assembly will regain substantially full strength with some loss in ductility. This is the best method of providing maximum weld strength, but it is usually uneconomical or impractical. The cost of heat treating can be high, especially if a large furnace is required, and the quenching operation may result in unacceptable distortion of the product.

It may be practical, at times, to weld a heat-treatable alloy in the solution-treated condition and age after welding. This can increase the strength over that in the as-welded condition while avoiding the distortion problem associated with solution heat treating.

There is no method of overcoming softening in nonheat-treatable alloys, other than further cold working of the parts after welding, and this is seldom practical. The weakest location in an as-welded assembly is the annealed (or partially annealed) heat affected zone.

STRESS DISTRIBUTION

WHERE WELDS ARE located in critical areas but do not cover the entire cross section, the strength of the section depends on the percentage of the cross-sectional area affected by the heat of welding. When members must be joined at locations of high stress, it is desirable that the welds be parallel to the principal member and to the main stress in that member. Transverse welds in tension members should be avoided.

Frequently, welds are more highly stressed at the ends than in the central portions. To avoid using thicker sections, areas of high stresses in welds can be minimized by sniping. This consists of beveling the end of a member to limit stress concentration in the weld at that end. The weld, however, should wrap around the end of the member. This type of member termination is illustrated in Figure 5.52.

In many weldments, it is possible to locate the welds where they will not be subjected to high stresses. It is frequently possible to make connections between a main member and accessories such as braces by welding at the neutral axis or other point of low stress.

SHEAR STRENGTH OF FILLET WELDS

THE SHEAR STRENGTH of fillet welds is controlled by the composition of the filler metal. Typical shear strengths of longitudinal and transverse fillet welds made with several aluminum filler metals are shown in Figures 5.53 and 5.54, respectively. The highest strength filler metal is alloy 5556. Use of a high-strength filler metal permits smaller welds. For example, assume a longitudinal fillet weld having a strength of 4000 lb/in. (700 kN/m) is desired. If 5356 filler metal is used, a 1/4 in. (6 mm) fillet weld can be applied in a single pass. However, if 4043 filler metal is used, it would require a 3/8 in. (9 mm) fillet weld that would probably require three passes to deposit. The use of the stronger filler metal has obvious economic advantage as it results in lower labor and material costs.

The minimum practical fillet weld sizes depend on the thickness of the base metal and the welding processes and procedure. Minimum recommended fillet weld sizes are shown in Table 5.15. Where minimum weld sizes must be used, a filler metal with the lowest suitable strength for the applied load should be selected to take advantage of the weld metal ductility.

By applying the appropriate safety factor to the shear strength of weld metal, the designer can determine the allowable shear stress in a fillet weld. Appropriate factors of safety and allowable shear stresses in fillet welds for aluminum structures are given in the *Specification for Aluminum Structures* published by The Aluminum Association.

FATIGUE STRENGTH

THE FATIGUE STRENGTH of welded aluminum structures follows the same general rules that apply to fabricated assemblies of other metals. Fatigue strength is governed by the peak stresses at points of stress concentration, rather than by nominal stresses. Eliminating stress raisers to reduce the peak stresses tends to increase the fatigue life of the assembly.

Average fatigue strengths of as-welded joints in small scale specimens of four aluminum alloys are shown in Figure 5.55. These are average test results for butt joints

Table 5.14
Tensile Strength of Welded Aluminum Alloys* (GTAW or GMAW with No Postweld Heat Treatment)

Base Metal Group No.	Alloy and Temper	Product and Thickness Range (in.)	(mm)	Minimum Tensile Strength (ksi)
1	1060 - H12, H14, H16, H18 - H112	All	All	8
1	1100 - H12, H14, H16, H18 - H112	All	All	11
1	3003 - H12,H14, H16, H18 - H112	All	All	14
1	Alclad 3003 - H12,H14, H16, H18 - H112	0.125-0.499 0.500-3.000	3-19 19-75	13 14
2	3004 - H32, H34, H36, H38 - H112	All	All	22
2	Alclad 3004 - H32,H34, H36, H38 - H112	0.125-0.499 0.500-3.000	3-19 19-75	21 22
1	5005 - H12, H14, H16, H18 - H32, H34, H36, H38 - H112	All		14
1	5050 - H32, H34, H36, H38 - H112	All		18
2	5052 - H32, H34, H36, H38 - H112	All		25
4	- H111,H112 5083 - H111 - H112, H116, H321 - H323, H343 - H112, H116, H321	Forgings—up to 4.000 Extrusions Sheet and plate 0.125-1.500 Plate 1.500-3.000	up to 100 3-38 38-75	38 39 40 39
4	5086 - H111, H112, H116 - H32, H34, H36, H38	0.125 to 2.000 2.001 to 3.00	3-50 50-75	35 34
2	5154 - H32, H34, H36, H38 - H112	All	All	30
2	5254 - H32, H34, H36, H38 - H112	All	All	30
2	5454 - H111, H112, H32, H34	All	All	31
4	5456 - H112, H116, H321 - H323, H343 - H112, H116, H321 - H116	Sheet and plate 0.125-1.500 Plate 1.501-3.000 Plate 3.001-4.000	 3-38 38-75 75-100	 42 41 40
3	6006 - T1, T5	Extrusions up to 1.000	up to 25	24
3	6061 - T4, T6, T651	All	All	24
3	Alclad 6061 - T4, T6, T62, T651	All	All	22
3	6063 - T4, T5, T6	All	All	17
3	6351 - T54	All	All	24
5	7005 - T53	All	All	40
6	356.0 - T51, T6, T7	Castings	Castings	23
6	443.0 - F	Castings	Castings	17

*AWS D1.2-83, Structural Welding Code Aluminum.
a. To convert to MPa, multiply the dsi value by 6.895

in 3/8 in. (9 mm) plate welded by the gas metal arc welding process. Specimens were welded on one side, back gouged, and then back welded. The stress ratio of zero means that the tensile stress went from zero to the plotted value and back to zero during each cycle. The effect of the stress ratio on fatigue life is shown in Figure 5.56. This behavior is typical for all metal.[23]

The fatigue strengths of several aluminum alloys are markedly different, and below 10^4 cycles, the designer may prefer one alloy over another for a particular application. However, beyond 10^6 cycles, the differences among various alloys are small. A solution to fatigue problems beyond 10^6 cycle range is primarily found in a change of design rather than by a change of alloy.

The designer should utilize symmetry in the assembly for balanced loading, and should avoid sharp changes in direction, notches, and other stress raisers. The fatigue strength of a groove weld may be significantly increased by removing weld reinforcement or by peening the weld-

23. Dieter, G. *Mechanical metallurgy.* New York: McGraw-Hill; 1961.

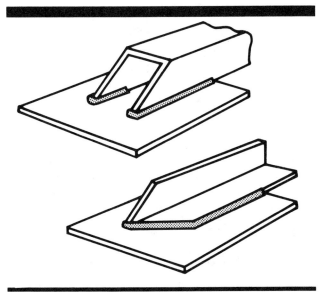

Figure 5.52—Beveling or Sniping the end of a Member to Reduce Stress Concentration at End of Connection

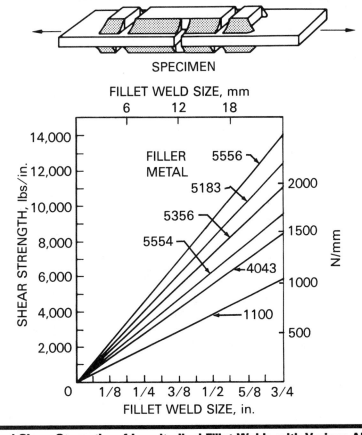

SPECIMEN

Figure 5.53—Typical Shear Strengths of Longitudinal Fillet Welds with Various Aluminum Filler Metals

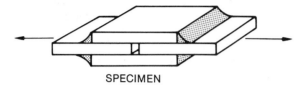

SPECIMEN

Table 5.15
Minimum Fillet Weld Size Aluminum

Base Metal Thickness of Thicker Part Joined (T)		Minimum Size of Fillet Weld*		
in.	mm	in.	mm	
T ≤ 1/4	T ≤ 6.4	1/8**	3	
1/4 < T ≤ 1/2	6.4 < T ≤ 12.7	3/16	5	Single-pass
1/2 < T ≤ 3/4	12.7 < T ≤ 19.0	1/4	6	welds must
3/4 < T	19.0 < T	5/16	8	be used.

* Except that the weld size need not exceed the thickness of the thinner part joined. For this exception, particular care should be taken to provide sufficient preheat to ensure weld soundness.
** Minimum size for dynamically loaded structures is 3/16 in. (5 mm).

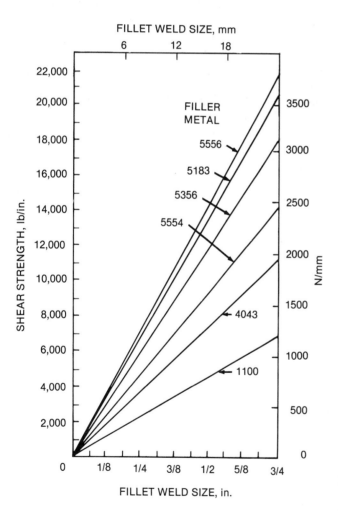

Figure 5.54—Typical Shear Strengths of Transverse Fillet Welds with Various Aluminum Filler Metals

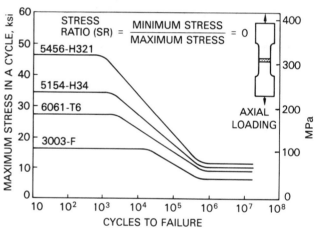

Figure 5.55—Fatigue Test Results of Butt Joints in Four Aluminum Alloys, 3/8 in. Thick Gas Metal Arc Welded Plates

ment. If such procedures are not practical, the weld reinforcement should blend smoothly into the base metal to avoid abrupt changes in thickness. With welding processes that produce relatively smooth weld beads, there is little or no increase in fatigue strength gained by smoothing the weld faces. The benefit of smooth weld beads can be nullified by excessive spatter during welding. Spatter marks sometimes create severe stress raisers in the base metal adjacent to the weld.

While the residual stresses from welding are not considered to affect the static strength of aluminum, they can be detrimental in regard to fatigue strength. Several

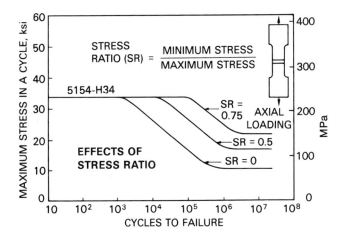

Figure 5.56—Effect of Stress Ratio on Fatigue Life of 5154-H34 Aluminum Weldment

methods can be employed to reduce residual welding stresses, including shot peening, multiple-pin gun peening, thermal treatments, and hydrostatic pressurizing of pressure vessels beyond the yield strength. Shot peening or hammer peening is beneficial when it changes the residual stresses at the weld face from tension to compression for a depth of 0.005 to 0.030 in. (0.1 to 0.8 mm).

Thermal treatments to relieve residual stresses are beneficial. They increase fatigue resistance and provide dimensional stability during subsequent machining. Thermal treatments for nonheat-treatable alloys, such as the 5000 series, can relieve up to 80 percent of the residual welding stresses with little decrease in the static strength of the base metal. Heat-treatable alloys are not so well suited to thermal treatments for relieving residual stresses because temperatures that are high enough to cause significant reductions in residual stress may also substantially diminish strength properties. However, a reduction in residual welding stresses of about 50 percent is possible if a decrease in strength of about 20 percent can be tolerated.

EFFECT OF SERVICE TEMPERATURE

THE MINIMUM TENSILE strengths of aluminum arc welds at various temperatures are shown in Table 5.16. The performance of welds in nonheat-treatable alloys closely follows those of the annealed base metals.

Most aluminum alloys lose a substantial portion of their strength at temperatures above 300°F (150°C). Certain alloys, such as 2219, have better elevated temperature properties, but their applications have definite limitations. The 5000 series alloys with magnesium content of 3.5 percent or higher are not recommended for use at sustained temperatures above 150°F (65°C). Alloy 5454 with its comparable filler metal ER5554 is the strongest of the 5000 series alloys recommended for such applications as hot chemical storage and tank trailers.

Table 5.16
Minimum Tensile Strengths at Various Temperatures of Arc-Welded Butt Joints in Aluminum Alloys

Base Alloy Designation	Filler Metal	Ultimate Tensile Strength, ksi[a]					
		−300°F	−200°F	−100°F	100°F	300°F[b]	500°F[b]
2219-T37[c]	2319	48.5	40.0	36.0	35.0	31.0	19.0
2219-T62[d]	2319	64.5	59.5	55.0	50.0	38.0	22.0
3003	ER1100	27.5	21.5	17.5	14.0	9.5	5.0
5052	ER5356	38.0	31.0	26.5	25.0	21.0	10.5
5083	ER5183	54.5	46.0	40.5	40.0	–	–
5083	ER5356	48.0	40.5	35.5	35.0	–	–
5454	ER5554	44.0	37.0	32.0	31.0	26.0	15.0
5456	ER5556	56.0	47.5	42.5	42.0	–	–
6061-T6[c]	ER4043	34.5	30.0	26.5	24.0	20.0	6.0
6061-T6[d]	ER4043	55.0	49.5	46.0	42.0	31.5	7.0

a. To convert to MPa, multiply ksi by ksi value by 6.895.
b. Alloys not listed at 300°F and 500°F are not recommended for use at sustained operating temperatures of over 150°F.
c. As welded.
d. Heat treated and aged after welding.

Aluminum is an ideal material for low temperature applications. Most aluminum alloys have higher ultimate and yield strengths at temperatures below room temperature. The 5000 series alloys have good strength and ductility at very low temperatures together with good notch toughness. Alloys 5083 and 5456 have been used extensively in pipelines, storage tanks, and marine vessel tankage for handling cryogenic liquids and gases.

SUPPLEMENTARY READING LIST

The Aluminum Association. *Specification for aluminum structures*. Washington, DC: The Aluminum Association, latest edition.

American Institute of Steel Construction, Inc. *Manual of steel construction*. Chicago: American Institute of Steel Construction, Inc. (AISC), latest edition.

American Welding Society. ANSI/AWS D1.1, *Structural welding code—steel*. Miami, FL: American Welding Society, latest edition.

American Welding Society. ANSI/AWS D1.2, *Structural welding code—aluminum*. Miami, FL: American Welding Society, latest edition.

Australian Welding Research Association. *Economic design of weldments*. AWRA Technical Note 8. Australian Welding Research Association (March 1979).

Blodgett, O. W. *Design of welded structures*. Cleveland, OH: Lincoln Arc Welding Foundation (1966).

Canadian Welding Bureau. *Welded structural design*. Toronto, Canada: Canadian Welding Bureau (1968).

Cary, H. B. *Modern welding technology*. Englewood Cliffs, NJ: Prentice-Hall, 1979.

International Institute of Welding. Prevention of lamellar tearing in welded steel fabrication. Prepared by Commission XV of IIW, Great Britain. *Welding World*. Vol. 23, No. 7/8, 1985.

Kaiser Aluminum and Chemical Sales, Inc. *Welding Kaiser aluminum*, 2nd Ed. Oakland, CA: Kaiser Aluminum and Chemical Sales, Inc. (1978).

The Lincoln Electric Company. *Procedure handbook of arc welding*, 12 Ed. Cleveland, OH: The Lincoln Electric Co. (June 1973).

The Lincoln Electric Company. *Solutions to the design of weldments*, D810.17. Cleveland, OH: The Lincoln Electric Co. (Jan. 1975).

Marshall, P. W. *Welding of tubular structures, proceedings of second international conference*. Boston, MA: International Institute of Welding (IIW) (1984).

Rolfe, S. T. Fatigue and fracture control in structures. *AISC Engineering Journal*. 1977: 14 (1).

Rolfe, S. T. and Barsom, J. M. *Fracture and fatigue control in structures: applictions of fracture mechanics*. Englewood Cliffs, NJ: Prentice-Hall (1977).

Sandus, W. W. and Day, R. H. Fatigue behavior of aluminum alloy weldments. *WRC Bulletin*: No. 286, Aug. (1983).

SYMBOLS FOR WELDING AND INSPECTION

PREPARED BY A COMMITTEE CONSISTING OF:

W. L. Green, Chairman
Ohio State University

J. T. Biskup
Canadian Welding Bureau

G. B. Coates
General Electric Company

M. D. Cooper
Hobart School of Technology

E. A. Harwart
Consultant

WELDING HANDBOOK COMMITTEE MEMBER
E. H. Daggett
Babcock & Wilcox

D. R. Spisiak, P. E.
Gaymar Industries Incorporated

CHAPTER 6

SYMBOLS FOR WELDING AND INSPECTION

PURPOSE

STANDARD SYMBOLS ARE used universally to indicate desired welding and brazing information on engineering drawings. They convey the design requirements to the shop in a concise manner. A welding symbol, for example, can be used to specify the type of weld, groove design, weld size, welding process, face and root contours, sequence of operations, length of weld, and other information. However, there are cases where all information cannot be conveyed by a symbol alone. Supplementary notes or dimensional details, or both, are sometimes required to provide the shop with complete requirements. The designer must be sure that the requirements are fully presented on the drawing or specifications.

Nondestructive examination requirements for welded or brazed joints can also be called out with symbols. The specific inspection methods[1] to be used are indicated on the symbols. The appropriate inspection methods depend upon the quality requirements with respect to discontinuities in welded or brazed joints.

The complete system of symbols is described in *ANSI/ AWS A2.4, Standard Symbols for Welding, Brazing, and Nondestructive Examination*, latest edition, published by the American Welding Society. This publication should be referred to when actually selecting the appropriate symbols for describing the desired joint and the inspection requirements. In practice, most designers will use only a few of the many available symbols. The information presented here describes the fundamentals of the symbols and how to apply them.

---

1. Nondestructive testing methods, procedures, and the type of discontinuities that each method will reveal are discussed in: *ANSI/AWS B1.10, Guide for the Nondestructive Inspection of Welds*. Miami, Florida: American Welding Society; latest edition.

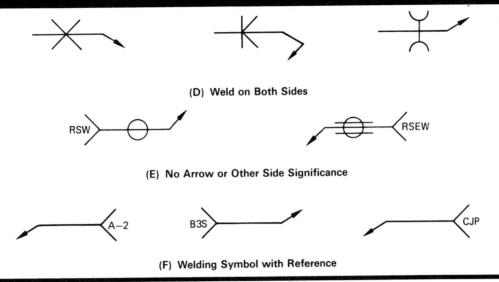

(D) Weld on Both Sides

(E) No Arrow or Other Side Significance

(F) Welding Symbol with Reference

Figure 6.3 (continued)—Significance of Arrow

Some weld symbols have no arrow or other side significance. However, supplementary symbols used in conjunction with these weld symbols may have such significance. For example, welding symbols for resistance spot and seam welding have no side significance, Figure 6.3(E), but GTAW, EBW, or other spot and seam welds may have arrow and other side significance.

References

WHEN A SPECIFICATION, process, test, or other reference is needed to clarify a welding symbol, the reference is placed in a tail on the welding symbol, as shown in Figure 6.3(F). The letters CJP may be used in the tail of the arrow to indicate that a complete joint penetration groove weld is required, regardless of the type of weld or joint preparation. The tail may be omitted when no specification, process, or other reference is required with a welding symbol.

Dimensions

DIMENSIONS OF A weld are shown on the same side of the reference line as the weld symbol. The size of the weld is shown to the left of the weld symbol, and the length of the weld is placed on the right. If a length is not given, the weld symbol applies to that portion of the joint between abrupt changes in the direction of welding or between specified dimension lines. If a weld symbol is shown on each side of the reference line, dimensions are required to be given for each weld even though both welds are identical.

Either US Customary or SI units may be used when specifying dimensions. However, only one of the two should be used for a product or project. Examples of dimensioning for typical fillet welds are shown in Figure 6.4.

If a weld in a joint is to be intermittent, the length of the increments and the pitch (center-to-center spacing) are placed to the right of the weld symbol, as shown in Figure 6.5.

The location on the symbol for specifying groove weld root opening, groove angle, plug or slot weld filling depth, the number of welds required in a joint, and other dimensions are shown in Figure 6.2.

SUPPLEMENTARY SYMBOLS

FIGURE 6.6 SHOWS supplementary symbols that may be used on a welding symbol. They complement the basic symbols and provide additional requirements or instructions.

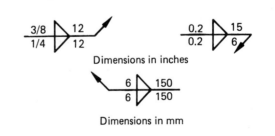

Dimensions in inches

Dimensions in mm

Figure 6.4—Weld Size and Length

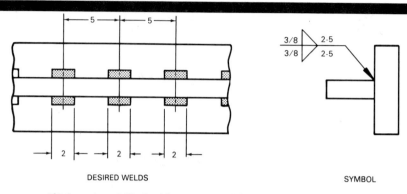

DESIRED WELDS SYMBOL

(A) Length and Pitch of Increments of Chain Intermittent Welds

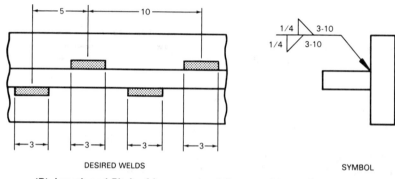

DESIRED WELDS SYMBOL

(B) Length and Pitch of Increments of Staggered Intermittent Welds

Figure 6.5—Dimensioning Intermittent Fillet Welds

Weld-All-Around Symbol

A WELD THAT extends completely around a joint is indicated by the weld-all-around symbol. Figure 6.7(A), (B), and (C) shows examples of its use. The weld can be in more than one plane, as in Figure 6.7(C).

Field Weld Symbol

FIELD WELDS ARE made at the erection site, not in the shop or at the place of initial construction. Each of these welds is designated by a field weld symbol (flag) that is always placed above and at a right angle to the reference line at the junction with the arrow (see Figure 6.8).

Melt-Through Symbol

THE MELT-THROUGH SYMBOL is used to show complete joint penetration (CJP) with root reinforcement on the back side of a weld to be made from one side only. The reinforcement is shown by placing the melt-through symbol on the side of the reference line opposite the weld symbol, as shown in Figure 6.9(A). The height of root reinforcement can be specified to the left of the symbol, as shown in Figure 6.9(B). Control of the root reinforcement height should be consistent with the specified joint design and welding process. The melt-through symbol

differs from the back or backing weld symbol, shown in Figure 6.1, in that the melt-through symbol is filled in.

Consumable Insert Symbol

THE CONSUMABLE INSERT symbol (a square) is placed on the side of the reference line opposite the groove weld symbol. The AWS classification and the class and style of the insert are placed in the tail of the welding symbol (see the latest edition of *AWS A5.30, Specification for Consumable Inserts*). A welding symbol for a typical joint with a consumable insert is shown in Figure 6.10(A).

Backing and Spacer Symbols

A BACKING SYMBOL is placed above or below the reference line to indicate that a backing ring, strip, or bar, is to be used in making the weld. It is used in combination with a groove weld symbol to avoid interpretation as a plug or slot weld. A welding symbol for a typical joint with backing is shown in Figure 6.10(B). It is a combination of a groove weld symbol on one side of the reference line and a backing symbol on the opposite side. The letter *R* may be placed within the backing symbol if the backing is to be removed after welding. The backing

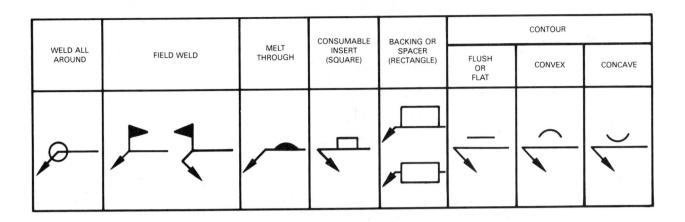

Figure 6.6—Supplementary Symbols

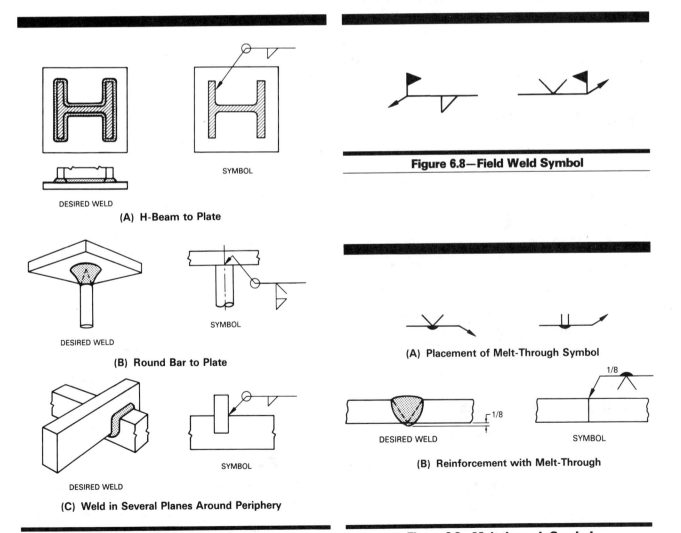

SYMBOL

DESIRED WELD

(A) H-Beam to Plate

SYMBOL

DESIRED WELD

(B) Round Bar to Plate

SYMBOL

DESIRED WELD

(C) Weld in Several Planes Around Periphery

Figure 6.7—Weld-All-Around Symbol

Figure 6.8—Field Weld Symbol

(A) Placement of Melt-Through Symbol

DESIRED WELD

SYMBOL

(B) Reinforcement with Melt-Through

Figure 6.9—Melt-through Symbol

type, material, and dimensions should be specified in the tail of the weld symbol, or elsewhere on the drawing, such as in the Bill of Material.

A welding symbol for a typical joint with a spacer strip inserted in the root of the joint is shown in Figure 6.10(C). It is a modified groove weld symbol having a rectangle within it. The material and dimension of the spacer strip should be specified in the tail of the weld symbol or elsewhere on the drawing, such as in the Bill of Material.

Contour Symbol

A CONTOUR SYMBOL is used on a welding symbol to indicate the shape of the finished weld. Welds that are to be made approximately flat (fillet welds), flush (groove welds), convex, or concave without subsequent finishing are represented by adding the flat, flush, convex, or concave contour symbol to the weld symbol, as shown in Figure 6.11(A). Welds that are to be finished by mechanical means are depicted by adding both the appropriate contour symbol and the user's standard finish symbol to the weld symbol, as in Figure 6.11(B).

CONSTRUCTION OF SYMBOLS

BEVEL-, J-, AND flare-bevel-groove, fillet, and corner-flange weld symbols are constructed with the perpendicular leg always to the left. When only one member of a joint is to be prepared for welding, the arrow is pointed with a definite break toward that member unless the preparation is obvious. The arrow need not be broken if either member may be prepared. These features are illustrated in Figure 6.12. Suggested size dimensions for welding symbol elements are given in *ANSI/AWS A2.4, Standard Symbols for Welding, Brazing, and Nondestructive Testing*, latest edition.

When a combination of welds is to be specified to make a joint, the weld symbol for each weld is placed on the welding symbol. Examples of such symbols are shown in Figure 6.13.

MULTIPLE REFERENCE LINES

TWO OR MORE reference lines may be used with a single arrow to indicate a sequence of operations, as shown in Figure 6.14. The first operation is shown on the reference line nearest the joint, as in Figure 6.14(A). Subsequent operations are shown sequentially on other reference lines joining the arrow. Reference lines may also be used to show data supplementing the welding symbol or to specify inspection requirements.

TYPES OF JOINTS

A JOINT IS the junction of members or the edges of members that are to be joined or have been joined. The five basic joints used in welding and brazing design are butt, corner, lap, edge, and T-joints. These joints are shown in *Appendix A*, Figure A1.

PROCESSES

LETTER DESIGNATIONS ARE used in the tail of a welding symbol to indicate the appropriate welding or brazing process. The more frequently used welding process designations are listed in Table 6.1. A complete listing of designations for welding, brazing, and allied processes is given in *ANSI/AWS A2.4, Standard Symbols for Welding, Brazing, and Nondestructive Examination*, latest edition.

EXAMPLES

AFTER THE JOINT is designed, a welding symbol can generally be used to specify the required welding. Figures 6.15 through 6.21 show examples of welded joints and the proper symbols to describe them. When

Table 6.1
Frequently Used Welding Process Designations

Letter Designation	Welding Process
SMAW	Shielded metal arc welding
SAW	Submerged arc welding
GMAW	Gas metal arc welding
FCAW	Flux cored arc welding
GTAW	Gas tungsten arc welding
PAW	Plasma arc welding
OFW	Oxyfuel gas welding
EBW	Electron beam welding
LBW	Laser beam welding
RSW	Resistance spot welding
RSEW	Resistance seam welding

the desired weld cannot be adequately described with welding symbols, the joint preparation and welding should be detailed on the drawing. Reference is made to the detail in the tail of the welding symbol.

Groove Welds

FOR A SINGLE-V-GROOVE weld, both members are beveled equally to form a groove at the joint. Figure 6.15(A) shows such a weld and the appropriate welding symbol.

If a V-shaped groove is required on both sides of the joint, the weld is a double-V-groove type. The symbol

(A) Consumable Insert Symbol

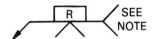

R — BACKING REMOVED
AFTER WELDING

NOTE: MATERIALS AND DIMENSIONS
OF BACKING AS SPECIFIED

(B) Backing Symbol

DOUBLE-BEVEL-GROOVE

DOUBLE-V-GROOVE

NOTE: MATERIAL AND DIMENSIONS OF SPACER AS SPECIFIED

(C) Spacer Symbol

Figure 6.10—Consumable Insert, Backing, and Spacer Symbols

(A) Contour Without Finishing

 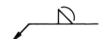

G — GRINDING   M — MACHINING
C — CHIPPING   U — UNISPECIFIED METHOD

(B) Contour With Finishing

Figure 6.11—Contour Symbols

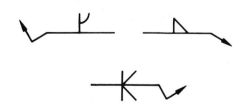

Figure 6.12—Weld Symbol Construction

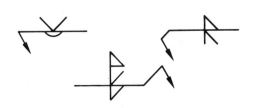

Figure 6.13—Combined Weld Symbols

for a double-V-groove with symmetrical preparation is shown in Figure 6.15(B). The depth of preparation (S) is not shown in the symbol in Figure 6.15(B), and therefore, a joint symmetrical about the plate mid or thickness is required. If an unsymmetrical V-groove geometry is desired, then the depth of preparation (S) must be specified as shown in Figure 6.14(B).

When a round member is placed on a flat surface and a weld is made lengthwise along one side, the weld is a single-flare-bevel-groove type. The weld and the appropriate symbol are shown in Figure 6.16(A). If two round members are placed side by side and welded together lengthwise, the weld is a single-flare-V-groove type. The weld and the applicable symbol are shown in Figure 6.16(B). The round shapes may be bent or rolled plates, pipes, or tubes.

Fillet Welds

JOINTS THAT CAN be joined by fillet welds are lap, corner, and Tee types.[2] Fillet welds are also used in conjunction with groove welds as reinforcement in corner and T-joints. Examples of fillet weld symbols are shown in Figure 6.17.

Plug and Slot Welds

PLUG AND SLOT welds are similar in design but are different in shape. In either case, a hole or slot is made in only one member of the joint. These welds are not to be confused with a fillet weld in a hole. Plug and slot welds require definite depths of filling. An example of a plug weld and the welding symbol is shown in Figure 6.18(A), and of a slot weld and the symbol in Figure 6.18(B). The weld size (diameter at faying surface), angle of countersink, depth of fill, and the center-to-center distance of plug welds may be dimensioned on the welding symbol. Only the depth of fill may be dimensioned on the welding symbol of slot welds. The details of the slot including the orientation and distance between the slots should be in a separate detail and referenced in the tail of the weld symbol.

Flange Welds

A FLANGE WELD is made on the edges of two or more members that are usually light-gage sheet metal. At least one of the members is required to be flanged by bending

it approximately 90 degrees. Examples of flange welds and welding symbols are shown in Figure 6.19(A) and (B).

Spot Welds

A spot weld is made between or upon overlapping members. Coalescence may start and continue over the faying surfaces, or may proceed from the surface of one member. The weld cross section (plan view) is approximately circular. Fusion welding processes that have the capability of melting through one member of a joint and fusing with the second member at the faying surface may be used to make spot welds. Resistance welding equipment is also used. Examples of arc and resistance spot welds are shown in Figure 6.20, together with the proper welding symbols.

Projection Welds

THE WELD SYMBOL for projection welds is the same as that for spot welds except that it is placed above or below the reference line to specify the member to be embossed. The process is indicated in the tail of the welding symbol. A resistance projection weld with the proper welding symbol is shown in Figure 6.21.

Seam Welds

A SEAM WELD is a continuous weld made between or upon overlapping members. Coalescence may start and occur on the faying surfaces, or may proceed from the surface of one member. The continuous weld may be a single weld bead or a series of overlapping spot welds. Seam welds are made with processes and equipment that are similar to those used for spot welding. A means of moving the welding head along the seam must be provided. Examples of arc and resistance seam welds and the appropriate welding symbols are shown in Figure 6.22.

Stud Welds

THE WELD SYMBOL for stud welds is similar to that for spot welds except that the circle contains a cross. The symbol is placed below the reference line and the arrow is pointed to the surface to which the stud is to be welded. The size of the stud is specified to the left of the weld symbol, the pitch is indicated to the right, and the number of stud welds is placed below the symbol in parentheses. Spacing of stud welds in any configuration other than a straight line must be dimensioned on the drawing. Figure 6.23 illustrates the use of the stud weld symbol.

---

2. A fillet weld has an approximate triangular cross section and joins two surfaces at about 90 degrees to each other. When the surfaces are at a greater or lesser angle, the weld should be specified with appropriate explanatory details and notes.

(A) Multiple Reference Lines

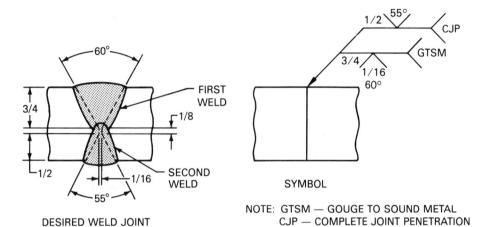

DESIRED WELD JOINT

SYMBOL

NOTE: GTSM — GOUGE TO SOUND METAL
CJP — COMPLETE JOINT PENETRATION

DEPTH OF PERPARATION (ARROW SIDE) — 3/4 in.
DEPTH OF PREPARATION (OTHER SIDE) — 1/2 in.
ROOT OPENING — 1/16 in.
GROOVE ANGLE (ARROW SIDE) = 60°
GROOVE ANGLE (OTHER SIDE) = 55°

(B) Application of Multiple Reference Lines

Figure 6.14—Multiple Reference Lines

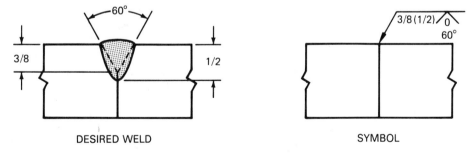

DESIRED WELD

SYMBOL

DEPTH OF PREPARATION — 3/8 in
GROOVE WELD SIZE (ALWAYS SHOWN IN PARENTHESES) — 1/2 in.
ROOT OPENING — 0
GROOVE ANGLE — 60°

(A) Single-V-Groove Weld From Arrow Side

Figure 6.15—Single-V- and Double-V-Groove Welds

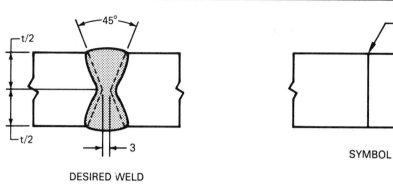

DESIRED WELD

DEPTH OF PREPARATION (EACH SIDE) — t/2
GROOVE ANGLE (EACH SIDE) — 45°
ROOT OPENING — 3 mm

(B) Double-V-Groove Weld (SI Units)

Figure 6.15 (continued)—Single-V- and Double-V-Groove Welds

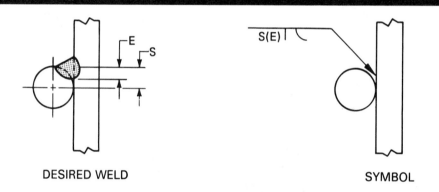

DESIRED WELD   SYMBOL

NOTE: DEPTH OF PREPARATION (S) EQUALS THE RADIUS OF THE ROUND MEMBER

(A) Single-Flare-Bevel-Groove Weld

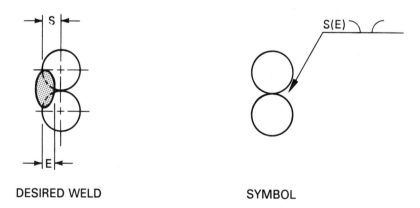

DESIRED WELD   SYMBOL

(B) Single-Flare-V-Groove Weld

Figure 6.16—Flare-Bevel- and Flare-V-Welds

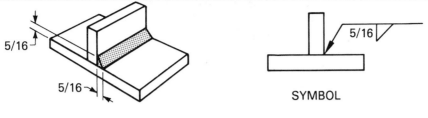

DESIRED WELD

SYMBOL

SIZE OF WELD — 5/16 in.

(A) Fillet Weld with Equal Legs

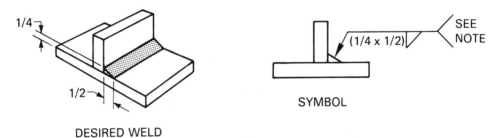

DESIRED WELD

SYMBOL

SIZE OF VERTICAL LEG — 1/4 in.
SIZE OF HORIZONTAL LEG — 1/2 in.

NOTE: VERTICAL LEG TO BE
1/2 HORIZONTAL LEG

(B) Fillet Weld with Unequal Legs

Figure 6.17—Fillet Welds

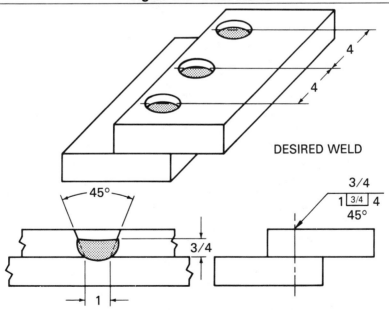

DESIRED WELD

SECTION OF DESIRED WELD

SYMBOL

SIZE — 1 in.
ANGLE OF COUNTERSINK — 45°

DEPTH OF FILLING — 3/4 in.
PITCH (CENTER-TO-CENTER SPACING) — 4 in.

Figure 6.18A—Plug Welds

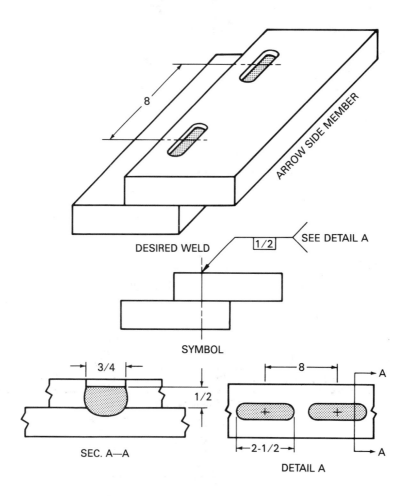

DESIRED WELD

SYMBOL

SEC. A—A

DETAIL A

Figure 6.18B—Slot Welds

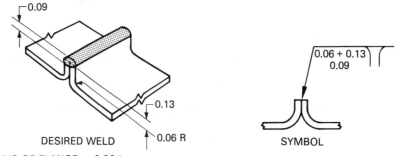

DESIRED WELD

SYMBOL

RADIUS OF FLANGE — 0.06 in.

HEIGHT OF FLANGE ABOVE POINT OF TANGENCY — 0.13 in.

WELD THICKNESS — 0.09 in.

(A) Edge-Flange Weld

Figure 6.19—Flange Welds

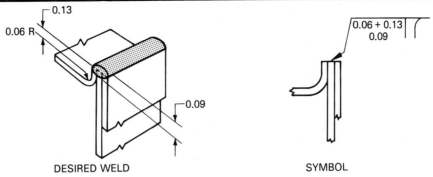

RADIUS OF FLANGE — 0.06 in.

HEIGHT OF FLANGE ABOVE POINT OF TANGENCY — 0.13 in.

WELD THICKNESS — 0.09 in.

(B) Corner-Flange Weld

Figure 6.19 (continued)—Flange Welds

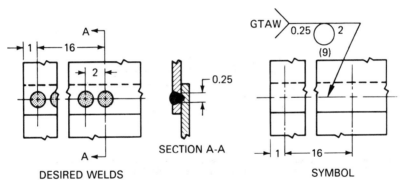

DESIRED WELDS

SIZE (AT FAYING SURFACE) — 0.25 in.

NUMBER OF SPOT WELDS — 9

PITCH (CENTER-TO-CENTER SPACING) — 2 in.

NOTE: SIZE CAN BE GIVEN IN POUNDS OR NEWTONS PER SPOT RATHER THAN THE DIAMETER.

(A) Arc Spot Welds

DESIRED WELDS

SIZE (AT FAYING SURFACE) — 0.25 in.

NUMBER OF SPOT WELDS — 5

PITCH (CENTER-TO-CENTER SPACING) — 1 in.

DISTANCE FROM CENTER OF FIRST SPOT WELD TO EDGE — 1/2 in.

(B) Resistance Spot Welds

Figure 6.20—Spot Welds

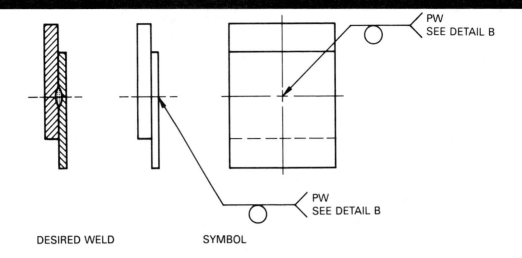

DESIRED WELD

SYMBOL

NOTE: SYMBOL REQUIRES THE
ARROW-SIDE MEMBER TO
BE EMBOSSED.

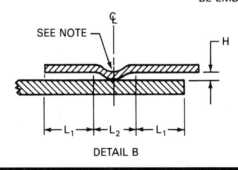

DETAIL B

Figure 6.21—Projection Weld

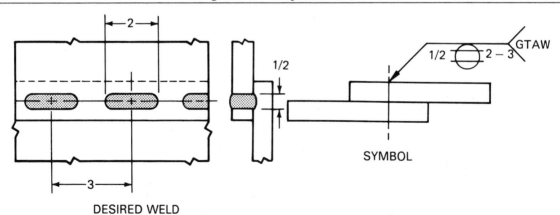

DESIRED WELD

SYMBOL

SIZE (AT FAYING SURFACE) — 1/2 in.

LENGTH — 2 in.

PITCH (CENTER-TO-CENTER SPACING) — 3 in.

NOTE: SIZE CAN BE GIVEN IN POUNDS PER
LINEAR IN. OR NEWTONS PER MILLIMETER.

(A) Arc Seam Weld

Figure 6.22—Seam Welds

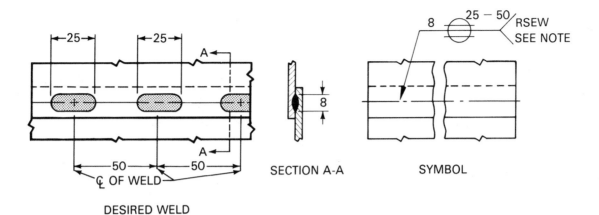

SIZE (AT FAYING SURFACE) — 8 mm

LENGTH — 25 mm

PITCH (CENTER-TO-CENTER SPACING) — 50 mm

NOTE: IF REQUIRED BY ACTUAL LENGTH OF THE JOINT, THE LENGTH OF THE INCREMENT OF THE WELDS AT THE END OF THE JOINT SHOULD BE INCREASED TO TERMINATE THE WELD AT THE END OF THE JOINT.

(B) Resistance Seam Weld (SI Units)

Figure 6.22 (continued)—Seam Welds

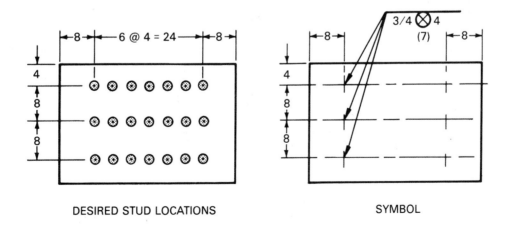

DESIRED STUD LOCATIONS

SYMBOL

Figure 6.23—Stud Weld Symbol for Multiple Rows

BRAZING SYMBOLS

BRAZED JOINTS

WHEN NO SPECIAL preparation other than cleaning is required for making a brazed joint, the arrow and reference line are used with the brazing process designation indicated in the tail, as shown in Figure 6.24(A). Application of conventional weld symbols to brazed joints is illustrated in Figure 6.24(B) through (H). Joint clearance can be indicated on the brazing symbol.

BRAZING PROCESSES

BRAZING PROCESSES TO be used in construction can be designated by letters. The brazing processes and their letter designations are given in Table 6.2.

Table 6.2
Letter Designations for Brazing Processes

Process	Designation
Arc Brazing	AB
Block Brazing	BB
Carbon Arc Brazing	CAB
Diffusion Brazing	DFB
Dip Brazing	DB
Electron Beam Brazing	EBB
Exothermic Brazing	EXB
Flow Brazing	FLB
Furnace Brazing	FB
Induction Brazing	IB
Infrared Brazing	IRB
Laser Beam Brazing	LBB
Resistance Brazing	RB
Torch Brazing	TB

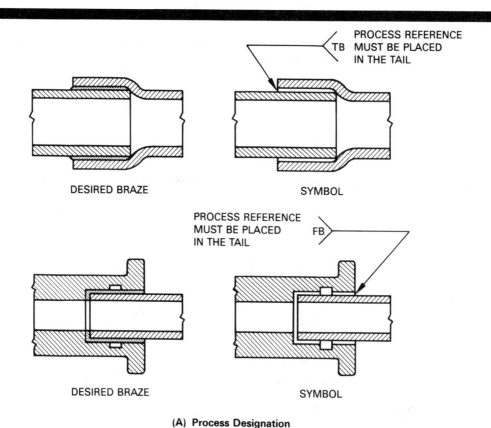

(A) Process Designation

Figure 6.24—Application of Brazing Symbols

CI — CLEARANCE
L — LENGTH OF OVERLAP
S — FILET SIZE

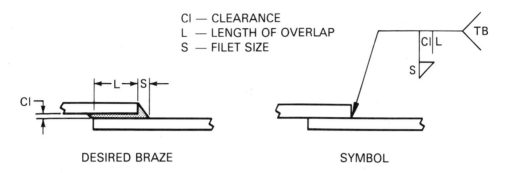

DESIRED BRAZE   SYMBOL

(B) Location of Elements of Brazing Symbol

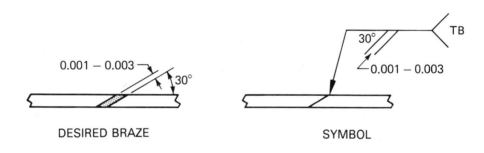

DESIRED BRAZE   SYMBOL

(C) Scarf Joint

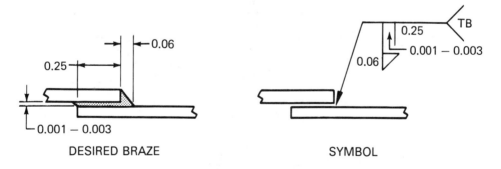

DESIRED BRAZE   SYMBOL

(D) Lap Joint with Fillet

Figure 6.24 (continued)—Application of Brazing Symbols

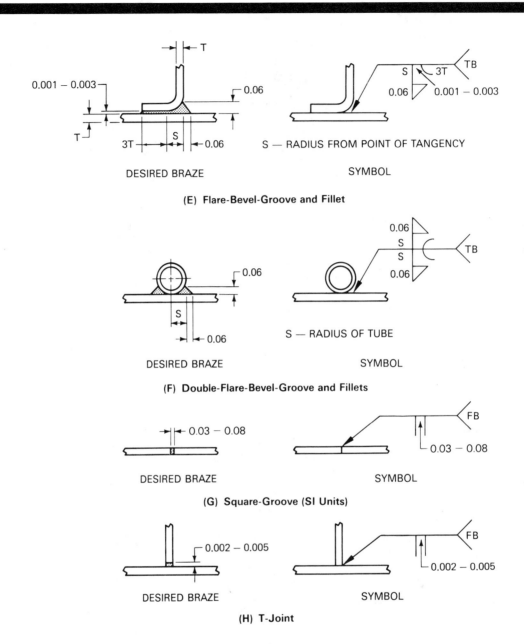

(E) Flare-Bevel-Groove and Fillet

(F) Double-Flare-Bevel-Groove and Fillets

(G) Square-Groove (SI Units)

(H) T-Joint

Figure 6.24 (continued)—Application of Brazing Symbols

NONDESTRUCTIVE EXAMINATION SYMBOLS

SYMBOLS FOR NONDESTRUCTIVE examinations (NDE) provide means for specifying on engineering drawings the method of examination to be used. Nondestructive examination symbols may be combined with welding symbols by using an additional reference line or by specifying the examination method in the tail of the welding symbol.

EXAMINATION SYMBOLS

THE SYMBOLS FOR the various nondestructive examination processes are shown in Table 6.3. The elements of a nondestructive examination symbol are as follows:

(1) Reference line
(2) Arrow
(3) Examination method letter designations
(4) Dimensions, areas, and number of examinations
(5) Supplementary symbols
(6) Tail
(7) Specifications or other references

The standard locations of these elements with respect to each other are shown in Figure 6.25.

Table 6.3
NDE Method Letter Designations

Examination Method	Letter Designation
Acoustic emission	AET
Electromagnetic	ET
Leak	LT
Magnetic particle	MT
Neutron radiographic	NRT
Penetrant	PT
Proof	PRT
Radiographic	RT
Ultrasonic	UT
Visual	VT

Significance of Arrow Location

THE ARROW CONNECTS the reference line to the part to be examined. The side of the part to which the arrow points is the arrow side. The side opposite from the arrow side is the other side.

Arrow Side Examination

EXAMINATIONS TO BE made on the arrow side of a joint are indicated on the NDE symbol by placing the basic examination symbol below the reference line; i.e., toward the reader. Figure 6.26(A) illustrates this type of symbol.

Other Side Examination

EXAMINATIONS TO BE made on the other side of the joint are indicated on the NDE symbol by placing the basic examination symbol above the reference line; i.e., away from the reader. This position is illustrated in Figure 6.26(B).

Examinations on Both Sides

EXAMINATIONS THAT ARE to be made on both sides of the joint are indicated by basic NDE symbols on both sides of the reference line. See Figure 6.26(C).

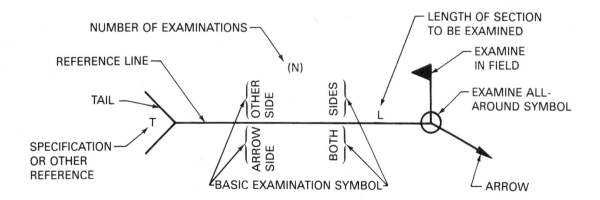

Figure 6.25—Standard Location of NDE Symbol Elements

No Side Significance

WHEN THE EXAMINATION may be performed from either side or has no arrow or other side significance, the basic examination symbols are centered in the reference line. Figure 6.26(D) shows this arrangement.

(A) Examine Arrow Side   **(B) Examine Other Side**

(C) Examine Both Sides   **(D) No Side Significance**

Figure 6.26—Significance of Symbol Placement

Radiographic Examination

THE DIRECTION OF radiation may be shown in conjunction with radiographic (RT) and neutron radiographic (NRT) examination symbols. The direction of radiation may be indicated by a special symbol and line located on the drawing at the desired angle. Figure 6.27 shows the symbol together with the NDE symbol.

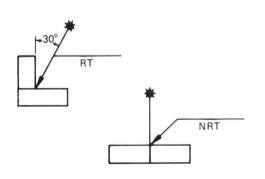

Figure 6.27—Specifying the Direction of Radiographic Examination

Combined Symbols

NONDESTRUCTIVE EXAMINATION SYMBOLS may be combined with welding symbols, as shown in Figure 6.28(A), or with each other if a part is to be examined by two or more methods, as in Figure 6.28(B).

Where an examination method with no arrow or other side significance and another method that has side significance are tobe used, the NDE symbols may be combined as shown in Figure 6.28(C).

REFERENCES

SPECIFICATIONS OR OTHER references need not be used on NDE symbols when they are described elsewhere. When a specification or other reference is used with an NDE symbol, the reference is placed in the tail, as shown in Figure 6.28(D).

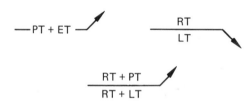

(A) Combined Symbols for NDE and Welding

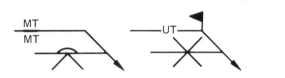

(B) Combined Symbols for Multiple NDE

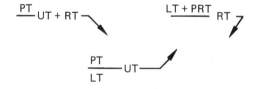

(C) Combined Symbols with Side and No-Side Significance

(D) NDE Symbol with Reference

Figure 6.28—Combined Nondestructive Examination Symbols

EXTENT OF EXAMINATION

Length

TO SPECIFY THE extent of examinations of welds where only the length of a section need be considered, the length is shown to the right of the NDE symbol [see Figure 6.29(A)].

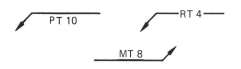

(A) Length of Examination

(B) Exact Location of Examination

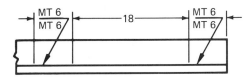

(C) Partial Examination

(D) Number of Examinations

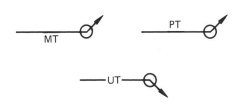

(E) All-Around Examination

Figure 6.29—Extent of Examination

Exact Location. To show the exact location of a section to be examined as well as its length, appropriate dimensions are shown on the drawing [see Figure 6.29(B)].

Partial Examination. When a portion of the length of a weld or part is to be examined with the locations determined by a specified procedure, the percentage of the length to be examined is indicated on the symbol, as illustrated in Figure 6.29(C).

Full Length. The full length of a part is to be examined when no length dimension is shown on the NDE symbol.

Number of Examinations

WHEN SEVERAL EXAMINATIONS are to be made on a joint or part at random locations, the number of examinations is given in parentheses, as shown in Figure 6.29(D).

All-Around Examination

FIGURE 6.29(E) SHOWS the use of the examine-all-around symbol to indicate that complete examination is to be made of a continuous joint, such as a circumferential pipe joint.

Area

NONDESTRUCTIVE EXAMINATION OF areas of parts is indicated by one of the following methods.

Plane Areas. To indicate a plane area to be examined on a drawing, the area is enclosed by straight broken lines with a circle at each change of direction. The type of nondestructive examination to be used in the enclosed area is designated with the appropriate symbol, as shown in Figure 6.30(A). The area may be located by coordinate dimensions.

Areas of Revolution. For nondestructive examination of areas of revolution, the area is indicated by using the examine-all-around symbol and appropriate dimensions. In Figure 6.30(B), the upper right symbol indicates that the bore of the hub is to be examined by magnetic particle inspection for a distance of three inches from the flange face. The lower symbol indicates an area of revolution is to be examined radiographically. The length of the area is shown by the dimension line.

The symbol shown in Figure 6.30(C) indicates that a pipe or tube is to be given an internal proof examination and an external eddy current examination. The entire length is to be examined because no limiting dimensions are shown.

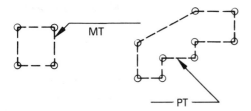

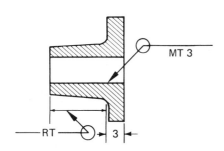

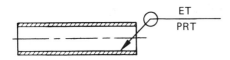

(A) Plane Areas

(B) Area of Revolution — One Side

(C) Area of Revolution — Both Sides

Figure 6.30—Examining Specified Areas

Acoustic Emission

ACOUSTIC EMISSION EXAMINATION is generally applied to all or a large portion of a component, such as a pressure vessel or a pipe. The symbol shown in Figure 6.31 indicates acoustic emission examination of the component without specific reference to locations of the sensors.

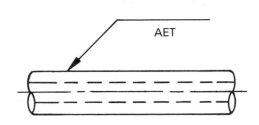

Figure 6.31—Acoustic Emission

RESIDUAL STRESSES AND DISTORTION

PREPARED BY A COMMITTEE CONSISTING OF:

K. Masubuchi, Chairman
Massachusetts Institute of Technology

O. W. Blodgett
Lincoln Electric Company

S. Matsui
Kawasaki Heavy Industries, Ltd.

F. P. Ross
Western Stress, Inc.

C. O. Rudd
Pennsylvania State Univ.

C. L. Tsai
Ohio State University

WELDING HANDBOOK COMMITTEE MEMBER:
L. J. Provoznik
Westinghouse Electric Corporation

CHAPTER 7

RESIDUAL STRESSES AND DISTORTION

RESIDUAL STRESSES

A WELDMENT IS locally heated by most welding processes, therefore, the temperature distribution in the weldment is not uniform, and structural and metallurgical changes take place as welding progresses along a joint. Typically, the weld metal and the heat-affected zone immediately adjacent to the weld are at temperatures substantially above that of the unaffected base metal. As the weld pool solidifies and shrinks, it begins to exert stresses on the surrounding weld metal and heat-affected zones. When it first solidifies, the weld metal is hot, relatively weak, and can exert little stress. As it cools to ambient temperature, however, the stresses in the weld area increase and eventually reach the yield point of the base metal and the heat-affected zone.

When a weld is made progressively, the already solidified portions of the weld resist the shrinkage of later portions of the weld bead. As a result, the portions welded first are strained in tension in a direction longitudinal to the weld, that is, down the length of the weld bead as shown in Figure 7.1. In the case of a butt joint, there commonly is little motion of the weld permitted in the transverse direction because of the weld joint preparation or the stiffening effect of underlying passes. As a result of shrinkage in the weld, there will also be transverse residual stresses, as shown in Figure 7.1. For fillet welds, the shrinkage stresses will be tensile along the length and across the face of the weld, as shown in Figure 7.2. The types and distribution of the stresses in

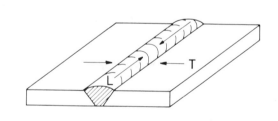

Figure 7.1—Longitudinal (L) and Transverse (T) Shrinkage Stresses in a Butt Weld

welds are complex. A detailed analysis of stresses in single and multiple pass welds and various factors which interact to increase or decrease their magnitude of stresses are discussed here.

Residual stresses in weldments can have two major effects: namely, they may produce distortion or cause premature failure in weldments, or both. Distortion is caused when the heated weld region contracts nonuniformly causing shrinkage in one part of a weld to exert eccentric forces on the weld cross section. The weldment strains elastically in response to these stresses, and detectable distortion occurs as a result of this nonuniform strain. The distortion may appear in butt joints as both longitudinal and transverse shrinkage or contraction, and as angular change (rotation) when the face of the weld shrinks more than the

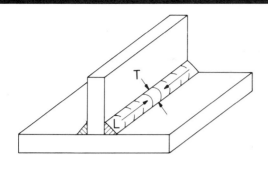

Figure 7.2—Longtudinal and Transverse Shrinkage Stresses in a T-Joint

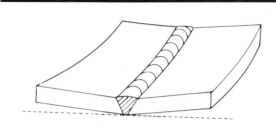

Figure 7.3—Distortion in a Butt Joint

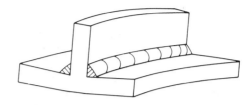

Figure 7.4—Distortion in a T-Joint

root. The latter change produces transverse bending in the plates along the weld length. These effects are illustrated in Figure 7.3.

Distortion in fillet welds is similar to that in butt welds: namely, transverse and longitudinal shrinkage as well as angular distortion result from the unbalanced nature of the stresses in these welds. This behavior is shown in Figure 7.4. Fillet welds are often used in combination with other welds in a weldment. Therefore, the specific resulting distortion may be complex. Distortion in weldments is an important factor in their use, and the procedures used to predict distortion in weldments are discussed subsequently.

Control of distortion can be achieved by a number of techniques. The most commonly used techniques control the geometry of the weld joint, either before or during welding. Examples of these techniques include prepositioning the workpieces prior to welding so that weld distortion leaves them in the desired final geometry, or restraining the workpieces so they cannot distort during welding. Designing and welding the joint so that weld deposits are balanced on each side of the weld centerline is another useful technique. Welding process selection and weld sequence can also influence distortion and residual stress. Some distorted weldments can be straightened after welding, if necessary, by mechanical means or by thermal or flame straightening. These procedures are discussed later in the Chapter.

Residual stresses and distortion affect the fracture behavior of materials by contributing to buckling and brittle fracture when such failures occur at low applied stress levels. When residual stresses and the accompanying distortion are present, buckling may occur at lower compressive loads than would otherwise be predicted. In tension, residual stresses may lead to high local stresses in weld regions of low notch toughness and, as a result, may initiate brittle cracks that can be propagated by low overall stresses that may be present. In addition, residual stresses may contribute to fatigue or corrosion failures.

Residual stress may be reduced or eliminated by both thermal and mechanical means. During thermal stress relief, the weldment is heated to a temperature at which the yield point of the metal is low enough for plastic flow to occur and thus allow relaxation of stress. As a result of thermal stress relief, the mechanical properties of the weldment are usually affected. For example, the brittle fracture resistance of many steel weldments is often improved by thermal stress relief because residual stresses in the weld are reduced and the heat affected zones are tempered. The toughness of the heat affected zones is improved by this procedure. Mechanical stress relief treatments can also reduce residual stresses, but they will not significantly change the microstructure or hardness of the weld or heat affected zone.

Improving the reliability of welded metal structures is very important. Engineers must consider during the design phase the effects of residual stresses and distortion, the presence of discontinuities, the mechanical properties of the weldment, the requirements for nondestructive testing, and the total fabrication and eviction costs.

Increasing knowledge of residual stresses and distortion will not in itself achieve the goal of reducing their effects. Reducing residual stresses and distortion may be achieved through a number of ways including:

(1) Selecting appropriate processes, procedures, welding sequence and fixturing

(2) Selecting the best methods for stress relieving and removing distortion

(3) Selecting design details and materials to minimize the effects of residual stresses and distortion

CAUSES OF RESIDUAL STRESSES

RESIDUAL STRESSES ARE those stresses that would exist in a weldment after all external loads are removed. Various technical terms have been applied to residual stresses such as internal stresses, initial stresses, inherent stresses, reaction stresses, and locked-in stresses. Residual stresses that occur when a structure is subjected to nonuniform temperature change are usually called thermal stresses.

Residual stresses develop in metal structures for many reasons during various manufacturing stages. Such stresses may be produced in many structural components including plates, bars, and sections during casting or mechanical working (rolling, forging, bending). They may also occur during fabrication by welding, brazing, and thermal cutting.

Heat treatments at various stages in manufacture can also influence residual stresses. For example, quenching from elevated temperature can cause residual stresses, while stress relieving heat treatments reduce them.

MACROSCOPIC AND MICROSCOPIC RESIDUAL STRESSES

AREAS IN WHICH residual stresses exist vary greatly from a large portion of a metal structure to areas on the atomic scale. Examples of macroscopic residual stresses are shown in Figure 7.5. When a structure is heated by solar radiation from one side, thermal distortions and thermal stresses are produced in the structure, as shown in Figure 7.5(A). Residual stresses produced by welding are illustrated Figure 7.5(B). The residual stresses are confined to areas near the weld. Figure 7.5(C) shows residual stresses produced by grinding; they are highly localized in a thin layer near the surface.

Residual stresses also occur on a microscopic scale. For example, residual stresses are produced in steels during the martensitic transformation[1] because it takes place at a low temperature and results in the expansion of the metal.

---

1. Martensitic transformation in steel is described in Chapter 4 of this volume.

(A) Thermal Distortion in a Structure Due to Solar Heating

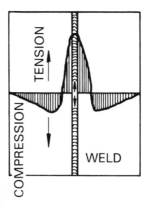

(B) Residual Stresses Due to Welding

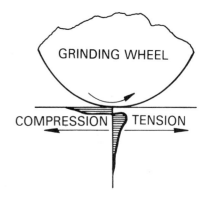

(C) Residual Stresses Due to Grinding

Figure 7.5—Macroscopic Residual Stresses on Various Scales

FORMATION OF RESIDUAL STRESSES

RESIDUAL STRESSES ARE classified according to the mechanisms that produce them: namely, structural mismatching and the uneven distribution of nonelastic strains, including plastic and thermal strains.

Residual Stresses Resulting from Mismatch

A SIMPLE CASE in which residual stresses are produced when bars of different lengths are forcibly connected is shown in Figure 7.6.[2] Tensile stresses are produced in the shorter bar, Q, and compressive stresses are produced in the longer bars, P and P'.

Figure 7.7 shows how a heating and cooling cycle can cause mismatch resulting in residual stresses. Three carbon steel bars of equal lengths and cross sectional areas are connected at the ends by two rigid members. The members and the middle bar are heated to 1100°F (595°C) and then cooled to room temperature while the two outside bars are kept at the room temperature. The stress in the middle bar is plotted against temperature in Figure 7.7 to show how residual stresses are produced. The two side bars resist the deformation of the middle

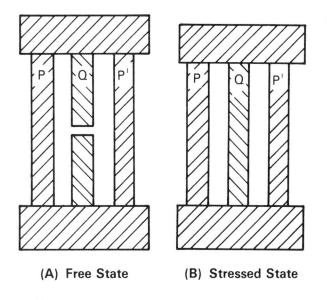

(A) Free State   **(B) Stressed State**

Figure 7.6—Residual Stresses Produced When Bars of Different Lengths are Forcibly Connected

bar. Therefore, the stress in each side bar is equal to half the stress in the middle bar and opposite in direction to the stress in the middle bar.

When the middle bar is heated, compressive stresses are produced in the bar because its expansion is restrained by the side bars. As the temperature of the middle bar increases, the compressive stress increases, as shown by Line AB. The yield strength in compressions is reached at a temperature of approximately 340°F (170°C), as indicated by Point B. As the temperature rises, beyond Point B, the stress in the middle bar is limited to the yield strength which decreases with temperature, as shown by Curve BC. When the middle bar reaches 1100°F (595°C), heating is stopped at Point C.

As the temperature decreases below 1100°F (595°C) on cooling, the action in the middle bar changes to elastic behavior. The stress in the middle bar drops rapidly, changes to tension, and soon reaches the yield strength in tension at Point D. Then, as the temperature decreases further, the stress in the middle bar is again limited to the yield strength at temperature, as shown by Curve DE. Thus, a residual tensile stress equal to the yield strength at room temperature is set up in the middle bar. The residual stresses in the side bars are compressive and equal to one-half of the tensile stress in the middle bar. If the heating of the middle bar is stopped between Points B and C, and allowed to cool to room temperature, tensile stresses would develop elastically along a line parellel to B'E until the yield stress level was reached on curve DE. At room temperature the final residual stress condition would be the same (i.e., residual tensile stress equal to the yield stress in the middle bar and compressive stresses are half that magnitude in the side bars).

Residual Stresses Produced by Unevenly Distributed Nonelastic Strains

WHEN A METAL is heated uniformly, it expands uniformly, and no thermal stresses are produced. On the other hand, if it is heated unevenly, thermal stresses and strains develop in the metal.[3] Residual stresses may also be produced when the object is plastically deformed. The following are fundamental relationships for a plane stress residual stress field[4] ($\sigma_z = 0$):

---

3. Thermal strains are related to the coefficient of linear thermal expansion and the change in temperature.

$$\epsilon''_x = \epsilon''_y = \alpha\Delta T, \ \gamma''_{xy} = 0$$

where α is the coefficient of linear thermal expansion and ΔT is the change of temperature from the initial temperature.

4. In a general three-dimensional stress field ($\sigma \neq O$) six stress components exist:

$\sigma_x, \ \sigma_y, \ \sigma_z, \ \tau_{xy}, \ \tau_{zy}, \ $ and τ_{zx}

2. Masubuchi, K., *Analysis of welded structures*. Pergamon Press (1980); Masubuchi, K., Residual stresses and distortion, *Metals Handbook*, 9th Ed., Vol. 6. Metals Park, OH: American Society for Metals (1983): 856-894.

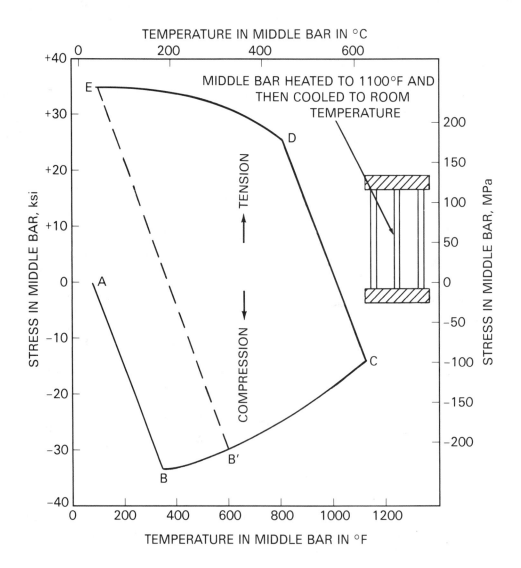

Figure 7.7—Effect of Heating Restrained Bar on the Residual Stresses

(1) Strains are composed of elastic strain and plastic strain:

$$\epsilon_x = \epsilon_x' + \epsilon_x''$$
$$\epsilon_x = \epsilon_y' + \epsilon_y''$$
$$\gamma_{xy} = \gamma_{xy}' + \gamma_{xy}'' \tag{7.1}$$

where

ϵ_x, ϵ_y, γ_{xy} are components of the total strain
ϵ_x', ϵ_y', γ_{xy}' are components of the elastic strain
ϵ_x'', ϵ_y'', γ_{xy}'' are components of the plastic strain

(2) A Hooke's Law relationship exists between stress and elastic strain, thus:

$$\epsilon_x' = \frac{1}{E}(\sigma_x - \nu\sigma_y)$$
$$\epsilon_y' = \frac{1}{E}(\sigma_y - \nu\sigma_x)$$
$$\gamma_{xy} = \frac{1}{G}\tau_{xy} \tag{7.2}$$

(3) The stress must satisfy the equilibrium conditions:

$$\frac{\partial \sigma_x}{\partial x} + \frac{\partial \tau_{xy}}{\partial y} = 0$$

$$\frac{\partial \tau_{xy}}{\partial x} + \frac{\partial \sigma_y}{\partial y} = 0 \qquad (7.3)$$

(4) The total strain must satisfy the condition of compatibility:

$$\left[\frac{\partial^2 \epsilon'_x}{\partial y^2} + \frac{\partial^2 \epsilon'_y}{\partial x^2} - \frac{\partial^2 \gamma'_{xy}}{\partial x \partial y}\right] + \left[\frac{\partial^2 \epsilon''_x}{\partial y^2} + \frac{\partial \epsilon''_y}{\partial x^2} - \frac{\partial^2 \gamma''_{xy}}{\partial x \partial y}\right] = 0 \quad (7.4)$$

Equations 7.3 and 7.4 indicate that residual stresses exist when the value of R, which has been called incompatibility, is not zero.[5] R is determined by the plastic strain using the following equation:

$$R = -\left[\frac{\partial^2 \epsilon''_x}{\partial y^2} + \frac{\partial \epsilon''_y}{\partial x^2} - \frac{\partial \gamma''_{xy}}{\partial x \partial y}\right] \qquad (7.5)$$

The incompatibility can be considered the cause of residual stresses.[6]

Several equations have been proposed to calculate stress components, σ_x, σ_y, and τ_{xy} for given values of plastic strain, ϵ''_x, ϵ''_y, and γ''_{xy}. Conclusions obtained from these mathematical analyses include the following:

(1) Residual stresses in a body cannot be determined by measuring the stress change that takes place when external load is applied to the member.

(2) Residual stresses σ_x, σ_y, and τ_{xy} can be calculated from equation 7.2 when elastic strain components ϵ'_x, ϵ'_y, and γ'_{xy} are determined. However, components of plastic strain ϵ''_x, ϵ''_y, and γ_{xy} which have caused residual stresses, cannot be determined without knowing the history of residual stress formation.

Equilibrium Condition of Residual Stresses

SINCE RESIDUAL STRESSES are not a result of external forces, they cannot produce a resultant force or a resultant moment. Expressed mathematically,

$$\int \sigma dA = 0 \text{ on any plane section} \qquad (7.6)$$

$$\int dM = 0 \text{ about any point} \qquad (7.7)$$

where A = area, and M = the moment.

5. If the plastic strain components are linear functions of the position,
$\epsilon''_x = a + bx + cy$
$\epsilon''_y = d + fx + gy$
$\gamma''_{xy} = k + mx + ny$
then $R = 0$. Consequently residual stresses will not occur.
6. Masubuchi, K., *Analysis of welded structures.* Pergamon Press (1980); Masubuchi, K., Residual stresses and distortion, *Metals Handbook*, 9th Ed., Vol. 6. Metals Park, OH: American Society for Metals (1983): 856-894; Masubuchi, K., Control of distortion and shrinkage in welding. *Welding Research Council Bulletin* 149 (1970).

THERMAL STRESSES AND RESULTING RESIDUAL STRESSES

THE CHANGES IN temperature and stresses during welding are shown schematically in Figure 7.8. A bead-on-plate weld is being deposited along line x-x. The welding arc is moving at a velocity, v, and is presently located at the point, O, as shown in Figure 7.8(A).

Figure 7.8(B) shows the temperature distributions transverse to line x-x at locations A, B, C, and D. Across Section A-A, which is ahead of the welding arc, the temperature change T due to welding is essentially zero. However, the temperature distribution is very steep across Section B-B, through the welding arc. At some distance behind the welding arc, along Section C-C, the temperature distribution is much less severe. At a farther distance from the welding arc, the temperature across Section D-D, has returned to a uniform distribution.

The distribution of stresses σ_x, in the x direction at Sections A-A, B-B, C-C, and D-D is illustrated in Figure 7.8(C). Stress in the y direction, σ_y, and shearing stress, τ_{xy}, also exist in a two-dimensional stress field but are not shown in Figure 7.8.

At Section A-A, thermal stresses due to welding are almost zero. Stresses in regions below the weld pool at section B-B are nearly zero because the hot metal cannot support a load. Stresses in the heat-affected zones on both sides of the weld pool are compressive because the expansion of these areas is restrained by surrounding metal that is at lower temperatures. The metal temperature near the arc is high, and the resulting yield strength is low. The compressive stresses will reach yield level at the temperature of the metal. The magnitude of compressive stress reaches a maximum with increasing distance from the weld (i.e., with decreasing temperature). At some distance away from the weld pool, tensile stresses must balance with those compressive stresses in the heat-affected zones from equilibrium conditions. This balance satisfies equation 7.6. The instantaneous stress distribution along Section B-B is shown in Figure 7.8(C)2.

At Section C-C, the weld metal and heat-affected zones have cooled. As they try to shrink, tensile stresses are induced in the weld metal. The tensile stresses are balanced by compressive stresses in the cooler base metal. The stress distribution is illustrated in Figure 7.8(C)3.

The final condition of residual stress for a bead-on-plate weld is shown in Section D-D. Along Section D-D, high tensile stresses exist in the weld and heat-affected zones, while compressive stresses exist in base metal away from the weld. The final stress distribution of the member is shown in Figure 7.8(C)4.

It is obvious that thermal stresses during welding are produced by a complex series of mechanisms that involve plastic deformation at a wide range of tempera-

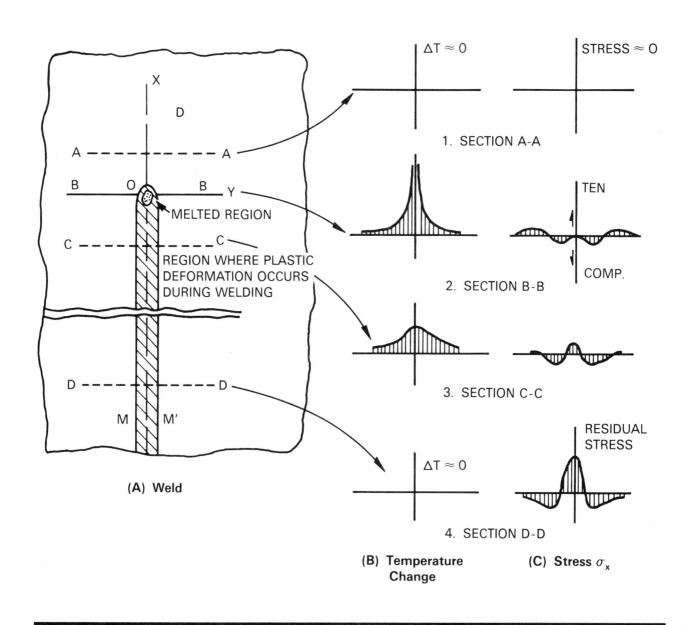

Figure 7.8—Distribution of Temperature and Stress

tures from ambient temperature to melting temperature. Because of the difficulty in analyzing plastic deformation, especially at elevated temperatures, mathematical analyses are presently limited to very simple cases.

Metal Movement During Welding

DURING WELDING, THE weldment undergoes shrinkage and deformation. The transient deformation, or metal movement, is most evident when the weld line is away from the neutral axis of the weldment causing a bending moment.[7] Figure 7.9 shows the deflection of a rectangular metal bar when a longitudinal edge is heated by a moving welding arc or an oxyfuel gas heating torch. The metal near the heat source (the upper regions of the bar)

7. Masubuchi, K., Residual stresses and distortion, *Metals Handbook,* 9th Ed., Vol. 6. Metals Park, OH: American Society for Metals (1983): 856-894.

is heated to higher temperatures than the metal away from the heat source (the lower regions of the bar). The hotter metal expands, and the bar first deforms as shown by Curve AB.

If all of the material remained completely elastic during the entire thermal cycle, any thermal stresses produced during the heating and cooling cycle would disappear when the bar returns to ambient temperature. Then the deflection of the bar would follow Curve AB'C'D', and the bar would be straight after the thermal cycle.

However, in most cases, plastic strains are produced by the thermal cycle, and after cooling, the bar will contain residual stresses. The transient deformation of the bar during the heating and cooling cycle is shown by Curve ABCD. After the bar cools to the ambient temperature, a final deformation δ_f remains. This deformation is called *distortion*.

An interesting characteristic is that the metal movement during welding and the distortion after welding is completed are in opposite directions and generally of the same order of magnitude.

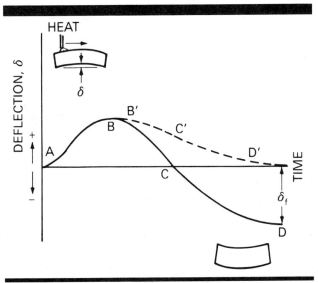

Figure 7.9—Center Deflection of a Metal Bar Under the Influence of a Longitudinal Moving Heat Source

RESIDUAL STRESSES AND REACTION STRESSES IN WELDMENTS

RESIDUAL STRESSES THAT occur during the fabrication of welded structures are classified into (1) residual welding stresses that are produced in the welding of unrestrained members and (2) reaction stresses that are caused by external restraint.

A typical distribution of both longitudinal and transverse residual stresses in a single-pass weld in a butt joint is shown in Figure 7.10. The stresses of concern are those parallel to the welding direction, designated σ_x and those transverse to it, designated σ_y, as shown in Figure 7.10(A).

Figure 7.10(B) shows the distribution of the longitudinal residual stress, σ_x. Residule stresses of high magnitude in tension are produced in the region near the weld; they decrease rapidly to zero over a distance several times the width of the weld metal. Farther away, the residual stresses are compressive in nature. The stress distribution is characterized by two variables: the maximum stress in weld region, σ_m, and the width of the tension zone of residual stress, b. In weldments made in low carbon steel, the maximum residual stress, σ_m, is usually as high as the yield strength of the weld metal. The distribution of longitudinal residual stress is shown in Figure 7.10B, and may be approximated by the following equation.[8]

$$\sigma_x(y) = \sigma_m \left\{ 1 - \left(\frac{y}{b}\right)^2 \right\} \exp\left\{ -\frac{1}{2}\left(\frac{y}{b}\right)^2 \right\} \qquad (7.8)$$

The distribution of the transverse residual stress, σ_y, along the length of the weld is represented by Curve 1 in Figure 7.10(C). Tensile stresses of relatively low magnitude are produced in the middle section of the joint and compressive stresses at both ends of the joint.

If the lateral contraction of the joint is restrained by an external constraint, tensile stresses approximately uniform along the length of the weld are added to the residual stress as reaction stress. This is illustrated by Curve 2, Figure 7.10(C). An external constraint affects the magnitude of the residual stress, but has little influence on the stress distribution.

Residual stresses in the thickness direction, σ_z, can become significant in weldments over about 1 in. (25 mm) thick.

8. Uhlig, H. H. *Corrosion and corrosion control and introduction to corrosion science and engineering.* New York: John Wiley and Sons, 1963.

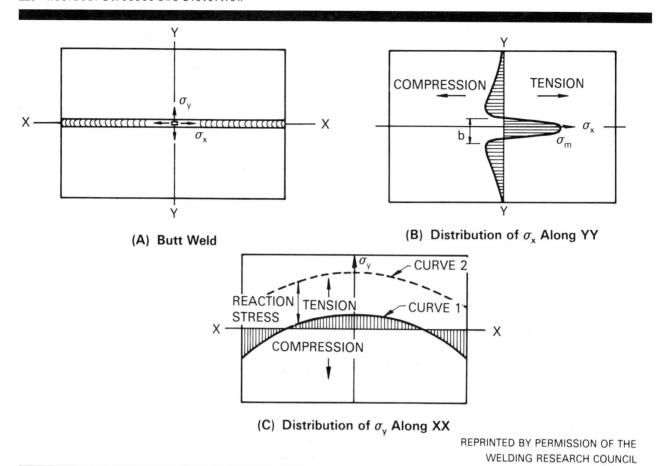

(A) **Butt Weld**

(B) **Distribution of σ_x Along YY**

(C) **Distribution of σ_y Along XX**

Figure 7.10—Typical Distribution of Residual Stresses in a Butt Joint

EFFECTS OF RESIDUAL STRESSES

RESIDUAL STRESSES DECREASE the fracture strength of welded structures only when certain conditions exist; however, the loss of strength can be drastic under these conditions. In general, the effects of residual stresses are significant if fractures can take place under low applied stress.

FRACTURE UNDER TENSILE STRESSES

Changes of Residual Stresses in Weldments Subjected to Tensile Loading

CHANGES IN RESIDUAL stresses when a welded butt joint is subjected to tensile loading is shown in Figure 7.11. Curve A shows the lateral distribution of longitudinal residual stress in the as-welded condition. When uniform tensile stress σ_1 is applied, the stress distribution will be shown by Curve B. The stresses in areas near the weld reach yield stress, and most of the stress increase occurs in areas away from the weld. When the tensile applied stress increases to σ_2, the stress distribution will be as shown by Curve C. As the level of applied stress increases, the stress distribution across the weld becomes more even (i.e., the effect of welding residual stress on the stress distribution decreases.)

When the level of applied stress is further increased, general yielding takes place (i.e., yielding occurs across the entire cross section.) The stress distribution at general yielding is shown by Curve D. Beyond general yielding, the effect of residual stresses on the stress distribution virtually disappears.

The next consideration is the distribution of residual stresses after the tensile loads are released. Curve E shows the residual stress which remains after unloading when the tensile stress σ_1 is applied to the weld and then released. Curve F shows the residual stress distribution

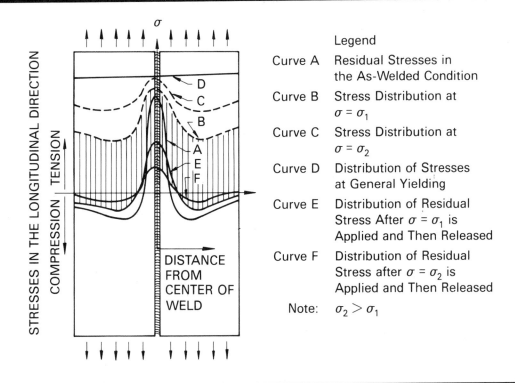

Figure 7.11—Effect of Uniform External Loads on Residual Stress Distribution of Welded Butt Joint

when the tensile stress σ_2 is applied and then released.

Compared to the original residual stress distribution Curve A, residual stress distributions after loading and unloading are less severe. As the level of loading increases, the residual stress distribution after unloading becomes more nearly uniform (i.e., the effect of welding residual stress on the overall stress distribution across the welded joint decreases.)

Based on this analysis, the effects of residual welding stresses may be summarized as follows:

(1) The effect of residual welding stresses on the performance of welded structures is significant only on phenomena which occur under low applied stress, such as brittle fracture, fatigue, and stress corrosion cracking.

(2) As the level of applied stress increases, the effect of residual stresses decreases.

(3) The effect of residual stress is negligible on the performance of welded structures under applied stresses greater than the yield strength.

(4) The effect of residual stress tends to decrease after repeated loading.

Brittle Fracture or Unstable Fracture Under Low Applied Stress

EXTENSIVE STUDIES HAVE been conducted on the effects of residual stresses on brittle fracture of welded steel structures.

Data obtained from brittle fractures in ships and other structures differ from experimental results obtained with notched specimens. Actual fractures have occurred at stresses far below the yield strength of the material. However, the nominal fracture stress of a notched specimen is as high as the yield strength, even when the specimen contains very sharp cracks. Under certain test conditions, complete fracture of a specimen has occurred even though the magnitude of applied stress was considerably below the yield stress of the material.[9]

Figure 7.12 shows the general fracture strength tendencies of welded carbon steel specimens at various tem-

9. Hall, W. S., Kihara, H., Soete, W., and Wells, A. A. *Brittle fracture of welded plate.* Englewood Cliffs, New Jersey: Prentice-Hall, 1967.

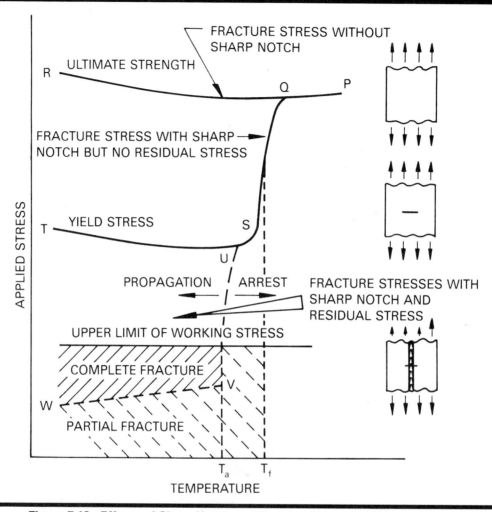

Figure 7.12—Effects of Sharp Notch and Residual Stress on Fracture Strength

peratures, and the effects of a transverse sharp notch and residual stress on fracture strength.[10]

When a specimen does not contain a sharp notch, fracture will occur at the ultimate strength of the material as represented by Curve PQR. When a specimen contains a sharp notch (but no residual stress), fracture will occur at the stresses represented by Curve PQST. When the test temperature is higher than the fracture transition temperature, T_f, a high energy (shear-type) fracture occurs at high stress. When the test temperature is below T_f, the fracture appearance changes to a low energy (cleavage) type, and the stress at fracture decreases to near the yield strength. When a notch is located in areas where high residual tensile stresses exist, these various types of fracture can occur:

(1) At temperatures higher than T_f, fracture stress is the ultimate strength (Curve PQR). Residual stress has no effect on fracture stress.

(2) At temperatures lower than T_f but higher than the crack-arresting temperature, T_a, a crack may initiate at a low stress, but it will not propagate.

(3) At temperatures lower than T_a, one of two phenomena can occur, depending upon the stress level at fracture initiation:

    (a) If the stress is below a critical stress, VW, the crack will stop after running a short distance.

    (b) If the stress is higher than the critical stress, VW, complete fracture will occur.

The effect of residual stress on unstable fracture has been analyzed using fracture mechanics concepts.[11] Such

10. Kihara, H. and Masubuchi, K. Effect of residual stress on brittle fracture. *Welding Journal* 38(1959): 159s-168s.

11. These concepts are described in more detail in Chapter 11 of this volume.

an analysis shows that an unstable fracture can develop even from small cracks that would normally be stable if residual stresses were not present. The sequence of events follows.

When a small subcritical flaw lies in a region free of or containing compressive residual stresses, these stresses do not contribute to the stress intensity at the tip of the discontinuity. If the flaw tip should occur in a region of tensile residual stress, then the residual stress would add to the applied stress and increase the stress intensity around the tip of the discontinuity. The increased stress intensity may cause the flaw to crack, and the crack to extend until the tip is outside the residual stress region. At this point, the crack may arrest or continue to grow, depending on the crack length and the stress intensity. Thus, residual stresses are local stresses and only affect fracture performance within the region of the residual stress field.

BUCKLING UNDER COMPRESSIVE LOADING

FAILURES DUE TO instability or buckling sometimes occur in metal structures composed of slender bars, beams, or thin plates, when they are subjected to compressive axial loading, bending, or torsional loading. Residual compressive stresses decrease the buckling strength of a metal structure and in addition, initial distortions caused by residual stresses also decrease the buckling strength.

Columns Under Compressive Loading

RESIDUAL STRESSES CAN significantly reduce the buckling strength of columns, fabricated by welding, particularly when made from universal mill plate.[12] However, the effects are similar to comparable hot rolled columns. Such columns also have very high residual tensile stresses at the intersections of the flanges and webs, and residual compressive stresses at the outer ends of the flanges.

On the other hand, the columns made from oxygen cut plates normally have residual tensile stresses in the outer areas of the flange. Columns fabricated from flame cut plates and stress relieved columns exhibit better buckling performance, Figure 7.13, than columns fabricated from universal mill plates. In this case, the residual stresses that result from preparing plate by oxy-

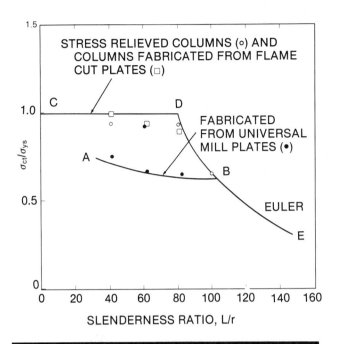

Figure 7.13—Effects of Residual Stresses on Buckling Strength of Columns

gen cutting and from subsequent welding into the column section will combine to counter-balance each other from a buckling strength viewpoint. The performance of these columns should be about equivalent to hot rolled ones. Almost all welded fabricated columns use flame cut plate rather than universal mill plate.

Plate and Plate Structures Under Compressive Loading

RESIDUAL STRESSES IN the direction parallel to the weld line are compressive in regions away from the weld, while they are tensile in regions near the weld [see Figure 7.10(B)]. Consequently, the effects of residual stresses on the buckling strengths of welded structures are complex. Residual stresses may reduce buckling strengths under some conditions and increase them under other conditions.

Both residual stresses and distortion exist in many structures. They create complex effects on the buckling strengths of welded structures.[13] In general, out-of-plane distortion causes considerable reduction in the buckling strength.

12. Kihara, H. and Fujita, Y. The influence of residual stresses on the instability problem. In colloquium on the *Influence of Residual Stresses on Stability of Welded Structures and Structural Members*. London: International Institute of Welding (distributed in the United States by the American Welding Society) (1960); Tall, L., Huber, A. W. and Beedle, L. S. Residual stress and the instability of axially loaded columns. In colloquium on the *Influence of Residual Stresses on Stability of Welded Structures and Structural Members*. London: International Institute of Welding (distributed in the United States by the American Welding Society) (1960).

13. Masubuchi, K. *Analysis of welded structures*. Pergamon Press, 1980.

FATIGUE STRENGTH

FATIGUE STRENGTH (THE number of cycles to fracture under a given load or the endurance limit) increases when a specimen has compressive residual stresses, especially on the specimen surface. For example, experimental studies have shown that local spot heating of certain types of welded specimens increases the fatigue strength.[14] On the other hand, it is possible that residual stresses are relieved during cyclic loading, and their effects on fatigue strength of weldments are negligible.

Surface smoothness is an important chracteristic of fatigue fracture because most fatigue cracks originate at the surface.[15] For example, removing weld reinforcement and smoothing surface irregularities, including undercut, are effective in reducing stress concentrations and increasing the fatigue strength of weldments.[16]

EFFECTS OF ENVIRONMENT

IN THE PRESENCE of hostile environments, residual stresses can cause cracking in metals without any applied loads. Both ferrous and nonferrous alloys can be susceptible to stress corrosion cracking. Examples of sensitive combinations of metals and environments are shown in Table 7.1.

Table 7.1
Environments Producing Stress Corrosion Cracking

Alloy	Sensitive Environment
Low alloy steels	Nitrates, hydroxides, hydrogen sulfide
Chromium stainless steels (greater than 12% Cr)	Halides, hydrogen sulfides, steam
Austenitic stainless steels (18 Cr-8 Ni)	Chlorides, hydroxides
Aluminum alloys	Sodium chloride, tropical environments
Titanium alloys	Red fuming nitric acid, chlorinated hydrocarbons

The susceptibility of high strength steels to hydrogen-induced cracking has been used to indicate the distribu-

tion of residual stresses in welds.[17] The crack pattern along a welded butt joint fabricated from oil quenched-and-tempered SAE 4340 steel plates is shown in Figure 7.14. Longitudinal and transverse cracking are apparent. The longitudinal cracks are in the heat-affected zone, and the transverse cracks are in the heat-affected zone and the base metal. The tensile residual stresses were transverse to the weld in the heat-affected zone, as indicated by the longitudinal cracks. The residual tensile stresses were longitudinal further from the weld, as indicated by the transverse cracks. The transverse cracks are longer and more numerous near the midlength of the weld than near the ends of the weld. This shows that the residual stresses were lower near the ends.

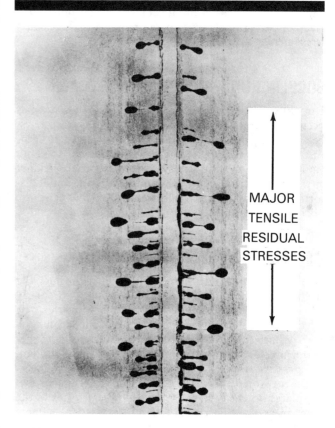

MAJOR TENSILE RESIDUAL STRESSES

Figure 7.14—Crack Pattern in a Simple Butt Joint

14. Gurney, T. R. *Fatigue of welded structures,* 2nd Ed. Cambridge: Cambridge University Press, 1979.

15. Masubuchi, K. *Analysis of welded structures.* Pergamon Press, 1980.

16. Fatigue life is also discussed in Chapters 5, 11, and 12.

17. Masubuchi, K. and Martin, D. C. Investigation of residual stressed by use of hydrogen cracking. *Welding Journal* Part I. 40(12): 1961; 553s–563s; Part II 45(9): 1966; 401s–418s.

MEASUREMENT OF RESIDUAL STRESSES IN WELDMENTS

CLASSIFICATION OF TECHNIQUES FOR MEASURING RESIDUAL STRESSES

MANY TECHNIQUES ARE used for measuring residual stresses in metals. Table 7.2 classifies presently available techniques for measuring residual stresses into the following groups: (1) stress-relaxation, (2) x-ray diffraction, and (3) cracking.[18]

In the stress-relaxation techniques (Group A-1) residual stresses are determined by measuring elastic strain release. This takes place when residual stresses are relaxed by cutting the specimen into pieces or by removing a piece from the specimen. In most cases, electric or mechanical strain gages are used for measuring the strain release. A variety of techniques exists depending upon the methods of sectioning specimens to determine residual stresses. Some techniques are applicable primarily to plates, while others are applicable to cylinders, tubes, or three-dimensional solids.

Strain release during stress relaxation (Group A-2) can be determined using a grid system, brittle coatings, or photoelastic coatings. An inherent disadvantage of the stress-relaxation techniques is that they are destructive; the specimen must be sectioned either entirely or partly. Nevertheless, the stress-relaxation techniques are most widely used for measuring residual stresses in weldments primarily because they provide reliable, quantitative data.

Elastic strains in metals can be determined by measuring the lattice parameter using x-ray diffraction techniques. Since the lattice parameter of a metal in the unstressed state is known, elastic strains in the metal can be determined nondestructively without machining or drilling. Two techniques are available at present: the x-ray film technique and the x-ray diffraction technique.

With the x-ray diffraction techniques, surface strains can be determined in a small area, say to a depth of 0.004 in. (0.01 mm) and an area of several square millimeters. These techniques are the only ones suitable for measuring residual stresses as in ball bearings and gear teeth and also surface residual stresses after machining or grinding.

X-ray diffraction techniques have several disadvantages. First, they are rather slow processes. At each measuring point, measurements must be made in two directions for the film technique.[19] Secondly, measurement is not very accurate, especially when applied to heat treated materials in which the atomic structures are distorted.

---

18. Masubuchi, K. *Analysis of welded structures.* Pergamon Press, 1980.

19. Ruud, C. O. Position-sensitive detector improves x-ray powder diffraction. *I. R. and D. Magazine* (June 1983): 84–88.

Table 7.2
Classification of Techniques for Measuring Residual Stresses

A-1 Stress-relaxation using electric and mechanical strain gages	Techniques applicable primarily to plates	1. Sectioning technique using electric resistance strain gages 2. Gunnert technique 3. Mathar-Soete drilling technique 4. Stablein successive milling technique
	Techniques applicable primarily to solid cylinders and tubes	5. Heyn-Bauer successive machining technique 6. Mesnager-Sachs boring-out technique
	Techniques applicable primarily to three-dimensional solids	7. Gunnert drilling technique 8. Rosenthal-Norton sectioning technique
A-2 Stress-relaxation using apparatus other than electric and mechanical strain gages		9. Grid system-dividing technique 10. Brittle coating-drilling technique 11. Photoelastic coating-drilling technique
B X-ray deffraction		12. X-ray film technique 13. X-ray diffractometer technique
C Cracking		14. Hydrogen-induced cracking technique 15. Stress corrosion cracking technique

Table 7.3
Stress-Relaxation Techniques for Measuring Residual Stresses

Technique	Application	Advantages	Disadvantages
1. Sectioning Technique Using Electric-Resistance Strain Gages	Relative all-around use, with the measuring surface placed in any position.	Reliable method, simple principle, high measuring accuracy.	Destructive. Gives average stresses over the area of the material removed from the plate; not suitable for measuring locally concentrated stresses. Machining is sometimes expensive and time consuming.
2. Mathar-Soete Drilling Technique	Laboratory and field work use and on horizontal, vertical, and overhead surfaces.	Simple principle, causes little damage to the test piece; convenient to use on welds and base metal.	Drilling causes plastic strains at the periphry of the hole which may displace the measured results. The method must be used with great care.
3. Gunnert Drilling Technique	Laboratory and field work use. The surface must be substantially horizontal.	Robust and simple apparatus. Semi-destructive. Damage to the object tested can be easily repaired.	Relatively large margin of error for the stresses measured in a perpendicular direction The underside of the weldment must be accessible for the attachment of a fixture. The method entails manual training.
4. Rosenthal-Norton Sectioning Technique	Laboratory measurements.	Fairly accurate data can be obtained when measurements are carried out carefully.	Troublesome, time-consuming, and completely destructive method.
5. Photoelastic Coating-Drilling Technique	Primarily a laboratory method, but it can also be used for field measurements under certain circumstances.	Permits the measurement of local stress peaks. Causes little damage to the material.	Sensitive to plastic strains which sometimes occur at the edge of the drilled hole.

Techniques are available for studying residual stresses by observing cracks in the specimen (Figure 7.14). The cracks may be induced by hydrogen or stress corrosion. The cracking techniques are useful for studying residual stresses in complex structural models which have complicated residual stress distributions. However, they provide qualitative rather than quantitative data.

MEASUREMENT BY STRESS-RELAXATION TECHNIQUES

THE STRESS-RELAXATION techniques are based on the fact that strains taking place during unloading are elastic even when the material has undergone plastic deformation. Therefore, it is possible to determine residual stresses without knowing the history of the material.

There are five techniques for measuring residual stresses based on stress-relaxation techniques that can be used for weldments. The first two techniques apply primarily to thin plates; the next two techniques apply primarily to solids. The first four techniques employ electric or mechanical strain gages, while the fifth one employs a photoelastic coating.[20] The range of applica-

tion and the advantages and disadvantages of each of the five techniques are summarized in Table 7.3.

Sectioning Technique for a Plate Using Electric-Resistance Strain Gages

WITH THIS TECHNIQUE, electric-resistance strain gages are mounted on the surface of the test structure or specimen.[21] A small piece of metal containing the gages is then removed from the structure, as shown in Figure 7.15. The strains that take place during the removal of the piece are determined. The piece removed is small enough so that the remaining residual stress may be neglected, and therefore, the measured strains are the elastic strains resulting from the residual stress distribution of the specimen. Stated mathematically

20. Masubuchi, K. Nondestructive measurement of residual stresses in metals and metal structures. RSIC—410, Redstone Arsenal, Ala-bama: Redstone Scientific Information Center. U. S. Army Missile Command, April 1965.

21. In the resistance-type bonded strain gage technique, gages made from either metallic wire or foil materials are bonded on the specimen. As the specimen is strained, the resistance of the gages changes, and the magnitude of strain is determined by measuring the resistance change. Information on electric strain gages is available from various sources including the *Handbook of Experimental Stress Analysis*.

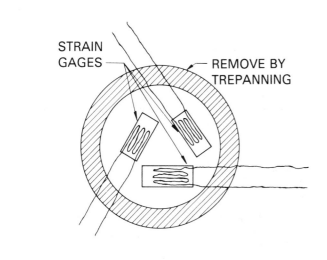

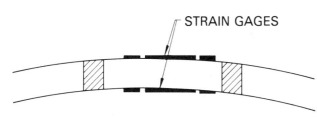

**Figure 7.15—Complete Stress-Relaxation Technique
Applied to a Plate**

It is advisable to make strain measurements on both surfaces of the plate because residual stresses may be caused by bending. The mean value of strains measured on both surfaces represents the normal stress component, while the difference between the strains on both surfaces represents the bending stress component.

The Mathar-Soete Drilling Technique

WHEN A SMALL circular hole is drilled in a plate containing residual stresses, those stresses in areas outside the hole are partially relaxed. It is possible to determine residual stresses that existed in the drilled area by measuring stress relaxation in areas outside the drilled hole.[22] The hole method of measuring stress was first proposed and used by Mathar and was further developed by Soete.

A common way to determine residual stresses is to place strain gages at 120 degrees from each other, and drill a hole in the center, as shown in Figure 7.16. The magnitudes and directions of the principle stresses are calculated by measuring strain changes at the three gages.

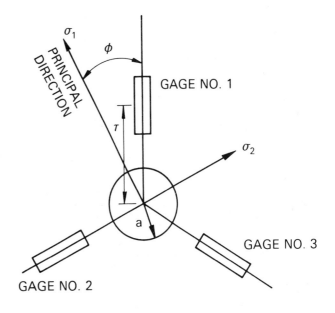

**Figure 7.16—120 Degree Star Arrangement of Strain
Gages for the Mather-Soete Drilling Technique**

$$\bar{\epsilon}_x = -\epsilon'_x$$
$$\bar{\epsilon}_y = -\epsilon'_y$$
$$\bar{\gamma}_{xy} = -\gamma'_{xy} \qquad (7.9)$$

where, ϵ'_x, ϵ'_y, and γ_{xy} are the elastic strain components of the residual stress and $\bar{\epsilon}_x$, $\bar{\epsilon}_y$, and $\bar{\gamma}_{xy}$ are the measured values of the strain components. The negative sign indicates that the strains are in opposite directions. Then residual stresses can be determined by substituting equations 7.9 into equations 7.3 and solving the simultaneous equations for σ_x, σ_y, and τ_{xy}. Thus:

$$\sigma_x = \frac{-E}{1 - \nu^2}\left[\bar{\epsilon}_x + \nu\bar{\epsilon}_y\right]$$
$$\sigma_y = \frac{-E}{1 - \nu^2}\left[\bar{\epsilon}_y + \nu\bar{\epsilon}_x\right]$$
$$\tau_{xy} = -G\bar{\gamma}_{xy} \qquad (7.10)$$

22. Soete, W. Measurement and relaxation of residual stresses. *Welding Journal* 28(8): 1949; 354s–364s.

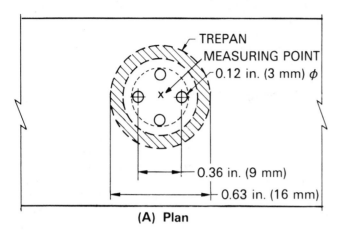

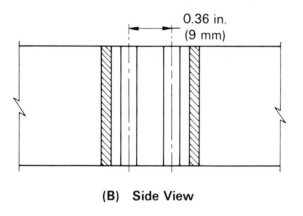

Figure 7.17—Gunnert Drilling Method

The Gunnert Drilling Technique

FOUR 0.12 IN. (3 mm) holes located on a circle with a 0.35 in. (9 mm) diameter are drilled through the plate perpendicular to the surface at the measuring point, as shown in Figure 7.17. The diametrical distance between these holes at different levels below the surface of the plate is measured by means of a specially designed mechanical gage. The perpendicular distance between the plate surface an the gage location at different levels below the surface is also measured. A groove 0.63 in. (16 mm) from the measuring point is then trepanned around the holes in steps, and the same measurements are made. Formulas are given to calculate residual stresses in the interior of the specimen from the measurement data.

The Rosenthal-Norton Sectioning Technique

ROSENTHAL-NORTON PROPOSED A technique for determining residual stresses in a thick weldment.[23] Two narrow blocks with full thickness of the plate, one parallel to the weld and the other transverse to the weld, are cut from the weld. Then any residual stresses still left in the narrow blocks are measured. Formulas are used to estimate residual stresses in the interior of the weldments from (1) strain changes which take place while cutting the narrow blocks and (2) residual stresses which are left in the blocks.

TYPICAL RESIDUAL STRESSES IN WELDMENTS

DISTRIBUTION IN TYPICAL WELDS

Welded Butt Joint

A TYPICAL DISTRIBUTION of both longitudinal and transverse residual stresses in a butt joint welded in a single pass has been discussed previously and is shown in Figure 7.10.

Plug Weld

FIGURE 7.18 SHOWS the distribution of residual stresses in a circular plug weld. In the weld and surrounding areas, tensile stresses as high as the yield strengths of the materials were produced in both radial and tangential directions.[24] In areas away from the weld, radial stresses, σ_r, were tensile and tangential stresses, t_ϕ, were

23. Rosenthal, D. and Norton, T. "A Method for measuring residual stresses in the interior of a meterial". *Welding Journal 24(5)*, Research Supplement, 295s–307s; 1945.

24. Kihara, H., Watanabe, M., Masubuchi, K., and Satoh, K. *Researches on welding stress and shrinkage distortion in Japan*. 60th Anniversary Series of the Society of Naval Architects of Japan, Vol. 4. Tokyo: Society of Naval Architects of Japan, 1959.

compressive; both stresses decreased with increasing distance r from the weld.

Welded Beam and Column Shapes

TYPICAL DISTRIBUTIONS OF residual stresses in welded shapes are shown in Figure 7.19.

Figure 7.19(A) shows residual stresses and distortion produced in a welded T-section. In a section some distance from the end of the column, Section X-X, high tensile residual stresses in the direction parallel to the axis are produced in areas near the weld. In the flange, stresses are tensile in areas near the weld and compressive in areas away from the weld. Tensile stresses in areas near the upper edge of the web are a result of the longitudinal bending distortion of the shape caused by the longitudinal shrinkage of the weld. Angular distortion may also take place.

Figures 7.19(B) and (C) show typical distributions of residual stresses in H Sections and a welded box section, respectively. Residual stresses shown are parallel to the axis; they are tensile in areas near the welds and compressive in areas away from the welds.

Welded Pipes

THE DISTRIBUTION OF residual stresses in a welded pipe is complex. In a girth-welded pipe, for example, shrinkage of the weld in the circumferential direction induces both shearing force, Q, and bending moment, M, to the pipe, as shown in Figure 7.20. The angular distortion caused by welding also induces a bending moment. Distribution of residual stresses is affected by the following:

(1) Diameter and wall thickness of the pipe
(2) Weld joint design
(3) Welding procedure and sequence

Figure 7.21 shows the residual stresses resulting from a circumferential weld in a low carbon steel pipe.[25] The pipe was 30 in. (760 mm) diameter by 7/16 in. (11 mm) wall thickness. Residual stresses were determined using the Gunnert technique.

Residual Stresses in Aluminum and Titanium Weldments

Aluminum Alloy Weldments. The distribution of residual stress in a 36 in. (914 mm) wide by 1/2 in. (13 mm) thick by 48 in. (1219 mm) long aluminum gas

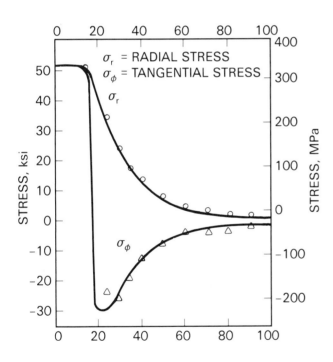

Figure 7.18—Effect of Distance from Center of Plug Weld on Radial and Tangential Residual Stresses

25. Burdekin, F. M. Local stress relief of circumferential butt welds in cylinders. *British Welding Journal.* 10(9): 1963; 483–490.

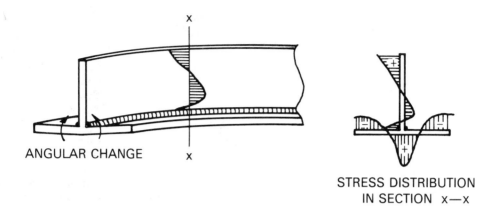

x

ANGULAR CHANGE x

STRESS DISTRIBUTION
IN SECTION x—x

(A) Residual Stresses and Distortion of a Welded T-Shape

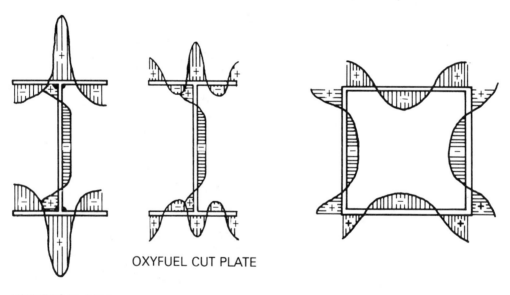

OXYFUEL CUT PLATE

UNIVERSAL MILL

Note: + Indicates Tensile Stress
      – Indicates Compressive Stress

**(B) Residual Stresses in
Welded H-Shapes**

**(C) Residual Stresses in a
Welded Box Shape**

Figure 7.19—Typical Residual Stresses in Welded Shapes

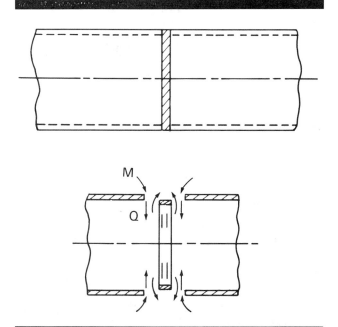

Figure 7.20.—Reaction Forces in a Girth Weld of a Pipe

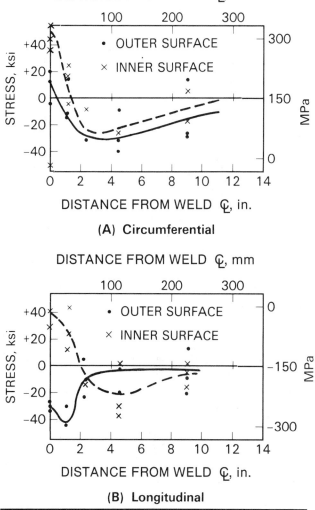

(A) Circumferential

(B) Longitudinal

Figure 7.21—Residual Stresses in a Girth Weld in Low Carbon Steel Pipe

metal arc welded plate is shown in Figure 7.22. The results are typical of longitudinal residual stresses in aluminum welds. The low residual stress at the weld centerline results because the weld and heat affected zone are annealed by the welding heat, and therefore, the low strength material relieves the stresses by plastically yielding.

Titanium Alloy Weldments. The distribution of residual stress in an edge welded titanium alloy is shown in Figure 7.23. The base metal was 56 in. (1,442 mm) long by 7.5 in. (191 mm) wide by 1/2 in. (12.7 mm) thick and was welded on one edge with the gas metal arc process. The experimental results agreed reasonably well with analytical predictions and may therefore be considered typical for titanium alloys.

EFFECTS OF SPECIMEN SIZES ON RESIDUAL STRESSES

WHEN STUDYING RESIDUAL stresses in a welded specimen, it is important that the specimen is large enough to contain residual stresses as high as those which exist in actual structures.

Effect of Specimen Length

TO STUDY THE effect of weld length on the residual stresses in unrestrained low carbon steel welded butt joints, two series of welds were prepared by the sub-

merged arc process and the shielded metal arc process. The welding conditions are summarized in Figure 7.24. In each series, the only variable was the length of the weld as given in Figure 7.25. The width of each specimen was sufficient to assure that full restraint was applied. Figure 7.25 shows the distribution of longitudinal and transverse residual stresses along the welds made by the two processes.[26]

Longitudinal residual stresses must be zero at both ends of the welds, and high tensile stresses exist in the central regions of the welds. The peak stress in the central region increases with increasing weld length. This

26. DeGarmo, E. P., Meriam, J. L. and Jonassen, F. The effect of weld length upon the residual stresses of unrestrained butt welds. *Welding Journal* 25(8): 1946; 485s–486s.

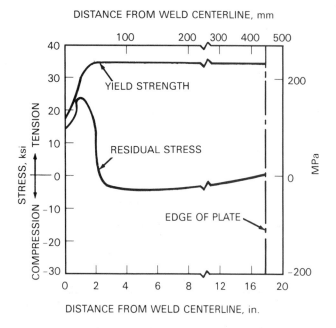

Figure 7.22—Distribution of Yield Strength and Longitudinal Residual Stresses in an Aluminum Alloy Weldment

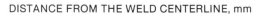
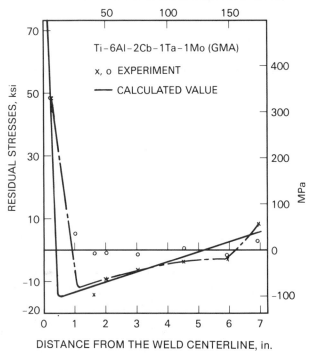

Figure 7.23—Distribution of Residual Stresses in a Titanium Alloy Weldment

effect is clearly shown in Figure 7.26 in which the peak stress for each panel is plotted versus the weld length. The figure indicates that welds longer than 18 in. (457 mm) are needed to produce maximum residual tensile stresses in the longitudinal direction. Longitudinal residual stresses become uniform in the central region for welds longer than 18 in. (457 mm).

Concerning the transverse residual stresses shown in Figure 7.24, stresses were tensile in central areas and compressive in areas near the plate ends. The weld length had little effect on the maximum tensile stresses in the central area and the maximum stress in areas near the plate ends.

Residual stress distributions were similar in welds made by the submerged arc process and the shielded metal arc process. Smooth stress distributions were obtained in welds made by the submerged arc process, while stress distributions in welds made by the shielded metal arc process were somewhat uneven.

Effect of Specimen Width

COMPARED WITH THE effect of specimen length discussed above, the effect of the specimen width on residual stresses is very small, as long as the specimen is long enough. In fact, the effect of specimen width is negligible when the width is greater than several times the width of the weld and heat affected zones.

RESIDUAL STRESSES IN HEAVY WELDMENTS

WHEN A WELDMENT is made in plate over 1 in. (25 mm) thick, residual stresses can vary significantly through the plate thickness.[27] Figure 7.27 shows distributions of residual stress in the three directions in the weld metal of a butt joint 1 in. thick, (25 mm) low-carbon steel plates. Welds were made with covered electrodes 1/8 to 3/16 in. (3 to 5 mm) in diameter; welding was sequenced alternately on both sides so that angular distortion Would be minimized. Residual stress measurements were obtained by using the Gunnert drilling technique. As shown in Figure 7.27(A) and (B), longitudinal and transverse stresses were tensile in areas near both surfaces of the weld. Compressive stresses in the interior of the weld apparently were produced during the welding of top and bottom passes.

Figure 7.27(C) shows the distribution of stresses (σ_z) normal to the plate surface. At both surfaces, σ_z must be zero. Residual stresses were primarily compressive below the surface.

27. Gunnert, R. Method for measuring tri-axial residual stresses. *Welding Research Abroad* 4(10): 1958; 17–25.

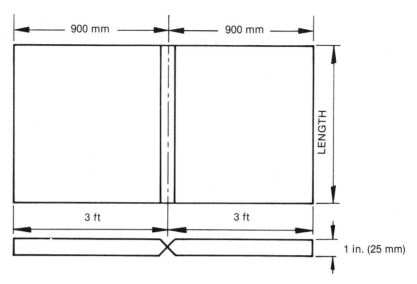

(A) Submerged Arc Process

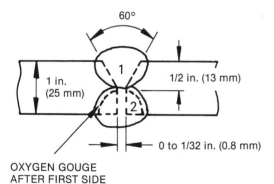

60°

1 in.
(25 mm)

1/2 in. (13 mm)

0 to 1/32 in. (0.8 mm)

OXYGEN GOUGE
AFTER FIRST SIDE
WELDED

(B) Shielded Metal Arc Process, (Double-V)

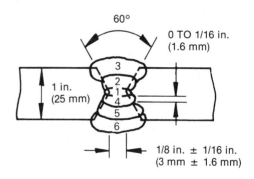

60°

0 TO 1/16 in.
(1.6 mm)

1 in.
(25 mm)

1/8 in. ± 1/16 in.
(3 mm ± 1.6 mm)

Electrode:	EH14 1/4 in. (6 mm) F62-EH14-200
Voltage:	32
Current:	1050 A
Speed of Arc Travel:	~12.5 in./min (~5.2 mm/s)
Details:	Pass 1: Oxygen Gouge Back Side
	Pass 2.

Electrode:	Passes 1 & 4:	5/32 in. (4 mm) E6010
	Passes 2, 3, 5, 6:	1/4 in. (6 mm) E6012
Current:	Root Passes:	150–165 A
	Other passes:	300–320 A
Details:	Passes 1, 2, 3:	Back Chip
	Passes 4, 5, 6.	

Figure 7.24—Welding Conditions for Residual Stress Measurements Shown in Figure 7.25 and 7.26

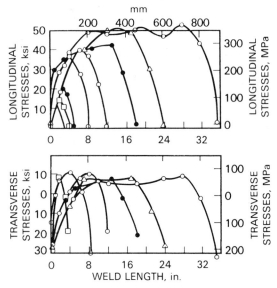

(A) Submerged Arc Welds

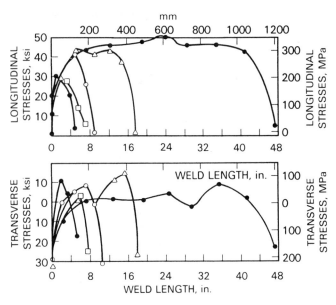

(B) Shielded Metal Arc Welds

Figure 7.25—Effect of Length on Residual Stress Distribution in Weldments

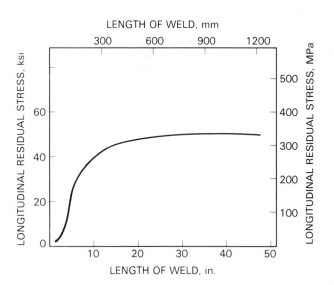

Figure 7.26—Effect of Length of Weld on Maximum Longitudinal Residual Stress

stresses and distortion. The selection of a proper welding sequence is important, especially when welding joints with high restraint, such as repair welds.

The effects of welding sequence on residual stresses and shrinkage in restrained butt welds and circular patch welds may be summarized as follows:[28]

(1) Welding sequence has little effect on the residual stresses along the weld. However, longitudinal tensile stresses will likely be relatively high.

(2) Differences in welding sequence can result in considerable differences in transverse shrinkage, the total elastic energy stored in restrained joints, and the reaction stresses in the inner plates of circular welds.

(3) Block welding generally results in less shrinkage, less strain energy, and less reaction stress than multilayer sequences.

28. Kihara, H., Watanabe, M., Masubuchi, K., and Satoh, K. *Researches on welding stress and shrinkage distortion in Japan.* 60th Anniversary Series of the Society of Naval Architects of Japan. Vol. 4 Tokyo: Society of Naval Architects of Japan (1959); Jonassen, F., Meriam, J. L. and DeGarmo, E. P. Effect of certain block and other special welding procedures on residual welding stresses. *Welding Journal* 25 (9): 492s–496s (1946); Weck, R. Transverse contractions and residual stresses in butt welded mild steel plates. Report R4, London: Admiralty Ship Welding Committee (Jan. 1947); Kihara, H., Masubuchi, K., and Matsuyama, Y. Effect of welding sequence on transverse shrinkage and residual stresses. Report 24. Tokyo: Transportation Technical Research Institute. (Jan. 1957).

EFFECTS OF WELDING SEQUENCE ON RESIDUAL STRESSES

WHEN WELDING A long butt joint, various types of welding sequences, such as backstep, block, built-up, and cascade, may be used in an attempt to reduce residual

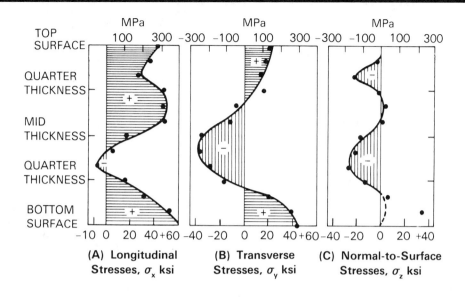

Figure 7.27—Distribution of Longitudinal, Transverse, and Short Transverse Residual Stresses Through the Thickness of a 1 in. (25 mm) Thick Steel Plate

RESIDUAL STRESSES OF WELDS MADE BY DIFFERENT WELDING PROCESSES

IT IS LIKELY that residual stresses in welds made by various processes, including the shielded metal arc, submerged arc, gas metal arc, and gas tungsten arc processes, are similar. For example, Figures 7.24(A)

and (B) show that the residual stresses in welds made by shielded metal arc welding and submerged arc welding, are similar. However, this is true when the design and relative size of the weldments are similar. When the joint design and weld size are significantly different, magnitudes and distributions of residual stresses may also change.

WELD DISTORTION

FUNDAMENTAL TYPES OF DISTORTION

THE DISTORTION FOUND in fabricated structures is caused by three fundamental dimensional changes that occur during welding:

(1) transverse shrinkage that occurs perpendicular to the weld line, (2) longitudinal shrinkage that occurs parallel to the weld line, and (3) an angular change that consists of rotation around the weld line. These dimensional changes are illustrated in Figure 7.28.

Figure 7.28(A) shows transverse shrinkage in a simple welded butt joint. The distribution of the longitudinal residual stress, σ_x, discussed earlier, is shown again in Figure 7.28(B). This stress causes longitudinal shrinkage. This shrinkage of the weld metal and base metal regions adjacent to the weld is restrained by the

surrounding base metal.

Figure 7.28(C) shows the angular change that occurs in a butt joint. Nonuniformity of transverse shrinkage in the thickness direction causes this rotation. Figure 7.28(D) shows the angular change that occurs in a fillet weld. Here, the angular change is caused by the unbalance of shrinkage on opposite sides of the flange member.

Shrinkage and distortion that occur during the fabrication of actual structures are far more complex than illustrated in Figure 7.28. When longitudinal shrinkage occurs in a fillet welded joint, the joint will bend longitudinally unless the weld line is located along the neutral axis of the joint (this problem will be discusssed later). Whether or not a joint is constrained externally also affects the magnitude and form of distortion.

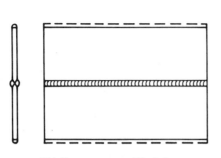

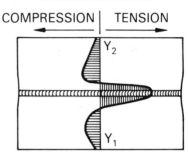

RESIDUAL STRESS

COMPRESSION | TENSION

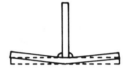

(A) Transverse Shrinkage in a Butt Joint

(B) Longitudinal Shrinkage in a Butt Joint (Distribution of Longitudinal Residual Stress, σ_x, is Also Shown)

(C) Angular Change in a Butt Joint

(D) Angular Change in a Fillet Weld

REPRINTED BY PERMISSION OF THE WELDING RESEARCH COUNCIL

Figure 7.28—Fundamental Dimensional Changes that Occur in Weldments

TRANSVERSE SHRINKAGE IN BUTT JOINTS

IN FIGURE 7.28(A), the shrinkage is uniform along the weld. However, transverse shrinkage that occurs in welded butt joints, in actual structures, is usually much more complex. The major factors causing nonuniform transverse shrinkage in welded butt joints are (1) rotational distortion during welding and (2) joint restraint.

The amount of transverse shrinkage of carbon and low alloy steel welds may be estimated using the following equation:

$$S = 0.2 \frac{A_w}{t} + 0.05\, d \qquad (7.11)$$

where

S = transverse shrinkage, in.
A_w = cross sectional area of weld, in.
t = thickness of plates, in.
d = root opening, in.

Rotational Distortion of Butt Joints

ROTATIONAL DISTORTION OF butt joints during welding is affected by both heat input and welding speed.[29] When mild steel plates are joined by shielded metal arc welding, the unwelded portion of the joint tends to close because of the relatively low welding speeds of the process. This is shown in Figure 7.29(A). When similar steel plates are joined with the submerged arc welding process, the unwelded portion of the joint tends to open, as shown in Figure 7.29(B). This is a result of thermal expansion of the heated edges ahead of the arc. To maintain desired root opening, tack welds for a submerged arc welded joint must be large enough to withstand stresses caused by the rotational distortion as welding progresses along the joint.

29. Masubuchi, K. Analytical investigation of residual stresses and distortions due to welding. *Welding Journal* 39 (12): 525s–537s (1960).

The amount of transverse shrinkage that occurs in welds is affected by the degree of restraint applied to the joint. External restraint may be considered to act like a system of transverse springs; the degree of restraint is expressed by the rigidity of the system of springs. The amount of shrinkage decreases as the degree of restraint increases. In many joints, the degree of restraint is not uniform along the weld. For example, in the slit weld specimen, shown in Figure 7.30, the degree of restraint varies along the length of the slit, and is highest at the ends.[30] Consequently, the amount of transverse shrinkage after welding is greater at the center of the slit than at the ends. When a long butt joint is welded using a back-step sequence, the transverse shrinkage is not uniform along the weld.

The rotational distortion is affected by welding heat input and the location of tack welds. The welding sequence produces complex effects on the rotational distortion and the distribution of restraint along the weld.

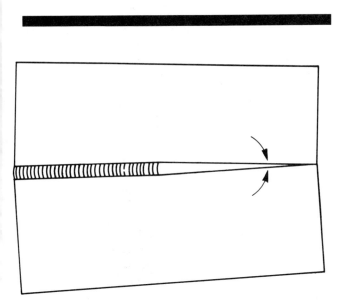

(A) Unwelded Portion of the Joint Closes (in Shielded Metal Arc Welds)

Mechanisms of Transverse Shrinkage in a Single-Pass Butt Joint

THE MAJOR PORTION of transverse shrinkage of a butt joint welded in a single pass is a result of contraction of the base metal. The base metal expands during welding. When the weld metal solidifies, the expanded base metal must shrink, and this shrinkage accounts for the major part of transverse shrinkage. Shrinkage of the weld itself is only about 10 percent of the actual shrinkage.[31]

Figures 7.31(A) and (B) show experimental results obtained on butt joints in low-carbon steel.[32] The curves labeled T show the temperature changes, and those labeled S show the changes of transverse shrinkage. Most of the shrinkage occurs after the weldment has cooled down to relatively low temperatures. The figure shows that in thicker plate, transverse shrinkage starts earlier, but the final value of the shrinkage is smaller. It should be noted, however, that this is true only when the same amount of heat input is always used regardless of the joint thickness. Welding thicker plates normally requires more than one pass.

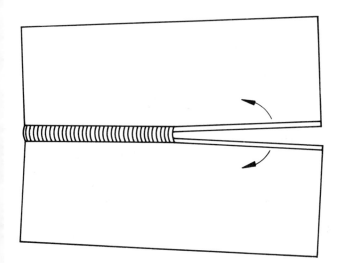

(B) Unwelded Portion of the Joint Opens (Submerged Arc Welds)

30. Welds similar to the slit weld are frequently made in repairs. Transverse shrinkage in the slit weld is constrained by the base metal surrounding the weld.

31. Naka, T. *Shrinkage and cracking in welds.* Tokyo: Komine Publishing Co., (1950); Matsui, S. "Investigation of shrinkage, restraint stress, and cracking in arc welding". Ph. D. Thesis. Osaka University (1964); Iwamura, Y. "Reduction of transverse shrinkage in aluminum butt welds". M.S. Thesis. M.I.T. (1974).

32. Matsui, S. "Investigation of shrinkage, restraint stress, and cracking in arc welding". Ph. D. Thesis. Osaka University (1964).

Figure 7.29—Rotational Distortion

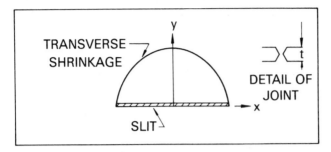

SLIT-WELD SPECIMEN AND TYPICAL
DISTRIBUTION OF SHRINKAGE

REPRINTED BY PERMISSION OF
THE WELDING RESEARCH COUNCIL

**Figure 7.30—Transverse Shrinkage in a Restrained
Butt Joint**

Transverse Shrinkage During Multipass Welding

DURING MULTIPASS WELDING of constrained butt joints
in carbon steel transverse shrinkage increases with each
weld pass.[33] Shrinkage is relatively pronounced during
the first weld passes but diminishes during later passes,
as shown in Figure 7.32(A). The resistance to shrinkage
increases as the weld metal cross section increases. A linear relationship exists between total transverse
shrinkage and the logarithm of the total weight of the
weld metal deposited, as shown in Figure 7.32(B). The
equation for this relationship is

$$u = u_o + b \, (log \, w - log \, w_o) \qquad (7.12)$$

where,

u = total transverse shrinkage
u_o = transverse shrinkage after first pass
w = total weight of weld metal
w_o = weight of first pass weld metal
b = a coefficient

Figure 7.33 suggests three methods for reducing transverse shrinkage:

(1) Decrease the total weight of weld metal, as shown
by line B-C.
(2) Decrease the tangent (b in Equation 7.12) which
decreases the amount of shrinkage from B to D.
(3) Deposit a larger first pass changing the shrinkage
from A to A'. The amount of shrinkage after the completion of the weld changes from B to E.

The effects of various factors on transverse shrinkage,
including joint design, root opening, type and size of
electrodes, degree of constraint, peening, and oxygen
gouging, are summarized in Table 7.4.[34] The relationship of these factors to the above methods for reducing
shrinkage is also given in the table. Root opening and
joint design produce the greatest effects.

Figure 7.34(A) shows the effect of root opening on
transverse shrinkage. As the root opening decreases, the
shrinkage and the total amount of weld metal decreases.

Figure 7.34(B) shows the effect of electrode size on
transverse shrinkage. Shrinkage decreases as the electrode size increases. However, reduction of shrinkage
cannot be obtained with large size electrodes unless they
are also used for the first pass.

Regarding the effect of welding heat input on transverse shrinkage, shrinkage decreases as the total heat
input required to weld a certain joint is decreased. However, when a weld is completed in several passes,
shrinkage decreases when the first pass is welded with a
greater heat input. In this case, an increase in shrinkage
is produced after welding the first pass by using a greater
heat input. (Compare points A' and A on Figure 7.33.)
However, the amount of shrinkage after the completion
of the weld decreases from B to E on Figure 7.33.

As to the effect of chipping and gouging on transverse
shrinkage, removal of the weld metal by chipping produces little effect on shrinkage, and shrinkage increases
due to rewelding. Since heat is applied to the weld area
during oxygen gouging, it increases shrinkage. Then,
shrinkage is increased further during repair welding.[35]

33. Kihara, H. and Masubuchi, K. Studies on the shrinkage and
residual welding stress of constrained fundamental joint. *Journal of
the Society of Naval Architects of Japan.* Part I, 1954—95: 181–195;
Part II, 1955—96: 99–108; Part III, 1955—97: 95–104 (in Japanese).

34. Kihara, H. and Masubuchi, K. Studies on the shrinkage and
residual welding stress of constrained fundamental joint. *Journal of
the Society of Naval Architects of Japan.* Part I, 1954—95: 181–195;
Part II, 1955—96: 99–108; Part III, 1955—97: 95–104 (in Japanese).

35. Kihara, H. and Masubuchi, K. Studies on the shrinkage and
residual welding stress of constrained fundamental joint. *Journal of
the Society of Naval Architects of Japan. Part I, 1954—95: 181–195;
Part II, 1955—96: 99–108; Part III, 1955—97: 95–104 (in Japanese).*

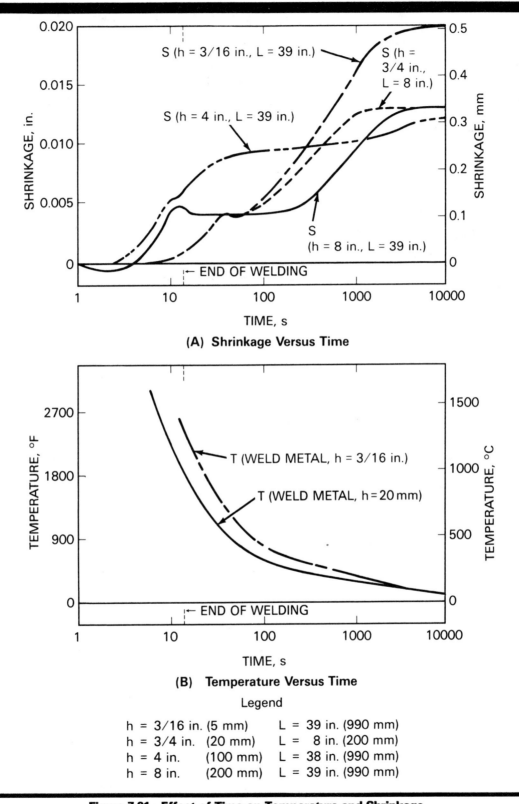

(A) Shrinkage Versus Time

(B) Temperature Versus Time

Legend

h = 3/16 in. (5 mm)	L = 39 in. (990 mm)
h = 3/4 in. (20 mm)	L = 8 in. (200 mm)
h = 4 in. (100 mm)	L = 38 in. (990 mm)
h = 8 in. (200 mm)	L = 39 in. (990 mm)

Figure 7.31—Effect of Time on Temperature and Shrinkage

Table 7.4
Effects of Several Procedure Variables on Transverse Shrinkage in Butt Joints

Procedures	Effects
Root opening	Shrinkage increases as root opening increases. See Fig. 7.33. Effect is great. (Methods 1 and 2.)*
Joint design	A single-V-joint produces more shrinkage than a double-V-joint. Effect is great. (Methods 1 and 2.)
Electrode diameter	Shrinkage decreases by using larger sized electrodes. See Fig. 7.33. Effect is medium.
Degree of constraint	Shrinkage decreases as the degree of constraint increases. Effect is medium. (Method 2.)
Electrode type	Effect is minor. (Method 2.)
Peening	Shrinkage decreases by peening. Effect is minor. (Method 2.)

Effects of Joint Restraint on Transverse Shrinkage of Butt Joints

IN ORDER TO study quantitatively the effect of joint restraint on the transverse shrinkage of a welded butt joint, it is first necessary to define analytically the degree of restraint of a joint. As a simple example, Figure 7.35 shows a butt joint under restraint. Points A and A' are fixed unmoved during welding. When transverse shrinkage occurs, it causes reaction stress as follows:

$$\sigma = E \left(\frac{u}{B} \right) \tag{7.13}$$

where
   σ = reaction stress, psi (Pa)
   E = Modulus of elasticity, psi (Pa)
   u = transverse shrinkage, in. (mm)
   B = width of the joint in. (mm)

The degree of restraint of the joint, ks, may be defined as the amount of reaction stress caused by a unit amount of transverse shrinkage, as follows:

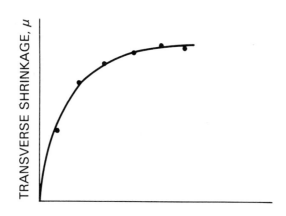

(A) Increase of Transverse Shrinkage in Multipass Welding

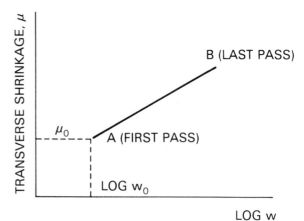

(B) Relationship Between Log w and μ

Figure 7.32—Effect of Weld Metal Deposit Weight on Transverse Shrinkage in Multipass Butt Joints

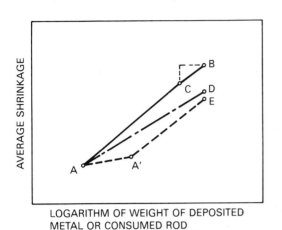

Figure 7.33—Methods for Reducing Transverse Shrinkage

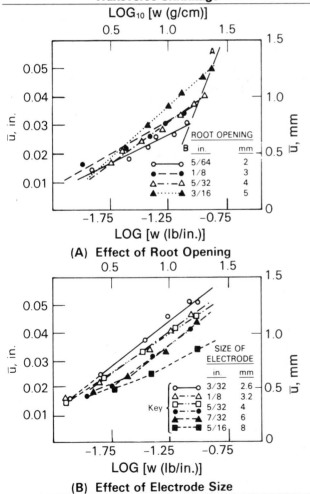

(A) Effect of Root Opening

(B) Effect of Electrode Size

Figure 7.34—Effect of Root Opening and Electrode Diameter on Shrinkage of Ring Specimen (Figure 7.36)

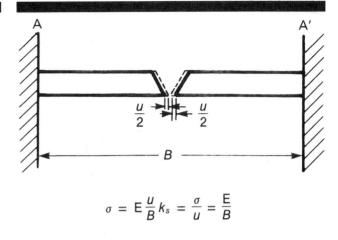

$$\sigma = E\frac{u}{B} \quad k_s = \frac{\sigma}{u} = \frac{E}{B}$$

Figure 7.35—Definition of Degree of Restraint, k_s, of a Simple Butt Joint

$$k_s = \frac{\sigma}{u} = \frac{E}{B} \tag{7.14}$$

By definition, the joint restraint has the dimension of psi/in. (Pa/mm). The joint restraint is inversely proportional to the size of deformable portions of the joint. For example, when the distance between the fixed points A and A′ are increased from B to 2B, the degree of restraint will be decreased to one half.

Figure 7.36 shows definitions of the degree of restraint of three types of butt joints, including the circular-ring type specimen, the slit-type specimen, and the H-type constraint specimen. Figure 7.37 shows the effects of degree of restraint on the transverse shrinkage obtained with the three types of butt joints.[36] The ordinate is the ratio of the transverse shrinkage of a restrained joint, S_t, and the transverse shrinkage of a unrestrained joint, S_{tf}.[37] By using k_s and the S_t/S_{tf} experimental data obtained by different investigators using different types of joints are aligned on one curve.

TRANSVERSE SHRINKAGE OF FILLET WELDS

THE TRANSVERSE SHRINKAGE across a fillet weld is much less than across a butt joint. Shrinkage of transverse fillet welds in carbon and low alloy steels may be estimated by the following simple formulas:

36. Watanabe, M. and Satoh, K. Effect of welding conditions on the shrinkage and distortion in welded structures. *Welding Journal* 40(8): 1961; 377s–384s.

37. Masubuchi, K., *Analysis of Welded Structures*. Pergamon Press, 1980.

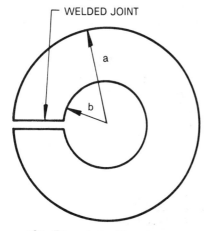

(A) Ring-type Specimen

$$k_s = \frac{E}{4\pi} \cdot \frac{1}{b-a}\left(\text{LOG}_e\frac{b}{a} - \frac{b^2-a^2}{b^2+a^2}\right)$$

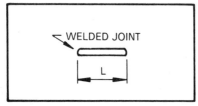

(B) Slit-type Specimen

$$k_s = \frac{2E}{\pi L}$$

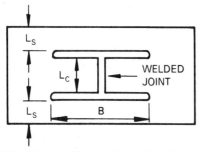

(C) H-type Restrained Specimen

$$k_s = \frac{E}{B\left(1+\dfrac{L_c}{2L_s}\right)}$$

Figure 7.36—Effect of Specimen Type on the External Constraint (k_s)

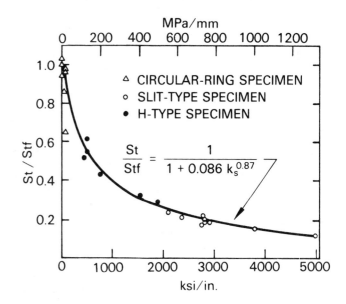

Figure 7.37—Effect of External Constraint on the Transverse Shrinkage of Butt-Welded Joints

(1) For T-joints with two continuous fillets:

$$S = C_1\left(\frac{D_f}{t_b}\right) \qquad (7.15)$$

where

S = shrinkage, in. or mm
D_f = fillet leg length, in. or mm
t_b = thickness of bottom plate, in. or mm
C_1 = 0.04 and 1.02 when S, L, and t_b are in inches and millimeters, respectively.

(2) For intermittent fillet welds, a correcting factor of proportional length of fillet weld to total length of joint should be used.

(3) For fillet welds in a lap joint between plates of equal thickness (two welds):

$$S = C_2\left(\frac{D_f}{t}\right) \qquad (7.16)$$

where

S = shrinkage, in or mm
D_f = fillet leg length, in. or mm
t = plate thickness, in. or mm
C_2 = 0.06 and 1.52 when S, L, and t are in inches and millimeters, respectively.

LONGITUDINAL SHRINKAGE OF BUTT JOINTS

THE AMOUNT OF longitudinal shrinkage in a butt joint is about 1/1000 of the weld length, much less than the transverse shrinkage. The following equation has been proposed for the longitudinal shrinkage in a butt weld.[38]

$$\Delta L = \frac{C_3 I L}{t} 10^{-7} \qquad (7.17)$$

where

ΔL = the longitudinal shrinkage
I = welding current, A
L = length of weld, in. or mm
t = plate thickness in. or mm
C_3 = 12 and 305 when L and t are in inches and millimeters, respectively.

ANGULAR CHANGE OF BUTT JOINTS

ANGULAR CHANGE OFTEN occurs in a butt joint when transverse shrinkage is not uniform in the thickness direction.

Figure 7.38 shows experimental data gathered on ring-type specimens, discussed earlier. A radial groove was cut and then welded using 1/4 in. (6 mm) diameter covered electrodes. Five groove designs ranging from symmetrical double-V to single-V were used. Welding was first completed on one side; then the specimen was turned over and the other side was back chipped and welded. Angular change was measured after welding each pass. The back chipping did not affect the angular change.

Angular change in the reverse direction resulted during welding of the second side of double V-grooves. The angular change remaining after completion of welding depended on the ratio of weld metal deposited on the two sides of the plates. Since the angular change increased more rapidly during welding of the second side, the minimum angular change was obtained when the first V-groove welded was slightly larger than the V-groove on the opposite side. The angular change in a double V-groove weld may be a minimum when

$$\frac{t_1 + 0.5\, t_3}{t_1 + t_2 + t_3} = 0.6 \qquad (7.18)$$

where

t_1 = depth of V-groove welded first, in. or mm
t_2 = depth of V-groove welded last, in. or mm
t_3 = width of the root face, in. or mm

The variables t_1, t_2, and t_3 are illustrated in Figure 7.38.

38. King, C. W. R. In *Transactions of the Institute of Engineers and Shipbuilders in Scotland.* 87: 1944; 238–255.

Figure 7.39 shows the groove design most suitable to minimizing angular changes in butt joints of various thicknesses. Curves are shown for weldments with and without strongbacks.

To achieve a minimum angular change in a double V-grove weld in a 3/4-inch-thick (19 mm) plate with no root face, the groove depth on the first side should be about 7/16 in. (11 mm). For the same conditions using the results shown in Figure 7.39 without restraint the groove depth on the first side should be about 1/2 in. (13 mm).

ANGULAR DISTORTION OF FILLET WELDS

IF A JOINT is free from external restraint, during fillet welding, contraction of the weld metal causes angular distortion about the joint axis as shown in Figure 7.40(A). However, if the members are restrained by some means, the distortion depends on the degree of restraint. For example, when the movement of stiffeners welded to a plate is prevented, the plate, distorts in a wave pattern as shown in Figure 7.40(B). The problem of analyzing wavy distortion and the associated stresses can be handled as a problem of stress in a rigid frame.[39] In the simplest case of a uniform distortion, the relationship between angular change and distortion at the weld is given as follows:

$$\frac{\delta}{L} = 0.25\, \phi - \left[\left(\frac{x}{L}\right) - 0.5\right]^2 \phi \qquad (7.19)$$

where

δ = distortion
L = length of span
ϕ = angular change
x = distance from the center line of frame to the point at which δ is measured. (see Figure 7.40B.)

When $x = 0.5L$, $\delta = 0.25\phi L$ and is a maximum value.

The angular change (ϕ) is related to the angular change (ϕ_o) that would occur in an unrestrained weld [see Figure 7.40(A)] under the same welding conditions. Thus

$$\phi = \frac{\phi_o}{1 + (2\frac{R}{L}) \cdot (\frac{1}{C})} \qquad (7.20)$$

39. Masubuchi, K., Ogura, Y., Ishihara, Y., and Hoshino, J. Studies on the mechanism of the origin and the method of reducing the deformation of shell plating in welded ships. *International Shipbuilding Progress* 3 (19): 1956; 123-133.

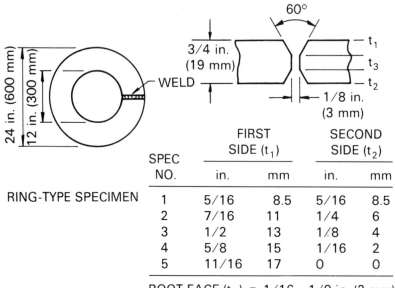

SPEC NO.	FIRST SIDE (t_1)		SECOND SIDE (t_2)	
	in.	mm	in.	mm
1	5/16	8.5	5/16	8.5
2	7/16	11	1/4	6
3	1/2	13	1/8	4
4	5/8	15	1/16	2
5	11/16	17	0	0

ROOT FACE (t_3) = 1/16 – 1/8 in. (2 mm)

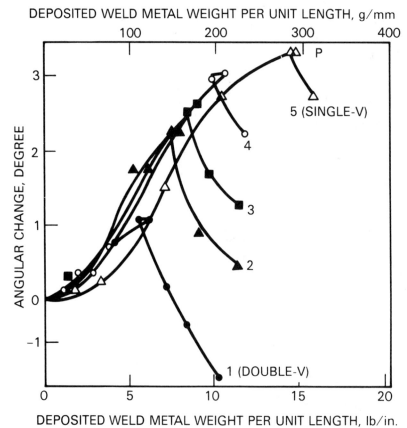

Figure 7.38—Experimental Data Gathered on Ring-Type Specimens

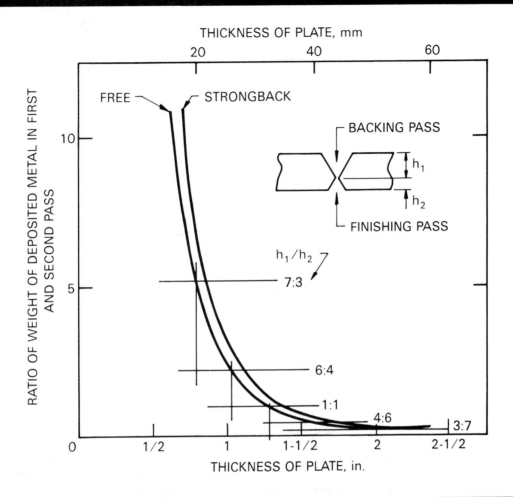

Figure 7.39—Groove Design Most Suitable to Minimizing Angular Changes in Butt Joints of Various Thicknesses

where

ϕ_o = the angular change of an unrestrained weld, degrees

L = the distance between stiffeners, in. (mm)

C = a coefficient of rigidity for angular changes, lb·in/in (kg·mm/mm)

R = the rigidity of the flange or bottom plate, ksi in.³ (MPa mm³)

The rigidity R is a function of the plate thickness and the elastic constants of the material. Thus

$$R = \frac{E t^3}{12(1 - \nu^2)} \qquad (7.21)$$

where

R = rigidity of flange plate, ksi in.³ (MPa mm³)

E = modulus of elasticity, ksi (MPa)

t = thickness of the flange plate, in. (mm)

ν = Poisson's ratio

For most metals ν is approximately 0.3 and equation 7.21 reduces to

$$R = 0.09 \, E t^3 \qquad (7.21)$$

The value of C in equation 7.23 can be interpolated from Tables 7.5 and 7.6 for steel and aluminum respectively.

The final step is to determine ϕ_o, the angular distortion of an unrestrained fillet weld of leg D_f. To estimate angular distortion ϕ_o from Figure 7.41 (7.42 in S.I. Units) for steel and aluminum, the parameter, w, is the weight of electrode deposited per foot of weld, must be determined. The parameter, w, is related to the deposition efficiency of the process (DE), the density of the

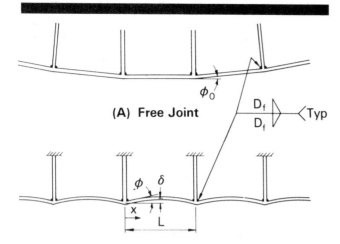

(A) Free Joint

$D_f \triangleright$ ⟨Typ
D_f

(B) Restrained Joint
Reprinted by permission of the Welding Research Council

Figure 7.40—Distortion Caused by Angular Change in Two Types of Fillet Welded Structures

weld metal (ρ), and the leg of the fillet.[40] Thus, for continous fillets on both sides of the web plate:

$$w = 12(D_f)^2 \left(\frac{\rho}{DE} \right) \tag{7.22}$$

The value of, ϕ_o, can be estimated from Figure 7.41 or 7.42, and δ, the height of the distortion, may be determined from equations 7.20 and 7.19.

COMPARISON OF THE ANGULAR DISTORTION OF STEEL AND ALUMINUM FILLET WELDS

THE ABOVE ANALYSIS was performed on steel and aluminum designs within the ranges shown in Table 7.7. The calculated values of δ_m for steel and aluminum fillet weld distortions are shown in Figure 7.43. These results indicate that angular distortion of fillet welds in aluminum is significantly lower than in steel.

BENDING DISTORTIONS INDUCED BY LONGITUDINAL SHRINKAGE

WHEN THE WELD line does not coincide with the neutral axis of a welded structure, longitudinal shrinkage of the weld metal induces bending moments, and longitudinal distortion of the structure results. A theory similar to the bending beam theory has been developed for the analysis of longitudinal distortion caused by welding of a long slender beam as shown in Figure 7.44. Longitudinal

residual stress, σ_x, and curvature of longitudinal distortion, $1/R$, are given in the following equations:

$$\sigma_x = -E\epsilon_x'' + \frac{M_y}{I_y}z + \frac{P_x}{A}$$

$$\frac{1}{R} = \frac{M_y''}{EI_y} = \frac{P_x^* L^*}{EI_y}$$

$$P_x^* = \int E\epsilon_x'' dy\, dz \tag{7.23}$$

where

ϵ_x'' = incompatible strain
A = sectional area of the joint
I_y = moment of inertia of the joint around the neutral axis
P_x = apparent shrinkage force,
  = $\int E\epsilon_x'' dy\, dz$
M_y = apparent shrinkage moment,
  = $\int\int E\epsilon_x\, z\, dy\, dz = P_x^* L^*$
L^* = distance between the neutral axis and the acting axis of apparent shrinkage force
z = distance

Equation 7.23 shows that it is necessary to know the distribution of incompatible strain, ϵ_x'', in order to know the distribution of residual stress, σ_x; however, information about moment, M_y^*, is only sufficient for determining the amount of distortion, $1/R$. Moment, M_y^*, is determined when the magnitude of apparent shrinkage force, P_x^*, and the location of its acting axis are known. Through experiments, it was found that the acting axis of P_x^* is located somewhere in the weld metal. The apparent shrinkage force, P_x^*, is the origin of residual stress and distortion is produced as the result of the existance of P_x^*. More information can be obtained when the value of P_x^* rather than the value of distortion itself is used in the analysis of experimental results. For example, in discussing the influence of various factors on the magnitude of distortion, it is possible to separate them into those attributable to the change in geometry (A, I_y, or L^*) or those attributed to the change in the value of P_x^* itself.

The increase of longitudinal distortion (apparent shrinkage force P_x^*) during multipass welding is as shown in Figure 7.45. The specimens were 48 in. long (1200 mm) by 1/2 inch (13 mm) thick mild steel plates. After the first layer, the values of P_x^* increased proportionally with the weight per weld length of electrode consumed. More distortion occurred in the first layer than in subsequent layers.

Practically no distortion was produced by intermittent welding. This is probably due to the fact that longitudinal residual stress does not reach a high value in a short intermittent weld.[41]

---

40. Densities of various metals and typical deposition efficiencies of several welding processes are shown in Chapter 8 of this volume.

41. The reduction in angular change by the use of intermittent welding may not be as great as that obtained in longitudinal distortion.

Table 7.5A
Coefficient of Angular Rigidity for Low Carbon Steel**

Fillet Weld Size in.	Weight of Electrode* Consumed lb/ft	Logrithm of Electrode Weight	Coefficient of Angular Rigidity, lb · in/in			
			t = 3/8	t = 1/2	t = 3/4	t = 1
1/4	0.16	− 0.79	11 900	43 900	167 700	375 000
5/16	0.25	− 0.60	9 700	37 700	133 600	300 500
3/8	0.33	− 0.44	7 900	31 500	121 000	266 500
1/2	0.65	− 0.19	7 100	25 600	82 700	242 700
5/8	1.01	0.01	6 600	20 500	77 200	228 200

*Deposition efficiency (DE) is 0.657
$w = 1/2D_f^2(12)\rho/DE = 2.59 \, D_f^2$
where ρ is the density of steel (0.28 lb/in^3)

Table 7.5B
Coefficient of Angular Rigidity C for Low-Carbon Steel

Fillet Weld Size mm	Weight of Electrode Consumed g/cm	Logrithm of Electrode Weight log$_{10}$w	Coefficient of Angular Rigidity, kg · mm/mm			
			t = 10 mm	t = 13 mm	t = 18 mm	t = 25.4 mm
6.58	2.51	0.4	5400	19 900	76 100	170 100
7.38	3.16	0.5	4700	18 000	65 200	142 400
8.29	3.98	0.6	4100	16 300	56 100	130 200
9.30	5.01	0.7	3800	15 000	48 800	125 000
10.45	6.31	0.8	3500	13 600	43 000	116 800
12.20	7.95	0.9	3300	12 200	38 900	112 000
13.15	10.00	1.0	3100	11 000	36 100	108 200
14.80	12.60	1.1	3000	9 800	35 200	105 000
16.55	15.85	1.2	2950	8 800	34 800	102 000

$w = 1/2 \, D_f^2(x10^{-2}) \, \rho/DE = 0.058 \, D_f^2$
where ρ is the density of steel (7.85 g/cm^3) and DE is the deposition efficiency (0.657)

Table 7.6A
Coefficient of Angular Rigidity for Aluminum

Fillet Weld Size in.	Weight of Electrode Consumed lb/ft	Logrithm of Electrode Weight	Coefficient of Angular Rigidity, lb · in/in					
			t = 1/8	t = 1/4	t = 3/8	t = 1/2	t = 5/8	t = 3/4
3/8	0.08	− 1.07	121	1680	30 000	45 900	55 800	159 800
7/16	0.12	− 0.94	115	1600	17 400	37 500	39 700	70 100
1/2	0.15	− 0.82	105	1480	11 000	27 300	24 500	43 200
5/8	0.24	− 0.63	88	1260	7500	18 400	14 300	28 000

*Deposition efficiency (DE) is 0.95
$w = 1/2D_f^2(12)\rho/DE = 0.605 \, D_f^2$
where ρ is the density of aluminum (0.1 lb/in^3)

Table 7.6B
Coefficient of Angular Rigidity for Aluminum

Fillet Weld Size mm	Weight of Electrode* Consumed g/cm	Logrithm of Electrode Weight $\log_{10} w$	Coefficient of Angular Rigidity, kg · mm/mm					
			t = 3.18 mm	t = 6.4 mm	t = 9.5 mm	t = 12.7 mm	t = 15.9 mm	t = 19.1 mm
8.969	1.122	0.05	57	782	14 390	22 800	31 000	78 400
9.567	1.259	0.1	55	762	13 600	20 800	25 300	72 500
10.660	1.585	0.2	52	725	7 900	17 000	18 000	31 800
11.960	1.995	0.3	49	686	5 600	13 800	12 900	22 200
13.420	2.512	0.4	46	645	4 300	11 000	9 200	17 000
15.057	3.162	0.5	43	608	3 600	8 900	6 900	13 500

*$w = 1/2 \, D_f{}^2 \, (\text{x}10^{-2}) \, \rho/DE$
where ρ is the density of Aluminum (2.65 g/cm³) and DE is the deposition efficiency (0.95)

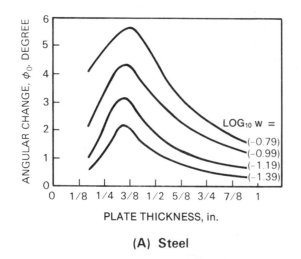

(A) Steel

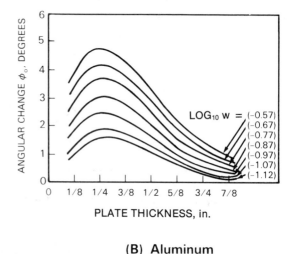

(B) Aluminum

Figure 7.41A—Effect of Plate Thickness and Filler Metal Weight per Unit Length of Weld on the Angular Change of Unrestrained Steel and Aluminum Fillet Welds—US Customary Units

Comparison Between Steel and Aluminum Weldments

FIGURE 7.46 COMPARES values of longitudinal bending distortion expressed in terms of the radius of curvature of built-up beams in steel and aluminum.[42] Aluminum weldments distorted less than steel weldments, perhaps due to the fact that the temperature distribution in the

42. Yamamoto, G. Study of longitudinal distortion of welded beams, M.S. thesis. Massachusetts Institute of Technology (May 1975).

z-direction is more uniform in an aluminum weldment than in a steel weldment.

BUCKLING DISTORTION

WHEN THIN PLATES are welded, residual compressive stresses occur in areas away from the weld and cause buckling. Buckling distortion occurs when the specimen length exceeds a critical length for a given thickness of a specific specimen. In studying weld distortions in thin-plate structures, it is important to first determine

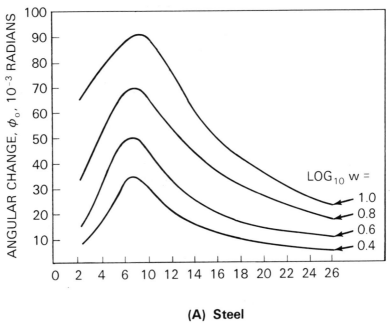

(A) Steel

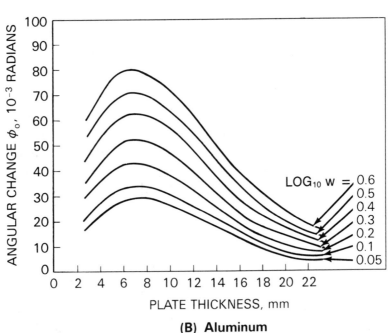

PLATE THICKNESS, mm

(B) Aluminum

Figure 7.42—Effect of Plate Thickness and Filler Metal Weight per Unit Length of Weld on the Angular Distortion of Unrestrained Steel and Aluminum Fillet Welds—SI Units

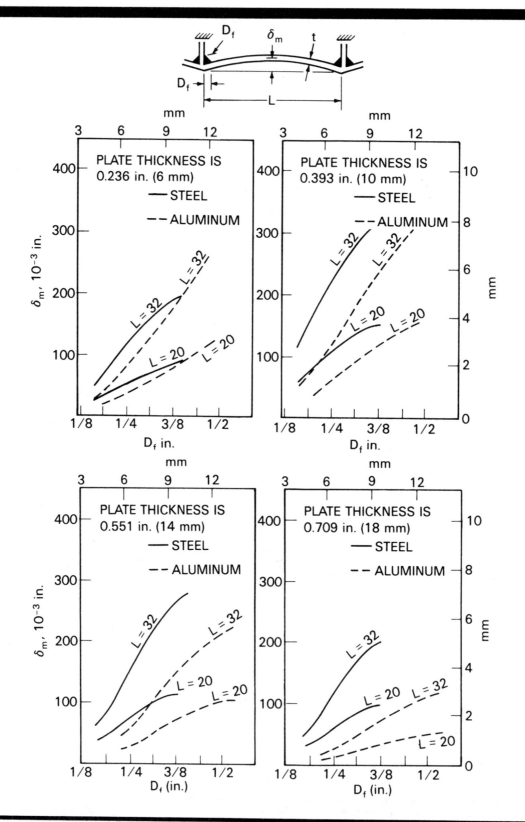

Figure 7.43—Out-of-Plane Distortion, δ_m, as a Function of Plate Thickness, T, Span Length, L, and the Size of Fillet Weld, D_f, for Steel and Aluminum

Table 7.7
Ranges of Designs for Calculated Angular Distortion
of Steel and Aluminum Fillet Welds

	Inch		Millimeter	
	min	max	min	max
Plate Thickness (t)	0.24	0.70	6	18
Distance Between Stiffeners (l)	20	32	500	800
Weld size (D_f)	3/16	1/2	5	12

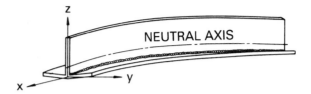

(A) General View

whether the distortion is produced by buckling or by bending. Buckling distortion differs from bending distortion in that there may be more than one stable deformed shape, and the amount of deformation in buckling distortion is much greater.

Buckling distortion of welded plates resulting from residual stresses has been the subject of several studies. One of these addressed the buckling distortion of 0.09 in. (2.3 mm) thick by 280 in. (7 M) long steel strips.[43] The strip width varied from 4 to 16 in. (100 to 400 mm). The strips were submerged-arc welded along the centerline, and the buckling distortions were observed. One specimen exhibited eight different stable deformation patterns along the weld centerline. These patterns are shown in Figure 7.47.

Experiments were made on thin plates with stiffeners fillet welded to the plate edges to investigate buckling in stiffened thin-wall structures.[44] The central deflection (out-of-plane distortion) of low-carbon steel panels was measured for several values of plate thickness, panel dimensions, and heat input, using both shielded metal arc and gas tungsten arc welding. The test conditions are summarized in Table 7.8.

Distortion increased with increasing time after welding until a stable condition of buckling was reached. This is illustrated in Figure 7.48 for a 20 × 20 × 1/4 in. (500 × 500 × 6 mm) panel. The final condition of the panel was reached approximately 30 minutes after the start of welding. The time was apparently required for the panel to reach thermal equilibrium. There was a sudden increase in the relative deflection when the heat input increased from 21 000 J/in. (8400 J/cm) to 23 000 J/in. (9240 J/cm).[45] This is illustrated in Figure 7.49 in which the heat input divided by the plate thickness is shown as a function of the relative deflection. The effect is also shown in Figure 7.50. In this figure, the

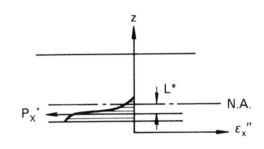

(B) Incompatible Strain, ε_x''

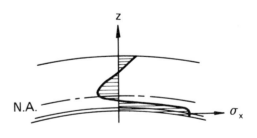

(C) Residual Stress, σ_x

43. Masubuchi, K. Buckling-type deformation of thin plate due to welding, in *Proceedings fo Third International Congress for Applied Mechanics of Japan* (1954): p107–111.
44. Terai, K., Matsui, S., and Kinoshita, T. Study on prevention of welding deformation in thin plate structures. Kawasaki Technical Review. Kawasaki Heavy Industries, Ltd., Iobe, Japan: No. 61 (Aug 1978): p 61–66.
45. The relative deflection is the ratio of the central deflection to the plate thickness.

Figure 7.44—Analysis of Longitudinal Distortion in a Fillet Joint

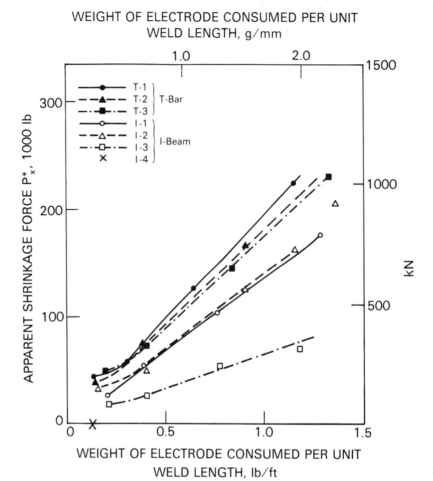

Figure 7.45—Effect of Filler Metal Consumed on Apparent Shrinkage Force

heat input is divided by the cube of the thickness and multiplied by the panel width. From Figure 7.50, it was concluded that buckling distortion occurred when the critical heat input for a panel was defined as

$$H_{cr} > 33 \text{ MJ/in.}^3 \text{ (2 MJ/cm}^3\text{)} \qquad (7.24)$$

where

$$H_{cr} = \frac{Q}{t^3} \cdot b$$

and Q is the heat input kJ/in. (J/cm)
t is the thickness, in. (cm)
b is the panel width, in. (cm)

COMPARISON OF DISTORTION IN ALUMINUM AND STEEL WELDMENTS

ALTHOUGH THE MAJORITY of information on weld

distortion pertains to steel weldments, some information on distortion in aluminum weldments is available.[46]

Compared with steel, aluminum has the following physical characteristics:

(1) The value of thermal conductivity of aluminum is about five times that of steel.
(2) The coefficient of linear thermal expansion of aluminum is about two times that of steel.
(3) The modulus of elasticity of aluminum is about one third that of steel.

46. Masubuchi, K., Residual stresses and distortion in welded aluminum structures and their effects on service performance. Welding Research Council Bulletin 174 (July 1972); Masubuchi, K. and Papazoglou, V. J. Analysis and control of distortion in welded aluminum structures. *Transactions of the Society of Naval Architects and Marine Engineers.* 86(1978): 77–100.

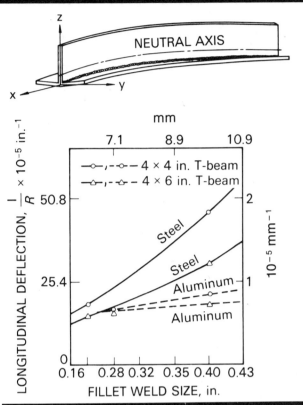

Figure 7.46—Effect of Fillet Weld Size on Longitudinal Deflection in T-Section Beams

Table 7.8
Summary of Conditions for Buckling Tests

	Customary		SI	
	min	**max**	**min**	**max**
Plate thickness	0.18 (in)	0.40 (in)	4.5 (mm)	10 (mm)
Panel Dimensions	20 × 20 (in)	40 × 40 (in)	500 × 500 (mm)	1000 × 1000 (mm)
Heat Input	19 200 (J/in)	34 300 J/in	7560 J/cm	13 500 J/cm

*Experiments were conducted using the SMAW and the GTAW Processes.

(A) Buckling Distortion

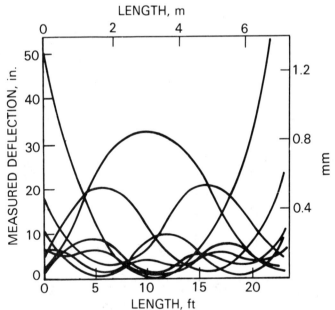

(B) Eight Stable Distortions Observed in One Specimen

NOTE: Distortions along the weld line are shown.

Figure 7.47—Buckling Distortion of a Bead-on-Plate Weld

Transverse Shrinkage of butt Joint

THE TRANSVERSE SHRINKAGE of a butt joint in aluminum is considerably greater than that of a steel joint of similar dimensions. This is mainly due to the fact that the welding heat is conducted into wider regions in an aluminum weld than in a steel weld. Since the major part of the transverse shrinkage comes from the shrinkage of the base metal, considerably more transverse shrinkage is produced in an aluminum weld than in a steel weld.

Angular Change of a Fillet Weld

AS SHOWN IN Figure 7.43, angular changes of aluminum fillet welds are less than those of steel fillet welds. Angular change is caused by temperature differences between the top and the bottom surfaces of the flange plate to which the web is welded. The temperature distribution in the thickness direction is more uniform in an aluminum weld than in a steel weld because of the higher thermal conductivity. Therefore, the amount of angular change of an aluminum fillet weld is less than that of a steel fillet weld.

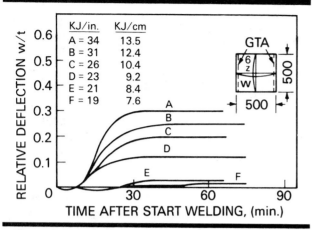

Figure 7.48—Effect of Time After Welding Start on Ratio of Central Deflection to Plate Thickness (Relative Deflection)

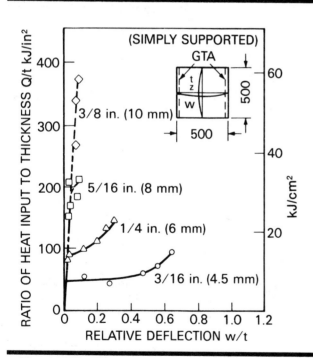

Figure 7.49—Effect of Relative Deflection on Heat Input to Thickness Ratio

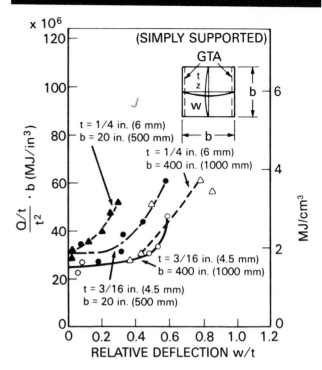

PARAMETER Qb/t^3

Q = HEAT INPUT
w = CENTRAL DEFLECTION
t = THICKNESS
b = PANEL WIDTH

Figure 7.50—Effect of Relative Deflection (W/t) on the Parameter Qb/t^3

Longitudinal Distortion

THE LONGITUDINAL DISTORTION expressed in terms of the radius of curvature of built-up beams in steel and aluminum is compared in Figure 7.46. The aluminum beams distorted less than the steel beams.

REDUCING RESIDUAL STRESSES AND DISTORTION

TO DEVELOP EFFECTIVE means for reducing residual stresses and distortion, it is essential to understand how residual stresses and distortions are formed and how they are affected by design and welding procedures. Some of the previous discussions should be helpful in determining appropriate designs and procedures. Some practical approaches to the reduction of residual stress and control of distortion follow.

REDUCTION OF RESIDUAL STRESSES

THE EFFECTS OF residual stresses on the service behavior of welded structures varies significantly, depending upon a number of conditions. It is generally preferable to keep residual stresses to a minimum. Therefore, it is advisable to take precautions to reduce residual stresses. This is particularly important when welding thick sections.

Effect of Amount of Weld and Welding Sequence

SINCE RESIDUAL STRESSES and distortion are caused by thermal strains due to welding, a reduction in the volume of weld metal usually results in less residual stress and distortion. For example, the use of a U-groove instead of a V-groove should reduce the amount of weld metal. It is desirable to use the smallest groove angles and root openings that will result in adequate accessibility for welding and the production of sound weldments.

CONTROL OF DISTORTION

Design

THE MOST ECONOMIC design for a welded fabrication is the one that requires the fewest number of parts and a minimum of welding. Such a design also assists in reducing distortion.

The type of joint preparation is important, particularly for unrestrained butt joints, because it can influence the amount of angular distortion of the joint.

The smallest fillet welds that meet the shear strength requirements will also produce the lowest residual stresses and distortions. This will avoid overwelding and excessive distortion.

Distortion cannot be controlled solely by good design and proper joint details, but they serve to reduce the magnitude of the problem.

Assembly Procedure

DISTORTION IN SOME form or another cannot be avoided. Therefore, it is necessary to take appropriate steps to minimize distortion. Distortion may best be controlled using one of the following assembly methods.

(1) Estimate the amount of distortion likely to take place during welding, and then preset the members to compensate for the distortion.

(2) Assemble the job so that it is nominally correct before welding, and then use some form of restraint to minimize the distortion from welding.

The first method is attractive because the parts have almost complete freedom to move during welding and residual stress will be lower than with the second method. However, the first method is difficult to apply except on relatively simple fabrications. A good approach is to fabricate subassemblies using the first method. The subassemblies can be welded without restraint. This approach is especially attractive when the weldment is made up of a large number of parts. The welded subassemblies may then be assembled together and welded to complete the job. Often this final welding has to be carried out under conditions of restraint.

Figure 7.51 shows simple examples of the presetting method applied to fillet and butt welds. The amount of preset required varies somewhat according to plate thickness, plate width, and the welding procedure. For this reason, it is advisable to establish the correct preset using welding tests rather than by judgment.

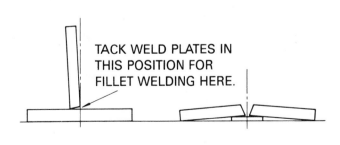

TACK WELD PLATES IN THIS POSITION FOR FILLET WELDING HERE.

Figure 7.51—Presetting for Fillet and Butt Welds

The restrained assembly method is the one generally preferred because of its comparative simplicity. The restraint may be applied by clamps, the use of fixtures, or simply by adequate tack welding. While this method minimizes distortion, it can result in high residual stresses. High residual stresses and the risk of cracking can often be minimized by the use of a suitable welding sequence and, with thick sections, by preheating. Where service requirements demand the removal of residual stresses, a stress relieving heat treatment must be applied after welding.

To attempt to impose complete restraint can be undesirable, but by restraining movement in one direction and allowing freedom in another, the overall effect can usually be controlled. An example is shown in Figure 7.52 where the clamping arrangement is designed to prevent angular distortion in a single-V-groove weld while permitting transverse shrinkage.

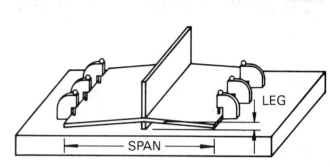

Figure 7.53—Apparatus for Welding T-Joints Submitted to Elastic Prestrain by Bolting Down Both Free Ends

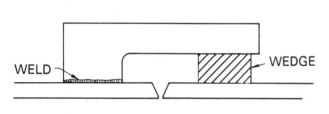

Figure 7.52—Arrangement of Clamp to Prevent Angular Distortion While Permitting Transverse Shrinkage

Elastic Prestraining

ELASTIC PRESTRAINING IS the method in which a plate is bent elastically before stiffeners are fillet welded to it, as shown in Figure 7.53. Angular changes after the removal of the restraint can be reduced significantly by this method.

Preheating

PREHEATING MAY BE used as a method of reducing distortion in a weldment. As shown in Figure 7.54, preheat on the top of the flange plate of a T-section, increases the angular distortion for some combinations of thickness and welding conditions and decreases it for other combinations. However, preheating the bottom of the flange plate, which helps to balance the heat of welding, reduces the angular distortion for all combinations of thickness and welding conditions.

In Figure 7.53, the parameter Z is determined using the following equation:

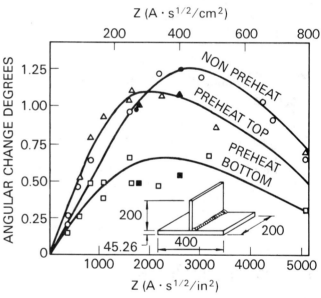

Figure 7.54—Effect of Preheat and Welding Variables on Angular Change of Steel Fillet Welded T-Joints

$$Z = \frac{I}{t\sqrt{vt}} \qquad (7.25)$$

where
I = welding current
t = thicknesses of horizontal plate
v = welding speed

The parameter Z, indirectly includes the fillet weld size because the weld area is proportional to the ratio of the amperage to the travel speed for consumable electrode welding.

CORRECTION OF DISTORTION

IT IS NOT always possible to control distortion within acceptable limits, especially with a new design or fabrication. In such circumstances, it is usually possible to remove distortion using one or more processes.

Distortion can be removed by producing adequate plastic deformation in the distorted member or section. The required amount of plastic deformation can be obtained by thermal or mechanical methods.

Thermal Straightening

THERMAL OR FLAME straightening can be used to remove distortion. The distorted area is straightened by heating spots to 1100-1200°F (600-650°C) and quenching with air or water. This procedure will cause the material to upset during heating and then shrinkage stresses will tend to straighten the plate or beam. Repeated applications result in additional shrinkage, but each successive application may be less effective. The best procedure is to apply the heat to a pattern of spots or lines in the distorted member.

This technique can also be used to provide desired shape. A bridge girder may require a fixed amount of camber or sweep. Camber can be induced by applying the heat to the outside of one flange along the length of the girder. Sweep can be induced by applying the heat to one edge of each flange along the length of the girder. Initially the girder will bend toward the torches as the flange lengthens due to thermal expansion. When the flange temperature returns to ambient the flange will have shortened and the girder will be bent away from the heated surface. This principle is illustrated in Figure 7.9.

Mechanical Straightening

DISTORTED MEMBERS CAN be straightened with a press or jacks. Heat may or may not be required for straightening.

Thermal Methods of Controlling Residual Stresses

THERMAL TREATMENTS ARE often necessary to maintain or restore the properties of base metal affected by the heat of welding. Thermal treatment may also affect the properties of the weld metal. The extent of the changes in the properties of the base metal, weld metal, and heat-affected zone are determined by the soaking temperature, time, and cooling rate, as well as the material thickness, grade, and initial temper.

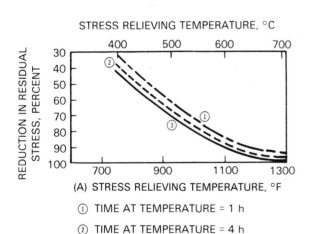

(A) STRESS RELIEVING TEMPERATURE, °F

① TIME AT TEMPERATURE = 1 h

② TIME AT TEMPERATURE = 4 h

③ TIME AT TEMPERATURE = 6 h

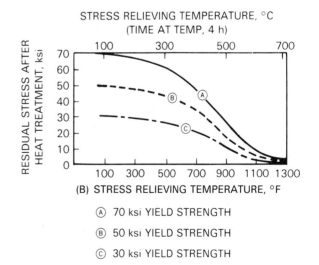

(B) STRESS RELIEVING TEMPERATURE, °F

Ⓐ 70 ksi YIELD STRENGTH

Ⓑ 50 ksi YIELD STRENGTH

Ⓒ 30 ksi YIELD STRENGTH

Figure 7.55A—Effect of Time at Temperature on the Reduction of Residual Stresses

Figure 7.55B—Effect of Stress Relieving Temperature on Residual Stresses

Preheat. The most common thermal treatment applied to weldments is preheat. As discussed previously, the proper use of preheat can minimize the distortion that would normally occur during welding. Distortion and the residual stresses are reduced as a result of lower thermal gradients around the weld. Preheat has the beneficial effect in steels of reducing the tendency for heat-affected zone or weld metal cracking.

Postweld Thermal Treatments. A properly executed postweld heat treatment results in uniform mechanical properties and reduced residual stresses. The effects of time at temperature and the stress relieving temperature on residual stresses are shown in Figures 7.55A and B, respectively. When thick weldments require a postweld machining operation, a stress relief heat treatment is usually necesary to achieve normal machining tolerances.

SUPPLEMENTARY READING LIST

American Bureau of Shipping. *Rules for building and classing steel vessels* (latest edition). New York, NY: American Bureau of Shipping.

American National Standards Institute. *Code for pressure piping*, ANSI B31.1 to B31.8 (latest edition). New York, NY: American National Standards Institute.

American Society of Mechanical Engineers. *ASME boiler and pressure vessel code,* Sections I, III, and VIII Divs. 1 and 2 (latest edition). New York, NY: American Society of Mechanical Engineers.

American Welding Society. *ANSI/AWS D1.1—Structural Welding Code—Steel*, Miami, FL: American Welding Society (Latest Edition).

Cullity, B. D. *Elements of x-ray diffraction, Second Edition..* Addison-Wesley Publishing Co.; 1978: 467–472.

Gurney, T. R. *Fatigue of welded steel structures.* Cambridge: Cambridge University Press, Second Edition, (1979).

Hall, W. J., Kihara, H., Soete, W. and Wells, A. A. *Brittle fracture of welded plates.* Englewood Cliffs, N.J.: Prentice-Hall, Inc. (1967).

Hwang, J. S. *Residual stresses in weldments in high-strength steels*, M.S. thesis. Massachusetts Institute.

Klug, H. P. and Alexander, L. E., *X-Ray diffraction procedures*, Second Edition. J. Wiley & Sons (1974): 768–770.

Masubuchi, K. Control of distortion and shrinkage in welding. *Welding Research Council Bulletin* 149: (1970).

———. Nondestructive measurement of residual stresses in metals and metal structures. RSIC-410. April 1965. Redstone Arsenal, Alabama: Redstone Scientific Information Center, U. S. Army Missile Command.

———. Residual stresses and distortion in welded aluminum structures and their effects on service performance. *Welding Research Council Bulletin* 174: 1974.

———. *Analysis of Welded Structures—Residual Stresses, Distortion and their Consequences,* Pergamon Press. (1980).

Masubuchi, K., Nishida, M., Yamamoto, G., Kitamura, K. and Taniguchi, C., *Analysis of thermal stresses and metal movements of weldments: a basic study toward computer-aided analysis and control of welded structures,* Transactions of the Society of Naval Architects and Marine Engineers, Vol., 83, 1975, 143–167.

Munse, W. H. *Fatigue of welded steel structures.* New York: Welding Research Council (1964).

Navy Department. *Fabrication welding and inspection; and casting inspection and repair for machinery, piping and pressure vessels in ships of the united states navy.* MIL-STD-278 (Ships) (latest edition). Washington, D.C.: Navy Department.

Navy Department. *General specification for ships of the united states navy*, Spec. S9-1 (latest edition). Washington, D.C.: Navy Department.

Ruud, C. O. and Farmer, G. D., "Residual Stress Measurement by X-Rays: Efforts, Limitations, and Applications", *Nondestructive Evaluation of Materials*, edited by J. J. Burke and V. Weiss, Plenum Press, 1979, 101–116.

Tall, L., ed. *Structural steel design.* 2d ed. New York, NY: Ronald Press Co. (1974).

Treuting, R. G., Lynch, J. J., Wishart, H. B., and Richards, D. G. *Residual stress measurements.* Metals Park, Ohio: American Sociey for Metals (1952).

United states coast guard marine engineering regulations and materials, Spec. CG-115 (latest edition). Washington, D.C.: United States Coast Guard.

WELDING AND CUTTING COSTS: ESTIMATES AND CONTROLS

PREPARED BY A COMMITTEE CONSISTING OF:

A. Lesnewich, Chairman
Consultant

C. W. Case
Inco Alloys International

L. P. Connor
American Welding Society

WELDING HANDBOOK COMMITTEE MEMBER:
C. W. Case
Inco Alloys International

CHAPTER 8

WELDING AND CUTTING COSTS: ESTIMATES AND CONTROLS

SCOPE

THE COST ELEMENTS of a product are material, labor, and overhead. Overhead costs are not addressed in this chapter because the amount of overhead varies from industry to industry. The method of distributing overhead costs also varies. Only welding materials such as filler metals, gases, and fluxes are considered in the chapter, and only welding labor is specifically addressed. Thus, information on base metal costs as well as layout, forming, fitting, and other metalworking operations is included.

Factual information and guidance in developing welding cost standards to suit each individual enterprise are also given. The information provided will assist in the development of welding production standards. The production standards can then be used to:

(1) Estimate welding costs
(2) Manage production planning
(3) Forecast personnel, inventory, and equipment requirements
(4) Justify new equipment
(5) Analyze job performance
(6) Manage cost reduction programs
(7) Set up incentive programs

To preserve the usefulness of the information and to address the greatest number of industries, material cost units will be in pounds and cubic feet, and labor cost units will be in manhours. Users can convert the cost units to dollar values using their specific labor and overhead rates and the current costs of consumables.

ESTIMATING WELDING COSTS

MATERIAL ESTIMATE

THE FUNDAMENTAL BASIS of any cost estimate is the quantity of material and the operations necessary to perform the required task.

The labor hours required can be estimated from an accurate summary of materials and a list of the operations required on each piece. Many fabricators develop standard manufacturing practices so that the labor allowances can be determined directly from the material requirements.

For a welding cost estimate, the material estimate is a list of each weld in the assembly, including the weld type, size, and length. For enterprises that produce a small range of similar products with a few standard manufacturing practices, less judgment is required by the estimator to reduce the material requirements to a labor estimate and to prepare a purchase bill of material. However, for custom fabricators that produce a large variety of products, estimators should know that certain procedures will be followed in the shop. Further, they must forecast, based on the nature of the work, whether the efficiencies normally achieved in the shop procedures will be achieved on this project. If the normal efficiencies are not anticipated by the estimators, they must assume different values based on the complexity or simplicity of the work.

When the material estimate is prepared and a welding procedure and joint geometry are assigned to each weld, the weight of deposited metal per foot of weld can be estimated from tables such as those shown in Tables 8.1 through 8.8.[1] The data in the tables are for steel, but these data can be used to determine the weight of any deposited metal. The equation is as follows:

$$W = \rho DV \tag{8.1}$$

where

W = weight of the deposited metal in question lb/ft (kg/m)

ρ = density of the deposited metal lb/in.^{3} (g/mm^3)

DV = deposited metal volume, in.^{3}/ft (mm^3/mm) from Tables 8.1 through 8.8

The densities of some common metals and alloys are shown in Table 8.9.

The deposited metal weight is the fundamental information needed to determine all welding costs. The welding process and the welding procedure affect the quantity of filler metal, flux, gas, and labor required to fabricate each weld. All of these quantities are derived from the deposited metal weight.

To determine the quantities of filler metal, gas or flux, and labor from the deposited metal, the estimator needs the filler metal deposition rate, the deposition efficiency, and the operator factor.

The deposition rate is the weight of weld metal deposited per unit of time. Typical deposition rates for welding steel with several consumable electrode processes as a function of welding current are shown in Figure 8.1.

The deposition efficiency is the ratio of the deposited metal weight to the weight of filler metal used, expressed as a percent. For estimating purposes, the weight of filler metal used includes stub weight. The deposition efficiency ranges for several welding processes are shown in Table 8.10.

The operator factor is the ratio of arc time or actual weld deposition time to the total work time of the welder or welding operator. The operator factor is expressed as a percentage, and some typical ranges are shown in Table 8.11.

In assigning values to these variables, the estimator should use data based on shop experience. If such data are not available, or the work is different than normal shop work, then the estimator should make an educated judgment on each factor. To guide the estimator, typical ranges of deposition efficiencies and operator factors are shown in Tables 8.10 and 8.11, respectively, and approximate deposition rates of several welding processes are shown in Figures 8.1 through 8.5.

Estimating Filler Metal Requirements

THE AMOUNT OF filler metal required depends on the deposition efficiency as well as the deposited metal.

The required weight of filler metal for each weld is as follows:

$$FM = \frac{100(DW)(L)}{DE} \tag{8.2}$$

where

FM = weight of filler metal, lb (kg)

DW = deposit metal, lb/ft (kg/m)

L = weld length, ft (m)

DE = deposition efficiency, %

Supplementary Requirements for Welding Consumables

THE SHIELDED METAL arc and self-shielded flux cored arc welding processes do not require additional consum-

---

1. Deposited metal is the filler metal that has been added during welding.

Table 8.1
Volume and Weight of Steel Fillet Welds

Fillet Weld Size		Deposited Metal			
		Volume		Weight	
in.	mm	in.³/ft	mm³/mm	lb/ft	kg/m
3/16	5	0.34	18.2	0.10	0.15
1/4	6	0.43	21.1	0.12	0.18
5/16	8	0.68	36.6	0.19	0.28
3/8	10	0.96	51.2	0.27	0.40
7/16	11	1.3	69.9	0.36	0.54
1/2	13	1.7	91.4	0.48	0.71
5/8	16	2.5	13.4	0.71	1.06
3/4	19	3.6	19.4	1.0	1.5
7/8	22	5.0	26.9	1.4	2.1
1	25	6.4	34.4	1.8	2.9

Table 8.2
Volume and Weight of Square-Groove Butt Joints in Steel, Welded Both Sides

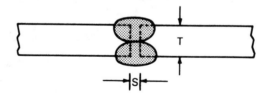

Joint Dimensions				Deposited Metal			
T		S		Volume		Weight	
in.	mm	in.	mm	in.³/ft	mm³/mm	lb/ft	kg/m
1/8	3	0	0	0.43	23.1	0.12	0.18
		1/32	1	0.46	24.7	0.13	0.19
3/16	5	1/32	1	0.71	38.2	0.20	0.29
		1/16	2	0.79	42.5	0.22	0.33
1/4	6	1/16	2	0.93	50.0	0.26	0.39
		3/32	2	1.0	53.8	0.29	0.43

Table 8.3
Volume and Weight of Square-Groove Butt Joints in Steel with Backing Strip

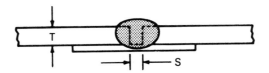

Joint Dimensions				Deposited Metal			
T		S		Volume		Weight	
in.	mm	in.	mm	in.³/ft	mm³/mm	lb/ft	kg/m
1/8	3	0	0	0.21	11.3	0.06	0.09
		1/16	2	0.32	17.2	0.09	0.13
3/16	5	1/16	2	0.46	24.7	0.13	0.19
		3/32	2	0.53	28.5	0.15	0.22
1/4	6	3/32	2	0.64	33.4	0.18	0.27
		1/8	3	0.75	40.3	0.21	0.31

Table 8.4
Volume and Weight of Single-V-Groove Butt Joints in Steel with a Back Weld

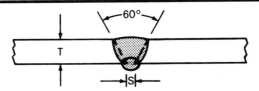

Joint Dimensions				Deposited Metal			
T		S		Volume		Weight	
in.	mm	in.	mm	in.³/ft	mm³/mm	lb/ft	kg/m
1/4	6	1/16	2	0.81	43.5	0.23	0.34
5/16	8	3/32	2	1.2	64.5	0.35	0.52
3/8	10	1/8	3	2.0	10.8	0.57	0.85
1/2	13	1/8	3	3.5	18.8	1.0	1.5
5/8	16	1/8	3	4.8	25.8	1.4	2.1
3/4	19	1/8	3	5.7	30.6	1.6	2.4
1	25	1/8	3	10.0	53.8	2.8	4.2

able supplies. However, submerged arc welding requires a flux, and gas metal arc or gas shielded flux cored arc welding require a shielding gas. Flux consumption will vary, but an average value of a pound of flux per pound of filler metal is an accepted amount. Users should analyze their results to determine the appropriate flux consumption. Gas consumption for gas metal arc and flux cored arc welding is about 10-to-15 ft³/(100 A)·(h) [0.28-to-0.42 m³/(100 A)(h)] depending on the gas, equipment and other local conditions. A sample calculation of shielding gases required to deposit 300 lb (136 kg) of filler metal is shown in Table 8.12.

Gas consumption depends on the arc time, which is a function of deposition rate. Thus, the higher deposition

Table 8.5
Volume and Weight of Double-V-Groove Joints in Steel

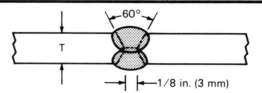

Joint Dimensions		Deposited Metal			
T		Volume		Weight	
in.	mm	in.³/ft	mm³/mm	lb/ft	kg/m
5/8	16	3.0	160	0.86	1.3
3/4	19	3.9	210	1.1	1.6
1	25	6.0	320	1.7	2.5
1-1/4	32	8.5	460	2.4	3.6
1-1/2	38	11.5	620	3.3	4.9
1-3/4	44	14.9	800	4.2	6.2
2	50	18.8	1000	5.3	7.9
2-1/4	57	23.0	1240	6.5	9.7
2-1/2	64	27.8	1500	7.9	11.8
3	75	38.5	2070	10.9	16.2

Table 8.6
Volume and Weight of Single-U-Groove Butt Joints in Steel

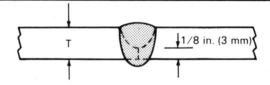

Joint Dimensions		Deposited Metal			
T		Volume		Weight	
in.	mm	in.³/ft	mm³/mm	lb/ft	kg/m
1/2	13	3.0	160	0.84	1.3
5/8	16	3.9	210	1.1	1.6
3/4	19	5.4	290	1.5	2.2
1	25	7.9	420	2.2	3.3
1-1/4	32	10.7	580	3.0	4.5
1-1/2	38	13.9	750	3.9	5.8
1-3/4	44	17.1	910	4.8	7.1
2	50	20.0	1070	6.0	8.3
2-1/4	57	25.4	1370	7.1	10.6
2-1/2	64	30.0	1610	8.4	12.5
2-3/4	70	34.6	1860	9.7	14.4
3	75	40.0	2150	11.2	16.6
3-1/2	89	51.1	2750	14.3	21.2
4	100	63.9	3450	17.9	26.6

Table 8.7
Volume and Weight of Single-Bevel Groove Joints in Steel

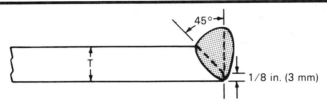

Joint Dimensions		Deposited Metal			
T		Volume		Weight	
in.	mm	in.³/ft	mm³/mm	lb/ft	kg/m
1/4	6	0.21	11.3	0.06	0.09
5/16	8	0.39	21.0	0.11	0.16
3/8	10	0.61	32.8	0.17	0.25
1/2	13	1.2	64.5	0.34	0.51
5/8	16	2.0	108	0.56	0.83
3/4	19	3.0	160	0.84	1.25
1	25	5.7	310	1.6	2.4

Table 8.8
Volume and Weight of Single-J-Groove Joints in Steel

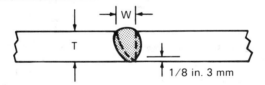

Joint Dimensions		Deposited Metal			
T		Volume		Weight	
in.	mm	in.³/ft	mm³/mm	lb/ft	kg/m
1	25	5.7	310	1.6	2.4
1-1/4	32	7.9	420	2.2	3.3
1-1/2	38	10.4	560	2.9	4.3
1-3/4	44	13.2	710	3.7	5.5
2	50	15.7	840	4.4	6.5
2-1/4	57	18.6	1000	5.2	7.7
2-1/2	64	22.1	1190	6.2	9.2
2-3/4	70	25.7	1380	7.2	10.7
3	75	29.6	1590	8.3	12.4
3-1/2	89	38.2	2050	10.7	15.9
4	100	47.5	2550	13.3	19.8

Table 8.9
Approximate Density for Some Common Engineering Alloys

Alloy Group	Density	
	lb/in.^{3}	g/mm^3
Carbon steel	0.28	7.8×10^3
Stainless steels	0.29	8.0×10^3
Copper alloys	0.31	8.6×10^3
Nickel alloys	0.31	8.6×10^3
Aluminum alloys	0.10	2.8×10^3
Magnesium alloys	0.065	1.8×10^3
Titanium alloys	0.17	4.7×10^3

Table 8.10
Deposition Efficiency for Welding Processes and Filler Metals

Filler Metal Form and Process	Deposition Efficiency,%[a]
Covered electrodes (SMAW)	
14 in. long	55 to 65
18 in. long	60 to 70
28 in. long	65 to 75
Bare solid wire	
Submerged arc welding (SAW)	95 to 99
Gas metal arc welding (GMAW)	90 to 97
Flux cored electrodes (FCAW)	80 to 90

a. Includes stub loss.

Table 8.11
Operator Factor for Various Welding Methods

Method of Welding	Operator Factor Range, Percent
Manual	50-30
Semiautomatic	10-60
Machine	40-90
Automatic	50-100

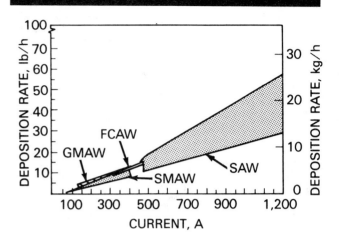

NOTE: Deposition rate will vary with electrode extension.

Figure 8.1—Effect of Welding Process and Current on Deposition Rate

rate flux cored arc process requires less shielding gas than the gas metal arc process. The deposition rate is an important variable in welding economics. It is approximately dependent on the welding current regardless of welding process, as shown in Figure 8.1. The major differences in deposition rates with various welding processes are related to the usability of the processes at high welding currents. However deposition rate is not the only factor in the choice of a welding process. Other factors include deposition efficiency, welding position, weld quality, required penetration, and availability of equipment and qualified personnel.

Methods of improving deposition rates with a given process, and factors affecting the choice of higher deposition rate processes will be discussed later.

ESTIMATING LABOR COSTS

TO DETERMINE LABOR manhours required, the estimator should judge the complexity of the work. If the work involves either frequent relocation of the welder and the welding equipment or repositioning of the work, then a low operator factor should be expected. When the number of supplementary activities for the welder is low, the operator factor should be high because the welder can accomplish more actual welding. The operator factor is used to determine labor costs. A sample calculation to determine the labor required to deposit 300 ft of 1/4 in. (90 m of 6 mm) fillet weld in the horizontal position using SMAW and FCAW is shown in Table 8.13. Both deposition rate and operator factor contribute to the labor estimate. This method can be used to estimate the required labor to weld any joint.

STANDARDS PREPARATION

MANY ORGANIZATIONS HAVE developed standard welding procedures to fabricate welds frequently encountered in their normal products. They also have predetermined the quantity of materials and labor required to make each type of weld. Such standards can be used to estimate welding costs, measure shop production performance, and set production incentive standards.

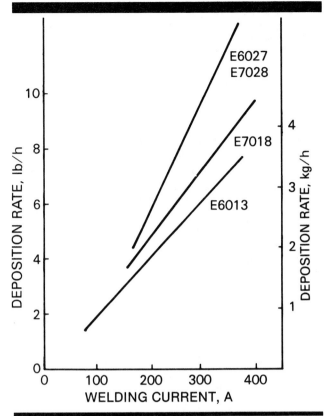

**Figure 8.2—Effect of Current and Electrode
Classification on Deposition Rate**

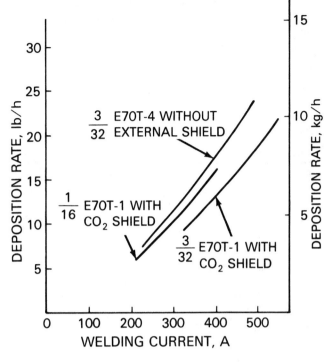

NOTE: Deposition rate will vary with electrode extension.

**Figure 8.4—Effect of Current on Deposition Rates of
1/16 Diameter Steel and Aluminum Welding Wires**

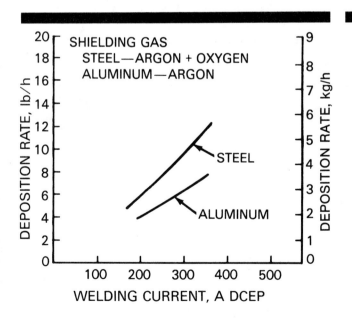

Note: Deposition rates will vary with electrode extension.

**Figure 8.3—Effect of Current on Deposition Rates of
1/16 Diameter Steel and Aluminum Welding Wires**

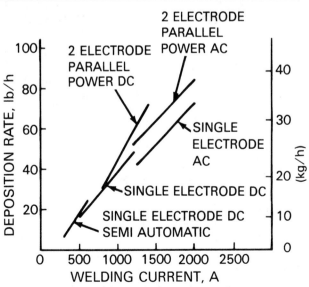

Note: Deposition rate will vary with electrode extension.

**Figure 8.5—Effect of Current on Deposition Rate on
Submerged Arc Welding**

Published values of deposition rate, deposition efficiency, and operator factor are industry averages, and should only be used as starting values for the development of cost standards. Enterprises that use welding should develop standard practices and procedures with associated cost standards. The standards can be refined by including production costs on actual jobs. In fact, when using actual production costs, the standards can be expanded to include sawing, punching, fitting, and other metalworking standards.

A sample format for welding cost standards is shown in Table 8.14. The data used for the calculations are also given in Table 8.14. The data shown in the Table were developed in accordance with the procedures discussed here.

Table 8.12
Sample Calculation of Shielding Gas Requirements

	Process	
	GMAW	FCAW
Deposited weld metal required, lb (kg)[a]	300 (136)	300 (136)
Deposition rate, lb/h (kg/h)[a]	10 (4.5)	15 (6.8)
Arc time, h	30	20
Gas flow rate, ft³/h (m³/h)	40 (1)	40 (1)
Total shielding gas required, ft³ (m³)	1200 (34)	800 (23)

Note: Arc time = $\dfrac{\text{Deposited metal}}{\text{Deposition Rate}}$

$\left\{\begin{array}{l}\text{Gas flow rate}\\\text{used for calculation}\end{array}\right\}$ $\left\{\begin{array}{l}\text{10-to-15 ft}^3/(100\text{A})\cdot(\text{h})\\\text{0.3-to-0.4 m}^3/(100\text{A})\cdot(\text{h})\end{array}\right\}$

Gas required = Gas flow rate × Arc time
a. Assumed for this example

Table 8.13
Sample Calculation of Labor Hours Required for 300 ft of Fillet Weld

	SMAW	FCAW
Weld size, in. (mm)	1/4 (6)	1/4 (6)
Weld Length, ft (m)	300 (91)	300 (91)
Deposited weld metal weight, lb (kg)[a]	36 (16)	36 (16)
Deposition rate, lb/hr (kg/h)[b]	6 (2.7)	9 (4.1)
Operating factor, %[c]	30	40
Labor, h[d]	20	10
Gas required, ft³ (m³)[e]	NA	160 (4.5)

a. From Table 8.1
b. From weld procedure (not shown) and Figures 8.2 and 8.4
c. From Table 8.11
d. Labor calculation

Arc time, (h) $= \dfrac{\text{Deposited weld metal}}{\text{Deposition rate}}$

Labor, (h) $= \dfrac{100X \text{ arc time}}{\text{Operator factor}}$

e. Gas required (at 40 ft³/h) = Gas flow rate × arc time

Table 8.14
Typical Welding Material and Labor Standards

Joint Configuration	Plate Thickness, in.[b]	Root Opening, in.[b]	Deposited Metal Weight, lb/ft[c]	Submerged Arc[a]			Shielded Metal Arc[a]	
				Filler Metal Required, lb/ft[c]	Flux Required, lb/ft[c]	Labor Required, h/ft[d]	Filler Metal Required, lb/ft[c]	Labor Required, h/ft[d]
Table 8.4	1/4	1/16	0.23	0.23	0.23	0.02	0.4	0.18
Table 8.4	1/2	1/8	1.0	1.0	1.0	0.07	1.5	0.8
Table 8.4	3/4	1/8	1.6	1.6	1.6	0.11	2.5	1.3
Table 8.3	1/8	1/16	0.09	0.09	0.09	0.012	0.14	0.072
Table 8.3	3/16	3/32	0.15	0.15	0.15	0.02	0.23	0.12
Table 8.3	1/4	1/8	0.21	0.21	0.21	0.028	0.32	0.17

Welding Process	Electrode Diameter, in.	Welding Current, A	Deposition Efficiency, %	Flux Ratio	Deposition rate, lb/h	Operator Factor, %
Submerged Arc	5/32	500	100	1	15	65
Shielded Metal Arc	7/32	350	65	N.A.	5	25

a. Process data.
b. To Convert from in. to mm, multiply inches by 25.4.
c. To Convert from lb/ft to kg/m, multiply lb/ft by 1.49.
d. To Convert from h/ft to h/m, multiply h/ft by 3.28.

CONTROL OF WELDING COSTS

PREPRODUCTION FUNCTIONS

Joint Design

THE FUNDAMENTAL FACTOR in welding cost is the weight of deposited metal. All other welding costs can be related to the deposited metal weight. Therefore, any change that results in less deposited metal will also reduce each welding cost item. Deposited metal weights can be reduced in several ways. The simplest is to reduce the cross-sectional area of the joint by decreasing the root opening, using a root face on groove welds,

decreasing the groove angle, and using double V- or U-grooves, as shown in Figure 8.6. To achieve a reduction in cross section, the parts must be cut and fit accurately so that the overall dimensions of the assembly meet the requirements. In addition, if the groove angle is too small or the root face is too wide, the possibility of incomplete joint penetration or other unacceptable weld discontinuities may be increased. Defective welds are very costly because they must be removed and rewelded. Changes in groove geometry should be carefully analyzed and tested before being implemented to assure that overall dimensions meet the specification requirements and that weld quality will be acceptable.

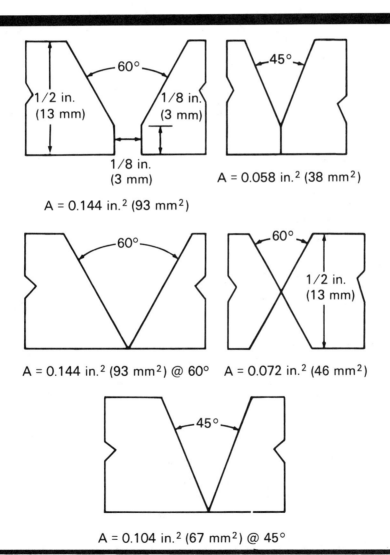

Figure 8.6—Effects of Minor Changes in Joint Design on Weld Cross-sectional Area

Process Selection

EACH WELDING PROCESS in commercial use has areas of application where it has economic advantages. The areas are broad and there is considerable overlap, especially with the consumable electrode arc welding processes.

Many fabricators have production capacity for several welding processes. Choosing the most suitable welding process for each application is vital to minimizing welding costs.[2] Some of the characteristics and advantages of the consumable electrode arc welding processes are shown in Table 8.15.

The choice of welding processes is the responsibility of each enterprise. To a degree, the choice of processes will control the type of products that can be produced competitively. To fabricate weldments at the lowest cost, processes that produce acceptable quality at high deposition rates and with high operator factors are necessary. The best process for a job depends on the job, and is the responsibility of the enterprise.

The deposition rates of the SMAW, GMAW, FCAW, and SAW processes are shown in Figures 8.2 through 8.5. When large quantities of weld metal are required, the operator factor will normally increase with increasing mechanization, as shown in Figure 8.7.

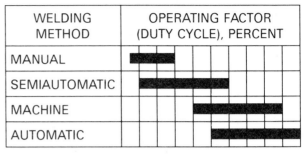

MECHANIZATION RAISES OPERATING FACTOR

Figure 8.7—Effect of Mechanization on Operator Factor

Assuming the work can be properly positioned and the equipment guided, the most efficient fusion welding process in plate and structural work, such as shipbuilding, bridge building, and pressure vessel fabrication, is the submerged arc welding process. However, when welds must be made out-of-position or when several short welds are required on many pieces involving frequent moves of the welder or the work, a flexible welding process should be used, such as SMAW, GMAW, or FCAW. The best choice is a compromise between welding speed (deposition rate), versatility (all position), and portability (operator factor).

PRODUCTION FUNCTIONS

Production Planning

MAINTAINING PRODUCTION SCHEDULES is vital to controlling manufacturing costs in general and welding costs in particular. Therefore, the materials should be delivered in a timely manner. The following rules help to prevent work delays:

(1) Have the proper equipment for the specified welding process.
(2) Provide the welding materials specified in the welding procedures.
(3) Provide accurately positioned and aligned parts using fixtures where possible, and inspect the fit-up before welding.
(4) Use positioning equipment when possible. Positioning for welding in the flat position greatly increases efficiency and reduces costs.
(5) Provide power tools to remove slag and finish the weld surfaces.
(6) Supervise work to verify that procedures are followed, consumables are not wasted, and workmanship is satisfactory.

Table 8.15
Advantages of Consumable Electrode Arc Welding Processes

Process	Advantages
SMAW	Flexible, all position process; low initial cost; portable; large variety of filler metals available with special characteristics (high deposition, fast travel speed, good fillet weld contour, deep penetration, etc.). Slag requires removal.
GMAW	Relatively flexible; requires wire feeder and external gas; need a special power source for all position capability; higher deposition rates than SMAW; no slag; can be adapted to mechanized, automatic, and robotic welding.
FCAW	Relatively flexible; requires wire feeder, and most electrodes require external gas; all position capability without special power source; higher deposition rate than SMAW and GMAW; can be adapted to mechanized, automatic, and robotic welding; requires slag removal.
SAW	Flat and horizontal position only, but very high deposition rate process. Must be mechanized for highest deposition rates. High current power sources, heavy-duty wire feeders, and welding head or workpiece manipulators are high capitol cost items. Mechanized SAW is high quality, low-cost process. Flux is required. Slag and excess flux requires removal.

2. Welding processes discussed in Chapter 1 are discussed in detail in the *Welding Handbook, Vols. 2 and 3, 7th Ed.* Welding processes will be covered in the *Welding Handbook,* Vol. 2, 8th Ed.

Welding Procedures

WELDING PROCEDURES ARE written instructions for shop personnel to follow in fabricating production weldments. The procedures should be thoroughly tested to verify that weldments of desired quality will be produced when the procedure is followed.

Supervision should select welding procedures from a portfolio of qualified procedures that best meet the requirements of the job. When selecting the welding process, the following should be considered:

(1) Product quality
(2) Manufacturing schedule
(3) Manufacturing cost

Supporting Activities

PRODUCTION WELDING IS normally preceded by preassembly and fitting, and often followed by cleaning, machining, painting, or all three. Cost control is a cooperative effort, and the support of all departments is required. Implementation of the following rules can help to reduce overall manufacturing costs:

(1) Prepare parts as accurately as needed, particularly those parts that are bent or formed to shape.
(2) Prepare parts by shearing or blanking where possible; it may be more economical than thermal cutting.
(3) Accurately and clearly mark all cut parts with job and part identity.
(4) Use automatic equipment when available and suited to the job.
(5) Inspect all parts for accuracy before delivery to the fitting area.
(6) Fit parts accurately.
(7) Avoid excessive use of temporary welded strongbacks or other fitting aids that have to be removed by gouging and grinding.
(8) Place tack welds in grooves or in fillet locations. (Tack welds should be small enough to be consumed by the production weld.)
(9) Schedule delivery of materials so that all parts of an assembly are available as needed at the fitting area.
(10) After welding, remove all slag, spatter, and other matter from the weldment.
(11) Inspect each operation prior to sending the assembly to the next operation.

Overwelding

OVERWELDING SIGNIFICANTLY CONTRIBUTES to excessive welding costs. The dramatic increase in weld cross

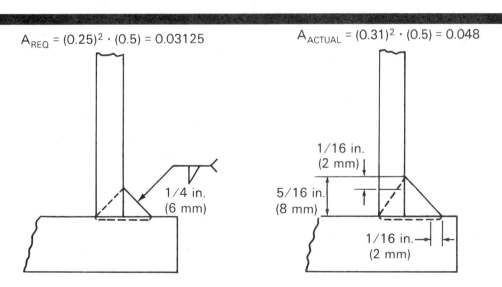

$A_{REQ} = (0.25)^2 \cdot (0.5) = 0.03125$

$A_{ACTUAL} = (0.31)^2 \cdot (0.5) = 0.048$

1/4 in. (6 mm)

1/16 in. (2 mm)

5/16 in. (8 mm)

1/16 in. (2 mm)

**Overwelding 1/4 in. (6 mm) Fillet by 1/16 in. (2 mm)
Increases Area by 56%**

Figure 8.8A — Effect of Overwelding on Weld Cross Section

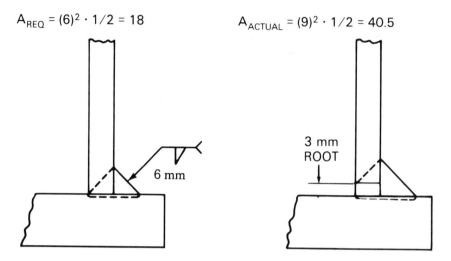

$$A_{REQ} = (6)^2 \cdot 1/2 = 18 \qquad A_{ACTUAL} = (9)^2 \cdot 1/2 = 40.5$$

6 mm

3 mm
ROOT

**Fitting Web with 3 mm Root Opening Requires Additional 3 mm Fillet Leg
and Increases Area by 125%**

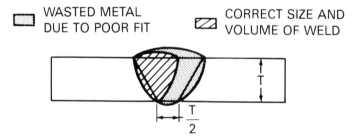

WASTED METAL
DUE TO POOR FIT

CORRECT SIZE AND
VOLUME OF WELD

$$\frac{T}{2}$$

T

Increased Root Opening of Groove Weld Doubles Cross Section

Figure 8.8B—Effect of Poor Fitup on Weld Cross Section

section as a result of overwelding is shown in Figures 8.8A and 8.8B. Overwelding results from inaccurate cutting and fitting, poor supervision, and insufficient training or lack of confidence in the strength of welds. Two joint configurations that often result in overwelding are full or partial penetration welds in T-joints fabricated in the horizontal position, and butt joints between plates of unequal thickness. The desired welds and the common overwelded conditions are shown in Figure 8.9 and 8.10.

Weld Quality

WORK OF POOR quality will also adversely affect welding costs. The cost of weld repairs can be two-to-three times that of the original weld. Because the repairs take time as well as workers and material, valuable shop space is lost and the overall production schedule is delayed. Poor quality work may adversely affect the reputation of the manufacturer, and that may be detrimental to future sales.

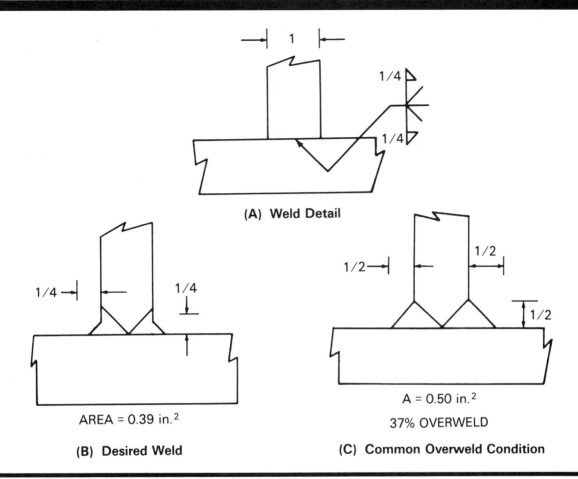

(A) Weld Detail

AREA = 0.39 in.²

(B) Desired Weld

A = 0.50 in.²

37% OVERWELD

(C) Common Overweld Condition

Figure 8.9—Common Overwelded Full Penetration T-Joint Fabricated in Horizontal Position

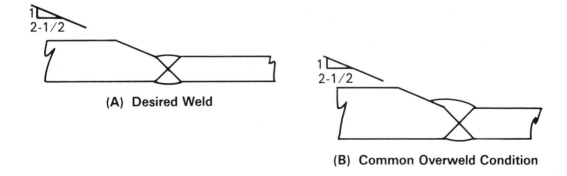

(A) Desired Weld

(B) Common Overweld Condition

Figure 8.10—Potential Overwelding of Transition Butt Joints

CAPITAL INVESTMENTS

JUSTIFICATION OF CAPITAL EQUIPMENT

INVESTMENTS IN CAPITAL equipment can reduce costs and improve a manufacturing operation. Such investments can be low cost and simple, such as replacing shielded metal arc welding with a continuous wire welding process (gas metal arc or flux cored arc welding). They can be costly and complex such as a computer-controlled, robotic, flexible manufacturing cell to replace several operations. (Flexible manufacturing cells are discussed in Chapter 10 of this volume.)

Before committing investment capital, verify that the intended facility is economically viable by conducting a financial analysis in which the costs and potential benefits of the investment are forecast.

The costs should include the following:

(1 The engineering, purchase, and installation cost of the proposed production equipment

(2) The cost of any peripheral equipment necessary to use the proposed equipment

(3) Annual operating costs of the production and support equipment

(4) Maintenance cost of the production and support equipment

The potential benefits include:

(1) Cost savings on discontinued operations
(2) Savings from improved quality
(3) Profit from increased production
(4) Identifiable savings from safety improvements
(5) A reduction in work-in-process inventory

The time periods when the costs are incurred and the benefits achieved are needed to prepare a discounted cash flow analysis. A discounted cash flow analysis provides the annual and cumulative financial results of the capital investment. The analysis uses current or projected interest rates, and calculates the time for capital recovery.

Several factors should be considered to determine the best capital investments and to develop the cost and benefits of the investments. They are discussed below.

SALES VOLUME ESTIMATES

FIRST AND FOREMOST in planning for welding shop improvements is a firmly established production volume forecast. A production volume of some minimum number of units will be required. Generally, the cost of manufacturing equipment is proportional to its production

capacity. For long production runs, dedicated automation is frequently more efficient to use. A pipe mill is an extreme example of dedicated automation, and the manufacturing cost of transmission pipe is low. However, such mills are high capital facilities that can only be used to manufacture pipe.

For relatively short production runs, flexible, multipurpose automation can be used to advantage even though expensive. An example of the latter is a microprocessor-controlled robot. In still other cases, investment in semiautomatic welding equipment and additional fixtures may result in significant cost benefits.

Thus, the types of product and the production quantities should be forecast to determine if dedicated or flexible automatic or semiautomatic equipment should be purchased. The relative unit cost of manual or semiautomatic, dedicated, and flexible manufacturing is a function of volume. The relationship is different for every plant, but the relative productivity of the three approaches to manufacturing is illustrated in Figure 8.11.

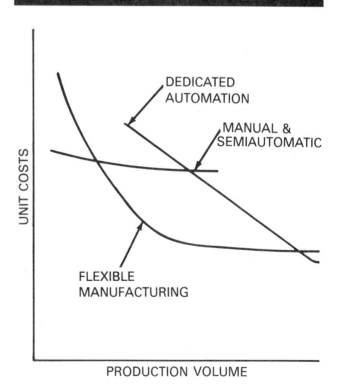

Figure 8.11—Effect of Production Volume on Unit Cost for Manual, Dedicated, and Flexible Manufacturing Stations

PART GEOMETRY

SPECIAL PURPOSE AUTOMATION may require geometric simplicity, such as straight or circular welds. For example, tubes and pipes can be welded relatively fast and at low cost because of the simplicty of their geometry. Parts having complicated weld paths may have to be welded manually unless the plant can invest in special purpose equipment that (1) can be preprogrammed to track the weld path, (2) is equipped with a seam tracking device, or (3) is designed with permanent mechanical motion paths using cams, gears, and templates. The thickness and size of the part, as well as a requirement for mobility may favor special purpose equipment.

ACCURACY OF PARTS

AUTOMATED WELDING EQUIPMENT generally requires better component accuracy than would be needed for manual welding. A welder visually senses variations in parts and instinctively makes corrections to compensate for them. Sensory systems and adaptive feedback controls can be added to a welding machine to correct for joint location and width, but the equipment is expensive and primitive relative to the skilled response of a human welder. A better solution would be to improve the component accuracy and assembly fitup. The inherent accuracy of the welding machine is also an important factor. Component accuracy is meaningless if the tracking of the welding head and the welding current controls are not accurately coordinated to deposit the correct weld size in the correct location.

MATERIAL HANDLING

THE CONTINUOUS CHARACTER of automated welding requires associated systems for material handling that frequently are integrated to work in sequence with the welding machine. Manually-loaded welding machines often are paced by the operator. In many cases, automatic loading and ejection of parts is incorporated in the welding machine. Conveyors or large containers are used to bring parts to the machine and to transfer them to the next manufacturing operation. Material handling equipment can also improve the productivity of shops using manual and semiautomatic welding equipment.

SAFETY

THE CONSIDERATION FOR safety requires more care than usual in operating, setting up, and maintaining an automated welding machine. Most automatic welding machines have rapidly moving components, and they do not sense the presence of operating and maintenance personnel. Typical safety requirements of those machines are a fail-safe system and easy access to emergency stop buttons that stop the machine immediately. In normal production operations, the operator and others must be kept out of the operating envelope of the machine. This requires guards that may be interlocked with the machine to prevent careless operation of the machine if a guard is removed for maintenance.

DEPOSITION EFFICIENCY

ONE OF THE benefits of automation is the precision with which the equipment will repeat each activity. Welding current and travel speed can be set to give the exact weld size required and to minimize weld spatter. Automation thus improves the deposition efficiency even if the welding process is unchanged. Significant improvements in deposition efficiency are achieved over manual welding.

OPERATOR FACTOR

IMPROVEMENTS IN THE operator factor can be significant with automatic welding equipment, as indicated previously in Figure 8.7. However, operator factor improvements depend on the production volume. When large weld deposits are required on a single weldment, a significant amount of continuous welding is necessary. For example, the girth seam in a 15 ft (4.6 m) diameter pressure vessel with a two-inch (50 mm) wall thickness may provide eight hours of continuous work for an automatic submerged arc station. On the other hand, a GMAW robotic station may require only two or three minutes to weld a frame for a furniture manufacturer, and a steady flow of frames is required to take advantage of the productive capacity of the robot.

WELD JOINT DESIGN

ANOTHER BENEFIT THAT can be realized by the use of automation is the capability for higher filler metal deposition rates which, in turn, permits the use of more efficient weld joint designs. For example, it might be possible with automation to change from a 60 degree V-groove to a square-groove joint design. Such a change may permit higher welding speed, lower deposited metal requirements, and smaller variation in weld size. Smaller variations in weld size will result in less filler metal being used and will minimize overwelding. Overwelding has been explained to be a negative factor in both labor and material costs.

OVERALL IMPROVEMENT FACTOR

SEVERAL FACTORS HAVE been considered that may result in increased productivity with automated equipment.

The overall improvement factor is the product of those factors, including operator factor, a factor for avoidance of overwelding, and the welding process improvement factor.

Typically, substitution of an automated GMAW system for semiautomatic welding to produce fillet welds can result in improvement by a factor of about 1.4 for the welding process; the reduction in overwelding may be about 2.25; and the operator factor improvement may be about 2.5. Overall, automatic GMAW may be nearly eight times more productive than semiautomatic GMAW, and even more productive than an SMAW station. The productivity improvements are illustrated in Figure 8.12.

COMBINED OPERATIONS

MECHANIZATION OF WELDING equipment can allow the simultaneous operation of two or more welding heads. Welding both fillets of a double fillet weld simultaneously is a common practice in both shipbuilding and girder fabrication. Shown in Figure 8.13 is a multiple-head gantry-type side beam that illustrates the efficiencies of mechanization.

Computer controlled flexible manufacturing cells can extend this concept of combined operations to include nonwelding activities such as fitting, punching, or gouging. The cost savings can be leveraged significantly by integrating several operations within a cell.

INSTALLATION COSTS

THERE ARE SEVERAL resource requirements which are unique to automated equipment. Planning must include provisions for adequate floor space, foundations, special installation equipment, and utilities.[2] It is not uncommon for the cost of installing automated equipment to be

5 to 10 percent of the initial cost of the machine. It is, therefore, important to plan for these costs and include them in the initial economic analyses.

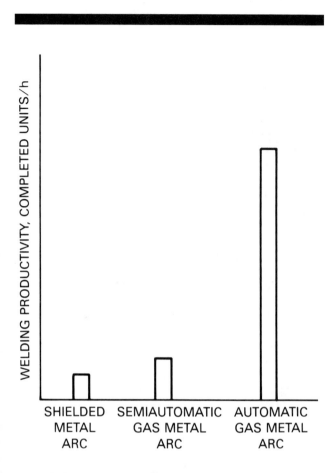

Figure 8.12—Effect of Mechanization on Welding Productivity

---

2. "Planning for Automation" is discussed in Chapter 10, Automation and Control.

Figure 8.13—High Productivity Multiple Head Stiffener Welder

ECONOMICS OF BRAZING AND SOLDERING

BRAZING AND SOLDERING processes require the use of filler metal. Many filler metals for these processes are costly because they contain noble or rare elements to provide desired flow and joint properties. Every effort must be made to minimize the amount of filler metal for each joint. This improves both cost and quality. The amount of filler metal required is dependent on a number of factors. Among the factors are joint designs, fabrication techniques, procedures, and operations used to prepare and finish parts. Changes in any of these factors might improve quality and reduce costs.

JOINT DESIGN

MANY VARIABLES MUST be considered in designing brazed and soldered joints. From the mechanical standpoint, the joints must be capable of carrying the service loads. Rules applying to concentrated loads, stress concentration, static loading, and dynamic loading are more complex than with other machined or fabricated parts.

The design of a brazed or soldered joint has specific requirements that should be met. Some of the important factors follow.

Chemical Compositions

THE COMPOSITIONS OF the base and filler metals and the fluxes must be compatible. The properties of the filler metal in the joint as well as the cost of the filler metal must be considered when designing for a given service condition.

Type and Design of Joint

THERE ARE TWO basic types of joints used in brazing and soldering operations, the lap joint and the butt joint. Excessive overlap in a lap joint leads to unnecessarily high costs. Although butt joints are cheaper in terms of filler metal, they may not only lack strength but also require costly preparation and fixturing to achieve the required critical fitup.

Service Requirements

MECHANICAL PERFORMANCE, ELECTRICAL conductivity, pressure tightness, corrosion resistance, and service temperature are design considerations that affect costs. For example, brazing an electrical connection requires a high conductivity joint, and high cost filler metal may be needed to satisfy the requirement.

Joint Clearance

IMPROPER JOINT CLEARANCE can affect costs by requiring excessive amounts of filler metal or repair of defective joints.

PRECLEANING AND SURFACE PREPARATION

CLEAN, OXIDE-FREE surfaces are imperative to ensure acceptable brazed joints. The costs of precleaning and surface preparation generally include cleaning materials and labor. Selection of an appropriate cleaning method depends on the nature of the contaminant, the base metal, the required surface condition, and the joint design.

ASSEMBLY AND FIXTURING

THE COST OF assembly fixturing and the added processing times for fixturing of assemblies make it advantageous to design parts that are easily assembled and self-fixturing. Some methods employed to avoid fixtures are resistance tack welding; arc tack welding; interlocking tabs and slots; and mechanical methods, such as staking, expanding, flaring, spinning, swaging, knurling, and dimpling. Some self-fixturing assemblies are shown in Figure 8.14.

COSTS OF BRAZING AND SOLDERING

MANY VARIABLES AFFECT the cost of making sound brazed and soldered joints, as shown in Table 8.16. Of fundamental importance is the technique selected, based on the following factors:

(1) Base metal
(2) Joint design
(3) Surface preparation
(4) Fixturing
(5) Flux or atmosphere
(6) Filler metal
(7) Method of heating

These factors, singly or in combination, have a direct bearing on the costs of the operation. A correctly applied technique based on sound engineering will result in a quality product at minimum cost. The desired goal is to heat the joint or assembly to brazing temperature as uniformly and as quickly as possible, while avoiding localized overheating. Mechanized equipment must be suitably constructed to provide proper control of temperature, time, and atmosphere. In many cases, consideration must be given to thermal expansion of the base metal to preserve correct joint clearance while the assembly is raised to the brazing temperature.

Table 8.16
Typical Variables for Estimating Brazing Costs

Brazing process	Brazing temperature range
Method of application	Brazing flux or atmosphere
Type of joint	Labor
Filler metal	Postbraze cleaning
Type	Inspection
Form and size	Consumable costs
Method of application	Power or fuel costs

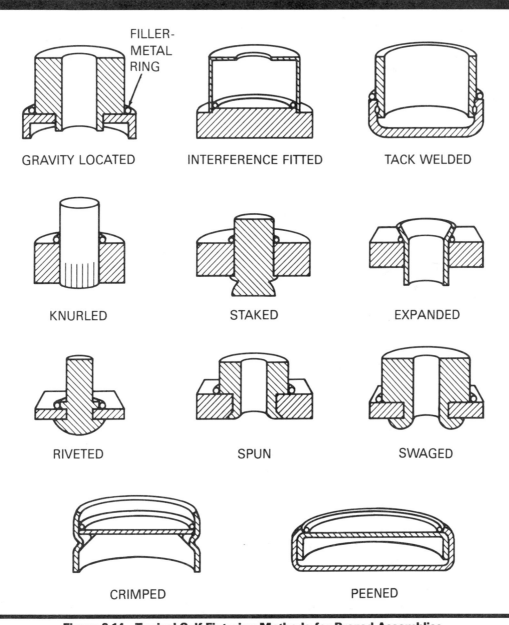

Figure 8.14—Typical Self-Fixturing Methods for Brazed Assemblies

ECONOMICS OF THERMAL CUTTING

OXYFUEL GAS AND plasma arc cutting are commonly used for shape cutting of parts from sheet and plate.[3] Oxyfuel gas cutting (OFC) is generally limited to carbon and low alloy steels; special process modifications are required to cut high alloy steels. Plasma arc cutting (PAC) can be used to cut any metal. Most PAC applications involve carbon steel, aluminum, and stainless steel. Both processes can be used for plate beveling, shape cutting, and piercing.

---

3. For information on oxyfuel gas and plasma arc cutting, refer to the *Welding Handbook, Vol. 2, 7th Ed.*

Both processes can be operated manually or used with portable machine cutting equipment. However, the PAC process requires a special power source, and PAC cutting torches are usually more bulky than OFC torches. As a result, when either OFC or PAC can be used, OFC is preferred in both manual and portable machine cutting applications.

Shape cutting machines for PAC and OFC are similar in design. Generally, plasma arc shape cutting machines can cut at higher speeds than similar OFC machines.

Carbon steel plate can be cut faster with PAC than with OFC in thicknesses under 3 in. (75 mm) when the appropriate equipment is used. For thicknesses under 1 in. (25 mm), PAC speeds can be up to five times those for OFC. Over 1-1/2 in. (38 mm) thickness, the selection of PAC or OFC will depend on other factors such as equipment costs, load factor, and applications for cutting thinner steel plates and nonferrous metals.

Capital costs for OFC equipment are relatively low compared to PAC equipment. The highest cost item for PAC equipment is the power source, and large, high voltage power sources are required for cutting thick plates. Because the cutting speed advantage of PAC over OFC decreases with increasing plate thickness, PAC stations are usually delegated to plate thickness less than 1 in. and OFC is used on thicker plates.

The economic advantages of PAC over OFC are exploited best in multiple torch, numerically or optically controlled cutting machines. These are high volume stations used in large fabrication plants, service centers, shipyards, and other facilities where large quantities of cut material are required.

High capacity plasma-arc cutting machines are often equipped with water tables. The use of water tables improves the cut quality, reduces the smoke and glare associated with PAC in air, but further increases the initial capital expenditure.

SUPPLEMENTARY READING LIST

An easy way to compute cost of weld metal, *Welding Design and Fabrication,* July 1982: 54-56.

Cary, H. *Modern welding technology,* Englewood Cliffs, NJ: Prentice Hall (1979): 541-59.

Hines, W. G. Jr. Selecting the most economical welding process. *Metal Progress.* 102: Nov. 1972; 42-44.

Lesnewich, A. The real cost of depositing a pound of weld metal. *Metal Progress.* 121: Apr. 1982; 52-55.

Mahler, V. Designer's guide to effective welding animation—Part II: flexibility and economics. *Welding Journal.* 65(6): June 1986; 43-52.

Necastro, N. P. How much time to weld that assembly. *Welding Design and Fabrication.* Feb. 1983: 50-53.

Oswald, P. Cost estimating for engineers and managers. Englewood Cliffs, NJ: Prentice-Hall, (1974).

Pandjiris, A. K., Cooper, N. C., and Davis, W. J. Know costs—then weld. *Welding Journal.* 47(7): July 1968; 561-68.

Pavone, V. J. Methods for economic justification of an arc welding robot installation. *Welding Journal.* 62(11): Nov. 1983; 40-46.

The Lincoln Electric Company. *Procedure handbook of arc welding.* Cleveland, OH: The Lincoln Electric Co. (1973): Section 12.

Rudolph, H. M. The hidden costs of thermal cutting. *Welding Design and Fabrication.* Aug. 1979: 76-80.

American Welding Society. Shipbuilder beats the elements while cutting energy costs. *Welding Journal.* 61(12): Dec. 1982; 46-47.

Smith, D., and Layden, L. Figure the way to cutting economy. *Welding Design and Fabrication.* Mar. 1978: 106-7.

Sullivan, M. J. Application considerations for selecting industrial robots for arc welding. *Welding Journal.* 59(4): Apr. 1980; 28-31.

Penton Publishing, Inc. *Welding & Fabricating Data Book 1986/1987.* Cleveland, OH: Penton Publishing, Inc. (1986).

FIXTURES AND POSITIONERS

PREPARED BY A COMMITTEE CONSISTING OF:

D. R. Spisiak, P. E., Chairman
Gaymar Industries Incorporated

L. P. Connor
American Welding Society

P. H. Galton
Big Three Industries, Inc.

E. B. Gempler
United Aircraft Products, Inc.

L. M. Layden
The BOC Group Tech. Ctr.

P. K. Wadsworth
Ohio State University

WELDING HANDBOOK COMMITTEE MEMBER:
E. H. Daggett
Babcock and Wilcox Company

CHAPTER 9

FIXTURES AND POSITIONERS

FIXTURES

GENERAL DESCRIPTION

IN ENGINEERING, THE terms *jig* and *fixture* have essentially the same meaning. The function of a fixture is to facilitate assembly of parts and to hold the parts in a fixed relationship. The parts may be partially or completely welded in the fixture or may be tack welded only and removed prior to welding. If the assembly is only tack welded together it is called a *fitting jig*. The use of a fixture also promotes good fitting tolerances in the final product. Quality weldments can be produced in fitting-and-welding fixtures at low costs.

The design and manufacture of the fixture should reflect the number of parts to be produced. Small quantities may be produced on temporary jigs assembled specifically for the product. For large quantities, the fixture could be an integral part of the whole production system. In this case, the fixture may include automatic clamping on a positioner, and the fixture may accommodate welding by an industrial robot. This type of fixture is expensive to build, but if many parts are produced on it, the fixture is cost effective.

Fixtures are used for three major purposes: (1) assembly fixtures, (2) precision fitting and welding fixtures, and (3) robotic welding fixtures.

Design

THERE ARE FEW standard commerical fixtures, but many light and heavy duty clamping devices are available. These devices can be incorporated into dedicated fixtures for large production runs, or into adjustable fixtures that can be easily modified for several short run products. For the most part, fixtures are designed and built by plant operations personnel to facilitate the production of one or more assemblies.

A production fixture should be designed with the following desirable features:

(1) Weld joints must be accessible in the fixture.
(2) The fixture must be more rigid than the assembly.
(3) Holddowns, clamps, and threads of bolts and nuts should be protected from weld spatter.
(4) The fixture should allow assembly of the work with a minimum of temporary welds that are visible on completion.
(5) The workpiece must be easy to remove from the fixture after the welding is complete.

The designer must decide how many welds are to be made while the work is in the jig. For example, the second side of a complete joint penetration weld may be deposited after the weldment is removed from the fixture. Sufficient welds should be completed in the jig to restrain the assembly from distortion during the completion of welding outside the jig. Since most weldments are fabricated as subassemblies, the tolerances are critical. However, often the intermediate dimensions are less important than the end and edge dimensions that control the fit in the final assembly.

Applications

THERE ARE MANY types and variations of fixtures in metal fabrication, but most fall into three broad categories. These categories are: assembly or fitting fixtures, precision jigs designed to produce accurate fabrications, and fixtures for robotic welding.

Assembly Fixtures. The fabrication of a light weight dump trailer is illustrated in Figure 9.1. Internal jigs, shown in Figure 9.1(A), illustrate the ease of assembly of the trailer shell. The external box beams shown in Figures 9.1(A) and (B) are pushed against the shell plate by bolts screwed through the heavy vertical angles. The jig produces a satisfactory fit-up quickly without any temporary welds to the workpiece. The completed assembly is shown in the jig in Figure 9.1(C).

Precision Jigs. A jig designed to produce a fabrication to close tolerances is shown in Figures 9.2A and B. The completed assembly is an aluminum air duct for a jet aircraft. The duct passes through a fuel tank, and therefore, close face plate tolerances are required. In addition to a precise jig, a detailed step-by-step assembly and welding procedure was required to meet the fabrication tolerances of 0.030 inches (8 mm).

Robotic Welding. Fixtures for robotic welding have several specific requirements. They must allow access for the robot. Most robots are not equipped with sensing systems and, furthermore, robotic sensing and vision systems are primitive in comparison to the vision of a human being. Therefore, the fixture should have low profile clamps, and the clamps should be located away from the joint. The fixture should contain at least two reference points that are in a fixed relationship to the weld seams of the workpiece. The robot is then programmed to locate the reference points on the fixture. The reference points establish a coordinate system for the robot to use to find its way along the joint on the workpiece.

The fixture should be easy to use so that the workpieces can be loaded and removed rapidly. The robotic cell shown in Figure 9.3 includes a turntable so that the operator can load and remove a small electronic chassis while the robot makes the production welds behind the partition. At the completion of the cycle, the turntable rotates 180 degrees, and the process is repeated.

(A) Internal

Figure 9.1—Assembly Jig for Production of Dump Trailers

(B) External

(C) Finished Assembly

Figure 9.1 (Continued)—Assembly Jig for Production of Dump Trailers

Figure 9.2A—Fitting and Welding Jig to Produce Tight Tolerance Assembly

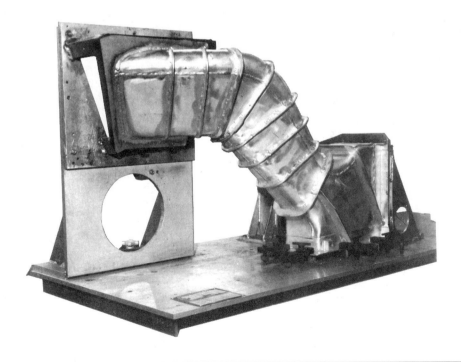

Figure 9.2B—Alternate View of Fitting and Welding Fixture

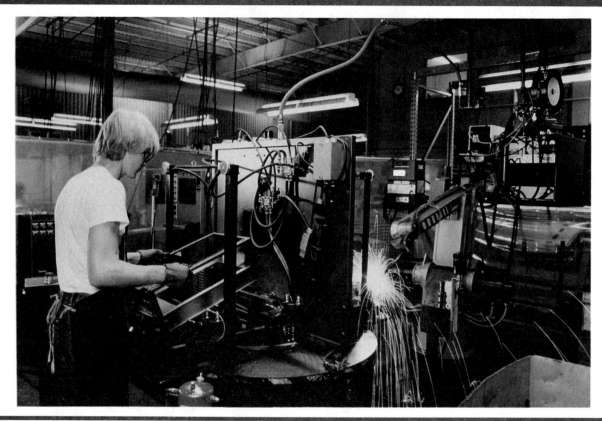

Figure 9.3—Operator Loads Assembly on Turntable in Robotic Welding Station

POSITIONERS

GENERAL DESCRIPTION

A POSITIONER IS a mechanical device that supports and moves a weldment to the desired position for welding and other operations. In some cases, a positioner may move a weldment as welding progresses along a joint. A welding fixture may be mounted on a positioner to place the fixture and the weldment in the most advantageous positions for loading, welding, and unloading.

Some assemblies may be fixtured on the floor and the joints tack welded to hold the assembly together. Then, the weldment is removed from the fixture and mounted on a positioner for welding the joints in the best positions for economical production.

A weldment on a positioner may be repositioned during welding and upon completion for cleaning, machining and nondestructive inspection of the weld.

ECONOMIC CONSIDERATIONS

THERE ARE BOTH positive and negative considerations in the use of positioners for welding. They are primarily governed by welding and handling costs. On the positive side are the high deposition rates, high operating factors and high product quality that can be achieved using positioners to orient the work for flat position welding. On the negative side, there is the handling cost to load and unload the positioner.

Deposition Rates

THE HIGHEST DEPOSITION rates in arc welding can be obtained when welding is done in the flat position because gravity keeps the molten metal in the joint. The next best position is the horizontal. The overall result of positioning of weldments should be lower welding costs.

Welder Skill

IT TAKES LESS skill to arc weld in the flat position than in other welding positions because it is easier for the welder to control the molten weld pool. Therefore, welding labor costs should be lower when the assembly can be easily manipulated for welding joints in the flat position.

Operator Factor and Set-Up Costs

OPERATOR FACTOR IS the ratio of arc time to total time that a welder applies to a weldment. When the welder must manually reposition a weldment, wait for a crane operator to move it, or weld in positions other than flat, the operator factor will be lower and welding costs higher than when a positioner is used. Operator factor should be higher when a weldment can be rapidly positioned for welding. However, labor costs for safely loading and unloading a heavy weldment on a positioner, or for repositioning a weldment with a crane or other lift, must also be considered.

For relatively short, small welds, it may be more economical to weld the joints in fixed positions than to reposition the weldment for ease of welding. The welding costs may be somewhat higher but the overall labor costs may be lower because of savings in handling costs. The cost of positioning equipment and material handling must be offset by labor savings.

Weld Quality

IN GENERAL, A qualified welder produces welds in the flat position with fewer defects than in other welding positions. The result is fewer weld repairs. Also, a joint can be filled with fewer passes because of the high deposition rates in the flat position. Welds made with a minimum of passes generally have lower welding stresses and associated distortion. However, heat input limitations must be observed with some alloy steels.

TYPES OF POSITIONING

POSITIONING CAN BE done with one, two, or three different motions. One motion is rotation about one axis. This is normally accomplished with turning rolls, headstock and tailstock arrangements, or turntables, all of which rotate the assembly about a single axis.

Two motion positioning is a combination of rotation and tilting. It is normally accomplished with a positioner that has a tilting table as well as rotation. Three motion positioning is accomplished by adding vertical movement with an elevating device in the machine base, thus providing rotation, tilt, and elevation.

TURNING ROLLS

TURNING ROLLS ARE used in sets, as shown in Figure 9.4. Each set consists of one powered roll and one or more idler rolls, each roll having two or more wheels. Turning roll design is quite simple. A set of rolls normally consists of a fabricated steel frame, wheels, drive train, drive motor, and controls. Simplicity of design offers low initial cost as well as low maintenance and repair costs. Standard models can be used for many applications.

Turning rolls can be manufactured to specific requirements, such as wheel construction and surface composition, weldment weight and diameter, special motions, and unitized frames.

Turning rolls are normally limited to rotation of a cylindrically shaped weldment about its horizontal axis. Noncylindrical assemblies can also be rotated on turning rolls using special round fixturing to hold the assembly. The fixtures rest on the turning rolls.

Turning rolls can position seams for manual, semiautomatic, and machine welding in the flat position. With machine welding, circumferential welds may be rotated under a fixed welding head. Longitudinal welds can be made with a welding head mounted on a traveling carriage or manipulator.

HEADSTOCK-TAILSTOCK POSITIONERS

General Description

A HEADSTOCK IS a single-axis positioning device providing complete rotation of a vertical table about the horizontal axis. This device provides easy access to all sides of a large weldment for welding in the flat position and for other industrial operations.

A headstock is sometimes used in conjunction with a tailstock, as shown in Figure 9.5. A tailstock usually has the same configuration as the headstock, but it is not powered. A simple trunnion, an outboard roller support, or any other free-wheeling support structure best suited to the configuration of the weldment may be used in place of a tailstock.

In all instances, precise installation and alignment of a headstock-tailstock positioner is essential so that the axes of rotation are aligned. Otherwise, the equipment or the weldment may be damaged.

Generally, the concept and application of headstock-tailstock positioners are much the same as lathes used in machine shop operations. However, in welding fabrication, the applications can be more versatile with the use of special fixtures and the addition of horizontal or vertical movement, or both, to the bases of the headstock and tailstock.

A headstock may be used independently for rotating weldments about the horizontal axis, provided the overhanging load capacity is not exceeded with this type of usage. An example is the welding of elbows or flanges to short pipe sections. A headstock can also be used in combination with idler rolls.

Applications

HEADSTOCK-TAILSTOCK POSITIONERS are well suited for handling large assemblies for welding. Typical assem-

Figure 9.4—Turning Rolls

blies are structural girders, truck and machinery frames, truck and railroad tanks and bodies, armored tank hulls, earthmoving and farm equipment components, pipe fabrication, and large transformer tanks. A headstock-tailstock positioner is sometimes used in conjunction with one or more welding head manipulators or with other machine welding equipment. Appropriate fixtures and tooling are needed with a headstock-tailstock positioner.

Sound engineering principles and common sense must be applied in the joining of an assembly held in a head-stock-tailstock positioner. Most long and narrow weldments supported at the ends tend to sag under their own weight or distort slightly as a result of fabrication. Rigid mounting of a weldment between the two faceplates is to be avoided when the headstock and tailstock are installed in fixed locations. Distortion from welding may damage the equipment when the weldment is rotated.

A flexible connection must be provided at each end of a clamping and holding fixture designed to accommo-

date a weldment between the headstock and tailstock tables. A universal joint or pivot trunnion in the fixture is recommended to compensate for lack of tolerance in the weldment components.

Consideration must be given to the inertia of a large rotating weldment when selecting the proper capacity of the headstock. The larger the polar moment of inertia, the greater is the required torque to start and stop rotation.[1] When one or more components of a weldment are at considerable distances from the rotational axis, the flywheel effect created by this condition can have serious

---

1. The unbalanced torque required to accelerate or decelerate a rotating body is related to the polar moment of inertia of the body and the angular acceleration by the following equation:

$$T_o = J_M \alpha$$

where

T_o = unbalanced torque
J_M = polar moment of inertia of the mass about the axis of rotation
α = acceleration or deceleration, radians /s/

Figure 9.5—Headstock and Tailstock

effects on the headstock drive and gear train. Under such circumstances, rotation speed should be low to avoid excessive starting, stopping, and jogging torque, and possible damage to the drive train.

Where usages involve weldments with a significant portion of the mass located at some distance from the rotation centerline, it is advisable to select a headstock capacity greater than that normally needed for symmetrical weldments of the same mass.

FEATURES AND ACCESSORIES

TO PROVIDE VERSATILITY to headstock-tailstock positioners, many features and accessories can be added. The following are the more common variations used with this equipment:

(1) Variable speed drive
(2) Self-centering chucks

(3) Automatic indexing controls

(4) Powered elevation

(5) Power and idler travel carriages with track and locking devices

(6) Through-hole tables

(7) Powered tailstock movement to accommodate varying weldment lengths and to facilitate clamping

(8) Digital tachometers

TURNTABLE POSITIONERS

Description

A TURNTABLE IS a single-axis positioning device that provides rotation of a horizontal table or platform about a vertical axis, thus allowing rotation of a workpiece about that axis. They may range in size from a small bench unit to large floor models. A typical turntable is shown in Figure 9.6.

Applications

TURNTABLE POSITIONERS ARE available in capacities from 50 lb (20 kg) to over 500 tons (250 M tons). They are used extensively for machine welding, scarfing and cutting, cladding, grinding, polishing, assembly, and nondestructive testing. Turntables can be built in a "lazy susan" concept to index several assemblies on a common table. Several smaller turntables are mounted on the main table together with an indexing mechanism. The turntable shown in Figure 9.3 allowed an operator to load and unload the work for a welding robot.

Features and Accessories

A TURNTABLE POSITIONER can be equipped with a number of optional features and accessories. The following are some popular adaptations:

(1) Constant or variable speed drive

(2) Manual rotation of the table

(3) Indexing mechanism

(4) Tilting base

(5) Travel carriage and track

(6) Through hole in the center of the table

(7) Digital tachometer

TILTING-ROTATING POSITIONERS

General Description

WHERE VERSATILITY IN positioning is required for shop operations, a tilting-rotating positioner is recommended. In most cases, the positioner table can be tilted through an angle of 135 degrees, including the horizontal and vertical positions. The turntable mechanism is mounted on an assembly that is pivoted in the main frame of the machine to provide a tilt axis. The tilt angle is limited by the design of the frame. Two common configurations of gear-driven tilting positioners differ by the limitation imposed by the tilt mechanism. A flat-135° positioner can tilt the table from the horizontal through the vertical to 45 degrees past the vertical. A typical positioner of this type is shown in Figure 9.7.

Figure 9.6—Typical Turntable

Figure 9.7—Typical Flat-135° Tilting Positioner

To provide clearance for the table and weldment when tilted past the vertical position, the base may be mounted on legs which enable the positioner to be raised to an appropriate height for the load to clear the floor.

The other common configuration for gear-driven tilting positioners is the 45°-90° design. This type can tilt the table from 90 degrees forward through the horizontal to 45 degrees backward. A typical 45°-90° positioner is shown in Figure 9.8. This design is most common in large sizes, and is available in capacities from about 20 tons (18 M tons) to several thousand tons (M tons).

The basic capacity rating of a positioner is based on (1) the total load (combined weight of the workpiece and the fixturing) and (2) the location of the center of gravity[2] (CG) of the combined load. Two dimensions are used to locate the CG of the combined load: (1) the distance

from the positioner table surface to the CG of the combined load, and (2) the distance from the axis of rotation to the CG of the combined load, referred to as eccentricity (ECC). Thus, it is common practice to rate a positioner for a given combined load at a given CG distance and ECC. The load that can be placed on a positioner will be less than the rated basic capacity if the CG of the load is more than 12 in. (305 mm) from either axis.

There is a broad overlap in basic capacity of the two common types of positioners. The selection depends upon the application.

A positioner can perform two functions. It may be used to position the weldment so that the welds can be made in the flat position and are easily accessible for welding. It may also be used to provide work travel for welding with a stationary welding head.

If work travel is intended, the drive system for rotation must have variable speed. Commonly, travel about the rotation axis is used to make circumferential welds. The tilt axis is less frequently used for work travel and is usually equipped with a constant speed drive.

2. The center of gravity of a body is that point where the body would be perfectly balanced in all positions if it were suspended from that point.

Figure 9.8—Typical 45°-90° Positioner

Drop-Center Tilting Positioners

FOR SPECIAL WORK where it is advantageous to use the tilt drive for work travel, a tilting positioner of special design can be used. An example of using work travel about the tilt axis is the fabrication of hemispherical pressure vessel heads from tapered sections. The welding fixture is mounted on the positioner table so that the center of a fixtured head coincides with the positioner tilt axis. Rotation around the tilt axis must be provided for welding the meridian seams.

The drop-center variation of the 45°-90° positioner is particularly useful for this application. In a convention-ally designed positioner, the surface of the table is not on the tilt axis but is offset by a distance called inherent overhang. This distance may vary up to 18 in. (450 mm) depending on positioner size.

For rotation about the tilt axis, the center of the hemispherical head must coincide with that axis. This often poses problems in mounting such a weldment on a standard tilting positioner because of its inherent overhang.

To avoid this difficulty, a drop-center positioner is used. It is usually of the 45°-90° configuration, but the trunnion assembly is modified so that the tilt axis lies in or below the plane of the table face. Then, a hemispheric head can be mounted so the center is on the tilt axis. The arrangement of a hemisphere on a typical drop-center positioner is shown in Figure 9.9.

The maximum size of weldments that can be handled on a drop center positioner is limited by the distance between the vertical side members of the base. Also, this type of positioner requires more floor space than a conventional positioner with the same table size because the tilt journals must be outside the table rather than beneath it.

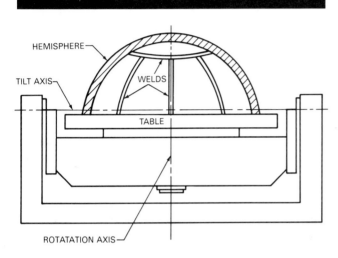

HEMISPHERE

TILT AXIS

WELDS

TABLE

ROTATION AXIS

Figure 9.9—Fabricated Hemispheric Head on Drop Center Positioner

Powered Elevation Positioners

WHEN A POSITIONER is intended for general purpose application in a welding shop, a flat-135° positioner with powered elevation generally is recommended because it is more versatile than a 45°-90° positioner. The ability to raise the chassis of a flat-135° positioner to permit full tilting of large weldments is particularly useful. A standard positioner must be raised and lowered with a shop crane while the positioner is unloaded. This limitation is avoided when a powered elevation positioner is used for the application. The positioner, loaded or unloaded, can be raised and lowered under power with integral rack and pinion or jack screw drives. A typical powered elevator positioner is shown in Figure 9.10. Powered elevator positioners are the most versatile of the gear driven types, but they are the most costly. They are available with capacities up to about 60 tons (55 M tons).

A powered elevator positioner used in the fabrication of a flanged cylinder is illustrated in Figure 9.11. In Figure 9.11(A), the axis of the spool is horizontal and the girth welds are made. The operator then tilts the table up 45 degrees and makes the first fillet weld in the optimum

position, shown in Figure 9.11(B). In Figure 9.11(C), the chassis is elevated and the table tilted down 45 degrees to make the second fillet weld.

Special Positioners

THERE ARE A number of variations of the basic tilt-plus-rotation positioners that are useful for specific applications. A *sky hook* positioner, shown in Figure 9.12, is basically a turntable mounted on the end of a long arm. The arm is attached to a headstock that provides tilting capability. This configuration provides unlimited rotation about both the tilt and the rotation axes.

A positioner can be part of an automatic welding station to provide movement of the weldment. Positioners can be integrated with welding robots to provide additional programmable axes for the robot system.

SAFETY

THE USE OF mechanical positioners rather than manual labor to position weldments during welding should reduce the likelihood of injury during the operation. For safety, a weldment must be firmly mounted on a positioner so that it will not move or fall off during welding and allied operations.

The weldment or the fixture may be bolted or welded to the positioner table. In either case, the fastening means must be strong enough to hold the work secure under any condition of tilt or rotation. The reaction loads at the fastening locations must be calculated or correctly estimated to ensure that the load is secured by fasteners of proper design. Methods for attaching a weldment or fixture to a positioner table are discussed later.

Positioners are designed to be stable when loaded within their rated capacity, but may be unstable if overloaded. Tilting positioners may be unstable during tilting because of the inertia of the load. Injury to personnel or damage to equipment could be serious if an overloaded positioner suddenly tipped during operation. For this reason, positioners must be fastened to the floor or other suitable foundation for safe operation. The instructions provided by the manufacturer concerning foundation design and fastening methods must be carefully followed to assure a safe installation.

Overloading of positioners, which is frequently practiced by users, cannot be condoned. It is possible for a user to mount an overload on a positioner table, tilt it to the 90 degree position, and be unable to return the table to the flat position because of insufficient power. Shop management must make every effort to ensure that the equipment is operated within its rated capacity to assure safe operation and long life. It is imperative to accurately determine the load, center of gravity, and eccentricity to avoid overload. The safety features needed in position-

Figure 9.10—Typical Powered Elevator Positioner with Weldment in the 135° Position

ing equipment include thermal-limiting or current-limiting overload protection of the drive motor, low voltage operator controls, a load capacity chart, and emergency stop controls.

The area surrounding the work station must be clear to avoid interference between the weldment and other objects in the area. Sufficient height must be provided for weldments to clear the floor. Manual or powered elevation in positioner bases is recommended for this purpose.

When using powered weldment clamping fixtures, a safety electrical interlock should prevent rotation (or elevation where applicable) until the clamping mechanism is fully locked to the weldment components. The use of proximity and limit switches is also advisable, especially in the positioning of oversized loads to avoid overtravel

or collision of the weldment with the floor or other obstructions.

TECHNICAL CONSIDERATIONS

Center of Gravity

CORRECT USE OF a positioner depends on knowing the weights of the weldment and the fixture, if used, and the location of the center of gravity (CG) of the total load. The farther the center of gravity of the load is from the axis of rotation, the greater is the torque required to rotate the load. Also, the greater the distance from the table surface to the CG, the larger is the torque required to tilt the load.

Consider a welding positioner that has the worktable tilted to the vertical position. Figure 9.13(A) shows the worktable with no load. In Figure 9.13(B), a round, uniform weldment has been mounted on the worktable concentric with the axis of rotation. The center of gravity coincides with the axis of rotation. The torque required to maintain rotation of the weldment is only that needed to overcome friction in the bearings. However, torque is required to accelerate or decelerate the weldment.

In Figure 9.13(C), an irregular weldment is mounted with its center of gravity offset from the rotation axis. In this case, torque is required to rotate the weldment, and is equal to the weight multiplied by the offset distance (moment arm). If a 5000 lb (2200 kg) weldment has its center of gravity located 10 in. (250 mm) from the rotation axis, it will require a minimum torque of 50 000 lb·in. (5600 N·m) to rotate it. The positioner must have enough drive capacity to provide this torque.

The same consideration would apply to a vessel being rotated by turning rolls. If the center of gravity of the vessel is not on the rotation axis, the turning rolls must provide enough traction to transmit the required torque for rotation.

Location of the center of gravity of the load is also important in the tilting action of welding positioners. When the loaded table is in the vertical position, the load exerts a torque around the tilt axis, as shown in Figure 9.14. The torque, T, is equal to the weight, W, of the load multiplied by the distance from its center of gravity to the tilt axis of the positioner. The tilt axis of most conventional positioners is located behind the table face for a length called the inherent overhang, IO. This length must be added to the distance, D, between the center of gravity of the load and the table face when calculating the tilt torque. The torque is given by the following equation:

$$T = W(IO + D)$$

where
T = tilt torque, lb·in.
W = weight of the load, lb
IO = inherent overhang, in.
D = distance from the table to center of gravity of the load, in.

Manufacturers of positioners usually attach a rating table to each positioner that specifies the torque ratings in tilt and rotation, the inherent overhang, and the load capacity at various distances between the center of gravity of the load and the table. A rating table is shown in Figure 9.15(A) and (B).

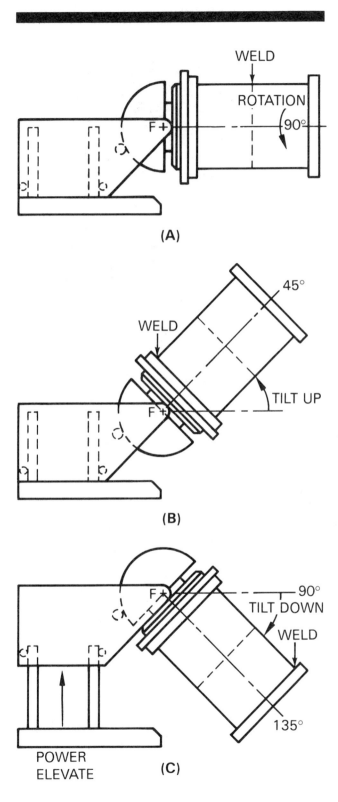

Figure 9.11—Power Elevator in Three Positions

Figure 9.12—Skyhook Positioner

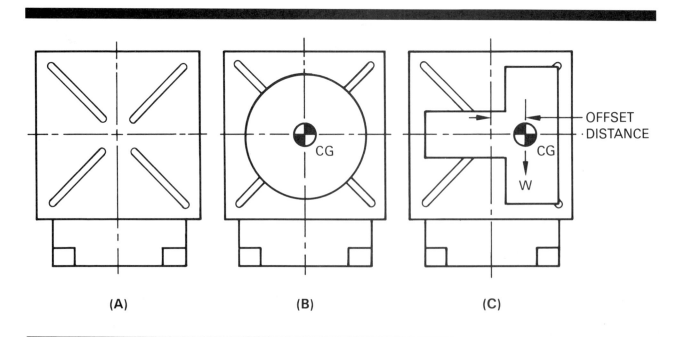

(A) (B) (C)

Figure 9.13—Effect of Location of Center of Gravity of Workpiece on Rotational Torque

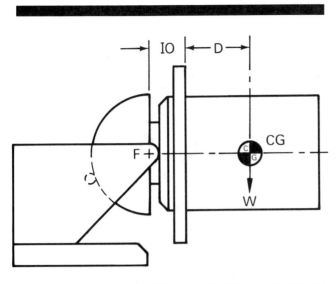

Figure 9.14—Effect of Location of Center of Gravity of Workpiece on Tilting Torque

For turning rolls, the tractive effort required to be supplied by the rolls can be calculated with the following formula:

$$TE = \frac{WD}{R}$$

where
  TE = tractive effort, lb
  W = weight of the load, lb
  D = the distance from the axis of rotation of the load to the center of gravity, in.
  R = radius of the load or holding fixture, in.

The tractive effort rating of the rolls should be about twice the calculated value to allow for drag and misalignment.

To select a proper positioner for a given weldment, the location of the center of gravity has to be known to determine the required torque. The center of gravity can be calculated, or it can be determined experimentally from a sample weldment.

The center of gravity of a symmetrical weldment of uniform density is at the geometric center. For example, the center of gravity of a cylinder is located on the center axis at the mid-point of the length.

Calculating the Center of Gravity. The center of gravity with respect to mutually perpendicular X, Y, and Z axes of a nonsymmetrical weldment can be determined by dividing it into simple geometric shapes. The center of gravity of each shape is calculated by formulas found in standard mechanical engineering or machinery handbooks. The location of the center of gravity of the total

weldment with respect to selected X, Y, and Z axes is determined by a transfer formula of the form:

$$y = \frac{Aa_1 + Bb_1 + Cc_1 + ...}{A + B + C + ...}$$

where
  y = distance from the X axis to the common center of gravity
  A = weight of shape A
  B = weight of shape B
  C = weight of shape C
  a_1 = distance from X axis to center of gravity of A
  b_1 = distance from X axis to center of gravity of B
  c_1 = distance from X axis to center of gravity of C

Additional calculations as required are made to determine the location of the common center of gravity from the Y and Z axes. The shop must be able to physically locate the positions of the selected X, Y, and Z axes with respect to the weldment.

Finding the Center of Gravity by Experimentation. There are two practical methods of finding the center of gravity without the need for calculations in the shop. The first method can be used to find the center of gravity of small box-shaped weldments. The weldment is balanced on a rod under one flat surface, as shown in Figure 9.16A. A vertical line is made on the weldment above the rod. The weldment is rotated to an adjacent flat surface and the procedure repeated, as shown in Figure 9.16B. The center of gravity lies on a line passing through the intersection of the two lines. The location of the center of gravity in a third direction can be found by repeating the procedure with a flat surface perpendicular to the other two.

In the second method of experimentally finding the center of gravity, a hoist can be used. The center of gravity of the weldment will settle under a single lifting point. When a plumb bob is suspended from the lift hook, a chalk mark along the line will pass through the center of gravity, as shown in Figure 9.17. A second lift from a different point on the weldment will establish a second line that pinpoints the center of gravity at the intersection of the two lines, Figure 9.17. (The marks from the balancing bar method were left on the weldment to show that both methods indicate the center of gravity at the same location.) A scale might also be used during lifting to obtain the weight of the weldment.

Attachment of the Weldment

WITH ANY TYPE of positioning equipment involving a rotating or tilting table, the weldment must be fastened to the table. In the case of horizontal turntables, fasten-

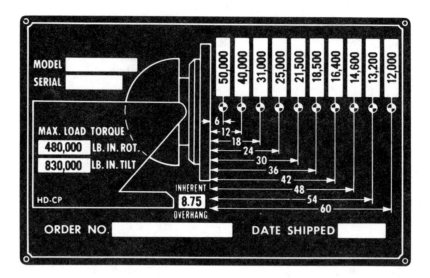

(A) U.S. Customary Units

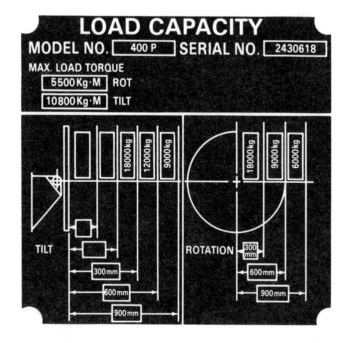

(B) S.I. Units

Figure 9.15—Typical Nameplates for Tilting Positioner

(A)

(B)

Figure 9.16—Method of Finding Center of Gravity by Balancing on Rod

(A)

Figure 9.17—Method of Finding Center of Gravity with a Hoist

ing serves only to prevent the weldment from being accidentally dislodged from the table. Where the positioner table will be used in positions other than horizontal, the fastening means must firmly hold the weldment to the table in any work position.

When the positioner table is horizontal, as in Figure 9.18(A), the only force tending to move the weldment from the table is the centrifugal force of rotation. Attachment requirements are minimal at low rotation speeds. As the table is tilted, the weldment tends to slide off the table because a component of the weights acts parallel to the table, Figure 9.18(B). The attachments must have sufficient strength in shear to prevent sliding

(B)

Figure 9.17 (Continued)—Method of Finding Center of Gravity with a Hoist

of the weldment. The shearing force increases as the table tilt angle increases, and is equal to the weight of the weldment at 90° tilt, Figure 9.18(C).

During tilting, a point is reached where the weldment would rotate about lower fasteners if not restrained by upper fasteners. The restraint results in tensile force on the upper fasteners [F1 and F2 in Figure 9.18(C) and (D)]. The amount of tensile force depends on the weight and geometry of the weldment.

The moment of the weldment about the lower fasteners P is the product of the weight W and the horizontal distance x between the center of gravity and the lower fasteners. As the tilt angle increases, the horizontal distance increases. The weight moment is balanced by the moment of the tensile force on the upper fasteners.

Assume that the weldments in Figure 9.18(C) and (D) have equal weights, W, but the one in (D) is longer and narrower than the one in (C). The force, F, in both cases, can be determined by the following equation:

$$F = \frac{Wx}{y}$$

where
  x = horizontal distance between P and CG
  y = vertical distance between P and F

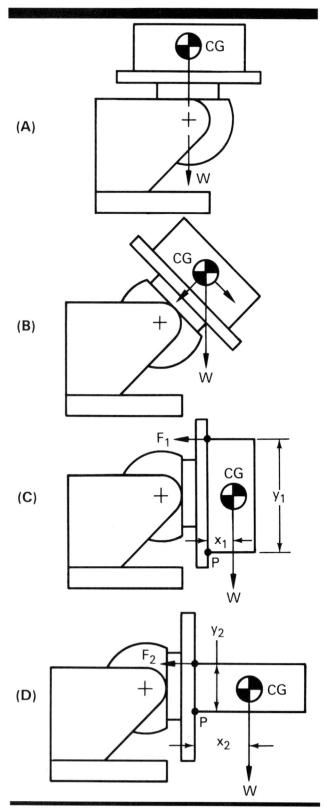

(A)

(B)

(C)

(D)

Figure 9.18—Effects of Tilting and Weldment Geometry on Positioner Fastener Requirements

In the two cases, x_2 is greater than x_1, and y_2 is less than y_1. Therefore, F_2 is greater than F_1. Accordingly, the attachments for the weldment in Figure 9.18(D) must be stronger than those in Figure 9.18(C). In both cases, all attachments must have sufficient strength to withstand both the shearing force and the tensile force.

There are two common methods employed to hold a weldment to a positioner table: bolting and welding. Most tables are provided with slots for mounting bolts and some manufacturers supply T-nuts which match the slots. The largest bolts that will fit the slots should always be used to avoid failure of the attachment. As a general rule, a weldment with similar base dimensions and height can be safely mounted using four bolts of a diameter that is slightly less than the slot width in the table. If a weldment is unusually tall with respect to its base dimensions, the force, F, on the mounting bolts should be calculated as described previously. The distance y should be the minimum bolt spacing to be used. The calculation assumes that all the load is on a single bolt, and is therefore conservative. Generally, neither the table nor the workpiece is flat, which makes this assumption valid.

The mounting bolts, properly tightened, will normally provide adequate shear resistance to prevent the weldment from sliding off the table. If there is any doubt that the load is adequately contained against sliding, steel stop blocks should be placed against the base of the weldment and tack welded to the table.

A weldment can be fastened to the table with temporary welds. The required size and length of each tack weld to carry the applied force, F, or shear load, or both should be determined as described in Chapter 5 of this volume. When the job is completed, the tack welds are cut loose after the table is returned to the flat position or after the weldment is adequately supported by shoring or a hoist.

If welding directly to the positioner table is objectionable, a waster plate can be securely bolted to the table and the weldment welded to the plate. The waster plate can be replaced, when necessary.

Work Lead Connection

WELDING POSITIONERS THAT support a weldment on rotating members should provide some means of carrying the welding current from the weldment to a point on the chassis where the work lead is connected. It would be impractical to continually relocate the work lead on the weldment as the table is rotated.

Two common methods for conducting welding current from the table to the base are sliding brushes and bearings. Either method can be satisfactory if properly designed, constructed, and maintained.

Sliding brushes are usually made of copper-graphite matrix, graphite, or copper. Several units are employed in parallel to provide adequate current capacity. The brushes are spring-loaded to bear against a machined surface on a rotating element. For proper operation, the brushes and rotating surface must be clean, the brushes must have the proper length, and the spring load must be adequate. Brushes should not be lubricated.

Conducting bearings must be properly designed and adequately preloaded. If the bearing preload is lost by mechanical damage, improper adjustment, or overheating of the spindle, the bearings can be damaged by internal arcing between bearing surfaces. In normal usage, this system requires little maintenance.

Poor connections in the welding circuit will cause unexplainable fluctuations in arc length, varying joint penetration, or changes in bead shape. If the positioner rotating connection is suspected, it can be checked by measuring the voltage drop under actual arc welding conditions. The voltage drop should be low (0.1 to 0.25 V), but more important, it should be constant. The cause of excessive voltage drop or fluctuation across the connection must be determined and corrected.

Positioners For Welding Robot Applications

THE ADAPTATION OF positioners to robotic welding involves the coordination of the motions of the robot with that of the positioner.

The simplest adaptation is the turn table shown in Figure 9.3 where an operator merely indexes the work into the robot work envelope by a 180 degree rotation of the turntable, and starts the robot on its welding sequence. The operator then removes the welded assembly and sets up an unwelded assembly. When the robot completes its weld sequence, the robot arm returns to its "home" position. The operator then rotates the turntable 180 degrees to locate the work in the robot work envelope, and initiates the robot weld sequence. As indicated earlier, the robot weld sequence begins with finding the reference points that establish the coordinate system. Because the robot is "blind," the location of the weld seams or points must always be in the same relationship to the reference points on the fixture.

For this kind of system, the coordination between the positioner and the robot is provided by a human operator. The operator recognizes when to rotate the turntable, and when to activate the robot sequence.

To adapt a tilting rotating positioner for use with a robot, as shown in Figure 9.19, more sophisticated controls are necessary. There are two types of systems in use. One system involves a robot with a relatively large programmable controller (PC) that has capacity to control devices in addition to the robot. The capacity includes adequate memory to hold the motion program of the robot and the positioner, and additional input/output (I/O) lines to send the motion commands to each

Figure 9.19—Robotic Work Station Using Positioner Interfaced with Robot

device in sequence, and to receive feedback from the device of the current status of the command execution.

The other system is one in which the robot controller can manage only the robot motions, and signal another controller to manage the positioner. The robot shown in Figure 9.3 can be adapted to service as shown in Figure 9.19. However, the positioner would have to include a programmable controller (PC) to store and manage the positioner motions, and a simple communication system between the robot controller (master) and positioner controller (slave) would be required. Two motion programs would be written. One program would be for the positioner and it would be stored in the slave PC of the positioner, and the other program would be for the robot and would be stored in the master PC. The two programs would have to be integrated so that all of the

moves would be coordinated. The program in the slave PC would be a complete instruction set which would issue commands, accept feedback from the positioner on command execution, and make any required adjustments. However, the slave PC would issue a motion command only when directed to do so by the master PC.

The slave PC then monitors the motion and when completed the slave advises the master PC to proceed with the program. The program in the master PC would contain commands to the slave PC intermingled with the commands for the robot. A flow chart illustrating the program sequence is shown in Figure 9.20.

The technology exists to have the positioners manipulate the work to provide weld travel, but except for specific application requirements, the usual approach has been to vest the weld motion function with the robot.

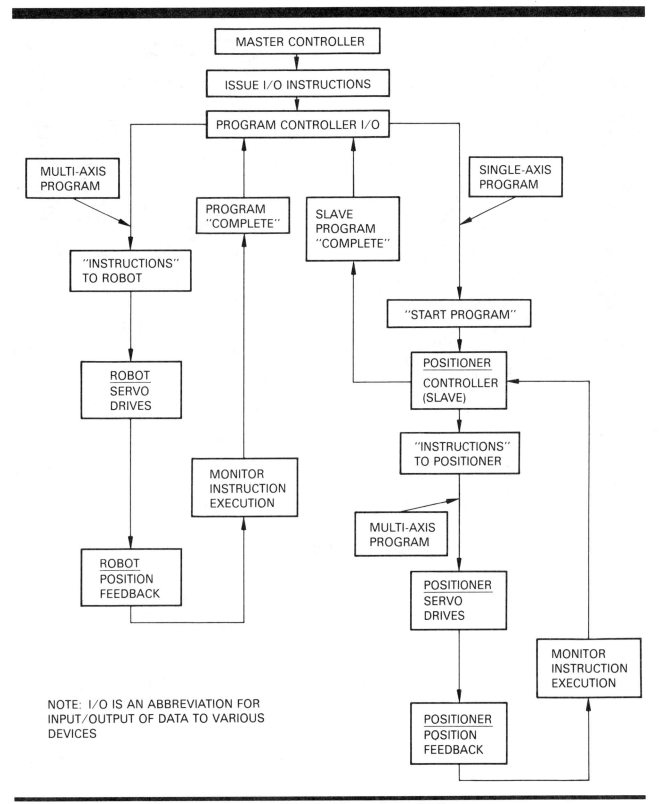

Figure 9.20—Flow Chart for Master Controller Program

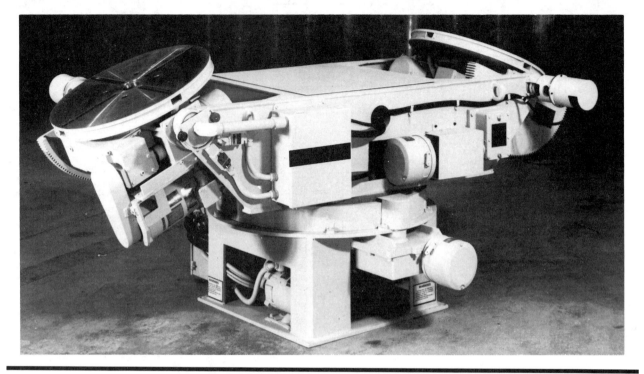

Figure 9.21—Indexing Turntable

A combination turntable and coordinated motion positioner is shown in Figure 9.21. The two tables rotate about a center axis, and each table tilts and rotates about its own axis. An operator can load work onto one table, rotate it to the robot, and initiate a programmed motion sequence for the positioner and the robot. Double-ended turntables of the kind shown in Figure 9.21 are not normally equipped with a PC. However, if the robot does not have adequate capacity to manage the table program, a slave PC can be added to the system.

SUPPLEMENTARY READING LIST

American Welding Society. *Brazing manual*, 3rd Ed. Revised. Miami, FL: American Welding Society 1976.

Boyes, W. E., ed. *Jigs and fixtures*. Dearborn, MI: Society of Manufacturing Engineers 1979.

Canadian Welder and Fabricator. Fixtures, positioners, and manipulation. *Canadian Welder and Fabricator.* 66(10): Oct. 1975; 14-16.

Cary, H. B. *Modern arc welding*. Englewood Cliffs, NJ: Prentice-Hall 1979.

Gage, J. H. Identification of a successful robotic welding application. *Welding Journal.* 62(11): Nov. 1983; 19-20.

Nissley, L. Adapting positioners for robotic welding. *Welding Design and Fabrication.* 56(8): Aug. 1983; 45-47.

———. Understanding positioner errors in your robotic arc welding system. *Welding Journal.* 62(11): Nov. 1983; 30-37.

Payne, S. The engineering of arc welding fixtures. *Manufacturing Engineering Management.* 64(1): Jan. 1970; 47-53.

Pierce, M. A guide to robotic arc welding positioners. *Welding Journal.* 64(11): Nov. 1985; 28-30.

Still, J. R. Positioning for welding. *Metal Construction.* 9(6): June 1977; 246-247.

The Welding Institute. *Development in mechanized automation and robotic welding.* Cambridge, England: The Welding Institute (1980).

Westinghouse Electric Company. *The robotics handbook.* Pittsburgh: Westinghouse Electric Company 1984.

AUTOMATION AND CONTROL

PREPARED BY A COMMITTEE CONSISTING OF:

R. C. Reeve, Chairman
R. C. Reeve, Incorporated

F. Prinz
Carnegie-Mellon University

J. Tincher
General Motors Corporation

J. Speranza
General Motors Corporation

R. Bruggeman
Caterpillar Tractor Company

S. Babcock
Rockwell International Corporation

T. Nevitt
McDermott Fabricators Division

WELDING HANDBOOK COMMITTEE MEMBER:
J. R. Condra
E. I. DuPont de Nemours and Company

CHAPTER 10

AUTOMATION AND CONTROL

INTRODUCTION

MEANING OF AUTOMATION

IN MANUFACTURING, THE term automation means that some or all of the functions or steps in an operation are performed in sequence by mechanical or electronic means. Automation may be partial, with certain functions or steps performed manually (partial automation), or it may be full, meaning that all functions and steps are performed by the equipment in proper sequence without adjustment by an operator (total automation). Automation may include the loading and unloading of the components of the operation.

Automation can be applied to many welding, brazing, soldering, and thermal cutting processes, as well as ancillary operations. Automatic equipment may be designed to accommodate a single assembly or family of similar assemblies (fixed automation), or automation equipment may be flexible enough to be quickly modified to perform the similar operations on different components or assemblies (flexible automation).

WHEN TO USE AUTOMATION

REGARDLESS OF THE degree of automation, its objective is to reduce manufacturing costs by increasing productivity, and improving quality and reliability. This is possible by the reduction or elimination of human errors. Other benefits of automation include lower floor space requirements, lower in-process inventory, and increased throughput. When automation systems are integrated into production scheduling systems, improved production flows result in better deliveries to customers.

Automation can succeed or fail depending on the application. A successful application requires careful planning, economic justification, and full cooperation and support of management, product designers, manufacturing engineers, labor, and maintenance. Major factors that should be carefully analyzed to determine if automation is feasible include: the product, the plant and the equipment, and costs.

FUNDAMENTALS OF WELDING AUTOMATION

AUTOMATION INVOLVES MORE than equipment or computer controls. It is a way of organizing, planning, and monitoring a production process. The procedures, systems, and production controls used for the manual welding cannot usually be adapted to automatic welding.

Most manual welding methods rely heavily upon the knowledge, skill, and judgement of the welder.

To successfully produce sound welds automatically, the equipment must perform in much the same manner as a welder or welding operator. Automatic welding equipment is not necessarily intelligent, nor can it make complex judgements about welds. However, adaptive control of arc voltage, filler feed rate, travel speed, gas flow rate and other variables are possible with the introduction of microprocessors and the ability to analyze large quantities of real time data. Methods and procedures for automatic welding should anticipate and deal with judgmental factors. Many of the topics discussed here should be considered when acquiring automatic welding equipment and when developing automatic welding procedures, methods, and production controls.

MACHINE WELDING

IN MACHINE WELDING, the operation is performed under the constant observation and control of a welding operator. The equipment may or may not perform the loading and unloading of the work. The work may be stationary while a welding head is moved mechanically along the weld joint, or the work may be moved under a stationary welding head.

Machine welding can increase welding productivity by reducing operator fatigue and by increasing the consistency and quality of the welds. The equipment should control the following variables when welding:

(1) Initiation and maintenance of the welding arc
(2) Feeding of the filler metal into the joint
(3) Travel speed along the joint

In machine welding, the operator must be near to the point of welding to closely observe the operation. The operator interacts continuously with the equipment to assure proper placement and quality of the weld metal. Machine welding should provide the welding operator with sufficient time to monitor and control the guidance aspects of the operation, as well as the welding process variables. Weld quality can be enhanced as a result of proper control of process variables. When a change in part production requires a new setup, human errors in setting up the new procedure may result in lower quality and lost production. Using microprocessors to set up the production equipment according to a preset procedure will often eliminate human setup errors. Additionally, a smaller number of weld starts and stops with machine welding reduces the likelihood of certain types of weld discontinuities.

Machine welding devices generally fit into one of the following categories:

(1) Machine carriages
(2) Welding head manipulators
(3) Side-beam carriages

(4) Specialized welding machines
(5) Positioners, turntables, and turning rolls[1]

These travel devices provide means for moving an automatic welding head relative to the part being welded or vice versa. Machine welding can be used with most fusion welding processes.

Machine Carriages

A WELDING MACHINE carriage provides relatively inexpensive means for providing arc motion. A typical carriage rides on a linear or curved track of the same contour as the joint to be welded, as shown in Figure 10.1. Some carriages are specifically designed to ride on the surface of the material being welded in the flat position, and some use the material for weld joint guidance. The travel speed of the carriage is a welding variable, and uniformity is important to weld quality. Tractor type carriages are used primarily for welding in the flat or horizontal position.

Figure 10.1—Welding Machine Carriage with a Single Submerged Arc Welding Head

Welding carriages may be employed for welding in the horizontal, vertical, or overhead positions. Others may be designed to follow irregular joint contours. They employ a special track upon which the welding carriage

1. Positioners, turntables, and turning rolls are discussed in Chapter 9 of this volume.

is mounted. Since the welding operator must continually interact with the controls, the carriage controls and the welding controls are typically placed within easy reach of the operator. This type of equipment is most productive in producing long flat position groove and fillet welds such as those encountered in ship and barge building. It is also useful in field work such as bridge or storage tank construction. The rigidity with which the welding carriage is held to the track and the uniformity of travel are critical to good performance.

Welding Head Manipulators

A WELDING HEAD manipulator typically consists of a vertical mast and a horizontal boom that carries an automatic welding head. A large welding head manipulator carrying a twin submerged arc welding unit is shown in Figure 10.2. Manipulators usually have power for moving the boom up and down the mast, and in most units, the mast will swivel on the base. In some cases, the welding head may move by power along the boom, while in others the boom itself may move horizontally on the mast assembly. Most manipulators also have slow-speed vertical and transverse motion control capabilities. This allows the operator to adjust the position of the welding head to compensate for variations along the weld joint.

It is esential during operation that the boom or welding head move smoothly at speeds that are compatible with the welding process being used. The carriage itself must also move smoothly and at constant speeds if the manipulator is designed to move along tracks on the shop floor. In selecting and specifying a welding manipulator, it is important to determine the actual weight to be carried at the end of the boom. The manipulator must be rigid and the deflection minimized during the welding operation.

Welding manipulators are used to position the welding head for longitudinal, transverse, and circular welds. Heavy duty manipulators are designed to support the weight of the operator as well as that of the welding equipment. All of the welding and manipulating controls are placed at the operator station as shown in Figure 10.2.

Side-Beam Carriage

A SIDE-BEAM CARRIAGE consists of a welding carriage mounted on a horizontal beam, as shown in Figure 10.3. It provides powered linear travel for the welding head.

The powered welding carriage supports the welding head, filler wire feeder, and sometimes, the operator controls. Typically, the welding head on the carriage is adjustable for vertical height and transverse position.

During operation, the welding process is monitored by the welding operator. Welding position and travel speed of the side-beam carriage are adjusted by the oper-

ator to accommodate different welding procedures and variations in part fit-up.

Special Welding Machines

SPECIAL WELDING MACHINES are available with custom clamping devices, torch travel mechanism, and other specific features. Applications include equipment for making circumferential and longitudinal welds on tanks and cylinders, welding piping and tubing, fabricating flanged beams, welding studs or bosses to plates, and for performing special maintenance work, such as the rebuilding of track pads for crawler type tractors.

A special welding machine is shown in Figure 10.4 that welds covers on light truck axle housings. This machine may be quickly adapted to different size axles.

AUTOMATIC WELDING

AUTOMATIC WELDING IS performed with equipment that manages an entire welding operation without adjustment of the controls by a welding operator. An automatic welding system may or may not perform the loading and unloading of the components. An automatic arc welding machine that welds a brake flange on an axle housing is shown in Figure 10.5.

The equipment and techniques currently used in automatic welding may not be capable of completely controlling the operation. A welding operator is required to oversee the process. An important aspect of automatic welding is that the operator need not continuously monitor the operation resulting in increased productivity, improved quality, and reduced operator fatigue.

Welding operators must have certain skills to operate and interact with automatic equipment. They are responsible for the proper operation of a complex electromechanical system, and must recognize variations from normal operation. In addition, deductive skills may be required if the welding operator cannot directly view the actual welding process.

Automatic welding uses the same basic elements of machine welding plus a welding cycle controller. Continuous filler wire feeding, where used, arc movement, and workpiece positioning technology are basic to automatic welding. Automatic welding requires welding fixtures to hold the component parts in position with respect to each other.[2] The welding fixtures are usually designed to hold one specific assembly and, by minor variation, may allow for a family of similar assemblies. Because of the inherent high cost of design and manufacture, the use of welding fixtures is economical only in high volume production where large numbers of identical or similar parts are produced on a continuous basis.

---

2. Welding fixtures are discussed in Chapter 9 of this volume.

Figure 10.2—Welding Head Manipulator Carrying a Tandom-Arc Submerged Arc Welding Unit

Figure 10.3—Sidebeam Carriage with GMA Welding Head

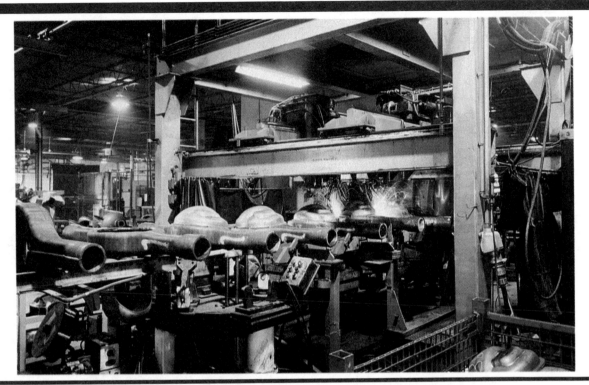

Figure 10.4—Automatic Welding of Covers on Axle Housings

Figure 10.5—Automatic Arc Welding Joins Brake Flanges to Axle Housing

Automatic welding also requires that the parts to be joined be prepared in a consistent and uniform manner. Therefore, parts preparation for automatic welding is often more expensive than for manual or machine welding. Part preparation costs can be reduced and fitting tolerance maintained or improved by utilizing numerically controlled equipment in the cutting and forming areas. Part preparation costs can often be partially offset by design changes including weld joint configuration.

Automatic welding requires a detailed sequential plan of each motion and operation. This includes motions and operations of the handling equipment, the welding head, and the welding equipment (wire feeders and power sources). Each operation of each device must be coordinated to begin at the right time, at the correct location in space, execute using the correct variables, and stop when the operation is complete. The devices in the system are all subordinate to the welding machine controller. Each device executes its function as commanded by the controller, and feeds back the status reports to the controller which compares the feedback data with the plan and sends adjustments to the device.

The welding machine controller is the primary element of an automtic welding system. Microprocessor technology is used in all new automatic welding system controls. A microprocessor panel is shown in Figure 10.6.

Successful application of automatic welding offers the following advantages:

(1) Consistent weld quality
(2) Reduced variable welding costs
(3) Predictable welding production rates
(4) Integration with other automatic operations on the weldment
(5) Increased productivity as a result of increased welding speeds and filler metal deposition rates
(6) Increased arc time
(7) Lower part costs

Automatic welding has limitations, including the following:

(1) Capital investment requirements are much higher than for manual or machine welding equipment.
(2) Dedicated fixturing is required for accurate part location and orientation.
(3) Elaborate arc movement and control devices with predetermined welding sequence may be required, depending on the complexity of the part.
(4) Production requirements must be large enough to justify the costs of equipment and installation, the training of programmers, and the maintenance of the equipment.

Figure 10.6—Microprocessor Panel for a Welding Machine

FLEXIBLE AUTOMATED WELDING

IN FLEXIBLE AUTOMATED welding, a numerical control or computer control program replaces complex fixturing and sequencing devices of fixed automated welding. The welding program can be changed to accommodate small lot production of several designs that require the same welding process, while providing productivity and quality equivalent to automatic welding. Small volumes of similar parts can be produced automatically in a short time. A flexible system can be reprogrammed for a new part.

An industrial robot is the most flexible of the automated systems used in manufacturing operations. A robot is essentially a mechanical device that can be programmed to perform a task under automatic control. A welding gun or torch can be mounted on a "wrist"[3] at the end of the robot arm.

Robots come in a variety of motion configurations;

3. A wrist is a combination of two mutually perpendicular revolute axes.

two common systems, articulated (jointed) and rectilinear, are shown in Figure 10.7. The choice of robot configuration depends upon the applications for which a robot is to be used.

A robot work station is a combination of various pieces of equipment integrated by a single control. A typical station would incorporate a robot arm, a robotic work positioner, a welding process package, and fixturing to hold the parts for welding.

A welding robot can be programmed to trace a path which is a welding seam. Some robots are programmed by moving the welding torch along the weld joint. Position signals are sent to the robot controller where they are recorded in memory. After programming, some robot controllers provide a means to make a permanent record of the program for future production of that specific weldment. Some robot controllers can also set and change welding variables in addition to controlling the welding torch path.

If distortion is a significant factor, or the piece parts are not accurately prepared or located, the weld joint may not be at the preprogrammed location. To compensate for mislocation, the flexible or robotic welding system may require sensory feedback to modify the program in real time. Sensors can be used to provide weld joint location information to certain intelligent robot controllers to make real-time adaptive corrections to the programmed path. Currently, a limited number of sensors and adaptive controls are available for use with certain welding processors. Most of these are adaptive for joint tracking only.

Flexible automation for welding can eliminate the need for complicated customized fixtures, and it is finding increased acceptance in low volume production. The inherent flexibility of robotic systems can accommodate several different types of weldments at the same work station. Furthermore, flexible welding automation can be interfaced with other automatic manufacturing operations.

EXTENT OF WELDING AUTOMATION

THE EXTENT TO which automation should be employed is governed by four factors: product quality, production levels, manpower, and investment.

The importance of these factors will change with the application and with the type of industry. The decisions on how much and when to automate can best be evaluated by examining each factor.

PRODUCT QUALITY

Process Control

A WELD IS usually considered a critical element of a fabricated component if failure of the weld would result

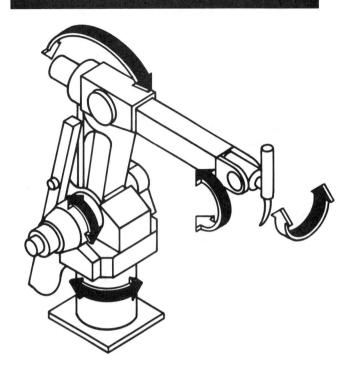

(A) Articulated robot

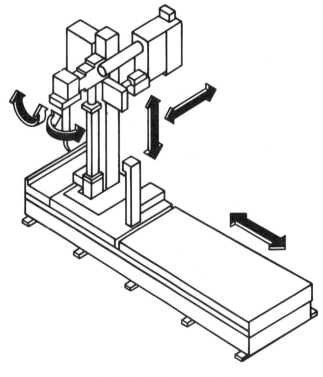

(B) Rectilinear robot

Figure 10.7—Robotic Motion Systems

in serious injury or death, or extensive damage to and loss of service of a welded structure. Thicker sections, additional consumables, welding process development, proper part preparation, adequate supervision, or extensive inspection and testing may be specified for critical welds to assure the required product quality.

Automated welding can significantly improve product quality while reducing costs. By regulating and controlling process variables, the possibility of human error is reduced. Automation also allows the welding operator to devote attention to other factors that can improve weld quality. Automation of critical welds is common in the aerospace, nuclear, and construction industries. Welding automation is also being applied in many other industries.

Product Improvement

AS MARKET PRESSURES dictate the need for higher quality, and often at lower cost, the long term viability of a manufacturer can hinge upon welding automation. Often, reductions in repair and scrap made possible through automation can financially justify conversion to automation. A reduction in the repair and scrap costs from about 10 percent to about 1 percent is not uncommon. In addition, microprocessors provide the ability to maintain data on the entire process in a convenient format. This can be used in quality control or maintenance troubleshooting. The financial benefits of quality products and customer satisfaction are more difficult to measure. At best, they can translate into some increase in market share. At worst, they can still mean the survival of the business in the midst of increased competition.

PRODUCTION LEVELS

Production Volume And Changeover Time

A MAJOR FACTOR in measuring utilization of an automated welding system is the percentage of time the system is kept productive. Where high production levels are required for a single or a limited number of parts, fixed automation is appropriate. Flexible automation should be considered if frequent model or engineering changes are anticipated. Where production lots are small, flexible automation should be considered.

Answers to the following questions are useful when analyzing the potential cost savings of an automatic welding installation:

(1) How many welders are currently working or projected to be working on the assemblies of interest?
(2) How many shifts are involved?
(3) What is the extent of welding, and what sizes of welds are needed on each assembly?
(4) How many assemblies are required per year?

(5) What is the projected frequency of changeovers (based on production batch size, model changes, engineering changes, and other factors)?

Floor Space

PLANT FLOOR SPACE is a part of product costs. Frequently, manual welding fabrication requires a relatively large amount of floor space because of the number of workers required for the operations. A reduction in floor space for a welding operation or an increase in output in the same floor space may be an important consideration. Therefore, automation of welding operations should be considered where floor space is limited.

Inventory

FLEXIBLE WELDING AUTOMATION can generate considerable savings through increased rate of internal inventory turn-over. Savings can be realized through the work-in-process and finished-goods inventory. These reductions can lead to additional savings in floor space. Flexible automation allows scheduling of smaller lot sizes as a result of reduced tooling and setup costs.

MANPOWER

Human And Environmental Factors

WELDERS ARE FREQUENTLY required to perform a skilled trade in a noisy, dirty, and hot environment. Welding fumes and gases are a potential health hazard, as is the possibility of being burned. Arc welding automation permits the welding operator to work outside of the hazardous environment. Additionally, welding automation can relieve the operator of repetitive, monotonous, and tiring tasks. Automation directly improves welding productivity.

Availability Of Manpower

WELDING SHOPS ARE often limited in skilled manpower. Skilled welders are normally in demand and receive relatively high labor rates. Operation of automated welding equipment by semiskilled operators can lead to large savings in labor cost.

Education And Training

THE INTRODUCTION OF automation and technology to the manufacturing operation requires education and training of personnel. Introduction of machine welding usually does not materially change the skills required in the shop. However, the technology of automatic and flexible welding systems may require planned education and training programs. These programs should cover the required skills to set up, operate, and maintain the equipment and a basic understanding of the technology being introduced. The availability of trained personnel or personnel capable of being trained materially affects the cost of automation. The operator of an automated welding system should have a welding background.

INVESTMENT

A COST-BENEFIT ANALYSIS should be made of potential welding automation projects. The discounted cash-flow technique is generally accepted in the metal working industry. This technique uses a proforma annual cash flow for the expected life of the system. Typically, a ten-year life is more than adequate to assess the financial impact of a welding automation system.

The estimated savings and costs that will result from the project should be clearly identified. Welding usually affects manufacturing costs both upstream and downstream from the actual welding step. Financial justification requires a review of the financial condition of the company, including availability and cost of capital.

PLANNING FOR AUTOMATION

PLANNING IS A necessary prerequisite to successful application of automation. Such planning includes the identification and in-depth analysis of potentially successful applications, as well as planning for the implementation of the selected automated welding systems. Companies often form automation teams to manage and perform the required planning and implementation tasks. The team should study all pertinent factors concerning planning, procurement, installation, and startup of the automation project.

PRODUCT DESIGN

Designing For Automation

MOST WELDMENTS ARE not designed for efficient automatic welding. Considerable savings can be realized by reviewing a weldment design to optimize it for automation. The costs and benefits of redesign should be included in the financial analysis of a project. Component parts, weld joints, and assemblies should be specifically designed for robotic welding.

Tolerances And Part Fit-Up

PART TOLERANCES AND joint fit-up are important in automatic welding. The dimensional accuracy of the parts, so long as they are within functional tolerances, is not nearly so important as the consistency. If the weld joint geometry is consistent, automatic welding can be done without the need for adaptive feedback controls. The manufacturing processes used in the preparation of component parts must be reviewed for consistency. Costs of improving component consistency must be included in the financial analysis.

Processing And Scheduling

THE PROCEDURES AND scheduling in most shops were initially developed for manual welding. Conversion to automation requires new procedures and scheduling to maximize productivity. A detailed review of subassembly designs and component scheduling is warranted. Often, an entire family of assemblies can be fabricated with one flexible automatic welding system.

Fixturing

WHEN ASSEMBLIES ARE redesigned or procedures are changed for automation, new welding fixtures are often required. Flexible automation can significantly reduce the required investment in new fixturing.

There is a substantial difference between the fixturing required for fixed automatic welding and that required for flexible automatic welding. Fixed automatic welding requires customized fixtures that provide exact weld joint locations. In flexible automation, the software program can be designed to avoid complex customized fixturing. Often all that is required for flexible automatic welding is a tack welding fixture.

PROCESS DESIGN

Welding Process Selection And Welding Procedures

AUTOMATION OF A welding operation often requires a change in the welding process. The selection of an appropriate welding process and associated welding consumables is an important step. Procedures for automatic welding should be qualified according to the applicable welding standards, such as *ANSI/AWS D1.1, Structural Welding Code—Steel*, or in the absence of a specific standard *AWS B2.1, Standard for Welding Procedure and Performance Qualification*.

Selection Of Welding Equipment

AUTOMATIC WELDING SYSTEMS are normally required to operate at high duty cycles to justify the investment. High duty cycles require the use of rugged heavy duty welding equipment designed for continuous operation. Water cooling should be specified where applicable to avoid overheating of the welding equipment.

Safety

THE WELDING OPERATOR should be prevented from entering the work envelope of an automatic welding system. However, if this is not possible, safety interlocks are necessary to prevent injury when the equipment is operational. Safe practices for the welding process and equipment must be followed when installing and operating automatic equipment (see Chapter 18).

Operator Intervention

UNDER FULLY AUTOMATIC welding conditions, the operator's responsibility may be limited to loading and unloading the machine and stopping the machine activity if it malfunctions. This will normally let the operator perform other activities such as inspection of incoming parts and finished products, supervision of other automatic equipment, material handling, or housekeeping.

FEASIBILITY TESTING

WELDING TESTS ARE normally conducted to prove the feasibility of automatic welding. The test parts may be actual production components or simulations. When designing weld test specimens, many factors, such as part weight, moments of inertia, and surface conditions, may be important to the welding process. Some of the more common test objectives are the following:

(1) Optimization of weld quality and consistency
(2) Determination of welding conditions and production cycle times (includes set-up, indexing, welding, and unloading times)
(3) Establishment of working relationships with equipment manufacturers
(4) Identification of critical operator skills and safety requirements
(5) Evaluation of the effects of distortion, fit-up tolerances, heat buildup, weld spatter, and other factors
(6) Provision of welded specimens for testing and evaluation
(7) Evaluation of product design for automatic products

Feasibility tests can be conducted by the user, the equipment manufacturer, or a subcontractor. The tests should be conducted in two or more phases. Initial testing should be performed by the user to determine the optimum balance between welding conditions, weld quality, and component fit-up tolerances. The results should be provided to the equipment manufacturer for subsequent testing of the automatic equipment. After installation in the user's plant, the equipment should be tested to verify its production capability.

Testing And Weld Development Costs

WHEN AUTOMATION IS being evaluated, a welding development and testing program should be undertaken to ensure that optimum welding performance is obtained from the proposed investment in capital equipment. The investment in development and testing programs should be proportional to the expected capital outlay.

FACILITIES

Utilities

ELECTRICAL POWER, AIR, and water requirements for automatic welding equipment are usually within the ranges of typical machinery used in a fabrication shop. One exception may be resistance welding equipment that has high power demands. Voltage stability in the power lines should be measured with a recording voltmeter or oscillograph. If voltage fluctuations exceed the range recommended by the equipment manufacturer, heavier service lines or a constant voltage transformer should be installed.

Where cooling water is required, a recirculating cooling system is recommended to conserve water. A compressed air supply may require filters, driers, or oilers.

Location

THE FOLLOWING FACTORS should be considered in evaluating the location of an automatic welding system:

(1) Safety—clearances, arc radiation exposure, fume control, spatter
(2) Work area conditions—lighting, temperature, humidity, cleanliness, vibrations.

The effects of lighting, noise, and smoke on the operator should be considered. Many automatic welding operations require some observation by the operator. Therefore, working conditions in the area should be conducive to stable, reliable performance.

Safety requirements, such as a clear area around the machine, can normally be determined from the manufacturer's specifications. Collision hazards include possible contacts with material handling equipment, fixtures, hand tools, and robot movement, and should receive special consideration.

Specifications

USER REQUIREMENTS ARE generally written for custom designed machines, and the specifications are made a part of the purchase (or lease) order. Manufacturer specifications serve the following purposes:

(1) Define the performance expectations of the purchaser and the supplier.

(2) Establish minimum quantitative requirements for defining the equipment.

(3) Provide uniform criteria to allow comparison of several vendor proposals against the purchaser's needs.

(4) Provide support groups (maintenance, programming, and others) with an information base in preparing for delivery of the equipment.

Environment

STRICT ENVIRONMENTAL REQUIREMENTS are not normally specified for automatic welding equipment. However, extended exposure to a hostile environment can reduce equipment life. When installing automatic welding equipment, it is best to provide as much protection as possible. Unlike conventional welding machines, automatic equipment often contains vulnerable electronic and mechanical devices. A clean environment will help assure good performance and long machine life. Potential environmental problems with automatic welding equipment are listed in Table 10.1.

The environment for automatic welding operations should meet the minimum requirements specified by the equipment manufacturer. A harsh environment can rapidly degrade the performance of the machine, create costly maintenance problems, and shorten the overall useful life of the system. Considering the cost of automatic welding equipment, investment in providing a suitable environment is often justified.

Table 10.1
Potential Environmental Problems with Automatic Welding Equipment

Environmental Factor	Potential Problem Areas
Temperature	Electronic and mechanical accuracies
Humidity	
Common moisture	Electronic components and rust
Salt air	Electronic components and severe corrosion
Vibration	Stability of electronic and servoparts
Dust and dirt	
Abrasive	Gears, bearings, and other mechanical devices
Nonabrasive	Air-cooled components
Magnetic	Collection on solid state devices
Lighting, noise, fumes	Operator efficiency

PROCUREMENT SCHEDULING

SCHEDULING THE ACQUISITION and installation of an automatic welding machine can be straightforward for machines of standard design. These machines include those that are inventoried and have been tested with actual production parts by the manufacturer. On the other hand, custom designed systems that automate new welding processes or involve unique and difficult applications may require a lengthy debugging period that can cause scheduling difficulties. A suggested checklist for procurement scheduling is given in Table 10.2.

TRAINING AND EDUCATION

Skills Required

THE BASIC JOB skills for operators of automatic welding equipment are a good knowledge of the welding process and training in the programming of automatic equipment.

Knowledge of the Welding Process. Fixed automatic or flexible automatic welding systems may not require a trained welder. However, a skilled welding operator may be required by the governing welding standard.

Table 10.2
Procurement Scheduling Checklist

1. Identify potential welding applications.
2. Estimate potential costs savings.
3. Research literature for similar application case histories.
4. Identify potential equipment suppliers.
5. Perform welding tests to determine feasibility and identify limitations.
6. Obtain approvals from engineering, manufacturing, quality control, and other departments, as required.
7. Prepare equipment and fixturing specifications.
8. Obtain budgetary proposals from potential equipment suppliers.
9. Appropriate procurement funds.
10. Select an equipment supplier.
11. Monitor design, construction, and test of the equipment by the supplier.
12. Install and check out the equipment.
13. Train operators, engineers, and maintenance personnel.
14. Perform pilot production welding tests.
15. Turn the equipment over to production.

Welding skills are an asset for applications where the operator performs a decision making role. Bead placement in multiple-pass arc welded joints and in-process adjustments of welding variables are examples of duties that may require basic welding skills.

Knowledge of Programming. The type and level of skill required to program automatic welding machines vary with the machine. Many robots are programmed on-line in a "teach-learn" mode by jogging through the desired motions and entering the welding conditions (speed, current, voltage). No numerical programming skills and only a nominal amount of training are required. This method uses an expensive production tool to "write a program." Some of the newer equipment available allows programming of the weld process parameters and weld motions off line on a microcomputer. The program is then transferred to the work station. Operator training in this area is required.

Other types of automatic welding machines, in particular those integrated into a computer or computer numerical control (CNC) systems, may require numerical control (NC) programming skills. Programmers may need experience in computer languages to make robot programs. A knowledge of welding is also an important factor.

Maintenance. Maintenance technicians must be skilled in electronics, mechanical apparatus, and often computer controls.

Training Required

TRAINING IS USUALLY required to prepare operators for automatic welding, and it should be provided by the equipment manufacturer. Other production factors, such as part preparation, processing procedures, welding procedures, quality control, and remedial measures may also require special training.

The following factors should be considered in estimating the scope of training required:

(1) Number of controls, gauges, meters, and displays presented to the operators
(2) Degree of difficulty and the criticalness of the setup
(3) Number of steps in the operating procedure
(4) Degree of operator vigilance required
(5) Responsibility for evaluation of results (weld quality and distortion)
(6) Corrective actions that may need to be taken or reported
(7) Required quality level
(8) Potential product liability
(9) Value of parts and completed weldments

The operator of a simple automatic welding machine, which has little more than ON and OFF switches and no variations in procedures, may not require formal training but only a few minutes of instruction by a foreman. An elaborate computer-controlled electron beam welding machine, such as the one shown in Figure 10.8, may require formal training for the operator, the responsible engineer, and the maintenance personnel. On-the-job training can provide the operator with a functional knowledge of the machine. However, such training is usually more costly than classroom training.

PRODUCTION CONSIDERATIONS

Space Allocation

AUTOMATIC WELDING EQUIPMENT will rarely require the same floor space as that occupied by the manual welding operations it replaces. It usually requires less floor space. To obtain good production flow, floor space requirements for the entire operation should be studied in context. Additional space savings may be realized by automating other operations as well.

Material Handling

ESTABLISHED METHODS FOR material and parts handling may not be suitable for automatic welding operations. For example, an overhead crane that is suitable for transferring parts to work tables for manual welding may be unsatisfactory for moving parts to an automatic welding machine.

Automated material handling should be considered for improved efficiency, orderly parts flow, and reduced risk of parts damage. Robotic material handling offers versatility but may exceed the capacity of a simple automated welding machine. Custom designed material handling systems are usually justifiable on high volume, dedicated production lines.

Flexibility and Expansion

THE ABILITY TO change or add to the capability of an automatic welding system may be needed by some users. Examples of changes and additions that may be considered are as follows:

(1) A change of welding process
(2) An increase in welding power
(3) A change in the program storage medium, for example, from paper tape to floppy disk
(4) Addition to the system of new operating features or enhancements
(5) Addition of other controlled equipment, such as positioners, material handling devices, and markers

Figure 10.8—Computer Controlled High Vacuum Electron Beam Welding Machine

(6) Retrofit of new or different adaptive control[4] sensors (TV cameras or tracking systems)

Complications can arise when changing or adding to automatic welding stations. Hardware, software, and electronic components of different models, and equipment from different manufacturers are usually not compatible. Manufacturers and users of computer equipment are active in the development of communication standards to assure that equipment from different vendors is compatible. The newly organized nonprofit Corporation for Open Systems (COS) is a group of computer vendors and users formed to promote computer communications standards.[5]

Shop Acceptance

IMPROVED PRODUCTIVITY IS best realized from a new automatic welding machine planned for the system. Shop personnel should be advised in advance that an automatic welding machine is being considered for a particular operation. Welding supervisors, engineers, welders, maintenance personnel, and welding inspectors can play an important role in making a transition from manual to automatic welding proceed smoothly. They can often formulate requirements for the machine.

Maintenance

AUTOMATIC WELDING MACHINES are often complex electromechanical systems that include precision mechanisms and components. The equipment can be extremely reliable, when it is treated properly.

A preventive maintenance program is essential for long-term reliability without degradation of performance. Preventive maintenance activities may include the following:

(1) Calibration of speeds and other welding variables (voltage, current, welding wire feed)
(2) Routine replacement of finite life components (e.g., guides, hoses, belts, seals, contact tips)
(3) Cleaning of critical surfaces (bearings, optics, and similar items)
(4) Oil changes and lubrication of bearings, seals, screws, etc.
(5) Replacement of filters
(6) Monitoring of temperature indicators, flow meters, line voltage, fluid levels, etc.
(7) Inspection of structural components and mechanisms for cracks, loose bolts, worn bearings, evidence of impact or abrasion, etc.
(8) Discussion of operation with the operator

A good maintenance program should include:

(1) A file of all available documentation (wiring diagrams, manuals), located near to machine
(2) A step-by-step troubleshooting procedure with clearly marked test points for instrumentation
(3) A generous supply of spare parts, such as circuit boards, fuses, and valves, located close to the machine
(4) A log of performed maintenance with details of problems encountered, the symptoms, and the actions taken to overcome them
(5) A planned schedule of daily, weekly, monthly, and yearly maintenance procedures, inspections and other maintenance activities from minor housekeeping to major overhaul

RESISTANCE WELDING AUTOMATION

AUTOMATION OF RESISTANCE welding operations may range from an indexing fixture that positions assemblies between the electrodes of a standard resistance welding machine to dedicated special machines that perform all welding operations on a particular assembly.[6] An example of the latter type is shown in Figure 10.9.

When flexibility is required in an automatic resistance welding operation to handle various assemblies, a robotic installation is a good choice. An example is the robotic spot welding line for automobile bodies shown in Figure 10.10. Such an installation can be program-

---

4. An adaptive control system is one that monitors certain independent and dependent variables of a process and continually adjusts the independent variables to maintain a preset level for each variable, and to provide a correct output. If the adjustments are made instantaneously during the process, they are said to be made in "real time."

5. The American Welding Society formed a committee on Automatic and Robotic Welding. Among the responsibilities of the new AWS committee is the development of standards for interfacing among automatic welding devices.

6. Resistance welding processes and equipment are discussed in the *Welding Handbook*, Volume 3, 7th Edition and are planned for Volume 2, 8th Edition.

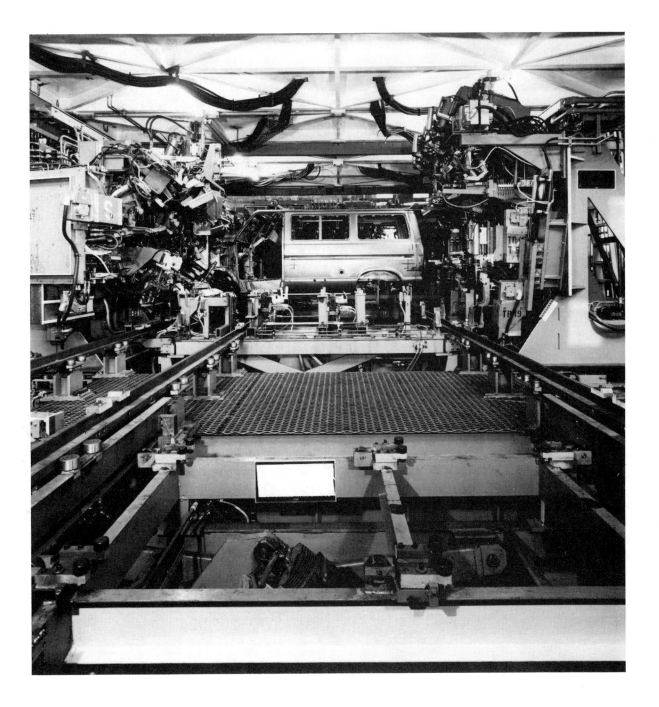

Figure 10.9—A Dedicated Automatic Resistance Welding Line

Figure 10.10—Automotive Robotic Resistance Welding Line

med to spot weld many different locations on the same car body, and similar car bodies for various models.

MATERIAL HANDLING

WHEN APPLYING AUTOMATION to a resistance spot welding application, it may be practical to move either the assembly or the welding gun to each weld location. Obviously, weight has to be considered in this decision. A typical resistance spot weld gun weighs 50 to 100 lb (23 to 45 kg), and the welding cable another 50 lb (23 kg). A typical portable resistance spot welding gun, cable, and power source are shown in Figure 10.11.

Moving The Assembly

JOINING RELATIVELY THICK sheets of low resistance alloys, such as aluminum, may require large welding current conductors. Heavy welding cables may be required to carry such high currents from the power source to the welding gun. Such conductors may exceed the load carrying capability of a robot. One solution is to move the assembly into position and weld with a stationary or indexing spot welding machine. Alternatively, the assembly may be indexed while single spot welds are made with a stationary welding machine.

Assemblies that weigh less than 150 lb (68 kg) may be moved with most industrial robots. Supporting tooling may be required to prevent damage to the assembly during movement.

Moving The Welding Gun

IN HIGH PRODUCTION spot welding, the time between welds must be minimized. At high welding rates, moving a light weight welding gun may be the better approach for welding relatively large, bulky assemblies.

PRODUCTION REQUIREMENTS

Quantity Of Production

IN HIGH VOLUME applications, such as automotive or appliance manufacturing, each welding gun in an automatic installation makes the same spot welds repeatedly. In low volume applications, such as aerospace work, a welding robot can be used to make a number of different spot welds before repeating an operation. A flexible, programmable industrial robot can be used for both types of applications.

Production Rates

TYPICALLY, PRODUCTION RATES of 15 to 30 spot welds per minute can be achieved when the welds are in close proximity. As distances between spot welds increase, production rates will depend on the maximum practical travel speed for the welding gun or the assembly, whichever is moved.

Production rate computations should include the following factors:

(1) Transfer time between weld locations
(2) Time for the gun or work to stabilize in position
(3) Welding sequence time, comprising:
    (a) Squeeze time (typically 0 to 10 cycles)
    (b) Weld time (typically 5 to 20 cycles)
    (c) Hold time (0 to 10 cycles)
(4) Fixturing sequence time
(5) Material handling time

Welding cycle time can be optimized using special programming, such as upsloping welding current during the latter portion of squeeze time. Likewise, welding current downslope can be applied during a portion of hold time. Hold time can be minimized by initiating the release of the gun before the time is completed, to take advantage of gun retraction delay. A properly designed welding cycle can significantly reduce the total welding cycle time for the assembly.

On a typical high production application, a single welding gun may make 50 spot welds on an assembly. Increasing the welding cycle time by so little as 10 cycles (60 Hz) will increase the total cycle time for 50 welds by 8 seconds. Therefore, production rates can be significantly affected by minute increases in the cycle time.

Quality Control

QUALITY CONTROL IS an important benefit of automation. However, weld quality is dependent upon process control. An automatic spot welding machine will repeat bad welds as well as good ones.

An automatic spot welding machine can maintain good spot weld quality with feedback controls and appropriate electrode maintenance. For example, a robot is normally programmed to always position the plane of the welding gun perpendicular to the weld seam. In manual spot welding, an operator can easily misalign the welding gun as a result of fatigue, boredom, or carelessness. This condition may result in improper welding pressure and weld metal expulsion from the seam.

When different thicknesses of metal are to be spot welded with the same welding gun, an automatic spot welding machine can be programmed to use a different welding schedule for each thickness. Weld quality is improved when the proper welding schedule can be used for every joint thickness.

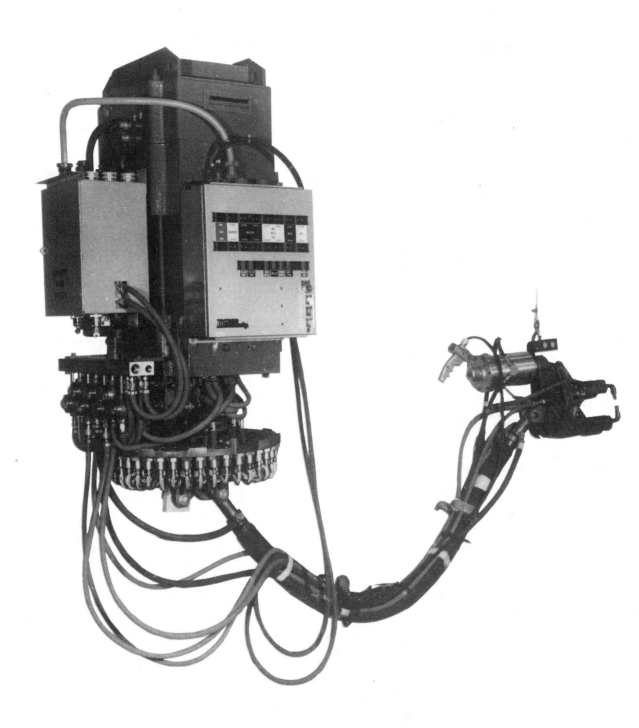

Figure 10.11—Typical Portable Resistance Spot Welding Machine

WELDING ROBOTS

Types Of Robots

THERE ARE FOUR general geometric classifications of industrial robots:

(1) Rectilinear (cartesian coordinate) robots have linear axes, usually three in number, which move a wrist in space. Their working zone is box shaped.

(2) Cylindrical coordinate robots have one circular axis and two linear axes. Their working zone is a cylinder.

(3) Spherical coordinate robots employ two circular axes and one linear axis to move the robot wrist. Their working zone is spherical.

(4) Articulating (jointed arm) robots utilize rotary joints and motions similar to a human arm to move the robot wrist. The working zone has an irregular shape.

All four robot geometries perform the same basic function; the movement of the robot wrist to a location in space. Each geometry has advantages and limitations under certain conditions. (Articulating and rectilinear robots are favored designs for arc welding.)

Load Capacity

MOST SPOT WELDING robots are designed to manipulate welding guns weighing between 75 and 150 lb (34 to 68 kg). Manual spot welding guns are often counterbalanced using an overhead support that can limit positioning of the gun. Robots of sufficient load capacity can move spot welding guns without counterbalances. Counterbalances may be needed to support heavy welding cables unless integral spot welding gun-transformer units of low weight are used.

Accuracy And Repeatability

A ROBOT CAN repeatedly move the welding gun to each weld location and position it perpendicular to the weld seam. It can also replay programmed welding schedules. A manual welding operator is less likely to perform as well because of the weight of the gun and monotony of the task. However, an operator can detect and repair poor spot welds; spot welding robots cannot.

Robot accuracy and repeatability vary with the manufacturer. Some robot models require periodic recalibration or reprogramming.

Number Of Axes

SPOT WELDING ROBOTS should have six or more axes of motion and be capable of approaching points in the work envelope from any angle. This permits the robot to be flexible in positioning a welding gun to weld an assembly. In the case of an automobile, a door, a window, and a roof can all be spot welded at one work station. Some movements that are awkward for an operator, such as positioning the welding gun upside down, are easily performed by a robot.

Reliability

INDUSTRIAL ROBOTS ARE currently capable of 2000 to 4000 hours of operation between failures with duty cycles of 98 percent or better. Robots are capable of operating continuously, as long as proper maintenance procedures are followed. On continuous lines with multiple robots, interruption of production can be minimized by (1) installing backup units in the line, (2) distributing the work of an inoperative robot to other nearby robots, or (3) quick replacement of the inoperable robot.

Maintenance

IN ADDITION TO the recommended maintenance of the robot, maintenance of the resistance welding equipment is necessary. Typically, the following actions are required during periodic maintenance of an automatic spot welding machine:

(1) Weekly or monthly
 (a) Tighten or replace electrical shunts.
 (b) Replace frayed or worn welding cables.
 (c) Repair or replace leaking air or hydraulic pressure lines.
 (d) Replace leaking or damaged air or hydraulic system regulators.
 (e) Replace welding transformers that show signs of deterioration.
(2) Daily
 (a) Inspect electrical cables for shorts.
 (b) Tighten loose bolts.
 (c) Inspect for oil or water leaks.
 (d) Clean the robot surroundings.
 (e) Report maintenance findings to supervisor.
(3) Hourly or after a specific number of welds, regrind, repair, or replace the spot welding electrodes.

Safety

THE OPERATOR OF a robot can easily avoid close proximity to jagged edges of parts, moving conveyors, weld metal expulsion, and other welding hazards. However, the movement of the robot arm creates a dangerous environment. The workers in the area must be completely prevented from entering the working envelope of the robot. Protective fences, power interlocks, and detection devices should be installed to assure worker safety.

RESISTANCE WELDING EQUIPMENT

AUTOMATIC WELDING IMPOSES specific demands on resistance spot welding equipment. Often, equipment must be specially designed and welding procedures developed to meet automatic welding requirements.

Welding Electrodes

ELECTRODE LIFE IS influenced by electrode force, electrode material, welding schedule, welding current, and the type of material being welded. Upslope of welding current can greatly reduce electrode wear. Progressively increasing the current after a fixed number of welds can compensate for mushrooming of the electrodes, thereby extending electrode life. Electrode life is also extended by proper positioning of the electrodes on the seam.

Spot Welding Guns

PORTABLE SPOT WELDING guns are normally designed to fit the assembly. Two basic designs are scissor and C-types.

Pneumatic guns are usually preferred because they are faster, and they apply a uniform electrode force. Hydraulic spot welding guns are normally used where space is limited or where high electrode forces are required. Unfortunately, hydraulic guns do not have the consistent electrode follow-up characteristics of pneumatic guns.

Integral transformer-gun units with up to 480 V primary can be safely used for automatic welding. Power consumption is lowered with these units by closely coupling the transformer and the welding gun.

CONTROLS

Multiple Schedule Controls

MICROCOMPUTER CONTROLS ARE available to program a number of different welding schedules. An automatic welding machine can be interfaced to such a control, and programmed to use the appropriate welding schedule for each spot weld or group of welds. Certain applications can benefit from multiple schedule controls, such as those shown in Figure 10.12. Multiple schedule controls are commonly used for spot welding such as automobile bodies and home appliances which involve welding various gauges of sheet metal in various combinations.

Supervisory Controls

Off-Flange Detectors. Automatic spot weld machines can be equipped to detect faulty or deteriorating process conditions. When teamed with optional sensory devices, such as off-flange detectors, automatic spot welding machines can monitor process conditions.

An off-flange detector measures either the voltage between the electrodes or the power factor. If the voltage or the power factor is below a set threshold value, the system signals the absence of material between the electrodes. An off-flange detector can be used to sense that a particular spot weld was not made.

Steppers. Steppers can be used to automatically increase the welding heat after a set number of spot welds to compensate for electrode mushrooming, to signal the need to replace electrodes, or to indicate the need for preventive maintenance functions on the machine. The devices can also be used to measure production rates.

Current Monitoring. A current monitoring device can indicate contaminated welding electrodes by detecting a predetermined change in welding current.

Energy Monitoring. Weld energy monitoring is a variation of current monitoring. The monitor is used to indicate trends or detect deterioration in equipment performance.

Resistance Monitoring. Resistance monitoring devices measure the change in electrical resistance between the electrodes as an indication of spot weld formation. When a resistance change occurs early in the welding cycle, the welding current can be terminated to avoid making an unacceptable spot weld.

Interface Between Components

THE CONVENTIONAL INTERFACE between an automatic machine control and the spot welding equipment is through input-output (I/O) channels. These channels enable the machine control to send and receive discrete signals. For example, an output signal can be sent from the machine to a welding control unit to initiate a weld. An input signal from the control unit to the machine can signal the machine to continue the operation sequence. A typical input/output communication link is as follows:

(1) The machine control sends an output signal to the welding control unit to initiate the electrode force and the welding sequence.

(2) The machine control sends an output signal to select the programmed weld schedule.

(3) The machine control receives an input signal from the welding control unit signifying that the control initiated a sequence.

Figure 10.12—Master Control Panel for Computer-Controlled Robotic Welding of Automobile Bodies

(4) The machine control receives an input signal from the welding control unit at the completion of the weld.

(5) The machine control receives an input signal from a supervisory control to turn on an alarm or interrupt the machine operation.

(6) The machine control sends an output signal to the welding control unit to confirm that it has received the weld completion signal and is moving to the next weld location.

(7) The machine control sends an output signal to a supervisory control to signal when the welding electrodes need to be redressed.

High level communication between automatic machine controls and welding controls is possible through standard communication links. These communication links permit long distance communications between central controls, welding controls, machine controls, and data acquisition centers for the following purposes:

(1) In-process maintenance, such as detection of a malfunctioning silicon controlled rectifier (SCR) or a missed weld
(2) Continuous in-process monitoring of welding current, power, or electrical resistance for quality control
(3) Monitoring of equipment status and condition
(4) Coordination of equipment activities
(5) Monitoring or reporting of production output
(6) In-process modification of programs or schedules

ACCESSORY EQUIPMENT

Monitors

WELD MONITORS CAN provide nondestructive quality assurance information. If weld quality data are difficult to obtain through random or periodic destructive testing, weld monitors may be an alternative. Data can be recorded for review at a later date or as a permanent record.

Monitors can reduce the amount of destructive testing required. Only welds of questionable quality need to be examined. Monitoring data and periodic destructive testing can provide an effective quality control program.

Monitoring equipment can perform some or all of the following functions:

(1) Sense position for detecting the presence of a part
(2) Measure welding current or power to detect long-term changes to indicate the need for cable and gun maintenance
(3) Measure welding voltage to detect the absence or presence of a part
(4) Measure resistance or impedance to detect the formation of a weld

(5) Sense electrode position to detect movement of electrodes signifying weld formation
(6) Measure ultrasonic reflection from a weld to verify weld size for quality control
(7) Perform acoustic emission detection and analysis of sound emitted from a weld as it forms to determine weld size or the presence of flaws

Adaptive Process Control

A WELDING PROCESS is said to be adaptively controlled when changes in weld integrity are measured and corrected in real time. The sensors listed previously for monitors can be used to detect certain properties of spot welds. These inputs can be analyzed and acted upon by an adaptive control program to optimize or correct the welding schedule variables. Users of adaptive controls are cautioned to thoroughly test and evaluate the controls before abandoning established quality control programs.

PROGRAMMING

Assembly Program Development

THE FIRST STEP in the development of a program for an assembly is to establish welding schedules for each proposed material and thickness combination. A robot can be "taught" the sequence of welding positions with the plane of the welding gun perpendicular to the seam. The sequence of weld positioning should be chosen to minimize the repositioning and index time for the robot. Finally, a travel speed between weld locations is selected. It should not be so high that oscillation or overshooting of the electrodes takes place. Appropriate welding schedules are selected and programmed at each welding position.

Touch-up Programs

TOUCH-UP PROGRAMS ARE the editing changes that are made in welding programs to improve or correct automatic operation. Typical reasons for editing include the following:

(1) Part dimensional changes
(2) Optimization of welding variables to improve weld quality
(3) Optimization of gun movements and speeds to increase production rate
(4) Changes in fixturing location
(5) Correction for movement, drift, or calibration changes

ARC WELDING AUTOMATION

TYPES OF AUTOMATION

VARIOUS ARC WELDING processes are used for automatic welding operations. Included are gas metal arc, flux cored arc, gas tungsten arc, plasma arc, and submerged arc welding. In most cases, automatic welding is best confined to the flat or horizontal positions for easy control of the molten weld pool. Nevertheless, automatic welding can be done in other positions using special arc control programming. An example of an automatic arc welding machine that is programmed to weld continuously around a pipe point is shown in Figure 10.13. This machine will orbit the pipe as the weld is deposited.

As with resistance welding, either the work or the welding gun or torch may be moved during automatic arc welding, depending on the shape and mass of the weldment and the path of the joint to be welded. It is easier to make circumferential welds in piping, tubing, and tanks if the work can be rotated under a stationary head for welding in the flat position. Long seams in plate, pipe sections, tanks, and structural members are easier to weld by moving the welding gun or torch along the seam. Another example of an automatic arc welding machine where the welding gun is moved while the work remains stationary is shown in Figure 10.14.

Figure 10.13—Automatic Arc Welding Machine Making Pipe Joints in the Horizontal Fixed Position (5G)

The automatic welding machines shown in Figures 10.4, 10.5, 10.13, and 10.14 represent fixed automation because they are designed to accommodate families of similar assemblies. Where flexible automation is desirable for arc welding several different assemblies, industrial robots are commonly used.

ARC WELDING ROBOTS

ROBOTIC ARC WELDING is applicable to high, medium, and low volume manufacturing operations under certain conditions. It can be applied to automation of medium and low volume production quantities where the total volume warrants the investment. Even job shop quantities can be welded where the investments in the equipment and the programming time can be justified.

An arc welding robot requires a number of peripheral or supporting devices to achieve optimum productivity. The basic elements of a robotic work cell are shown in Table 10.3. Many variations are possible, and each device could contain its own controller that would execute instructions from its program on command from the robot or host controller. All robot stations can be enhanced by one or more of the components listed in Table 10.4. These components help to quickly "teach the robot," minimize times for scheduled and unscheduled maintenance, and assure operator and equipment safety. Also included in Table 10.4 are several features or components that are not necessary for efficient robot cell use but can enhance the productivity of the cell.

Robot Features

ROBOTS REQUIRE SPECIAL features and capabilities to successfully perform arc welding operations. Arc welding robots are generally high precision machines containing electric servomotor drives and special interfaces with the arc welding equipment.

An articulating (jointed arm) robot is favored for arc welding small parts where the travel distances between welds are large. The arm of this type of robot is capable of quick motion. This robot design also is preferred for nonmoveable assemblies that require the robot to reach around or inside a part to position the robot wrist.

Rectilinear robots are favored for most other arc welding applications for safety reasons. They are particularly suited for applications where a welding operator is required to be in close proximity to the welding arc. Rectilinear robots move slower and in a more predictable path than articulating robots.

Figure 10.14—An Automatic Arc Welding Machine for Fillet Welding Stiffeners to Panels

Table 10.3
Elements of a Robot Work Cell

Component	Function
Host controller	Manages robot motions, welding process functions, and safety interlocks according to stored program. May also manage or direct motions of positioners, tooling, fixtures and material handling devices. (For most systems, the host controller is part of the robot.)
Robot	Manipulates welding torch as directed by controller.
Process package	Performs welding process as directed by host or robot controller. Process functions include shielding gas flows, wire feed, and arc voltage and current.
Positioner	Manipulates workpiece to fixed position so that robot can perform weld sequence. May also manipulate workpiece during welding to provide weld motion. May be activated by robot controller or by a human operator.
Fixture and clamping tools mounted on positioner	Hold workpiece components in a fixed position relative to two or more orientation points on the fixture. Clamping tools may be activated by the robot controller or by a human operator.
Material handling	Moves components into work cell, and removes welded assemblies from work cell. May be manual, machine or automatic.

Table 10.4
Supplementary Equipment for Welding Robot Work Cell

Component	Recommended Function
Master part (component)	Teaches the robot. Then tests and verifies program and work cell accuracy, and identifies changes.
Robot program listing	Programs the robot. Then verifies work cell performance.
Program save device	Permits permanent storage of robot program produced by "Teach Method" for future use.
Torch set-up jig	Allows quick set-up after torch maintenance or replacement. Establishes the position of the point of welding arc with respect to the robot.
Torch cleaning station	Keeps welding torch operating properly. May be performed manually by an operator or automatically.
Safety screens and interlocks	Provide operator protection from arc flash, smoke and fume, burns and heat. Prevent physical harm from robot, tooling, or material handling equipment.

Component	Desirable Function
Inspection jig	Enables quick inspection of products, helps to identify set-up or program problems.
Cell set-up and troubleshooting guide	Permits quick detection of problems. Provides orientation and training for new operators, maintenance and supervisory personnel.
Work cell tool kit	Keeps at hand special critical tools used in work cell set-up, adjustment and repair.
Offline programming station	Enables reprogramming, program debugging, and program editing without lost cell time. Relatively new technology still under development.
Download link	Permits off-line programming station to send programs to work cell. May allow work cell to send programs for storage.

Number Of Axes

ARC WELDING ROBOTS usually have five or six axes (see Figure 10.7), and some may be equipped with seven or eight axes. A complete robotic work station may contain as many as eleven axes of coordinated motion.

Six axes are required to position and orient an object in space. At least three of these axes must be revolute. The others may be linear. Many robot models have as many as six revolute joints.

Arc welding processes that use a consumable electrode, such as gas metal arc and flux cored arc welding, have one built-in degree of freedom because the welding gun can weld in any direction. Hence, a robot with five axes is capable of positioning the welding gun with these processes.

Robots can be equipped with additional axes to extend their working volume. These axes will be redundant degrees of freedom, and may or may not be usable while the robot is in motion.

Arc welding often involves manipulation of both the assembly and the welding torch. The axes needed to manipulate the assembly are equivalent to those needed to manipulate the welding gun, and additional axes will expand the flexibility of the system. If there are more than six unique degrees of freedom, a point in space can be approached from multiple positions. This flexibility is important in arc welding because gravity affects the behavior of the molten weld pool.

Accuracy And Repeatability

THE PRECISION WITH which a robot can approach an abstract point in space is called accuracy. Accuracy is required in robots where the control programs are developed numerically. It is a measure of the ability of the robot to reproduce a program that was generated by a computer or a digitizer. Robots which are "taught" the programs do not have to be as accurate because they depend upon memorization and replay.

Repeatability is a measure of the ability of a robot to repeatedly approach a point in space. Arc welding robots must have repeatability, but accuracy is not mandatory if the robot is "taught the path." The more repeatable a robot is, the larger are the allowable component tolerances. The total allowable variation in arc welding is the sum of the robot repeatability, the component parts variations, and the part positioning equipment repeatability.

Process Control

TO BE AN efficient arc welding machine, an industrial robot must be capable of controlling the arc welding process. As a minimum, the robot must be able to turn shielding gas on and off, initiate and terminate the welding sequence, and select the programmed welding conditions. Some robots control the welding process by selecting preset values. Other, more sophisticated robots are capable of directly controlling filler wire feeder and power supply, travel speed, and establishing the process conditions as a part of the robot program.

Robots that can select only preprogrammed welding conditions place more responsibility upon the welding equipment manufacturer, the user, and the operator. Such robots increase the costs of the peripheral equipment and the cost of interfacing it to the robot. Also, the welding process variables must be programmed separately from the robot program. Robots with built-in process control capabilities usually offer more flexibility.

ARC WELDING EQUIPMENT

IN GENERAL, WELDING equipment for automatic arc welding is designed differently from that used for manual arc welding. Automatic arc welding normally involves high duty cycles, and the welding equipment must be able to operate under those conditions. In addition, the equipment components must have the necessary features and controls to interface with the main control system. The main controller is actually the welding machine. It manages the functions of the peripheral equipment in the system.

Arc Welding Power Sources

AUTOMATIC ARC WELDING machines may require power sources more complex than those used for semiautomatic welding.[7] An automatic welding machine usually electronically communicates with the power source to control the welding power program for optimum performance.

The main control system can be designed to program welding power settings as a part of the positioning and sequence program, to set the appropriate welding power for each weld. The automatic welding program becomes the control of quality for the weld.

Welding power sources for automatic welding must be accurate and have a means for calibration. Manual welders often adjust the welding machine controls at the mid-point of a range that is suitable for several different welds. An automatic welding machine merely replays an established program, and the elements of that program must be accurately reproduced every time it is used.

Welding Wire Feeders

AN AUTOMATIC ARC welding machine is easily designed to directly control a motorized welding wire feed system. This allows flexibility in establishing various welding wire feed rates to suit specific requirements for an assembly. An automatic machine additionally can insure that the wire feed rate is properly set for each weld through "memorization" of the setting in the program. Special interfaces are required to allow the welding machine to control the wire feed rate. Typically, an analog output is provided from the welding machine control to interface to the wire feed system. If the welding machine is not equipped to receive a signal of actual wire feed rate, the wire feeder itself must have speed regulation. A means to calibrate the wire feeder is also required. Wire feed systems for automatic welding should have a closed-loop feedback control for wire speed. An automatic arc welding machine is normally run at high welding rates and duty cycles. The wire consumption rate is typically three to four times that of manual welding and the arc on-time can be two to three times that of manual welding.

Because the wire consumption rates are large, wire conduit liners and guides can become clogged with residual lubricants from the wire drawing process and small debris on the wire surface. This can result in rough wire feeding and voltage fluctuations. Therefore, conduit liners and guides should be inspected often and cleaned or replaced as required.

Arc Welding Guns And Torches

THE HIGH DUTY cycle of automatic arc welding (usually between 75 percent and 95 percent) requires a welding gun or torch that can withstand the heat generated by the welding process. Watercooling systems require maintenance, and suitable flow controls must be provided.

Flexible welding torch mounts having positive locating and emergency stop capabilities are preferred for robotic welding. A welding robot does not have the ability to avoid collisions with objects in the programmed travel path. During robot programming, the welding gun or torch may collide with the work. Collisions can cause damage to the robot and the welding gun or torch. Flexible mounts with emergency stop capabilities minimize the possibility of damage. Typically, the mounts are springloaded with two or more tooling pins to assure proper positioning. The emergency stop switch is activated when a collision knocks the gun or torch out of position.

---

7. Arc welding power sources are discussed in the *Welding Handbook*, Vol.2, 7th Ed., 1978.

Arc Voltage Control

THE GAS TUNGSTEN arc welding process requires real time adjustment of the electrode-to-work distance to control arc voltage. Automatic gas tungsten arc welding machines should be equipped with an adaptive control to adjust arc voltage. These controls usually sense the actual arc voltage and reposition the welding torch as required to maintain a preset voltage.

With gas tungsten arc welding, the arc is usually initiated with superimposed high frequency voltage or high voltage pulses. Application of the high voltage must be incorporated in the welding program, and welding travel must not start until the arc has been initiated.

MATERIAL HANDLING

FOR AN AUTOMATIC welding system to efficiently perform a task, the component parts must be reliably brought to the work station. Material handling systems are often used to move component parts into position. Material handling may be as simple as hand loading of multiple welding fixtures that are successively moved into the welding station, or as complicated as a fully automated loading system.

Part Transfer

A COMMON MATERIAL handling procedure is to transfer the component parts into the working envelope of an automatic welding machine. This task can be performed by turntables, indexing devices, or conveying systems. Such a procedure is used for small assemblies where parts are relatively light in weight.

Part transfer usually requires two or more fixturing stations, increasing fixturing costs. The number of transfer stations required is determined by the relationship between the loading time, the transfer time, and the welding time.

Robot Transfer

LARGE ASSEMBLIES AND assemblies that require significant arc time can be welded by moving or rotating a robotic arc welding machine into the welding area. This procedure is usually faster than moving the assembly, but it requires a relatively large working area.

A welding robot can be transferred between multiple work stations. This allows production flexibility, and reduces inventory and material handling costs. Work stations can be left in place when not in use while the robot is utilized at other locations.

Some robots can access multiple welding stations located in a semicircle. Rectilinear robots can move to welding stations that are organized in a straight line.

Positioner

A POSITIONER CAN be used to move an assembly under an automatic arc welding head during the welding operation, or to reposition an assembly, as required, for robotic arc welding.[8] The assembly can be moved so that the welding can be performed in a favorable position, usually the flat position.

There are two types of positioners for automatic arc welding. One type indexes an assembly to a programmed welding position. The other type is incorporated into the welding system to provide an additional motion axis. A positioner can provide continuous motion of the assembly while the machine is welding, to improve cycle time. Positioners can have fully coordinated motion for positioning a weld joint in the flat position in a robot cell.

CONTROL INTERFACES

AN AUTOMATIC ARC welding system requires control interfaces for component equipment. Two types of interfaces are usually provided: contact closures and analog interfaces.

Welding Process Equipment

WELDING POWER SOURCES and welding wire feeders are controlled with both electrical contact closure and analog interfaces. Contact closures are used to turn equipment on and off. Analog interfaces are used to set output levels.

Fixtures And Positioners

FIXTURES FOR AUTOMATIC welding are often automated with hydraulic or pneumatic clamping devices. The welding machine often controls the operation of the clamps. Clamps can be opened to permit gun or torch access to the weld joint. Most clamping systems and positioner movements are activated by electrical contact closures.

Other Accessories

ACCESSORIES, SUCH AS torch cleaning equipment, water cooling systems, and gas delivery equipment, are also controlled by electrical contact closures. Flow and pressure detectors are installed on gas and water lines. Signals from these detectors are fed back to the automatic welding machine control through contact closures.

---

8. Positioners are discussed in Chapter 9 of this volume.

PROGRAMMING

AN AUTOMATIC ARC welding system must be programmed to perform the welding operation. Programming is the establishment of a detailed sequence of steps that the machine must follow to successfully weld the assembly to specifications.

Each assembly to be welded requires an investment in programming. Programming costs vary widely depending upon the welding system being used, the experience of the programmers, and the complexity of the welding process. Investment in programming must be taken into account when determining the economics of automatic welding. Once an investment is made for a specific weldment, the program can usually be stored for future use.

A welding program development involves the following steps:

(1) Calibrate the automatic welding system. Calibration insures that future use of the program will operate from a known set point.

(2) Establish the location of the assembly with respect to the welding machine. Often, simple fixturing is sufficient.

(3) Establish the path to be followed by the welding gun or torch as welding progresses. Some robots can be "taught" the path while other automatic welding systems have to be programmed off-line.

(4) Develop the welding conditions to be used. They must then be coordinated with the work motion program.

(5) Refine the program by checking and verifying performance. Often, a program requires editing to obtain the desired weld joint.

Offline Programming and Interfacing with Computer Aided Design

THE PROCESS OF "teaching" the robot can be time consuming, utilizing productive robot time. If the need of only a few parts is forecast, robotic welding may not be economical. However, off-line programming using computer aided design (CAD) systems can be used to program the sequence of motions of the robot and the positioner. Graphic animation programs help to visualize and debug the motion sequence. The CAD model cannot always duplicate the actual conditions at the actual welding station. It is often necessary to edit any motion program generated off line.

ACCESSORIES FOR AUTOMATION

Welding Gun Cleaners

PERIODIC CLEANING OF gas metal arc and flux cored arc welding guns is required for proper operation. The high duty cycle of an automatic operation may require automated gun cleaning. Systems are available that spray an antispatter agent into the nozzle of the gun. Additionally, tools that ream the nozzle to remove accumulated spatter are available. The cleaning system can be activated at required intervals by the welding control system.

Handling Of Fumes And Gases

THE HIGH DUTY cycle of an automatic welding system can generate large volumes of fumes and gases. An appropriate exhaust system is highly recommended to safely remove fumes and gases from the welding area for protection of personnel.

Seam Tracking Systems

ONE PROBLEM WITH automatic arc welding operations is proper positioning of the welding gun or torch with respect to the weld joint to produce welds of consistent geometry and quality in every case. Dimensional tolerances of component parts, variations in edge preparation and fit-up, and other dimensional variables can affect the exact position and uniformity of the weld joints from one assembly to the next. Consequently, some adjustment of the welding gun or torch position may be desirable as welding proceeds along a joint.

There are several systems available for guiding a welding gun or torch along a joint. The simplest one is a mechanical follower system which utilizes spring-loaded probes or other devices to physically center the torch in the joint and follow vertical and horizontal part contours. These systems are, of course, limited to weld joints with features of sufficient height or width to support the mechanical followers.

An improvement over these units are electromechanical devices utilizing lightweight electronic probes. Such probes operate motorized slides that adjust the torch position to follow the joint. These devices can follow much smaller joint features and operate at higher speeds than purely mechanical systems. However, they are limited in their ability to trace multiple pass welds and square-groove welds. Also, they are adversely affected by welding heat. The welding head of the automatic welding machine shown previously in Figure 10.14 is equipped with electro-mechanical seam-following and video-monitoring, as shown in Figures 10.15(A) and (B).

There are seam tracking systems that utilize arc sensing. The simplest form of these systems is an arc voltage control used with the gas tungsten arc welding process. This control maintains consistent torch position above the work by voltage feedback directly from the arc.

More sophisticated versions of the arc seam tracking

Figure 10.15A—An Automatic Welding Machine Equipped with Video Monitoring Cameras and Electromechanical Seam Tracking Devices

systems employ a mechanism to oscillate the arc and interpret the variations in arc characteristics to sense the location of the joint. This information can then be fed back to an adjustment slide system or directly into a machine control to adjust the welding path. These systems, which depend on arc oscillation, may or may not be desirable with a particular welding process, and they can be limited in travel speed by the oscillation requirements. Also, arc characteristics can vary from process to process, and some tuning may be required to make an arc seam tracking system work in a given situation. The manufacturers should be consulted for details.

The most sophisticated seam tracking systems are optical types. These systems utilize video cameras or other scanning devices to develop a two- or three-dimensional image of the weld joint. This image can be analyzed by a computer system that determines such information as the joint centerline, depth, width, and possibly weld volume. These systems can be used for adaptive control of welding variables as well as seam tracking.

There are several types of optical tracking systems presently available. One is a two-pass system in which a camera first is moved along the nominal weld path with the arc off. The system performs an analysis and correction. Then, a second pass is made for actual welding. This system cannot correct for any distortion that occurs during welding.

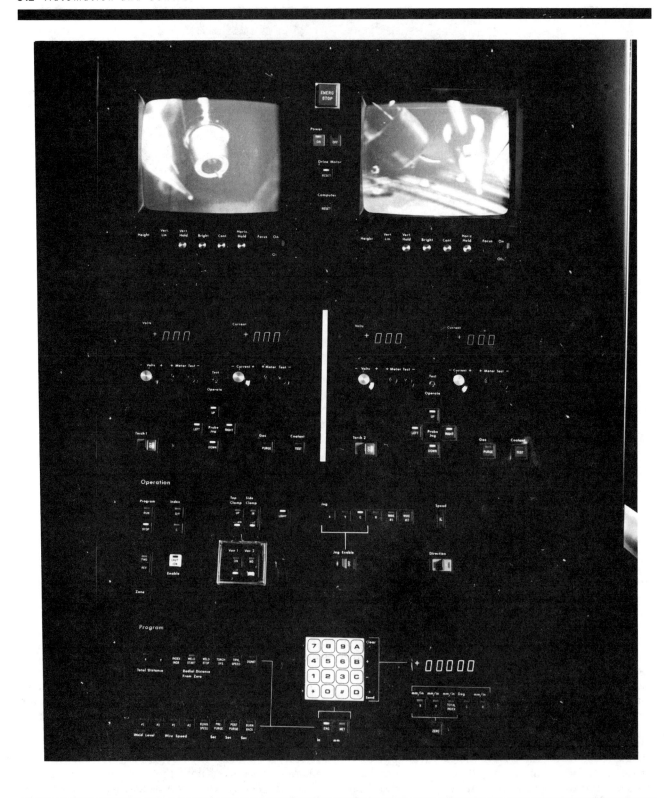

Figure 10.15B—Control Panel for the Automatic Welding Machine Showing Video Screens for Viewing the Welding Operation

A more sophisticated version is a single-pass or real-time system. Presently, these units look slightly ahead of (preview) the actual operating arc and feed back data for correction of tracking torch path and welding variables. These systems have some limitations, such as tracking sharp corners, and they can be influenced by smoke and arc heat. Furthermore, real-time "preview" systems frequently create interference problems between the sensor and the workpiece, especially when welding in tight corners.

The most sophisticated seam tracking systems are fully adaptive to compensate for volume changes in groove welds by varying either travel speed or wire feed speed accordingly, as well as tracking the joint.

Welding Wire Supply

PRODUCTION RATES OF automatic arc welding machines put heavy demands on welding wire supply systems. Welding wire spools of 60 lb (27 kg) or less used for semiautomatic arc welding are not satisfactory for automatic arc welding. Spools and drums of 250 lb (113 kg) or more are better choices.

Powered payoff packs that support the spool and reel the wire from the spool are desirable. These prevent slippage of the welding wire in the wire feed system, and resultant defects that can occur in the welds. In some cases, a push-pull wire feed system may be necessary if long distances or bends are involved.

BRAZING AUTOMATION

THE IMPORTANT VARIABLES involved in brazing operations include brazing temperature, time at temperature, filler metal, and brazing atmosphere. Other variables are joint fit-up, amount of filler metal, and rate and mode of heating.

Many brazed joints are made automatically using welding torches. Other processes that may be automated include furnace brazing (e.g., vacuum and atmosphere), resistance brazing, induction brazing, dip brazing, and infrared brazing.[9] Generally, the amount of heat supplied to the joint is automated by controlling temperature and the time at temperature. Brazing filler metal can be preplaced at the joints during assembly of components or automatically fed into the joints while at brazing temperature.

Further automation of brazing can be achieved by inclusion of automatic fluxing, in-line inspection and cleaning (flux removal), simultaneous brazing of multiple joints in an assembly, and continuous brazing operations.

Generally, the more automated a process becomes, the more rigorous must be its economic justification. Frequently, the increased cost of automation is justified by increased productivity. However, in the case of brazing, a further justification may well be found in the energy savings associated with the joint heating requirements.

FUNDAMENTALS

BRAZING TEMPERATURE IS generally dictated by the base metal characteristic and the type of filler metal. Gener-

ally speaking, filler metals with higher melting temperatures correlate with higher joint strengths. However, the actual brazing temperature must be related to the melting temperature and heat treatment response of the base metal.

It is standard practice to minimize the time at brazing temperature. Brazing times of less than a minute are common. Minimal times at brazing temperatures are desirable to avoid excessive filler metal flow, base metal erosion, and oxidation.

A major consideration for filler metal type is the cost. Obviously, compatibility of the base metal and filler metal must be considered. Compatibility factors include such things as corrosion potentials, brittle compound formation, and base metal wettability.

EQUIPMENT

WHILE MANUAL TORCH brazing represents the simplest brazing technique, there is some economic justification for this process. First, the braze joint is visible to the operator who can adjust the process based on observation. Consequently, the operator represents an in-line inspection control. Second, this process directs heat only to the joint area versus heating of an entire assembly when furnace brazing techniques are used. When energy costs represent a large percentage of the cost of a braze joint, this is an important consideration.

The major drawback to torch brazing is, of course, the fact that it is labor intensive and consequently represents low productivity. A continuous belt furnace can alleviate this problem; however, the joint is not visible during the brazing operation, and energy efficiency is low because the entire assembly is heated.

Automatic brazing machines can provide the attributes of both torch and furnace brazing, that is, energy

---

9. Brazing is covered in the *Welding Handbook*, Vol. 2, 7th Ed., 1980 and also the *Brazing Manual*, 3rd Ed., 1976, published by the American Welding Society.

economy and continuous operation. Typically, heat is directed to the joint area by one or more torches. Similar effects can be obtained by induction heating. A typical machine has provisions for assembly and fixturing, automatic fluxing, preheating (if needed), brazing, air or water quenching, part removal, and inspection.

Temperature Control

PROCESS CONTROL AND consistency that can be accomplished with automatic equipment are of primary importance. Temperature control is critical to successful brazing. When joining components of different masses, all components must attain brazing temperature at the same time. Overheating of a thinner component can cause molten filler metal to flow onto that component before it wets the thicker component. This can reduce the amount of filler metal available to fill the joint.

Automatic brazing equipment should provide a correct and predetermined heating pattern. This can be accomplished effectively with either induction or torch heating. Induction heating works best on ferrous metal parts, especially those of symmetrical shape. Induction heating coils must be designed for efficient localized heating of the joint area. Fixturing and removal of the brazed part from the heating coil must also be considered during coil design.

Torch heating involves the use of a preset heating pattern by appropriately designed torches located at one or more machine stations. Torches of various sizes, utilizing different fuel gases mixed with air or oxygen, may be used. Gas-air torches, with gentle wrap-around flames, can be used on many applications. When a pinpoint flame is required or when hotter and faster heating is desired, oxygen may be used in place of or in combination with air. The use of oxygen represents an additional operating cost, however, and should be considered only when its cost can be justified by increased production rates.

Heating Time

TIME AT TEMPERATURE is an important control feature of automatic brazing equipment. Heating time is a function of the desired production rate and can be determined in advance. Once the heating time of an assembly is determined, the number of heat stations and dwell time at each can be allocated among the total number of work stations. Digital timing equipment can provide exact control of the dwell or heating time.

To maximize production, assemblies should be heated only to the temperature at which filler metal melts and flows into the joint. Holding the assembly at temperature beyond this point is not recommended in most cases because it can promote oxidation and cause other problems, such as volatilization of filler metal constituents and base metal erosion.

Filler Metal Application

CONTROL OF FILLER metal amounts and consequent joint quality is also possible with automatic brazing equipment. Brazing filler metal application can be effectively controlled with the use of preplaced shims or wire rings, or with timed automatic feed of wire to a joint from a filler wire feeder. Volume per joint can be controlled to provide consistency or repeatability that is not attainable with manual rod feeding.

Flux Application

CONTROL OF THE amount and coverage of brazing flux can be provided by automatic equipment. Dispensers are available that can automate this operation. Dispensable fluxes are available for use in these units, and flow can be adjusted to handle a variety of applications.

ADVANTAGES OF BRAZING AUTOMATION

THE MAIN ADVANTAGES of automatic brazing are as follows:

(1) High production rates
(2) High productivity per worker
(3) Filler metal savings
(4) Consistency of results
(5) Energy savings
(6) Adaptability and flexibility

Production requirements essentially determine the level or amount of automation required. Low production requires only basic equipment, such as a motorized conveyor transporting assemblies through a furnace. High production requirements may involve a fully automated system utilizing parts feeding and automatic unloading of completed assemblies. This type of equipment is able to produce in excess of 1000 parts per hour.

Automatic brazing equipment can dramatically increase the productivity per worker. As an example, one unskilled brazing operator running an automatic brazing machine is able, in many cases, to produce as much as four or five highly skilled manual brazers. This increase in productivity alone is often enough to justify the purchase of automatic brazing equipment.

Another feature of automatic brazing equipment that reflects a major cost advantage and economic justification for purchase is savings in filler metal. Instead of feeding extra filler metal for assurance, as can be the case with manual brazing, automatic equipment can provide an exact and correct amount each and every time.

The advantage of consistent quality production is one that should not be overlooked. Rework or repairs of brazements are expensive, and can require extra operations. The best way to avoid rework is to produce a good brazement the first time. This is expected with automatic equipment.

Automatic brazing equipment should be energy efficient. Localizing heat input only to the joint area is good energy conservation. Utilizing low cost fuels mixed with compressed or shop air is effective energy efficiency.

PROBLEMS OF AUTOMATION

RISK FACTORS

Factory Integration

THE INTRODUCTION OF an automatic welding or brazing machine can create many problems in a manufacturing shop. For example, there can be incompatibilities with other shop operations, such as machining, cutting, or forming. Material handling for automatic operations can present problems. Tasks that are often overlooked in preparation for welding automation include the following:

Part Programming. The operator may be required to develop the part program using the "teach method." This is new to most factory operators, unlike CNC machine tools and other automatic machines where programs are written off-line by a trained programmer.

Program Proofing. The program for automatic welding or brazing equipment is usually evaluated using actual production parts. Simulated production parts cannot duplicate the effects of distortion, heat input, and clamping, and such parts should only be used in the initial stages of process development. Decisions regarding the use of simulations and the time to switch to production parts can be difficult. An inexperienced user should solicit help from the equipment manufacturer or other experts.

Documentation. A system of numbering, filing, revising, and maintaining operating programs is required for programmable automatic welding and brazing machines. Systems for numerically controlled machine tools may not be suitable for the following reasons:

(1) Welding and brazing are operations joining two or more parts, whereas machining is done on a single part.

(2) Welding programs, which are usually small in terms of data storage, are often formulated into a single tape for fast, convenient access. Machine tool programs are usually generated individually for each operation.

(3) Programs may be edited by personnel who are not familiar with welding processes.

(4) Several different programs may be required to compensate for joint fit-up or part variations with welding and brazing operations. Machine tools usually need only one program for each part or operation.

(5) The costs of mishandling program documentation can be large. Expensive materials can be ruined using an old or incorrect program.

System Integration

THE INTRODUCTION OF automatic welding or brazing may have widespread impact on company operations. Several aspects should be evaluated.

Design. Special design considerations are required for automatic welding and brazing. Joint designs and placements should be selected to optimize repeatability of joint location and fit-up. Component strength with automatic welding or brazing may be different from that with manual welding or brazing. Appropriate tests should be made to determine actual mechanical properties using the proposed automatic joining equipment. Corrosion resistance may also be affected. Savings in materials and consumables may be possible through design changes that recognize the increased consistency and reliability of automatic welding or brazing.

Manufacturing Planning. Component sequence may need to be changed for automatic welding or brazing. A few potential changes are

(1) Periodic destructive or nondestructive testing and inspection may be required for quality control.

(2) Additional quality control tests, such as joint fit-up inspection, may be required.

(3) Multiple welding operations may be combined at a single work station.

(4) Welding or brazing conditions, machine control programs, and other formulated information may need to accompany components or be attached to planning sheets.

(5) New process specifications and acceptance criteria may be required.

(6) NC networking and integration with other manufacturing processes must be considered.

Fixturing and Positioning. The fixturing and positioning of parts for automatic welding or brazing can be different from any previous experience of the average fabrication shop. Fixtures and positioners that are not furnished by the machine manufacturer with automatic welding or brazing equipment should be planned and designed by personnel experienced with the specific joining process.

Manufacturing. Automatic welding and brazing operations should be integrated into other factory operations. The following items should be considered:

(1) Operator job classification
(2) Operator skills and training
(3) Production backup capability
(4) Component part repeatability
(5) Repair procedures
(6) Intershift communications
(7) Safety procedure
(8) Consumables and spare parts
(9) Test parts and testing procedures
(10) Maintenance capability
(11) Material handling
(12) Calibration of equipment and procedures
(13) Production support personnel

Inspection and Quality Control. Automatic welding often complicates inspection. Fillet welds are preferred for automatic welding, but are difficult to inspect. Multiple welding operations performed at one station often restrict access for inspection. In-process inspection may not be feasible from operational and economical standpoints. New quality control procedures are usually required.

Reliability of Equipment

GENERALLY, THE RELIABILITY of a system is inversely proportional to its complexity, and automatic welding and brazing machines are complex systems. Nevertheless, the inherent precision and durability of equipment components usually exceeds the duty requirements because welding and brazing operations impose light dynamic machine loads and accuracy requirements compared to machining. The reliability problems in welding machines are usually related to environmental conditions of heat, spatter, fume, and collision. Electronic components require consideration of the electromagnetic and electric power environment.

CONSISTENCY AND PREDICTABILITY

SEVERAL OF THE many variables that can affect the quality of welded or brazed joints include the following:

(1) Material variations (thickness, geometry, composition, surface finish)
(2) Welding or brazing conditions (voltage, current, gas flow rates)
(3) Fixturing (part location, chill bars, materials)
(4) Procedures (alignment, preheat, heating rate)
(5) Consumables (filler metals, gases)
(6) Machine design and operation (bearings, guides, gear trains)

VENDOR ASSISTANCE

RELIABLE SUPPLIERS OF automatic welding and brazing systems can assist the users in several ways.

Evaluation and Consulting Services

IN GENERAL, IT is prudent to consult with other knowledgeable persons when possible. Equipment manufacturers and engineering consultants are often more capable of determining the feasibility of a proposed automatic welding or brazing system than the user. They may be able to cite examples of similar installations and arrange for plant visits, point out potential problems, and indicate the need for special equipment. Discussions with plant management of similar installations are especially valuable.

Testing and Prototype Production

WELDING OR BRAZING tests can be performed on simple shapes, such as flat plate or pipe, when evaluating automation feasibility. Prototype production parts should be welded or brazed during equipment acceptance tests to establish production readiness. These tests can reveal unexpected problems, which are best solved before production commitments are made.

Training and Education

AN APPROPRIATE SCHOOL or training program should be located for training of the equipment operators. A manufacturer of automatic welding or brazing equipment can be a source of training programs. Other sources are technical societies, vocational schools, and colleges or universities. Training should be considered for the engineers and maintenance personnel also.

Installation Services

INSTALLATION IS NORMALLY performed by or with the guidance of the equipment manufacturer. The installation of foundations, utilities, and interfaces with other equipment is usually performed by the user. The installation project should be thoroughly planned in advance of delivery of the equipment. Engineering drawings and installation requirements should be obtained from the manufacturer as early as possible.

Debugging the system can be a complex process and consume more time than expected. The equipment manufacturer should provide the needed test equipment and instrumentation to accomplish the job. The user can help expedite the task by providing required assistance to the manufacturer's start-up personnel.

SUPPLEMENTARY READING LIST

Computer Control and Monitoring

Bollinger, J.G. and Ramsey, P.W. Computer controlled self programming welding machine. *Welding Journal.* 58(5): May 1979; 15-21.

Hanright, J. Robotic arc welding under adaptive control—a survey of current technology. *Welding Journal.* 65(11): Nov. 1986; 19-24.

———. Selecting your first arc welding robot—A guide to equipment and features. *Welding Journal.* 63(11): Nov. 1984; 40-45.

Homes, J.G., and Resnick, B.J. Human combines with robot to increase welding versatility. *Welding and Metal Fabrication.* 48(1): Jan./Feb. 1980; 13-14, 17-18, 20.

Kuhne, A.H., Frassek, B., and Starke, G. Components for the automated GMAW process. *Welding Journal.* 63(1): Jan. 1984; 31-34.

Paxton, C.F. Solving resistance (spot) welding problems with mini- and micro-computers. *Welding Journal.* 58(8): Aug. 1979; 27-32.

Scott, J.J. and Brandt, H. Adaptive feed-forward digital control of GTA welding. *Welding Journal.* 61(3): Mar. 1982; 36-44.

Automatic Systems

Bennington, R. J. Speed doubled on robotic arc welding of complicated automotive assembly. *Welding Journal.* 65(11): Nov. 1986; 27-30.

Fuller, J. H. Outdoor equipment manufacturer installs robots, reaps long-term benefits. *Welding Journal.* 65(4): Apr. 1986; 54-56.

Flanigan, L. Factors influencing design and selection of GTAW robotic welding machines for the space shuttle main engine. *Welding Journal.* 65(11): Nov. 1986; 62-67.

Henschel, C. and Evans, D. Automatic brazing systems. *Welding Journal.* 61(10): Oct. 1982; 29-32.

Jones, S.B. Automan '83 —state of the art and beyond. *Metal Construction.* 15(5): May 1983; 283-285.

Lee, J. Robotic welding in the factory-of-the-future. *Welding Journal.* 63(11): Nov. 1984; 35-37.

Malin, V. Designers guide to effective welding animation—part I: analysis of welding operations as objects for automation. *Welding Journal.* 64(11): Nov. 1985; 17-27.

———. Problems in design of integrated welding automation—part I: analysis of welding-related operations as objects for welding automation. *Welding Journal.* 65(11): Nov. 1986; 53-60.

Miller, L. Development of the "apprentice" arc welding robot. *Metal Construction.* 12(11): Nov. 1980; 615-619.

Nozaki, T. et al. Robot sees, decides, and acts. *Welding and Metal Fabrication.* 47(9): Nov. 1979; 647, 649-651, 653-655, 657-658.

Soroka, D.P. and Sigman, R.D. Robotic arc welding: what makes a system? *Welding Journal.* 61(9): Sept. 1982; 15-21.

Sullivan, M.J. Application considerations for selecting industrial robots for arc welding. *Welding Journal.* 59(4): Apr. 1980; 28-31.

Wolke, R.C., Integration of a robotic welding system with existing manufacturing processes. *Welding Journal.* 61(9): Sept. 1982; 23-28.

Seam Tracking

Agapakis, J. E., Katz, J. M., Koifman, M., Epstein, J. M., Eyring D. O., and Rutishauser, H. J. Joint tracking and adaptive robotic welding using vision sensing of the weld joint geometry. *Welding Journal.* 65(11): Nov. 1986; 33-41.

Bollinger, J.G. and Harrison, M.L. Automated welding using spatial seam tracking. *Welding Journal.* 50(11): Nov. 1972; 787-792.

Goldberg, F. and Karlen, R. Seam tracking and height sensing—inductive systems for arc welding and thermal cutting. *Metal Construction.* 12(12): Dec. 1980; 668-671.

Hanright, J. Robotic arc welding under adaptive control technology. *Welding Journal.* 65(11): Nov. 1986; 19-24.

Lacoe, D. and Seibert, L. 3-D vision-guided robotic welding system aids railroad repair shop. *Welding Journal.* 63(3): Mar. 1984; 53-56.

Pan, J.L. et al. Development of a two-directional seam tracking system with laser sensor. *Welding Journal.* 62(2): Feb. 1983; 28-31.

Pandjuris, A.K. and Weinfurt, E.J. Tending the arc. *Welding Journal.* 51(9): Sept. 1972; 633-637.

Richardson, R.W. et al. Coaxial arc weld pool viewing for process monitoring and control. *Welding Journal.* 63(3): 1984; 43-50.

———. Robotic weld joint tracking systems—theory and implementation methods. *Welding Journal.* 65(11): Nov. 1986; 43-51.

CHAPTER 11

WELD QUALITY

PREPARED BY A COMMITTEE CONSISTING OF:

James C. Papritan,
Chairman
Ohio State University

Karl R. Anderson
Newport News Shipbuilding

James R. Hannahs
Midmark Corporation

Jack W. Lee
AVCO-Lycoming Division

Allan Lemon
Welding Institute of Canada

Carl D. Lundin
University of Tennessee

David R. Miller
Newport News Shipbuilding

Alan W. Pense
Lehigh University

Les Sandor
Widener University

James P. Snyder
Bethlehem Steel Corporation

WELDING HANDBOOK COMMITTEE MEMBER
James C. Papritan
Ohio State University

WELD QUALITY

INTRODUCTION

MEANING OF WELD QUALITY

IF A WELDMENT (or brazement) is to have the required reliability throughout its life, it must have a sufficient level of *quality* or *fitness for purpose*. Quality includes design considerations, which means that each weldment should be:

(1) Adequately designed to meet the intended service for the required life
(2) Fabricated with specified materials and in accordance with the design concepts
(3) Operated and maintained properly

Quality is a relative term, and to have higher quality than is needed for the application is not only unnecessary, it is often costly. Thus, quality levels may be permitted to vary among different weldments and individual welds, depending on service requirements.

Quality considerations are often narrowly confined to physical features normally examined by inspectors, but quality should also include such factors as hardness, chemical composition, and mechanical properties. All of these characteristics contribute to the fitness for purpose of a weld. The quality level required to provide the desired reliability depends on the expected modes of failure under the anticipated service conditions.

Weld quality relates directly to the integrity of a weldment. It underlies all of the fabrication and inspection steps necessary to ensure that a welded product will be capable of serving the intended function for the desired life. Both economics and safety influence weld quality considerations. Economics require that a product must be competitive, and safety requires that the product function without hazard to personnel or property.

The majority of welded fabrication standards define quality requirements to insure reasonably safe operation in service.[1] The requirements of such standards are to be considered minimums, and the acceptance criteria for welds should not be encroached upon without sound engineering judgement. For critical applications, more stringent requirements than those specified in the fabrication standard may be necessary to ensure safety.

Weld quality is verified by nondestructive examination. The acceptance standards for the welds are generally related to the method of nondestructive examination. All deviations are evaluated, and the acceptance or rejection of a weld is usually based on well defined conditions. Repair of unacceptable or defective conditions is normally permitted so that the quality of the weld may be brought up to acceptance standards. Many standards relating to weld quality do not govern product usage, and leave maintenance of the fabricated product to the owner. It is the owner or a designated engineering representative who must modify, amplify, or impose additional weld quality standards during maintenance activities to ensure that the product continues to function properly.

Weld quality, while sometimes difficult to define precisely, is often governed by codes, specifications, or regulations based on rational assessments of economics and safety. The documents may be modified by the owner or his engineering representative to reflect additional concerns of usage related to safety or economics, or both. Welds are examined in regard to size, shape, contour, soundness, and other features. Thus, at the heart of

---

1. Standards are discussed in Chapter 13 of this volume.

weld quality is the understanding of the occurrence, examination, and correction of any defects.[2]

SELECTION OF WELD QUALITY

DETERMINATION OF THE overall quality requirements for a weldment is a major consideration involving design teams and quality groups. Specifying excessive quality can lead to high costs with no benefits, while low quality weldments lead to high maintenance costs and excessive loss of service. Thus, the aim is to specify features which lead to fitness for purpose.

Fortunately, valuable guidance is provided in fabrication standards, which often indicate allowable levels of stress and discontinuities, and minimum levels of mechanical properties. These standards are based on experience, and they have proven to be safe.

Quality selection also involves consideration of those major factors that can be analyzed by fracture mechanics (discussed later), as well as a number of other factors. The main considerations affecting quality selection are design, fabrication, operation, and maintenance.

Design Considerations

FITNESS FOR SERVICE requires that the design of a weldment or structure be consistent with sound engineering practices. Components of adequate size should be specified to ensure that stresses from anticipated service loads are not excessive. The intended service should be carefully analyzed to determine whether cyclic loading might result in fatigue failure in highly stressed members. If the anticipated number of loading cycles exceeds 10 000 then fatigue of critical members should be considered.

Environmental conditions during service should be forecast, and the ramifications of those conditions should be considered in the design. Brittle fracture is a possiblility in low temperature service; creep is a consideration in high temperature service. Furthermore, corrosion and wear can reduce the section size and increase service stresses, as well as create sites for fatigue cracks to initiate and further reduce fitness for service of the assembly.

Designers should select materials of proper size with chemical and mechanical properties that will ensure satisfactory service. If material that can meet all of the requirements is not available, other means of providing fitness for service should be specified. For example, in a highly corrosive environment, suitable solutions may be painting the metal surfaces or cathodic protection.

Fabriction Considerations

THE CONTRACTOR should select fabrication procedures and practices that ensure that the weldments meet the design specifications. The procedures should be rigorously followed, and the weldments should be properly inspected to verify that the base metal and the weld joint are free of unacceptable discontinuities.

All parties should recognize that only the exposed surfaces of certain welds may be inspected. In that case, a sufficient margin of safety should be incorporated to allow for the possibility of undetected discontinuities.

Operating and Maintenance Considerations

OPERATING A FACILITY safely requires periodic inspection. If failure at a facility would result in a public hazard or the destruction of property, then inspections should be more frequent and more rigorous. Power plants, chemical plants and refineries, dams, and bridges are examples of fixed facilities that may endanger facility employees, the public, and the surrounding neighborhoods. Automobiles, trucks, trains, and ships are mobile facilities that may also damage individuals and properties. Maintenance of these facilities is a requirement for continued safe operation.

Some facilities can partially continue operations during inspection and maintenance, and others may require a complete shutdown. Often the loss of production is far more expensive than the direct cost of the repair. In such cases, premium materials and costly fabrication practices can easily be justified, provided these precautions permit extended operating periods between inspections or improve the likelihood that extensive repairs will not be required as a result of periodic inspections.

Finally, the basis for selection of overall quality is a combination of design, fabrication, and inspection which will provide the lowest total cost over the full life of the weldment. Low initial cost, minimum weight, least welding, fewest imperfections, and other factors taken individually should not be the basis for quality selection.

Optimum cost quality is based on (1) costs of design, materials, fabrication, quality assurance, and money; (2) costs of possible failure multiplied by the probability of failure; and (3) service costs (including maintenance). The lowest fabrication cost seldom represents the lowest total cost.

---

2. Inspection methods and procedures for detecting discrepancies in welded and brazed joints are discussed in Chapter 15.

TERMINOLOGY

EXCEPT FOR THOSE terms defined here, the terms used in this chapter are defined in AWS A3.0, *Standard Welding Terms and Definitions.*

Excessive melt-through. A hole through the weld metal, usually occurring in the first pass.

Inadequate joint penetration. Joint penetration that is less than that specified.

Incomplete fusion. A weld discontinuity in which fusion did not occur between weld metal and joint fusion faces or between adjoining weld beads.

Incomplete joint penetration. Joint penetration that is unintentionally less than the thickness of the weld joint.

Joint misalignment. In plate, offset or mismatch in a direction perpendicular to the plate surface and weld axis. In pipe, offset or mismatch in the radial direction at a butt joint or a T-joint.

Lamination. A type of discontinuity with separation or weakness generally aligned parallel to the worked surface of a metal. It may be the result of piping blisters, seams, inclusions, or segregations that became elongated and made directional by working.

Lap. A surface imperfection, caused by folding over hot metal, fins, or sharp corners and then rolling or forging them into the surface.

Seam. An unwelded fold or lap that appears as a crack on the surface of a metal product, usually resulting from a discontinuity formed during casting or rolling.

Surface irregularities. Any number of surface conditions that result in notches or abrupt changes in thickness. Some are excessive reinforcement, convexity, concavity, undercut, and overlap. Figure 11.1 illustrates eleven weld surface irregularities.

Tungsten inclusion. Droplets from the tungsten electrode entrapped in weld metal or along the weld interfaces.

Warpage. Buckling of sheet or plates parallel or transverse to the weld axis.

DISCONTINUITIES IN FUSION WELDED JOINTS

CLASSIFICATION

DISCONTINUITIES IN FUSION welded joints may be classified under three major headings, related to (1) procedure and process, (2) design, and (3) metallurgical behavior. Those discontinuities usually considered in each of the three major groups are shown in Table 11.1. The groups should be applied loosely because discontinuities listed in each group may have secondary origins in other groups. Discontinuities related to process, procedure, and design are, for the most part, those that alter stresses in a weld or heat-affected zone. Metallurgical discontinuities may also alter the local stress distribution, and in addition, may affect the mechanical or chemical (corrosion resistance) properties of the weld and heat-affected zone.

Discontinuities alter stresses in two ways. The least detrimental effect is stress amplification. Discontinuities amplify stresses by reducing the cross-sectional area. The average stress is amplified in direct proportion to the reduction in area. Stress concentration is more detrimental. Stresses are concentrated at notches, which occur at sudden changes in weldment geometry. The more severe the change in geometry, the greater the notch effect and the accompanying stress concentration. Tensile stresses perpendicular to a notch and shear stresses parallel to a notch are concentrated at the notch tip. Extremely high concentrations of stress can develop at sharp notches, such as cracks.

Discontinuities should be characterized not only by their nature, but also by their shape. Planar type discontinuities, such as cracks, laminations, incomplete fusion, and inadequate joint penetration, create serious notch effects. Three-dimensional discontinuities create almost no notch effect, but amplify stresses by reducing the weldment area. Therefore, the following characteristics of discontinuities should always be considered:

(1) Size
(2) Acuity or sharpness
(3) Orientation with respect to the principal working stress and residual stress
(4) Location with respect to the weld, the exterior surfaces of the joint, and the critical sections of the structure

Table 11.1
Classification of Weld Joint Discontinuities

Welding Process or Procedure Related
- A. Geometric
  - Misalignment
  - Undercut
  - Concavity or convexity
  - Excessive reinforcement
  - Improper reinforcement
  - Overlap
  - Burn-through
  - Backing left on
  - Incomplete penetration
  - Lack of fusion
  - Shrinkage
  - Surface irregularity
- B. Other
  - Arc strikes
  - Slag inclusions
  - Tungsten inclusions
  - Oxide films
  - Spatter
  - Arc craters

Metallurgical
- A. Cracks or fissures
  - Hot
  - Cold or delayed
  - Reheat, stress-relief, or strain-age
  - Lamellar tearing
- B. Porosity
  - Spherical
  - Elongated
  - Worm-hole
- C. Heat-affected zone, microstructure alteration
- D. Weld metal and heat-affected zone segregation
- E. Base plate laminations

Design Related
- A. Changes in section and other stress concentrations
- B. Weld joint type

The significance of these characteristics is described in detail later.

Procedure and Process Discontinuities

CERTAIN DISCONTINUITIES FOUND in welded joints are related to specific welding processes. Slag inclusions are generally associated with shielded metal arc, flux cored arc, submerged arc, and electroslag welding. Tungsten inclusions result from improper gas tungsten arc welding practices. The types of discontinuities that may result with various welding processes are shown in Table 11.2. Weld discontinuities encountered with a specific welding

process are covered in detail in the appropriate volume of this handbook.[3]

Design Discontinuities

SOME DISCONTINUITIES FOUND in welded joints are the result of design decisions. Two examples are abrupt changes in weldment shape or cross-sectional area which concentrate stresses and partial penetration joints which leave a recognized mechanical discontinuity. Proper design of a weldment will ensure that stress concentrations from design decisions do not adversely affect the weldment performance.

Design can also affect the occurrence of mechanical and metallurgical discontinuities. Conditions of high restraint increase the likelihood of cracks, and limited access for welding (any condition that makes it difficult for the welder to deposit a sound weld) increases the likelihood of procedure- and process-related discontinuities.

Metallurgical Discontinuities

METALLURGICAL DISCONTINUITIES ARE changes in the properties of the weld or base metals.[4] A crack is usually the result of a metallurgical discontinuity. For example, cold cracks are often the result of localized embrittlement. Most metallurgical discontinuities are closely related to the welding process and the design of the weldment. High heat input may, for example, cause segregation of alloying element in some metals. Welding together of incompatible materials can cause severe embrittlement. Metallurgical discontinuities may occur in the base metal prior to fabrication. Laminations, for example, are base metal defects.

LOCATION AND OCCURRENCE OF DISCONTINUITIES

DISCONTINUITIES IN WELDMENTS may be found in the weld metal, the weld heat-affected zone, and the base metal. The common weld discontinuities, general locations, and specific occurrences are presented in Table 11.3.

The discontinuities listed in Table 11.3 are depicted in butt, lap, corner, and T-joints in Figures 11.1 through 11.6. Where the list indicates that a discontinuity is generally located in the weld, it may be expected to appear in almost any type of weld. However, there are excep-

---

3. Welding processes are covered in Vols. 2 and 3 of the 7th Ed. of this Handbook. They will be included in Vol. 2, 8th Ed.
4. Typical discontinuities associated with specific base metals are discussed in Vol. 4 of the 7th Ed. of the *Welding Handbook* and are planned for Vol. 3 of the 8th Ed.

Table 11.2
Discontinuities Commonly Associated with Welding Processes

Welding Process	Type of discontinuity						
	Porosity	Slag Inclusions	Incomplete Fusion	Inadequate Joint Penetration	Undercut	Overlap	Cracks
Stud welding	–	–	X	–	–	–	X
Plasma arc welding	X	–	X	X	–	–	X
Submerged arc welding	X	X	X	X	X	X	X
Gas tungsten arc welding	X	–	X	X	–	–	X
Gas metal arc welding	X	X	X	X	X	X	X
Flux cored arc welding	X	X	X	X	X	X	X
Shielded metal arc welding	X	X	X	X	X	X	X
Carbon arc welding	X	X	X	X	X	X	X
Resistance spot welding	–	–	X	–	–	–	X
Resistance seam welding	–	–	X	–	–	–	X
Projection welding	–	–	X	–	–	–	X
Flash welding	–	–	X	–	–	–	X
Upset welding	–	–	X	–	–	–	X
Percusion welding	–	–	X	–	–	–	X
Oxyacetylene welding	X	–	X	X	–	–	X
Oxyhydrogen welding	X	–	X	X	–	–	X
Pressure gas welding	X	–	X	–	–	–	X
Cold welding	–	–	X	–	–	–	X
Diffusion welding	–	–	X	–	–	–	X
Explosion welding	–	–	X	–	–	–	–
Forge welding	–	–	X	–	–	–	X
Friction welding	–	–	X	–	–	–	–
Ultrasonic welding	–	–	X	–	–	–	–
Electron beam welding	X	–	X	X	–	–	X
Laser beam welding	X	–	X	–	–	–	X
Electroslag welding	X	X	X	X	X	X	X
Induction welding	–	–	X	–	–	–	X
Thermit welding	X	X	X	–	–	–	X

X indicates that the type of discontinuity may occur in welds produced by the process.
– indicates the occurrence of this type defect in these welds is very rare.

tions. For example, tungsten inclusions are only found in welds made by the gas tungsten arc welding process, and microfissures normally occur only in electroslag and electrogas welds.

Each general type of discontinuity is discussed below. The term *fusion type discontinuity* is sometimes used inclusively to describe slag inclusions, incomplete fusion, inadequate joint penetration, and similar generally elongated discontinuities in fusion welds. Many codes consider fusion type discontinuities less critical than cracks. However, some codes specifically prohibit cracks as well as fusion type defects. Spherical discontinuities, almost always gas pores, can occur anywhere within the weld. Elongated discontinuities may be encountered in any orientation. Specific joint types and welding procedures have an effect on type, location, and incidence of discontinuities. The welding process, joint details, restraint on the weldment, or a combination of these may have an effect on the discontinuities to be expected.

FUSION WELD DISCONTINUITIES

Porosity

POROSITY IS THE result of gas being entrapped in solidifying weld metal. The discontinuity is generally spherical, but it may be elongated.

Table 11.3
Fusion Weld Discontinuity Types

Type of Discontinuity	Discontinuity Identification	Appearing in Figure Number	Location[a]	Remarks
Porosity			W	
Uniformly scattered	1a	11.2, 11.6		
Cluster	1b	11.1, 11.2, 11.3, 11.5, 11.6		Weld only, as discussed herein
Linear	1c	11.2, 11.6		
Piping	1d	11.1, 11.2, 11.6		
Inclusions			W	
Slag	2a	11.1, 11.2, 11.3, 11.4, 11.5, 11.6		
Incomplete fusion	3	11.1, 11.2, 11.3, 11.4, 11.5	W	At joint boundaries or between passes
Inadequate joint penetration	4	11.1, 11.2, 11.3, 11.4, 11.5	W	Root of weld preparation
Undercut	5	11.1, 11.2, 11.3, 11.4, 11.5, 11.6	HAZ	Junction of weld and base metal at surface
Underfill	6	11.1, 11.2, 11.3, 11.5	W	Outer surface of joint preparation
Overlap	7	11.1, 11.2, 11.3, 11.4, 11.5, 11.6	W/HAZ	Junction of weld and base metal at surface
Laminations	8	11.1, 11.2, 11.3, 11.4, 11.5, 11.6	BM	Base metal, generally near midthickness of section
Delamination	9	11.1, 11.2, 11.3, 11.4, 11.5, 11.6	BM	Base metal, generally near midthickness of section
Seams and laps	10	11.1, 11.3, 11.4, 11.5, 11.6	BM	Base metal surface, almost always longitudinal
Lamellar tears	11	11.3, 11.5	BM	Base metal, near weld HAZ
Cracks (includes hot cracks and cold cracks)				
Longitudinal	12a	11.1, 11.2, 11.3, 11.4, 11.5, 11.6	W, HAZ	Weld or base metal adjacent to weld fusion boundary
Transverse	12b	11.1, 11.2, 11.3, 11.4, 11.5, 11.6	W, HAZ, BM	Weld (may propagate into HAZ and base metal)
Crater	12c	11.1, 11.2, 11.3, 11.4, 11.5, 11.6	W	Weld, at point where arc is is terminated
Throat	12d	11.1, 11.2, 11.3, 11.4, 11.5, 11.6	W	Weld axis
Toe	12e	11.1, 11.4, 11.5, 11.6	HAZ	Junction between face of weld and base metal
Root	12f	11.1, 11.2, 11.3, 11.4, 11.5, 11.6	W	Weld metal, at root
Underbead and heat-affected zone	12g	11.1, 11.2, 11.4, 11.5, 11.6	HAZ	Base metal, in HAZ
Fissures			W	Weld metal

a. W—weld, BM—base metal, HAZ—weld heat-affected zone

Uniformly scattered porosity may be distributed throughout single pass welds or throughout several passes of multiple pass welds. Whenever uniformly scattered porosity is encountered, the cause is generally faulty welding technique or defective materials, or both.

Cluster porosity is a localized grouping of pores that may result from improper initiation or termination of the welding arc.

Linear porosity may be aligned along (1) a weld interface, (2) the root of a weld, or (3) a boundary between weld beads. Linear porosity is caused by gas evolution from contaminants along a particular boundary.

Piping porosity is a term for elongated gas pores. Piping porosity in fillet welds normally extends from the root of the weld toward the face. When one or two pores are seen in the surface of the weld, it is likely that many

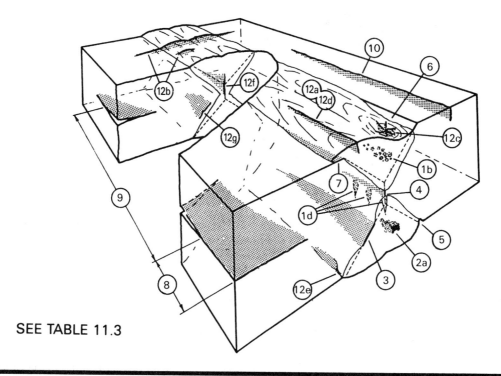

SEE TABLE 11.3

Figure 11.1—Double-V-Groove Weld in Butt Joint

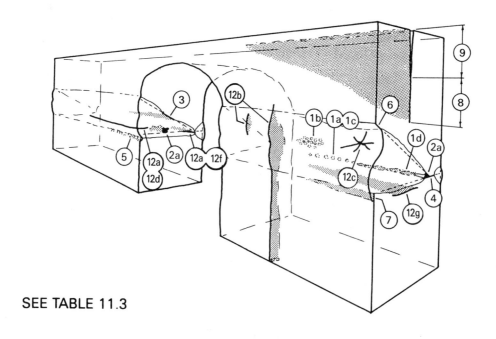

SEE TABLE 11.3

Figure 11.2—Single-Bevel-Groove Weld in Butt Joint

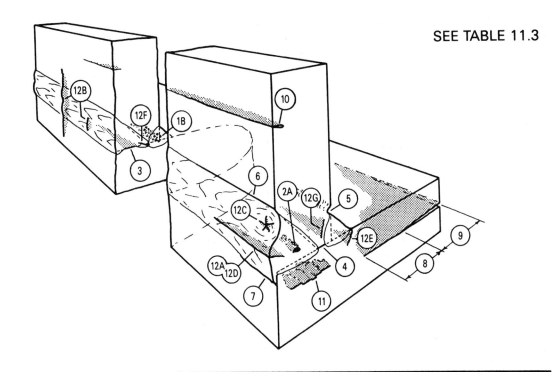

SEE TABLE 11.3

Figure 11.3—Welds in Corner Joint

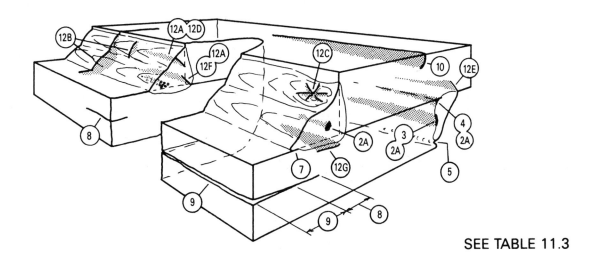

SEE TABLE 11.3

Figure 11.4—Double Fillet Weld in Lap Joint

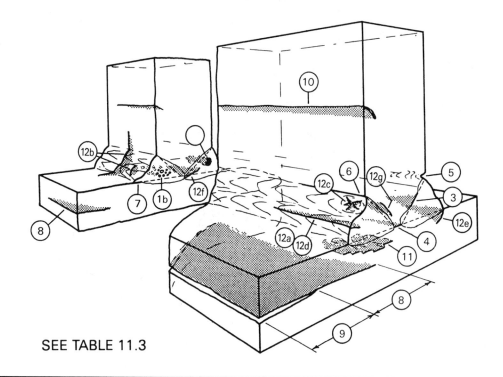

SEE TABLE 11.3

Figure 11.5—Combined Groove and Fillet Welds in T-Joint

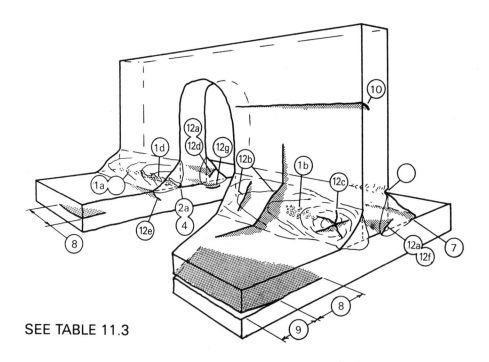

SEE TABLE 11.3

Figure 11.6—Single Pass Fillet Welds in a T-Joint

subsurface piping pores are interspersed among the exposed pores. Much of the piping porosity found in welds does not extend to the surface. Piping porosity in electroslag welds may be relatively long.

Inclusions

SLAG INCLUSIONS ARE nonmetallic solid materials trapped in the weld metal or at the weld metal interfaces. They may be present in welds made by most arc welding processes. In general, slag inclusions result from faulty welding techniques, improper access to the joint for welding, or both. With proper welding techniques, molten slag will float to the surface of the molten weld metal. Sharp notches in joint boundaries or between weld passes promote slag entrapment in the weld metal.

Tungsten inclusions are particles trapped in weld metal deposited with the gas tungsten arc welding process. These inclusions may be trapped in a weld if the tungsten electrode is dipped into the molten weld metal, or if the welding current is too high and causes melting and transfer of tungsten droplets into the molten weld metal. Tungsten inclusions appear as light areas on radiographs because tungsten is more dense than the surrounding metal and absorbs larger amounts of x-rays or gamma radiation. Almost all other weld discontinuities are indicated by dark areas on radiographs.

Incomplete Fusion

INCORRECT WELDING TECHNIQUES, improper preparation of the materials for welding, or wrong joint designs promote incomplete fusion in welds. The welding conditions that principally contribute to incomplete fusion are insufficient welding current and lack of access to all faces of the weld joint that should be fused during welding. Insufficient preweld cleaning may contribute to incomplete fusion, even if the welding conditions and technique are adequate. Preweld cleaning is more critical in certain metals.

Inadequate Joint Penetration

WHEN THE ACTUAL root penetration of a weld is less than specified in the design, the discontinuity at the root is called inadequate penetration. This condition may result from insufficient welding heat, improper joint design (too much metal for the welding arc to penetrate), incorrect bevel angle, or poor control of the welding arc. Some welding processes have great penetrating ability, and that characteristic is often used to advantage. However, the process must be matched to the joint preparation or incomplete fusion will result.

Many welding procedures for double groove welds require backgouging of the root of the first weld to expose sound metal prior to depositing the first pass on the second side. This procedure is used to ensure that there are no areas of inadequate joint penetration.

Undercut

VISIBLE UNDERCUT IS generally associated with either improper welding techniques or excessive welding currents, or both. It is generally located parallel to the junction of weld metal and base metal at the toe or root of the weld. Undercut discontinuities create a mechanical notch at the weld interface. If examined carefully, many welds have some undercut. Often, the undercut may only be seen in metallographic tests where etched weld cross sections are examined under magnification. When undercut is controlled within the limits of the specifications and does not constitute a sharp or deep notch, it is usually not deleterious.

The term *undercut* is sometimes used in the shop to describe melting away of the groove face of a joint at the edge of a layer or bead of weld metal. This "undercut" forms a recess in the joint face where the next layer or bead of weld metal must fuse to the base metal. If the depth of fusion at this location is too shallow when the next layer of weld metal is applied, voids may be left in the fusion zone. These voids would more correctly be identified as incomplete fusion. This undercut is usually associated with incorrect manipulation of the welding electrode while depositing a weld bead or layer next to the joint face.

Underfill

UNDERFILL RESULTS SIMPLY from the failure of the welder or welding operator to fill the joint with weld metal as called out in the welding procedure specification or on the design drawing. Normally, the condition is corrected by adding one or more additional layers of weld metal in the joint prior to subsequent processing.

Overlap

OVERLAP IS USUALLY caused by incorrect welding procedures, wrong selection of welding materials, or improper preparation of the base metal prior to welding. If tightly adhering oxides on the base metal interfere with fusion, overlap will result along the toe, face, or root of the weld. Overlap is a surface discontinuity that forms a severe mechanical notch parallel to the weld axis.

Cracks

CRACKS WILL OCCUR in weld metal and base metal when localized stresses exceed the ultimate strength of the metal. Cracking is often associated with stress amplification near discontinuities in welds and base metal, or near mechanical notches associated with the weldment design. Hydrogen embrittlement often contributes to crack formation in steel. Plastic deformation at the crack edges is very limited.

Cracks can be classified as either hot or cold types. Hot cracks develop at elevated temperatures. They commonly form during solidification of the weld metal. Cold cracks develop after solidification of a fusion weld as a result of stresses. Cold cracks in steel are sometimes called *delayed cracks*, and are often associated with hydrogen embrittlement. Hot cracks propagate between the grains while cold cracks propagate both between grains and through grains.

Cracks may be longitudinal or transverse with respect to the weld axis. Longitudinal weld metal cracks and the heat-affected zone cracks are parallel to the axis of the weld; transverse cracks are perpendicular to the weld axis.

Throat cracks. Throat cracks run longitudinally in the face of the weld and extend toward the root of the weld. They are generally, but not always hot cracks.

Root cracks. Root cracks run longitudinally and originate in the root of the weld. Both hot cracks and cold cracks can form in the root of the weld.

Longitudinal cracks. Longitudinal cracks in automatic submerged arc welds are commonly associated with high welding speeds, and are sometimes related to internal porosity. Longitudinal cracks in small welds between heavy sections are often the result of high cooling rates and high restraint.

Transverse cracks. Transverse cracks are nearly perpendicular to the axis of the weld. They may be limited in size and completely within the weld metal, or they may propagate from the weld metal into the adjacent heat-affected zone, and the base metal. Transverse cracks are generally the result of longitudinal shrinkage strains acting on weld metal of low ductility. Transverse cracks in steel weld metals are almost always related to hydrogen embrittlement.

Crater cracks. Crater cracks are formed by improper termination of a welding arc. They are usually shallow hot cracks and sometimes are referred to as *star cracks* when they form a star-like cluster.

Toe cracks. Toe cracks are generally cold cracks that initiate approximately normal to the base material surface and then propagate from the toe of the weld where residual stresses are higher. These cracks are generally the result of thermal shrinkage strains acting on a weld heat-affected zone that has been embrittled. Toe cracks sometimes occur when the base metal cannot accommodate the shrinkage strains that are imposed by welding.

Underbead cracks. Underbead cracks are generally cold cracks that form in the heat-affected zone. They may be short and discontinuous, but may also extend to form a continuous crack. Underbead cracking can occur in steels when three elements are present:

(1) Hydrogen in solid solution
(2) A crack-susceptible microstructure
(3) High residual stresses

When present, these cracks are usually found at regular intervals under the weld metal, and do not normally extend to the surface. They cannot be detected by visual inspection, and may be difficult to detect by ultrasonic and radiographic examinations.

Fissures are crack-like separations of small or moderate size along internal grain boundaries. They can occur in fusion welds made by most welding processes, but they are more frequent in electroslag welds because of the larger grain sizes in the weld. The separations can be either hot or cold cracks. Their effects on the performance of welded joints are the same as cracks of similar sizes in the same location and orientations.

Surface Irregularities

Surface Pores. Occasionally, pores form in the face of a weld bead. The pattern can vary from a single pore every few inches to many pores per inch. The pores are caused by improper welding conditions such as excessive current, inadequate shielding, or use of the wrong polarity. Unsatisfactory gas shielding may also adversely affect the weld surface. Gas shielding is usually better at the bottom of a weld groove than near the top of the groove. Improvement in weld bead appearance may be achieved by changing such welding conditions as polarity or arc length.

It is important to eliminate surface pores because they can result in slag entrapment during subsequent passes. Sound multiple pass welds are not normally achieved unless surface pores are removed prior to depositing the next weld layer.

Other Surface Irregularities. Varying widths of weld surface layers, depressions, variations in weld height or reinforcement, nonuniformity of weld ripples, and other surface irregularities are not classified as weld discontinuities. However, poor surface appearance indicates that the specified welding procedure was not followed or that a satisfactory welding technique was not used. Surface conditions may not affect the integrity of a completed weld, but they frequently are covered by specification requirements and are subject to inspection.

Magnetic disturbances, poor welding technique, and improper electrical conditions can account for certain surface irregularities. Such conditions might be caused by lack of experience, inaccessibility to the weld joint, or other factors peculiar to a specific job. Generally speaking, surface appearance reflects the ability and experience of the welder, and the presence of gross surface irregularities may be deleterious. Welds with uniform surfaces are desirable for structural as well as cosmetic reasons.

A fillet weld with poor surface appearance is shown in Figure 11.7. A satisfactory weld is shown in Figure 11.8. Acceptable weld surfaces are best judged by comparison with workmanship samples.

Figure 11.7—Single-Pass Horizontal Fillet Weld with Poor Surface Appearance Caused by Improper Welding Technique

Figure 11.8—Single-Pass Horizontal Fillet Weld with the Proper Welding Technique

BASE METAL DISCONTINUITIES

NOT ALL DISCONTINUITIES are a result of improper welding procedures. Many difficulties with weld quality may be traced to the base metal. Base metal requirements are usually defined by an ASTM specification.[5] Departure from these requirements should be considered cause for rejection.

5. The *Annual Book of ASTM Standards* is available from ASTM, Philadelphia, PA.

Base metal properties that should meet specification requirements include chemical composition, cleanliness, laminations, stringers, surface conditions (scale, paint, oil), mechanical properties, and dimensions. The inspector should keep such factors in mind when evaluating the sources of indications in welded joints that have no apparent cause. Several flaws that may be found in base metal are also shown in Figures 11.1 through 11.6.

Laminations

LAMINATIONS IN PLATE and other mill shapes are flat, generally elongated discontinuities found in the central zone of wrought products. Laminations may be completely internal and only detectable by ultrasonic tests, or they may extend to an edge or end where they may be visible at the surface. They may also be exposed when the base metal is cut.

Laminations are flattened discontinuities that run generally parallel to the surfaces of rolled products, and are most commonly found in mill shapes and plates. Some laminations are partially welded together during hot rolling operations. Other laminations may be so tight that they may not be detected by ultrasonic tests. Structural wrought shapes containing laminations cannot be relied upon to carry tensile stress in the through-thickness direction.

Delamination in the base metal may occur when the laminations are subject to transverse stresses. The stresses may be residual from welding or they may result from external loading. Delamination may be detected visually at the edges of pieces or by ultrasonic testing with longitudinal waves through the thickness. Delaminated metal should not be used to transmit tensile loads.

Lamellar Tears

SOME ROLLED STRUCTURAL shapes and plates are susceptible to a cracking defect known as *lamellar tearing*. Lamellar tearing, a form of fracture resulting from high stress in the through-thickness direction, may extend over long distances. Lamellar tears are generally terrace-like separations in base metal typically caused by thermally-induced shrinkage stresses resulting from welding. The tears take place roughly parallel to the surface of rolled products. They generally initiate either in regions having a high incidence of coplanar, stringer-like, nonmetallic inclusions or in areas subject to high residual (restraint) stresses, or both. The fracture usually propagates from one lamellar plane to another by shear along planes that are nearly normal to the rolled surface.

Laps and Seams

LAPS AND SEAMS are longitudinal flaws at the surface of the base metal that may be found in hot-rolled mill products. When the flaw is parallel to the principal stress, it is not generally a critical defect; if the lap or seam is perpendicular to the applied or residual stresses, or both, it will often propagate as a crack. Seams and laps are surface discontinuities. However, their presence may be masked by manufacturing processes that have subsequently modified the surface of a mill product. Welding over seams and laps can cause porosity, incomplete fusion, and cracking. Seams and laps may be harmful in applications involving welding, heat treating, or upsetting, and also in certain components that will be subjected to cyclic loading. Mill products can be produced with special procedures to control the presence of laps and seams. Open laps and seams can be detected by magnetic particle, penetrant, and ultrasonic inspection methods, but those tightly closed may be missed during inspection.

Arc Strikes

STRIKING AN ARC on base metal that will not be fused into the weld metal should be avoided. A small volume of base metal may be momentarily melted when the arc is initiated. The molten metal may crack from quenching, or a small surface pore may form in the solidified metal. These discontinuities may lead to extensive cracking in service. Any cracks or blemishes caused by arc strikes should be ground to a smooth contour and reinspected for soundness.

DIMENSIONAL DISCREPANCIES

THE PRODUCTION OF satisfactory weldments depends upon, among other things, the maintenance of specified dimensions. These may be the size and shape of welds or the finished dimensions of an assembly. Requirements of this nature will be found in the drawings and specifications. Departures from the requirements in any respect should be regarded as dimensional discrepancies that, unless a waiver is obtained, must be corrected before acceptance of the weldment. Dimensional discrepancies can be largely avoided if proper controls are exercised when the base metals are cut to size.

Warpage

WARPAGE OR DISTORTION is generally controllable by using suitable jigs or welding sequences, or by presetting of joints prior to welding. The exact method employed should be dictated by the size and shape of the parts as well as by the thickness of the metal.

Incorrect Joint Preparation

ESTABLISHED WELDING PRACTICES require proper dimensions for each type of joint geometry consistent with the base metal composition and thickness and the welding process. Departure from the required joint geometry may increase the tendency to produce weld discontinuities. Therefore, joint preparation should meet the requirements of the shop drawings, and be within the specified limits.

Joint Misalignment

THE TERM *MISALIGNMENT* is often used to denote the amount of offset or mismatch across a butt joint between members of equal thickness. Many codes and specifications limit the amount of allowable offset because misalignment can result in stress raisers at the toe and the root of the weld. Excessive misalignment is a result of improper fit-up, fixturing, tack welding, or a combination of these factors.

Incorrect Weld Size

Fillet Welds. The required size of fillet welds should be on the detail drawings. Fillet weld size can be measured with gages designed for this purpose. Undersize fillet welds can be corrected by adding one or more weld passes. Oversize fillet welds are not harmful if they do not interfere with subsequent assembly. However, they are uneconomical and can cause excessive distortion.

Groove Welds. The size of a groove weld is its joint penetration, which is defined as the depth of the joint preparation plus root penetration. Underfilled groove welds can be repaired by adding additional passes.

Repair of inadequate joint penetration depends upon the requirements of the joint design. When the weld is required to extend completely through the joint, repair may be made by back-gouging to sound metal from the back side and applying a second side weld. When only partial joint penetration is required, or when the back side of the weld is inaccessible, the weld must be removed, and the joint rewelded using a modified welding procedure that will provide the required weld size.

Weld Profiles

THE PROFILE OF a finished weld may affect the service performance of the joint. The surface profile of an internal pass or layer of a multiple pass weld may contribute to the formation of incomplete fusion or slag inclusions when the next layer is deposited. Requirements concerning discontinuities of this nature in finished welds are usually included in the specifications. Figure 11.9 illustrates various types of acceptable and unacceptable weld profiles in fillet and groove welds.

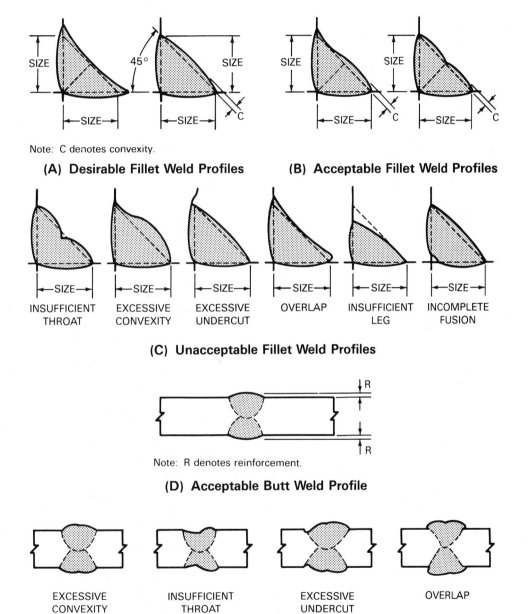

Note: C denotes convexity.

(A) Desirable Fillet Weld Profiles   **(B) Acceptable Fillet Weld Profiles**

INSUFFICIENT THROAT   EXCESSIVE CONVEXITY   EXCESSIVE UNDERCUT   OVERLAP   INSUFFICIENT LEG   INCOMPLETE FUSION

(C) Unacceptable Fillet Weld Profiles

Note: R denotes reinforcement.

(D) Acceptable Butt Weld Profile

EXCESSIVE CONVEXITY   INSUFFICIENT THROAT   EXCESSIVE UNDERCUT   OVERLAP

(E) Unacceptable Butt Weld Profiles

Figure 11.9—Weld Profiles of Acceptable and Unacceptable Surface Conditions

CAUSES AND REMEDIES FOR FUSION WELD DISCONTINUITIES

POROSITY IN FUSION WELDS

Causes of Porosity

THE COMMON CAUSES of porosity, and suggested methods of preventing it, are summarized in Table 11.4.

Dissolved gases are usually present in molten weld metal. Porosity is formed as the weld metal solidifies if the dissolved gases are present in amounts greater than their solid solubility limits. The causes of porosity in weld metal are related to the welding process and the welding procedure, and in some instances, to the base metal type and chemistry. The welding process, welding procedure, and base metal type (including the manufacturing method) directly affect the quantities and types of gases that are present in the molten weld pool. The welding process and welding procedure control the solidification rate, which in turn affects the amount of weld metal porosity. Proper welding procedures for a given combi-

Table 11.4
Common Causes and Remedies of Porosity

Cause	Remedies
Excessive hydrogen, nitrogen, or oxygen in welding atmosphere	Use low-hydrogen welding process; filler metals high in deoxidizers; increase shielding gas flow
High solidification rate	Use preheat or increase heat input
Dirty base metal	Clean joint faces and adjacent surfaces
Dirty filler wire	Use specially cleaned and packaged filler wire, and store it in clean area
Improper arc length, welding current, or electrode manipulation	Change welding conditions and techniques
Volatization of zinc from brass	Use copper-silicon filler metal; reduce heat input
Galvanized steel	Use E6010 electrodes and manipulate the arc heat to volatilize the zinc ahead of the molten weld pool*
Excessive moisture in electrode covering or on joint surfaces	Use recommended procedures for baking and storing electrodes
	Preheat the base metal
High sulfur base metal	Use electrodes with basic slagging reactions

* Note: See Fume Control, Chapter 16.

nation of welding process and base metal should produce welds that are essentially free of porosity.

Dissolved Gases in Weld Metal

THE GASES WHICH may be present in the molten weld pool include the following:

(1) Hydrogen
(2) Oxygen
(3) Nitrogen
(4) Carbon monoxide
(5) Carbon dioxide
(6) Water vapor
(7) Hydrogen sulphide
(8) Argon
(9) Helium

Of these, only hydrogen, oxygen, and nitrogen are soluble to any significant extent in a molten weld pool, and the solubility of these gases in solidified metal is significantly less than in liquid metal.

Hydrogen is the major cause of porosity in the welding of metals. It may enter the molten weld pool from many sources. For example, it may be present in the gas atmosphere surrounding the arc zone or in hydrogen-forming constituents, such as cellulose in the flux or electrode covering. Hydrogen may also be introduced into the molten weld pool by the dissociation of water. Moisture may be present in fluxes, electrode coverings, ambient atmosphere, or on the base metal surfaces. Residual lubricant from wire drawing can remain on filler wire surfaces and can become a significant contributer to weld metal hydrogen content, particularly with small diameter electrodes. Hydrogen dissolved in the base metal itself or in surface oxides may remain in the weld metal. Filler metals may also contain dissolved hydrogen. Sulfur or selenium in the base metal may combine with hydrogen to form other gases.

Nitrogen may cause porosity in steel and nickel alloy welds. This gas may enter the molten weld pool from the atmosphere or from contaminated shielding gas. It may also be present in the base metal or filler metal in the form of dissolved nitrogen or nitrides.

Oxygen dissolved in the molten weld metal may also cause porosity. When present in molten steel, oxygen will react with carbon to form CO or CO_2. Oxygen may enter the molten weld pool as oxides on filler wire or base metal, or both, in the form of compounds in a flux or an electrode covering, and from the atmosphere. Insufficient amounts of deoxidizers in steel base metals, filler metals, flux, or electrode coverings may result in incomplete deoxidation of the molten weld pool.

Significance to Weld Integrity

POROSITY HAS BEEN extensively evaluated as a discontinuity in welds. Tests have been conducted to determine the effects of porosity on both static and dynamic behaviors of welded joints, and for virtually all types of base metals. In amounts less than 3 percent by volume, porosity has an insignificant effect on static tensile or yield strength. This level is generally higher than that permitted by industry fabrication standards.

The effect of porosity on ductility is slightly more pronounced. The higher the yield strength of the metal, the greater is the adverse effect of porosity on ductility.

The gas or other contaminant causing porosity may influence the properties of the weld metal by dissolving in it. Also, the gas in the pores or cavities may influence the metal surrounding the pore to the extent that the pore acts as a crack initiator when loaded in service. (Hydrogen has this effect in steel, but other gases may not.)

The influence of porosity on the dynamic toughness of weld metal is less certain. The designer should investigate the effects of porosity on welds subjected to the expected type of loading before specifying acceptable porosity limits for a weldment.

The effect of porosity in ferrous metal welds is mitigated to a great extent by a post-weld treatment.

In face-centered cubic alloys (Al, Cu, Ni), the influence of porosity is small. At high temperatures (creep range), the reduction in properties is in proportion to the loss in cross-sectional area.

The most significant studies of porosity have addressed its effect on the fatigue properties of fusion welded butt joints with and without weld reinforcement. In any amount, the effect of porosity on the fatigue strengths of reinforced welds was overshadowed by stress concentrations on the surfaces. However, when the weld reinforcement was removed, exposed porosity contributed to failure by fatigue.

For fillet welds, the stress concentration effects of the weld toes and the start and stop locations are great, and they override all porosity considerations. Internal porosity in fillet welds does not appear to affect service performance.

The effect of surface porosity in butt and fillet welds is slightly different. Surface porosity is considered more injurious than buried or internal porosity, but certainly no worse than a crack. Surface porosity may reduce the effective throat of the weld below that needed to support the desired load. Surface porosity also often indicates that the process is out-of-control.

SLAG INCLUSIONS

Causes and Remedies

COMMON CAUSES AND remedies of inclusion-type discontinuities are shown in Table 11.5.

Entrapped slag discontinuities typically occur only with the flux shielded welding processes: shielded metal arc, flux cored arc, submerged arc, and electroslag welding. Entrapped slag is a reaction product of the flux and the molten weld metal. Oxides, nitrides and other impurities may dissolve in the slag to refine the weld metal. The slag density is less than the weld metal density and therefore slags normally float to the surface.

During welding, slag is formed and forced below the surface of the molten weld metal by the stirring action of the arc. Slag may also flow ahead of the arc, and metal may be deposited over it. The latter is especially true when multipass welds are made without proper interpass cleaning. In any case, slag tends to rise to the surface because of its lower density.

A number of factors may prevent the release of slag and result in the slag being trapped in the weld metal. Some of these factors are

(1) High viscosity weld metal
(2) Rapid solidification
(3) Insufficient welding heat
(4) Improper manipulation of the electrode
(5) Undercut on previous passes

Geometric factors such as poor bead profile, sharp undercuts, or improper groove geometry promote

Table 11.5
Common Causes and Remedies of Slag Inclusions

Cause	Remedies
Failure to remove slag	Clean surface and previous weld bead
Entrapment of refractory oxides	Power wire brush the previous weld bead
Tungsten in the weld metal	Avoid contact between the electrode and the work; use larger electrode
Improper joint design	Increase groove angle of joint
Oxide inclusions	Provide proper gas shielding
Slag flooding ahead of the welding arc	Reposition work to prevent loss of slag control
Poor electrode manipulative technique	Change electrode or flux to improve slag control
Entrapped pieces of electrode covering	Use undamaged electrodes

entrapment of slag by providing places where it may accumulate beneath the weld bead. In making a root pass, if the electrode is too large and the arc impinges on the groove faces instead of the root faces, the slag may roll down into the root opening and be trapped under the weld metal.

The factors that contribute to slag entrapment may be controlled by welding technique.

Significance to Weld Integrity

THE INFLUENCE OF slag inclusions on weld behavior is similar to that of porosity, as described previously. The effect of slag inclusions on static tensile properties is significant principally to the extent it influences the cross-sectional area available to support the load. The toughness of the weld metal seems generally to be unaffected by isolated slag with volumes of 4 percent or less of the weld zone. In weld metals of less than 75 ksi (517 MPa) tensile strength, ductility is generally unaffected. As the tensile strength increases, however, ductility drops in proportion to the amount of slag present.

Slag inclusions may be elongated with tails that may act as stress raisers. Therefore, slag can influence the fatigue behavior of welds, particularly when the weld reinforcement is removed and the weld is not postweld heat treated. As with porosity, slag at or very near to the weld surface (face or root) influences fatigue behavior to a considerably greater extent than similarly constituted slag buried within the weld metal. Slag, together with hydrogen dissolved in the weld metal, may influence fatigue strength by reducing the critical slag particle size for the initiation of a fatigue crack.

INCOMPLETE FUSION

INCOMPLETE FUSION RESULTS from the failure to fuse the weld metal to the base metal, or to fuse adjacent beads or layers of weld metal to each other. Failure to obtain fusion may occur at any point in a groove or fillet weld, including the root of the weld.

Causes and Remedies

INCOMPLETE FUSION ALMOST always occurs as a result of improper welding techniques for a given joint geometry and welding process. It may be caused by failure to melt the base metal or the previously deposited weld metal, or both. The presence of oxides or other foreign materials, such as slag, on the surfaces of the metals may also promote the occurance of incomplete fusion. In some cases, unsuitable combinations of joint design and welding process may lead to incomplete fusion. The causes and remedies for incomplete fusion are summarized in Table 11.6.

Table 11.6
Common Causes and Remedies of Incomplete Fusion

Cause	Remedies
Insufficient heat input, wrong type or size of electrode, improper joint design, or inadequate gas shielding	Follow correct welding procedure specification
Incorrect electrode position	Maintain proper electrode position
Weld metal running ahead of the arc	Reposition work, lower current, or increase weld travel speed
Trapped oxides or slag on weld groove or weld face	Clean weld surface prior to welding

Significance to Weld Integrity

INTERMITTANT INCOMPLETE FUSION affects weld joint integrity in much the same manner as porosity and slag inclusions. The degree to which intermittent incomplete fusion can be tolerated in a welded joint for various types of loading is similar to the limits for porosity and slag inclusions. Continuous incomplete fusion has the same affect as inadequate joint penetration discussed below.

INADEQUATE JOINT PENETRATION

Causes and Remedies

INADEQUATE JOINT PENETRATION is generally associated with groove welds. Complete joint penetration is not always required in all welded joints. Some joints are designed with partial joint penetration welds. However, such welds can have inadequate joint penetration when the effective throat of the weld is less than that specified in the welding symbol. The occurrence of inadequate joint penetration in welds is a function of groove geometry as well as welding procedure. The causes and remedies of inadequate joint penetration are as shown in Table 11.7.

Significance to Weld Integrity

INADEQUATE JOINT PENETRATION is undesirable. Furthermore, shrinkage stresses can cause distortion of the parts during further welding and may cause a crack to initiate at the unfused area. Such cracks may propagate,

Table 11.7
Common Causes and Remedies of Inadequate Joint Penetration

Causes	Remedies
Excessively thick root face or insufficient root opening	Use proper joint geometry
Insufficient heat input	Follow welding procedure
Slag flooding ahead of welding arc	Adjust electrode or work position
Electrode diameter too large	Use small electrodes in root or increase root opening
Misalignment of second side weld	Improve visibility or backgouge
Failure to backgouge when specified	Backgouge to sound metal if required in welding procedure specification
Bridging of root opening	Use wider root opening or smaller electrode in root pass

as successive beads are deposited, and if undetected could result in the presence of a dangerous flaw in service.

In welds deposited from one side of the joint, those with inadequate joint penetration may be loaded in bending at the root, and the stress concentration there may cause failure without appreciable deformation. If the joint is welded from both sides, and the inadequate joint penetration is at the neutral axis, the bending stresses will be lower, but they will still be concentrated at the ends of the discontinuity. Furthermore, buried inadequate joint penetration is more difficult to detect than a discontinuity at the surface.

Inadequate joint penetration is undesirable in any groove weld that will be subjected to cyclic tension loading in service. The discontinuity can initiate a crack that may propagate and result in catastrophic failure.

CRACKS

CRACKING IN WELDED joints results from localized stresses that exceed the ultimate strength of the metal. When cracks occur during or as a result of welding, they usually do not show evidence of deformation.

Weld metal or base metal that has considerable ductility under uniaxial stress may fail without appreciable deformation when subjected to biaxial or triaxial stresses. Shrinkage occurs in all welds, and if a joint or any portion of it (such as the heat-affected zone) cannot accommodate the shrinkage stresses by plastic deformation, then high stresses will develop. These stresses can cause cracking.

The chemical compositions of the base metal and the weld metal affect crack susceptibility. An unfused area at the root of a weld may result in cracks without appreciable deformation if this area is subjected to tensile or

bending stresses. When welding two plates together, the root of the weld is subjected to tensile stress as successive layers are deposited, and, as already stated, incomplete fusion in the root will promote cracking.

After a welded joint has cooled, cracking is more likely to occur if the weld metal or heat-affected zone is either hard or brittle. A ductile metal, by localized yielding, may withstand stress concentrations that might cause a hard or brittle metal to fail. The causes and remedies of weld metal and heat-affected-zone cracking are shown in Table 11.8.

Table 11.8
Common Causes and Remedies of Cracking

Causes	Remedies
Weld Metal Cracking	
Highly rigid joint	Preheat
	Relieve residual stresses mechanically
	Minimize shrinkage stresses using backstep or block welding sequence
Excessive dilution	Change welding current and travel speed
	Weld with covered electrode negative; butter the joint faces prior to welding
Defective electrodes	Change to new electrode; bake electrodes to remove moisture
Poor fit-up	Reduce root opening; build up the edges with weld metal
Small weld bead	Increase electrode size; raise welding current; reduce travel speed
High sulfur base metal	Use filler metal low in sulfur
Angular distortion	Change to balanced welding on both sides of joint
Crater cracking	Fill crater before extinguishing the arc; use a welding current decay device when terminating the weld bead
Heat-Affected Zone	
Hydrogen in welding atmosphere	Use low-hydrogen welding process; preheat and hold for 2 h after welding or postweld heat treat immediately
Hot cracking	Use low heat input; deposit thin layers; change base metal
Low ductility	Use preheat; anneal the base metal
High residual stresses	Redesign the weldment; change welding sequence; apply intermediate stress-relief heat treatment
High hardenability	Preheat; increase heat input; heat treat without cooling to room temperature
Brittle phases in the microstructure	Solution heat treat prior to welding

Weld Metal Cracking

THE ABILITY OF weld metal to remain intact under a stress system imposed during a welding operation is a function of the composition and structure of the weld metal. In multiple layer welds, cracking is most likely to occur in the first layer (root bead) of weld metal. Unless such cracks are repaired, they may propagate through subsequent layers as the weld is completed. Weld metal cracking resistance can be improved by one or more of the following:

(1) Changing electrode manipulation or electrical conditions to improve the weld face contour or the composition of the weld metal

(2) Changing filler metal to develop a more ductile weld metal

(3) Increasing the thickness of each weld pass by decreasing the welding speed and providing more weld metal to resist the stresses

(4) Using preheat to reduce thermal stresses

(5) Using low hydrogen welding procedures

(6) Sequencing welds to balance shrinkage stresses

(7) Avoiding rapid cooling conditions.

Three types of cracks that can occur in weld metal are as follows:

Transverse weld cracks. These cracks are perpendicular to the axis of the weld and, in some cases, extend beyond the weld into the base metal. This type of crack is more common in joints that have a high degree of restraint.

Longitudinal weld cracks. These cracks are found mostly within the weld metal, and are usually confined to the center of the weld. Such cracks may be the extension of cracks formed at the end of the weld. They may also be the extension, through successive layers, of a crack that started in the first layer.

Crater cracks. Whenever the welding operation is interrupted, there is a tendency for a crack to form in the crater. These cracks are usually starshaped and progress only to the edge of the crater. However, these may be starting points for longitudinal weld cracks, particularly when cracks occur in a crater formed at the end of the weld.

Crater cracks are found most frequently in metals with high coefficients of thermal expansion, such as austenitic stainless steel. However, the occurrence of such cracks can be minimized or prevented by filling craters to a slightly convex shape prior to breaking the welding arc.

Base Metal Cracking

HEAT-AFFECTED-ZONE CRACKING may be longitudinal or transverse and is almost always associated with hardenable base metals. High hardness and low ductility in the heat-affected zone result from the metallurgical response to the weld thermal cycles.[6] These two conditions are among the principal factors that contribute to crack susceptibility.

In ferritic steels, the maximum attainable hardness increases and the ductility decreases with increasing carbon content and increasing cooling rates from welding temperature. The rate of cooling will depend upon a number of physical factors such as:

(1) The peak temperature produced in the heat-affected zone

(2) The initial base metal (preheat) temperature

(3) The thickness and thermal conductivity of the base metal

(4) The heat input per unit time at a given section of the weld

(5) The ambient temperature

As indicated in Chapter 4, Welding Metallurgy, the heat-affected-zone hardness is related to the hardenability of the base metal, which in turn is dependent on the base metal chemical composition. Carbon has the strongest effect on the hardenability of steel and in addition, increases the hardness of the transformation products. Nickel, manganese, chromium, and molybdenum also contribute to the hardenability of steel, but unlike carbon, these elements only moderately increase the base metal hardness.

High alloy steels include the austenitic, ferritic, and martensitic stainless steels. The martensitic stainless steels behave similarly to medium carbon and low-alloy steels, but they are more crack susceptible. Austenitic and ferritic stainless steels do not undergo a phase transformation that would harden the heat-affected zone. The heat-affected zone ductility of ferritic stainless steels may be adversely affected by welding.

The metallurgical characteristics of the base metal affect the heat-affected zone crack susceptibility. Small changes in chemical composition of the base metal and filler metal (hydrogen content), and added joint restraint can appreciably increase the cracking tendency. There can be significant differences in crack susceptibility among several heats of the same grade of low alloy steel.

The primary base metal cracking problem encountered when welding many steels is induced by soluble hydrogen.[7] Such cracking is known by various other

---

6. The metallurgical response to weld thermal cycles is discussed in Chapter 4 of this volume.

7. Hydrogen induced cracking is discussed in detail in Chapter 4 of this volume.

names, including underbead, cold, and delayed cracking. It generally occurs at some temperature below 200 °F (100 °C) immediately upon cooling or after a period of several hours. The time delay depends upon the type of steel, the magnitude of the welding stresses, and the hydrogen content of the weld and heat-affected zones. In any case, it is caused by diffusible hydrogen trapped in the weld metal or the heat-affected zone. Weld metal may crack, but this seldom occurs when the yield strength is below about 90 ksi (620 MPa).

Diffusion of hydrogen into the heat-affected zone from the weld metal during welding contributes to cracking in this zone. The microstructures of the weld metal, heat-affected zone, and base metal are also contributing factors.

Hydrogen-induced cracking can be prevented by using (1) a low-hydrogen welding process (2) a combination of welding and thermal treatments that promotes the escape of hydrogen by diffusion (this may also produce a microstructure that is more resistant to hydrogen induced cracking), or (3) welding procedures that result in low welding stresses.

Significance to Weld Integrity

CRACKING, IN ANY form, is an unacceptable discontinuity, and is considered most detrimental to performance. A crack, by nature, is sharp at its extremities, and thus acts as a stress concentrator. The stress concentration effect produced by cracks is greater than that of most other discontinuities. Therefore, cracks, regardless of size, are not normally permitted in weldments governed by most fabrication codes. They must be removed regardless of location and the excavation filled with sound weld metal.

SURFACE IRREGULARITIES

THE FOLLOWING SURFACE irregularities may be observed on welds:

(1) Sharp, irregular surface ripples
(2) Excessive spatter
(3) Craters
(4) Protrusions (such as an overfilled crater)
(5) Arc strikes

The welder or welding operator is directly responsible for these discontinuities since they result from incorrect welding technique or improper machine settings. Poor workmanship should not be accepted, even though the joint may be adequate for its intended service. Unsatisfactory workmanship indicates that proper procedures are not being followed. Failure to correct the situation can lead to more serious quality problems.

In some cases, faulty or wet electrodes and unsuitable base metal chemistry may cause discontinuities and unsatisfactory weld appearance.

Gross bead irregularities are discontinuities inasmuch as they constitute an abrupt change of section. Such changes of section are potential sources of high stress concentration, and should be carefully evaluated with respect to service requirements.

Spatter in itself is not necessarily a defect, but likely indicates improper welding technique or other associated problems.

Arc strikes with either the electrode or the holder can initiate failure of the weldment in bending or cyclic loading. They can create a hard and brittle condition in alloy steels, and are inadvisable even on mild steel when high static or normal cyclic loading may be encountered.

INADEQUATE WELD JOINT PROPERTIES

SPECIFIC MECHANICAL PROPERTIES or chemical compositions, or both, are required of all welds in a weldment. These requirements depend on the codes or specifications covering the weldment, and departure from specified requirements is unacceptable. The required properties are generally determined with specially prepared test plates, but may be determined from sample weldments taken from production. Where test plates are used, the inspector should verify that standard production procedures are followed. Otherwise, the results obtained will not represent the properties of production weldments.

Mechanical properties that may not be satisfactory are tensile strength, yield strength, ductility, hardness, and toughness. The chemical composition of the weld metal may be improper because of incorrect filler metal composition or excessive dilution or both. This condition may result in lack of corrosion resistance in the weld zone.

DISCONTINUITIES IN RESISTANCE AND SOLID STATE WELDS

SPOT, SEAM, AND PROJECTION WELDING

THE REQUIRED WELD quality depends primarily upon the application. Weld quality may be affected by the

chemical composition and condition of the base metal, joint and part designs, electrode condition, and welding equipment. In some applications, each weld must meet the minimum requirements of a rather stringent specification. This is true for aircraft and space vehicles. Other applications may have standards for satisfactory welds but may also permit some percentage of undersize or defective welds. Automotive components are examples.

Unfortunately, there are no reliable nondestructive test methods for resistance spot and seam welds. In addition, the commonly used destructive testing of sample welds has limitations inherent in such tests. The designer should be aware of these facts when considering resistance spot or seam welding for an application.

Design requirements may include surface appearance, minimum strength and, with some seam welding applications, leak tightness. These should be monitored by a system of quality control that includes visual inspection and destructive examination of test samples or actual weldments. The most important indicators of weld quality are the following:

(1) Surface appearance
(2) Weld size
(3) Penetration
(4) Strength and ductility
(5) Internal discontinuities
(6) Sheet separation and expulsion
(7) Weld consistency

Surface Appearance

THE SURFACE APPEARANCE of a resistance weld is not an infallible indication of weld strength, size, or internal soundness. It is an indication of the conditions under which the weld was made, but it should not be used as the sole criterion for qualifying production welds. For example, a group of spot welds in a joint may have identical surface appearance. However, the second and succeeding welds may be undersized at the faying surface because of the shunting of current through the previous spot welds. Adjacent spot welds of similar surface size are shown in Figure 11.10 and 11.11, but the weld size

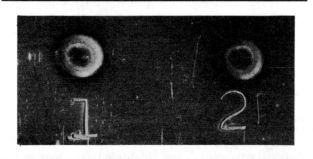

Figure 11.10—Surface Appearance of Two Succeeding Spot Welds in 0.040 in. (1 mm) Stainless Steel Sheet

Figure 11.11—Cross Section of Spot Welds in Figure 11.10 Showing the Effect of Current Shunting

at the faying surface of the first weld is greater than that of the second weld, Figure 11.11. The second weld is smaller than the first because part of the welding current passed through the first weld.

Both welds have identical surface appearance because the welding current enters through the outside surface from the electrodes. The size difference is greater for closely spaced welds, welds in metals of low electrical resistivity, and welds in thicker sheets.

Normally, the surface appearance of a spot, seam, or projection weld should be relatively smooth; it should be round or oval in the case of contoured work, and free from surface fusion, electrode deposit, pits, cracks, deep electrode indentation, or any other condition that would indicate improper electrode maintenance or equipment operation. The causes of some common undesirable spot weld surface conditions and their effects on weld quality and cost are shown in Table 11.9.

Weld Size

THE DIAMETER OR width of the fused zone must meet the requirements of the appropriate specifications or the design criteria. In the absence of such requirements, either accepted shop practices or the following general rules should be used.

(1) Spot welds that are reliably reproduced under normal production conditions should have a minimum nugget diameter of 3.5 to 4 times the thickness of the thinnest outside part of the joint. In cases of three or more dissimilar thicknesses, the nugget diameters between adjacent parts can be adjusted somewhat by the selection of the electrode design and materials.

(2) The individual nuggets in a pressure-tight seam weld should overlap a minimum of 25 percent. The width of the nugget should be at least 3.5 to 4 times the thickness of the thinnest part.

(3) Projection welds should have a nugget size equal to or larger than the diameter of the original projection.

Table 11.9
Undesirable Surface Conditions for Spot Welds

Type	Cause	Effect
1. Deep electrode indentation	Improperly dressed electrode face; lack of control of electrode force; excessively high rate of heat generation due to high contact resistance (low electrode force)	Loss of weld strength due to reduction of metal thickness at the periphery of the weld area; bad appearance
2. Surface fusion (usually accompanied by deep electrode identation)	Scaly or dirty metal; low electrode force; misalignment of work; high welding current; electrodes improperly dressed; improper sequencing of pressure and current	Undersize welds due to heavy expulsion of molten metal; large cavity in weld zone extending through to surface; increased cost of removing burrs from outer surface of work; poor electrode life and loss of production time from more frequent electrode dressings
3. Irregularly shaped weld	Misalignment of work; bad electrode wear or improper electrode dressing; badly fitting parts; electrode bearing on the radius of the flange; skidding; improper surface cleaning of electrodes	Reduced weld strength due to change in interface contact area and expulsion of molten metal
4. Electrode deposit on work (usually accompanied by surface fusion)	Scaly or dirty material; low electrode force or high welding current; improper maintenance of electrode contacting face; improper electrode material; improper sequencing of electrode force and weld current	Bad appearance; reduced corrosion resistance; reduced weld strength if molten metal is expelled; reduced electrode life
5. Cracks, deep cavities, or pin holes	Removing the electrode force before welds are cooled from liquidus; excessive heat generation resulting in heavy expulsion of molten metal; poorly fitting parts requiring most of the electrode force to bring the faying surfaces into contact	Reduction of fatigue strength if weld is in tension or if crack or imperfection extends into the periphery of weld area; increase in corrosion due to accumultion of corrosive substances in cavity or crack

There is a maximum limit to the nugget size of a spot, projection, or seam weld. This limitation is based on the economical and practical limitations of producing a weld together with the laws of heat generation and dissipation. The maximum useful nugget size is difficult to specify in general terms. Each user should establish this limit in accordance with the design requirements and prevailing shop practices.

Depth of Fusion

DEPTH OF FUSION is the distance to which the nugget extends into the pieces that are in contact with the electrodes. The minimum depth of fusion is generally accepted as 20 percent of the thickness of the thinner piece. If the depth of fusion is less than 20 percent, the weld is said to be "cold" because the heat generated in the weld zone was too low. Normal variations in welding current, time, electrode force, and other changes will cause undesirable changes in the strength of "cold" welds. In extreme cases, no weld nugget may be formed. The depth of fusion should not exceed 80 percent of the thickness of the thinner piece. Greater depth will result in expulsion, excessive indentation, and rapid electrode wear. Normal, excessive, and insufficient depth of fusion are shown in Figure 11.12.

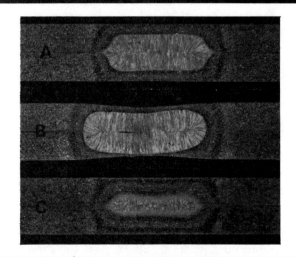

Figure 11.12—Penetration in Spot Welds; (A)—Normal, (B)—Excessive, (C)—Insufficient

The depth of fusion into each piece should be approximately uniform for equal or nearly equal sheet thicknesses. For dissimilar thickness ratios of three-to-one and greater, the depth of fusion into the thicker piece need not exceed that in the thinner piece.

Strength and Ductility

STRUCTURES EMPLOYING SPOT, seam and projection welds are usually designed so that the welds are loaded in shear when the parts are exposed to tension or compression loading. In some cases, welds may be loaded in tension when the direction of loading is normal to the plane of the joint, or the welds may be loaded in combinations of tension and shear. In the case of flanged tank sections that are seam welded along the flanges, the seam welds may be subjected to peeling action when the tank is pressurized.

The strength requirements for spot and projection welds are normally specified in pounds per weld. For seam welds, the strength is usually specified in pounds per inch of joint length. It is good practice to specify a weld strength that is greater than that of welds of minimum recommended nugget size, but not more than 150 percent of such welds.

The strength of spot and projection welds increases as the diameter becomes larger, even though the average unit stress decreases. The unit stress decreases because of the increasing tendency for failure to occur at the edge of the nugget as its size increases. In low carbon steel, for example, the calculated average shear stress in good welds at rupture will vary from 10 to 60 ksi (69–414 MPa). Low values apply to relatively large welds, and high values to relatively small welds. In both instances, the actual tensile stress in the sheet at the weld periphery is at or near the ultimate tensile strength of the base metal. This factor tends to cause the shear strength of circular welds to vary linearly with diameter.

Single spot and projection welds are not strong in torsion where the axis of rotation is perpendicular to the plane of the welded parts. This strength tends to vary with the cube of the diameter (weld size). Little torsional deformation is obtained with brittle welds prior to failure. Angular displacements may vary from 5 to 180 degrees depending upon weld metal ductility. Torsion is normally used to shear welds across the interface to measure the nugget diameter. Periodic testing of production spot welds verifies that weld schedules are producing adequate weld sizes.

The ductility of a resistance weld is determined by the composition of the base metal and the effect of high temperatures and subsequent rapid cooling rates on the weld and base metal. Unfortunately, the standard methods of measuring ductility are not adaptable to spot, seam, and projection welds. The nearest thing to ductility measurement is a hardness test, becaue the hardness of a metal is usually an inverse indication of its ductility. For a given alloy, ductility decreases with increasing hardness, but different alloys of the same hardness do not necessarily possess the same ductility.

Another method of indicating the ductility of spot or projection welds of equal size is to determine the ratio of direct tensile strength to tension-shear strength. A weld with good ductility has a high ratio; a weld with poor ductility has a low ratio.

There are various methods which can be used in production welding to minimize the hardening effect of rapid cooling. Some of these are

(1) Use long weld times to put heat into the work.
(2) Preheat the weld area with a preheat current.
(3) Temper the weld and heat-affected zones using a temper cycle at some interval after the weld time.
(4) Furnace anneal or temper the welded assembly.

These methods are not always practical. For instance, the first will produce greater distortion of the assembly and reduce production rates; the second and third methods require welding machine controls that provide these features; the fourth method involves an additional operation that may reduce the strength of cold worked base metal. If a welded assembly is quenched from the annealing temperature, it may cause excessive distortion.

Internal Discontinuities

INTERNAL DISCONTINUITIES INCLUDE cracks, porosity or spongy metal, large cavities, and, in the case of some coated metals, metallic inclusions in the nugget. Generally speaking, these discontinuities will have no detrimental effect on the static or fatigue strength of a weld if they are located entirely in the central portion of the weld nugget. On the other hand, it is extremely important that no defects occur at the periphery of a weld where the load stresses are highly concentrated.

Spot, seam, and projection welds in metal thicknesses of approximately 0.040 in. (1 mm) and greater may have small shrinkage cavities in the center of the weld nugget as illustrated in Figure 11.13(A). These cavities are less pronounced in some metals than in others due to the difference in forging action of the electrodes on the hot metal. Such shrinkage cavities are generally not detrimental in the usual applications. However, the cavity that results from heavy expulsion of molten metal, as shown in Figure 11.13(B), may take up a very large part of the fused area and is detrimental.

A certain number of expulsion cavities is to be expected in the production welding of most commercial steels. Heavy expulsion of molten metal is a result of improper welding conditions, and the number of such welds that can be accepted should be limited by specifications. The best method of assuring satisfactory adherence to specified spot welding schedules is through a structured statistical quality control program with regular production sampling and destructive testing.

Internal defects in spot, seam, and projection welds are generally caused by low electrode force, high weld-

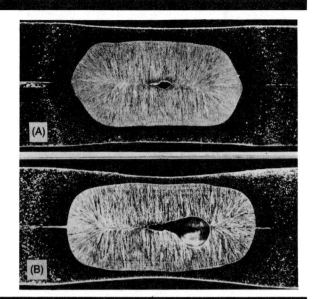

Figure 11.13—Shrinkage Cavities in Spot Welds: (A)—Small, (B)—Large

ing current, or any other conditions that produce excessive welding heat. They are also caused by removing the electrode force too soon after the welding current stops. When this occurs, the weld nugget is not properly forged during cooling. Such action may result during high-speed seam and roll spot welding.

Sheet Separation and Expulsion

SHEET SEPARATION OCCURS at the faying surfaces as a result of the expansion and contraction of the weld metal and the forging effect of the electrodes on the hot nugget. The amount of separation varies with the thickness of the base metal, increasing with greater thickness. Normal separation is shown in Figure 11.12(A).

Excessive sheet separation results from the same causes as surface indentation to which it is related. Improperly dressed electrode faces can act as punches under high electrode force. This tends to decrease the joint thickness, upset the weld metal radially, and force the sheets up around the electrodes. Excessive sheet separation is illustrated in Fig. 11.14 (note that one sheet is laminated).

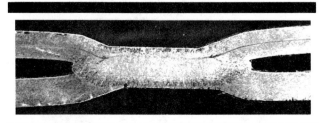

Figure 11.14—Excessive Sheet Separation

Expulsion is the result of overheating which generally results from excessive welding current. This is particularly true when the high current is combined with inadequate electrode force, improperly faced electrodes, or inadequate follow-up of the electrodes. Expulsion results in internal cavitation, and usually reduces weld strength. This tendency is so pronounced that the maximum current is normally limited to a value where expulsion will not occur.

Weld Quality and Consistency

CONSISTENT RESISTANCE WELD quality can be maintained with proper control of the factors that tend to produce variations in the final product. The factors include:

(1) Joint design and fit-up
(2) Material thickness tolerance
(3) Composition, temper, and surface condition of the base metal
(4) Electrode material and shape
(5) Electrode and weldment cooling
(6) Welding cycle variables
(7) Postweld thermal treatments

The importance of joint design and fit-up is discussed elsewhere in this chapter. Large variations in part thickness, particularly with three or more thicknesses, may produce inconsistent fit-up which, in turn, can affect weld quality. Changes in base metal composition, temper, or surface conditions may require revision of the welding schedule to produce acceptable welds.

Consistent quality resistance welds are obtained by use of correct welding settings and techniques, and maintaining them for the duration of a particular production run. Factors such as welding current, weld time, and electrode force must be controlled within the limits of the resistance welding schedule. The best control method is periodic testing of workpieces or test samples.

The number of workpieces tested as well as the test method may vary. The test pieces may be examined nondestructively or a certain number may be tested destructively. Statistical methods are then used to predict the quality of the production lot. In any case, an inspector must be able to (1) recognize conditions that may cause variations in weld quality, (2) make certain that the test pieces are representative samples, and (3) verify that production pieces are made under the same welding conditions as the test samples.

The application of statistical control to production quality has three prime objectives:[8]

---

8. The advantages, methods, and procedures for statistical quality control of resistance welding are described in *AWS C1.1-66, Recommended Practices for Resistance Welding*. Miami: American Welding Society, 1966.

(1) To reduce the number of rejections and machine shutdowns because of poor performance

(2) To assist in establishing the optimum procedure limits for satisfactory quality

(3) To provide a reasonably reliable measure of actual production quality

If these objectives are achieved, the product will have high quality at low cost, and minimum scrap.

The basic principles of statistical control are widely used in industry. Briefly these principles are to:

(1) Select samples of actual production and test them for performance to specifications.

(2) Estimate the probable quality or conformance of all production by analysis of samples.

(3) Predict the future quality by considering the trend of past and present quality.

The methods of sampling, extracting data from the samples, and deciding whether to permit further use of the welding procedure constitute the quality control system.

SOLID STATE PROCESSES

CERTAIN RESISTANCE AND solid state welding processes are done at elevated temperatures essentially below the melting point of the base metal being joined without the addition of filler metal. When the joint reaches the desired welding temperature, force is applied to the joint to consummate the weld. As welding is being accomplished, some molten metal may be generated between the surfaces being joined, but it is expelled as a result of pressure applied on the joint during the welding cycle. Typical resistance and solid state welding processes of this nature are flash welding, friction welding, high frequency resistance welding, and upset welding.[9]

9. Explosion welding and ultrasonic welding are described in *The Welding Handbook*, Vol. 3, 7th Ed.

The resulting welds are all characterized by a flat weld interface. The most common discontinuities are two dimensional and lie in the plane of the joint. However cracks may occur with other orientations. Further, some inspection methods applicable to fusion welds, such as radiographic and ultrasonic testing, require special techniques when applied to these welds and the results of the tests are difficult to interpret.

Weld Discontinuities

NORMALLY, RESISTANCE AND solid state welding processes are completely automated with good reproducibility. Discontinuities encountered in solid state welded joints may be classified as either mechanical or metallurgical.

Mechanical problems are the dominent cause for rejection of solid state welds. The discrepancies caused by mechanical problems are easily found by visual inspection, and can usually be corrected by machine adjustments. An acceptable weld and an unacceptable weld caused by misalignment or offset of the workpieces are illustrated in Figure 11.15.

In the case of flash and upset welds, the shape and contour of the upset metal is a good indicator of weld quality. The geometry of the weld in Figure 11.16(A) indicates proper heat distribution as well as proper upset. Insufficient upset, as in Figure 11.16(B), may indicate trapped oxides or flat spots at the weld interface and possibly an incomplete weld.

The condition and shape of the flash on friction welds is an indicator of possible discontinuities along the weld interface. Figure 11.17 shows the effect of axial shortening on weld quality. These inertia friction welds were made with the same speed and inertial mass but with a decreasing heating pressure from left to right. Two of the welds exhibited center discontinuities because the axial shortening (pressure) was insufficient.

Welding may be incomplete at the center of the joint in continuous drive friction welds when inadequate speed,

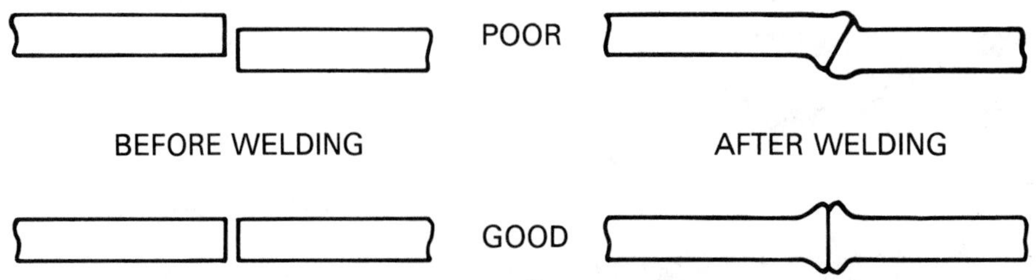

Figure 11.15—Effect of Part Alignment on Joint Geometry

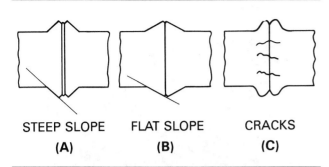

Figure 11.16—Visual Indications of Flash Weld Quality: (A) Satisfactory Heat and Upset, (B) Insufficient Heat or Upset or Both, (C) Cracks Due to Insufficient Heat

STEEP SLOPE **(A)**

FLAT SLOPE **(B)**

CRACKS **(C)**

EXCESSIVE   DECREASING   INSUFFICIENT

AXIAL SHORTENING

Figure 11.17—The Effect of Axial Shortening on the Bonding and Flash of Friction Welds

heating time, or pressure is used. Depressions in the faying surfaces that prevent uniform contact during the early stages of friction welding often limit center heating and entrap oxides.

Metallurgical discontinuities are usually associated with material defects or heterogeneities. Such discontinuities are much harder to find by nondestructive examination methods.

Various types of metallurgical discontinuities found in resistance and solid state welds are as follows:

(1) Cracks
(2) Intergranular oxidation
(3) Decarburization
(4) Voids
(5) Oxides and other inclusions
(6) Flat spots
(7) Cast metal at the interface
(8) Out-turned fibrous metallurgical structure at the weld

Cracking in welds can be divided into two categories depending upon the temperature of formation: cold cracking and hot cracking. Cold cracking, illustrated in Figure 11.16(C), can be caused by insufficient heating prior to or during upsetting. Excessive cooling rates in hardenable steels can cause cold cracking, but slow cooling rates will eliminate it. The most common form of hot cracking in upset welds occurs as microfissures in the heated zone, and is known as *break-up*.

In the case of flash welds, a form of intergranular oxidation known as die-burn can occur at clamp locations. This discontinuity is caused by localized overheating of the workpiece where it contacts the clamping dies. Precleaning of the workpiece surfaces in the clamping area to bright metal will usually eliminate this problem. Excessive initial spacing between the clamping dies can result in overheating of workpieces near the faying surfaces during flashing. This may result in intergranular oxidation as well as nonuniform upsetting and joint misalignment.

Occasionally, another type of solid state welding discontinuity results from elemental redistribution during welding. In carbon steel, this may be manifested as decarburization. This discontinuity appears as a bright band on a polished and etched surface of an upset welded steel specimen which is cut transverse to the weld interface. Chemical heterogeneity affects the mechanical properties of a welded joint. In particular, a variation in carbon concentration greatly affects hardness, ductility, and strength.

Oxides and other inclusions, voids, and cast metal along the interfaces of weld joints are related by the fact that they can usually be eliminated by increasing the upset distance. In the case of flash welds, craters are formed on the faying interfaces by the expulsion of molten metal during flashing. If the flashing voltage is too high or the platen motion is incorrect, violent flashing causes deep craters in the faying surfaces. Molten metal and oxides may be trapped in these deep craters and not be expelled during upsetting.

The cause of and ways to prevent flat spots in welds are not well understood. In fact, the term itself is vague and is often used to describe several different features. Flat spots may be the result of a number of metallurgical phenomena having inherently different mechanisms. The smooth, irregular areas indicated by arrows on the fractured surfaces of a flash weld in Figure 11.18 are typical flat spots. Zones of low ductility (flat spots) along the weld interface in flash and friction welds apparently are associated with base metal characteristics as well as welding process variables.

The inherent fibrous structure of wrought mill products may cause anisotropic mechanical behavior. An out-turned fibrous structure at the weld interface often results in some decrease in mechanical properties as compared with the base metal, particularly ductility.

The decrease in ductility is not normally significant unless one or both of the following conditions are present:

(1) The base metal is extremely inhomogeneous. Examples are severely banded steels, alloys with excessive stringer type inclusions, and mill products with seams and cold shuts produced during the fabrication process.

(2) The upset distance is excessive. When excessive upset distance is employed, the fibrous structure may be completely reoriented transverse to the original structure.

Figure 11.18—Flat Spots, Indicated by Arrow, on the Mating Fractured Surfaces of a Flash Weld

DISCONTINUITIES IN BRAZED AND SOLDERED JOINTS

BRAZING AND SOLDERING are joining processes that rely on capillary action to draw liquid filler metal into a controlled joint clearance and also on wetting of the faying surfaces by the liquid filler metal. The two processes differ only in filler metals used and temperatures employed. Discontinuities found in soldered joints and their causes and remedies are similar to those in brazed joints. Therefore, the discussion will be limited to brazed joints.[10]

COMMON DISCONTINUITIES IN BRAZED JOINTS

Lack of Fill

LACK OF FILL in the form of voids and porosity can be the result of improper cleaning of the base metal, excessive joint clearance, insufficient filler metal, entrapped gas, or movement of the mating parts before the filler metal solidifies. Lack of fill reduces the strength of the joint by reducing the load-carrying area, and it may provide a path for leakage in joints designed for pressure- or liquid-containing applications.

Flux Entrapment

ENTRAPPED FLUX MAY be found in a brazed joint where a flux is used to prevent and remove oxidation during the heating cycle. Entrapped flux prevents flow of filler metal into that particular area, thus reducing the joint area. It may also cause false leak or proof test acceptance. The entrapped flux, if corrosive, may reduce service life.

Discontinuous Fillets

SEGMENTS OF JOINTS containing undersized or no fillets are usually found by visual inspection. They may or may not be acceptable depending upon the specification requirements of the brazed joint.

Base Metal Erosion

EROSION OF THE base metal by the filler metal is caused by the filler metal alloying with the base metal during brazing. It may cause undercut or the disappearance of the mating surface and reduce the strength of the joint by

---

10. Additional information on discontinuities in soldered joints and inspection procedures maybe found in the *Soldering Manual,* 2nd Ed. Miami: American Welding Society; Revised, 1978.

changing the filler metal composition and by reducing the base metal cross-sectional area.

Unsatisfactory Surface Appearance

EXCESSIVE FLOW OF brazing filler metal onto the base metal, surface roughness, and excessive filler metal may be detrimental for several reasons. In addition to cosmetic considerations, these may act as stress concentrations, corrosion sites, or may interfere with inspection of the brazement.

Cracks

CRACKS REDUCE BOTH the strength of the brazement and its service life. They act as stress raisers that may cause premature failure under cyclic loading, as well as lower the static strength of the brazement.

ACCEPTANCE LIMITS

THE SHAPE, ORIENTATION, and location of the discontinuity in a brazement should be considered when defining the acceptance limit for a particular type of brazing discontinuity. The relationship of the discontinuity to other imperfections and whether the discontinuity is on the surface or not, should be considered.

Judgments for disposition of components containing discontinuities should be made by persons competent in the fields of brazing metallurgy and quality assurance, and who fully understand the function of the component. Such dispositions should be documented.

TYPICAL DISCONTINUITIES AND THEIR CAUSES

DISCONTINUITIES FOUND IN brazed joints are an indication that the brazing procedure is out of control, or that improper techniques were used. Several typical discontinuities are shown here so that they may be recognized when they occur in production. The possible cause is given in each case.

A large void in the fillet of a section through a brazed copper lap joint is shown in Figure 11.19. The flawed joint was detected by bubble-testing the assembly with compressed air in a water tank. The discontinuity may have been caused by underheating or improper fluxing procedures, or both. The filler metal would not exhibit irregular flow into the joint if the brazing temperature and fluxing procedure were correct.

A section through a brazed joint in which severe erosion of the base metal occurred is shown in Figure 11.20. This erosion resulted from overheating of the joint during brazing. Such erosion may not be serious in thick sections, but cannot be tolerated in relatively thin

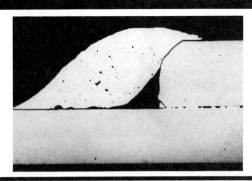

Figure 11.19—Void Under the Filler Metal Fillet in a Brazed Copper Lap Joint

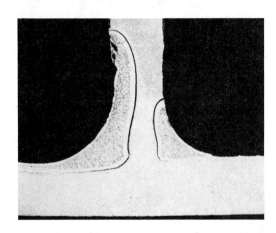

Figure 11.20—Severe Alloying and Erosion of the Base Metal in a Brazed Joint

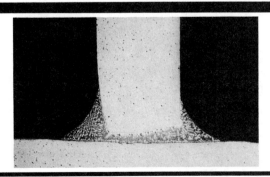

Figure 11.21—Brazed Joint with Essentially No Base Metal Erosion

sections. A joint with essentially no erosion is shown in Figure 11.21.

A brazed copper socket joint is shown in Figure 11.22A. A radiograph of the joint, Figure 11.22B, shows that large areas of the joint are void of filler metal.

A macrograph of a cross section through the braze fillet showing an extensive void in the capillary of the joint is shown in Figure 11.22C. The voids throughout the joint were caused by insufficient heating. If a joint leaks, voids may be detected by a pressure test.

A brazed lap joint, a radiograph of the joint, and cross sections through the joint are shown in Figure 11.23A, B, and C. The assembly shown in Figure 11.23A is of two flat pieces of low carbon steel brazed with a silver brazing filler metal. The radiograph taken through the joint, Figure 11.23B, shows large voids as dark areas. Cross sections 1-1 and 2-2 through the joint are shown in Figure 11.23C. The voids in the joint capillary are evident.

Figure 11.22A—Brazed Copper Socket Joint

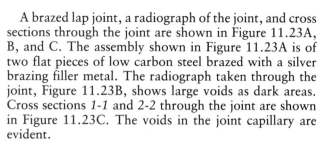

Figure 11.23A—Brazed Lap Joints

Figure 11.22B—Radiograph of Joint

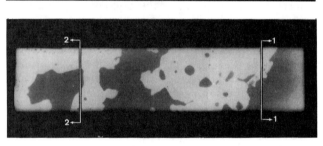

Figure 11.23B—Radiograph Showing Voids (Dark Areas in the Joint)

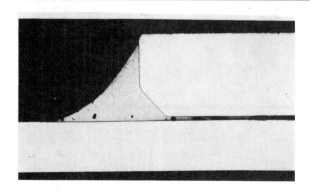

Figure 11.22C—Cross Section Through the Braze Fillet and Capillary

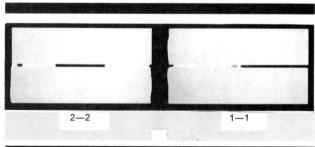

Figure 11.23C—Cross Sections Through the Joint Showing Voids in the Capillary

SIGNIFICANCE OF WELD DISCONTINUITIES

INCREASING DESIGN DEMANDS are resulting in more critical inspection methods and acceptance standards. Ideally, these acceptance standards should represent the minimum weld (or braze) quality that can be tolerated to assure satisfactory performance of the product. They should be based on tests of welded specimens containing the particular discontinuity under consideration. Correlation of these test results with allowable results should be the basis for acceptance of these particular discontinuities. Of course, a safety factor would be added to determine a final acceptance standard. The study of the fracture performance of materials containing defects is called fracture mechanics. The application of fracture mechanics to acceptance standards is often referred to as fitness for service (or fitness for purpose).

Fitness-for-service acceptance standards are not yet a reality. Current welding standards are based on the soundness that can be achieved with good workmanship. The standards are not based on engineering principles, and therefore, may be either inadequate or overly conservative. Overly conservative quality standards can result in late schedules and high production costs. Despite these drawbacks, the current standards have withstood the test of time, and a designer can be relatively certain the standards will provide a satisfactory structural performance in the future.

DISCONTINUITY–MATERIAL RELATIONSHIPS

THE EFFECT OF a particular weld discontinuity on structural integrity, economics, and safety depends to a certain extent on the metals being welded. Metals with high fracture toughness are more resistant to failure in the presence of a discontinuity than those with low fracture toughness.[11] The relationship between the metal mechanical properties and the allowable discontinuities is not addressed in current fabrication codes or inspection standards. Fracture mechanics provides a quantitative relationship to assess the significance of a discontinuity in terms of maximum flaw size and the material fracture toughness under conditions of plane strain.

DISCONTINUITY–MECHANICAL PROPERTY RELATIONSHIPS

THE SIGNIFICANCE OF a particular discontinuity in a weld depends on the intended service. A discontinuity may be innocuous under normal operation conditions,

but may grow to critical size in a hostile corrosive or fatigue environment.

The heat-affected zone often plays a major role in the determination of mechanical properties of the joint and can render the presence of weld discontinuities less significant. Heat-affected zones of strain-hardened and precipitation-hardened base metals experience recrystallization and resolution annealing, respectively, and both exhibit some grain growth near the fusion line. In such metals, both the strength and hardness of the heat-affected zone are diminished.

The effects of discontinuities on mechanical properties are governed by the shape, size, quantity, interspacing, distribution, and orientation of the discontinuities within the weld. Because of the number of variables, any qualitative correlation between a specific discontinuity class and mechanical properties is extremely difficult. Therefore, trends rather than direct relationships are presented here.

As the strength of a base metal is increased, a corresponding reduction in toughness and ductility often occurs. This increases the sensitivity of the base metal to discontinuities. Ultra-high-strength steels, as well as hard heat-affected zones in relatively soft mild steels, may be extremely sensitive to small discontinuities. When welding ultra-high-strength steels, it is also difficult to match the strength and toughness of the weld to that of the base metal.

Tensile Strength

FOR THE PURPOSE of characterizing static tensile performance, welded joints fall into two categories:

(1) Those having weld metal strengths that closely match or undermatch the base metal strength
(2) Those having weld metal strengths that overmatch the base metal strength

In laboratory tensile tests, weld discontinuities decrease the strength of welded joints in the first category more than those in the second category, because there is no reserve of additional weld metal strength to counteract the decrease in cross-sectional area. Because of the additional strength of overmatching weld metal, a certain amount of discontinuity can be tolerated before transverse tensile properties are adversely affected. The loss in transverse tensile strength is roughly proportional to the loss in cross-sectional area. Additional cross-sectional area provided by any weld reinforcement may compensate for some of this loss.

In welded joints of the overmatching type, the transverse strength degradation is usually accompanied by a

---

11. Determining fracture toughness is discussed in Chapter 12 and, the relation between toughness and design is discussed in Chapter 5 of this volume.

change in location of the fracture from base metal to weld metal when the number of discontinuities present changes from a few to a significant quantity. Usually, the ductility will be reduced roughly proportionally to the volume of the discontinuities present. The yield strength is not significantly affected.

Nevertheless, in production weldments, cracks in high strength weld metal are much more serious than cracks in low strength, ductile weld metal. One solution for welding high strength heat treated steels is to deposit a weld metal with lower strength but better ducility to accommodate strains during welding. Weld metal stronger than the base metal containing cracks has no value.

Fatigue

FATIGUE FAILURE at normal working stresses is invariably associated with stress concentrations. Fatigue is probably the most common cause of failure in welded construction.[12] The discontinuities most significant in promoting fatigue failures, except for the obvious defect of gross cracking or extensive incomplete fusion, are those which affect the weld faces. The combination of excessive weld reinforcement, Figure 11.24A, and slight undercutting is one of the most serious discontinuities affecting fatigue life. Fabrication codes usually specify the maximum permissible height of the reinforcement, and the condition shown in Figure 11.24B with the excess metal ground off without tapering the contact angle with the base metal may meet some code requirements for maximum reinforcement. However, the condition of Figure 11.24B does not improve the fatigue life. The reinforcement should blend smoothly into the base metal at the edges of the weld as shown in Figure 11.24C for service in fatigue applications.

The effect of the weld reinforcement contact angle on fatigue properties is illustrated in Figure 11.25. These data indicate how abrupt changes in section size may affect service life. For example, using a fillet welded lap joint in place of a butt joint can reduce the fatigue strength by a factor of up to three.

Tests of welds containing porosity have shown that fatigue cracks initiate at the toe of the weld reinforcement, and that porosity has little effect on the fatigue life. If the reinforcement is removed, porosity on or near the surface will adversely affect the fatigue strength more dramatically than porosity deep below the surface.

Similar tests conducted on welds with very large tungsten inclusions showed similar results except that when

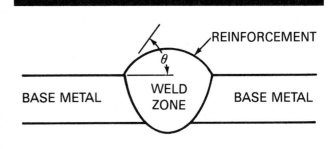

Figure 11.24A—Weld with Excessive Reinforcement

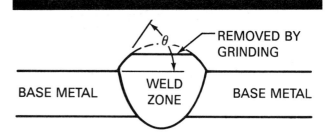

Figure 11.24B—Improper Treatment of Weld Reinforcement

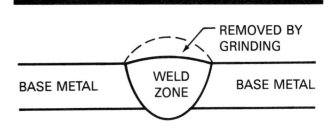

Figure 11.24C—Acceptable Weld Reinforcement Profile for Fatigue Applications

the reinforcement was removed, fatigue failure resulted from small oxide inclusions associated with the tungsten inclusions.

Increasing slag inclusion length initially leads to decreasing fatigue strength. However, when the discontinuity length becomes large relative to its depth through the thickness, no further reduction in fatigue strength occurs.

Discontinuities which are located in the middle of the weld can be blanketed by compressive residual stress so that other smaller discontinuities nearer to the surface control the fatigue strength. Such discontinuities may be below the size detectable by radiography.

---

12. The design of welded structural members for fatigue applications joints is discussed in Chapter 5, and fatigue testing of weld joints is discussed in Chapter 12 of this volume.

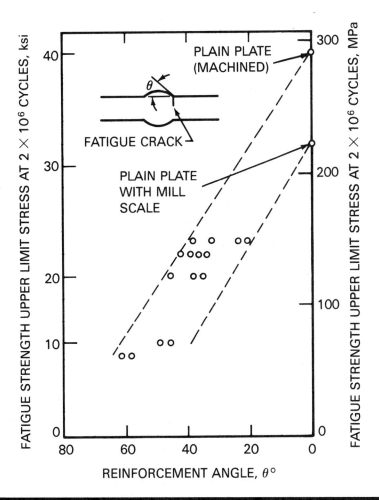

Figure 11.25—Effect of Weld Reinforcement Angle on Fatigue Strength of Steel

Toughness

FRACTURE MECHANICS PROVIDES a relationship between stresses and a critical discontinuity size that will propagate under plane strain conditions. This critical discontinuity size is inversely proportional to the square of the applied stress. The plane strain fracture toughness of metals generally decreases as the yield strength increases.

The fracture-mechanic equations that provide relationships among the plane strain fracture toughness, applied stress, and the critical crack length depend on the geometry and location of the discontinuity with respect to the stress field. The location and shape (depth-to-length ratio) of the discontinuity must be known or accurately predicted, particularly in those instances where discontinuities can grow by fatigue or stress corrosion. With this information and a valid plane-strain or elastic-plastic fracture toughness value, whichever is appropriate, one can estimate the combination of stress and discontinuity size at which a structure can be operated safely.

The effectiveness of a discontinuity as an initiator of fracture in a given weld metal depends on the existence of plane strain at the discontinuity tip. The edge radius of discontinuities will have an effect on performance. Sharp natural cracks are the most severe discontinuities. Incomplete fusion is the next most severe discontinuity; inadequate joint penetration is not as severe as incomplete fusion if the edges of the inadequately penetrated joint are less sharp. Slag inclusions and porosity are relatively harmless in initiating brittle fracture.

The resistance of weld metal to crack propagation under impact loading is not significantly affected by porosity.

FRACTURE MECHANICS EVALUATION OF DISCONTINUITIES

FRACTURE MECHANICS PROVIDES an analytical method to determine the critical crack size for unstable fracture. The most widely used model is a sharp crack in a member with the crack tip loaded under conditions of plane strain. If the stress intensity factor, K, exceeds the critical stress intensity factor for the material, K_{IC}, then the crack will become unstable and the member will fail.

The plane strain stress intensity factor K, the stress σ, and the crack length a are related by the equation

$$K_I = C\sigma(\pi a)^{1/2}$$

where

K_I = a plane strain stress intensity factor
σ = the applied or residual stress magnitude acting on a discontinuity
C = a constant depending on discontinuity size and shape
a = the discontinuity size or depth
π = 3.1416

In its simplest form, the application of linear elastic fracture mechanics to fitness for service is as follows:

(1) The worst case is assumed: the weld discontinuity is a crack.
(2) The plane strain fracture toughness of the material, K_{IC}, is determined.
(3) The vector sum of the applied and residual stresses is estimated.
(4) The critical size for unstable fracture is determined from the above equation.

(5) A margin of safety is applied and the maximum allowable crack-like flaw size is selected.
(6) Acceptance criteria for more innocuous discontinuities are defined.

There are two published documents that use linear elastic fracture mechanics to determine fitness for service. One document is *American Society for Mechanical Engineers Boiler and Pressure Vessel Code, Section XI, Inservice Inspection.* Fracture mechanics is used in this document to develop acceptance criteria for weld inspections. Inspection reports contain flaw dimensions determined from nondestructive examination, and these results are compared to discontinuity standards for evaluation.

In-service inspection of utilities under *Section XI* of the *ASME Boiler and Pressure Vessel Code* will provide facts on the performance of weldments containing discontinuities that will lead to further use of the concepts of fitness for service.

The other document is *British Standards Institute Document PD 6493, Guidance on some methods for the Deviation of Acceptance Levels for Defects in Fusion Welded Joints.* This document is not a fabrication-and-inspection code. It provides a procedure of *Engineering Critical Assessment* (ECA) that may be used to establish acceptance criteria for combinations of materials; welding process, procedure, and consumables; and also stress and environmental factors.

Clearly the ECA can only be used in applications where an existing fabrication code is not required by local law. Furthermore, the application of the ECA rules is to be agreed upon by all contracting parties including local authorities, when applicable.

SUPPLEMENTARY READING LIST

American Welding Society. Microprocessor-controlled welding yields consistent quality. *Welding Journal.* 65(11): Nov. 1986; 81-82.

Boulton, C.F. Acceptance levels of weld defects for fatigue service. *Welding Journal.* 56(1): Jan. 1977; 13s-22s.

Burdekin, F.M. Some defects do - some defects don't (lead to the failure of welded structures). *Metal Construction.* 14(2): Feb. 1982; 91-94.

Cox, E.P. and Lamba, E.P. Cluster porosity effects on transverse fillet weld strength. *Welding Journal.* 63(1): Jan. 1984; 1s-7s.

British Standards Institute. *Guidance on some methods for the derivation of acceptance levels for defects in fusion welded joints.* London: British Standards Institute (1980): PD6493.

Lundin, C.D. Fundamentals of weld discontinuities and their significance. *Welding Research Council Bulletin.* No. 295: June 1984.

———. Review of worldwide discontinuity acceptance standards. *Welding Research Council Bulletin.* No. 268: June 1981.

———. The significance of weld discontinuities—a review of current literature. *Welding Research Council Bulletin.* No. 222: Dec. 1976.

Lundin, C.D. and Pawel, S.J. An annotated bibliography on the significance, origin, and nature of discontinuities in welds, 1975-1980. Welding Research Council Bulletin. No. 263: Nov. 1980.

Masubuchi, K. *Analysis of welded Structures—residual stresses, distortion, and their consequences.* Oxford, England: Pergamon Press (1980).

Pellini, W. S. Principles of fracture-safe design. *Welding Journal.* 50(1): Mar. 1971; 91s-109s; 50(2): Apr. 1971; 147s-162s

Pfluger, A.R. and Lewis, R.E., eds. *Weld imperfections.* Addison-Wesley Co. (1968).

Reed, R.P., McHenry, H.I., and Kasan, M.B. A fracture mechanics evaluation of flaws in pipeline girth welds. *Welding Research Council Bulletin.* No. 245: Jan. 1979.

Sandor, L.W. A perspective on weld discontinuities and their acceptance standards in the U.S. maritime industry. *Fitness-for-Purpose in Welded Constructions.* Cambridge CB1 6AL England: The Welding Institute. (1982) (Available from AWS).

Sandor, L.W. et al. Weld discontinuities. *ASM Metals Handbook,* Vol. 6, 9th Ed. Metals Park, OH: American Society for Metals (1983).

Tsai, C.L. and Tsai, M. J. Significance of weld undercut in design of fillet welded T-joints. *Welding Journal.* 63(2): Feb. 1984; 64s-70s.

Wells, A.A. Fitness for purpose and the concept of defect tolerance. *Metal Construction.* 13(11): Nov. 1981; 677-81.

Wilkowski, G.M. and Eiber, R.J. Review of fracture mechanics approaches to defining critical size girth weld discontinuities. *Welding Research Council Bulletin.* No. 239: July 1978.

Will, W. Technical basis for acceptance standards for weld discontinuities. *Naval Engineers Journal.* April 1979; 60-70.

TESTING FOR EVALUATION OF WELDED JOINTS

PREPARED BY A COMMITTEE CONSISTING OF:

A. G. Portz, Chairman
Cleveland Pneumatic

H. Hahn
ARTECH Corporation

H. W. Mishler
Edison Welding Institute

H. S. Sayre
Consultant

B. W. Schaaf Jr.
E. I. duPont de Nemours and Company

WELDING HANDBOOK COMMITTEE MEMBER:
J. R. Condra
E. I. duPont de Nemours and Company

TESTING FOR EVALUATION OF WELDED JOINTS

INTRODUCTION

ALL TYPES OF welded structures, from steel bridges to high-temperature components of jet engines, serve a function. Likewise, the welded joints in these structures or components are designed for service-related capabilities or properties. To assure that they will fulfill their intended function, a test of some type is usually performed. The ideal test, of course, is the observance of the structure in actual service, but actual service tests are expensive and time consuming. Therefore, standardized tests and testing procedures are used that give results which can be related to metals and structures that have performed satisfactorily in service.

Various testing methods that are regularly used to evaluate the expected performance of welded joints are described here. The property being tested, the test methods which may be used, and the application of results with special consideration of their relationship to welded joints are covered. Reference is frequently made to *ANSI/AWS B4.0, Standard Methods for Mechanical Testing of Welds*[1] and to *ASTM A370, Standard Methods and Definitions for Mechanical Testing of Steel Products*.[2]

The problem of predicting the performance of structures from a laboratory type test is a complex one, because the size, configuration, environment, and type of loading normally differ. In welded joints, the complexity is further increased by the nature of the joint, which is both metallurgically and chemically heterogeneous. In addition to the unaffected base metal, the welded joint consists of weld metal and a heat-affected zone. Those regions are composed of a multitude of metallurgical structures as well as chemical heterogeneities. A variety of properties is thus to be expected throughout the welded joint. Any test will measure either the properties of some discrete portion of the joint or some composite average of all or a portion of the joint. When testing a welded joint, the investigator not only has the problem of relating the test to the service of the actual structure, but also of determining whether the true properties are measured by the limited region tested. The investigator therefore should use care in the interpretation and application of any test results.

When selecting a test method, the function to be performed by that test must be considered and balanced against the time and cost. Tension and hardness tests, for example, both provide a measure of strength, yet the hardness test is the simpler and less costly of the two. A hardness test is not adequate to establish the strength of a welded joint, but it is sufficient to verify that a maximum heat-affected zone hardness is not exceeded. Each laboratory test provides a limited amount of information on the properties of welded joints. Accordingly, most weldments are evaluated by several tests. Each test provides specific data on the serviceability of the weldment.

1. Available from American Welding Society, Miami, Florida.
2. Available from ASTM, Philadelphia, PA.

TENSILE PROPERTIES—STRENGTH AND DUCTILITY

TENSION AND BEND tests are frequently used to evaluate the breaking strength and ductility of a metal and to determine that the metal meets applicable specification requirements. Welded joints contain metallurgical and compositional differences that result from the welding process. It is important to know the effects of these changes on mechanical properties. Therefore, tension and bend tests are frequently made to determine the suitability of the welded joint for service. They also are often used to qualify welding procedures or welders according to specific code requirements.

TENSION TESTS

Base Metal Tension Test

Longitudinal or Transverse Test. The strength and ductility of metals are generally obtained from a simple uniaxial tensile test in which a machined specimen is subjected to an increasing load while simultaneous observations of extension are made. Longitudinal speci-mens are oriented parallel to the direction of rolling, and transverse specimens are oriented perpendicular to the rolling direction. The load can be plotted against the elongation, customarily as shown in Figure 12.1. The stress (load divided by original area) is plotted against the strain (elongation divided by the original gage length). Several important engineering properties can be determined from the tensile test. These properties are the yield strength, tensile strength, and ductility, expressed in percentage elongation, and percentage reduction in cross sectional area.

The yield strength shown by the engineering stress-strain curve is generally the strength at some arbitrary amount of extension under load or a permanent plastic strain (offset). For certain steels that exhibit such a phenomenon, yielding takes place at the point where plastic extension occurs with no increase in load (upper yield point, Figure 12.1). The ultimate tensile strength is the maximum stress on the stress-strain curve. It can be calculated by dividing the maximum load by the original cross-sectional area. Typical stress-strain curves for several commercial steels are shown in Figure 12.2.

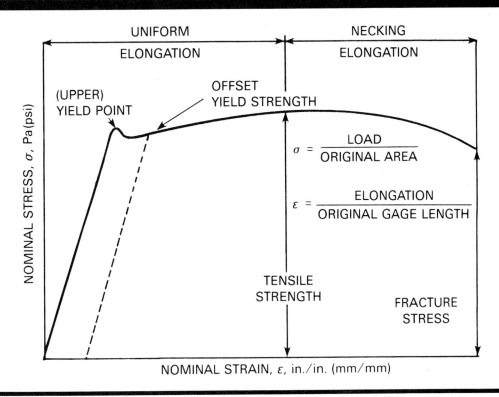

Figure 12.1—Engineering Stress-Strain Curve for Low Carbon Steel

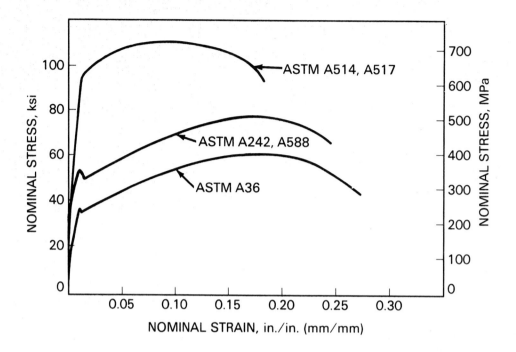

Figure 12.2—Nominal Stress-Nominal Strain Curves for Various Constructional Steels

Two standard measurements of ductility are derived from an engineering tension test—elongation and reduction of area expressed in percent. The percent elongation is the ratio of the increase in gage length to its original gage length. Because a measurement of percent elongation includes both uniform strain and local or necking strain, this ductility value is affected by the choice of original gage length and by the geometry of the specimen cross section. Therefore, comparisons of percent elongations should be made only for identical specimens. The percent reduction of area is the ratio of the decrease in cross-sectional area after fracture to the original cross-sectional area.

The details of specimen preparation and test procedures for smooth tension tests for base metals are described in ASTM A370, *Standard Methods and Definitions for Mechanical Testing of Steel Products.*

Weld Tension Test

THE TENSION TESTING of welds is somewhat more involved than testing base metal because a weld test section is heterogeneous, containing weld metal, heat-affected zone, and unaffected base metal. To obtain an accurate assessment of the strength and ductility of welds, several different specimens and orientations may

be used as shown in Figure 12.3. In some cases, the weld reinforcement is left intact on the test specimen.

All-Weld-Metal Test.   To determine the tensile properties of weld metal, the test specimen orientation is parallel to the axis of the weld, and the entire specimen is machined from the weld metal. The chemical composition of the weld metal will be affected by the joint penetration.

If the purpose of the test is to qualify a filler metal, then melting of the base metal should be minimized when making the test weld. This procedure is described in *AWS A5.1, Specification for Covered Carbon Steel Arc Welding Electrodes* and other filler metal specifications. If the purpose of the test is to determine properties of the weld metal in a particular weldment, the welding process and procedure used in actual fabrication should be employed to make the test weld.

The mechanical properties measured and reported in an all-weld-metal tension test are tensile strength, yield strength, elongation, and reduction in area.

Transverse Weld Test.   Interpretation of test results for a welded joint as a whole is not possible in the transverse weld specimen. The reduced section of this specimen contains base metal, heat-affected zones, and weld

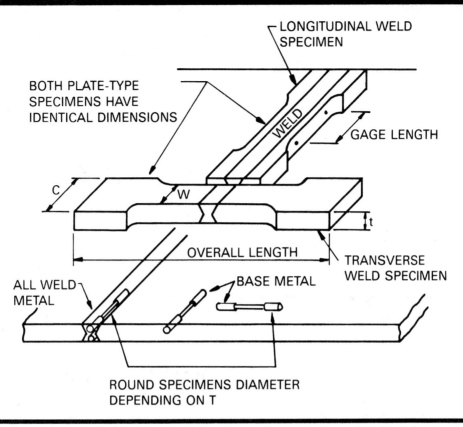

Figure 12.3—Typical Tension Test Specimens for Evaluation of Welded Joints

metal. When all of these materials are simultaneously subjected to the same stress, the one with the lowest strength will elongate and break. For example, if the weld metal strength is higher then that of the unaffected base metal, then failure will occur outside of the weld area, and no quantitative information about the weld metal strength is provided by the test. Thus, this test should not be used to make quantitative comparisons of weld metals.

The most common use of transverse weld tensile tests is to qualify welding procedures. The purpose of the test is to verify that the welding procedure will produce welds that equal or exceed design strength requirements. Only the tensile strength and the location of the fracture are normally reported for transverse weld tension tests.

Longitudinal Weld Test. In the longitudinal weld test, the direction of loading of the specimens is parallel to the weld axis, but the test differs from an all-weld-metal test specimen in that the reduced cross section contains weld, heat-affected zone, and base metal. During testing, all of these zones must strain equally and simultaneously. Weld metal, regardless of strength, elongates with the base metal until failure occurs. Low weld or heat-

affected zone ductility may initiate fracture at strength levels below that of the base metal. Only the tensile strength of longitudinal weld tests is normally reported. Elongation may be measured as an indication of minimum ductility of the joint.

Tension-Shear Test

Fillet Weld Shear Test. Tension-shear tests determine the shear strength of fillet welds. The test specimens are usually intended to represent actual weldments, and therefore, the welds are prepared using production procedures. Two specimen types, transverse and longitudinal, are shown in Figure 12.4.

To avoid rotation and bending stresses during testing, transverse-shear specimens are tested as double lap joints.

The shear strength of the weld metal, calculated as shown in Figure 12.5, and the location of fracture are the test results normally reported.

The longitudinal shear test measures the shear strength of fillet welds when the specimen is loaded par-

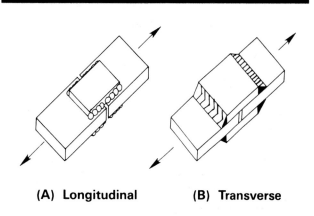

(A) Longitudinal   **(B) Transverse**

Figure 12.4—Longitudinal and Transverse Fillet Weld Shear Stress Specimens

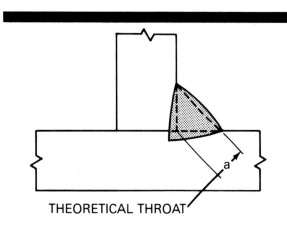

THEORETICAL THROAT

$$\tau = \frac{P}{\ell \times a}$$

where:

P = load
ℓ = total length of fillet weld shared
a = theoretical throat dimension
τ = shear strength of weld

Figure 12.5—Method of Determining Shear Strength of Fillet Welds

allel to the axis of the welds. To avoid bending during testing, two identical welded specimens are machined, and then tack welded together, as indicated in Figure 12.6. The results are calculated and reported as shown for the transverse shear tests, Figure 12.5.

Fillet weld shear test results are sensitive to specimen preparation procedures. The stress concentration at the root of the transverse fillet welds increases with increas-

ing root opening. Variations in root opening can cause inconsistent test results. Test specimens are sensitive to heat-affected zone cracking, undercut, and bead surface contour. It is recommended that the longitudinal edges of transverse specimens be machined to eliminate crater effects and to provide smooth surfaces. The corners should be rounded slightly.

Tension-Shear Test for Brazed Joints. The tension-shear test is used to determine the strength of the filler metal. There are various specimen types and joint designs used for this test, as indicated in Figure 12.7. The test is used primarily as a research tool for the development of filler metals and brazing procedures. However, the test has been standardized for control testing of samples from production brazing cycles and for comparison of filler metals produced by various manufacturers.

Two single 1/8 in. (3 mm) thick ferrous or nonferrous sheets are joined by brazing with a filler metal. The shear strength of the filler metal is calculated from the quotient of the tensile load at failure to the brazed area. Such test specimens require suitable fixturing during brazing to maintain accurate specimen alignment.

Strength Tests for Spot Welds

Tension-Shear Test. This test is widely used for determining the strength of arc and resistance spot welds.[3] It is also used to evaluate spot welding schedules for ferrous and nonferrous alloys.

The test specimen in Figure 12.8 is made by overlapping suitable size coupons and making a spot weld in the center of the overlapped area. The specimen is tested in a standard tensile test machine. In sheet thicknesses less than about 0.040 in. (1 mm) the eccentric load on the weld causes bending and rotation of the weld that results in failure around the edge of the nugget. In thicker sheets, the base metal tends to resist bending, and the spot weld may fail in shear through or around the nugget.

When the specimen thickness becomes as large as 0.19 in. (4.8 mm) or greater, the wedge grips of the test machine should be offset to reduce the eccentric loading on the weld.

Tension-shear tests are commonly used in quality assurance testing of production welds because they are easy and inexpensive to perform. Test specimen dimensions and test fixtures as well as statistical methods for evaluating resistance weld test results may be found in AWS C1.1, *Recommended Practices for Resistance Welding.*

---

3. Dimensions are different for ferrous and nonferrous metals. Refer to the applicable specification for appropriate specimen dimensions. See also *AWS C1.1, Recommended Practices for Resistance Welding.* Miami: American Welding Society, latest Ed.

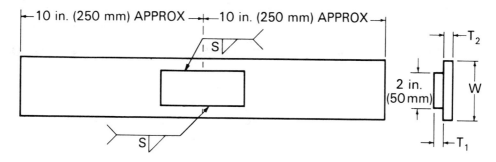

(A) Step 1: Deposit Weld "S"

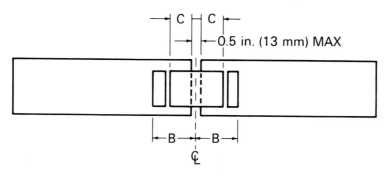

(B) Step 2: Machine Groove in Base Plate

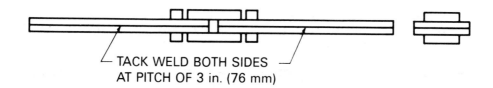

TACK WELD BOTH SIDES
AT PITCH OF 3 in. (76 mm)

(C) Step 3: Tack Two Pieces As Indicated

S	T_1	T_2	W	B	C
1/8	1/4	1/4	3		1-1/2
1/4	1/2	3/8	3		1-1/2
3/8	3/4	1/2	3		1-1/2
1/2	1	5/8	3-1/2*		1
ALL				3	
ALL				2	

in.	mm
1/8	3.0
1/4	6.5
3/8	9.5
1/2	12.5
5/8	16.0
3/4	19.0
1	25.5
1-1/2	38.0
2	51.0
3	76.0
3-1/2	89.0
7	178.0
10	254.0

S = specified size of fillet weld
* For 7 in. (178.0 mm) on each end of the base plates, the 3-1/2 in. (89.0 mm)
  width may have to be reduced to 3 in. (76.0 mm) to accommodate the jaws
  of the test machine.

Figure 12.6—Procedure for Preparation of Longitudinal Fillet Weld Shear Test Specimen

Direct-Tension Test. The direct-tension spot weld test is used to measure the strength of welds for loads applied in a direction normal to the spot weld interface.

This test is used mostly for welding schedule development and as a research tool for the weldability of new alloys. The tension test can be applied to ferrous and

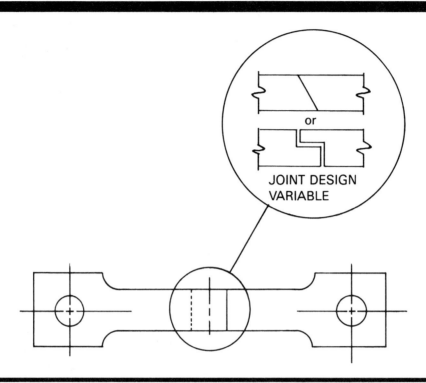

Figure 12.7—AWS Standard Shear Test Specimen for Brazed Joints (Refer to *Standard Method for Evaluating and Strength of Brazed Joints*, AWSC3.)

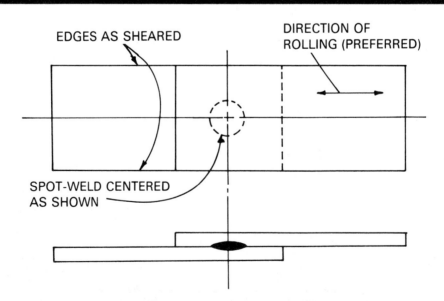

Figure 12.8—Tension-Shear Test Specimens

nonferrous metals of all thicknesses. The direct-tension test specimen is also used to determine the relative notch sensitivity of spot welds.

There are two types of specimens used for the direct-tension test. The cross-tension specimens of Figure 12.9 can be used for all alloys and all thicknesses. When the metal thickness is less than 0.04 in. (1 mm), it is necesary to reinforce the specimen to prevent excessive bending. The test jig shown in Figure 12.10 is used for thicknesses to 0.19 in. (4.9 mm); the jig in Figure 12.11 is used for greater thicknesses.

The U-specimen, Figure 12.12, is another type of direct-tension test. Two U-channels are spot welded back-to-back to form the test specimens. The specimens are pinned to filler blocks with pull tabs for applying a tensile load to the spot weld through the pin connections. The maximum load, which causes the weld to fail either by pulling a plug (tearing around the edge of the spot weld) or by tensile failure across the weld metal, is measured and reported. The direct-tension load is normally less than the tension-shear load for the same size weld and alloy.

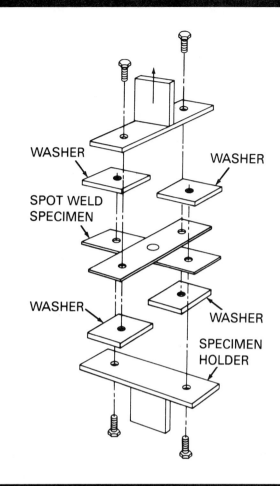

Figure 12.10—Jig for Cross-Tension for Thicknesses up to 0.19 in. (4.8 mm)

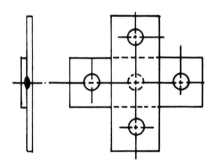

(A) Thickness Up to 0.19 in. (4.8 mm)

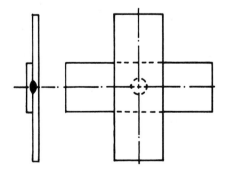

(B) Thickness Over 0.19 in. (4.8 mm)

Figure 12.9—Cross-Tension Test Specimens

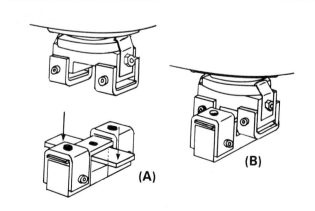

Figure 12.11—Jig for Cross-Tension Test for Thickness 0.19 in. (4.8 mm) and over

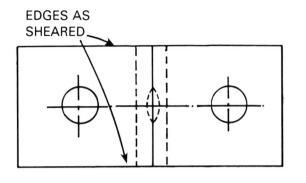

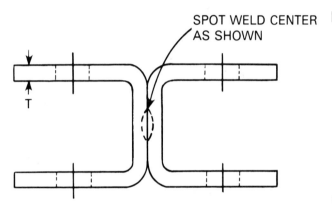

Figure 12.12—U-Test Specimen

Direct-tension testing of production welds is not normally used for quality control.

Peel Test. A variation of the direct-tension test that is commonly used as a production control test is the peel test shown in Figure 12.13. The size of the nugget is measured and compared to a standard weld size determined by tension-shear and direct tension tests. If the nugget size is equal to or greater than the standard size for the design, then the production welds are acceptable. This test is fast and inexpensive to perform. However, it may not be suitable for high-strength base metals or thicker sheets. Dimensions for the peel test vary according to sheet thickness. *AWS C1.1, Recommended Practices for Resistance Welding*, should be consulted for specimen dimensions.

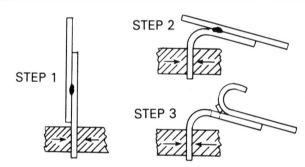

Figure 12.13—Peel Test. Step 1: Grip in Vise or Other Suitable Device. Step 2: Bend Specimen. Step 3: Peel Pieces Apart with Pincers or Other Suitable Tool

The ratio of the direct-tension load to the tension-shear load is frequently referred to as the ductility of the weld, and is a measure of the notch sensitivity of the weld. A ratio greater than 0.5 is considered ductile. Typical ratio ranges for spot welds in several base metals are listed in Table 12.1. Ratios less than 0.30 indicate notch sensitive welds.

Table 12.1
Ratio of Direct-Tension to Tension-Shear Test Specimen Loads

Material	Typical Ratio Range
Low carbon steel	0.60 to 0.99
Medium carbon steel (0.2C)	0.18 to 0.21
Low alloy high strength steel	0.21 to 0.28
Austenitic stainless steel	0.55 to 0.82
Ferritic stainless steel	0.25 to 0.33
Aluminum-base	0.37 to 0.43
Nickel-base	0.71 to 0.81
Titanium-base	0.27 to 0.52

HARDNESS TESTS

HARDNESS TESTING MAY be used in weld evaluations, either alone or to complement information gained through other tests. Hardness test methods for metals include Brinell, Vickers, Knoop and Rockwell. The first three tests use the area of indentation under load as the measure of hardness. The Rockwell hardness test relates hardness to the depth of indentation under load.

Hardness measurements can provide information about the metallurgical changes caused by welding. In alloy steels, a high hardness could indicate presence of untempered martensite in the weld heat-affected zone, while a low hardness may indicate an over-tempered condition. Welding may result in significantly lower heat-affected zone hardness of cold worked or age-hardened metal as a result of recrystallization or overaging, respectively.

Hardness tests should be made on ground, polished, and polished-and-etched cross sections of a weld joint. Measurements can be made on any specific area of the weld or base metal depending on the test method. Frequently, hardness indentations are made at regular intervals across an entire cross section, as shown in Figure 12.14.

Hardness test selection depends primarily on the hardness or strength of the metal, the size of the welded joint, and the type of information desired. The Brinell test produces a large indentation, typically 2 to 6 mm (0.08 to 0.24 in.) in diameter. The Rockwell test produces a much smaller indentation more suited for hardness traverses, but the indentations are macroscopic in size and may be larger than the precise areas of interest (for instance, a fusion zone or a coarse grain region in the heat-affected zone). The Vickers and Knoop microhardness tests make relatively small indentations that are well-suited for hardness measurements of the various regions of the heat-affected zone and for closely spaced traverses.

Each test measures the average hardness of all of the material under the indication. Microhardness tests, such as Vickers and Knoop, are used with a metallograph and can measure the hardness of individual grains and inclusions in the metal. Tests that make larger indentations are more representative of the bulk properties of the metal. The Brinell test makes the largest indentation and so gives an average hardness for the largest sample of metal. Several tests should be conducted on each sample, and the hardness scale and the average hardness should be reported.

There is an approximate interrelationship among the different hardness test results and the tensile strength of some metals. One relationship for nonaustenitic steels is shown in Table 12.2. A relationship for austenitic steels is shown in Table 12.3. Such correlations should be used with caution when applied to welded joints or any metal with a heterogeneous structure.

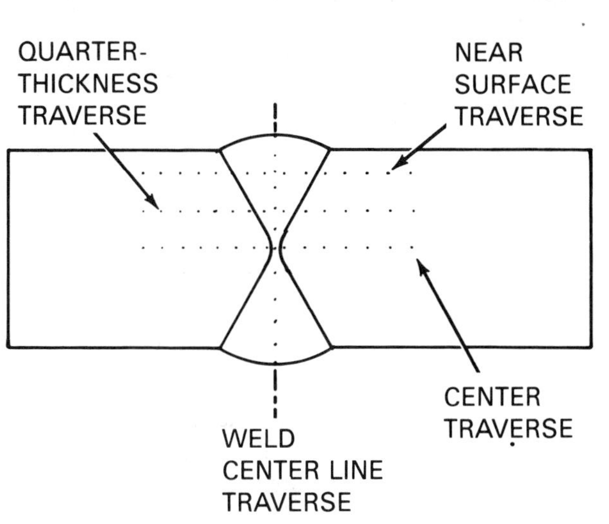

BEND TESTS

VARIOUS TYPES OF bend tests are used to evaluate the ductility and soundnes of welded joints. Guided bend specimens may be longitudinal or transverse to the weld axis and may be bent in tensile test machines or in wraparound bend test jigs. Typical guided bend testing fixtures are illustrated in Figure 12.15.

Face bend tests are made with the weld face in tension; root bend tests are made with the weld root in tension. When bend testing thick plates, transverse slices or side bend test specimens are usually cut from the welded joint and bent with the weld cross section in tension. The relative orientations of these specimens are illustrated in Figure 12.16.

The guided bend test is most commonly used in welding procedure and welder performance qualification. The required specimen dimensions are specified in the relevant code. The strain on the outside fiber of the bend specimen can be approximated from the following formula:

$$\epsilon = \frac{100t}{(2R+t)} \tag{12.1}$$

where
  ϵ = strain, %
  t = bend test specimen thickness, in. (mm)
  R = inside bend radius, in. (mm)

In qualification testing, the specimen thickness and bend radius are chosen according to the ductility of the metal being tested. Sound mild steel welds can easily achieve an outside fiber elongation of 20 percent. However, in the presence of weld defects, the bend test specimens will consistently fail.

Bend tests suffer from the same weakness as the transverse weld tension test in that nonuniform properties along the length of the specimen can cause nonuniform bending. The bend test is sensitive to the relative strengths of the weld metal, the heat-affected zone, and the base metal.

Several types of problems can develop in transverse bend tests. An overmatching weld strength may prevent the weld zone from conforming exactly to the bending die radius, and thus may force the base metal to deform to a smaller radius. The desired elongation of the weld will not be achieved. With an undermatching weld strength, the specimen may bend in the weld to a radius smaller than the bending die. Failure in this case may result when the weld metal ductility is exceeded, and not because the weld metal contained a defect.

Figure 12.14—Typical Hardness Traverses for a Double-V Groove Welded Joint

Table 12.2
Approximate Hardness Conversion Numbers for Nonaustenitic Steels

| Rockwell B Scale, 100-kgf Load 1/16-in. (1.588 mm) Ball | Vickers Hardness Number | Brinell Indentation Diameter, mm | Brinell Hardness, 3000-kgf Load, 10-mm Ball | Knoop Hardness, 500-gf Load and Over | Rockwell A Scale, 60-kgf Load, Diamond Penetrator | Rockwell F Scale, 60-kgf Load, 1/16-in. (1.588-mm) Ball | Rockwell Superficial Hardness | | | Approximate Tensile Strength ksi (MPa) |
							15T Scale, 15-kgf Load, 1/16-in. (1.588-mm) Ball	30T Scale, 30-kgf Load, 1/16-in. (1.588-mm) Ball	45T Scale, 45-kgf Load, 1/16-in. (1.588-mm) Ball	
100	240	3.91	240	251	61.5	...	93.1	83.1	72.9	116 (800)
99	234	3.96	234	246	60.9	...	92.8	82.5	71.9	114 (785)
98	228	4.01	228	241	60.2	...	92.5	81.8	70.9	109 (750)
97	222	4.06	222	236	59.5	...	92.1	81.1	69.9	104 (715)
96	216	4.11	216	231	58.9	...	91.8	80.4	68.9	102 (705)
95	210	4.17	210	226	58.3	...	91.5	79.8	67.9	100 (690)
94	205	4.21	205	221	57.6	...	91.2	79.1	66.9	98 (675)
93	200	4.26	200	216	57.0	...	90.8	78.4	65.9	94 (650)
92	195	4.32	195	211	56.4	...	90.5	77.8	64.8	92 (635)
91	190	4.37	190	206	55.8	...	90.2	77.1	63.8	90 (620)
90	185	4.43	185	201	55.2	...	89.9	76.4	62.8	89 (615)
89	180	4.48	180	196	54.6	...	89.5	75.8	61.8	88 (605)
88	176	4.53	176	192	54.0	...	89.2	75.1	60.8	86 (590)
87	172	4.58	172	188	53.4	...	88.9	74.4	59.8	84 (580)
86	169	4.62	169	184	52.8	...	88.6	73.8	58.8	83 (570)
85	165	4.67	165	180	52.3	...	88.2	73.1	57.8	82 (565)
84	162	4.71	162	176	51.7	...	87.9	72.4	56.8	81 (560)
83	159	4.75	159	173	51.1	...	87.6	71.8	55.8	80 (550)
82	156	4.79	156	170	50.6	...	87.3	71.1	54.8	77 (530)
81	153	4.84	153	167	50.0	...	86.9	70.4	53.8	73 (505)
80	150	4.88	150	164	49.5	...	86.6	69.7	52.8	72 (495)
79	147	4.93	147	161	48.9	...	86.3	69.1	51.8	70 (485)
78	144	4.98	144	158	48.4	...	86.0	68.4	50.8	69 (475)
77	141	5.02	141	155	47.9	...	85.6	67.7	49.8	68 (470)
76	139	5.06	139	152	47.3	...	85.3	67.1	48.8	67 (460)
75	137	5.10	137	150	46.8	99.6	85.0	66.4	47.8	66 (455)
74	135	5.13	135	147	46.3	99.1	84.7	65.7	46.8	65 (450)
73	132	51.8	132	145	45.8	98.5	84.3	65.1	45.8	64 (440)
72	130	5.22	130	143	45.3	98.0	84.0	64.4	44.8	63 (435)
71	127	5.27	127	141	44.8	97.4	83.7	63.7	43.8	62 (425)
70	125	5.32	125	139	44.3	96.8	83.4	62.1	42.8	61 (420)
69	123	5.36	123	137	43.8	96.2	83.0	62.4	41.8	60 (415)
68	121	5.4	121	135	43.3	95.6	82.7	61.7	40.8	59 (405)
67	119	5.44	119	133	42.8	95.1	82.4	61.0	39.8	58 (400)
66	117	5.48	117	131	42.3	94.5	82.1	60.4	38.7	57 (395)
65	116	5.51	116	129	41.8	93.9	81.8	59.7	37.7	56 (385)
64	114	5.54	114	127	41.4	93.4	81.4	59.0	36.7	...
63	112	5.58	112	125	40.9	92.8	81.1	58.4	35.7	...

Problems with weld strength mismatch can be avoided by using longitudinal bend specimens in which the weld runs the full length of the specimen; the bend axis is then perpendicular to the weld axis. In this test, all zones of the welded joint (weld, heat-affected zone, and base metal) are strained equally and simultaneously. This test is generally used for evaluations of joints in dissimilar metals.

In the longitudinal bend test, weld flaws that are inherently oriented parallel to the weld axis—such as incomplete fusion, inadequate joint penetration, or undercut—are only moderately strained and may not cause failure.

Side-bend specimens strain the entire weld cross section, and thus are especially useful for exposing defects near the midthickness that might not contribute to fail-

Table 12.3
Approximate Hardness Conversion Numbers for Austenitic Steels

Rockwell C Scale, 150-kgf Load, Diamond Penetrator	Rockwell A Scale, 60-kgf Load, Diamond Penetrator	Rockwell Superficial Hardness		
		15N Scale, 15-kgf Load, Diamond Penetrator	30N Scale, 30-kgf Load, Diamond Penetrator	45N Scale, 45-kgf Load, Diamond Penetrator
48	74.4	84.1	66.2	52.1
47	73.9	83.6	65.3	50.9
46	73.4	83.1	64.5	49.8
45	72.9	82.6	63.6	48.7
44	72.4	82.1	62.7	47.5
43	71.9	81.6	61.8	46.4
42	71.4	81.0	61.0	45.2
41	70.9	80.5	60.1	44.1
40	70.4	80.0	59.2	43.0
39	69.9	79.5	58.4	41.8
38	69.3	79.0	57/5	40.7
37	68.8	78.5	56.6	39.6
36	68.3	78.0	55.7	38.4
35	67.8	77.5	54.9	37.3
34	67.3	77.0	54.0	36.1
33	66.8	76.5	53.1	35.0
32	66.3	75.9	52.3	33.9
31	65.8	75.4	51.4	32.7
30	65.3	74.9	50.5	31.6
29	64.8	74.4	49.6	30.4
28	64.3	73.9	48.8	29.3
27	63.8	73.4	47.9	28.2
26	63.3	72.9	47.0	27.0
25	62.8	72.4	46.2	25.9
24	62.3	71.9	45.3	24.8
23	61.8	71.3	44.4	23.6
22	61.3	70.8	43.5	22.5
21	60.8	70.3	42.7	21.3
20	60.3	69.8	41.8	20.2

Copyright ASTM. Reprinted with permission.

ure in face- or root-bend tests. They are normally used for relatively thick sections [over 3/4 in. (19mm)].

Codes generally specify a maximum allowable size for cracks in bend tests made for procedure or welder qualification. More detailed descriptions of bend test procedures and requirements should be obtained from applicable codes, such as AWS or ASME codes. The latest edition of *AWS B4.0, Standard Methods for Mechanical Testing of Welds* provides requirements for specimen preparation, testing conditions, and testing procedure, but does not give acceptance criteria.

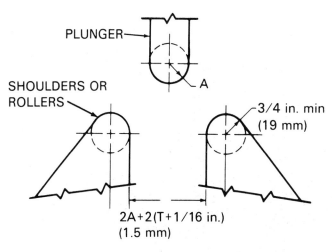

(A) Guided Bend Test Jig

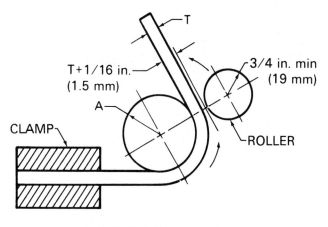

(B) Wrap-Around Bend Test Jig

Figure 12.15—Guided Bend Test Jigs

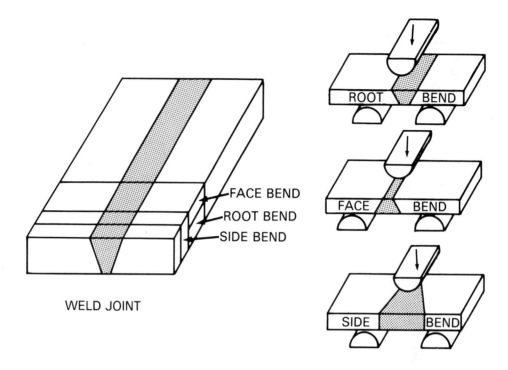

WELD JOINT

Figure 12.16—Triaxial Stress Distribution. Stress in the X- and Z-Directions Results from a Non-Uniform Elongation of the Elements in the Direction of Applied Stress (Y-Direction)

FRACTURE TOUGHNESS

FRACTURE TOUGHNESS IS a generic term for measures of resistance to extension of a crack.[4] The common methods of measuring the fracture toughness of welded joints are the Charpy V-notch, the Dynamic Tear, the Plane Strain Fracture Toughness (K_{IC}) and the Drop Weight tests. The test methods and the test rationale are described here. The use of test results in designing weldments is covered in Chapter 5 of this volume.

Crack propagation requires an energy source. In fracture toughness tests, the energy is produced by the testing machine. In service, the stored elastic strain energy in the structure is the driving force for crack propagation.

The stored elastic strain energy of a stressed member is a product of the stress and strain. The units of stress multiplied by strain are *energy per unit volume*. Thus, high strength materials can store more elastic strain than can low strength materials when both are loaded to the

same fraction of the yield strength. At the same stress level, the stored elastic energy of high and low strength materials is the same.

CHARPY V-NOTCH IMPACT TEST

THE CHARPY V-NOTCH impact test is the most common fracture toughness test. The test procedure is shown in *ASTM A370 Standard Methods and Definitions for Mechanical Testing of Steel Products*[5] and *AWS B4.0 Standard Methods for Mechanical Testing of Welds*.[6] The test is often used in specifiying minimum acceptance criteria for base metal and filler metal manufacturing, and for welding procedure qualifications.

The Charpy V-notch specimen is shown in Figure 12.17. For metals such as carbon and low alloy steels

---

4. Refer to *ASTM E616-82, Standard terminology relative to fracture testing.* Philadelphia: ASTM.

5. Available from ASTM, Philadelphia, PA.
6. Available from American Welding Society, Miami, Florida.

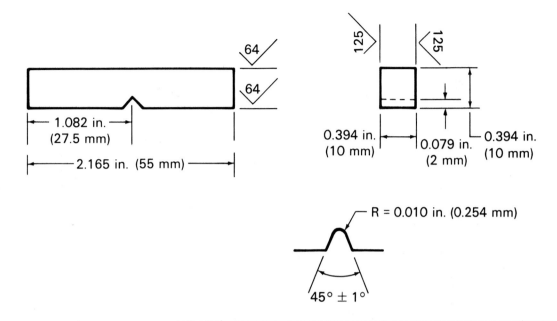

Figure 12.17—Charpy V-Notch Impact Specimen

that exhibit a change in failure mode with decreasing temperature, it is common to conduct the test at several temperatures. The most common reported result is the absorbed energy, but the percent shear fracture and the lateral expansion may also be reported. A typical transition temperature curve is shown in Figure 12.18. Metals exhibiting high Charpy V-notch energy absorption are resistant to brittle fracture in service. There is no threshold value of energy absorption that will assure ductile behavior in service. As previously indicated, high strength materials can store more elastic energy than low strength materials. Thus, to provide the same ductile fracture behavior found in lower strength metals, high strength metals should exhibit a higher Charpy V-notch energy absorption.

Specifications for most metals that include Charpy V-notch impact tests normally require a minimum energy absorption at a particular temperature.

Some fabrication codes may require that the Charpy V-notch energy absorption of the weld metal and the heat-affected zone of the weld procedure qualification test plate be determined. These codes also prescribe conditions that require making the test, the test temperature, and the minimum acceptable result. In addition, some fabrication codes require impact testing of production welds. The production test plates may be prepared as runoff plates from a straight seam weld, or may be separate test plates, but they should be fabricated at the same time as the production weld and follow the same welding procedure specification.

PLANE STRAIN FRACTURE TOUGHNESS TESTS

PLANE STRAIN FRACTURE toughness (K_{IC}) is the resistance to crack extension under conditions of plane strain at a crack tip. The tests require a specimen of sufficient size that the strain in one of the orthogonal directions (usually the thickness direction) is small in comparison to the crack size and the specimen dimensions. The minimum test specimen dimensions to develop a plane strain condition usually declines with increasing strength and with decreasing temperature. The plane strain fracture toughness tests are unique in that the specimen dimensions and configuration are not standard, and the validity of the test is determined only after the results are calculated. The test method is described in *ASTM E399, Standard Test Method for Plane-Strain Fracture Toughness of Metallic Materials.*[7] The specification should be consulted for the specific details of the test procedure. The test sequence can be summarized as follows:

(1) Select a specimen configuration and test method from ASTM E399. The compact tension specimen, recommended for testing welded joints in *AWS B4.0, Stan-*

---

7. Available from ASTM, Philadelphia, PA.

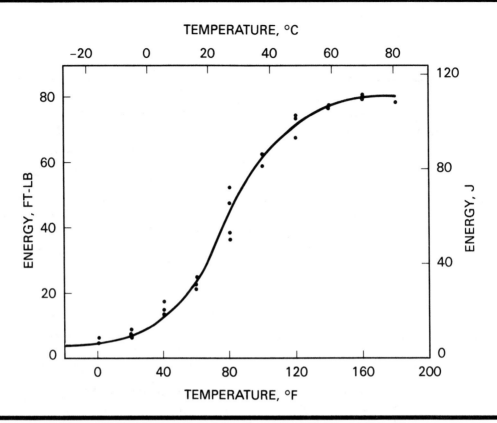

TEMPERATURE, °C

Figure 12.18—Typical Transition Curve for Mild Steel Ship Plate

dard Method for Mechanical Testing of Welds, is shown in Figure 12.19.[8]

(2) Assume a value of K_{IC}. It is recommended that the assumed value of K_{IC} be overestimated so the specimen will be conservatively large. The minimum specimen dimensions can then be determined as follows:

Crack length, $a \geq 2.5\ (K_{IC}/\sigma_{ys})^2$
Specimen width, $B \geq 2.5\ (K_{IC}/\sigma_{ys})^2$
Specimen depth, $W \geq 5\ (K_{IC}/\sigma_{ys})^2$

(3) The test specimen is machined and precracked by fatigue at a low stress level.

(4) The specimen is loaded in bending or tension, and the load and the crack opening are recorded on an X-Y recorder. The specimen is tested to failure and the maximum load, P_{max}, is recorded.

(5) A load P_Q on the load-displacement curve is found following the rules of E399, and the ratio P_{max}/P_A is determined. If the ratio exceeds 1.10, the test is invalid. If the ratio does not exceed 1.10, a conditional plain-strain fracture toughness K_Q is calculated, by using the equations given in ASTM E399 for the test specimen used.

8. Available from the American Welding Society, Miami, Florida.

(6) Calculate a_{min} and B_{min} as follows
$a_{min} = B_{min} = 2.5\ (K_{IC}/\sigma_{ys})^2$
If
$a \geq a_{min}$ and $B \geq B_{min}$
where a and B are the crack length and specimen width respectively, then
$K_Q = K_{IC}$ and the test is valid.

The test procedure is expensive and requires considerable time to perform. It is not normally used as an acceptance test for production material or for welding procedure qualification.

Applications

PLANE-STRAIN FRACTURE TOUGHNESS tests are preproduction or pilot plant type tests that provide a rational means for designers and engineers to estimate the effects of new designs, materials, or fabrication practices on the fracture-safe performance of structures.

The plane strain fracture toughness K_{IC} is measured under conditions most susceptible to unstable crack propagation. Therefore, valid K_{IC} values are the lowest

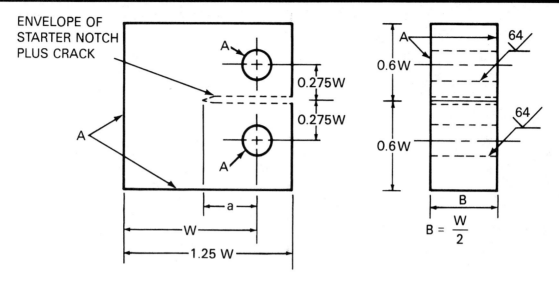

Notes:

1. Dimensions a, B, and W are to be determined in accordance with ASTM E399.

2. Surfaces marked A shall be perpendicular and parallel as applicable to within 0.002W total indicator reading (TIR).

3. The intersection of the crack starter tips with the two specimen faces shall be equally distant from the top and bottom edges of the specimen within 0.005W.

4. Integral or attachable knife edges for clip gage attachment to the crack mouth may be used.

Figure 12.19—Compact Tension Specimen to Determine Plane Strain Fracture Toughness in Weldments

conservative measures of resistance to crack propagation. When the conditions of plane strain are not met, K_Q is always less than the valid K_{IC}. Thus, reporting a K_Q that does not meet the validity tests of ASTM E399 results in an unnecessarily conservative approach to fracture control. Values of K_Q can be useful, but the conditions of the test, the calculated values of the test results (including the results of the validity tests), and the specimen dimensions should be reported. A sample report form is provided in ASTM E399.

Effects of Temperature and Strain Rate

THE MAGNITUDE OF K_{IC} decreases with decreasing temperature, and with increasing strain rates. The minimum specimen size to achieve plane strain conditions at the crack tip also decreases with decreasing temperature and increasing strain rates.

DROP WEIGHT TESTS

DROP WEIGHT TESTS were developed to determine the temperature above which a dynamic crack will be arrested. The test specimen and test procedure are described in *ASTM E208 Standard Method for Conducting Drop-Weight Test to Determine Nil-Ductility Transition Temperature of Ferritic Steels*.[9] The application of the test to welded specimens is described in *AWS B4.0, Standard Methods for Mechanical Testing of Welds*.

The procedure for drop-weight weld testing involves welding a "crack starter" bead on a test specimen, and then notching the weld bead. The specimen is chilled to a designated test temperature and tested with the notched bead in tension by applying a load from a falling weight. "Break" and "no break" performances are illustrated in Figure 12.20.

Application

THE DROP-WEIGHT TEST may be used as an acceptance test for steel plates and shapes. It is often used in the procurement of material for marine applications. The minimum thickness of applicability is 5/8-in. (16 mm),

9. Available from ASTM, Philadelphia, PA.

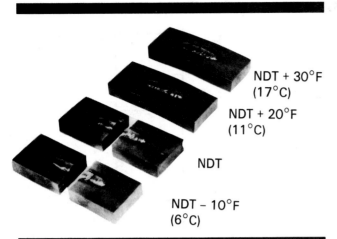

NDT + 30°F
(17°C)

NDT + 20°F
(11°C)

NDT

NDT − 10°F
(6°C)

**Figure 12.20—Typical Break and No Break
Performance**

but a nonstandard 1/2-in. (13 mm) thick plate has also been employed in research and development work.

When the Drop Weight test is used as an acceptance criterion, a temperature is usually specified, at which a "no-break" performance is required.

SPECIAL PROBLEMS OF WELDED JOINTS

THE WELDED JOINT presents a special problem in evaluating fracture toughness because it contains a variety of microstructures in the weld metal and weld heat-affected zone. Any test for evaluating the toughness of welded joints, therefore, should assess the effect of these varied microstructures and, if possible, assign a toughness index to each. This has the value of establishing which of these structures is the most critical, the "weak link," and what the relative behavior of each zone is. It is, however, not at all certain that a "weak link" will fail in actual service. For example, in some steels the heat-affected zone is known to be less tough than the base metal or weld metal. However this zone may be rather narrow and irregular and the fracture may not be restricted to it. As a result, fracture behavior may not reflect the toughness of this zone alone but may be affected by the portion of the crack front which passes through tougher regions.

The fracture toughness test specimens are suited to testing a specific zone because the notch can be located in the zone to be evaluated. However, locating the initial notch in a specific zone does not necessarily control the fatigue crack path. A test intended for the heat-affected zone may fail in the weld metal or the base metal.

Another problem in these tests is the complex nature of the zones themselves. The heat-affected zone is actually a series of zones.

Furthermore, the heat-affected zone does not exist by itself, as do the base metal and weld metal. The heat-affected zone must be tested either in the location where it exists, or in a separate specimen that was subjected to a simulated weld thermal cycle. Such simulated heat-affected zones can be adapted to the Charpy V-notch impact test, but they are not suitable for other tests.

For these reasons, most fracture toughness studies of weldments incorporate a number of different tests, to determine the resistance of the composite weld zone to fracture.

FATIGUE PROPERTIES OF WELDED STRUCTURAL JOINTS

THE TERM FATIGUE FAILURE denotes failure under cyclic loading. Fatigue[10] is a progressive failure. It originates at a notch that causes a stress concentration. In welded assemblies, the notch could be a structural detail such as an intersection or cover plate, or a discontinuity such as a bolt hole or a weld undercut. A fatigue crack develops as a result of the stress concentration and extends with each load cycle until failure occurs, or until the cyclic loads are transferred to redundant members. The fatigue performance of a member is more dependent on the localized state of stress than is the static strength of the base metal or the weld metal.

There are only two primary factors affecting the fatigue life of welded assemblies:

(1) The difference between the maximum and minimum stress—the stress range
(2) The design detail

These conclusions are demonstrated in Figures 12.21, 12.22, and 12.23. The minimum stress has no apparent influence on the cycles to failure, but the life is inversely

---

10. Fatigue is also discussed in Chapter 5 of this volume.

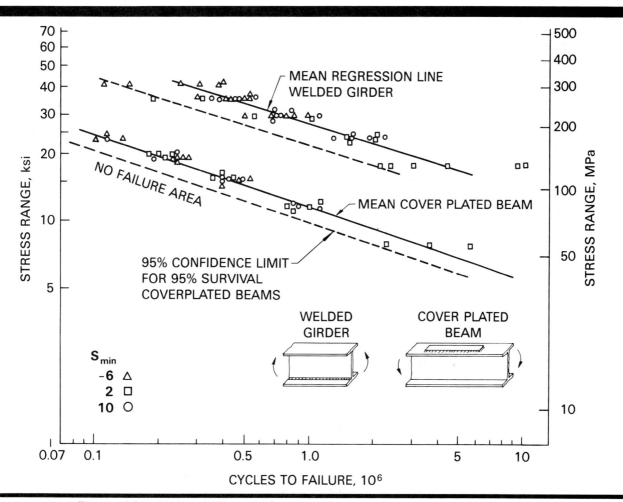

Figure 12.21—Effect of Minimum Stress on Stress Range-Cycle Life Relationship

proportional to the stress range, as shown in Figure 12.21. The steel grade is not a significant factor in determining the girder life, as shown in Figure 12.22. The life of girders with cover plates is less than girders without cover plates for the same stress range, as shown in Figures 12.21 and 12.22. The length of a welded attachment also affects fatigue life, see Figure 12.23.

Thus, in fatigue tests of welded joints, the specimen geometry significantly affects the results. Specimen geometry includes the surface condition of the weld and the internal soundness of the weld.

When welded fatigue specimens are tested with the weld reinforcement in place, the results are considerably scattered. The fatigue life of reinforced specimens at a given stress range is less than the life of welded specimens with the reinforcement ground smooth and feathered into the base plate.

Weld soundness is another geometry effect, because a stress concentration results at any discontinuity.

Thus, fatigue testing of weldments is performed primarily to develop design data for welded structures. The application of fatigue data to the design of welded girders is addressed in *ANSI/AWS D1.1, Structural Welding Code—Steel.*

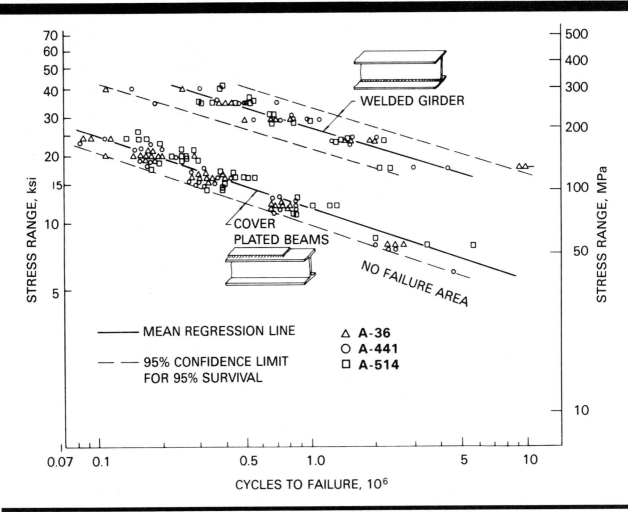

Figure 12.22—Effect of Type of Steel on Stress Range-Cycle Life Relationship

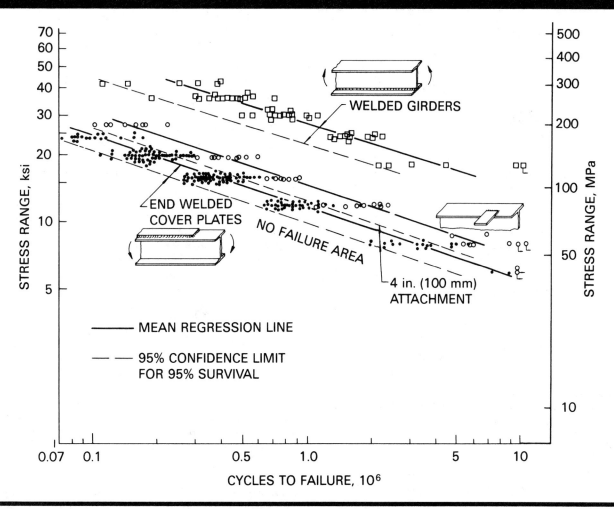

Figure 12.23—The Effect of Attachment Length

CORROSION FACTORS AFFECTING THE TESTING AND PERFORMANCE OF WELDED JOINTS

THE CORROSION RESISTANCE of welded joints may differ from that of the unwelded base metal, because the weld zone varies in chemical composition, metallurgical structure, and residual stress levels. These differences are brought about by the welding process itself. Fortunately, the welding process and filler metal can generally be selected and controlled to optimize the corrosion resistance of welded joints.

CORROSION TESTING OF WELDED JOINTS

A WELDED JOINT and the base metal may corrode uni-

formly over the entire surface. The welded joint itself may be susceptible to varying degrees of preferential attack, as shown in Figure 12.24. The weld metal may corrode more, Figure 12.24c, or less, Figure 12.24b, than the base metal as a result of differences in chemical composition or microstructure, or both. In addition, the heat-affected zone may be susceptible to corrosion attack in a specific region, Figures 12.24d and e, as a result of metallurgical reactions during welding.

More than one type of corrosion shown in Figure 12.24 may occur in the same welded joint. The figure represents a macroscopic view of the weldment. Microscopic attack, such as intergranular corrosion and pit-

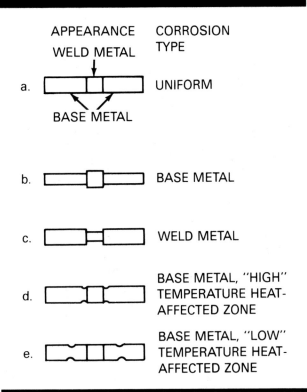

Figure 12.24—Types of Corrosion in a Welded Joint

ting corrosion, can occur at any location in a welded joint.

Some of the factors that influence the corrosion resistance of welded corrosion tests are the following:

(1) Chemical composition and structure of base metal and weld metal

(2) Welding process and procedure, especially shielding

(3) Dimensions of the welded test plate and the corrosion specimens removed from the plate

When reporting results of corrosion tests of welded joints, the items in the preceding list should also be reported.

The most common method of evaluating corrosion resistance is to measure the weight lost during exposure to the corrodent and convert this to an average corrosion rate using the formula:

$$R = \frac{KW}{ADt} \qquad (12.2)$$

where

$R =$ the average corrosion rate in depth of attack per unit time

$K =$ a constant, the value of which depends upon the units used

$W =$ the weight lost by the specimen during the test

$A =$ the total surface area of the test specimen

$D =$ the density of the specimen material

$t =$ the duration of exposure

This formula is ideally suited to general corrosion. Selective corrosion can occur in the weld metal or heat-affected zone. Since these areas are small in comparison to the total weld area, the average corrosion rate R may appear small. Therefore, corrosion test specimens should be visually examined for selective attack and any localization of attack should be noted in the report.

Another technique to determine if preferential corrosion occurs is to expose an unwelded sample of the same dimension as the welded specimen. If the corrosion rates of the welded and unwelded specimens are approximately equal, then no preferential attack is indicated.

If one rate is significantly greater than the other, then preferential attack should be suspected. Regardless of the ratio, a careful visual inspection should be conducted.

Test Specimens

Specimen Geometry. A corrosion test specimen design will vary with the size and type of product being evaluated. A butt joint with suitable dimensions is the simplest test specimen shape for general corrosion, Figure 12.25.

The overall dimensions of the test assembly should be representative of an actual welded assembly. To simulate the residual stresses formed in assemblies, the test plate should be at least 18 in. (450 mm) long and 12 in.

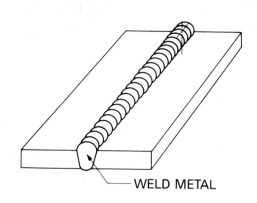

Figure 12.25—Typical Welded Joint

(300 mm) wide. The thickness should be representative of welded applications of the material tested.

Cross welds may be tested also. In some cases, it may be desirable to cut and test a sample from the weld metal only. For example, the corrosion resistance of a filler metal may be tested. However, this technique will not determine whether a galvanic corrosion couple will form between this weld metal and a base metal.

Surface Preparation. The surface preparation of a test specimen should fit the purpose of the test and the technique of evaluating the corrosion damage after the test. These steps should be followed:

(1) Test in the as-welded condition with no attempt to remove metal or welding scale. This is recommended if the weld will be used in this manner in service and if the test environment is the same or similar to that which the weld would experience in service.

(2) Keep the amount of exposed unaffected base metal to a minimum unless galvanic corrosion is suspected in the weld.

(3) Machine or grind all surfaces flush (except in case 1 above), leaving no undercut or surface imperfections, and then polish to 120 grit-finish. Avoid overheating during preparation. This method is recommended if surface examination is to be used for the evaluation. However, it is important to realize that the removal of large amounts of weld metal can significantly affect corrosion rates, particularly in the case of large sections and multipass welds.

In some cases where both faces of a weld are exposed to the test, a compromise method is to grind one side of the weld bead flush, but leave the other side as-welded and cleaned by one of the following methods:

(1) Clean the weld by mechanical means, such as by shot or vapor blasting, but do not remove the weld bead.

(2) Clean the weld by chemical means. This method may involve a descaling operation only or a pickling operation in which scale and metal are removed.

Normal corrosion test procedures prior to testing, such as degreasing, should be followed.

EVALUATION OF CORRODED WELDMENTS

Methods of Evaluation

Weight Loss Measurements. Allowing for their limitations, weight loss measurements are considered useful because they are generally easy to perform.

Macroscopic Examination. Low magnification examination of the tested sample is generally adequate to determine the existance of significant localized attack. Higher magnifications may also be used to advantage, provided the original surface finish of the test specimen permits this. Standard photographs may be used as a rating system for reference.

Electrical Resistance. Observing the change in electrical resistance across a weld as a result of corrosion is an accurate and effective method of detecting general and intergranular corrosion. Unfortunately, the specimen needs to be in the form of long, thin sections (preferably wires or strips) to detect small amounts of attack satisfactorily.

Measurement of Depth of Attack. Several devices are available to measure the depth of preferential attack in a corroded weld. They include (1) a fine-tipped micrometer which may be suitable for measuring a severe attack, such as that sketched in Figure 5.24c, (2) a microscope with calibrated vertical movement (with it, the distance between the focal point of the top and bottom of a locally corroded zone such as sketched in Figure 12.24e can be measured), and (3) a profilometer. Profilometers, which can be used to scan the surface of a tested sample, have possible application in the evaluation of welded joints. With this device, a fine needle moves over the surface of the tested sample and a transducer measures the movement of the needle relative to a flat-tipped shoe.

Ultrasonic Methods. Ultrasonic inspection techniques are applicable to the measurement of localized surface discontinuities and have been used to evaluate intergranular attack.

ELEVATED TEMPERATURE BEHAVIOR

CREEP-RUPTURE PHENOMENA

WHEN A LOAD that is lower than that required to produce rapid failure is applied to a metal at an elevated temperature, the metal will gradually elongate and ulti-

mately fracture. The gradual elongation or plastic deformation is called elevated temperature creep, while the eventual fracture is called rupture. This time-temperature phenomenon, illustrated in Figure 12.26, can be divided into three stages. In primary creep, the first

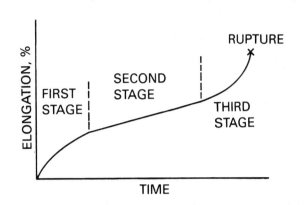

Figure 12.26—Typical Creep-Rupture Elongation Curve

stage, the rate of elongation is high but decreases with time. The second stage of creep is generally an extended period during which the rate of elongation is constant. In the third stage, the creep rate again increases, and culminates in rupture of the specimen. Increasing load or temperature, or both, shortens the time for completion of each stage.

TEST METHOD

THE TEST SPECIMEN used to determine creep-rupture strength may be either an all-weld-metal specimen or a transverse weld specimen. The advantages and limitations of each orientation are similar to those of the tension test. However, transverse creep tests are seldom conducted because the variations in elongation in each region of the welded joint make total elongation measurements meaningless.

Creep rupture tests are described in ASTM E 139—*Standard Practice for Conducting Creep, Creep-Rupture, and Stress-Rupture Tests in Metallic Materials.*[11]

In the rupture test, time to failure at a given applied load is the principal dependent-variable measured, although final elongation and reduction of area are also determined.

Reporting and Interpreting Creep Data

RUPTURE AND CREEP data are generally reported on log-log plots of stress versus time to failure, as illustrated in Figure 12.27. The time to failure of each test is plotted against the applied stress, and the best lower limit straight line is drawn for each test temperature.

As illustrated in Figure 12.27, the time to failure at a constant temperature increases with decreasing stress

11. Available from ASTM, Philadelphia, PA.

level and decreases with increasing temperatures. The linear relationship between the logarithm of stress and the logarithm of time is fairly reliable, provided that metallurgical reactions do not occur during the life of the test.

If a solid state reaction does occur, then a relatively distinct change in slope occurs in the curve, as illustrated in Figure 12.28 at points A, B, and C. Isothermal reactions in alloy systems are diffusion controlled; they occur more rapidly at a high temperature than they do at a lower temperature. Thus, the solid state reactions that occurred at points A, B, and C are the same reaction, and one should expect a similar reaction at temperature T_4. The time required for the reaction to begin at T_4 may

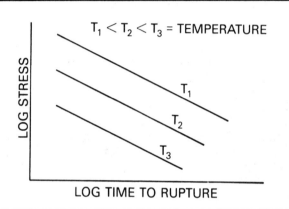

Figure 12.27—Typical Set of Rupture Data at Three Temperatures

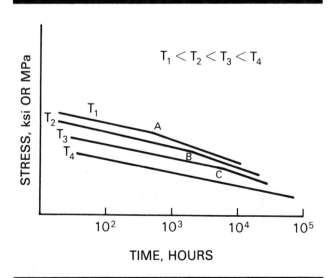

Figure 12.28—Effect of Metallurgical Reactions on the Slope of the Creep Rupture Curve

be greater than the life of the test. Therefore, the results at temperature T_4, should not be extrapolated to longer times.

The method now preferred when using high temperature tests to predict lower temperature creep performance is the use of parameters such as the Larson-Miller parameter. The Larson-Miller parameter and other parameters are discussed in ASTM E139—*Standard Practice for Conducting Creep, Creep-Rupture, and Stress-Rupture Tests of Metallic Materials.* The Larson-Miller parameter is determined as follows:

$$P = T (C + \log t_r) \tag{12.3}$$

where

   $P =$ the Larson Miller parameter
   $T =$ the absolute temperature, °R
   $C =$ a constant approximately equal to 20 for most alloys

$t_r =$ the time to rupture

Another time-temperature parameter discussed in ASTM E139 is

$$P = \frac{T - T_a}{(Log\ t - Log\ t_a)} \tag{12.4}$$

where

   $T =$ test temperature, °F
   $t =$ time to rupture,
   T_a and t_a are constants determined from the test data in accordance with ASTM E139

The use of these parameters, including guidelines for the minimum number of tests per thousand hours and the recommended maximum extrapolation, are addressed in ASTM E139.

TESTS OF THERMAL SPRAY APPLICATIONS

SEVERAL METHODS OF evaluation of thermal spray deposits have been devised. Most are summarized in the publication *Thermal Spraying Practice, Theory, and Application.*[12] Some of the tests recommended in this publication include tests covered by ASTM standards. The ASTM standard tests used in evaluating thermal spray deposits are listed in Table 12.4.

Several nonstandard tests to determine the shear strength, compressive strength, adhesion, and coating hardness are also described in *Thermal Spray, Theory, Practice and Application.*

Table 12.4
Standard Tests Used in the Evaluation of Thermal Spray Deposits

ASTM Standard		Type Test
C633	Standard Test Method for Adhesion or Cohesive Strength of Flame-Sprayed Coating	Tensile test
D3359	Standard Methods for Measuring Adhesion by Tape Test	Scratch test
G65	Practice for Conducting Dry Sand/Rubber Wheel Abrasion Test	Wear

SUPPLEMENTARY READING LIST

American Society for Mechanical Engineers. *ASME Boiler and pressure vessel code.* Section 8, Pressure Vessels. Divisions 1 and 2. New York, NY: American Society for Mechanical Engineers, (latest edition).

———. *ASME Boiler and pressure vessel code.* Section 9, Welding and Brazing Qualifications for Welders, Brazers, and Welding and Brazing Operators. New York, NY: American Society of Mechanical Engineers, (latest edition).

ASTM. Conducting drop-weight test to determine nil-ductility transition temperature of ferritic steels, E 208. *Annual Book of ASTM Standards.* Philadelphia, PA: ASTM (published annually).

Egan, G. R. An assessment of defects in welded carbon manganese steels—Part I. *Welding Research Abroad.* 18 (3): 1972; 2-32.

12. Available from the American Welding Society, Miami, Florida.

————. Free bend test for ductility of welds, E 16. *Annual Book of ASTM Standards*. Philadelphia, PA: ASTM (published annually).

————. Guided bend test for ductility of welds, E 190. *Annual Book of ASTM Standards*. Philadelphia. PA: ASTM (published annually).

Henthorne, M. Corrosion testing of weldments. *Corrosion* 30 1974: 39-46.

Manson, S. S. *Thermal stress and low cycle fatigue*. New York, NY: McGraw-Hill (1966).

————. Mechanical Testing of Steel Products, A 370. *Annual Book of ASTM Standards*. Philadelphia, PA: ASTM (published annually).

————. *Metals—Mechanical, Fracture, and Corrosion Testing; Fatigue; Erosion; Effect of Temperature*, Parts 10 and 11. Philadelphia, PA: ASTM (published annually).

Munse, W. H. and Grover, L. M. *Fatigue of welded steel structures*. New York, NY: Welding Research Council (1964).

————. Notched bar impact testing of metallic materials, E 23. *Annual Book of ASTM Standards*. Philadelphia, PA: ASTM (published annually).

————. Recommended practice for laboratory immersion corrosion testing of metals, G 31. *Annual Book of ASTM Standards*. Philadelphia, PA: ASTM (published annually).

————. Recommended practice for preparing, cleaning, and evaluating corrosion test specimens, G. 1. *Annual Book of ASTM Standards*. Philadelphia, PA: ASTM (published annually).

————. Standard method for conducting creep, creep-rupture, and stress-rupture tests in metallic materials, E 139. *Annual Book of ASTM Standards* 0301, Philadelphia, PA: ASTM (published annually).

————. Standard method of test for plane-strain fracture toughness of metallic materials, E 399. *Annual Book of ASTM Standards*. Philadelphia, PA: ASTM (published annually).

————. Tension Testing of Metallic Materials, E 8. *Annual Book of ASTM Standards*. Philadelphia, PA: ASTM (published annually).

American Welding Society. *Recommended practices for resistance welding*, C1.1. Miami: American Welding Society: (latest edition).

————. *Specification for bare mild steel electrodes and fluxes for submerged arc welding*, A5.17. Miami: American Welding Society: (latest edition).

————. *Standard method for evaluating the strength of brazed joints*, AWS C3.2. Miami: American Welding Society: (latest edition).

————. *Standard methods for mechanical testing of welds*, B4.0. American Welding Society: (latest edition).

————. *Structural welding code—steel*, D1.1. Miami: American Welding Society: (latest edition).

Thielsch, H. Thermal fatigue and thermal shock. *Welding Research Council Bulletin*. 1952: No.10.

Vagi, J. J., Meister, R. P., and Randall, M. D. *Weldment evaluation methods*. DMIC Report 244, Columbus, OH: Battelle Memorial Institute (Aug. 1968).

Yen, T. C. Thermal fatigue—a critical review. *Welding Research Council Bulletin*. 1961: No. 72.

CODES AND OTHER STANDARDS

PREPARED BY A COMMITTEE CONSISTING OF:

G. N. Fischer, *Chairman*
Fischer Engineering Company

W. L. Ballis
Colombia Gas Distribution Company

H. W. Ebert
Exxon Research and Engineering

P. D. Flenner
Consumers Power Company

J. W. McGrew
Babcock and Wilcox Company

WELDING HANDBOOK COMMITTEE MEMBER:
M. Tomsic
Plastronics Incorporated

CHAPTER **13**

CODES AND OTHER STANDARDS

DEFINITIONS

THIS CHAPTER IS intended to familiarize fabricators and purchasers of welded products with the basic documents that govern or guide welding activities. These documents serve to assure that (1) only safe and reliable welded products will be produced, and (2) those persons associated with welding operations will not be exposed to undue danger or other conditions that would be harmful to their health. Publications relating only to the manufacture of welding materials or welding equipment are not covered in this chapter. However, those publications may be referenced in the basic documents, and their relationship to safety and reliability should not be underestimated.

The American Welding Society uses the general term *standards* to refer to documents that govern and guide welding activities. Standards describe the technical requirements for a material, process, product, system, or service. They also indicate the procedures, methods, equipment, or tests used to determine that the requirements have been met.

Standards include codes, specifications, recommended practices, classifications, methods, and guides. These documents have many similarities, and the terms are often used interchangeably, but sometimes incorrectly.

Codes and *specifications* are similar types of standards that use the words *shall* and *will* to indicate the mandatory use of certain materials or actions, or both. Codes differ from specifications in that their use is generally applicable to a process. Specifications are generally associated with a product. Both become mandatory when specified by one or more governmental jurisdic-

tions or when they are referenced by contractural or other procurement documents.

Recommended practices and *Guides* are standards that are offered primarily as aids to the user. They use words such as *should* and *may* because their use is usually optional. However, if these documents are referenced by codes or contractural agreements, their use may become mandatory. If the codes or agreements contain non-mandatory sections or appendices, the use of referenced guides or recommended practices is at the user's discretion.

Classifications and *methods* generally provide lists of established practices or categories for processes or products. The most common example is a standard testing method.

The user of a standard should become acquainted with its scope and intended use, both of which are usually included within the *Scope* or *Introduction* section of the standard. It is equally important, but often more difficult, to recognize subjects that are not covered by the document. These omissions may require additional technical consideration. A document may cover the details of the product form without considering special conditions under which it will be used. Examples of special conditions would be corrosive atmospheres, elevated temperatures, and dynamic rather than static loading.

Standards vary in their method of achieving compliance. Some have specific requirements that do not allow for alternative actions. Others permit alternative actions or procedures, so long as they result in properties that meet specified criteria. These criteria are often given as minimum requirements; for example, the ultimate tensile strength of a welded specimen must meet or exceed the minimum tensile strength specified for the base material.

SOURCES

PRIVATE AND GOVERNMENTAL organizations develop, issue, and update standards that apply to their particular areas of interest. Table 13.1 lists those organizations of concern to the welding industry and their current addresses. The interests of many of these groups overlap with regard to welding, and some agreements have been made to reduce duplication of effort. Many standards that are concerned with welding, brazing, and allied processes are prepared by the American Welding Society (AWS) because these subjects are of primary interest to the members. Standards that apply to a particular product are usually prepared by the group that has overall responsibility. For example, those for railroad freight cars are published by the Association of American Railroads (AAR). However, freight cars are basically structures, and the applicable AAR specification currently refers to *ANSI/AWS D1.1, Structural Welding Code— Steel* for the qualification of welding procedures, welders, and welding operators. In 1986, the American Welding Society published *ANSI/AWS D15.1, Railroad Welding Specification*. Future revisions to the AAR standards will reference ANSI/AWS D15.1.

Each organization that prepares consensus standards has committees or task groups to perform this function. Members of these committees or task groups are specialists in their fields. They prepare drafts of standards that are reviewed and approved by a larger group. Each main committee is selected to include persons with diverse interests from producers, users, and government representatives. To avoid control or undue influence by one interest group, consensus must be achieved by a high percentage of all members.

The federal government develops or adopts standards for items and services that are in the public rather than the private domain. The mechanisms for developing federal or military documents are similar to those of private organizations. Standard-writing committees usually exist within a federal department or agency that has responsibility for a particular item or service.

The American National Standards Institute (ANSI) is a private organization responsible for coordinating national standards for use within the United States. ANSI does not actually prepare standards. Instead, it forms national interest review groups to determine whether proposed standards are in the public interest. Each group is composed of persons from various organizations concerned with the scope and provisions of a particular document. If there is consensus regarding the general value of a particular standard, then it may be adopted as an American National Standard. Adoption of a standard by ANSI does not, of itself, give it mandatory status. However, if the standard is cited by a governmental rule or regulation, it may then be backed by force of law.

Other industrial countries also develop and issue standards on the subject of welding. The following are examples of other national standards designations and the bodies responsible for them:

BS —British Standard issued by the British Standards Association

CSA—Canadian Standard issued by the Canadian Standards Association

DIN—West German Standard issued by the Deutsches Institute fuer Normung

JIS —Japanese Industrial Standard issued by the Japanese Standards Association

NF —French Standard issued by the Association Francaise de Normalisation

Of these, the Canadian Standards Association is discussed in a following section.

There is also an International Organization for Standardization (ISO). Its goal is the establishment of uniform standards for use in international trade. This organization is discussed at more length in a following section.

APPLICATIONS

THE MINIMUM REQUIREMENTS of a particular standard may not satisfy the special needs of every user. Therefore, a user may find it necessary to invoke additional requirements to obtain desired quality.

There are various mechanisms by which most standards may be revised. These are used when a standard is found to be in error, unreasonably restrictive, or not applicable with respect to new technological developments. Some standards are updated on a regular basis, while others are revised as needed. The revisions may be in the form of addenda, or they may be incorporated in superseding documents.

If there is a question about a particular standard regarding either an interpretation or a possible error, the user should contact the responsible organization.

When the use of a standard is mandatory, whether as a result of a government regulation or a legal contract, it is essential to know the particular edition of the document to be used. It is unfortunate, but not uncommon, to find that an outdated edition of a referenced document has been specified, and must be followed to be in compliance. If there is a question concerning which edition or revision of a document is to be used, it should be resolved before commencement of work.

Table 13.1
Sources of Standards of Interest to the Welding Industry

American Association of State Highway and Transportation Officials
(AASHTO)
    444 N. Capital St., N.W.
    Washington, DC 20001
    (202) 624-5800

American Bureau of Shipping (ABS)
    45 Eisenhower Drive
    Paramus, New Jersey 07652
    (201) 368-9100

American Institute of Steel Construction (AISC)
    400 N. Michigan Ave
    Chicago, Illinois 60611
    (312) 670-2400

American National Standards Institute (ANSI)
    1430 Broadway New York, New York 10018
    (212) 354-3300

American Petroleum Institute (API)
    1220 L Street, N.W.
    Washington, DC 20005
    (202) 682-8000

American Railway Engineering Association (AREA)
    50 F Street, N.W.
    Washington, DC 20001
    (202) 639-2190

American Society of Mechanical Engineers (ASME)
    345 East 47th Street
    New York, New York 10017
    (212) 705-7722

American Water Works Association (AWWA)
    6666 W. Quincy Avenue
    Denver, Colorado 80235
    (303) 794-7711

American Welding Society (AWS)
    550 N.W. LeJeune Road
    P. O. Box 351040
    Miami, Florida 33135
    (305) 443-9353

Association of American Railroads (AAR)
    50 F Street, N.W.
    Washington, DC 20001
    (202) 635-2100

ASTM
    1916 Race Street
    Philadelphia, Pennsylvania 19103
    (215) 299-5400

Canadian Standards Association (CSA)
    178 Rexdale Boulevard
    Rexdale, Ontario
    Canada M9W 1R3
    (416) 744-4000

Compressed Gas Association (CGA)
    1235 Jeff Davis Highway
    Arlington, Virginia 22202
    (703) 979-0900

International Organization for Standardization (ISO)
    (See American National Standards Institute)

National Board of Boiler and Pressure Vessel Inspectors (NBBPVI)
    1055 Crupper Avenue
    Columbus, Ohio 43229
    (614) 888-8320

National Fire Protection Association (NFPA)
    Batterymarch Park
    Quincy, Massachusetts 02269
    (617) 770-3000

Naval Publication and Forms Center[1]
    5801 Taber Avenue
    Philadelphia, Pennsylvania 19120
    (215) 697-2000

Pipe Fabrication Institute (PFI)
    612 Lenore Avenue
    Springdale, Pennsylvania
    (412) 274-4722

SAE
    400 Commonwealth Drive
    Warrendale, Pennsylvania 15096
    (412) 776-4841

Superintendent of Documents[2]
    U.S. Government Printing Office
    Washington, DC 20402
    (202) 783-3238

Underwriters Laboratories, Inc.
    333 Pfingsten Road
    Northbrook, Illinois 60062
    (312) 272-8800

Uniform Boiler and Pressure Vessel Laws Society (UBPVLS)
    2838 Long Beach Road
    Oceanside, New York 11572
    (516) 536-5485

---

1. Military Specifications
2. Federal Specifications

Organizations responsible for preparing standards that relate to welding are discussed in the following sections. The publications are listed without reference to date of publication, latest revision, or amendment. New publications relating to welding may be issued, and current ones may be withdrawn or revised. The responsible organization should be contacted for current information on the standards it publishes.

Some organizations cover many product categories while others may cover only one. Table 13.2 lists the organizations and the product categories covered by their documents. The National Fire Protection Association is not listed in the Table because its standards are concerned with safe practices and equipment rather than with products. The American Welding Society and the American Petroleum Institute also publish standards concerned with welding safety.

AMERICAN ASSOCIATION OF STATE HIGHWAY AND TRANSPORTATION OFFICIALS

THE MEMBER AGENCIES of the American Association of State Highway and Transportation Officials (AASHTO) are the U.S. Department of Transportation, and the Departments of Transportation and Highways of the fifty states, Washington DC, and Puerto Rico. AASHTO specifications are prepared by committees made up of employees of the member agencies. These documents are the minimum rules to be followed by all member agencies or others in the design and construction of highway bridges.

Standard Specifications for Highway Bridges

THIS AASHTO SPECIFICATION covers the design and construction requirements for all types of highway bridges. It refers to the welding fabrication requirements in the *AASHTO* standard, *Specifications for Welding of Structural Steel Highway Bridges* and the *ANSI/AWS D1.1, Structural Welding Code—Steel.*

Standard Specifications for Welding of Structural Steel Highway Bridges

THIS AASHTO SPECIFICATION provides modifications to the *ANSI/AWS D1.1, Structural Welding Code—Steel* deemed necessary for use by member agencies. These are referenced to the applicable sections of the AWS Code.

Guide Specifications for Fracture Critical Non-Redundant Steel Bridge Members

FRACTURE CRITICAL MEMBERS or components of a bridge are tension members or components, the failure of which would likely result in collapse of the structure. The document assigns the responsibility for specifying those bridge members or components, if any, that fall

Table 13.2
Products Covered by Standards of Various Organizations

Product	AAR	AASHTO	ABS	AISC	API	AREA	ASME	NBBPVI UBPVLS	ASTM	AWS	AWWA	FED	PFI	SAE	UL
Base metals			X		X		X		X			X		X	
Bridges		X		X	X					X		X			
Buildings				X						X					
Construction equipment										X		X		X	
Cranes, hoists							X			X					
Elevators, escalators							X								
Filler metals			X				X			X		X		X	
Food, drug equipment							X								
Machine tools										X					
Military equipment												X			
Power generation equipment			X				X		X			X			
Piping			X		X		X			X	X	X	X		
Presses										X					
Pressure vessels, boilers			X		X		X	X							
Railway equipment	X					X				X					
Sheet metal fabrication										X					
Ships			X							X		X			
Storage tanks					X					X	X				X
Structures, general				X						X					
Vehicles										X		X		X	

into the fracture-critical category. It requires that such members or components be fabricated to the required workmanship standards only by organizations having the proper personnel, experience, procedures, knowledge, and equipemnt. For example, all welding inspectors and nondestructive testing personnel must have demonstrated competency for assuring quality in compliance with the governing specifications. The document also contains requirements additional to those in the *Standard Specifications for Welding of Structural Steel Highway Bridges*.

AMERICAN BUREAU OF SHIPPING

THE FUNCTION OF the American Bureau of Shipping (ABS) is to control the quality of ship construction. Each year, ABS reissues the *Rules for Building and Classing Steel Vessels*. These rules are applicable to ships that are intended to have American registration.

To obtain American registration and insurance, a ship must be classed (approved) by ABS after inspections and reviews by its surveyors (inspectors). The surveys begin with a review of the proposed design. Reviews are also made during and after construction to verify that construction complies with the ABS rules. The process is completed with the assignment and registration of a class (numerical identification) for the ship.

One section of the ABS Rules addresses welding and is divided into the following parts:

Part 1—Hull Construction
Part 2—Boilers, Unfired Pressure Vessels, Piping, and Engineering Structures
Part 3—Weld Tests

The section addresses such topics as weld design, welding procedures, qualification testing, preparation for welding, production welding, workmanship, and inspection.

ABS also publishes a list of welding consumables, entitled *Approved Welding Electrodes, Wire-Flux, and Wire-Gas Combinations*. These consumables are produced by various manufacturers around the world. They are tested under ABS supervision and approved for use under the ABS rules.

AMERICAN INSTITUTE OF STEEL CONSTRUCTION

THE AMERICAN INSTITUTE OF STEEL CONSTRUCTION (AISC) is a non-profit trade organization for the fabricated structural steel industry in the United States. The Institute's objectives are to improve and advance the use of fabricated structural steel through research and engineering studies, and to develop the most efficient and economical design of structures. The organization also conducts programs to improve and control product quality.

Manual of Steel Construction

THE FIRST FOUR parts of the manual cover such topics as (1) the dimensions and properties of rolled structural steel shapes, (2) beam, girder, and column design, and, (3) welded connection design. Part 5 of the manual is the "Specification for the Design Fabrication and Erection of Structural Steel for Buildings." This part includes certain aspects of structural steel design that are not included in other parts of the manual.

Specification for the Design, Fabrication, and Erection of Structural Steel for Buildings

THIS DOCUMENT SPECIFIES, n detail, all principal steps required for the construction of structural steel buildings. It references the AWS filler metal specifications, and specifies the particular filler metal classification to be used with a welding process for each type of structural steel. Requirements for the types and details of fillet, plug, and slot welds are also included. The specification refers to *ANSI/AWS D1.1, Structural Welding Code—Steel* for welding procedure and welder performance qualifications.

Quality Criteria and Inspection Standards

THIS DOCUMENT COVERS such subjects as preparation of materials, fitting and fastening, dimensional tolerances, welding, surface preparation, and painting. It discusses the practical implementation of some of the requirements of other AISC specifications. Typical problems that may be encountered in steel construction and recommended solutions are presented. The welding section provides interpretations regarding AISC requirements for prequalification of welding procedures, preheating, control of distortion, and tack welding.

AMERICAN NATIONAL STANDARDS INSTITUTE

AS MENTIONED PREVIOUSLY, the American National Standards Institute (ANSI) is the coordinating organizaion for the U.S. voluntary standards system; it does not develop standards directly. The Institute provides the means for determining the need for standards, and ensures that organizations competent to fill these needs undertake the development work. The approval procedures of ANSI ensure that all interested persons have an opportunity to participate in the development of a standard or to comment on provisions of the standard prior to publication. ANSI is the U.S. member of non-treaty international standards organizations, such as the International Organization for Standardization (ISO), and the International Electrotechnical Commission (IEC).

American National Standards, which now number approximately 10 000 documents, encompass virtually every field and every discipline. They deal with dimensions, ratings, terminology and symbols, test methods, and performance and safety specifications for materials, equipment, components, and products in some two dozen fields. These fields include construction; electrical and electronics; heating, air conditioning, and refrigeration; information systems; medical devices; mechanical; nuclear; physical distribution; piping and processing; photography and motion pictures; textiles; and welding.

The American National Standards provide a common language that can be used confidently by industry, suppliers, customers, business, the public, government, and labor. Each of these interests has either participated in the development of the standards or has been given the opportunity to comment on their provisions. However, these standards are developed and used voluntarily. They become mandatory only when they are adopted or referenced by a governmental body.

The ANSI federation consists of companies, large and small, and trade, technical, professional, labor, and consumer organizations that total over 1000 members. The standards are primarily developed by the member organizations.

AMERICAN PETROLEUM INSTITUTE

THE AMERICAN PETROLEUM INSTITUTE (API) publishes documents in all areas related to petroleum production. Those documents that include welding requirements are discussed below.

Pipelines and Refinery Equipment

The Guide for Inspection of Refinery Equipment. This publication consists of 20 chapters and an appendix, all of which may be purchased separately from API. The appendix, entitled "Inspection of Welding," is the only part that applies specifically to welding. Its objective is to guide the user in determining whether welded joints are of acceptable quality and comply with both the requirements of the contract or job specifications and the prescribed welding procedure specifications.

Recommended Pipeline Maintenance Welding Practices, RP 1107. The primary purpose of this recommended practice is safety. It prohibits practices that are known to be unsafe, and also warns against practices for which caution is necessary. It also includes 18 methods for the inspection of repair welds, and for installing appurtenances on loaded piping systems being used for the transmission of natural gas, crude petroleum, and petroleum products.

The legal authority for RP 1107 comes from reference to it in *ASME B31.4, Liquified Petroleum Transporta-*

tion Piping Systems (described under the American Society of Mechanical Engineers). The latter publication, like *API Std 1104, Standard for Welding Pipelines and Related Facilities*, is also referenced by *Title 49, Part 195, Transportation of Liquids by Pipeline, of the United States Code of Federal Regulations.*

Recommended Practice for Welded, Plain Carbon Steel Refinery Equipment for Environmental Cracking Service, Publ 942. This publication proposes actions for protection against hydrogen stress cracking of welds in plain carbon steel that are exposed, under stress, to certain aqueous-phase acidic environments, such as moist hydrogen sulfide.

Standard for welding Pipelines and Related Facilities, API Std 1104. This standard applies to arc and oxyfuel gas welding of piping used in the compression, pumping, and transmission of crude petroleum, petroleum products, and fuel gases, and also to the distribution systems when applicable. It presents methods for the production of acceptable welds by qualified welders using approved welding procedures, materials, and equipment. It also presents methods for the production of suitable radiographs by qualified technicians using approved procedures and equipment, to ensure proper analysis of weld quality. Standards of acceptability and repair of weld defects are also included.

The legal authority for the use of API Std 1104 comes from *Title 49, Part 195, Transportation of Liquids by Pipeline*, of the *United States Code of Federal Regulations*.

Storage Tanks for Refinery Service

Inspection, Rating, and Repair of Pressure Vessels in Petroleum Refinery Service, RP 510. This recommended practice covers the inspection, repair, evaluation for continued use, and methods for computing the maximum allowable working pressure of existing pressure vessels. The vessels include those constructed in accordance with *Section VIII of the ASME Boiler and Pressure Vessel Code* or other pressure vessel codes.

Recommended Rules for Design and Construction of Large, Welded, Low-Pressure Storage Tanks, Std 620. These rules cover the design and construction of large, field-welded tanks that are used for storage of petroleum intermediates and finished products under pressure of 15 psig (103 kPa) and less.

Welded Steel Tanks for Oil Storage, Std 650. This standard covers the material, design, fabrication, erection, and testing requirements for vertical, cylindrical, welded steel storage tanks that are above ground and not subject to internal pressure.

Safety and Fire Protection

Repairs to Crude Oil, Liquified Petroleum Gas, and Products Pipelines, PSD 2200. This petroleum safety data sheet is a guide to safe practices for the repair of pipelines for crude oil, liquified petroleum gas, and petroleum products.

Safe Practices in Gas and Electric Cutting and Welding in Refineries, Gasoline Plants, Cycling Plants, and Petrochemical Plants, Publ 2009. This publication outlines precautions for protecting persons from injury and property from damage by fire that might result during the operation of oxyfuel gas and electric cutting and welding equipment in and around petroleum operations.

Welding or Hot Tapping on Equipment Containing Flammables, PSD 2201. This petroleum safety data sheet lists procedures for welding, as well as for making hot taps (connections while in operation), on pipelines, vessels, or tanks containing flammables. This data sheet and PSD 2200 are also requirements of *ASME B31.4, Liquified Petroleum Transportation Piping Systems.*

AMERICAN RAILWAY ENGINEERING ASSOCIATION

THE AMERICAN RAILWAY ENGINEERING ASSOCIATION (AREA) publishes the *Manual for Railway Engineering.* This manual contains specifications, rules, plans, and instructions that constitute the recommended practices of railway engineering. Two chapters specifically cover steel construction. One of these covers the design, fabrication, and erection of buildings for railway purposes. The other addresses the same topics for railway bridges and miscellaneous steel structures.

AMERICAN SOCIETY OF MECHANICAL ENGINEERS

TWO STANDING COMMITTEES of the American Society of Mechanical Engineers (ASME) are actively involved in the formulation, revision, and interpretation of standards covering products that may be fabricated by welding. These committees are responsible for preparing the *ASME Boiler and Pressure Vessel Code* and the *Code for Pressure Piping*, which are American National Standards.

Boiler and Pressure Vessel Code

THE *ASME BOILER AND PRESSURE VESSEL CODE* (BPVC) contains eleven sections. Sections I, III, IV, VIII, and X cover the design, construction, and inspection of boilers and pressure vessels. Sections VI, VII, and XI cover the care and operation of boilers or nuclear power plant components. The remaining Sections II, V, and IX cover material specifications, nondestructive examination, and welding and brazing qualifications, respectively.

Section I, Power Boilers, covers power, electric, and miniature boilers; high temperature boilers used in stationary service; and power boilers used in locomotive, portable, and traction service. Section III, Nuclear Power Plant Components, addresses the various components required by the nuclear power industry. Section IV, Heating Boilers, applies to steam heat and hot water supply boilers that are directly fired by oil, gas, electricity, or coal. Section VIII, Pressure Vessels, covers unfired pressure vessels. Unfired pressure vessels are containers for the containment of pressure either internal or external. All Code vessels not covered by Sections I, III, and IV are covered by Section VIII. These include towers, reactors and other oil and chemical refining vessels, heat exchangers for refineries, paper mills, and other process industries, as well as storage tanks for large and small air and gas compressors.

Section II, Material Specifications, contains the specifications for acceptable ferrous and nonferrous base metals, and for acceptable welding and brazing filler metals and fluxes. Many of these specifications are identical to and have the same numerical designation as ASTM and AWS specifications for base metals and welding comsumables, respectively. Section V, Nondestructive Examination, covers methods and standards for nondestructive examination of boilers and pressure vessels. Section IX, Welding and Brazing Qualifications, covers the qualification of (1) welders, welding operators, brazers, and brazing operators, and (2) the welding and brazing procedures that are to be employed for welding or brazing of boilers or pressure vessels. This section of the Code is often cited by other standards and regulatory bodies as the welding and brazing qualification standard for other types of welded or brazed products.

The *ASME Boiler and Pressure Vessel Code* is referenced in the safety regulations of most states and major cities of the United States, and also the provinces of Canada. A number of federal agencies include the Code as part of their respective regulations.

The Uniform Boiler and Pressure Vessel Laws Society (UBPVLS) has, as its objective, uniformity of laws, rules, and regulations that affect boiler and pressure vessel fabricators, inspection agencies, and users. The Society believes that such laws, rules, and regulations should follow nationally accepted standards. It recommends the *ASME Boiler and Pressure Vessel Code* as the standard for construction and the *Inspection Code* of the National Board of Boiler and Pressure Inspectors (NBBPVI), discussed in a following section, as the standard for inspection and repair.

The *ASME Boiler and Pressure Vessel Code* is unique in that it requires third party inspection independent of the fabricator and the user. The NBBPVI commissions inspectors by examination. These inspectors are employed either by authorized inspection agencies (usually insurance companies) or by jurisdictional authorities.

Prior to building a boiler or pressure vessel, a company must have a quality control system and a manual that describes it. The system must be acceptable to the authorized inspection agency and either the jurisdictional authority or the NBBPVI. Based on the results of an audit of the fabricator's quality system, ASME may issue the fabricator a Certificate of Authorization and a code symbol stamp. The authorized inspection agency is also involved in monitoring the fabrication and field erection of boilers and pressure vessels. An authorized inspector must be satisfied that all applicable provisions of the *ASME Boiler and Pressure Vessel Code* have been followed before allowing the fabricator to apply its code symbol stamp to the vessel name plate.

Code for Pressure Piping

THE ASME B31, CODE FOR PRESSURE PIPING, presently consists of seven sections. Each section prescribes the minimum requirements for the design, materials, fabrication, erection, testing, and inspection of a particular type of piping system.

B31.1, Power Piping, covers power and auxiliary service systems for electric generation stations; industrial and institutional plants; central and district heating plants; and district heating systems.

This section excludes boiler external piping which is defined by Section I of the *ASME Boiler and Pressure Vessel Code.* Boiler piping requires a quality control system and third party inspection similar to those required for boiler fabrication. Otherwise, the materials, design, fabrication, installation, and testing for boiler external piping must meet the requirements of section B31.1. A fabricator is not required to provide a quality control system and third party inspection for the other piping systems covered by B31.

B31.2, Fuel Gas Piping, covers piping systems for fuel gases including natural gas, manufactured gas, liquefied petroleum gas (LPG) and air mixtures above the upper combustible limits, LPG in the gaseous phase, or mixtures of these gases. These piping systems, both in and between buildings, extend from the outlet of the consumer's meter set assembly (or point of delivery) to and including the first pressure-containing valve upstream of the gas utilization device.

B31.3, Chemical Plant and Petroleum Refinery Piping, covers all piping within the property limits of facilities engaged in processing or handling of chemical, petroleum, or related products. Examples are chemical plants, petroleum refineries, loading terminals, natural gas processing plants (including liquefied natural gas facilities), bulk plants, compounding plants, and tank farms. This section applies to piping systems that handle all fluids, including fluidized solids, and to all types of service including raw, intermediate, and finished chemicals; oil and other petroleum products; gas; steam; air; water; and refrigerants, except as specifically excluded.

Piping for air and other gases, which is not now within the scope of existing sections of this code, may be designed, fabricated, inspected, and tested in accordance with the requirements of this section of the Code. The piping must be in plants, buildings, and similar facilities that are not otherwise within the scope of this section.

B31.4, Liquid Petroleum Transportation Piping Systems, covers piping for transporting liquid petroleum products between producers' lease facilities, tank farms, natural gas processing plants, refineries, stations, terminals, and other delivery and receiving points. Examples of such products are crude oil, condensate, gasoline, natural gas liquids, and liquefied petroleum gas.

B31.5, Refrigeration Piping, applies to refrigerant and brine piping for use at temperatures as low as $-320°F$, $(-196°C)$ whether erected on the premises or factory assembled. It does not include (1) self-contained or unit refrigeration systems subject to the requirements of Underwriters Laboratories or any other nationally recognized testing laboratory, (2) waterpiping, or (3) piping designed for external or internal pressure not exceeding 15 psig (103 kPa), regardless of size. Other sections of the Code may provide requirements for refrigeration piping in their respective scopes.

B31.8, Gas Transmission and Distribution Piping Systems, addresses gas compressor stations, gas metering and regulating stations, gas mains, and service lines up to the outlet of the customer's meter set assembly. Gas storage lines and gas storage equipment of the closed-pipe type that are either fabricated or forged from pipe or fabricated from pipe and fittings are also included.

B31.9, Building Services Piping, applies to piping systems for services in industrial, commercial, public, institutional and multi-unit residential buildings. It includes only those piping systems within the buildings or property limit.

All sections of the *Code for Pressure Piping* require qualification of the welding procedures and performance of welders and welding operators to be used in construction. Some sections require these qualifications to be performed in accordance with Section IX of the *ASME Boiler and Pressure Vessel Code,* while in others it is optional. The use of *API Std 1104, Standard for Welding Pipelines and Related Facilities* or *AWS D10.9, Specification for Qualification of Welding Procedures and Welders for Piping and Tubing* is permitted in some sections as an alternative to Section IX. Each section of the Code should be consulted for the applicable qualification documents.

ASTM

ASTM (FORMERLY THE AMERICAN SOCIETY FOR TESTING AND MATERIALS) develops and publishes specifications for use in the production and testing of materials. The committees that develop the specifications are comprised of producers and users, as well as others who have an interest in the subject materials. The specifications cover virtually all materials used in industry and commerce with the exception of welding consumables, which are covered by AWS specifications.

ASTM publishes an *Annual Book of ASTM Standards* that incorporates new and revised standards. It is currently composed of 15 sections comprising 65 volumes and an index. Specifications for the metal products, test methods, and analytical procedures of interest to the welding industry are found in the first three sections, comprising 17 volumes. Section 1 covers iron and steel products; Section 2, nonferrous metal products; and Section 3, metal test methods and analytical procedures. Copies of single specifications are also available from ASTM.

Prefix letters, which are part of each specification's alpha-numeric designation, provide a general idea of the specification content. They include A for ferrous metals, B for nonferrous metals, and E for miscellaneous subjects including examination and testing. When ASME adopts an ASTM specification for certain applications, either in its entirety or in a revised form, it adds an "S" in front of the ASTM letter prefix.

Many ASTM specifications include supplementary requirements that must be specified by the purchaser if they are desired. These may include vacuum treatment, additional tension tests, impact tests, or ultrasonic examination.

The producer of a material or product is responsible for compliance with all mandatory and specified supplementary requirements of the appropriate ASTM specification. The user of the material is responsible for verifying that the producer has complied with all requirements.

Some codes permit the user to perform the tests required by ASTM or other specification to verify that a material meets requirements. If the results of the tests conform to the requirements of the designated specification, the material can be used for the application.

Some products covered by ASTM specifications are fabricated by welding. The largest group is steel pipe and tubing. Some types of pipe are produced from strip by rolling and arc welding the longitudinal seam. The welding procedures generally must be qualified to the requirements of the *ASME Boiler and Pressure Vessel Code* or another code.

Other types of pipe and tubing are produced with resistance welded seams. There are generally no specific welding requirements in the applicable ASTM specification. The finished pipe and tubing must pass specific tests that should result in failure at the welded seam if the welding operation is out of control.

Two ASTM specifications cover joints in piping systems. These are *ASTM A422, Standard Specification for Butt Welds in Still Tubes for Refinery Service* and *ASTM F722, Standard Specification for Welded Joints for Shipboard Piping Systems*.

ASTM E190, Guided Bend Test for Ductility of Welds, is presently the only ASTM testing specification that is solely intended for welds.

AMERICAN WATER WORKS ASSOCIATION

THE AMERICAN WATER WORKS ASSOCIATION (AWWA) currently has two standards that pertain to the welding of water storage and transmission systems. One of these standards was developed jointly with and adopted by the American Welding Society.

Standard for Field Welding of Steel Water Pipe Joints C206

THIS STANDARD COVERS field welding of steel water pipe. It includes the welding of circumferential pipe joints as well as other welding required in the fabrication and installation of specials and accessories. The maximum wall thickness of pipe covered by this standard is 1.25 in. (31.8 mm).

Standard for Welded Steel Elevated Tanks, Standpipes, and Reservoirs for Water Storage, D100 (AWS D5.2)

THIS STANDARD COVERS the fabrication of water storage tanks. An elevated tank is one supported on a tower. A standpipe is a flat-bottomed cylindrical tank having a shell height greater than its diameter. A reservoir is a flat-bottomed cylindrical tank having a shell height equal to or smaller than its diameter. In addition to welding details, this standard specifies the responsibilities of the purchaser and the contractor for such items as the foundation plans, the foundation itself, water for pressure testing, and a suitable right-of-way from the nearest public road for on-site erection.

AMERICAN WELDING SOCIETY

THE AMERICAN WELDING SOCIETY (AWS) publishes numerous documents covering the use and quality control of welding. These documents include codes, specifications, recommended practices, classifications, methods, and guides. The general subject areas covered are

(1) Definitions and symbols
(2) Filler metals
(3) Qualification and testing
(4) Welding processes
(5) Welding applications
(6) Safety

Definitions and Symbols

ANSI/AWS A2.4, Symbols for Welding, Brazing, and Nondestructive Examination. This publication describes the standard symbols used to convey welding, brazing, and nondestructive testing requirements on drawings. Symbols in this publication are intended to facilitate communications between designers and fabrication personnel. Typical information that can be conveyed with welding symbols includes type of weld, joint geometry, weld size or effective throat, extent of welding, and contour and surface finish of the weld.

ANSI/AWS A3.0, Standard Welding Terms and Definitions. This publication lists and defines the standard terms that should be used in oral and written communications conveying welding, brazing, soldering, thermal spraying, and thermal cutting information. Nonstandard terms are also included; these are defined by reference to the standard terms. A glossary of terms and definitions from this publication appears in Appendix A of this volume.

Filler Metals

THE AWS FILLER metal specifications cover most types of consumables used with the various welding and brazing processes. The specifications include both mandatory and nonmandatory provisions. The mandatory provisions cover such subjects as chemical or mechanical properties, or both, manufacture, testing, and packaging. The nonmandatory provisions, included in an appendix, are provided as a source of information for the user on the classification, description, and intended use of the filler metals covered.

Following is a current listing of AWS filler metal specifications.

Aluminum and Aluminum Alloy Bare Welding Rods and Electrodes, Specification for, ANSI/AWS A5.10
Aluminum and Aluminum Alloy Covered Arc Welding Electrodes, Specification for, ANSI/AWS A5.3
Brazing Filler Metal, Specification for, ANSI/AWS A5.8

Composite Surfacing Welding Rods and Electrodes, Specification for, ANSI/AWS A5.21
Carbon Steel Electrodes for Flux Cored Arc Welding, Specification for ANSI/AWS A5.20
Carbon Steel Electrodes and Fluxes for Submerged Arc Welding, Specification for, ANSI/AWS A5.17
Carbon Steel Filler Metals for Gas Shielded Arc Welding, Specification for, ANSI/AWS A5.18
Consumable Inserts, Specification for, ANSI/AWS A5.30
Consumables Used for Electrogas Welding of Carbon and High Strength Low Alloy Steels, Specification for, ANSI/AWS A5.26
Consumables Used for Electroslag Welding of Carbon and High Strength Low Alloy Steels, Specification for, ANSI/AWS A5.25
Copper and Copper Alloy Bare Welding Rods and Electrodes, Specification for, ANSI/AWS A5.7
Copper and Copper Alloy Rods for Oxyfuel Gas Welding, Specification for, ANSI/AWS A5.27
Corrosion-Resisting Chromium and Chromium-Nickel Steel Bare and Composite Metal Cored and Stranded Arc Welding Electrodes and Welding Rods, Specification for, ANSI/AWS A5.9
Covered Carbon Steel Arc Welding Electrodes, Specification for, ANSI/AWS A5.1
Covered Copper and Copper Alloy Arc Welding Electrodes, Specification for, ANSI/AWS A5.6
Covered Corrosion-Resisting Chromium and Chromium-Nickel Steel Welding Electrodes, Specification for, ANSI/AWS A5.4
Flux-Cored Corrosion-Resisting Chromium and Chromium-Nickel Steel Electrodes, Specification for, ANSI/AWS A5.22
Iron and Steel Oxyfuel Gas Welding Rods, Specification for, ANSI/AWS A5.2
Low Alloy Steel Covered Arc Welding Electrodes, Specification for, ANSI/AWS A5.5
Low Alloy Steel Electrodes and Fluxes for Submerged Arc Welding, Specification for, ANSI/AWS A5.23
Low Alloy Steel Electrodes for Flux Cored Arc Welding, Specification for, ANSI/AWS A5.29
Low Alloy Steel Filler Metals for Gas Shielded Arc Welding, Specification for, ANSI/AWS A5.28
Magnesium Alloy Welding Rods and Bare Electrodes, Specification for, ANSI/AWS A5.19
Nickel and Nickel Alloy Bare Welding Rods and Electrodes, Specification for, ANSI/AWS A5.14
Nickel and Nickel Alloy Covered Welding Electrodes, Specification for, ANSI/AWS A5.11
Solid Surfacing Welding Rods and Electrodes, Specification for, ANSI/AWS A5.13
Titanium and Titanium Alloy Bare Welding Rods and Electrodes, Specification for, ANSI/AWS A5.16
Tungsten Arc Welding Electrodes, Specification for, ANSI/AWS A5.12

Welding Rods and Covered Electrodes for Cast Iron, Specification for, ANSI/AWS A5.15
Zirconium and Zirconium Alloy Bare Welding Rods and Electrodes, Specification for, ANSI/AWS A5.24

Most AWS filler metal specifications have been approved by ANSI as American National Standards and adopted by ASME. When ASME adopts an AWS filler metal specification, either in its entirety or with revisions, it adds the letters "SF" to the AWS alphanumeric designation. Thus, ASME SFA-5.4 specification would be similar, if not identical, to the AWS A5.4 specification.

AWS also publishes the following documents to aid users with the purchase of filler metals:

AWS A5.01, Filler Metal Procurement Guidelines provides methods for identification of filler metal components, classification of lots of filler metals, and specification of the testing schedule in procurement documents.

The *Filler Metal Comparison Charts*, provide lists of manufacturers that supply filler metals in accordance with the various AWS specifications and the brand names. Conversely, the AWS specification, classification, and manufacturer of a filler metal can be determined from the brand name.

Qualification and Testing

AWS C2.16, Guide for Thermal Spray Operator and Equipment Qualification. The guide provides for the qualification of operators and equipment for applying thermal sprayed coatings. It recommends procedural guidelines for qualification testing. The criteria used to judge acceptability are determined by the certifying agent alone or together with the purchaser.

AWS D10.9, Specification for Qualification of Welding Procedures and Welders for Piping and Tubing. This standard applies specifically to qualifications for tubular products. It covers circumferential groove and fillet welds but excludes welded longitudinal seams involved in pipe and tube manufacture. An organizaiton may make this specification the governing document for qualifying welding procedures and welders by referencing it in the contract and by specifying one of the two levels of acceptance requirements. One level applies to systems that require a high degree of weld quality. Examples are lines in nuclear, chemical, cryogenic, gas, or steam systems. The other level applies to systems requiring an average degree of weld quality, such as low-pressure heating, air-conditioning, sanitary water, and some gas or chemical systems.

AWS B2.2, Standard for Brazing Procedure and Performance Qualification. The requirements for qualification of brazing procedures, brazers, and brazing operators for furnace, machine, and automatic brazing are covered by this publication. It is to be used when required by other documents, such as codes, specifications, or contracts. Those documents must specify certain requirements applicable to the production brazement. Applicable base metals are carbon and alloy steels, cast iron, aluminum, copper, nickel, titanium, zirconium, magnesium, and cobalt alloys.

AWS B2.1, Standard for Welding Procedure and Performance Qualification. This standard provides requirements for qualification of welding procedures, welders, and welding operators. It may be referenced in a product code, specification, or contract documents. If a contract document is not specific, certain additional requirements must also be specified, as listed in this standard. Applicable base metals are carbon and alloy steels, cast irons, aluminum, copper, nickel, and titanium alloys.

ANSI/AWS C3.2, Standard Method for Evaluating the Strength of Brazed Joints in Shear. This standard describes a test method used to obtain reliable shear strengths of brazed joints. For comparison purposes, specimen preparation, brazing practices, and testing, procedures must be consistent. Production brazed joint strength may not be the same as test joint strength if the brazing practices are different. With furnace brazing, for example, the actual part temperature or time at temperature, or both, during production may vary from those used to determine joint strength.

ANSI/AWS B4.0, Standard Methods for Mechanical Testing of Welds. This document describes the basic mechanical tests used for evalution of welded joints, weldability, and hot cracking. The tests applicable to welded butt joints are tension, Charpy impact, drop-weight, dynamic-tear, and bend types. Tests of fillet welds are limited to break and shear tests.

For welding materials and procedure qualifications, the most commonly used tests are round-tension; reduced-section tension; face-, root-, and side-bend; and Charpy V-notch impact. Fillet weld tests are employed to determine proper welding techniques and conditions, and the shear strength of welded joints for design purposes.

AWS B1.10, Guide for the Nondestructive Inspection of Welds This standard describes the common nondestructive methods for examining welds. The methods included are visual, penetrant, magnetic particle, radiography, ultrasonic and eddy current inspection.

Welding Processes

AWS PUBLISHES RECOMMENDED practices and guides for arc and oxyfuel gas welding and cutting, brazing; resistance welding; and thermal spraying. The following is a list of processes and applicable documents.

ARC AND GAS WELDING AND CUTTING
Air Carbon-Arc Gouging and Cutting, Recommended Practices for, ANSI/AWS C5.3
Electrogas Welding, Recommended Practices for, AWS C5.7
Gas Metal Arc Welding, Recommended Practices for, AWS C5.6
Gas Tungsten Arc Welding, Recommended Practices for, AWS C5.5
Oxyfuel Gas Cutting, Operator's Manual for, AWS C4.2
Plasma Arc Cutting, Recommended Practices for, AWS C5.2
Plasma Arc Welding, Recommended Practices for, AWS C5.1
Stud Welding, Recommended Practices for, ANSI/AWS C5.4

BRAZING
Design, Manufacture, and Inspection of Critical Brazed Components, Recommended Practices for, AWS C3.3

RESISTANCE WELDING
Resistance Welding, Recommended Practices for, AWS C1.1
Resistance Welding Coated Low Carbon Steels, Recommended Practices for, AWS C1.3

THERMAL SPRAYING
Thermal Spraying: Practice, Theory, and Application, TS
Metallizing with Aluminum and Zinc for Protection of Iron and Steel, Recommended Practices for, AWS C2.2

Welding Applications

AWS PUBLISHES STANDARDS that cover various applications of welding. The subjects and appropriate documents are listed below.

AUTOMOTIVE
Automotive Portable Gun Resistance-Spot Welding, Recommended Practices for, AWS D8.5
Automotive Resistance Spot Welding Electrodes, Standard for, AWS D8.6
Automotive Welding Design, Recommended Practices for, AWS D8.4
Automotive Weld Quality-Resistance Spot Welding, Specification for, AWS D8.7

MACHINERY AND EQUIPMENT
Earthmoving and Construction Equipment Specification for Welding, AWS D14.3
Industrial and Mill Crane and Other Material Handling Equipment, Specification for Welding, ANSI/AWS D14.1
Machinery and Equipment, Classification and Application of Welded Joints for, AWS D14.4
Metal Cutting Machine Tool Weldments, Specification for, ANSI/AWS D14.2
Presses and Press Components, Specification for Welding of, AWS D14.5
Railroad Welding Specification, ANSI/AWS D15.1
Rotating Elements of Equipment, Specification for, AWS D14.6

MARINE
Aluminum Hull Welding, Guide for, ANSI/AWS D3.7
Steel Hull Welding, Guide for, ANSI/AWS D3.5
Underwater Welding, Specification for, ANSI/AWS D3.6

PIPING AND TUBING
Aluminum and Aluminum Alloy Pipe, Recommended Practices for Gas Shielded Arc Welding of, AMSI/AWS D10.7
Austenitic Chromium Nickel Stainless Steel Piping and Tubing, Recommended Practices for Welding, ANSI/AWS D10.4
Chromium-Molybdenum Steel Piping and Tubing, Recommended Practices for Welding of, ANSI/AWS D10.8
Local Heat Treatment of Welds in Piping and Tubing, AWS D10.10
Plain Carbon Steel Pipe, Recommended Practices and Procedures for Welding, AWS D10.12
Root Pass Welding and Gas Purging, Recommended Practices for, ANSI/AWS D10.11
Titanium Piping and Tubing, Recommended Practices for Gas Tungsten Arc Welding of, ANSI/AWS D10.6

Sheet Metal

ANSI/AWS D9.1, SPECIFICATION FOR WELDING OF SHEET METAL covers non-structural fabrication and erection of sheet metal by welding for heating, ventilating, and air-conditioning systems; architectural usages; food-processing equipment; and similar applications. Where differential pressures of more than 120 in. (30 kPa) of water or structural requirements are involved, other standards are to be used.

Structural

ANSI/AWS D1.2, STRUCTURAL WELDING CODE—ALU-MINUM, addresses welding requirements for aluminum alloy structures. It is used in conjunction with appropriate complementary codes or specifications for materials, design, and construction. The structures covered are tubular designs and static and dynamic nontubular designs.

ANSI/AWS D1.4, Structural Welding Code—Reinforcing Steel, applies to the welding of concrete reinforcing steel for splices (prestressing steel excepted), steel connection devices, inserts, anchors, anchorage details, and other welding in reinforced concrete construction. Welding may be done in a fabrication shop or in the field. When welding reinforcing steel to primary structural members, the provisions of *ANSI/AWS D1.1, Structural Welding Code—Steel* also apply.

ANSI/AWS D1.3, Structural Welding Code—Sheet Steel, applies to the arc welding of sheet and strip steel, including cold-formed members, that are 0.18 in. (5 mm) or less in thickness. The welding may involve connections of sheet or strip steel to thicker supporting structural members. When sheet steel is welded to primary structural members, the provisions of *ANSI/AWS D1.1, Structural Welding Code—Steel* also apply.

ANSI/AWS D1.1, Structural Welding Code—Steel, covers welding requirements applicable to welded structures of carbon and low alloy steels. It is to be used in conjunction with any complementary code or specification for the design and construction of steel structures. It is not intended to apply to pressure vessels, pressure piping, or base metals less than 1/8 in. (3 mm) thick. There are sections devoted exclusively to buildings (static loading), bridges (dynamic loading), and tubular structures.

Safety

ANSI/ASC Z49.1, SAFETY IN WELDING AND CUTTING, was developed by the ANSI Accredited Standards Committee Z49, Safety in Welding and Cutting, and then published by AWS. The purpose of the Standard is the protection of persons from injury and illness, and the protection of property from damage by fire and explosions arising from welding, cutting, and allied processes. It specifically covers arc, oxyfuel gas, and resistance welding, and thermal cutting, but the requirements are generally applicable to other welding processes as well. The provisions of this standard are backed by the force of law since they are included in the *General Industry Standards* of the U.S. Department of Labor, Occupational Safety and Health Administration.

Other safety and health standards published by AWS include the following:

Electron Beam Welding and Cutting, Recommended Safe Practices for, AWS F2.1

Evaluating Contaminants in the Welding Environments, A Sampling Strategy Guide, AWS F1.3

Measuring Fume Generation Rates and Total Fume Emission for Welding and Allied Processes, Laboratory Method for, ANSI/AWS F1.2

Preparation for Welding and Cutting Containers and Piping That Have Held Hazardous Substances, Recommended Safe Practices for the, AWS F4.1

Sampling Airborne Particulates Generated by Welding and Allied Processes, Method for, ANSI/AWS F1.1

Sound Level Measurement of Manual Arc Welding and Cutting Processes, Method for AWS F6.1

Thermal Spraying, Recommended Safe Practices for, AWS C2.1

ASSOCIATION OF AMERICAN RAILROADS

Manual of Standards and Recommended Practices

THE PRIMARY SOURCE of welding information relating to the construction of new railway equipment is the *Manual of Standards and Recommended Practices* prepared by the Mechanical Division, Association of American Railroads (AAR). This manual includes specifications, standards, and recommended practices adopted by the Mechanical Division. Several sections of the manual relate to welding, and the requirements are similar to those of ANSI/AWS D1.1 Structural Welding Code—Steel. This Code is frequently referenced, particularly with regard to weld procedure and performance qualification. In 1986, the American Welding Society published *AWS D15.1 Railroad Welding Specification.* The AWS Committee on Railroad Welding includes representatives of AAR, and the AWS D15.1 Railroad Welding Specification has been endorsed by the AAR. Therefore, it is expected that future revisions of the *Manual of Standards and Recommended Practices* will refer to AWS D15.1 for all welding requirements on construction and maintenance of steel and aluminum railcars.

The sections of the current Manual of Standards and Recommended Practices that relate to welding are summarized below.

Section C, Part II, Specifications for Design, Fabrication, and Construction of Freight Cars. This specification covers the general welding practices for freight car construction. Welding processes and procedures other than those listed in the document may be used. However, they must conform to established welding standards or proprietary carbuilder's specifications, and produce welds of quality consistent with design requirements and good manufacturing techniques. The welding requirements are similar to, though not so detailed as those in *ANSI/AWS D1.1, Structural Welding Code—Steel*. The qualification of welders and welding operators must be done in accordance with the AWS Code.

Section C, Part III, Specification for Tank Cars. This specification covers the construction of railroad car tanks used for the transportation of hazardous and non-hazardous materials. The requirements for fusion welding of the tanks, and for qualifying welders and welding procedures to be used are describedin one appendix. A second appendix describes the requirements for repairs, alterations, or conversions of car tanks. If welding is required, it must be performed by facilities certified by AAR in accordance with a third appendix. The rules for welding on the tanks are covered by the *ASME Boiler and Pressure Vessel Code*.

The U.S. Department of Transportation (DOT) issues regulations covering the transportation of explosives, radioactive materials, and other dangerous articles. Requirements for tank cars are set forth in the *United States Code of Federal Regulations, Title 49*, Sections 173.3l4, 173.316, and 179, which are included at the end of the AAR specifications.

Section D, Trucks and Truck Details. The procedures, workmanship, and qualification of welders employed in the fabrication of steel railroad truck frames are required to be in accordance with (1) the latest recommendations of the American Welding Society, (2) The Specifications for Design, Fabrication, and Construction of Freight Cars, and (3) the welding requirements of the Specifications for Tank Cars.

Field Manual of Association of American Railroads Interchange Rules

THIS MANUAL COVERS the repair of existing railway equipment. The U.S. railway network is made up of numerous interconnecting systems, and it is often necessary for one system to make repairs on equipment of another system. The repair methods are detailed and specific so that they may be used as the basis for standard charges between the various railroad companies.

CANADIAN STANDARDS ASSOCIATION

THE CANADIAN STANDARDS ASSOCIATION (CSA) is a voluntary membership organization engaged in standards development and also testing and certification. The CSA is similar to ANSI in the United States, but ANSI does not test and certify products. A CSA Certification Mark assures buyers that a product conforms to acceptable standards.

Examples of CSA welding documents are the following:

Aluminum Welding Qualification Code, CSA W47.2
Certification of Companies for Fusion Welding of Steel Structures, CSA W47.1
Code for Safety in Welding and Cutting (Requirements for Welding Operators), CSA W117.2
Qualification Code for Welding Inspection Organizations, CSA W178
Resistance Welding Qualification Code for Fabricators of Structural Members Used in Buildings, CSA W55.3
Welded Aluminum Design and Workmanship (Inert Gas Shielded Arc Processes), CSA S244
Welded Steel Construction (Metal Arc Welding), CSA W59
Welding Electrodes, CSA W48 Series
Welding of Reinforcing Bars in Reinforced Concrete Construction, CSA W186

COMPRESSED GAS ASSOCIATION

THE COMPRESSED GAS ASSOCIATION (CGA) promotes, develops, represents, and coordinate technical and standardization activities in the compressed gas industries, including end uses of products.

The Handbook of Compressed Gases, published by CGA, is a source of basic information about compressed gases, their transportation, uses, and safety considerations, and also the rules and regulations pertaining to them.

Standards for Welding and Brazing on Thin Walled Containers, CGA C-3, is directly related to the use of welding and brazing in the manufacture of DOT compressed gas cylinders. It covers procedure and operator qualification, inspection, and container repair.

The following CGA publications contain information on the properties, manufacture, transportation, storage, handling, and use of gases commonly used in welding operations:

Acetylene, G-1
Commodity Specification for Acetylene, G-1.1
Carbon Dioxide, G-6
Commodity Specification for Carbon Dioxide, G-6.2
Hydrogen, G-5
Commodity Specification for Hydrogen, G-5.3
Oxygen, G-4
Commodity Specification for Oxygen, G-4.3
The Inert Gases Argon, Nitrogen, and Helium, P-9
Commodity Specification for Argon, G-11.1
Commodity Specification for Helium, G-9.1
Commodity Specification for Nitrogen, G-10.1

Safety considerations related to the gases commonly used in welding operations are discussed in the following CGA pamphlets:

Handling Acetylene Cylinders in Fire Situations, SB-4
Oxygen-Deficient Atmospheres, SB-2
Safe Handling of Compressed Gases in Containers, P-1

FEDERAL GOVERNMENT

SEVERAL DEPARTMENTS OF the Federal Government, including the General Services Administration, are responsible for developing welding standards or adopting existing standards, or both. There are in excess of 48 000 standards adopted by the federal government.

Consensus Standards

THE U.S. DEPARTMENT of Labor, Transportation, and Energy are primarily concerned with adopting existing national consensus standards, but they also make amendments to these standards or create separate standards, as necessary. For example, the Occupational Safety and Health Administration (OSHA) of the Department of Labor issues regulations covering occupational safety and health protection. The welding portions of standards adopted or established by OSHA are published under Title 29 of the *United States Code of Federal Regulations*. Part 1910 covers general industry, while Part 1926 covers the construction industry. These regulations were derived primarily from national consensus standards of ANSI and the NFPA.

Similarly, the U.S.Department of Transportation is responsible for regulating the transportation of hazardous materials, petroleum, and petroleum products by pipeline in interstate commerce. Its rules are published under Title 49 of the *United Sates Code of Federal Regulations, Part 195*. Typical of the many national consensus standards incorporated by reference in these regulations are *API Standard 1104* and *ASME B31.4*, which were discussed previously.

The U. S. Department of Transportation is also responsible for regulating merchant ships of American registry. It is empowered to control the design, fabrication, and inspection of these ships by *Title 46 of the United States Code of Federal Regulations*.

The U.S. Coast Guard is responsible for performing the inspections of merchant ships. The *Marine Engineering Regulations* incorporate references to national consensus standards, such as those published by ASME, ANSI, and ASTM. These rules cover repairs and alterations that must be performed with the cognizance of the local Coast Guard Marine Inspection Officer.

The U.S. Department of Energy is responsible for the development and use of standards by government and industry for the design, construction, and operation of safe, reliable, and economic nuclear energy facilities. National consensus standards, such as the *ASME Boiler and Pressure Vessel Code, Sections III and IX*, and *ANSI/AWS D1.1, Structural Welding Code—Steel* are referred to in full or in part. These standards are supplemented by separate program standards, known as *RDT Standards*.

Military and Federal Specifications

MILITARY SPECIFICATIONS ARE prepared by the Department of Defense. They cover materials, products, or services specifically for military use, and commercial items modified to meet military requirements.

Military specifications have document designations beginning with the prefix MIL. They are issued as either coordinated or limited-coodination documents. Coordinated documents cover items or services required by more than one branch of the military. Limited coodination documents cover items or services of interest to a single branch. If a document is of limited coordination, the branch of the military which uses the document will appear in parentheses in the document designation. The Department of Defense has begun to replace military specifications with concension standards in the interest of economy.

Two current military specifications cover the qualification of welding procedures or welder peformance, or both. One is *MIL-STD-1595, Qualification of Aircraft, Missile, and Aerospace Fusion Welders*. The other, *MIL-STD-248, Welding and Brazing Procedure and Performance Qualification*, covers the requirements for the qualification of welding and brazing procedures, welders, brazers, and welding and brazing operators. It allows the fabricator to submit for approval certified records of qualification tests prepared in conformance with the standards of other government agencies, ABS, ASME, or other organizations. Its use is mandatory when referenced by other specifications or contractural documents.

MIL-STD-1595 establishes the procedure for qualifying welders and welding operators engaged in the fabrication of components for aircraft, missiles, and other aerospace equipment by fusion welding processes. This standard is applicable when required in the contracting documents, or when invoked in the absence of a specified welder qualification document.

MIL-STD-1595 superseded *MIL-T-5021, Tests; Aircraft and Missile Welding Operator's Qualification*, which is now obsolete. However, MIL-T-5021 is still referenced by other current government specifications and contract documents. When so referenced, a contractor has to perform the technically obsolete tests required by this standard.

Federal specifications are developed for materials, products, and services that are used by two or more Federal agencies, one of which is not a defense agency. Federal specifications are classified into broad categories. The *QQ* group, for example, covers metals and most welding specifications. Soldering and brazing fluxes are in the *O-F* group.

Some military and federal specifications include requirements for testing and approval of a material, process, or piece of equipment before its submission for use under the specification. If the acceptance tests pass the specification requirements, the material or equipment will be included in the applicable *Qualified Products List (QPL)*.

In other specifications, the supplier is responsible for product conformance. This is often the case for welded fabrications. The supplier must show evidence that the welding procedures and the welders are qualified in accordance with the requirements of the specification, and must certify the test report.

The following military and federal standards (currently listed in the Department of Defense Index) address welding, brazing, and soldering. Those standards covering base metals and welding equipment are not included.

Braze-Welding, Oxyacetylene, of Built-Up Metal Structures, MIL-B-12672

Brazing Alloy, Copper, Copper-Zinc, and Copper-Phosphorus, QQ-B-650

Brazing Alloy, Gold, QQ-B-653

Brazing Alloy, Silver, QQ-B-654

Brazing Alloy, 82 Gold-18 Nickel, Wire, Foil, Sheet and Strip, MIL-B-47043

Brazing Alloys, Aluminum and Magnesium, Filler Metal, QQ-B-655

Brazing, Aluminum, Process for, MIL-B-47292

Brazing and Annealing of Electromagnetic (Iron-Cobalt Alloy) Poles to Austenitic Stainless Steel, Process for, MIL-B-47291

Brazing, Nickel Alloy, General Specification for, MIL-B-9972

Brazing, Oxyacetylene, of Built-Up Metal Structures, MIL-B-12673

Brazing of Steels, Copper, Copper Alloys, Nickel Alloys, Aluminum, and Aluminum Alloys, MIL-B-78838

Brazing Sheet, Aluminum Alloy, MIL-B-20148

Electrode, Underwater Cutting, Tubular, Ceramic, MS-16857

Electrode, Welding, Bare, Aluminum Alloys, MIL-E-16053

Electrode, Welding, Bare, Copper and Copper Alloy, MIL-E-23765/3

Electrode, Welding, Bare, High Yield Steel, MIL-E-19822 Electrode, Welding, Bare, Solid, Nickel-Manganese-Chromium-Molybdenum Alloy Steel for Producing HY-130 Weldments for As-Welded Applications, MIL-E-24355

Electrode, Welding, Carbon Steel, and Alloy Steel, Bare, Coiled, MIL-E-18193

Electrode, Welding, Copper, Silicon-Deoxidized, Solid, Bare, MIL-E-45829

Electrode, Welding, Covered, Aluminum Bronze, MIL-E-278

Electrode, Welding Covered (Austenitic Chromium-Nickel Steel, for Corrossive and High Temperature Services), MIL-E-22200/2

Electrode, Welding, Covered, Austenitic Steel (19-9 Modified) for Armor Applications MIL-E-13080

Electrode, Welding, Covered, Bronze, for General Use, MIL-E-13191

Electrode, Welding, Covered, Coated, Aluminum and Aluminum Alloy, MIL-E-15597

Electrode, Welding, Covered, Copper-Nickel Alloy, MIL-E-22200/4

Electrode, Welding, Covered, Low Alloy Steel (Primarily for Aircraft and Weapons), MIL-E-6843

Electrode, Welding, Covered, Low-Hydrogen, and Iron Powdered Low-Hydrogen, Chromium-Molybdenum Alloy Steel and Corrosion-Resisting Steel, MIL-E-22200/8

Electrode, Welding, Covered, Low-Hydrogen, Heat-Treatable Steel, MIL-E-8697

Electrode, Welding, Covered, Mild Steel, QQ-E-450

Electrode, Welding, Covered, Molybdenum Alloy Steel Application, MIL-E-22200/7

Electrode, Welding, Covered, Nickel Base Alloy, and Cobalt Base Alloy, MIL-E-22200/3

Electrode, Welding, Mineral Covered; Iron-Powder, Low-Hydrogen, High Tensile Low Alloy Steel-Heat-Treatable Only, MIL-E-22200/5

Electrode, Welding, Mineral Covered, Iron-Powder, Low-Hydrogen Medium and High Tensile Steel, As-Welded or Stress-Relieved Weld Application, MIL-E-22200/1

Electrode, Welding, Mineral Covered, Low-Hydrogen, Chromium-Molybdenum Alloy Steel and Corrosion Resisting Steel, MIL-E-16589

Electrode, Welding, Mineral Covered, Low-Hydrogen or Hydrogen, Nickel-Manganese-Chromium-Molybdenum Alloy Steel for Producing HY-130 Weldments for As-Welded Application, MIL-E-22200/9

Electrode, Welding, Mineral Covered, Low-Hydrogen, Medium and High Tensile Steel, MIL-E-22200/6

Electrode, Welding, Surfacing, Iron Base Alloy, MIL-E-19141

Electrodes and Rods—Welding, Bare, Chromium and Chromium-Nickel Steels, MIL-E-19933

Electrodes and Rods—Welding, Bare, Solid, Low Alloy Steel, MIL-E-23765/2

Electrodes and Rods Welding, Bare, Solid, Mild and Alloy Steel, MIL-E-23765/1

Electrodes (Bare) and Fluxes (Granular), Submerged Arc Welding, High-Yield Low Alloy Steels, MIL-E-22749

Electrodes, Cutting and Welding, Carbon-Graphite, Uncoated and Copper-Coated, MIL-E-17777

Electrodes, Welding, Covered, General Specification for, MIL-E-22200

Electrodes, Welding, Mineral Covered, Iron-Powder, Low-Hydrogen-8 Nickel-Chromium-Molybdenum-Vanadium Alloy Steel for Producing HY-130 Weldments to be Heat Treated, MIL-E-22200/11

Electrodes, Welding, Mineral Covered, Low-Hydrogen, Iron-Powder for Producing HY-100 Steel Weldments for As-Welded Applications, MIL-E-22200/10

Fabrication Welding and Inspection, and casting Inspection and Repair for Machinery, Piping and Pressure Vessels in Ships of the United States Navy, MIL-STD-278

Fabrication Welding and Inspection of Hyperbaric Chambers and Other Critical Land Based Structures, MIL-STD-1693

Fabrication, Welding, and Inspection of HY-130 Submarine Hull, MIL-STD-1681

Flux, Aluminum and Aluminum Alloy, Gas Welding, MIL-F-6939

Flux, Brazing, Silver Alloy, Low Melting Point, O-F-499 Flux, Galvanizing, MIL-F-19197

Flux, Soldering, Liquid (Rosin Base), MIL-F-14256

Flux, Soldering, Paste and Liquid, O-F-506

Flux, Soldering, Rosin Base, General Purpose, MIL-F-20329

Flux, Soldering (Stearine Compound 1C-3), MIL-F-12784

Flux, Welding (for Copper-Base and Copper-Nickel alloys and Cast Iron), MIL-F-16136

Fluxes, Welding (Compositions), Submerged Arc Process With Type B Electrodes, Carbon and Low Alloy Steel Application, MIL-F-19922

Fluxes, Welding, Submerged Arc Process Carbon and Low Alloy Steel Application, MIL-F-18251

Rod, Welding, Copper and Copper Alloy, MIL-R-19631

Rod, Welding, Copper and Nickel Alloy, QQ-R-571

Rod, Welding, High Strength, MIL-R-47191

Solder Bath Soldering of Printed Wiring Assemblies, MIL-S-46844

Solder, Lead-Tin alloy, MIL-S-12204

Solder, Low-Melting Point, MIL-S-627

Solder, Tin alloy, Lead-Tin Alloy and Lead Alloy, QQ-S-571

Soldering, Manual Type, High Reliability, Electrical and Electronic Equipment, MIL-S-45743

Soldering of Electrical Connections and Printed Wiring Assemblies, Procedure for, MIL-STD-1460

Soldering of Metallic Ribbon Lead Materials to Solder Coated Conductors, Process for Reflow, MIL-S-46880

Soldering Process, General Specification for, MIL-S-6872

Welded Joint Design, MIL-STD-22

Welded Joint Designs, Armored-Tank Type, MIL-STD-21

Welder Performance Qualification, Aerospace, MIL-STD-1595

Welding, Aluminum Alloy Armor, MIL-W-45206

Welding, Arc and Gas, for Fabricating Ground Equipment for Rockets and Guided Missiles, MIL-W-47091

Welding and Brazing Procedure and Performance Qualification, MIL-STD-248

Welding, Flash, Carbon and Alloy Steel, MIL-W-6873

Welding, Flash, Standard Low Carbon Steel, MIL-W-62160

Welding, Fusion, Electron Beam, Process for, MIL-W-46132

Welding, Gas Metal-Arc and Gas Tungsten-Arc, Aluminum Alloys, Readily Weldable for structures, Excluding Armor, MIL-W-45205

Welding, Gas, Steels, Constructional, Readily Weldable, for Low Stressed Joints, MIL-STD-1183

Welding, High Hardness Armor, MIL-STD-1185

Welding, Metal Arc and Gas, Steels, and Corrosion and Heat Resistant Alloys, Process for, MIL-W-8611

Welding of Aluminum Alloys, Process for, MIL-W-8604

Welding of Armor, Metal-Arc, Manual, with Austenitic Electrodes, for Aircraft, MIL-W-41

Welding of Electronic Circuitry, Process for, MIL-W-46870

Welding of Homogeneous Armor by Metal Arc Processes, MIL-W-460-86 (MR)

Welding of Magnesium Alloys, Gas and Arc, Manual and Machine Processes for, MIL-W-18326

Welding Procedures for Constructional Steels, MIL-STD-1261 (MR)

Welding Process and Welding Procedure Requirements for Manufacture of Equipment Utilizing Steels, MIL-W-52574

Welding, Repair, of Readily Weldable Steel Castings (Other Than Armor) Metal-Arc, Manual, MIL-W-13773

Welding, Resistance, Electronic Circuit Modules (Asg), MIL-W-8939

Welding, Resistance, Spot and Projection for Fabricating Assemblies of Carbon Steel Sheets, MIL-W-46154

Welding, Resistance, Spot and Seam, MIL-W-6858

Welding, Resistance, Spot, Seam, and Projection, for Fabricating Assemblies of Low-Carbon Steel, MIL-W-12332

Welding, Resistance, Spot, Weldable Aluminum Alloys, MIL-W-45210

Welding rod and Wire, Nickel Alloy High Permeability, Shielding Grade, MIL-W-47192

Welding Rods and Electrodes, Preparation for Delivery of, MIL-W-10430

Welding, Spot, Hardenable Steels, MIL-W-45223

Welding, Spot, Inert-Gas Shielded Arc, MIL-W-27664

Welding, Stud, Aluminum, MIL-W-45211

Welding Symbols (ABCA-323), Q-STD-323

Welding Terms and Definitions (ABCA-324), Q-STD-324

Weldment, Aluminum and Aluminum Alloy, MIL-W-22248

Weldment, Steel, Carbon and Low Alloy (Yield Strength 30,000-60,000 psi), MIL-W-21157

INTERNATIONAL ORGANIZATION FOR STANDARDIZATION

THE INTERNATIONAL ORGANIZATION FOR STANDARDIZATION (ISO) promotes the development of standards to facilitate the international exchange of goods and services. It is comprised of the standards-writing bodies of more than 80 countries and has adopted or developed over 4000 standards.

ANSI is the designated U.S. representative to ISO. ISO standards and publications are available from ANSI.

The ISO standards that relate to welding are listed below. For convenience in locating specific standards they have been categorized into six groups.

GENERAL

Classification Groups for Fusion-Welded Joints in Steel, DP 58l7

Fundamental Welding Positions—Definitions and Values of Angles of Slope and Rotation for Straight Welds for These Positions, ISO 6947

Quality Control Levels for Final Assessment of Welded Joints, DP 6214

Weldability—Definition—Trilingual Edition, ISO 581

Welding—Factors to be Considered When Assessing Firms Using Welding as a Prime Means of Fabrication, ISO 3834

Welding—Items to be Considered for Ensuring Quality in Welded Structures, ISO 6213

ARC AND GAS WELDING AND CUTTING PROCESSES

Arc Welding with Consumable Electrode Wires (Mig—Mag process)—Specification of Power Sources, DIS 8172

Definitions of Welding Processes—Bilingual Edition—DP 857

Gas Welding Equipment—Acceptance Test for Oxygen Cutting Machines, DP 8206

Gas Welding Equipment—Flow Meter Regulators, DP 7292

Gas Welding Equipment—Hose Assembly Connections (end to end inclusive), DP 8207

Gas Welding Equipment—Pipeline Regulators, DP 7290

Gas Welding Equipment—Probe Couplings, DP 7288

Gas Welding Equipment—Probe Couplings with Shut-Off Valve, DP 7289

Gas Welding Equipment—Regulators for Manifolds, DP 7291

Graphic Symbols for Thermal Cutting, DP 7287

Hose Connections for Equipment for Welding, Cutting, and Related Processes, ISO 3253

Manual Blowpipes for Welding and Cutting, DP 5172

Oxygen/Fuel Gas Cutting Machine Blowpipe of Cylindrical Shaft type—General Requirements and Methods of Test, DP 5186

Power Sources for Manual Metal Arc Welding with Covered Electrodes and for the TIG Process, ISO 700

Pressure Gauges used in Welding, Cutting, and Related Processes, ISO 5171

Tungsten Electrodes for Inert Gas Shielded Arc Welding, and for Plasma Cutting and Welding, DIS 6848

Welding and Cutting Equipment and Allied Processes—Safety Devices for Fuel Gases, Oxygen, and Compressed Air—Part I General Specifications and Requirement, ISO 5175/1

Welding and Cutting Equipment and Allied Processes II—Test Characteristics for Safety Devices, DP 5175/2

Welding, Brazing, Braze Welding and Soldering of Metals—List of Processes, for Symbolic Representation on Drawings—Bi-Lingual Edition, ISO 4063

Welding—Flexible Hoses for Gas Welding and Allied Processes, ISO 3821

Welding—Regulators for Gas Cylinders used in Welding, Cutting and Related Processes, ISO 2503

RESISTANCE WELDING PROCESSES

Cylinders for Multiple Spot Welding Machines, DP 6210

Dimensions of Seam Welding Wheel Blanks, ISO 693

Electrode Taper Fits for Spot Welding Equipment—Dimensions, ISO 1089

Graphical Symbols on Resistance Welding Equipment, DP 7286

Insulation Cap and Brushes for Resistance Welding Equipment, DP 7931

Laminated Shunts for Resistance Welding Equipment, DP 6211

Materials for Resistance Welding Electrodes and Ancillary Equipment, ISO 5182

Most Important Details of Resistance Spot Welding Guns, DP 5831

Pneumatic Cylinders for Multiple Spot Welding Machines—Part I Cylinders Without Liners, DIS 7285/1

Rating of Resistance Welding Equipment, ISO 669

Resistance Spot Welding Electrode Adaptors, ISO 5183

Resistance Spot Welding Electrode Adaptors, Female Taper 1 : 10, ISO 5829

Resistance Spot Welding Electode Caps, ISO 5821

Resistance Spot Welding Electrode Holders—Part I: Taper Fixing 1:10, DP 8430/1

Resistance Spot Welding Electrode Holders—Part 2: Morse Taper Fixing, DP 8430/2

Resistance Spot Welding Electrode Holders—Part 3: Parallel Shank Fixing for End Thrust, DP 8430/3

Resistance Welding Equipment—Dimensions of Transformers for Portable Gun Stations, DP 8204

Resistance Spot Welding—Male Electrode Caps, ISO 5830

Resistance Welding Equipment—Multi-Spot Welding as Used in the Automobile Industry— Particular Specifications Applicable to Transformers with Two Separate Secondary Windings, DIS 7284

Resistance Welding Equipment—Secondary Cables for portable Spot Welders, DP 8205

Resistance Welding Equipment—Secondary Connecting Cables with Terminals Connected to Water-Cooled Lugs—Dimensions and Characteristics, ISO 5828

Slots in Platens for Projection Welding Machines, ISO 865

Spot Welding—Electrode Back-Ups and Clamps, ISO 5827

Spot Welding Equipment—Taper Plug Gauges and Taper Ring Gauges, ISO 5822

Straight Electrodes with Taper male, DP 6212

Straight Resistance Spot Welding Electrodes, ISO 5184

Taper Plug Gauges for Spot Electrode Fits—Type B, DP 5823

Taper Ring Gauges for Spot Electrode Fits—Type A, DP 5824

Taper Ring Gauges for Spot Electrode Fits—Type B, DP 5825

Transformers for Resistance Welding Machines— General Specifications Applicable to All Transformers, ISO 5826

Welding—Resistance Welding Equipment— Recommended Projections for Resistance Projection Welding, DP 8167

Welding—Resistance Welding Equipment—The Assessment and Testing of Resistance Spot Electrode Materials, DP 8166

FILLER METALS AND ELECTRODES

Code of Symbols for Covered Electrodes for Manual Metal Arc Welding of Cast Iron, ISO/R 1071

Covered Electrodes for Manual Arc Welding of Creep-Resisting Steels—Code of Symbols for Identification, ISO 3580

Covered Electrodes for Manual Arc Welding of Mild Steel and Low Alloy Steel—Code of Symbols for Identification, ISO 2560

Covered Electrodes for Manual Arc Welding of Stainless and Other Similar High Alloy Steels— Code of Symbols for Identification, ISO 3581

Drawn or Extruded Filler Rods for Welding Supplied in Straight Lengths—Lengths and Tolerance, ISO 546

Electrodes for Manual Arc Welding and Filler Metals for Gas Welding—Diameters and Tolerances, ISO 544

Electrodes for the Welding of Mild Steel and Low Alloy High Tensile Steel—Lengths and Tolerance, ISO 547

Filler Metals for Brazing and Soldering—Code of Symbols, DP 3677

Filler Products for Welding, Brazing and Braze Welding—Dimensional Criteria, DP 8706

Filler Rods for Braze Welding—Determination of Characteristics of Deposited Metal, ISO 688

Filler Rods for Braze Welding—Determination of Conventional Bond Strength on Steel, Cast Iron and Other Metals, ISO 698

Filler Rods for Gas Welding of Mild Steels and Low Alloy High Tensile Steels—Code of Symbols, DP 636

Filler Rods for Gas Welding of Mild Steels and Low Alloy High Tensile Steels—Determination of Mechanical Properties of Deposited Weld Metal, ISO 637

Filler Rods for Gas Welding—Test to Determine the Compatibility of Steel Filler Rods and the Parent Metal in the Welding of Steels, ISO/R 708

Filler Rods, Other Than Drawn or Extruded for Welding—Lengths and Tolerance, ISO 545

Manual Metal Arc Deposition of a Weld Metal Pad for Chemical Analysis, DIS 6847

Solid Wires for Gas Shielded Metal Arc Welding of Mild Steel—Dimensions of Wires, Spools, Rims and Coils, DP 864

Standardized Method Recommended for the Determination of the Ferrite Number in Weld Metal Deposited Chrome-Nickel Stainless Steel Electrodes, DIS 8249

DESIGN

Calculation of Rectangular Symmetrical Fillet Welds Statically Loaded in Such a Way That the Transverse Section Is Not Under Any Normal Stress, ISO/R 617

Simplifed Rules for the Verification of Statically Loaded Fillet Welds, DP 5185

Welding Requirements—Factors to be Considered in Specifying Requirements for Fusion Welded Joints in Steel (Technical Influencing Factors), ISO 3088

Welds—Symbolic Representation on Drawings, DIS 255

Welding Requirements—Categories of Service Requirements for Welded Joints, ISO 3041

TESTING AND EVALUATION

Classification of Imperfections in Metallic Fusion Welds, with Explanations—Bilingual Edition, ISO 6520

Covered Electrodes—Determintion of the Efficiency, Metal Recovery and Deposition Coefficient, ISO 2401

Fusion Welded Butt Joints in Steel—Transverse Root and Face Bend Test, ISO 5173

Fusion Welded Butt Joints in Steel—Transverse Side Bend Test, ISO 5177

Investigation of Brazeability Using a Varying Gap Test Piece, ISO 5179

Longitudinal Tensile Test on Cylindrical Weld Metal Specimens from Fusion Welded Butt Joints in Steel, DP 5178

Methods of Determining the Mechanical Properties of the Weld Metal Deposited by Electrodes 3.15 mm (1/8 in.) or More in Diameter, ISO/R 615

Qualification of Welders for General Purpose Welding—Manual Metal-Arc Welding with Covered Electrodes—Carbon Steels and Low-Alloy Steels, DP 4152

Radiographic Image Quality Indicators—Principles and Identification, ISO/R 1027

Radiographic Inspection of Resistance Spot Welds for Aluminum and Its Alloys—Recommended Practice, ISO 3777

Radiography of Welds and Viewing Conditions for Films—Utilization of Recommended Patterns of Image Quality Indicators (I Q I), DP 2504

Recommended Practice for Radiographic Inspection of Circumferential Fusion Welded Butt Joints in Steel Pipes Up to 50 mm (2 in.) Wall Thickness, DP 947

Recommended Practice for Radiographic Inspection of Fusion Welded Butt Joints for Steel Plates Up to 50 mm (2 in.) Thick, DIS 1106

Recommended Practice for Radiographic Inspection of Fusion Welded Butt Joints for Steel Plates 50 to 200 mm (2 to 4 in.) Thick, ISO 2405

Recommended Practice for the X-ray Inspection of Fusion Welded Butt Joints for Aluminum and Its Alloys and Magnesium and Its Alloys 5 to 50 mm (3/16 to 2 in.) Thick, DP 2437

Soft Soldered Joints—Determination of Shear Strength, ISO 3683

The Basis for a Proposed Acceptance Standard for Weld Defects—Part I—Porosity, DP 5818/1

The Basis for a Proposed Acceptance Standard for Weld Defects—Part II—Slag Inclusion, DP 5819

Transverse Tensile Test for Fusion Welded Butt Joints in Steel, DIS 4136

Welded Joints—Recommended Practice for Liquid Penetrant Testing, ISO 3879

Welding and Allied Processes—Assemblies Made with Soft Solders and Brazing Filler Metals—Mechanical Test Methods, DIS 5187

Welding—Determination of Hydrogen in Deposited Weld Metal Arising From the Use of Covered Electrodes for Welding Mild and Low Alloy Steels, ISO 3690

Welds in Steel—Calibration Block No. 2 for Ultrasonic Examination of Welds, DIS 7963

Welds in Steel—Calibration of Equipment for Ultrasonic Examination, Using a Reference Block, DIS 5180

Welds in Steel—Reference Block for the Calibration of Equipment for Ultrasonic Examination, ISO 2400

NATIONAL BOARD OF BOILER AND PRESSURE VESSEL INSPECTORS

THE NATIONAL BOARD OF BOILER AND PRESSURE VESSEL INSPECTORS (NBBPVI), often referred to as the National Board, represents the enforcement agencies empowered to assure adherence to the *ASME Boiler and Pressure Vessel Code*. Its members are the chief inspectors or other jurisdictional authorities who administer the boiler and pressure vessel safety laws in the various jurisdictions of the United States and provinces of Canada.

The National Board is involved in the inspection of new boilers and pressure vessels. It maintains a registration system for use by manufacturers who desire or are required by law to register the boilers or pressure vessels that they have constructed. The National Board is also responsible for investigating possible violations of the *ASME Boiler and Pressure Vessel Code* by either commissioned inspectors or manufacturers.

The National Board publishes a number of pamphlets and forms concerning the manufacture and inspection of boilers, pressure vessels, and safety valves. It also publishes the *National Board Inspection Code* for the guidance of its members, commissioned inspectors, and others. The purpose of this code is to maintain the integrity of boilers and pressure vessels after they have been placed in service by providing rules and guidelines for inspection after installation, repair, alteration, or rerating. In addition, it provides inspection guidelines for authorized inspectors during fabrication of boilers and pressure vessels.

In some states, organizations that desire to repair boilers and pressure vessels must obtain the National Board Repair *(R)* stamp by application to the National Board. The firm must qualify all welding procedures and welders in accordance with the *ASME Boiler and Pressure Vessel Code, Section IX*, and the results must be accepted by the inspection agency. The firm must also have and demonstrate a quality control system similar to, but not so comprehensive as that required for an ASME code symbol stamp.

NATIONAL FIRE PROTECTION ASSOCIATION

THE MISSION OF the National Fire Protection Association (NFPA) is the safeguarding of people and their environment from destructive fire through the use of scientific and engineering techniques and education. NFPA standards are widely used as the basis of legislation and regulation at all levels of government. Many are referenced in the regulations of the OSHA. The standards are also used by insurance authorities for risk evaluation and premium rating.

Installation of Gas Systems

NFPA PUBLISHES SEVERAL standards that present general principles for the installation of gas supply systems and the storage and handling of gases commonly used in welding and cutting:

Bulk Oxygen Systems at Consumer Sites, NFPA 50
Design and installation of Oxygen-Fuel Gas Systems for Welding and Cutting and Allied Processes, NFPA 51
Gaseous Hydrogen Systems at Consumer Sites, NFPA 50A
National Fuel Gas Code, NFPA 54
Storage and Handling of Liquefied Petroleum Gases, NFPA 58

Users should check each standard to see if it applies to their particular situation. For example, NFPA 51 does not apply to a system comprised of a torch, regulators, hoses, and single cylinders of oxygen and fuel gas. Such a system is covered by *ANSI/AWS Z49.1, Safety in Welding and Cutting.*

Safety

NFPA PUBLISHES SEVERAL standards which relate to the safe use of welding and cutting processes:

Cleaning Small Tanks and Containers, NFPA 327
Control of Gas Hazards on Vessels to be Repaired, NFPA 306
Fire Protection in Use of Cutting and Welding Processes, NFPA 51B
Installation of Blower and Exhaust Systems for Dust, Stock, and Vapor Removal or Conveying, NFPA 91
Standard on Aircraft Maintenance, NFPA 410

Again, the user should check the standards to determine those that apply to the particular situation.

PIPE FABRICATION INSTITUTE

THE PIPE FABRICATION INSTITUTE (PFI) publishes numerous documents for use by the piping industry. Some of the standards have mandatory status because they are referenced in one or more piping codes. The purpose of PFI standards is to promote uniformity of piping fabrication in areas not specifically covered by codes. Other PFI documents, such as technical bulletins, are not mandatory, but they aid the piping fabricator in meeting the requirements of codes. The following PFI standards relate directly to welding.

End Preparation and Machined Backing Rings for Butt Welds, ES1
Manual Gas Tungsten Arc Root Pass Welding End Preparation and Fit Up Tolerances, ES21
Minimum Length and Spacing for Welded Nozzles, ES7
Preheat and Postheat Treatment of Welds, ES19
Recommended Practice for Welding of Transition Joints Between Dissimilar Steel Combinations, ES28
Welded Load Bearing Attachments to Pressure Retaining Piping Materials, ES26
Visual Examination—The Purpose, Meaning, and Limitation of the Term, ES27

SAE

SAE (FORMERLY THE SOCIETY OF AUTOMOTIVE ENGINEERS) is concerned with the research, development, design, manufacture, and operation of all types of self-propelled machinery. Such machinery includes automobiles, trucks, buses, farm machines, construction equipment, airplanes, helicopters, and space vehicles. Related areas of interest to SAE are fuels, lubricants, and engineering materials.

Automotive Standards

SEVERAL SAE WELDING-RELATED automotive standards are written in cooperation with AWS. These are:

Automotive Resistance Spot Welding Electrodes, Standard for, HS J1156 (AWS D8.6)
Automotive Weld Quality-Resistance Spot Welding, Specification for, HS J1188 (AWS D8.7)
Automotive Frame Weld Quality-Arc Welding, Specification for, HS J1196 (AWS D8.8)

Aerospace Material Specifications

MATERIAL SPECIFICATIONS ARE published by SAE for use by the aerospace industry. The Aerospace Material Specifications (AMS) cover fabricated parts, tolerances, quality control procedures, and processes. Welding-related AMS specifications are listed below. The appropriate AWS filler metal classification follows some of the specifications, in parentheses, for clarification.

PROCESSES
2664 Brazing—Silver, For Use Up to 800°F (425°C)
2665 Brazing—for Use Up to 400°F (205°C)
2667 Brazing—Silver, For Flexible Metal Hose—600°F (315°C) Max Operating Temperature
2668 Brazing—Silver, For Flexible Metal Hose—400°F (200°C) Max Operating Temperature
2669 Brazing—Silver, For Flexible Metal Hose—800°F (425°C) Max Operating Temperature
2670 Brazing—Copper Furnace, Carbon and Low Alloy Steels
2671 Brazing—Copper Furnace, Corrosion and Heat Resistant Steels and Alloys
2672 Brazing—Aluminum
2673 Brazing—Aluminum Molten Flux (Dip)
2675 Brazing—Nickel Alloy
2680 Electron Beam Welding For Fatigue Critical Applications
2681 Electron Beam Welding
2685 Welding, Tungsten Arc, Inert Gas, Nonconsumable Electrode (GTAW Method)
2689 Fusion Welding, Titanium and Titanium Alloys
2690 Welding (Parallel Gap) of Microelectric interconnections to Thin Film Substrates
2694 Repair Welding of Aerospace Castings

FLUX
3410 Flux—Silver, Brazing
3411 Flux—Silver, Brazing, High Temperature
3412 Flux—Brazing, Aluminum
3414 Flux—Welding, Aluminum
3415 Flux—Aluminum Dip Brazing, 1030°F (555°C) or Lower Liquidus
3416 Flux—Aluminum Dip Brazing, 1090°F Fusion Point
3430 Paste, Copper, Brazing Filler Metal—Water Thinning

ALUMINUM ALLOYS
4184 Wire, Brazing—10Si 4Cu (R-84)
4185 Wire, Brazing—12Si (R-84)
4188 Welding Wire
4181 Welding Wire, 7.0Si 0.30Mg 0.10Ti (4008)
4189 Wire, Welding—4.1Si 0.2Mg (4643)
4190 Wire, Welding—5.2Si (4043)
4191 Wire, Welding—6.3Cu 0.3Mn 0.18Zr 0.15Ti 0.10V (2319)
4233 Welding Wire, 4.5Cu 0.70Ag 0.30MN 0.25Mg 0.25Ti (201)
4244 Welding Wire, 4.6Cu 0.35Mn 0.25Mg 0.22Ti (206)
4245 Welding Wire, 5.0Si l.2Cu 0.50Mg (355)
4246 Welding Wire, 7.0Si 0.52Mg (357)

MAGNESIUM ALLOYS
4395 Wire, Welding—9Al 2Zn (AZ92A)
4396 Wire, Welding—3.3Ce 2.5Zn 0.7Zr (EZ33A)

BRAZING AND SOLDERING FILLER METALS
4750 Solder—Tin-Lead 45Sn 55Pb (R-84)
4751 Solder—Tin-Lead, Eutectic 63Sn 37Pb
4764 Brazing Filler Metal—Copper, 52.5Cu 38Mn 9.5Ni, 1615°-1700°F (880°-925°C) Solidus-Liquidus Range (1600N)
4765 Brazing Filler Metal, Silver—56Ag 42Cu 2.0Ni, 1420°-1640°F (770°-895°C) Solidus-Liquidus Range, (BAg-13a)
4766 Brazing Filler Metal, Silver-85Ag 15Mn, 1760°-1780°F (960°-970°C), Solidus-Liquidus Range
4767 Brazing Filler Metal, Silver—92.5Ag 7.2Cu 0.22Li, 1435°-1635°F (780°-890°C) Solidus-Liquidus Range (BAg-19)
4768 Brazing Filler Metal, Silver—35Ag 26Cu 21Zn 18Cd, 1125°-1295°F (605°-700°C) Solidus-Liquidus Range (BAg-2)
4769 Brazing Filler Metal—Silver, 45Ag 24Cd 16Zn 15Cu, 1125°-1145°F (605°-620°C) Solidus-Liquidus Range (BAg-1)
4770 Brazing Filler Metal—Silver, 50Ag l8Cd 16.5Zn 15.5Cu, 1160°-1175°F (625°-635°C) Solidus-Liquidus Range (BAg-1A)

4771 Brazing Filler Metal—Silver, 50Ag 16Cd 15.5Zn 15.5Cu 3.0Ni, 1170°-1270°F (630°-690°C) Solidus-Liquidus Range (BAg-3)

4772 Brazing Filler Metal—Silver, 54Ag 40Cu 5.0Zn 1.0Ni, 1325°-1575°F (720°-855°C) Solidus-Liquidus Range (BAg-13)

4773 Brazing Filler Metal, Silver, 60Ag 30Cu 10Sn, 1115°-1325°F (600°-720°C) Solidus-Liquidus Range (BAg-18)

4774 Brazing Filler Metal, Silver, 63Ag 28.5Cu 6.0Sn 2.5Ni, 1275°-1475°F (690°-800°C) Solidus-Liquidus Range (BAg-21)

4775 Brazing Filler Metal, Nickel, 73Ni 4.5Si 14Cr 3.1B 4.5Fe, 1790°-1970°F (975°-1075°C) Solidus-Liquidus Range (BNi-1)

4776 Brazing Filler Metal, Nickel, 73Ni 4.5Si 14Cr 3.1B 4.5Fe (Low Carbon) 1790°-1970°F (975°-1075°C) Solidus-Liquidus Range (BNi-1A)

4777 Brazing Filler Metal, Nickel, 82Ni 4.5Si 7.0Cr 3.1B 3.0Fe, 1780°-1830°F (970°-1000°C) Solidus-Liquidus Range (BNi-2)

4778 Brazing Filler Metal, Nickel, 92Ni 4.5Si 3.1B, 1800°-1900°F (980°-1040°C) Solidus-Liquidus Range (BNi-3)

4779 Brazing Filler Metal, Nickel, 94Ni 3.5Si 1.8B, 1800°-1950°F (980°-1065°C) Solidus-Liquidus Range (BNi-4)

4780 Brazing Filler Metal, Manganese, 66Mn 16Ni 16Co 0.80B, 1770°-1875°F (965°-1025°C) Solidus-Liquidus Range

4782 Brazing Filler Metal, Nickel, 71Ni 10Si 19Cr, 1975°-2075°F (1080°-1135°C) Solidus-Liquidus Range (BNi-5)

4783 Brazing Filler Metal, High Temperature, 50Co 8.0Si 19Cr 17Ni 4.0W 0.80B, 2050°-2100°F (1120°-1150°C) olidus-Liquidus Range (BCo-1)

4784 Brazing Filler Metal, High Temperature, 50Au 25Pd 25Ni, 2015°-2050°F (1100°-1120°C) Solidus-Liquidus Range

4785 Brazing Filler Metal, High Temperature 30Au 34Pd 36Ni, 2075°-2130°F (1135°-1165°C) Solidus-Liquidus Range (BAu-5)

4786 Brazing Filler Metal, High Temperature 70Au 8Pd 22Ni, 1845°-1915°F (1005°-1045°C) Solidus-Liquidus Range

4787 Brazing Filler Metal, High Temperature 82Au 18Ni, 1740F (950°C) Solidus-Liquidus Temperature (BAu-4)

TITANIUM ALLOYS

4951 Wire, Welding, Commercially Pure
4953 Wire, Welding—5Al 2.5Sn
4954 Wire, Welding—6Al 4V
4955 Wire, Welding—8Al 1Mo 1V

4956 Wire, Welding—6Al 4V, Extra Low Intersitial, Environment Controlled Packaging

CARBON STEELS

5027 Wire, Welding—1.05Cr 0.55Ni 1.0Mo 0.07V (0.26-0.32C), Vacuum Melted, Environment Controlled Packaged (D6AC)

5028 Wire, Welding—1.05Cr 0.55Ni 1.0Mo 0.07V (0.34-0.40C), Vacuum Melted, Environment Controlled Packaged (D6AC)

5029 Wire, Welding—0.78Cr 1.8Ni 0.35Mo 0.20V (0.33-0.38C), Vacuum Melted, Environment Controlled Packaged

5030 Wire, Welding—Low Carbon

5031 Welding Electrodes, Covered, Steel—0.007-0.15C (E6013)

CORROSION AND HEAT RESISTANT STEELS AND ALLOYS

5675 Wire, Welding—70Ni 15.5Cr 7Fe 3.0Ti 2.4Mn

5676 Wire, Welding—80Ni 20Cr

5677 Electrodes, Covered Welding—75Ni 19.5Cr 1.6 (Cb+Ta)

5679 Wire, Welding—73Ni 15.5Cr 8Fe 2.2 (Cb+Ta)

5680 Wire, Welding—18.5Cr 11Ni 0.40 (Cb+Ta)

5681 Electrodes, Covered Welding—19.5Cr 10.5Ni 0.60(Cb+Ta) (E347)

5682 Rod or Wire, Coating Alloy—78Ni 20Cr

5683 Wire, Welding—75Ni 15.5Cr 8Fe

5684 Electrodes, Covered, Alloy, Welding—72Ni 15Cr 9Fe 2.8 (Cb+Ta) (ENiCrFe-1)

5691 Electrodes, Covered, Steel, Welding—18Cr 12.5Ni 2.2Mo (E316)

5694 Wire, Welding—27Cr 21.5Ni (ER310)

5695 Electrodes, Covered Welding—25Cr 20Ni (E310)

5696 Wire Welding, 19Cr 12.5Ni 2.5Mo

5774 Wire, Welding—16.5Cr 4.5Ni 2.9Mo 0.1N (AM 350)

5775 Electrodes, Welding, Covered—16.5Cr 4.5Ni 2.9Mo 0.10N

5776 Wire, Welding—12.5Cr (410)

5777 Electrodes, Covered Welding Steel—12.5Cr (E410)

5778 Wire, Welding—72Ni 15.5Cr 2.4Ti l(Cb+Ta) 0.7Al 7Fe

5779 Electrodes, Covered Welding—75Ni 15Cr 1.5(Cb+Ta) 1.9Ti 0.55Al 5.5Fe

5780 Wire, Welding—15.5Cr 4.5Ni 2.9Mo 0.10N

5781 Electrodes, Welding Covered, 15.5Cr 4.5Ni 2.9Mo 0.10N

5782 Wire, Welding—20.5Cr 9.0Ni 0.50Mo 1.5W 1.2(Cb+Ta) 0.20Ti (ER349)

5783 *Electrodes, Covered Welding—19.5Cr 8.8Ni 0.50Mo 1.5W 1.0 (Cb+Ta) (E349)*

5784 *Wire Welding—29Cr 9.5Ni (ER312)*

5785 *Electrodes, Covered Welding—28.5Cr 9.5Ni (E312)*

5786 *Wire, Welding—62.5Ni 5.0Cr 24.5Mo 5.5Fe (ERNiMo3)*

5788 *Coating Alloy—62Co 29Cr 4.5W*

5789 *Wire, Welding 54Co 25.5Cr 10.5Ni 7.5W*

5790 *Wire, Welding 20Cr 20Ni 0.75(Cb+Ta), High Ferrite Grade*

5791 *Powder, Plasma Spray, 56.5Co 25.5Cr 10.5Ni 7.5W*

5792 *Powder, Plasma Spray, 50(88W 12Co) + 35(70Ni 16.5Cr 4Fe 4Si 3.8B) + 15(80Ni 20Al)*

5793 *Powder, Plasma Spray, 95Ni 5Al*

5794 *Wire, Welding—31Fe 20Cr 20Ni 20Co 3Mo 2.5W 1(Cb+Ta)*

5795 *Electrodes, Covered, Welding—31Fe 21Cr 20Ni 20Co 3.0Mo 2.5W 1.0(Cb+Ta)*

5796 *Wire, Welding, Alloy, 52Co 20Cr 10Ni 15W*

5797 *Electrodes, Covered, Welding—51.5Co 20Cr 10Ni 15W*

5798 *Wire, Welding, Alloy 47.5Ni 22Cr 1.5Co 9.0Mo 0.60W 18.5Fe (ERNiCrMo-2)*

5799 *Electrodes, Covered Welding—48Ni 22Cr 1.5Co 9.0Mo 0.60W 18.5Fe (ENiCr Mo-2)*

5800 *Wire, Welding—54Ni 19Cr 11Co 10Mo 3.2Ti 1.5Al 0.006B Vacuum Melted*

5801 *Wire, Welding—39Co 22Cr 22Ni 14.5W 0.07La*

5804 *Wire, Welding—15Cr 25.5Ni 1.3Mo 2.2Ti 0.006B 0.30V*

5805 *Wire, Welding—15Cr 25.5Ni 1.2Mo 2.1Ti 0.004B 0.30V Vacuum Induction Melted, Environmntal Controlled Package*

5811 *Wire, Welding, 15Cr 30Ni 1.2Mo 2.2Ti 0.25Al 0.001B 0.30V (0.01-0.03C) Vacuum Induction Melted, Environmental Controlled Packaged*

5812 *Wire, Welding—15Cr 7.1Ni 2.4Mo 1Al Vacuum Melted*

5817 *Wire, Welding—13Cr 2Ni 3W*

5821 *Wire, Welding—12.5Cr, Ferrite Control Grade*

5822 *Wire, Welding—11.8Cr-2.8Ni-1.6Co-1.8 Mo-0.32V, Vacuum Melted*

5823 *Wire, Welding,—11.8Cr 2.8Ni-1.6Co- 1.8Mo-0.32V*

5824 *Welding—17Cr 7.1Ni 1.0Al*

5825 *Welding—16.5Cr 4.8Ni 0.22(Cb+Ta) 3.6Cu*

5826 *Wire, Welding—15Cr 5.1Ni 0.30(Cb+Ta) 3.2Cu*

5827 *Electrodes, Welding, Covered—16.4Cr 4.8Ni 0.22(Cb+Ta) 3.6Cu (E630)*

5828 *Wire, Welding—57Ni 19.5Cr 13.5Co 4.2Mo 3.1Ti 1.4Al 0.006B Vacuum Induction Melted*

5829 *Wire, Welding—56Ni 19.5Cr 18Co 2.5Ti 1.5Al Vacuum Induction Melted*

5832 *Wire, Welding, Alloy—52.5Ni 19Cr 3.0Mo 5.1(Cb+Ta) 0.90Ti 0.50Al 18Fe, Consumable Electrode or Vacuum Induction Melted*

5835 *Wire, Welding, Alloy, 72Ni 3.2Mo 20Cr 2.5(Cb+Ta) 0.48Ti, Vacuum Induction Melted, Environment Controlled Packaged*

5837 *Wire, Welding, Alloy 62Ni 21.5Cr 9.0Mo 3.7(Cb+Ta)*

5838 *Wire, Welding, Alloy—65Ni 16Cr 15Mo 0.30Al 0.06La*

5840 *Wire, Welding—13Cr 8.0Ni 2.3Mo 1.1Al, Vacuum Melted*

LOW ALLOY STEELS

6457 *Wire, Welding—0.95Cr 0.20Mo (0.28- 0.33C) Vac. Melted*

6458 *Wire, Welding—0.65Si 1.25Cr 0.50Mo 0.30V (0.28-0.33C), Vacuum Melted*

6459 *Wire, Welding—1.0Cr 1.0Mo 0.12V (0.18- 0.23C) Vacuum Induction Melted*

6460 *Wire, Welding—0.75Si 0.62Cr 0.20Mo 0.10Zr (0.10-0.17C)*

6461 *Wire, Welding—0.95Cr 0.20V Vacuum Melted (0.28-0.33C)*

6462 *Wire, Welding—0.95Cr 0.20V (0.28-0.33C) 83*

6463 *Wire, Welding—18.5Ni 8.5Co 5.2Mo 0.72Ti 0.10Al, Vacuum Melted, Environment Controlled Packaging*

6464 *Electrodes, Covered Welding—1.05Mo 0.20V (0.06-0.12C)*

6465 *Wire, Welding—2.0Cr 10Ni 8.0Co 1.0Mo 0.02Al 0.06V, (0.10-0.14C), Vacuum Melted, Environment Controlled 0)*

6466 *Wire, Welding Corrosion Resistant 5.2Cr 0.55Mo*

6467 *Electrodes, Covered Welding Steel—5Cr 0.55Mo*

6468 *Wire, Welding—1.0Cr 10Ni 3.8Co 0.45Mo 0.08V (0.14-0.17C) Vacuum Melted, Environment Controlled*

SAE Aerospace Recommended Practices of interest are

Electron Beam Welding, APR 1317
Welding of Structures for Ground Support Equipment, APR 1330
Wave Soldering Practice, APR 1332

Nondestructive Testing of Electron Beam Welded Joints in Titanium Base Alloys, APR 1333

Unified Numbering System

THE UNIFIED NUMBERING System (UNS) provides a method for cross referencing the different numbering systems used to identify metals, alloys, and welding filler metals. With UNS, it is possible to correlate over 3500 metals and alloys used in a variety of specifications, regardless of the identifying number used by a society, trade association, producer, or user.

UNS is produced jointly by SAE and ASTM, and designated SAE HSJ1086/ASTM DS56. It cross references the numbered metal and alloy designations of the following organizations and systems:

> AA (Aluminum Association)
> ACI (Steel Founders Society of America)
> AISI (American Iron and Steel Institute)
> ASME (American Society of Mechanical Engineers)
> ASTM (Formerly American Society for Testing and Materials)
> AWS (American Welding Society)
> CDA (Copper Development Association)
> Federal Specifications
> MIL (Military Specifications)
> SAE (Formerly Society of Automotive Engineers)
> AMS (SAE Aerospace Materials Specifications)

Over 500 of the listed numbers are for welding and brazing filler metals. Numbers with the prefix W are assigned to welding filler metals that are classified by deposited metal composition.

UNDERWRITER'S LABORATORIES, INC.

UNDERWRITER'S LABORTORIES, INC., (UL) is a not-for-profit organization which operates laboratories for the examination and testing of devices, systems, and materials to determine their relation to hazards to life and property. UL Standards for Safety are developed under a procedure which provides for participation and comment from the affected public as well as industry. This procedure takes into consideration a survey of known existing standards, and the needs and opinions of a wide variety of interests concerned with the subject matter of a given standard. These interests include manufacturers, consumers, individuals associated with consumer-oriented organizations, academicians, government officials, industrial and commercial users, inspection authorities, insurance interests, and others.

Examples of standards which contain welding requirements are the following:

Tanks, Steel Aboveground, for Flammable and Combustible Liquids, UL 58
Tanks, Steel Underground, for Flammable and Combustible Liquids, UL 142

Both of these standards include details in regard to the types of welded joints that are allowed to be used and how they are to be tested.

UL should be contacted if no standard can be found for a particular product. The UL *Standards for Safety* pertain to more than 11 000 product types in over 500 generic product categories.

MANUFACTURERS' ASSOCIATIONS

THE FOLLOWING ORGANIZATIONS publish literature which relates to welding. The committees that write descriptive literature are comprised of representatives of equipment or material manufacturers. They do not generally include users of the products. Although some bias may exist, there is much useful information that can be obtained from this literature. The organization should be contacted for further information.

> The Aluminum Association
> 900 19th Street, N.W.
> Suite 300
> Washington, DC 20006
> Phone (202) 862-5100

> American Iron and Steel Institute
> 1000 16th Street, N.W.
> Washington, DC 20036
> Phone (202) 452-7100

> Copper Development Association, Inc.
> Greenwich Office Park
> Building 2
> 51 Weaver Street
> Greenwich, Connecticut 06836
> Phone (203) 625-8210

> Electronic Industries Association
> 2001 Eye Street, N.W.
> Washington, DC 20006
> Phone (202) 457-4900

> National Electrical Manufacturers Association
> 2101 L Street, N.W.
> Washington, DC 20037
> Phone (202) 457-8400

> Resistance Welder Manufacturers Association
> 1900 Arch Street
> Philadelphia, Pennsylvania 19103
> Phone (215) 564-3484

QUALIFICATION AND CERTIFICATION

PREPARED BY A COMMITTEE CONSISTING OF:

M. J. Houle, Chairman
National Board of Boiler & Pressure Vessel Inspectors

R. A. Dunn
Stepson Limited

E. R. Holby
IFR Engineering

G. W. Oyler
Welding Research Council

H. A. Sosnin
Consultant

D. L. Sprow
McDermott, Incorporated

W. F. Urbick
The Boeing Company

J. H. Zirnhelt
Canadian Welding Bureau

WELDING HANDBOOK COMMITTEE MEMBER:
C. Case
Inco Alloys International Welding Products Company

QUALIFICATION AND CERTIFICATION

INTRODUCTION

MOST FABRICATING CODES and standards require qualification and certification of welding and brazing procedures and of welders, brazers, and operators who perform welding and brazing operations in accordance with the procedures.[1] Standards or contractual documents may also require that weldments or brazements be evaluated for acceptance by a qualified inspector. Nondestructive inspection of joints may be required. This should be done by qualified nondestructive testing personnel using specified testing procedures.

Technical societies, trade associations, and government agencies have defined qualification requirements for welded fabrications in standards generally tailored for specific applications such as buildings, bridges, cranes, piping, boilers, and pressure vessels. Welding procedure qualification performed for one standard may qualify for another standard provided the qualification test results meet the requirements of the latter. The objectives of qualification are the same in nearly all cases. Some standards permit acceptance of previous performance qualification by welders and welding operators having properly documented evidence.

A welding procedure specification (WPS) is a document that provides in detail required welding conditions for a specific application. The standard to which the product is being manufactured will normally identify which of the WPS variables are qualification variables.[2] Qualification variables are items in the WPS that if changed beyond specified limits, require requalification of the welding procedure. After requalification, a revised or new WPS should be prepared. Variables other than qualification variables are items in the WPS which may be changed, but the changes do not affect the qualification status. All changes in the procedure require a revision of the written WPS prior to using the revised procedure in production. Normally, a procedure specification must be qualified by demonstrating that joints made by the procedure can meet prescribed requirements. The actual welding conditions used to produce an acceptable test joint and the results of the qualification tests are recorded in a procedure qualification record (PQR).

Welders or welding operators are normally required to demonstrate their ability to produce welded joints that meet prescribed standards. This is known as *welder performance qualification.*

The results of welding procedure or performance qualification must be certified by an authorized representative of the organization performing the qualification tests. This is known as *certification.*

---

1. In the following discussion, the terms *weld, welder, welding*, and *welding operator* imply also *braze, brazer, brazing* and *brazing operator*, respectively, unless otherwise noted.

2. Certain codes including the *AWS D1.1 Structural Welding Code—Steel* and the *ASME Boiler and Pressure Vessel Code* use the expressions "essential variables" for qualification variables and "nonessential variables" for the balance of the WPS items. The *Welding Handbook* is using the terminology of the *AWS B2.1 Standard for Welding Procedure and Performance Qualification.*

STANDARDS REQUIRING QUALIFICATION

TYPICAL CODES AND specifications[3] for welded or brazed products that require performance or procedure qualification are

AWS D1.1, Structural Welding Code—Steel
AWS D1.2, Structural Welding Code—Aluminum
AWS D1.3, Structural Welding Code—Sheet Steel
AWS D1.4, Structural Welding Code—Reinforcing Steel
AWS D3.6, Specification for Underwater Welding
AWS D9.1, Specification for Welding Sheet Metal
AWS D14.1, Specification for Welding Industrial and Mill Cranes
AWS D14.2, Specification for Metal Cutting Tool Weldments
AWS D14.3, Specification for Earthmoving and Construction Equipment
AWS D14.4, Classification and Application of Welded Joints for Machinery and Equipment
AWS D14.5, Specification for Welding of Presses and Press Components
AWS D14.6, Specification for Rotating Elements of Equipment
AWS D15.1, Railroad Welding Specification
ASME Boiler and Pressure Vessel Code
ASME B31, Code for Pressure Piping
National Board Inspection Code
API STD 1104, Standard for Welding Pipelines and Related Facilities

The codes listed above generally include requirements for welding procedure and performance qualification. The *ASME Boiler and Pressure Vessel Code* specifically covers welding and brazing qualification in Section IX.

Other standards address only welding or brazing qualification, and may be referenced in contract documents or codes and specifications for welded or brazed products that do not address qualification. Examples are the following:

AWS B2.1, Welding Procedure and Performance Qualification
AWS B2.2, Brazing Procedure and Performance Qualification
AWS D10.9, Specification for Qualification of Welding Procedures and Welders for Piping and Tubing
MIL-STD-248, Welding and Brazing Procedure and Performance Qualification
MIL-STD-1595, Qualification of Aircraft, Missile, and Aerospace Fusion Welders

PROCESSES

QUALIFICATION OF PROCEDURES generally applies to all joining processes covered by a code or specification. The processes normally include arc, oxyfuel, electron beam, resistance, and electroslag welding, as well as brazing. Performance qualification is normally required of personnel who will do manual, semiautomatic, and machine welding in production.

PROCEDURE SPECIFICATIONS

ARC WELDING

MANY FACTORS CONTRIBUTE to the end result of a welding operation, whether it is manual shielded metal arc welding of plain carbon steel or gas metal arc welding of exotic heat-resistant alloys. It is always desirable and often essential that the vital elements associated with the welding of joints are described in sufficient detail to permit reproduction, and to provide a clear understanding of the intended practices. The purpose of a welding procedure specification (WPS) is, therefore, to define and document in detail the variables involved in welding a certain base metal. To fulfill this purpose efficiently, welding procedure specifications should be as concise and clear as possible, without extraneous detail.

Description and Details

TWO DIFFERENT TYPES of welding procedure specifications are in common use. One is a broad, general type that applies to all welding of a given kind on a specific base metal. The other is a narrower, more definitive type that spells out in detail the welding of a single size and type of joint in a specific base metal or part. Only the broader, more general type is usually required by codes, specifications, and contracts or by building, insurance, and other regulatory agencies.

The narrower, more definitive type is frequently used by employers for control of repetitive in-plant welding operations or by purchasers desiring certain specific metallurgical, chemical, or mechanical properties.[4] However,

---

3. Codes and other standards are discussed in Chapter 13 of this volume.

4. The term *employer* means the manufacturer or contractor who produces the weldment for which welding procedures and performance qualifications are required.

either type may be required by a customer or agency, depending upon the nature of the welding involved and the judgment of those in charge. In addition, the two types are sometimes combined to varying degrees, with addenda to show the exact details for specific joints attached to the broader, more general specification.

Arrangements and details of welding procedure specifications, as written, should be in accordance with the contract or purchaserequirements and good industry practice. They should be sufficiently detailed to ensure welding that will satisfy the requirements of the applicable code, rules, or purchase specifications.

Codes and other standards generally require that the employer prepare and qualify the welding procedure specifications. These should list all of the welding variables such as joint geometry, welding position, welding process, base metal, filler metal, preheat and interpass temperatures, welding current, arc voltage, shielding gas or flux, welding technique, and postweld heat treatment.

Some standards are very specific in defining the content of information to be included in a WPS. They may list the specific variables that are to be addressed, and they will also specify which of those variables are qualification variables. Other codes refer only to the welding variables of a specific process that affect qualification and leave it to the user to determine what other variables and information should be included in the WPS.

Some fabrication codes permit the use of prequalifed joint welding procedures. Under this system, the employer prepares a written welding procedure conforming to the specific requirements of that code for materials, joint design, welding technique, preheat, filler materials, etc. Weld qualification tests need not be made if the requirements are followed in detail. The employer must accept responsibility for the use of prequalified welding procedures in construction when permitted to use them. Deviations from the requirements negate the prequalified status and require qualification by testing. *AWS D1.1, Structural Welding Code—Steel* is an example of a code that permits prequalified joint welding procedures as an alternative to testing by each employer.

Use of prequalified and qualified welding procedures does not guarantee satisfactory production welds. The quality of welds should be verified by nondestructive testing applied during and after welding. Visual, magnetic particle, liquid penetrant, ultrasonic, and radiographic testing are commonly used on welded joints.

Most codes permit an organization other than the employer to prepare test coupons and specimens and to perform the required nondestructive or destructive testing. However, the employer must supervise or control the fabrication of the procedure qualification weldment, and the employer is responsible for the technical accuracy of the WPS and PQR, and for the implementation of the WPS in production.

Regardless of the differences in WPS requirements among fabrication standards, the WPS provides direction to the welder or welding operator and is an important control document. It should have a specific reference number and an approval signature[5] prior to release for production welding. Responsibility for the content, qualification status, and use of a WPS rests upon the employer.

Typical subjects that may be listed in a WPS are discussedbelow. They do not necessarily apply to every process or application. Also, there may be important variables for certain welding processes that are not covered. The applicable code or specification should be consulted.

Scope. The welding processes, the applicable base metal, and the governing specification should be stated clearly.

Base Metal. The base metal(s) should be specified. This may be done either by giving the chemical composition or by referring to an applicable specification. If required, special treatment of the base metal before welding also should be indicated (e.g., heat treatment, cold work, or cleaning). These factors may be important. A welding procedure that would provide excellent results with one base metal, might not provide the same results with another, or even with the same base metal treated differently. Thus, the fabricator should identify the base metal, condition, and thickness.

Welding Process. The welding process that is to be used and type of operation should be clearly defined.

Filler Metal. Composition, identifying type, or classification of the filler metal should always be specified to ensure proper use. Filler metal marking is usually sufficient for identification. In addition, sizes of filler metal or electrodes that can be used when welding different thicknesses in different positions should be designated. For some applications, additional details are specified. These may include manufacturer, type, heat, lot, or batch of the welding consumables.

Type of Current and Range. Whenever welding involves the use of electric current, the type of current should be specified. Some shielded metal arc welding electrodes operate with either ac or dc.If dc is specified, the proper polarity should be indicated. In addition, the current range for each electrode size when welding in different positions and for welding various thicknesses of base metal should be specified.

---

5. The employer may vest the authority for approval of a WPS, certification of a PQR, and certification of a record of performance qualification with a responsible individual in the organization.

Arc Voltage and Travel Speed. For most arc welding processes, it is common practice to list an arc voltage range. Ranges for travel speed are mandatory for automatic welding processes, and frequently for semiautomatic welding processes. If the properties of the base metal would be impaired by excessive heat input, permissible limits for travel speed or bead width are necessary.

Joint Design and Tolerances. Premissible joint design details should be indicated, as well as the sequence for welding. This may be done with cross-sectional sketches showing the thickness of the base metal and details of the joint, or by references to standard drawings or specifications. Tolerances should be indicated for all dimensions.

Joint and Surface Preparation. The methods that can be used to prepare joint faces and the degree of surface cleaning required should be designated in the procedure specification. They may include oxyfuel gas, air-carbon arc, or plasma arc cutting, with or without surface cleaning. Surface preparation may involve machining or grinding followed by vapor, ultrasonic, dip, or lint-free cloth cleaning. Methods and practices should conform to the application and metal.

Tack Welding. Tack welding can affect weld soundness, hence details concerning tack welding procedures should be included in the welding procedure specification. Tack welders must use the designated procedures.

Welding Details. All details that influence weld quality, in terms of the specification requirements, should be clearly outlined. These usually include the appropriate sizes of electrodes for different portions of joints and for different positions, the arrangement of weld passes for filling the joints, and pass width or electrode weave limitations. Such details can influence the soundness of welds and the properties of finished joints.

Positions of Welding. A procedure specification should always designate the positions in which welding can be done. In addition, the manner in which the welding is to be done in each position should be designated (i.e., electrode or torch size; welding current range; shielding gas flow; number, thickness, and arrangement of weld passes, etc.). Positions of welding are shown in Figures 14.1 through 14.4. The defined limits or boundries of each position are shown in Figures 14.5 and 14.6 for groove and fillet welds, respectively.

Preheat and Interpass Temperatures. Whenever preheat or interpass temperatures are significant factors in the production of sound joints or influence the properties of weld joints, the temperature limits should be specified. In many cases, the preheat and interpass temperatures must be kept within a well defined range to avoid degradation of the base metal heat-affected zone.

Peening. Indiscriminate use of peening should not be permitted. However, it is sometimes used to avoid cracking or to correct distortion of the weld or base metals. If peening is to be used, the details of its application and the appropriate tooling should be covered in the WPS.

Heat Input. Heat input during welding is usually of great importance when welding heat-treated steels and crack-sensitive ferrous and nonferrous alloys. Whenever heat input can influence final weld joint properties, details for its control should be prescribed in the WPS.

Second Side Preparation. When joints are to be welded from both sides, the methods that are to be used to prepare the second side should be described in the WPS. They may include chipping, grinding, and air-carbon arc or oxyfuel gas gouging of the root to sound metal. If the second side requires an inspection other than visual, it should be stated on the WPS. Frequently, this preparation is of primary importance in producing weld joints free from cracks and other unsound conditions.

Postheat Treatment. When welded joints or structures require heat treatment after welding to develop required properties, dimensional stability, or dependability, such treatment should be described in the WPS. The same heat treatment should be applied to all procedure qualification test welds. A full description of the heat treatment may appear in the WPS or in a separate fabrication document, such as a shop heat-treating procedure.

Records. If detailed records of the welding of joints are required, the specific requirements for these records should be included in the WPS.

Application

WELDING PROCEDURE SPECIFICATIONS are sometimes required by the purchaser to govern fabrication of a given product in an employer shop. More often, however, the purchaser will specify the properties desired in the weldment in accordance with a code or specification. The employer then develops a welding procedure that will produce the specified results. In other cases, the purchaser, through contract documents, may require specific properties of welded joints because of a critical function or novel design of a particular weldment. The employer must then either conduct additional tests using a standard practice WPS to prove that this practice meets the customer requirements, or devise a new WPS, and qualify it in accordance with the applicable code and the special contract requirements.

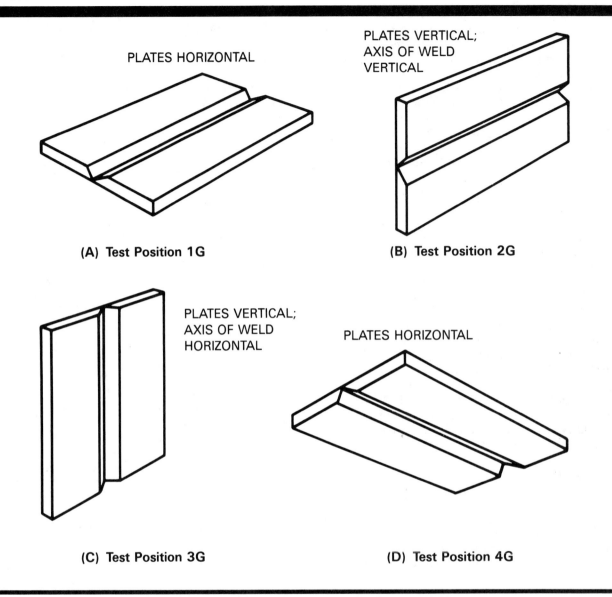

PLATES HORIZONTAL

(A) Test Position 1G

PLATES VERTICAL;
AXIS OF WELD
VERTICAL

(B) Test Position 2G

PLATES VERTICAL;
AXIS OF WELD
HORIZONTAL

(C) Test Position 3G

PLATES HORIZONTAL

(D) Test Position 4G

Figure 14.1—Positions of Test Plates for Groove Welds

Sample WPS

WELDING PROCEDURE SPECIFICATIONS may be prepared in many different ways and may be either brief or long and detailed. Some codes have suggested forms which provide sufficient information but will not cover complex welding conditions. Naturally, more complex and critical applications should have more detailed procedure specifications. These forms may be supplemented with additional notes or instructions or new forms devised to suit the specific requirement. A sample WPS is shown in Figure 14.7. The limits of qualification variables of this WPS are those of *AWS B2.1-84, Standard Welding Procedure and Performance Qualifica-* tion. A supporting PQR is shown in Figure 14.8. The data are not the results of actual tests, but they represent typical results for A515 Grade 70 steel weldments. Figure 14.8 illustrates the preparation of the WPS in Figure 14.7.

The Welding Research Council plans to publish an official collection of procedure qualifiction records, for use by the welding industry.

Oxyfuel Gas Welding

THE PRINCIPLES INVOLVED in the preparation of a WPS for oxyfuel gas welding are the same as for arc welding.

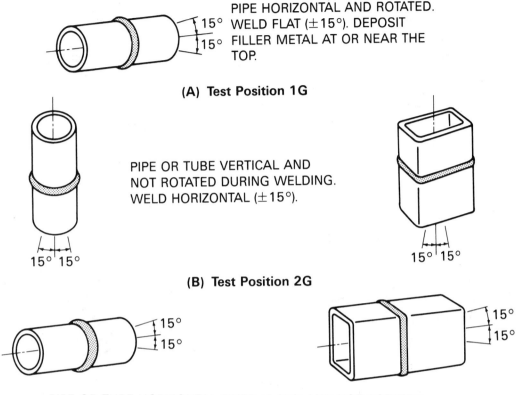

(A) Test Position 1G

PIPE HORIZONTAL AND ROTATED. WELD FLAT (±15°). DEPOSIT FILLER METAL AT OR NEAR THE TOP.

PIPE OR TUBE VERTICAL AND NOT ROTATED DURING WELDING. WELD HORIZONTAL (±15°).

(B) Test Position 2G

PIPE OR TUBE HORIZONTAL FIXED (±15°) AND NOT ROTATED DURING WELDING. WELD FLAT, VERTICAL, OVERHEAD.

(C) Test Position 5B

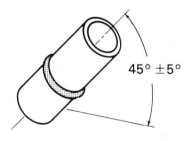

45° ±5°

PIPE INCLINED FIXED (45° ±5°) AND NOT ROTATED DURING WELDING.

(D) Test Position 6G

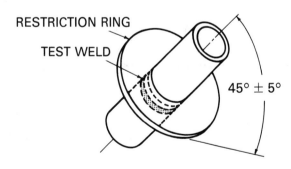

RESTRICTION RING

TEST WELD

45° ± 5°

(E) Test Position 6GR (T, K, or Y Connections)

Figure 14.2—Positions of Test Pipe or Tubing for Groove Welds

However, because of the nature of the process, there are additional qualification variables, and certain other variables may become nonqualification variables. The additional important variables to include in an oxyfuel WPS are as follows:

Type of Fuel Gas. The type fuel gas used is a qualification variable and should be specified. Oxygen is always used to support combustion, but it should be mentioned in the procedure. The flame type, either oxidizing, reducing, or neutral should be included in the WPS.

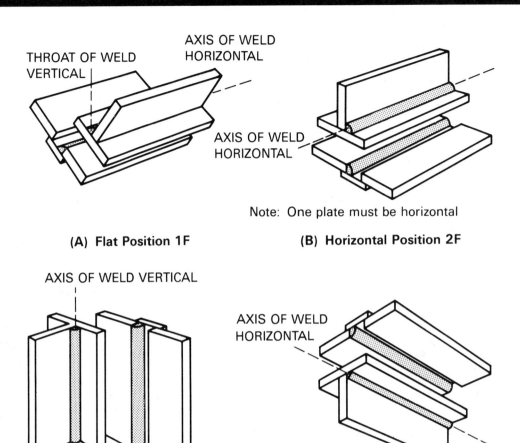

Figure 14.3—Positions of Test Plates for Fillet Welds

Tip Size. The tip size (or sizes) should be given in the WPS. These sizes are based on orifice size, and they control the gas consumption.

Fuel and Oxygen Pressures. Gas pressure ranges at the regulator should be stated.

Joint Design and Tolerances. Permissible joint design should be indicated. This may be done with sketches.

Joint and Surface Preparation. Methods of joint preparation and cleaning should be specified.

Tack Welding. Details of tack welds such as number, size and location should be included in the procedure.

Welding Details. All details which affect weld quality should be given in the procedure. The number of passes or layers, size of filler metal, rate of travel, width of weave (when applicable) should be recorded in the WPS.

Positions of Welding. Although oxyfuel gas Welding may be performed in any position, those positions for which the procedure was written should be stated.

BRAZING

BRAZING PROCEDURE SPECIFICATIONS (BPS) are similar to those for arc welding except for the process data. Pro-

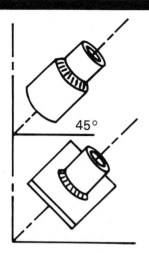

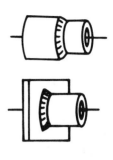

**(A) Test Position 1F
for Flat Position
(Rotated)**

**(B) Test Position 2F
for Horizontal Position
(Fixed)**

**(C) Test Position 2FR
for Horizontal Position
(Rotated)**

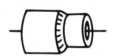

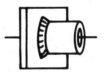

**(D) Test Position 4F
for Overhead Position
(Fixed)**

**(E) Test Position 5F
for Multiple Position
(Fixed)**

Figure 14.4—Positions of Test Pipes for Fillet Welds

cess information may include but not be restricted to the following:

(1) Type of brazing (torch, furnace, induction, etc.)
(2) Brazing filler metal and form
(3) Brazing temperature range
(4) Brazing flux or atmosphere
(5) Flow position
(6) Method of applying filler metal
(7) Time at brazing temperature
(8) Heating and cooling rates

Typical brazing flow positions are flat, vertical down, vertical up, and horizontal. In flat flow, the joint faces and capillary flow are horizontal. In vertical down and vertical up flow, the joint faces are vertical and capillary flow of filler metal is down and up respectively. With horizontal flow, the joint faces are also vertical, but capillary flow of filler metal is horizontal.

Typical codes and standards that address brazing qualification are *AWS B2.2, Standard for Brazing Procedure and Performance Qualification* and *Section IX, Welding and Brazing Qualification, ASME Boiler and Pressure Vessel Code.*

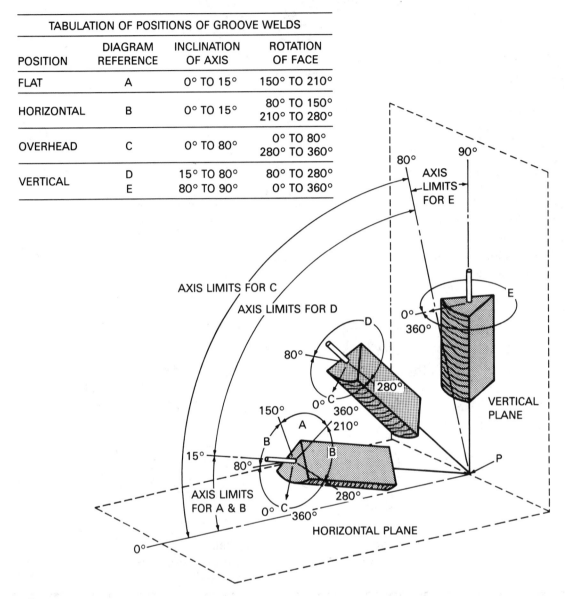

TABULATION OF POSITIONS OF GROOVE WELDS

POSITION	DIAGRAM REFERENCE	INCLINATION OF AXIS	ROTATION OF FACE
FLAT	A	0° TO 15°	150° TO 210°
HORIZONTAL	B	0° TO 15°	80° TO 150° 210° TO 280°
OVERHEAD	C	0° TO 80°	0° TO 80° 280° TO 360°
VERTICAL	D E	15° TO 80° 80° TO 90°	80° TO 280° 0° TO 360°

Notes:
1. The horizontal reference plane is always taken to lie below the weld under consideration.
2. The inclination of axis is measured from the horizontal reference plane toward the vertical reference plane.
3. The angle of rotation of the face is determined by a line perpendicular to the theoretical face of the weld which passes through the axis of the weld. The reference position (0°) of rotation of the face invariably points in the direction opposite to that in which the axis angle increases. When looking at point P, the angle of rotation of the face of the weld is measured in a clockwise direction from the reference position (0°).

Figure 14.5—Positions of Groove Welds

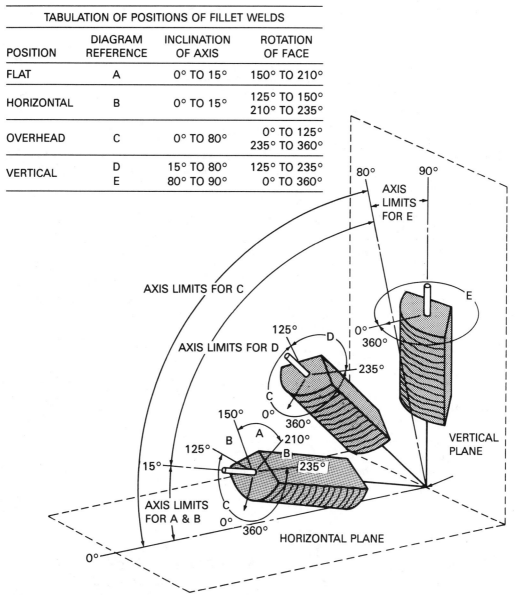

TABULATION OF POSITIONS OF FILLET WELDS			
POSITION	DIAGRAM REFERENCE	INCLINATION OF AXIS	ROTATION OF FACE
FLAT	A	0° TO 15°	150° TO 210°
HORIZONTAL	B	0° TO 15°	125° TO 150° 210° TO 235°
OVERHEAD	C	0° TO 80°	0° TO 125° 235° TO 360°
VERTICAL	D E	15° TO 80° 80° TO 90°	125° TO 235° 0° TO 360°

Note:

For groove welds in pipe the following definitions shall apply:

Horizontal Fixed Position: When the axis of the pipe does not deviate by more than 30° from the horizontal plane and the pipe is not rotated during welding.

Horizontal Rolled Position: When the axis of the pipe does not deviate by more than 30° from the horizontal plane, the pipe is rotated during welding, and the weld metal is deposited within an arc not to exceed 15° on either side of a vertical plane passing through the axis of the pipe.

Vertical Position: When the axis of the pipe does not deviate by more than 10° from the vertical position. (The pipe may or may not be rotated during welding.)*

*Positions in which the axis of the pipe deviates by more than 10° and less than 60° from the vertical shall be considered intermediate.

Figure 14.6—Positions of Fillet Welds

WELDING PROCEDURE SPECIFICATION (WPS)

Date _June 11, 1986_   Identification _WPS-001_

    Revision _0_

Company name _XYZ Fabricators_

Supporting PQR no.(s) _87-123_   Type - Manual (X)   Semi-Automatic ( )

Welding process(es) _SMA_   Machine ( )   Automatic ( )

Backing: Yes (X) No ( )

Backing material (type) _A36, (S)A516 Gr 70 Steel or E7018 WM_

Material number _1_ Group _2_ To material number _1_ Group _2_

Material spec. type and grade _(S) A516 Gr 70 Stl_ To material spec. type and grade _(S)A516 Gr 70 Stl_

Base metal thickness range: Groove _1/16 to 3/4_ Fillet _3/16_

Deposited weld metal thickness range _1/16 to 3/4_

Filler metal F no. _4_   A no. _1_

Spec. no. (AWS) _A5.1_   Flux tradename _N.A._

Electrode-flux (Class) _E7018_   Type _N.A._

Consumable insert: Yes ( ) No (X)   Classifications _N.A._

    Shape _N.A._

Position(s) of joint _All_   Size _N.A._

Welding progression: Up (X) Down ( )   Ferrite number (when reqd.) _____

PREHEAT: **GAS:**

   Preheat temp., min _60°F_   Shielding gas(es) _N.A._

   Interpass temp., max _450°F_   Percent composition _N.A._

   (continuous or special heating, where   Flow rate _N.A._

   applicable, should be recorded)   Root shielding gas _N.A._

POSTWELD HEAT TREATMENT: Trailing gas composition _N.A._

   Temperature range _None_   Trailing gas flow rate _N.A._

   Time range _None_

Tungsten electrode, type and size _N.A._

Mode of metal transfer for GMAW: _N.A._ Short-circuiting ( )   Globular ( )   Spray ( )

Electrode wire feed speed range: _N.A._

Stringer bead (X)   Weave bead (X)   Peening: Yes ( ) No (X)

Oscillation _N.A._

Standoff distance _N.A._

Multiple ( ) or single electrode (X)

Other _Visually inspect each pass for porosity, undercut, sharp corners, or other conditions that would promote inadequate fusion on the subsequent pass._

Weld layer(s)	Process	Class	Dia.	Type & polarity	Amp range	Volt range	Travel speed range	
1	SMA	E7018	1/8	DCEP	100/130	22/24	—	e.g., Remarks, comments, hot wire addition, technique, torch angle, etc.
Bal	SMA	E7018	5/32	DCEP	150/190	23/26	—	

Approved for Production by _To be signed by responsible manager._

Figure 14.7—Welding Procedure Specification

PROCEDURE QUALIFICATION RECORD (PQR)

Page 1 of 2

WPS no. used for test _87-123_ Welding process(es) _SMA_

Company _XYZ Fabricators_ Equipment type and model (sw) _Not available_

JOINT DESIGN USED (2.6.1)

Single V Groove with Backing

Single (X) Double weld ( )

Backing material _ASTM A 36_

Root opening _1/4_ Root face dimension _0_

Groove angle _60_ Radius (J-U) _None_

Back gouging: Yes ( ) No (X) Method _N.A._

BASE METALS (2.6.2)

Material spec. _SA-516_ To _SA-516_

Type or grade _70_ To _70_

Material no. _1_ To material no. _1_

Group no. _2_ To group no. _2_

Thickness _3/8_

Diameter (pipe) _N.A._

Surfacing: Material _N.A._ Thickness _N.A._

Chemical composition _SA-516 Gr 70_

Other _N.A._

FILLER METALS (2.6.3)

Weld metal analysis A no. _1_

Filler metal F no. _4_

AWS specification _A5.1_

AWS classification _E 7018_

Flux class _N.A._ Flux brand _N.A._

Consumable insert: Spec. _N.A._ Class. _N.A._

Supplemental filler metal spec. _N.A._ Class. _N.A._

Non-classified filler metals _N.A._

Consumable guide (ESW) Yes ( ) No (X)

Supplemental deoxidant (EBW) _N.A._

POSITION (2.6.4)

Position of groove _3G_ Fillet _N.A._

Vertical progression: Up (X) Down ( )

PREHEAT (2.6.5)

Preheat temp., actual min _60°F_

Interpass temp., actual max _150°F_

WELD INCREMENT SEQUENCE

POSTWELD HEAT TREATMENT (2.6.6): _None_

Temp. _N.A._

Time _N.A._

Other _N.A._

GAS (2.6.7)

Gas type(s) _N.A._

Gas mixture percentage

Flow rate _N.A._

Root shielding gas _N.A._ Flow rate _N.A._

EBW vacuum (_N.A._) Absolute pressure (_N.A._)

ELECTRICAL CHARACTERISTICS (2.6.8)

Electrode extension _N.A._

Standoff distance

Transfer mode (GMAW) _N.A._

Electrode diameter tungsten _N.A._

Type tungsten electrode _N.A._

Current: AC ( ) DCEP (X) DCEN ( ) Pulsed ( )

Heat input (J/in.) _69,000 min. 92,000 max._

EBW: beam focus current _N.A._ Pulse freq. _N.A._

Filament type _N.A._ Shape _N.A._ Size _N.A._

Other _N.A._

TECHNIQUE (2.6.9)

Oscillation frequency _Manual_ Weave width _Full_

Dwell time _Manual_

String or weave bead _Weave_ Weave width _Full_

Multi-pass or single pass (per side) _Multi_

Number of electrodes _One_

Peening _None_

Electrode spacing _N.A._

Arc timing (SW) _N.A._ Lift ( )

PAW: Conventional _N.A._ Key hole _N.A._

Interpass cleaning: _Grind root, wire brush each pass._

Pass no.	Filler metal size	Amps	Volts	Travel speed (ipm)	Filler metal wire (ipm)	Slope induction	Special notes (process, etc.)
1	1/8	120	24	2.5	N.A.	N.A.	Grind surface
2	1/8	120	23	1.8	N.A.	N.A.	Wire brush
3	1/8	120	23	1.8	N.A.	N.A.	wire brush

Note: Those items that are not applicable should be marked N.A.

Figure 14.8—Procedure Qualification Record

TENSILE TEST SPECIMENS: SUGGESTED PROCEDURE QUALIFICATION RECORD

Type: _B2.1-84 FigA4A_ Tensile specimen size: _15in. min._

Groove (X)   Reinforcing bar ( )   Stud welds ( )

Tensile test results: (Minimum required UTS _____ psi)

Specimen no.	Width, in.	Thickness, in.	Area, in.^{2}	Max load lbs	UTS, psi	Type failure and location
001A-1	1.508	0.384	0.579	43,425	75,000	S-BM
001A-7	1.506	0.378	0.569	43,244	76,000	S-BM

S=Shear; WM= Weld Metal

GUIDED BEND TEST SPECIMENS - SPECIMEN SIZE: _____

Type	Result	Type	Result
Root	No Cracks OK.	Face	No Cracks OK
Root	No Cracks OK	Face	No Cracks OK

MACRO-EXAMINATION RESULTS: Reinforcing bar (_NA_).   Stud (_NA_.)

1. _____   4. _____
2. _____   5. _____
3. _____

SHEAR TEST RESULTS - FILLETS: 1. _____   3. _____
2. _____   4. _____

IMPACT TEST SPECIMENS

Type: _N.A._   Size: _N.A._

Test temperature: _N.A._

Specimen location: WM = weld metal;   BM = base metal;   HAZ = heat-affected zone

Test results:

Welding position	Specimen location	Energy absorbed (ft.-lbs.)	Ductile fracture area (percent)	Lateral expansion (mils)

IF APPLICABLE   **RESULTS**

Hardness tests:_N.A._) Values _N.A._   Acceptable ( )   Unacceptable ( )
Visual (special weldments 2.4.2) (X) _Report No. XXX_   Acceptable (X)   Unacceptable ( )
Torque (_N.A._psi   Acceptable ( )   Unacceptable ( )
Proof test (_NA_) Method _____   Acceptable ( )   Unacceptable ( )
Chemical analysis (_N.A._   Acceptable ( )   Unacceptable ( )
Non-destructive exam (X) Process _X-Ray Report #XXX_ Acceptable (X)   Unacceptable ( )
Other _N.A._   Acceptable ( )   Unacceptable ( )
Mechanical Testing by (Company) _XYZ Fabricators_   Lab No. _XXX_

We certify that the statements in this Record are correct and that the test welds were prepared, welded, and tested in accordance with the requirements of the American Welding Society Standard for Welding Procedure and Performance Qualification (AWS B2.1-84).

Qualifier: _____   Reviewed by: _To be signed by_

Date: _____   Approved by: _responsible manager._

Figure 14.8—(Continued) Procedure Qualification Record

RESISTANCE WELDING

THE FOLLOWING ITEMS are ordinarily covered in a resistance welding procedure specification. Others may be required depending on the welding process.

Welding Process

A WPS SHOULD specify the specific resistance welding process to be used (i.e., spot, seam, flash welding, etc.) because the various processes are distinctly different in many respects.

Composition and Condition of Base Metal

THE BASE METAL to be welded should be specified either by reference to a specification or by chemical composition. The permissible chemical composition range may include base metals covered by more than one specification if they can be welded by the same procedure. The condition (temper) should also be stated. The WPS should include any specific cleaning requirements. The above information is of great importance because a welding procedure that produces excellent results with one base metal may not be satisfactory for another, or for even the same metal in a different condition of heat treatment or cleanliness.

Joint Design

A WPS SHOULD specify all details of the joint design including contacting overlap, weld spacing, type and size of projection, and other similar factors.

Type and Size of Electrode

THE WPS SHOULD state the type of electrode to be used including alloy, contour, and size. If plates, dies, blocks, or other such devices are used, properties that would affect the quality of welding should be specified.

Machine Settings

THE ELECTRODE FORCE, squeeze time, weld time, hold time, off time, welding speed, upsetting time, and other such factors controlled by machine settings should be specifically prescribed in the WPS.

Weld Size

THE SIZE OF each weld or the weld strength is generally an acceptance criterion and should be specified.

Surface Appearance

FACTORS SUCH AS indentation, discoloration, or amount of upset that affect the surface appearance of a weldment should be specified. These factors may be covered by a general requirement rather than by detailed requirements.

Inspection Details

THE PROPERTIES TO be checked—appearance, strength, tightness, etc.—should be specified as well as the method of testing, e.g., shear test, pillow test, peel test, or workmanship sample.

QUALIFICATION OF WELDING PROCEDURE SPECIFICATIONS

THE PURPOSE OF procedure qualification is to determine, by preparation and testing of standard specimens, that welding in accordance with the WPS will produce sound welds and adequate properties in a joint. The type and number of tests required are designated in the standard. The tests are selected to provide sufficient information on strength, ductility, toughness, or other properties of the joint. The qualification variables are specified to maintain desired properties. When significant changes are made to the WPS qualification variables, the properties of the weld joint may be affected, and requalification is required.

The mechanical and metallurgical properties of a welded joint may be altered by the WPS selected for the job. It is the responsibility of each employer to conduct the tests required by the applicable codes and contrac-

tual documents. It is the duty of the engineer or inspector to review and evaluate the results of such qualification tests. These qualification activities must be completed prior to production to assure that the selected combination of materials and methods is capable of achieving the desired results. Rules for the qualification of welding procedures are usually established by the fabrication standard. Individual codes handle procedures in different ways.

Employment of Prequalified Joint Welding Procedures. This concept is based on the reliability of certain proven procedures spelled out by the applicable code or specification. Any deviation outside specified limits voids prequalification.

Employment of Qualification Tests. Qualification tests may or may not simulate the actual conditions anticipated for a given project. Usually, such tests involve conventional butt joints on pipes or plates, or fillet welds between two plates. Base metals, welding consumables, and thermal treatments must follow production welding plans within specific ranges. However, certain other variables, such as joint geometry, welding position, and accessibility may not be considered to be qualification variables.

Employment of Mock-Up Tests. Mock-up tests should simulate actual production conditions to the extent necessary to ascertain that a sound plan with proper tooling and inspection has been selected. Generally, welding codes do not require preparation of welded mock-ups or sample joints unless they are needed to demonstrate that the welding procedures will produce the specified welds. Preparation of mock-ups or sample joints may be required to satisfy contractual conditions or to avoid problems in production. For the latter purpose, mock-ups can indicate the expected quality levels under difficult or restricted welding conditions.

WELDING PROCEDURES

THE BASIC STEPS in the qualification of a welding procedure are as follows:

(1) Prepare a preliminary WPS.
(2) Prepare and weld a suitable test weldment using the preliminary WPS.
(3) Conduct the required nondestructive and destructive tests.
(4) Evaluate the results of preparation, welding, and testing.
(5) Document results on PQR.
(6) Issue approved PQR.
(7) Issue approved WPS.

Preparation of Sample Joints

PLATE OR PIPE samples with a representative welding joint are usually prepared for procedure qualification testing. The size, type, and thickness of the sample are governed by the thickness and type of base metal to be welded in production, and by the type, size, and number of specimens to be removed for testing. The latter are usually prescribed by the applicable code or specification. The base and filler metals and other details associated with welding of sample joints should be in accordance with the particular welding procedure specification that is being qualified. Nondestructive examination of the sample joints is normally preferred,

and may be required by the fabrication code or specification.

Tests of Procedure Qualification Welds

TEST SPECIMENS ARE usually removed from the sample joints for examination to determine certain properties.[6] The type and number of specimens removed and the test details normally depend upon the requirements of the particular application or specification. Usually, the tests include tensile and guided bend specimens to determine strength, ductility, soundness, and adequacy of fusion. If only fillet welds are tested, tensile-shear tests and break or macroetch specimens are usually employed. Additional tests may be specified by the applicable codes or contract documents to meet specific needs. These may include the following:

(1) Fracture tests to determine the notch toughness of the weld and the heat-affected zone. Fracture tests are usually conducted at a specified temperature to minimize the risk of brittle fracture above that temperature. Charpy V-notch specimens are most commonly used for such tests, but many other tests, including drop weight and crack-opening-displacement (COD) tests, are sometimes employed.
(2) Nick-break tests to determine weld soundness at randomly selected locations.
(3) Free bend tests to determine the ductility of weld metal.
(4) Shear tests to determine the shear strength of fillet welds or clad bonding.
(5) Hardness tests to determine adequacy of heat treatment and suitability for certain service conditions. Such tests may be performed on surfaces or on cross sections of welds.
(6) All-weld-metal tenion tests to determine the mechanical properties of the weld metal in the diluted condition.
(7) Elevated temperature tests to determine mechanical properties at temperatures resembling service conditions.
(8) Restraint tests to determine crack susceptibility and the ability to achieve sound welds under restrained conditions.
(9) Corrosion tests to determine the properties needed to withstand aggressive environments.
(10) Nondestructive inspection and macroetch or microetch tests to determine the soundness of a weld.
(11) Delayed cracking tests to detect resistance to hydrogen cracking in high strength, low alloy steels and some other alloys.

---

6. Standard welding tests are covered in Chapter 12 of this volume, and in *AWS B4.0, Standard Methods for Mechanical Testing of Welds.* Miami: American Welding Society.

Recording Test Results

THE DETAILS OF the tests and the results of all tests and examinations are entered on a procedure qualification Record (PQR) as shown in Figure 14.8. (The data shown in Figure 14.8 are hypothetical). When the qualifier is satisfied that the records and results are accurate, they become certified when the qualifier signs the PQR. If the results meet the requirements of a job specification, then WPS may be prepared and issued for production welding. Since the PQR is a certified record of a qualification test, it should not be revised. If information needs to be added later, it can be in the form of a supplement or attachment; records should not be changed by revision. A PQR may support several WPS.

In evaluating a welding procedure or the test results, applicable codes provide general guidance and some specific acceptance-rejection criteria. For instance, the minimum tensile strength and the maximum number of inclusions or other discontinuities are specified by many documents. The acceptability of other properties should be based on engineering judgment. In general, it is desirable that the weld match the mechanical and metallurgical properties of the base metal, but this is not always possible. Not only are weld metal and base metal made of different product forms, but they often have somewhat different chemical compositions and mechanical properties. It requires engineering judgment to select the most important properties for each individual application. This is especially important for service at high or low temperature, or under corrosive conditions.

Changes in a Qualified Procedure

IF A FABRICATOR has qualified a welding procedure and desires at some later date to make changes in that procedure, it may be necessary to conduct additional qualifying tests. Those tests should establish that the changed welding procedure will produce satisfactory results.

Such requalification tests are not usually required when only minor details of the original procedure are changed. They are required, however, if the changes might alter the properties of the resulting welds. Reference should always be made to the governing code or specification to determine whether an essential variable has been changed. Typical procedure factors that may require requalification of the WPS are given in Table 14.1.

BRAZING PROCEDURES

PROCEDURE SPECIFICATIONS FOR brazing are qualified by brazing designated test specimens and evaluating the joints by appropriate tests. Typical tests, applicable joints, and purposes of the tests are given in Table 14.2.

The results of the tests are recorded on the Brazing Procedure Qualification Record, which is similar to a Welding Procedure Qualification Record. It lists the information on the BPS and the results of the appropriate tests. The record is then certified by the employer's representative who witnessed the tests.

RESISTANCE WELDING PROCEDURES

A PROPOSED WELDING procedure must be evaluated by appropriate tests to determine whether joints can be welded consistently and will satisfactorily meet the service requirements to which they will be exposed. Because of the varied nature of resistance welded products, qualification of resistance welding procedures is also varied. Generally, where the welded part is small in size, the procedure may be qualified by making a number of finished pieces and testing them to destruction under service conditions, either simulated or real. In other instances, welds can be made in test specimens that are tested in tension or shear, or inspected for other properties.[7] There are three main steps in qualifying a resistance welding procedure.

Preparation of Weld Specimens

SAMPLE WELD SPECIMENS are prepared and welded in accordance with the welding procedure specifications to be qualified. The number, size, and type of samples are governed by the nature of the tests to be performed for qualification. Where qualification is in accordance with a standard, these requirements are usually specified.

Testing of Specimens

THE NATURE OF the tests to be performed will vary according to the service requirements of the completed parts. Test specimens may actually be subjected to real or simulated service conditions. More frequently, however, tests will be made to determine specific properties of the weld, such as tensile strength, shear strength, surface appearance, and soundness.

Evaluation of Test Results

AFTER TESTING THE welded specimens, the results must be reviewed to determine whether they meet the specified requirements. If all of the requirements are met, the welding procedure is considered qualified.

---

7. Standard tests for resistance welds are given in the latest edition of *AWS C1.1, Recommended practices for resistance welding*. Miami: American Welding Society. Resistance welded butt joints can be evaluated by mechanical tests similar to those used for arc welded joints.

Table 14.1
Welding Procedure and Performance Qualification Variables That May Require Requalification

Procedure Variable	Procedure Requalification May be Required When the Welding Practices are Changed to the Extent Indicated Below	Welder or Welding Operator Requalification May be Required for the Practice Changes Indicated Below
Type, composition, or process condition of the base material	When the base metal is changed to one not conforming to the type, specification, or process condition qualified. Some codes and specifications provide lists of materials which are approximately equivalent from the standpoint of weldability and which may be substituted without requalification.	Usually not required.
Thickness of base metal	When the thickness to be welded is outside the range qualified. The various codes and specifications may differ considerably in this respect. Most provide for qualification on one thickness within a reasonable range; e.g., 3/16 in. to 2T, 1/2T to 2T, 1/2T to 1.1T, etc. Some may require qualification on the exact thickness or on the minimum and maximum thicknesses.	When the thickness to be welded is outside the range qualified. Most codes provide for an unlimited thickness test.
Joint design	When established limits of root openings, root face, and included angle of groove joints are increased or decreased; i.e., basic dimensions plus tolerances. Some codes and specifications prescribe definite upper and lower limits for these dimensions, beyond which requalification is necessary. Others permit an increase in the included angle and root opening and a decrease in the root face without requalification. Requalification is also often required when a backing or spacer strip is added or removed or the basic type of material of a backing or spacer strip is changed.	When changing from a double-welded joint or a joint using backing material to an open root, and vice versa. The addition or deletion of a consumable insert.
Pipe diameter	Usually not required. In fact, some codes permit procedure qualification on plate to satisfy the requirements for welding to be performed on pipe.	When the diameter of piping or tubing is reduced below specified limits. It is generally recognized that smaller pipe diameters require more sophisticated techniques, equipment, and skills.
Type of current and polarity (if dc)	Usually not required for changes involving electrodes or welding materials adapted for the changed electrical characteristics, although sometimes required for change from ac to dc, or vice versa, or from one polarity to the other.	Usually not required for changes involving similar electrodes or welding materials adapted for the changed electrical characteristics.
Electrode classification and size	When electrode classification is changed or when the diameter is increased beyond the allowed ranges specified in the relevant Code.	When the electrode classification grouping is changed and sometimes when the electrode diameter is increased beyond specified limits.
Welding current	When the current is outside the range qualified.	Usually not required.
Position, progression, or both	Usually not required, but desirable.	When the change exceeds the limits of the position(s) qualified or a change in progression.
Deposition of filler metal	When a marked change is made in the manner of filler metal deposition; e.g., from a small bead to a large bead or weave arrangement or from an annealing pass to a no-annealing-pass arrangement, or vice versa.	Usually not required.
Preparation of root of weld for second side welding	When method or extent is changed.	Usually not required.
Preheat and interpass temperatures	When the preheat or interpass temperature is outside the range for which qualified.	Usually not required.
Postheat treatment	When adding or deleting postheating or when the postheating temperature or time cycle is outside the range for which qualified.	Usually not required.

Note: This table is shown only for illustration to indicate the general requirements governing requalification of welding procedures, welders, and operators. It cannot be used by an inspector to determine whether requalification is required in a specific instance. For that information, reference must be made to the particular code or specification applicable to the work being inspected.

Table 14.2
Brazing Procedure Qualification Tests

Test	Applicable Joints	Properties Evaluated
Tension	Butt, scarf, lap, rabbet	Ultimate strength (tension or shear)
Guided bend	Butt, scarf	Soundness, ductility
Peel	Lap	Bond quality, soundness

Changes in a Qualified WPS

IF CHANGES BECOME necessary in an established and qualified WPS, it may be necessary to conduct additional qualifying tests to determine whether the modified WPS will yield satisfactory results. Such qualification tests should not be required when only minor changes have been made in the origianl WPS, but should be required when the changes might alter the properties of the resulting welds. Where a governing code or specification exists, reference should be made to it to determine if requalification is required.

PERFORMANCE QUALIFICATION

WELDER PERFORMANCE

WELDER, WELDING OPERATOR, and tack welder qualification tests determine the ability of the persons tested to produce acceptably sound welds with the process, materials, and procedure called for in the tests. Qualification tests are not intended to be used as a guide for welding during actual construction, but rather to assess whether an individual has a required minimum level of skill to produce sound welds. The tests cannot foretell how an individual will perform on a particular production weld. For this reason, complete reliance should not be placed on qualification testing of welders. The quality of production welds should be determiend by inspection during and following completion of the actual welding.

Various codes, specifications, and governing rules generally prescribe similar, though frequently somewhat different, methods for qualifying welders, welding operators, and tack welders. The applicable code or specification should be consulted for specific details and requirements. The types of tests that are most frequently required are described below.

Qualification Requirements

THE QUALIFICATION REQUIREMENTS for welding pressure vessels, piping systems, and structures usually state that every welder or welding operator shall make one or more test welds using a qualified welding procedure. Each qualification weld is tested in a specific manner (e.g., radiography or bend tests).

Qualification requirements for welding pressure pipe differ from those for welding plate and structural members chiefly in the type of test assemblies used. The test positions may also differ to some extent. As a rule, the tests require use of pipe assemblies instead of flat plate.

Space restrictions may also be included as a qualification factor if the production work involves welding in restricted spaces.

Limitation of Variables

CERTAIN WELDING VARIABLES that will affect the ability of welders or welding operators to produce acceptable welds are considered qualification variables. The specific variables and the limits of variation are called out in the applicable code. A welder or welding operator must be requalified before using a WPS that has qualification variables that are outside the limitation of variables of the individual's performance qualification test. The following are examples of qualification variables from *AWS D1.1, Structural Welding Code—Steel* that require requalification of welder performance if changed:

Welding Process. A change in the welding process requires requalification.

Filler Metal. Shielded metal arc welding electrodes are divided into groups and assigned "F" numbers. The grouping of electrodes is based essentially on those usability characerics that fundamentally determine the ability of a welder to make satisfactory welds with all electrodes in a given group.

Welding Position. Positions are classified differently for plate and pipe. (See Figures 14.1 through 14.4) A change in position may require requalification. For pipe welding, a change from rotated to fixed position requires requalification. The defined limits or boundries of each position are shown in Figures 14.5 and 14.6, respectively, for groove and fillet welds.

Joint Detail. A change in joint type, such as omission of backing on joints welded from one side only, requires requalification.

Plate Thickness. Test plates for groove welds vary in thickness according to the range required for production, generally 3/8 in. (9 mm) to 1 in. (25 mm). The 3/8 in. (9 mm) thickness is employed for qualifying a welder for limited thickness [3/4 in. (19 mm) max.] and the 1-in. (25 mm) test plate for unlimited thickness.

Technique. In vertical position welding, for example, a change in direction of progression of welding requires requalification.

Codes and specifications usually require that welder qualification tests be made in one or more of the most difficult positions to be encountered in production (e.g., vertical, horizontal, or overhead) if the production work involves other than flat position welding. Qualification in a more difficult position may qualify for welding in less difficult positions (e.g., qualification in the vertical, horizontal, or overhead position is usually considered adequate for welding in the flat position).

Test Specimens

TYPICAL GROOVE WELD qualification test plates for unlimited thickness are shown in Figures 14.9 and 14.10. The groove weld plates for limited thickness qualification are essentially the same except the plate thickness is reduced. The welder is usually then qualified to weld plates up to $2T$ where T is the thickness of the qualification plate.

Joint detail for groove weld qualification tests for butt joints on pipe or tubing should be in accordance with a qualified welding procedure specification for a single-welded pipe butt joint. As an alternative, the joint details shown in Figure 14.11 are frequently used. The *AWS D1.1, Structural Welding Code - Steel* provides special joint designs, Figure 14.12, for qualification of welders to weld T-, K-, or Y-connections in pipe, and also fillet and tack welds. A typical fillet weld test plate is shown in Figure 14.13.

Groove weld qualification usually qualifies the welder to weld both groove and fillet welds in the qualified positions. Fillet weld qualification limits the welder to fillet

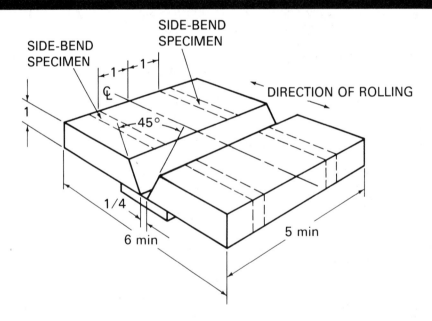

Notes:
1. When radiography is used for testing, no tack welds shall be in the test area.
2. The backing bar thickness shall be 1/4 in. min to 3/8 in. max; backing bar width shall be 3 in. min when not removed for radiography, otherwise 1 in. min.

Figure 14.9—Test Plate for Unlimited Thickness—Welder Qualification

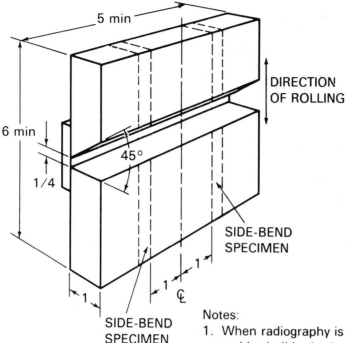

DIRECTION OF ROLLING

5 min

6 min

45°

1/4

SIDE-BEND SPECIMEN

SIDE-BEND SPECIMEN

1

1

1

℄

Notes:

1. When radiography is used for testing, no tack welds shall be in the test area.

2. The backing bar thickness shall be 1/4 in. min to 3/8 in. max; backing bar width shall be 3 in. min when not removed for radiography, otherwise 1 in. min.

Figure 14.10—Optional Test Plate for Unlimited Thickness—Horizontal Position—Welder Qualification

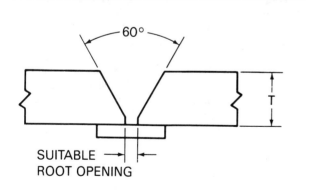

60°

T

SUITABLE ROOT OPENING

Figure 14.11—Tubular Butt Joint—Welder Qualification

Testing of Qualification Welds

ALL CODES AND specifications have definite rules for testing qualification welds to determine compliance with requirements. For groove welds, guided bend test specimens are cut from specific locations in the welded plates and bent in specified jigs. Fillet welds do not readily lend themselves to guided bend tests. In most cases, fillet weld break tests or marcoetch tests, or both, are required.

Radiographic testing may be permitted as an alternative to mechanical or other tests. The primary requirement for qualification is that all test welds be sound and thoroughly fused to the base metal. If the test welds meet the prescribed requirements, the welder or welding operator is considered qualified to weld with processes, filler metals, and procedures similar to those used in the test, within prescribed limits.

Qualification Records

RESPONSIBILITY FOR QUALIFICATION of welders, as for qualification of welding procedures, is with the

welding in only the position qualified and other specified positions of less difficulty. Common type and position limitations for welder qualification are shown in Table 14.3.

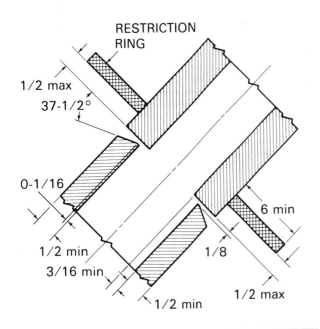

Figure 14.12A—Test Joint for T-, Y-, and K-Connections on Pipe or Square or Rectangular Tubing—Welder Qualification

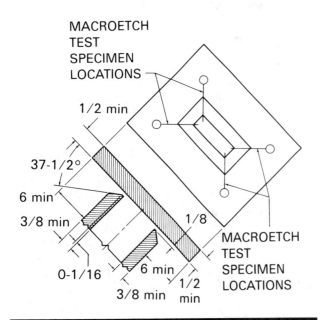

Figure 14.12B—Optional Test Joint for T-, Y-, and K-Connections on Square or Rectangular Tubing—Welder Qualification

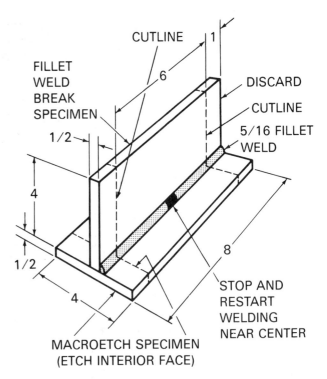

Note: Plate thickness and dimensions are minimum.

Figure 14.13—Fillet-Weld-Break and Macroetch Test Plate—Welder Qualification

employer. After successful qualification of a welder or welding operator, the employer is required to prepare a performance qualification record that provides the necessary details of qualification. The manufacturer then certifies the record by the affixed signature of a responsible individual within the organization. Hence, the term *certified welder* is sometimes referred to rather than qualified welder. In many fabrication codes a fperson that welds a test plate for a PQR becomes qualified within the limitations of variables in the WPS. A certification for a fictitious welder is shown in Figure 14.14. The certification is based on the hypothetical PQR illustrated in Figure 14.8.

Duration of Qualification

THE DURATION OF qualification for welders and welding operators varies from one code or specification to another. One code allows welders to be inactive with the welding process for three months and, if they have welded with another process, the period may be extended to six months. Other codes allow six months

Table 14.3
Welder Qualification—Type and Position Limitations

| Qualification Test | | Type of Weld and Position of Welding Qualified* | | | |
| | | Plate | | Pipe | |
Weld	Plate or pipe positions**	Groove	Fillet	Groove	Fillet
Plate-groove	1G	F	F,H	F (Note 1)	F,H
	2G	F,H	F,H	F,H (Note 1)	F,H
	3G	F,H,V	F,H,V	F,H,V (Note 1)	F,H
	4G	F,OH	F,H,OH		F
	3G & 4G	All	All		F,H
Plate-fillet (Note 2)	1F		F		F
	2F		F,H		F,H
	3F		F,H,V		
	4F		F,H,OH		
	3F & 4F		All		
Pipe-groove	1G	F	F,H	F	F,H
	2G	F,H	F,H	F,H	F,H
	5G	F,V,OH	F,V,OH	F,V,OH	F,V,OH
	6G	Note 3	Note 3	Note 3	Note 3
	2G & 5G	Note 3	Note 3	Note 3	Note 3
	6GR (Fig. 14.12A)	All	All	All	All
	6GR (Fig. 14.12B)		All	T-, Y-, K-Box Only	All
Pipe-fillet	1F		F		F
	2F		F,H		F,H
	2F Rolled		F,H		F,H
	4F		F,H,OH		F,H,OH
	4F & 5F		All		All

Notes:
1. Welders qualified to weld tubulars over 24 in. (600 mm) in diameter with backing or back gouging, for the test positions indicated.
2. Not applicable for fillet welds between parts having a dihedral angle ( ) of 60 deg. or less.
3. Qualified for all but groove welds for T-, Y-, and K-connections.
*Positions of welding: F = flat, H = horizontal, V = vertical, OH = overhead.
** See Figures 14.1 through 14.4.

of inactivity with the welding process before requalification is required. However, for most codes qualification may be extended indefinitely provided the welder or welding operator performs satisfactory work using the welding processes within the stated period of the standard.

BRAZER PERFORMANCE

THE PURPOSE OF brazer performance qualification tests is to verify the ability of brazers to make sound brazed joints following a brazing procedure specification (BPS) and under conditions that will be encountered in production applications. Brazing operators are tested to verify their ability to operate machine or automatic brazing equipment in accordance with a BPS.

Two standards that address brazing performance qualification are *A WS B2.2, Brazing Procedure and Per-* *formance Qualification* and *Section IX, ASME Boiler and Pressure Vessel Code.*

Acceptance Criteria

THE BRAZER OR brazing operator is required to make one or more test samples using a qualified brazing procedure. Acceptance of performance test brazements may be based on visual examination or specimen testing. Qualification by specimen testing qualifies the person to perform production brazing following a BPS that was qualified by either specimen testing or visual examination.

Qualification by Visual Examination

QUALIFICATION BY VISUAL examination is done with a workmanship test brazement representative of the

PERFORMANCE QUALIFICATION TEST RECORD

Name _I.M. Welder_   Identification _EZ_   Welder (X)   Operator ( )

Social security number: _116-26-1035_   Qualified to WPS no. _87-123_

Process(es) _SMA_   Manual (X)   Semi-Automatic ( )   Automatic ( )   Machine ( )

Test base metal specification _(S) A516 Gr 70_   To _(S) A516 Gr 70_

Material number _1 Group 2_   To _1 Group 2_

Fuel gas (OFW) _N.A._

AWS filler metal classification _E7018_   F no. _4_

Backing:	Yes	(X)	No	( )	Double ( ) or Single side (X)
Current:	AC	( )	DC	(X)	Short-circuiting arc (GMAW) Yes ( ) No (X)
Consumable insert:	Yes	( )	No	(X)	
Root shielding:	Yes	( )	No	(X)	

TEST WELDMENT   **POSITION TESTED**   **WELDMENT THICKNESS (T)**

GROOVE:

Pipe   1G ( )   2G ( )   5G ( )   6G ( )   6GR ( )   Diameter(s)_____ (T) _____

Plate   1G ( )   2G ( )   3G (X)   4G ( )   (T) _3/8_

Rebar   1G ( )   2G ( )   3G ( )   4G ( )   Bar size _____ Butt ( )

                                                        Spliced butt ( )

FILLET:

Pipe ( )   1F ( )   2F ( )   3F ( )   4F ( )   5F ( )   Diameter _____ (T) _____

Plate ( )   1F ( )   2F ( )   3F ( )   4F ( )   (T) _____

Other (describe) _Progression-Uphill_

Test results: Remarks

Visual test results	N/A ( )	Pass (X)	Fail ( )
Bend test results	N/A ( )	Pass (X)	Fail ( )
Macro test results	N/A ( )	Pass ( )	Fail ( )
Tension test	N/A ( )	Pass (X)	Fail ( )
Radiographic test results	N/A ( )	Pass (X)	Fail ( )
Penetrant test	N/A ( )	Pass ( )	Fail ( )

QUALIFIED FOR:

PROCESSES

GROOVE: **THICKNESS**

Pipe   1G ( )   2G ( )   5G ( )   6G ( )   6GR ( )   (T) Min _____ Max _____ Dia

Plate   1G (X)   2G ( )   3G (X)   4G ( )   (T) Min _3/16"_ Max _3/4"_

Rebar   1G (X)   2G ( )   3G (X)   4G ( )   _Flare_ Bar size _____ Min _____ Max

FILLET:

Pipe   1F (X)   2F ( )   4F ( )   5F ( )   (T) Min _____ Max _____

Plate   1F (X)   2F (X)   3F (X)   4F ( )   (T) Min _____ Max _____

Rebar   1F ( )   2F ( )   3F ( )   4F ( )   Bar size _____ Min _____ Max _____

Weld cladding ( ) _N.A._ Position(s) _____ T Min _____ Max _____ Clad Min _____

Consumable insert ( ) _N.A._ Backing type ( ) _N.A._

Vertical Up (X)   Down ( )

Single side ( )   Double side ( )   No backing ( ) _N.A._

Short-circuiting arc ( )   Spray arc ( )   Pulsed arc ( ) _N.A._

Reinforcing bar - butt ( ) or Spliced butt ( ) _N.A._

The above named person is qualified for the welding process(es) used in this test within the limits of essential variables including materials and filler metal variables of the AWS Standard for Welding Procedure and Performance Qualification (AWS B2.1).

To be signed by respon-

Date tested _Current Date_   Signed by _sible manager_

                                                             Qualifier

Figure 14.14—Performance Qualification

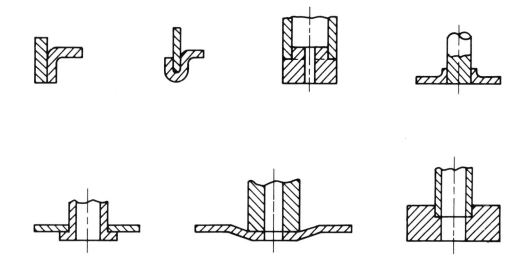

Figure 14.15—Typical Workmanship Test Brazements

design details of the joint to be brazed in production. Typical workmanship brazements are shown in Figure 14.15. The completed brazement must meet the requirements of the standard.

Qualification by Specimen Testing

EITHER A STANDARD test brazement or a workmanship test brazement is used for qualification by specimen testing. A standard test brazement may have a butt, scarf, lap, single- or double-spliced butt, or a rabbet joint in plate or pipe, as shown in Figure 14.16.

The test joint normally is sectioned, the exposed surfaces are polished and etched, and the brazed joint is examined at low magnification for discontinuities. Peel tests may be used in place of macroetch tests, or vice versa, using lap joints or spliced butt joints. In the latter case, the splice plates are peeled from the specimens. The macroetched cross sections or the peeled surfaces must meet the acceptance criteria of the appropriate standard.

Qualification Variables

TYPICAL BRAZING PERFORMANCE qualification variables that may require requalification when changed are as follows:

(1) Brazing process
(2) Base metal

(3) Base metal thickness
(4) Brazing filler metal composition
(5) Method of adding filler metal
(6) Brazing position
(7) Joint design

To minimize the number of brazing performance qualification tests, base metals are separated into groups that have similar brazeability. Brazing filler metals are grouped according to similarity of composition or melting range.

Performance Qualification Test Record

BRAZING OF THE test brazement and the results of the acceptance tests are reviewed by the person responsible for the test. The information is recorded on a Performance Qualification Test Record. A sample form similar to Figure 14.14 that addresses brazing variables is available in *AWS B2.2, Standard for Brazing Procedure and Performance Qualification*.

THERMAL SPRAY OPERATORS

OPERATORS OF EQUIPMENT for applying thermal sprayed coatings may be qualified using the recommended procedures in *AWS C2.16, Guide for Thermal Spray Operator and Equipment Qualification*. They should be qualified for a specific coating process and method of application.

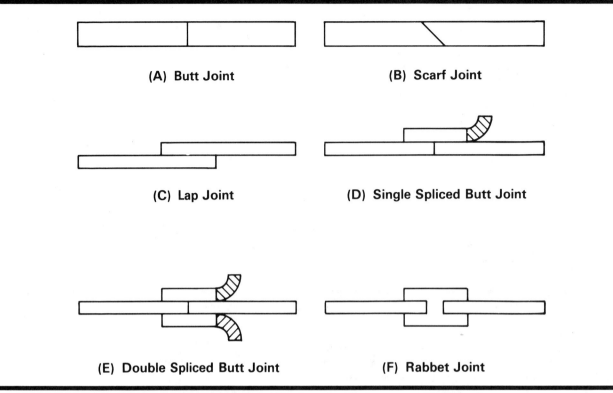

Figure 14.16—Brazed Joint Designs for Performance Qualification

Initial operator qualification should consist of the following:

(1) A written test covering all aspects of the coating process and application method
(2) Demonstration of ability to operate the equipment
(3) Testing to determine knowledge of proper masking procedures for both surface preparation and spraying

(4) Surface preparation of the thermal spray test specimens

Test specimens should be prepared from alloys specified by the purchaser and agreed to by the certifying agent. The type and number of test specimens to be prepared should be specified in the contract documents. The testing conditions and results of the test should be recorded and certified by the representative that observed the tests.

QUALITY CONTROL AND INSPECTION QUALIFICATION

NONDESTRUCTIVE TESTING PERSONNEL

QUALIFICATION OF PERSONNEL is one of the most important aspects of nondestructive testing. As in welding, qualification assures that the person performing the test method has proper knowledge and experience to apply the test and interpret the results. Most nondestructive testing personnel are qualified in accordance with the *American Society for Nondestructive Testing Recommended Practice No. SNT-TC-lA*. This document estab-

lishes guidelines for education, training, experience, and testing requirements for various levels of competence. The levels and the functions that can be performed by each level are as follows:

NDT Level I

LEVEL I PERSONNEL are qualified to properly perform specific calibrations, specific tests, and specific evaluations according to written instructions and to record the results. They receive the necessary guidance or supervision from a certified NDT Level II or III individual.

NDT Level II

LEVEL II PERSONNEL are qualified to set up and calibrate equipment, and to interpret and evaluate results with respect to applicable codes, standards, and specifications. They must (1) be thoroughly familiar with the scope and limitations of the method; (2) exercise responsibility for on-the-job training and guidance of trainers and NDT Level I personnel; and (3) organize and report nondestructive testing investigations.

NDT Level III

PERSONNEL CERTIFIED FOR Level III are capable of and responsible for (1) establishing techniques, (2) interpreting codes, standards, and specifications, and (3) designating the particular test method and technique to be used. They are accountable for the complete NDT operation and evaluation of results in terms of existing codes and specifications. It is desirable that they have general familiarity with all commonly used NDT methods. They may be responsible for the training and examination of Level I and Level II personnel for certification.

WELDING INSPECTOR QUALIFICATION

THE AMERICAN WELDING SOCIETY conducts examinations to determine the proficiency of welding inspectors and associate welding inspectors in accordance with *AWS QCI, Standard for Qualification and Certification of Welding Inspectors.* The objective of the welding inspector qualification and certification program is to verify the knowledge of welding inspectors who may be required to inspect weldments or welded products in accordance with codes or other specified requirements.

The examinations are designed to require a level of knowledge that can be determined by standardized tests,

and do not necessarily reflect the level of competence required for each and every code or specification used in the welding fabrication field. Employers must recognize their responsibility for additional training and education of their welding inspectors when needed to satisfy current production needs and the requirements of new standards and specifications.

A certified welding inspector (CWI) must be able to perform inspections and verify that the work and records conform to the applicable codes and specifications. A certified associate welding inspector (CAWI) performs inspections under the direction of a CWI. However, the CWI is responsible for determining if weldments conform to workmanship and acceptance standards.

The applicants must have specified minimum visual, educational, and work qualifications directly related to weldments fabricated to a code or specification. The applicant's work experience must include one or more of the following:

(1) Design of weldments
(2) Production welding
(3) Inspection of welds
(4) Repair of welds

Candidates are also required to pass written tests covering the following:

(1) One of four specified codes and specifications
(2) Principles of welding, nondestructive testing, materials, heat treatment, and other fundamentals
(3) Practical application of welding inspection

Upon successful completion of the examination, the American Welding Society issues a certificate to the applicant. Certification must be renewed every three years. This may be done by re-examination or by meeting certain requirements for renewal without examination.

STANDARDIZATION OF QUALIFICATION REQUIREMENTS

THE PRIMARY REASON for standardization in any field is economics, in particular as it affects the consumer. Increasing concern for the protection of consumer interest makes it imperative that standards be developed on a voluntary consensus basis with all affected interests represented. Standardization in the field of welding qualification could lead to exceptional savings. Most welding qualification standards have only minor differences, but now require separate qualification for each. The result-

ing duplication of effort and cost in the fabrication and construction industries is highly undesireable, and results in significant increases in cost to the consumer with no related benefits.

The American Welding Society has responded in the field of welding qualification in several progams. The Welding Insepctor Qualification and Certification program is one example of standardization of qualification that is receiving recognition and acceptance by industry and government agencies. The two general qualification standards, *AWS B2.1, Standard for Welding Procedure and Performance Qualification* and *AWS B2.2, Standard for Brazing Procedure and Performance Qualification*, are designed to provide single standards to replace several different standards that contain similar rules for qualification of procedures and performance.

SUPPLEMENTARY READING LIST

American Bureau of Shipping. *Rules for Building and Classing Steel Vessels*, New York: American Bureau of Shipping: (latest edition).

American Welding Society. *Standard for brazing procedure and performance qualification*, American Welding Society B2.2, Miami, FL: (1985) latest edition.

———. *Standard for Welding Procedurre and Performance Qualifiction*, B2.1, Miami, FL: American Welding Society (1984): latest edition.

———. *Welding inspection*, Miami, FL: American Welding Society (latest edition).

Cary, H. B., *Modern welding technology*. Englewood Cliffs, NJ: Prentice-Hall, (1979).

The Procedure handbook of arc welding, 12th Ed., Cleveland, OH: The Lincoln Electric Company (1973).

INSPECTION

PREPARED BY A COMMITTEE CONSISTING OF:

Philip I. Temple, Chairman
Detroit Edison

Harry W. Ebert
Exxon Research and Engineering Company

Saundra Walmsley
Westinghouse Electric

Leon A. Laime
Northeast Utilities

Philip A. Grimm
United McGill Corporation

WELDING HANDBOOK COMMITTEE MEMBER:
D. R. Amos
Westinghouse Electric Corporation

CHAPTER 15

INSPECTION

WELDING INSPECTORS

INSPECTORS OF WELDMENTS should know and understand (1) the requirements and the duties of welding inspectors, engineers, and others, and (2) the principles of those tests employed to evaluate the adequacies of weldments and their compliance with welding procedures, codes, and specifications.[1] In general, the information about the inspection of weldments also applies to brazements. However, some inspection methods are not suitable for brazed joints because of the thinness of the filler metal in a joint.

Welding inspectors must be familiar with all phases of the fabrication activities that apply to their product line, including a working knowledge of applicable codes, specifications, and laws governing the quality of specific components. Often this will encompass requirements for welding procedure specifications, qualification testing, and the application of mechanical, proof, and nondestructive tests. In this regard, it is beneficial if inspectors are certified under the Welding Inspector Qualification and Certification Program of the American Welding Society.[2]

REQUIREMENTS FOR INSPECTORS

IN CONJUNCTION WITH their duties, welding inspectors may represent (1) the manufacturer, the purchaser, or the insurer of welded components, or (2) a public interest or government organization that has no commercial involvement. Since the represented organization may subject inspectors to various pressures and influences, it is essential for inspectors to maintain highly ethical standards. Such pressures can be minimized by active management efforts to reduce conflicts and to support the inspectors in their work. Support of the inspectors is most effective when the inspection department reports directly to upper management. This frees the inspectors from pressures of costs and schedules.

Inspectors must be physically fit to observe welding and related activities. In some cases, this may involve ability to climb or to enter specific components. In all cases, the inspector's duties involve good vision. Some inspection tools also require adequate color vision for maximum effectiveness.

Good relationships with coworkers is a nontechnical trait possessed by effective inspectors. Inspectors serve as a link between departments with differing interests, and they may be called upon to make unpopular decisions. Inspectors need positive attitudes toward the job and other persons, and should strive to earn the cooperation and respect of coworkers by impartial, consistent, and technically correct decisions.

Inspectors must be knowledgeable in welding and welding related activities. Usually, this includes several welding processes and the associated welding consumables and preweld and postweld activities. The inspector's skills should include the ability to interpret drawings, specifications, code requirements, and fabrication and testing procedures. Inspectors must be able to communicate orally and in writing. They should

---

1. Additional information may be found in *Welding Inspection*, *ANSI/AWS B1.0, Guide for the Nondestructive Inspection of Welds*, and *AWS QC1, Standard for Qualification and Certification of Welding Inspectors*, published by the American Welding Society.

2. Refer to the latest edition of *AWS QC1, Standard for Qualification and Certification of Welding Inspectors, Including Guide to AWS Welding Inspector Qualification and Certification* for more information on this program.

understand testing methods and be able to perform or evaluate such tests. Finally, inspectors should maintain records that document their findings, acceptances, and rejections. The ability to analyze these findings and to recommend corrective actions is highly desirable. These skills can be learned through formal and on-the-job training. They can be demonstrated by successful completion of the certification program offered by the American Welding Society.

DUTIES OF INSPECTORS

INSPECTORS MUST BE familiar with the product, engineering drawings, specification requirements, and manufacturing and testing procedures. This includes the handling and disposition of any deviations from requirements or procedures.

The writing and qualifying of welding procedure specifications are usually an engineering function, while verification of welder and welding operator qualifications and the surveillance of performance records are inspection functions. Various codes and contractural documents have different testing and acceptance criteria. Decisions must be based on the detailed knowledge of those requirements. Knowledge of such requirements also applies when performing or monitoring applicable mechanical, nondestructive, and proof testing operations.

In many cases, actual inspection activities should not be limited to acceptance or rejection of the final product. Once a component has been completed, it may be difficult to evaluate its total quality and to take corrective actions that would ensure desired quality. Consequently, certain activities are appropriate both prior to and during fabrication. Typically, these may include:

(1) Inspection prior to welding:
   (a) Procedures and qualifications
   (b) Fabrication and testing plans
   (c) Base metal specifications and quality
   (d) Welding equipment and welding consumables
   (e) Joint designs and joint preparations
(2) Inspection during welding:
   (a) Proper fitup, procedures for control of distortion, and tack weld quality
   (b) Conformity to welding procedures and fabrication plans
   (c) Preheat and interpass temperature requirements and measurement methods
   (d) Control and handling of welding consumables
   (e) Use of welders qualified for specific operations
   (f) Interpass and final cleaning
   (g) Visual and, if required, nondestructive inspection
(3) Inspection after welding:
   (a) Conformity to drawings and specifications
   (b) Cleaning and visual inspection
   (c) Nondestructive, proof, and if required, mechanical testing
   (d) Repair activities
   (e) Postweld heat treatment, if required
   (f) Documentation of fabrication and inspection activities

When the above activities are applied to specific industries and products, the duties and requirements of welding inspectors must be tailored to meet specific needs. At times, these details are left to the discretion of individual inspectors; in other cases, they are documented as part of a detailed quality assurance plan.

INSPECTION PLAN

IN SOME CASES, 100 percent inspection of all production may be required. In other cases, sampling procedures are specified. Sampling is the selection of a representative portion of the production for inspection or test purposes. By statistically analyzing the inspection results of a representative sample of a production lot, valid conclusions can be drawn concerning the entire lot. The sampling technique, sample size, and inspection method should be part of a formal inspection procedure, or they should be reported with the inspection results.

SAMPLING METHODS

SOME OF THE terms and methods found in test procedures for sampling are *partial* sampling which may be specified or random, and *statistical sampling*.

Partial Sampling

PARTIAL SAMPLING INVOLVES the inspection of a certain number of weldments which is less than the total number in the lot or production run. The method of selection of the weldments to be inspected and the type of inspection should be prescribed. The rejection criteria and disposition for any substandard product found by the sampling inspection should also be specified on the procedure.

Specified Partial Sampling. A particular frequency or sequence of sample selection from the lot or production run may be prescribed on the drawing or specification. This is known as specified partial sampling. An example of this type of sampling is the selection of every fifth unit for inspection, starting with the fifth unit. It is usually specified for fully automated operations to monitor process deterioration.

Random Partial Sampling. When the units to be inspected are selected in a random manner, it is called *random partial sampling.* For example, one out of every five units from a production run may be inspected, and the specific selection is made by the inspector in a random manner. Since it is not generally known in advance which units will be selected for inspection, equal care must be taken in the production of all units and a more uniform product quality should result.

To improve the effectiveness of partial sampling, a progressive examination system may be employed. The frequency of sampling is increased if the percentage of rejections exceeds a specified figure. For example, two additional welds are inspected using the same inspection method when a weld is rejected. Additional constraints may be imposed, such as welds produced by a particular welder, or welds made by various welding operators of machine or automatic welding equipment at the same welding station. Selection criteria may also be based on the welding procedure. One result in invoking such criteria is that an increasingly large sample is required if the weld quality is low (i.e., two rejects may require two additional inspections, or two rejects require eight additional inspections.) All completed welds may eventually require examination. Also, the welds must be selected randomly without notifying the welder or welding operator of the welds to be examined.

Statistical Sampling

STATISTICAL SAMPLING USES one procedure to select probability samples, and another to summarize the test results so that specific inferences can be drawn and the risks calculated by probability. This method of sampling may miss some defective products. The percentage of defective products accepted is reduced by increasing the size of the sample. This method of sampling is used primarily for mass produced products.

Statistical sampling will function most economically after statistical control has been attained. Control charts are the usual criteria for establishing statistical control of a process or operation. When it is desirable to take full advantage of the various plans for sampling, a statistical control chart must first be established which provides criteria for expected tolerances and data trends for inspected items.

Plans for statistical sampling should be designed to maximize acceptance of good material and minimize acceptance of poor material. It is not possible to estimate the quality of a production lot from a small sample. The higher the percentage of production sampled, the lower the probability of error.

PLAN SELECTION

IF A SAMPLE plan is given in the procedure specification, the inspector need only follow the procedure.

Complete inspection is used when weldments of the highest quality are required for critical service. One or more methods of nondestructive testing, along with visual inspection, may be specified for critical joints.

For the average welding job, inspection generally involves a combination of complete examination and random partial sampling. All of the welds are inspected visually; random partial sampling is then applied using one or more of the other methods of nondestructive testing.

NONDESTRUCTIVE EXAMINATION

DEFINITIONS AND GENERAL DESCRIPTION

NONDESTRUCTIVE EXAMINATION (NDE) is a term used to designate those inspection methods that allow materials to be examined without changing or destroying their usefulness. *Nondestructive Testing* (NDT) and *Nondestructive Inspection* (NDI) are terms sometimes used interchangeably with NDE and are generally considered synonymous. Nondestructive examinations are per-

formed on weldments to verify that the weld quality meets the specification, and to determine if weld quality is degraded during service.

Nondestructive examination methods may include the following elements:

(1) A source of probing energy
(2) A component to be examined that is compatible with the energy source

(3) A detection device for measuring the differences in or the effects on the probing energy

(4) A means to display for record the results of the test

All NDE methods must include the following to render valid examination results:

(1) A trained operator
(2) A procedure for conducting the tests
(3) A system for reporting the results
(4) A standard to interpret the results

The forms of energy that can be used for nondestructive examination of a particular weldment depend on the physical properties of the base and weld metals, the joint designs, and the accessability with the energy sources. Thorough knowledge of each NDE method is needed for proper selection of the appropriate methods for each application. The commonly used NDE methods that are applicable to the inspection of weldments are

(1) Visual inspection, with or without optical aids (VT)
(2) Liquid penetrant (PT)
(3) Magnetic particle (MT)
(4) Radiography (RT)
(5) Eddy current (ET)
(6) Ultrasonic (UT)
(7) Acoustic emission (AET)

There are other NDE methods, such as heat transfer and ferrite testing, that are used for special cases. They are not discussed here. The considerations generally used in selecting an NDE method for welds are summarized in Table 15.1.

With respect to welding inspection, the meaning of the terms *discontinuity*, *flaw*, and *defect* are as follows:

A *discontinuity* is an interruption in the typical structure of a weldment. It may consist of a lack of homogeneity in the mechanical, metallurgical, or physical characteristics of the base metal or the weld metal. A discontinuity is not necessarily a defect.

A *flaw* is nearly synonymous with a discontinuity but has a connotation of undesirability.

A *defect* is a discontinuity which by nature or effect renders a weldment unable to meet specifications or acceptance standards. The term designates a rejectable condition.

VISUAL INSPECTION

VISUAL INSPECTION IS the most extensively used nondestructive examination on weldments. It is simple and relatively inexpensive; it does not normally require special equipment; and it gives important information about conformity to specifications. One requirement for

this method of inspection is that the inspector must have good vision.

Visual inspection should be the primary evaluation method of any quality control program. It can, in addition to flaw detection, discover signs of possible fabrication problems in subsequent operations, and can be incorporated in process control programs. Prompt detection and correction of flaws or process deviations can result in significant cost savings. Conscientious visual inspection before, during, and after welding can detect many of the discontinuities that would be found later by more expensive nondestructive examination methods.

Equipment

VISUAL AIDS AND gages are sometimes used to make detection of discontinuities easier and to measure the sizes of welds or discontinuities in welds. Lighting of the welded joint must be sufficient for good visibility. Auxiliary lighting may be needed. If the area to be inspected is not readily visible, the inspector may use mirrors, borescopes, flashlights, or other aids. Low power magnifiers are helpful for detecting minute discontinuities. However, care must be taken with magnifiers to avoid improper judgment of the discontinuity size.

Inspection of welds usually includes quantitative as well as qualitative assessment of the joint. Numerous standard measuring tools are available to make various measurements, such as joint geometry and fit-up, weld size, weld reinforcement height, misalignment, and depth of undercut. Typical gages for measuring welds are shown in Figure 15.1.

Some situations require special inspection gages to assure that specifications are met. Indicators, such as contact pyrometers and crayons, should be used to verify that the preheat and interpass temperatures called for in the welding procedure are being used. Proper usage of visual aids and gages requires proper inspector training.

Prior to Welding

EXAMINATION OF THE base metal prior to fabrication can detect conditions that tend to cause weld defects. Scabs, seams, scale, or other harmful surface conditions may be found by visual examination. Plate laminations may be observed on cut edges. Dimensions should be confirmed by measurements. Base metal should be identified by type and grade. Corrections should be made before work proceeds.

After the parts are assembled for welding, the inspector should check the weld joint for root opening, edge preparation, and other features that might affect the quality of the weld. Specifically, the inspector should check the following conditions for conformity to the applicable specifications:

Table 15.1
Nondestructive Testing Methods

Equipment Needs	Applications	Advantages	Limitations
Visual			
Magnifiers, color enhancement, projectors, other measurement equipment (i.e., rulers, micrometers, optical comparators, light source.)	Welds which have discontinuities on the surface.	Economical, expedient, requires relatively little training and relatively little equipment for many applications.	Limited to external or surface conditions only. Limited to the visual acuity of the observer/inspector.
Radiography (Gamma)			
Gamma ray sources, gamma ray camera projectors, film holders, films, lead screens, film processing equipment, film viewers, exposure facilities, radiation monitoring equipment.	Most weld discontinuities including incomplete fusion incomplete penetration, slag as well as corrosion and fit-up defects, wall thickness dimensional evaluations.	Permanent record—enables review by parties at a later date. Gamma sources may be positioned inside of accessible objects i.e., pipes, etc for unusual technique radiographs. Energy efficient source requires no electrical energy for production of gamma rays.	Radiation is a safety hazard—requires special facilities or areas where radiation will be used and requires special monitoring of exposure levels and dosages to personnel. Sources (gamma) decay over their half-lives and must be periodically replaced. Gamma sources have a constant energy of output (wavelength) and cannot be adjusted. Gamma source and related licensing requirements are expensive. Radiography requires highly skilled operating and interpretive personnel.
Radiography (X-Rays)			
X-ray sources (machines) electrical power source, same general equipment as used with gamma sources (above).	Same application as above.	Adjustable energy levels, generally produces higher quality radiographs than gamma soruces. Offers permanent record as with gamma radiography (above).	High initial cost of x-ray equipment. Not generally considered portable, radiation hazard as with gamma sources, skilled operational interpretive personnel required.
Ultrasonic			
Pulse-echo instrument capable of exciting a piezoelectric material and generating ultrasonic energy within a test piece, and a suitable cathode ray tube scope capable of displaying the magnitudes of the reflected sound energy. Calibration standards, liquid couplant.	Most weld discontinuities including cracks, slag, inadequate penetration, incomplete fusion; lack of bond, in brazing; thickness measurements.	Most sensitive to planar type discontinuities. Test results known immediately. Portable. Most ultrasonic flaw detectors do not require an electrical outlet. High penetration capability. Reference standards are required	Surface conditions must be suitable for coupling to transducer. Couplant (liquid) re-required. Small welds and thin materials may be difficult to inspect. Reference standards are required. Requires a relatively skilled operator/inspector. The results of the inspection are usually reported the operator on a preprinted form.
Magnetic Particle			
Prods, yokes, coils suitable for inducing magnetism into the test piece. Power source (electrical). Magnetic powders, some applilications require special facilities and ultraviolet lights.	Most weld discontinuities open to the surface—some large voids slightly sub-surface. Most suitable for cracks.	Relatively economical and expedient. Inspection equipment is considered portable. Unlike dye penetrants, magnetic particle can detect some near surface discontinuities. Indications may be preserved on transparent tape.	Must be applied to ferro-magnetic materials. Parts must be clean before and after inspection. Thick coatings may mask rejectable indiations. Some applications require parts to be demagnetized after inspection. Magnetic particle inspection requires use of electrical energy for most applications

Table 15.1 (Continued)

Equipment Needs	Applications	Advantages	Limitations
Liquid Penetrant			
Fluorescent or visible (dye penetrant, developers, cleaners, solvents, emulsifiers, etc.). Suitable cleaning gear. Ultraviolet light source if fluorescent dye is used.	Weld discontinuities open to surface (i.e., cracks, porosity, seams.)	May be used on all non-porous materials. Portable, relatively inexpensive equipment. Expedient inspection results. Results are easily interpreted. Requires no electrical energy except for light source. Indications may be further examined visually.	Suface films such as coatings, scale, smeared metal mask or hide rejectable defects. Bleed out from porous surfaces can also mask indications. Parts must be cleaned before and after inspection.
Eddy Current			
An instrument capable of inducing electromagnetic fields within a test piece and sensing the resulting electrical currents (eddy) so induced with a suitable probe or detector. Calibration standards.	Weld discontinuities open to the surface (i.e., cracks, porosity, incomplete fusion) as well as some subsurface inclusions. Alloy content, heat treatment variations, wall thickness.	Relatively expedient, low cost. Automation possible for symmetrical parts. No couplant required. Probe need not be in intimate contact with test piece.	Limited to conductive materials. Shallow depth of penetration. Some indications may be masked by part geometry due to sensitivity variations. Reference standard required.
Acoustic Emission			
Emission sensors, amplifying electronics, signal processing electronics including frequency gates, filters. A suitable output system for evaluating the acoustic signal (audio monitor, visual monitor, counters, tape recorders, X-Y recorder).	Internal cracking in weld during cooling, crack initiation and growth rates.	Real time and continuous surveilance inspection. May be inspected remotely. Portability of inspection apparatus.	Requires the use of transducers coupled on the test part surface. Part must be in "use" or stressed. More ductile materials yield low amplitude emissions. Noise must be filtered out of the inspection system.

(1) Joint preparation, dimensions, and cleanliness
(2) Clearance dimensions of backing strips, rings, or consumable inserts
(3) Alignment and fit-up of the pieces being welded
(4) Welding process and consumables
(5) Welding procedures and machine settings
(6) Specified preheat temperature
(7) Tack weld quality

Sometimes, examination of the joint fit-up reveals irregularities within code limitations. These may be of concern, and should be watched carefully during later steps. For example, if the fit-up for a fillet weld exhibits a root opening when none is specified, the adjacent leg of the fillet weld needs to be increased by the amount of the root opening.

Inspection During Welding

DURING WELDING, VISUAL inspection is the primary method for controlling quality. Some of the aspects of fabrication that can be checked include the following:

(1) Treatment of tack welds

(2) Quality of the root pass and the succeeding weld layers
(3) Proper preheat and interpass temperatures
(4) Sequence of weld passes
(5) Interpass cleaning
(6) Root condition prior to welding a second side
(7) Distortion
(8) Conformance with the applicable procedure

The most critical part of any weld is the root pass because many weld discontinuities are associated with the root area. Competent visual inspection of the root pass may detect a condition that would result in a discontinuity in the completed weld. Another critical root condition exists when second side treatment is required of a double welded joint. This includes removal of slag and other irregularities by chipping, arc gouging, or grinding to sound metal.

The root opening should be monitored as welding of the root pass progresses. Special emphasis should be placed on the adequacy of tack welds, clamps, or braces designed to maintain the specified root opening to assure proper joint penetration and alignment.

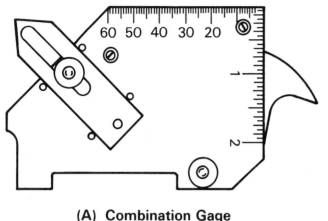

(A) Combination Gage

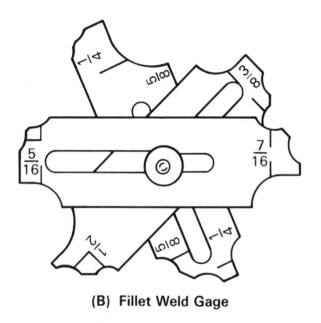

(B) Fillet Weld Gage

Figure 15.1—Typical Gages for Measuring Weld Size and Shape

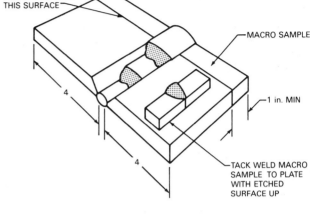

(A) Groove Weld Standard

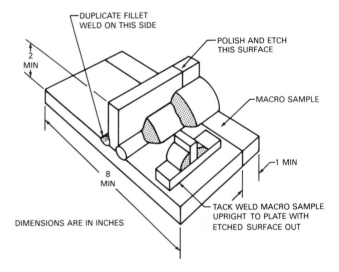

DIMENSIONS ARE IN INCHES

(B) Fillet Weld Standard

Figure 15.2—Workmanship Standards for Groove and Fillet Welds

Inspection of successive layers of weld metal usually concentrates on bead shape and interpass cleaning. Sometimes, it is carried out with the assistance of workmanship standards. Examples of such standards are shown in Figure 15.2. These examples show sections of joints similar to those in manufacture in which portions of successive weld layers are shown. Each layer of the production weld may be compared with the corresponding layer of the workmanship standard. In some situations, workmanship samples may be an actual product.

When preheat and interpass temperatures are specified, they should be monitored at the proper times with a suitable temperature measuring device (i.e., a temperature-indicating crayon or pyrometer). The amount of heat input and also the sequence and placement of each weld pass may be specified to maintain mechanical properties or limit distortion, or both.

To ensure the weld quality as work progresses, each weld layer should be visually checked, by the welder, for surface irregularities and adequate interpass cleaning to avoid subsequent slag inclusions or porosity.

After Welding

ITEMS THAT ARE checked by visual inspection after welding include the following:

(1) Final weld appearance
(2) Final weld size
(3) Extent of welding
(4) Dimensional accuracy
(5) Amount of distortion
(6) Postweld heat treatment

Most codes and specifications describe the type and size of discontinuities which can be accepted. Many of the following discontinuities on the surface of a completed weld can be found by visual inspection:

(1) Cracks
(2) Undercut
(3) Overlap
(4) Exposed porosity and slag inclusions
(5) Unacceptable weld profile
(6) Roughness of the weld faces

For detection and accurate evaluation of discontinuities, the weld surface should be thoroughly cleaned of oxide and slag. The cleaning operation must be carried out carefully to avoid masking discontinuities from view. For example, if a chipping hammer is used to remove slag, the hammer marks could mask fine cracks. Shot blasting may peen the surface of relatively soft weld metal and hide discontinuities.

Dimensional accuracy of weldments is determined by conventional measuring methods. The conformity of weld size and contour may be determined by the use of a suitable weld gage. The size of a fillet weld in joints whose members are at right angles, or nearly so, is defined in terms of the length of the legs. The gage should determine whether the leg size is within allowable limits, and whether there is excessive concavity or convexity. Special gages may be required where the members are at acute or obtuse angles.

For groove welds, the height of reinforcement should be consistent with specified requirements. When not specified, the inspector may rely on judgment, guided by what is considered good welding practice. Surface

appearance requirements differ widely. In general, the weld surface appearance should meet the requirements of the standard. Visual standards or sample weldments submitted by the fabricator and agreed to by the purchaser can be used as guides to appearance. Sometimes a smooth weld, strictly uniform in size, is required because the weld is part of the exposed surface of the product, and good appearance is desirable.

A fabrication standard may permit limited amounts of undercut, undersize, and piping porosity, but cracks, incomplete fusion, and unfilled craters are generally not acceptable. Undercut, overlap, and improper weld profile act as stress raisers under load, and cracks may develop at these locations under cyclic loading.

Some steels, such as ASTM A514 and A517, are susceptible to delayed cracking.[3] The applicable standard may specify a waiting period before visual inspection of welds in crack-sensitive steels.

When a postweld heat treatment is specified, the operation should be monitored and documented by an inspector. Items of importance in heat treatment may include the following:

(1) Area to be heated
(2) Heating and cooling rates
(3) Holding temperature and time
(4) Temperature measurement and distribution
(5) Equipment calibration

Care should be taken when judging the quality of a weld from the visible appearance alone. Acceptable surface appearance does not prove careful workmanship, and is not a reliable indication of subsurface weld integrity. However, proper visual inspection procedures before and during fabrication can increase product reliability over that based only on final inspection.

LIQUID PENETRANT TESTING

LIQUID PENETRANT TESTING (PT) is a method that reveals open discontinuities by bleedout of a liquid penetrant medium against a contrasting background developer.[4] The technique is based on the ability of a penetrating liquid to wet the surface opening of a discontinuity and to be drawn into it. If the discontinuity is significant, penetrant will be held in the cavity when the excess is removed from the surface. Upon application of a liquid-propelled or dry powder developer, blotter

---

3. Refer to *Welding Handbook*, Vol. 4, 7th Ed., (1982) 3-7.
4. Refer to *ASTM E165, Standard Recommended Practice for Liquid Penetrant Inspection* for more information on the subject. *ASTM E433, Standard Reference Photographs for Liquid Penetrant Inspection* may be useful for classifying indications. These standards are available from the American Society for Testing and Materials, 1916 Race Street, Philadelphia, PA, 19103.

action draws the penetrant from the discontinuity to provide a contrasting indication on the surface.

The basic steps involved in the application of a liquid penetrant test are relatively simple. It is a relatively inexpensive and reliable method for obtaining information on questionable welds. All inspectors should be trained and certified in this method.

The following sequence is normally used in the application of a typical penetrant test. When the order is changed or short cuts are taken, the validity of the test is suspect.

(1) Clean the test surface.
(2) Apply the penetrant.
(3) Wait for the prescribed dwell time.
(4) Remove the excess penetrant.
(5) Apply the developer.
(6) Examine the surface for indications and record results.
(7) Clean to remove the residue.

Liquid penetrant methods can be divided into two major groups: Method A—fluorescent penetrant testing, and Method B—visible penetrant testing. The two types differ in the penetrant used. For Method A, the penetrating medium is fluorescent, meaning that it glows when illuminated by ultraviolet or "black" light. This is illustrated in Figure 15.3. Method B uses a visible penetrant, usually red in color, that produces a contrasting indication against the white developer background as shown in Figure 15.4.

The sensitivity may be greater using the fluorescent method. However, both offer extremely good sensitivity when properly applied. The difference in sensitivity is primarily due to the fact that the eye can discern the contrast of a fluorescent indication under black light more readily than a color contrast under white light. In the latter case, the area must be viewed with adequate white light.

Applications

THE LIQUID PENETRANT test, when properly performed, is reliable for the inspection of welds. Except for visual examination, it is perhaps the most commonly used nondestructive test for surface examination of nonmagnetic parts. While the test can be performed on as-welded surfaces, the presence of weld bead ripples and other irregularities may hinder interpretation of indications. In the examination of welds joining cast metals with this method, the inherent surface imperfections in castings may also cause problems with interpretation. If the surface condition causes an excessive amount of irrelevant indications, it may be necessary to remove troublesome imperfections by light grinding prior to inspection. The visible penetrant test method is com-

Figure 15.3—Fluorescent Particles Illuminated Under "Black" Light

Figure 15.4—Red Penetrant Contrasted Against White Developer

monly used for field testing applications because of its portability.

Liquid penetrant testing may also be used to check the accuracy of results obtained by magnetic particle testing of ferromagnetic weldments.

Equipment

UNLIKE MANY OF the other types of nondestructive testing, liquid penetrant testing requires little equipment.

Most of the equipment consists of containers and applicators for the various liquids and solutions that are used. The testing materials are available in convenient aerosol spray cans, which can be purchased separately or in kits. Fluorescent penetrant testing requires a high intensity ultraviolet light source and facilities for reduction or elimination of outside lighting. Other related equipment, such as additional lighting, magnifiers, drying apparatus, rags, and paper towels, might be needed depending on the specific test application.

Materials

LIQUID PENETRANT TESTING materials consist of fluorescent and visible penetrants, emulsifiers, solvent base removers, and developers.[5] Intermixing of materials of different methods and manufacturers is not recommended. The inspection materials used should not adversely affect the serviceability of the parts tested.

Penetrants. Water-washable penetrants are designed to be directly washed from the surface of the test part after a suitable penetrant (dwell) time. It is important to avoid overwashing because the penetrants can be washed out of discontinuities if the rinsing step is too long or too vigorous.

Post-emulsifiable penetrants are insoluble in water and cannot be removed with water rinsing alone. They must first be treated with an emulsifier. The emulsifier combines with the penetrant to form a water-washable mixture, which can be rinsed from the surface of the part.

Solvent-removable penetrants can be removed by first wiping with clean, lint-free material, and repeating the operation until most traces of penetrant have been removed. The remaining traces are removed by wiping the surface with clean, lint-free material lightly moistened with a solvent remover. This type is intended primarily for portability and for localized areas of inspection. To minimize removal of penetrant from any discontinuities, the use of excess solvent must be avoided.

Emulsifiers. These are liquids used to emulsify the excess oily penetrant on the surface of the part, rendering it water washable.

Developers. Dry powder developers are free-flowing and noncaking as supplied. Care should be taken not to contaminate the developer with penetrant, as the specks can appear as indications.

Aqueous wet developers are normally supplied as dry powders to be suspended or dissolved in water, depending on the type of developer.

Nonaqueous suspendible developers are supplied as suspensions of developer particles in nonaqueous solvent carriers ready for use as supplied. They are applied to the area by conventional or electrostatic spray guns or by aerosol spray cans after the excess penetrant has been removed and the part has dried. Nonaqueous wet developers form a white coating on the surface of the part when dried, which serves as a contrasting background for visible penetrants and as developing media for fluorescent penetrants.

Liquid film developers are solutions or colloidal suspensions of resins in a suitable carrier. These developers will form a transparent or translucent coating on the surface. Some types of film developer may be stripped from the part and retained for record purposes.

Procedures

THE FOLLOWING GENERAL processing procedures, outlined in Figure 15.5, apply to both the fluorescent and visible penetrant testing methods. The temperature of the penetrant materials and the surface of the weldment should be between 60 ° and 125 °F (16 ° and 52 °C).

Satisfactory results can usually be obtained on surfaces in the as-welded condition. However, surface preparation by grinding or machining is necessary when surface irregularities might mask the indications of unacceptable discontinuities, or otherwise interfere with the effectiveness of the examination.

Precleaning. The success of any liquid penetrant testing procedure is greatly dependent upon the freedom of both the weld area and any discontinuities from contaminants (soils) that might interfere with the penetrant process. All parts or areas of parts to be examined must be clean and dry before the penetrant is applied. "Clean" is intended to mean that the surface must be free of any rust, scale, welding flux, spatter, grease, paint, oily films, dirts, etc., that might interfere with penetration. All of these contaminants can prevent the penetrant from entering discontinuities. If only a section containing a weld is to be inspected, the weld and adjacent areas within 1 in. (25 mm) of the weld must be cleaned.

Drying After Cleaning. It is essential that the parts be thoroughly dry after cleaning because any liquid residue will hinder the entrance of the penetrant. Drying may be accomplished quickly by warming the parts in drying ovens, with infrared lamps, or with forced hot air. Part temperatures must not exceed 125 °F (52 °C) prior to application of penetrant.

---

5. These materials can be flammable or emit hazardous and toxic vapors. Observe all manufacturer's instructions and precautionary statements.

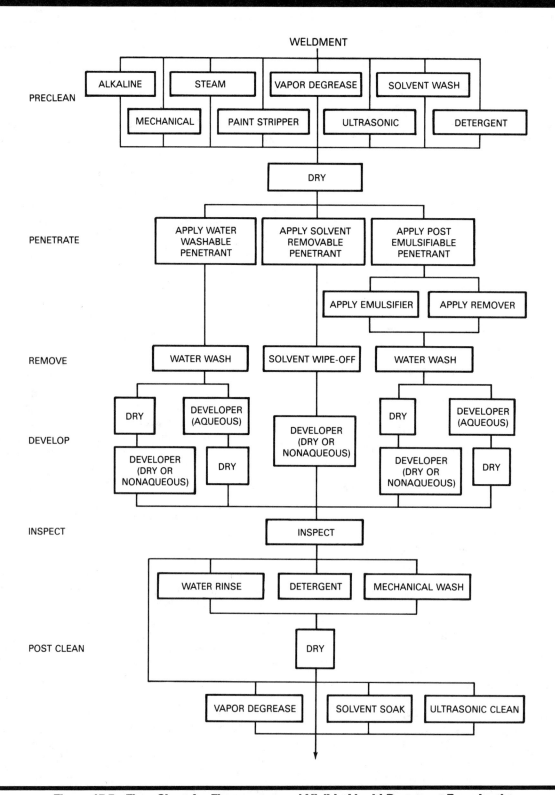

Figure 15.5—Flow Chart for Fluorescent and Visible Liquid Penetrant Examinations

Penetrant Application. After the part has been cleaned, dried, and cooled to approximate ambient temperatures [125 °F maximum (52 °C)], the penetrant is applied to the surface to be inspected so that the entire part or the weld area is completely covered with penetrant.

There are various methods for effective application of penetrant such as dipping, brushing, flooding, or spraying. Small parts are quite often placed in suitable baskets and dipped into a tank of penetrant. On larger parts, and those with complex geometries, penetrant can be applied effectively by brushing or spraying. Aerosol sprays are also a very effective and convenient means of application. With spray applications, it is important that there is proper ventilation. This is generally accomplished through the use of a properly designed spray booth and exhaust system.

After application, excess penetrant is drained from the part. Care should be taken to avoid pools of penetrant on the part.

The length of time that the penetrant remains on the part to allow proper penetration should be as recommended by the penetrant manufacturer or as required by the applicable code or procedure. If penetrant characteristics are materially affected by a prolonged dwell time, as evidenced by difficulty in removing the excess, reapply the penetrant for the original prescribed dwell time. The penetrant must remain wet for the entire dwell time to be effective.

Removal of Excess Penetrant. After the required dwell time, the excess penetrant is removed. The procedure depends on the type of penetrant.

Water-washable penetrants can be removed directly from the part surface using water spray or immersion equipment. Most water-washable penetrants can be removed effectively within a temperature range from 60 ° to 110 °F (16 to 45 °C), but for consistent results, the temperature recommended by the penetrant supplier should be used.

Excessive washing may cause penetrant to be washed out of discontinuities. The rinsing operation for fluorescent penetrants should be done under black light to show when the surface penetrant has been adequately removed.

In special applications where water rinse facilities are not available, penetrant may be removed by wiping the surface with a clean, absorbent material dampened with water until the excess surface penetrant is removed.

Post-emulsifiable penetrants are not directly water-washable; they require the use of an emulsifier (oil or water base). After the required penetration time, the excess penetrant on the part is emulsified by dipping, flooding, or spraying with the required emulsifier. After application of the emulsifier, the part should be drained to avoid pooling of the emulsifier. Emulsification dwell time begins as soon as the emulsifier has been applied. Nominal emulsification time should be as recommended by the manufacturer. Effective rinsing of the emulsified penetrant from the surface of the part can be accomplished in the same manner as for water-washable penetrants.

With *solvent-removable penetrants*, excess penetrant is removed by wiping with clean, lint-free material, repeating the operation until most traces of penetrant have been removed. Then, a lint-free material is lightly moistened with solvent and wiped over the surface until all remaining traces of excess penetrant have been removed. Excessive solvent must be avoided to minimize removal of penetrant from discontinuities.

For some critical applications, flushing of the surface with solvent should be prohibited. When effective penetrant removal cannot be done using the previously described procedure, flushing the surface with solvent may be necessary, but it may also jeopardize the accuracy of the test.

The solvent may be trichorethylene, perchlorethylene, acetone, or a volatile petroleum distillate. The former two are somewhat toxic; the latter two are flammable. Proper safety precautions must be observed. Producers of these testing materials also market compatible precleaners for their penetrant products, and their use is recommended.

Drying of Parts. During the preparation of parts for examination, drying is necessary following the removal of excess penetrant, and before the application of developers. Parts can be dried using the procedures described previously for drying after the initial cleaning operation.

Part temperature should never exceed 125 °F (52 °C). Excessive drying temperature or time can cause evaporation of the penetrant, which may impair the sensitivity of the inspection. Drying time will vary with the size, nature, and number of parts under inspection. When excess penetrant is removed with the solvent wipe-off technique, drying should be by normal evaporation.

Developing Indications. Indications are developed by drawing the penetrant back out of any discontinuities, through blotting action, which spreads it on the surface. This increases the visibility of the penetrant to the eye. Developers are used either dry or suspended in an aqueous or nonaqueous solvent that is evaporated to dryness before inspection. They should be applied immediately after the excess penetrant has been removed from the part surface, prior to drying in the case of aqueous developers, and immediately after the part has been dried for all other developer forms.

There are several methods for effective application of various types of developers, such as dipping, immersing, flooding, spraying, or dusting. The size, configuration, surface condition, and number of parts to be processed will influence the choice of developer.

Dry powder developers should be applied in a manner that assures complete coverage of the area being inspected. Excess powder may be removed by shaking or tapping the part gently, or by blowing with low-pressure (5 to 10 psi) dry, clean, compressed air.

Aqueous developers should be applied immediately after the excess penetrant has been removed from the part, but prior to drying. After drying, the developer appears as a white coating on the part. Aqueous developers should be prepared and maintained in accordance with the manufacturer's instructions and applied in a manner to assure complete and even coverage. The parts are then dried as described previously.

Nonaqueous wet developers are applied to the part by spraying, as recommended by the manufacturer, after the excess penetrant has been removed and the part has been dried. This type of developer evaporates very rapidly at normal temperature, and does not require the use of a dryer. It must be used with proper ventilation. The developer must be sprayed in a manner that assures complete coverage with a thin, even film.

The length of time before the coated area is examined visually for indications should not be less than about 7 minutes, or as recommended by the manufacturer. Developing time begins immediately after the application of dry powder developer or as soon as a wet (aqueous or nonaqueous) developer coating is dry. If bleedout does not alter the test results, development periods of over 30 minutes are permitted.

Examination. Although examination of parts is done after the appropriate development time, it is good practice to observe the surface while applying the developer as an aid in evaluating indications. Examination for fluorescent penetrant indications is done in a dark area. Maximum ambient light of about 3 footcandles (32 lux) is allowed for critical examination. Higher levels may be used for noncritical work. Black light intensity should be a minimum of 5000 W/in.² (800 W/cm²) on the surface of the part being examined.

Visible penetrant indications can be observed under natural or artificial white light. A minimum light intensity of 32.5 footcandles (350 lux) is recommended.

Postcleaning. Postcleaning is necessary in those cases where residual penetrant or developer could interfere with subsequent processing or with service requirements. It is particularly important where residual inspection materials might combine with other materials in service to produce corrosion products. A suitable technique, such as simple water rinsing, machine washing, vapor degreasing, solvent soaking, or ultrasonic cleaning may be employed. In the case of developers, it is recommended that cleaning be carried out within a short

time after examination so that the developer does not adhere to the part. Developers should be removed prior to vapor degreasing because the heat can bake the developer onto the parts.

Interpretation of Indications

PROBABLY THE MOST common defects found using this process are surface cracks. Because surface cracks are critical for most applications, this test capability is a valuable one. An indication of a crack is very sharp and well defined. Most cracks exhibit an irregular shape, and the penetrant indication will appear identical. The width of the bleedout is a relative measure of crack depth. A very deep crack will continue to produce an indication even after recleaning and redeveloping several times.

Surface porosity, metallic oxides, and slag will also hold penetrant and cause an indication. Depending on the exact shape of the pore, oxide, or slag pocket, the indication will be more or less circular. In any case, the length-to-width ratio usually will be far less than that of a crack.

Other discontinuities such as inadequate joint penetration and incomplete fusion can also be detected by penetrant inspection if they are open to the surface. Undercut and overlap are not easily detected by this type of testing; they can be evaluated more effectively using visual inspection.

MAGNETIC PARTICLE TESTING

Principles

MAGNETIC PARTICLE TESTING (MT) is a nondestructive method used to detect surface or near surface discontinuities in magnetic materials.[6] The method is based on the principle that magnetic lines of force in a ferromagnetic material will be distorted by a distinct change in material continuity. A discontinuity or a sharp dimensional change are examples of such a change. If a discontinuity in a magnetized material is open to or close to the surface, the magnetic flux lines will be distorted at the surface, a condition termed *flux leakage*. When fine magnetic particles are distributed over the area of the discontinuity while the flux leakage exists, they will accumulate at the discontinuity and be held in place. This principle is illustrated in Figure 15.6. The accumulation of particles will be visible under proper lighting conditions. While there are variations in the magnetic particle test method, they all are dependent on the prin-

---

6. Additional information is contained in *ASTM E709, Standard Recommended Practice for Magnetic Particle Examination*, and *ASTM E125, Magnetic Particle Indications on Ferrous Castings*, available from the American Society for Testing and Materials, 1916 Race Street, Philadelphia, PA 19103.

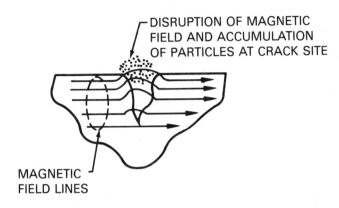

DISRUPTION OF MAGNETIC FIELD AND ACCUMULATION OF PARTICLES AT CRACK SITE

MAGNETIC FIELD LINES

Figure 15.6—Magnetic Particles Attracted to Discontinuities by Flux Leakage

ciple that magnetic particles will be retained at the locations of magnetic flux leakage.

The essential requirements of the test method are the following:

(1) The part must be magnetized.
(2) Magnetic particles must be applied while the part is magnetized.
(3) Any accumulation of magnetic particles must be observed and interpreted.

A ferromagnetic material can be magnetized either by passing an electric current through the material or by placing the material within a magnetic field originated by an external source. The entire component, or a portion of it, can be magnetized as dictated by size and equipment capacity or by need. As previously noted, the discontinuity must interrupt the normal path of the lines of force. If a discontinuity is open to the surface, the flux leakage will be a maximum for a given size and shape of discontinuity. When a discontinuity is below the surface, flux leakage will be lower. Discontinuities must be open to the surface or must be in the near subsurface to create flux leakage of sufficient strength to accumulate magnetic particles.

If a discontinuity is oriented nearly parallel to the lines of force, it will be essentially undetectable. Because discontinuities may occur in any orientation, it is usually necessary to magnetize the part at least twice so that induced magnetic lines of force are produced in different directions to perform an adequate examination.

The lines of force must be of sufficient strength to indicate those discontinuities which are unacceptable, yet not so strong that an excess of particles is accumulated locally, thereby masking relevant indications.

Applications

MAGNETIC PARTICLE TESTING is extensively used in nondestructive examination of weldments. The process can assist in determining the quality of welds in ferrous and other magnetic metals. Considerable magnetic particle testing is performed on finished weldments. Moreover, when used at prescribed intervals in the completion of a multiple pass weld, it can be an extremely valuable process control tool. When used in this manner, discontinuities are found while they are easily correctable rather than later when the difficulty and cost of repair are greater. One use is the examination of the root of a groove weld, or an excavation for weld repair prior to welding. This will ensure that all defects have been satisfactorily removed. Magnetic particle testing is also applied to weldments following, and sometimes prior to, stress relief. Most defects that might occur during this treatment would be surface-related.

The aerospace industry sometimes uses magnetic particle inspection on lightweight structural components. Fatigue behavior is a prime design consideration for many components where surface quality is extremely critical. With thin sections, many subsurface flaws are also detectable. Magnetic particle testing may be performed during routine maintenance. It provides a good check for potential structural problems.

Magnetic particle testing is frequently applied to plate edges prior to welding to detect cracks, laminations, inclusions, and segregations. It will reveal only those discontinuities that are near or extend to the edge being inspected. Not all discontinuities found on plate edges are objectionable. However, it is necessary to remove those that would affect either the soundness of the welded joint or the ability of the base metal to meet the designer requirements.

Magnetic particle inspection can be applied in conjunction with repair work or rework procedures on both new parts and parts that may have developed cracks in service. This applies not only to the repair of weldments but also to rework done by welding in the repair or salvage of castings and forgings. The completed repair should be tested for cracks or other objectionable discontinuities in the weld or in the adjacent metal before the part is placed in service. In general, the same inspection procedures should be used in connection with repair or rework procedures as would be used on the original parts.

Limitations

THE MAGNETIC PARTICLE method of inspection is applicable only to ferro-magnetic metals in which the deposited weld metal is also ferromagnetic. It cannot be used to inspect nonmagnetic metals such as austenitic steel.

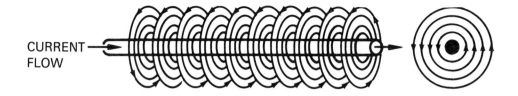

CURRENT FLOW

Figure 15.7—Magnetic Force Lines Around a Conductor

Difficulties may arise with weldments where the magnetic characteristics of the weld metal are appreciably different from those of the base metal. Joints between metals of dissimilar magnetic characteristics create magnetic discontinuities that may produce indications even though the joints themselves are sound.

Subsurface porosity and slag inclusions produce powder patterns that are not clearly defined. The degree of sensitivity in this method depends upon certain factors. Sensitivity decreases with a decrease in size of the discontinuity, and with an increase in depth below the surface. A decrease in sensitivity is evident when discontinuities are rounded or spherical rather than cracklike.

To be detected, a discontinuity must be sufficiently large to interrupt or distort the magnetic field and to cause external leakage. Fine elongated discontinuities, such as seams, inclusions, or fine cracks, will not interrupt a magnetic field that is parallel to the direction of the discontinuity. In this case, no indication will be apparent.

Surface conditions also influence the sensitivity of the process. The surface of the weld and surrounding areas should be clean, dry, and free from oil, water, excessive slag, or other accumulations that would interfere with magnetic particle movement. A rough surface decreases the sensitivity and tends to distort the magnetic field. It also interferes mechanically with the formation of powder patterns. Light grinding may be used to smooth rough weld beads, using care to avoid smearing the surface.

Orientation of the Magnetic Field

THE ORIENTATION OF the magnetic field has a great influence on the validity and performance of the test. If testing is done on a weld using only a single orientation of the magnetic field, some discontinuities may not be detected if they are aligned with the flux path. The direction of the magnetic field must be known so that it can be shifted to provide the necessary coverage. The best results are obtained when the magnetic field is perpendicular to the length of existing discontinuities.

Circular Magnetization

A MAGNETIC FIELD can be produced by passing an electrical current through a conductor. This method is referred to as *circular magnetization*. Most magnetic particle testing uses this principle to produce a magnetic field within the part. An electrically-induced magnetic field is highly directional. Also, the intensity of the field is proportional to the strength of the current. The direction of the magnetic lines of force for circular magnetism is shown in Figure 15.7.

When current is passed through a nonmagnetic conductor, a magnetic field is present on its surface as well as around it. However, when current is passed through a ferromagnetic conductor, such as carbon steel, most of the field is confined within the conductor itself. This behavior is illustrated in Figure 15.8.

When current is made to flow through a uniform steel section, the magnetic field will be uniform as well. Upon application of fine magnetic particles to the surface while the current is flowing, the particles will be uniformly distributed over the surface. If a discontinuity is present on the surface, the particles will tend to build up across the discontinuity because of flux leakage.

When the excess magnetic powder is removed, the outline of the discontinuity, with regard to dimensions and orientation, is well defined by the powder that remains, providing the discontinuity is nearly perpendicular to the flux path. This is illustrated in Figure 15.9. If the flux path is nearly parallel to the discontinuity, it is possible that no indication will appear. For a very large discontinuity parallel to the flux path, an indication might be visible, but it would likely be weak and indefinite.

If circular magnetization is utilized to detect longitudinal discontinuities lying on the inner surface of a hollow part, a slightly different technique is required because the inside surface is not magnetized when current passes directly through the part. First, a conductor is placed through the opening or hole in the hollow part. Then, current is passed through the conductor, and circular magnetic fields are induced at both the inner and outer surfaces of the hollow part.

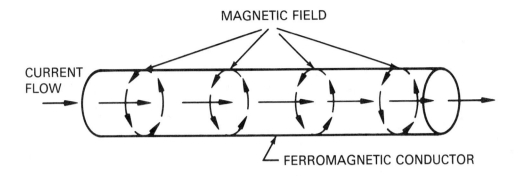

Figure 15.8—Magnetic Force Lines Within a Ferromagnetic Conductor

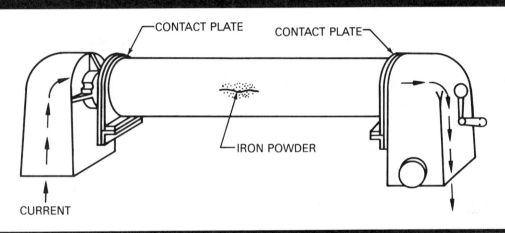

Figure 15.9—Accumulation of Magnetic Particles on a Discontinuity Using Circular Magnetization

Longitudinal Magnetization

SOMETIMES, DISCONTINUITIES ARE oriented such that they would be parallel to circular magnetic flux in a steel part. The detection of such discontinuities requires a different approach. A conductor is coiled, and the part to be tested is placed within the coil such that it becomes the core of a solenoid. This produces a magnetic field in line with the axis of the coil. Two or more poles are produced, usually at the ends of the part. This technique is referred to as *longitudinal* or *bipolar magnetization*. With longitudinal magnetization, it is possible to reveal those discontinuities lying nearly transverse to the long axis of the part, Figure 15.10. Flaws oriented at 45 degrees may be detected by circular or longitudinal magnetization. When magnetic particle testing equipment is not available at the test site, longitudinal magnetization can be induced in parts, such as shafts, pipes, and beams, by coiling a length of welding cable around the area to be tested and applying current.

Localized Magnetization

FOR LARGE PARTS, there are two types of equipment that can produce a magnetic field in a localized area. Both types can be utilized as portable methods for inspection on location.

The first of these techniques is referred to as *prod magnetization*. With this method, a localized area can be magnetized by passing current through the part by means of hand-held contacts or prods, as shown in Figure 15.11. Manual clamps or magnetic leeches may be used in place of prods.

The current creates local, circular magnetic fields in the area between the contact points. This method is used extensively for localized inspection of weldments where the area of interest is confined to the weld zone. The prods must be securely held in contact with the part to avoid arcing at the contact points even though a low open-circuit voltage (2 to 16 V) is used. This method provides only a unidirectional magnetic field. Therefore, it is necessary to reorient the contacts at about 90

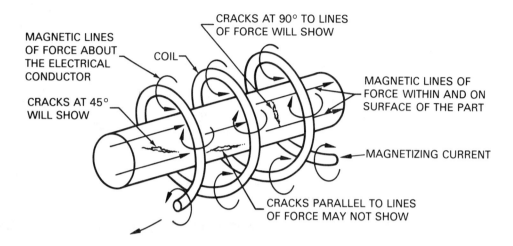

Figure 15.10—Accumulation of Magnetic Particles on Discontinuities With Longitudinal Magnetization

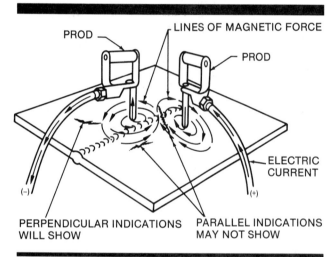

Figure 15.11—Local Magnetization of a Weld Using Prod Technique

degrees and remagnetize for complete inspection of the area.

Another way to induce a localized magnetic field is with a solenoid having flexible low carbon steel extensions of the core. When the two extensions make contact with the part and the coil is energized, the magnetic field of the solenoid is concentrated in the part between the contact points. Referred to as the *yoke method*, the equipment can be relatively small and lightweight. Another desirable feature of this technique is that electric current is not transferred to the part, as with the prod method. Thus, there is no tendency to arc or burn the part.

Types of Magnetizing Current

ALTERNATING OR DIRECT current may be used for magnetizing the parts. High amperage, low voltage power is usually employed.

Portable equipment that makes use of electromagnets or permanent magnets is occasionally used. These magnets are satisfactory for the detection of surface cracks only.

Alternating Current. The surface of the metal is magnetized by alternating current (ac). The method is effective for locating discontinuities that extend to the surface, such as cracks, but deeper discontinuities or incomplete fusion will not be detected. Alternating current may be used satisfactorily when subsurface evaluation is not required.

Direct Current. Direct current (dc) produces a magnetic field that penetrates throughout the part and is, therefore, more effective than alternating current for detecting subsurface discontinuities. Full-wave, three-phase rectified current produces results essentially comparable to uniform direct current obtained from a generator or batteries.

Half-wave rectified, single-phase current provides maximum sensitivity. The pulsating field increases particle mobility and enables particles to line up more readily in weak leakage fields.

With suitable equipment, two different types of current can be applied to a part. The characteristics of the same indications with different currents could provide useful information for more accurate interpretation of a discontinuity. Using both ac and dc magnetization, for example, the relative amounts of particle buildup might help to determine the depth of the discontinuity. If the particle buildup with dc magnetization is considerably greater than that with ac, it is highly probable that the discontinuity has a significant depth and is not just a surface irregularity.

Amount of Magnetizing Current

THE MAGNETIZING CURRENT should be of sufficient strength to indicate all detectable discontinuities that might affect the performance of the weldment in service. Excessive magnetizing currents should be avoided because they produce irrelevant patterns. Magnetizing currents should be specified in the test procedures or specifications. If they are unavailable, current requirements may be determined by experience or experiment. The applied voltage has no effect on the magnetic field strength, and should be kept low to prevent arcing and overheating.

The approximate amperage ranges for the various magnetizing methods are as follows:

(1) Longitudinal magnetization: 3 000 to 10 000 ampere-turns, depending on the ratio of the coil and part diameters.
(2) Overall circular magnetization: 100 to 1 000 amperes per inch of part diameter.
(3) Prod magnetization: 90 to 125 amperes per inch of prod spacing, depending on metal thickness.
(4) Yoke magnetization: The magnetizing current must be sufficient to lift 40 lb (18 kg) with dc magnetization and 10 lb (5 kg) with ac magnetization.

Inspection Media

MAGNETIC PARTICLES OF various colors, mobility, and luminescence are available. A suitable type can be selected to provide the greatest visual sensitivity for each specific test situation. The condition of the test surface as well as the types of discontinuities suspected are additional criteria to be considered.

Dry Method. With this technique, finely divided ferromagnetic particles in dry powder form are coated to provide for enhanced mobility, and then dyed various colors to create a distinct contrast with the background. They are applied uniformly to the part by means of a particle dispenser, an atomizer, or a spray gun. The dry method provides for the greatest portability of inspection. Dry powder is most satisfactory on rough surfaces. (Application of a wet medium would result in an increase in the extent of irrelevant indications, making interpretation more difficult.)

The powder should be applied in the form of a low velocity cloud with just enough motive force to direct the particles to the area of interest. This permits the particles to line up in indicating patterns as they are drawn to locations of leakage flux. The excess powder should be removed using a stream of air of low velocity but sufficient to carry the excess powder away while not disturbing lightly held powder patterns.

Wet Method. The magnetic particles for wet magnetic particle testing are smaller in size, and are suspended in a liquid bath of light petroleum distillate. Wet magnetic particle testing is better suited for the detection of fine surface discontinuities on smooth surfaces. Conversely, it is less likely to reveal a subsurface discontinuity than is the dry method.

The magnetic particles for liquid suspension are available in either paste or concentrate forms for use with either an oil or water bath. The manufacturer's recommendations should be used when making up the bath for proper testing sensitivity. Aerosol cans containing premixed particle bearing suspensions are commercially available. Continuous agitation of the solution is necessary to prevent the suspended particles from settling and reducing test sensitivity.

Both oil-based and water suspensions provide nearly equal sensitivity. A water suspension avoids any fire hazard due to arcing, but the presence of water near electrical apparatus creates a shock hazard that must be guarded against. Operators should be made aware of any potential hazards and be informed how to prevent them.

During testing with the wet technique, the magnetic particle solution is either flowed over or sprayed on the local area of interest, or the part is immersed in a tank containing the liquid bath. The smaller the particle size, the greater is the test sensitivity; very fine discontinuities can be revealed consistently. Magnetic particles are available in two colors, red and black. The red particles are better for dark surfaces. Particles coated with a dye that fluoresces brilliantly under ultraviolet (black) light increase the sensitivity of the test. Fluorescent particles can indicate very small or fine discontinuities and permit rapid inspection of irregular or dark sufaces.

Sequence of Operation

THE SEQUENCE OF operation in magnetic particle testing involves both the timing and the application of the particles and the magnetizing current. Two basic sequences, continuous and residual, are commonly employed.

Continuous Magnetization. This method of operation with either wet or dry particles is employed for most applications. The sequence of operation differs for wet and dry continuous magnetization techniques. The wet technique is generally used for those parts processed on a horizontal, wet-type testing unit. In practice, it involves

(1) bathing the part with the inspection medium to provide an abundant source of suspended particles on the surface, and (2) terminating the bath application simultaneously with the initiation of the magnetizing current. The duration of the magnetizing current is typically one-half second.

With the dry technique, the particles lose mobility when they contact the surface of a part. Therefore, it is imperative that the part be under the influence of the applied magnetic field while the particles are still airborne and free to migrate to leakage fields. The flow of magnetizing current must be initiated prior to the application of dry magnetic particles, and continued until the application of powder has been completed and any excess blown off. Half-wave rectified or alternating current provides additional particle mobility on the surface of the part, which can be an asset. Examination with dry particles is usually carried out in conjunction with prod type, localized magnetization.

Residual Magnetization. In this technique, the examination medium is applied after the magnetizing current has been discontinued. It can be used only if the weldment being tested has relatively high retentivity so that the residual magnetic field will be of sufficient strength to concentrate and hold particles at discontinuities. Unless experiments with typical parts show that the residual magnetic field has sufficient strength to produce satisfactory indications, it should not be used.

Equipment for residual magnetization must be designed to provide a consistent, quick interruption of the magnetizing current.

Equipment

THE BASIC EQUIPMENT for magnetic particle testing is relatively simple. It includes facilities for setting up magnetic fields of proper strengths and directions. Means are provided for adjusting current. An ammeter should be available so that the inspector can verify that the correct magnetizing current is being used for each test.

Most commercially available equipment has some degree of versatility. However, no single piece of magnetic particle equipment can perform all variations of testing in the most effective and economical manner. Therefore, the following factors should be considered when selecting the type of equipment for a specific task:

(1) Type of magnetizing current
(2) Size of part or weldment
(3) Specific purpose of test or the type of discontinuities anticipated
(4) Test media to be used
(5) Portable or stationary equipment
(6) Area to be examined and its location on the part
(7) Number of pieces to be tested

Record of Results

LITTLE IS GAINED by any inspection if a system for consistently and accurately recording indications is lacking. Sometimes, dimensions from reference locators can be recorded so that the exact location of the discontinuity can be determined later. If repair is to take place immediately following inspection, the powder build-up may remain to positively identify the affected area.

However, a permanent, positive test record is required in many cases. In such cases, the actual powder build-up can be preserved with the following technique. After discovery of a magnetic particle indication, a piece of transparent pressure-sensitive tape, sufficiently large to cover the entire area of interest, can be carefully applied to the surface over the indication. For location determination, it may be helpful to extend the tape to also cover other nearby reference locators such as holes, keyways, etc. Upon removal of the tape from the surface of the part, the magnetic particles will remain on the tape to provide an accurate record of the shape, extent, and location of the indication. This tape can then be applied to a piece of contrasting white paper. Sketches may be necessary to further clarify the exact location of the indication. Various photographic techniques also are used for the permanent recording of magnetic particle indications.

It may be helpful to provide better contrast of the indications on the part. Better contrast may be obtained by first spraying the part with white liquid penetrant developer prior to testing to provide a white background for a dark magnetic particle indication.

Demagnetization

FERROMAGNETIC STEELS EXHIBIT various degrees of residual magnetism after being magnitized. In some situations, a residual magnetic field remaining in a component would be detrimental in service, and demagnetization is necessary. An example is a component that will be located close to instruments that are affected by magnetic fields, such as a compass.

Demagnetization of small parts can be accomplished by inserting each one into the magnetic field of a strong ac solenoid, and then gradually withdrawing it from the field. Alternatively each part is to subjected to an alternating current field that is gradually reduced in intensity.

With a massive structure, alternating current does not work because the magnetic field cannot penetrate sufficiently to accomplish complete demagnetization. In such cases, direct current magnetization should be used, and the current should be gradually reduced to zero while undergoing cyclic reversals. Hammering on the component or rotating it in the magnetic field will sometimes assist demagnetization.

Annealing or stress relief heat treatment will partially demagnetize steel weldments, and total demagnetization is always accomplished when the weldment is heated above the Curie temperature [1414 °F for carbon steel (768 °C)].

Relevant Indications

MAGNETIC PARTICLE POWDER indications must be evaluated to judge compliance with a the governing standard. Indications will have a variety of configurations based on the flux leakage field caused by the discontinuity. Such indication characteristics as height, width, shape, and sharpness of detail provide information as to the type and extent of the discontinuity. Certain discontinuities exhibit characteristic powder patterns that can be identified by a skilled operator. Some of the typical discontinuity indications are discussed below.

Surface Cracks. The indication exhibited by a surface crack is well-defined and tightly held with heavy powder build-up. The amount of powder build-up is a relative measure of the depth of the crack.

Subsurface Cracks. Cracks that have not broken to the surface exhibit indications different from those of surface cracks. The powder build-up is slightly wider and not well defined.

Incomplete Fusion. The indication will be fairly well-defined, but it may only occur at the edge of a weld pass. Because incomplete fusion rarely is visible on the surface, the magnetic particle indication will not be sharp.

Slag inclusions and Porosity. Subsurface slag inclusions and porosity can be found, but the indications will be very vague unless the extent is severe. Powder build-up will not be clearly defined but can be distinguished from surface indications.

Inadequate Joint Penetration. Under certain conditions, inadequate joint penetration can be located with magnetic particle inspection. The powder indication will be wide and fuzzy, like a subsurface crack, but the pattern should be linear.

Laminations. When plate edges or weld preparations are inspected prior to welding, plate rolling laminations may be detected. The indications will be significant and very distinct. They may be continuous or intermittent.

Seams. Indications of seams in plate are straight, sharp, fine, and often intermittent. Powder build-up is less significant. A magnetizing current greater than that required for the detection of cracks may be necessary.

Undercut. A surface indication for undercut will be located at the toe of a groove or fillet weld. It will be slightly less well-defined than that from inadequate joint penetration. Visual inspection is probably a better method for evaluating this discontinuity.

Nonrelevant Indications Magnetic powder can collect at any location where there is a disturbance in the magnetic field. These indications are not associated with a mechanical discontinuity. They are not related to the soundness of the weld nor do they affect the service performance of the weld. Most nonrelevant indications can be properly identified with experience, but some codes require that all nonrelevant indications be regarded as relevant until their nonrelevance can be verified by re-examination with the same or other nondestructive testing methods. Once verified as nonrelevant, the standards allow repetitive nonrelevant indications of the same type to be ignored unless they can mask the detection of valid indications. Some of the more commonly encountered irrelevant indications are discussed below:

Surface Finish. When a weld has a rough or irregular surface, it is highly likely that the powder will build up and give false indications. Grinding the weld face smooth and retesting should determine the relevancy of any indications.

Magnetic Characteristics. Differences in magnetic characteristics in a welded joint can occur by several mechansims, but all result in irrelevant indications. Consequently, some cases may warrant the use of a different nondestructive test to check the weld quality. A change in magnetic characteristics can take place in the heat-affected zone. The resulting indications will run along the edge of the weld and be fuzzy. The pattern could be mistaken for undercut, but is less tightly held. Postweld thermal stress relief will eliminate this phenomenon.

Another change in magnetic characteristics is noted when two metals of differing magnetic properties are joined, or when the filler metal and the base metal have different magnetic properties. The most common example is the joining of carbon steel (magnetic) and austenitic stainless steel (nonmagnetic). Another common occurrence is when austenitic stainless steel filler metal is used to make a repair in carbon steel. A false indication will form at the junction of the two, sharp and well-defined, much like a crack.

Banding. A magnetic phenomenon occurs when the magnetizing current is too high for the volume of metal subjected to the magnetic field. When encountered, the current should be reduced or the prod spacing increased to prevent the possible masking of real flaws.

Residual Magnetism. Lifting of steel with electromagnets or contact with permanent magnets may give false indications from residual magnetism in the steel.

Cold Working. Sometimes, cold working of a magnetic steel will change the local magnetic characteristics such that the zone will hold powder. This can occur when the surface of a steel weld is marred by a blunt object. The track will appear as a fuzzy, lightly-held powder build-up.

Interpretation of Discontinuity Indications

WHEN MAGNETIC PARTICLE inspection is used, there is normally a standard that governs both the methodology, and the acceptance-rejection criteria. Most standards dictate the maximum permissible size of a discontinuity, and the minimum distance between otherwise acceptable discontinuities. Some standards, including the ANSI/AWS *D1 Structural Welding Codes,* also require that discontinuities be a minimum distance from the edge of the weld.

RADIOGRAPHIC TESTING

Description

RADIOGRAPHIC TESTING OF weldments or brazements employs x-rays or gamma rays to penetrate an object and detect any discontinuities by the resulting image on a recording or a viewing medium. The medium can be a photographic film, sensitized paper, a fluorescent screen, or an electronic radiation detector. Photographic film is normally used to obtain a permanent record of the test.

When a test object or welded joint is exposed to penetrating radiation, some of the radiation will be absorbed, some scattered, and some transmitted through the metal to a recording medium. The variations in amount of radiation transmitted through the weld depend upon (1) the relative densities of the metal and any inclusions, (2) through-thickness variations, and (3) the characteristics of the radiation itself. Nonmetallic inclusions, pores, aligned cracks, and other discontinuities result in more or less radiation reaching the recording or viewing medium. The variations in transmitted radiation produce optically contrasting areas on the recording medium.

The following are essential elements of radiographic testing:

(1) A source of penetrating radiation, such as an x-ray machine or a radioactive isotope

(2) The object to be radiographed, such as a weldment

(3) A recording or viewing device, usually photographic (x-ray) film enclosed in a light-tight holder

(4) A qualified radiographer, trained to produce a satisfactory exposure

(5) A means to process exposed film or operate other recording media

(6) A person skilled in the interpretation of radiographs

Sources

X-RAYS MOST SUITABLE for welding inspection are produced by high-voltage x-ray machines. The wavelengths of the x-radiation are determined by the voltage applied between elements in the x-ray tube. Higher voltages produce x-rays of shorter wavelengths and increased intensities, resulting in greater penetrating capability. Typical applications of x-ray machines for various thicknesses of steel are shown in Table 15.2. The penetrating ability of the machines may be greater or lesser with other metals, depending upon the x-ray absorption properties of the particular metal. X-ray absorption properties are generally related to metal density.

Table 15.2
Approximate Thickness Limitations of Steel for X-Ray Machines

Max Voltage, kV	Approx. Max Thickness	
	in.	mm
100	0.33	8
150	0.75	19
200	1	25
250	2	50
400	3	75
1000	5	125
2000	8	200

Gamma rays are emitted from the disintegrating nuclei of radioactive substances known as radioisotopes. Although the wavelengths of the radiation produced can be quite different, both x-radiation and gamma radiation behave similarly for radiographic purposes. The three radioisotopes in common use are cobalt-60, iridium-192, and cesium 137, named in order of decreasing energy level (penetrating ability). Cobalt-60 and iridum-192 are more widely used than cesium-137. The appropriate thickness limitations of steel for these radioisotopes are given in Table 15.3.

Table 15.3
Approximate Thickness Limitations of Steel for Radioisotopes

Radioisotope	Approx. Equivalent X-Ray Machine kV	Useful Thickness Range	
		in.	mm
Iridium-192	800	0.5-2.5	12-65
Cesium-137	1000	0.5-3.5	12-90
Cobalt-60	2000	2-9	50-230

The advantages and limitations of each source of radiation are listed in Table 15.4. The most significant aspect of a radiation source is usually related to its image quality producing aspects. However, other important considerations in selecting a source include its portability and costs.

All radiation producing sources are very hazardous. Special precautionary measures, discussed later, must be taken when entering or approaching a radiographic area.

Test Object

RADIOGRAPHIC TESTING DEPENDS upon the differential absorption of the radiation as it passes through the test object. The rate of absorption is determined by (1) the penetrating power of the source, (2) the densities of the materials subject to radiation, and (3) the relative thickness of materials in the radiation path.

Various discontinuities found in welds may absorb more or less radiation than the surrounding metal, depending on the density of the discontinuity and its thickness parallel to the radiation. Most discontinuities, such as slag inclusions, pores, and cracks are filled with material of relatively low density. They may appear as dark regions on the recording medium (film). Lighter regions usually indicate areas of greater thickness or density, such as weld reinforcement, spatter, and tungsten inclusions. These concepts are illustrated in Figure 15.12.

The thickest part of the test item allows the least amount of radiation to penetrate to the film. Therefore, the area of the film marked "A" would appear lightest on the film except in that location where discontinuity "B" is located. Assuming discontinuity "B" to be a gas or slag pocket, it will absorb less radiation than the more dense surrounding metal and result in a darker image on the film. (A tungsten inclusion in a gas tungsten arc weld will result in a lighter image on a radiographic film.) The film locations marked "C" and "D" would be progressively darker because the thinner sections allow more radiation to penetrate and expose the film. The boundaries between the film densities would be wider than indicated in Figure 15.12, and the density would gradually change from one condition to the adjoining condition. The film image is wider (and larger) than the test item. This phenomena is called *radiographic enlargement*, and is caused by the divergence of the radiation from the source. The degree of enlargement increases with decreasing source-to-object distance and with increasing object-to-film distance.

Table 15.4
Advantages and Limitations of Radiation Sources

Radioisotopes	X-Ray Machines
Advantages	
(1) Small and portable	(1) Radiation can be shut off
(2) No electric power required	(2) Penetrating power (kV) is adjustable
(3) No electrical hazards	(3) Can be used on all metals
(4) Rugged	(4) Radiographs have good contrast and sensitivity
(5) Low initial cost	
(6) High penetrating power	
(7) Access into small cavities	
(8) Low maintenance costs	
Limitations	
(1) Radiation emitted continuously by the isotope	(1) High initial cost
(2) Radiation hazard if improperly handled	(2) Requires source of electrical power
(3) Penetrating power cannot be adjusted	(3) Equipment comparatively fragile
(4) Radioisotope decays in strength requiring	(4) Less portable recalibration and replacement
(5) Radiographic contrast generally lower than with x-rays	(5) Tube head usually large in size
	(6) Electrical hazard from high voltage
	(7) Radiation hazard during operation

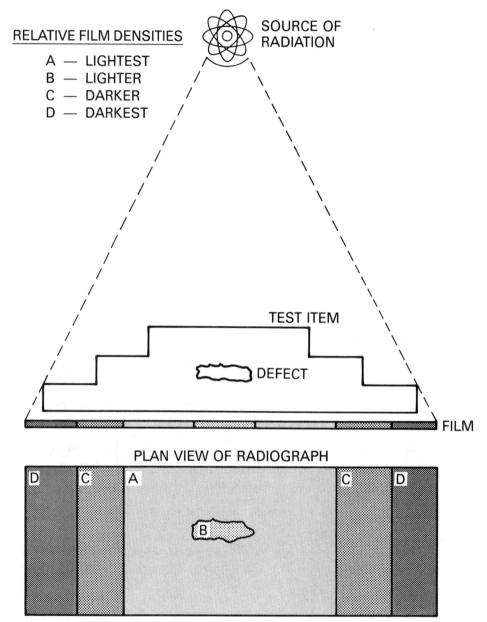

RELATIVE FILM DENSITIES

A — LIGHTEST
B — LIGHTER
C — DARKER
D — DARKEST

SOURCE OF RADIATION

TEST ITEM

DEFECT

FILM

PLAN VIEW OF RADIOGRAPH

Note: Relative film densities from light to dark are A, B, C, D.

Figure 15.12—Concept of Radiographic Testing

Recording or Viewing Means

THE COMMONLY USED recording means is radiographic film made expressly for the purpose. An industrial radiographic film is a thin transparent flexible plastic base on which a gelatin coating containing microscopic crystals of silver bromide has been deposited. Some film is coated on one side with this gelatin emulsion. Other films are coated on both sides. The emulsion is sensitive to both penetrating radiation and to visible light, and must be loaded in a darkroom into light-tight film cassettes (holders). Radiographic films are classified on the basis of speed, contrast, and grain size. Film selection depends on the nature of the inspection, the thickness and type of metal, required exposure time, and desired sensitivity.

Fluorescent screens or image amplifier systems may be used for direct viewing by the radiographer. Electronic devices can enhance the image, or convert it to electrical signals for further processing, display, or recording. However, most radiography of weldments is currently done with film.

Qualified Radiographer

THE RADIOGRAPHER PLAYS a key role in successful radiographic testing. The relative positioning of the source and film with the test object or weld affects the sharpness, density, and contrast of the radiograph. Many decisions must be made in choosing the procedure variables for specific test conditions. The radiographer must select the proper film type, intensifying screens, and filters.

Proper safety procedures must be followed to prevent dangerous radiation exposure to people in the area. Applicable federal and local safety regulations must be followed during handling and use of radiographic equipment.

Many fabrication codes and specifications require that radiographers be trained, examined, and certified to certain proficiency levels. The American Society of Nondestructive Testing publishes recommended procedures for certification of radiographers.[7]

Film processing

MANY TIMES, THE processing of exposed radiographic film can determine the success or failure of the method. Radiographs are only so good as the developing process. The processing is essentially the same as that for black and white photographic film. During film handling and processing steps, cleanliness and care are essential. Dust, oily residues, fingerprints, droplets of water, and rough handling can produce false indications or mask real ones.

Skilled Interpreter

A FINISHED RADIOGRAPH film must be evaluated or "read" to determine (1) the quality of the exposure, (2) the type and number of discontinuities present, and (3) the freedom of the weldment from unacceptable indications. This work requires a skilled film interpreter who can determine radiographic quality and knows the requirements of the applicable codes or specifications. The skills for film interpretation are acquired by a combination of training and experience, and require a knowledge of (1) weld and related discontinuities associated with various metals and alloys, (2) methods of fabrication, and (3) radiographic techniques. Most codes require film interpreters to meet the requirements of ASNT SNT-TC-1A.

Radiographic Image Quality

THE RADIOGRAPHIC IMAGE must provide useful information regarding the internal soundness of the weld. Image quality is governed by radiographic contrast and definition. The variables that affect contrast and definition are shown in Figure 15.13.[8] Control of these variables, except for the weld to be tested, is primarily with the radiographer and the film processor.

Image Quality Indicators

THERE ARE A number of variables that affect the image quality of a radiograph. Consequently, some assurance is needed that adequate radiographic procedures are used.[9] The tool used to demonstrate technique is an image quality indicator (IQI) or penetrameter. Typical penetrameters are shown in Figure 15.14. They consist of a piece of metal of simple geometric shape that has absorption characteristics similar to the weld under investigation.

When a weldment is to be radiographed, a penetrameter with a thickness equal to two percent of the weld

---

7. Refer to *SNT-TC-1A, Personnel Qualification and Certification in Nondestructive Testing.* Columbus, OH: American Society for Nondestructive Testing.

8. The factors affecting radiographic image quality are discussed in *Welding inspection,* 2nd Ed. Miami, FL: American Welding Society (1980) 128-34.

9. For more detailed information refer to *ASTM E142, Standard method for controlling quality of radiographic testing.* Philadelphia, PA: American Society for Testing and Materials.

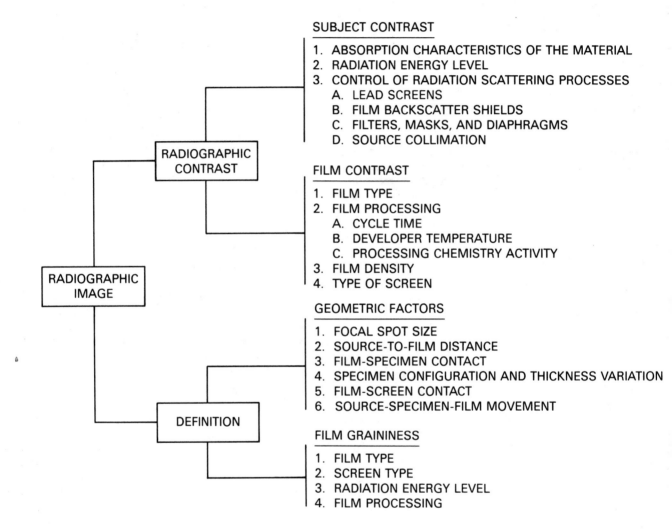

Figure 15.13—Factors Affecting Quality of Radiographic Image

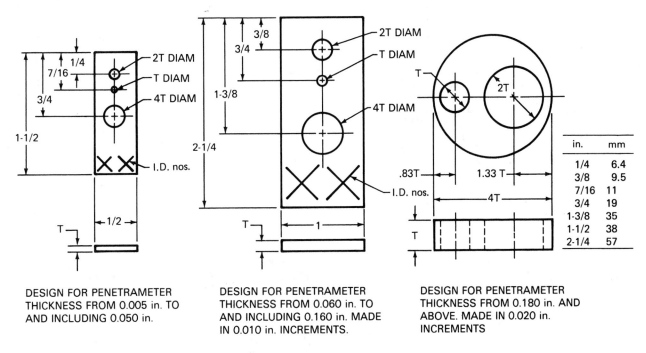

DESIGN FOR PENETRAMETER
THICKNESS FROM 0.005 in. TO
AND INCLUDING 0.050 in.

DESIGN FOR PENETRAMETER
THICKNESS FROM 0.060 in. TO
AND INCLUDING 0.160 in. MADE
IN 0.010 in. INCREMENTS.

DESIGN FOR PENETRAMETER
THICKNESS FROM 0.180 in. AND
ABOVE. MADE IN 0.020 in.
INCREMENTS

DIMENSIONS ARE IN INCHES

in.	mm
1/4	6.4
3/8	9.5
7/16	11
3/4	19
1-3/8	35
1-1/2	38
2-1/4	57

Figure 15.14—Typical Penetrameter Designs

thickness is generally selected. A lead identification number at one end shows the thickness of the penetrameter in thousandths of an inch or in millimeters.

Conventional penetrameters usually contain three holes, the diameters of which vary in size as multiples of the thickness. Most specifications and codes call for 1T, 2T, and 4T diameter holes, where T is the penetrameter thickness. Penetrameters are manufactured in standard sizes and increments of thickness. When an exact thickness penetrameter is not available for a particular weld thickness, the next closest one is normally used.

In addition, conventional penetrameters are usually manufactured in material groupings. Most codes and specifications organize commonly used metals and alloys into a minimum of five groups (absorption categories), ranging from light to heavy metals. A penetrameter of the appropriate grade is used when radiographic testing a weldment made of an alloy in the group.

Most radiographic image quality requirements are expressed in terms of penetrameter thickness and desired hole size. For example, the requirement might be 2-2T level of sensitivity. The first 2 requires the penetrameter

thickness to be two percent of the thickness of the specimens; the symbol 2T requires that the hole having a diameter twice the penetrameter thickness be visible on the radiograph. This image quality level is commonly specified for routine radiography. For more sensitive radiography, a 1-2T or 1-1T could be required. More relaxed image quality requirements would include 2-4T and 4-4T.

Most fabrication and inspection standards specify the exact penetrameter for a range of nominal base metal thicknesses. Table 15.5 shows the requirements of *ANSI/AWS D1.1-88, Structural Welding Code—Steel* which is a typical format for most fabrication standards.

A penetrameter placed on the source side of the weld gives a less well defined image than a film-side penetrameter. However, it is not always practical to place the penetrameter on the source side. In radiographing a circumferential weld in a long pipe section, for example, the pipe although accessible on the inside for the source is not accessible on the inside for accurate placement of the penetrameters. In this case, a thinner penetrameter may be located on the film side of the weldment.

Table 15.5
Penetrameter Requirements of the AWS D1.1 Structural Welding Code—Steel

Nominal Material Thickness Range		Source Side		Film Side*	
in	mm	Designation**	Essential Hole	Designation**	Essential Hole
Up to 0.25 incl	Up to 6.4 incl.	10	4T	7	4T
Over 0.25 to 0.375	Over 6.4 through 9.5	12	4T	10	4T
Over 0.375 to 0.50	Over 9.5 through 12.7	15	4T	12	4T
Over 0.50 to 0.625	Over 12.7 through 15.9	15	4T	12	4T
Over 0.625 to 0.75	Over 15.9 through 19.0	17	4T	15	4T
Over 0.75 to 0.875	Over 19.0 through 22.2	20	4T	17	4T
Over 0.875 to 1.00	Over 22.2 through 25.4	20	4T	17	4T
Over 1.00 to 1.25	Over 25.4 through 31.7	25	4T	20	4T
Over 1.25 to 1.50	Over 31.7 through 38.1	30	2T	25	2T
Over 1.50 to 2.00	Over 38.1 through 50.8	35	2T	30	2T
Over 2.00 to 2.50	Over 50.8 through 63.6	40	2T	35	2T
Over 2.50 to 3.00	Over 63.6 through 76.2	45	2T	40	2T
Over 3.00 to 4.00	Over 76.2 through 102	50	2T	45	2T
Over 4.00 to 6.00	Over 102 through 152	60	2T	50	2T
Over 6.00 to 8.00	Over 152 through 203	80	2T	60	2T

*The penetrameter designation is the penetrameter thickness in inches times 1000. Thus, a 40 penetrameter is 0.040 in. (1.02 mm) thick.
**To be used only when the source side of the joint is inaccessible for penetrameter location, such as on the tubular butt joints.

The appearance of the penetrameter image on the radiograph will indicate the quality of the radiographic technique. Even though a certain hole in a penetrameter may be visible on the radiograph, a discontinuity of the same approximate diameter and depth as the penetrameter may not be visible. Penetrameter holes have sharp boundaries and abrupt changes in dimensions, whereas voids or discontinuities may have gradual changes in dimension and shape. A penetrameter, therefore, is not used for measuring the size of a discontinuity or the minimum detectable flaw size.

Identification Markers

RADIOGRAPH IDENTIFICATION MARKERS of lead alloy are usually in the form of a coded series of letters and numbers. The markers are placed on the test piece at marked locations adjacent to the welded joint during setup. When a welded joint is radiographed, a distinct, clear image of the identification markers should be produced at the same time. Identification markers must be located so that their projected shadows do not coincide with the shadows of any regions of interest in the weldment.

The view identification and the test-piece identification almost always appear in coded form. View identification is usually a simple code (such as A, B, C, or 1, 2, 3) that relates some inherent feature of the weldment or some specific location on the weldment to the view used. The location of the view markers is handwritten in chalk or crayon directly on the piece so that correlation of the radiographic image with the test piece itself can be made

during interpretation and evaluation of the radiograph. As a minimum requirement, the identification code for a weld must enable each radiograph to be traced to a particular test piece or section of a test piece. The pertinent data concerning weld and test-piece identification should be recorded in a logbook opposite the corresponding identification number.

Exposure Techniques

RADIOGRAPHIC FILM EXPOSURES are performed in a number of different arrangements; typical ones are shown in Figure 15.15. The exposure arrangement is chosen with consideration of the following factors:

(1) The best coverage of the weld and the best image quality
(2) The shortest exposure time
(3) Optimum image of discontinuities that are most likely to be present in a particular type of weld
(4) The use of either multiple exposures or one or more exposures at some angle to fully cover all areas of interest
(5) Radiation safety considerations
(6) Whether single- or double-wall exposures should be used with a pipe weld

Typical exposure arrangements for radiography of pipe welds are shown in Figure 15.15(B) through (F). The appropriate arrangement depends upon the pipe diameter and wall thickness.

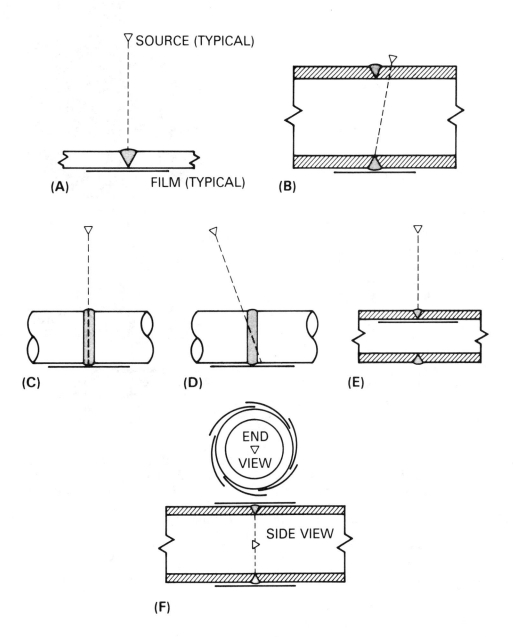

Figure 15.15—Typical Radiographic Exposure Arrangements for Pipe Welds

Interpretation of Radiographs

THE ESSENTIAL STEPS of radiograph interpretation are as follows:

(1) Determine the accuracy of the identification of the radiograph.

(2) Determine the weld joint design setup and the welding procedure.

(3) Verify the radiographic setup and procedure.

(4) Review the film under good viewing conditions.

(5) Determine if any false or irrelevant indications are present on the film. (Re-radiograph if necessary.)

(6) Identify any surface irregularities and verify their type and presence by visual or other NDT methods.

(7) Evaluate relevance of discontinuities with code or specification requirements.

(8) Prepare radiographic report.

Viewing Conditions

FILM VIEWING EQUIPMENT must be located in a space with subdued lighting to reduce interfering glare. A masking arrangement should be provided so that only the film itself is illuminated and the light is shielded from the viewer's eyes. Variable intensity lighting is usually desirable to accommodate film of various average densities.

Discontinuities in Welds

RADIOGRAPHIC TESTING CAN produce a visible image of weld discontinuities, either surface or subsurface, when the discontinuities have significant differences in radiographic density from the base metal and adequate thickness parallel to the radiation. The process will not reveal very narrow discontinuities, such as cracks, laps, and laminations that are not closely aligned with the radiation beam. Surface discontinuities are better identified by visual, penetrant, or magnetic particle testing unless the face and root of the weld are not accessible for examination.

Images of slag inclusions are usually irregularly shaped dark areas, and appear to have some width. Inclusion indications are most frequently found at the weld interfaces, as illustrated in Figure 15.16. In location, slag inclusions often are found between passes. Tungsten inclusions will appear as highly contrasted light areas (white spots).

Porosity (gas holes) indications appear as nearly round voids that are readily recognizable as dark spots with the radiographic contrast varying directly with diameter. These voids may be randomly dispersed, in clusters, or may even be aligned along the centerline of the fusion zone. An example of porosity is shown in Figure 15.17. A worm hole image appears as a dark rectan-

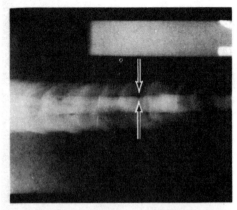

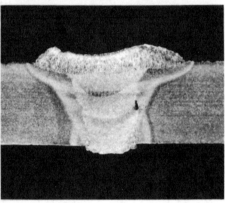

Figure 15.16—Elongated Slag at the Weld Interface Between Passes. Top: Radiograph; Bottom: Metallographic Section

gle if the long axis is perpendicular to the beam, and concentric circles if the long axis is parallel to the beam.

Crack indications often appear in a radiograph as fine dark lines of considerable length, but without great width. Even some fine crater cracks are readily detected. Cracks in weldments may be transverse or longitudinal, and may be either in the fusion zone or in the heat affected zone of the base metal. Examples of cracks are shown in Figures 15.18 and 15.19. Cracks may be undetected if they are very small or are not aligned with the radiation beam.

Incomplete fusion images appear as elongated dark lines or bands. They sometimes appear very similar to a crack or an inclusion, and could even be interpreted as such. Incomplete fusion occurs between weld and base metal and also between successive beads in multiple-pass welds.

Incomplete joint penetration shows on a radiograph as a very narrow dark line near the center of the weld, as shown in Figure 15.20. The narrowness can be a result of drawing together of the plates being welded. Slag

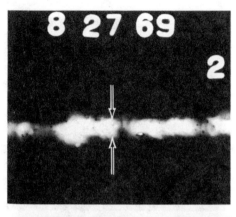

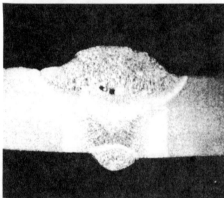

Figure 15.17—Weld Metal Porosity. Top: Radiograph; Bottom: Metallographic Section

Figure 15.18—Transverse Crack in Weld. Top: Radiograph; Bottom: Longitudinal Metallographic Section

inclusions and gas holes sometimes are found in connection with incomplete joint penetration, and cause the line to appear broad and irregular.

Undercutting appears as a dark zone of varying width along the edge of the fusion zone, as shown in Figure 15.21. The darkness or density of the line is an indicator of the depth of the undercut.

Concavity at the weld root occurs only in joints that are welded from one side, such as pipe joints. It appears on the radiograph as a region darker than the base metal, and running along the center of the weld.

If weld reinforcement is too high, the radiograph shows a lighter zone along the weld seam. There is a sharp change in image density where the reinforcement meets the base metal, and the edge of the reinforcement image is usually irregular.

As joint thickness increases, radiography becomes less sensitive as an inspection method. Thus, other nondestructive inspection methods are preferred for thick welds.

Safe Practices

Radiation Protection. Federal, state, and local governments issue licenses for the operation of radiographic facilities. The federal licensing program is concerned mainly with those companies that use radioactive isotopes as sources. However, in most localities, state and local agencies exercise similar regulatory prerogatives. To become licensed under any of these programs, a facility or operator must show that certain minimum requirements for protection of both operating personnel and the general public from excessive levels of radiation have been met. Although local regulations may vary in the degree and type of protection afforded, certain general principles apply to all.

The amount of radiation that is allowed to escape from the area over which the licensee has direct and exclusive control is limited to an amount that is safe for continuous exposure. In most instances less than 2 millirems (mrem) per hour, 100 mrem in seven consecutive

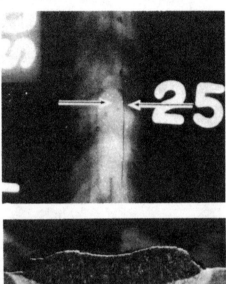

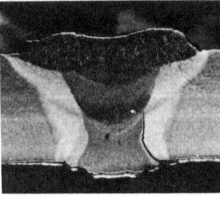

Figure 15.19—Longitudinal Crack in the Heat Affected Zone. Top: Radiograph; Bottom: Longitudinal Metallographic Section

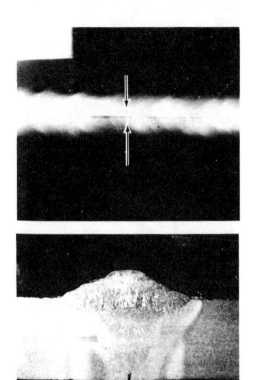

Figure 15.20—Incomplete Penetration in a Weld. Top: Radiograph; Bottom: Metallographic Section

days, and 500 mrem in a calendar year can be considered safe.

Radiation Monitoring. A radiation safety program must be controlled to ensure that both the facility itself and all personnel subject to radiation exposure are monitored. Facility monitoring generally is accomplished by periodically taking readings of radiation leakage during operation of each source under various conditions. Calibrated instruments can be used to measure radiation dose rates at various points within the restricted area, and at various points around the perimeter of the restricted area.

To guard against inadvertent leakage of large amounts of radiation from a shielded work area, interlocks and alarms are often required. Basically, an interlock disconnects power to an x-ray tube if an access door is opened, or prevents any door from being opened if the unit is turned on. Alarms are connected to a separate power source and activate visible or audible signals, or

both, whenever the radiation level exceeds a preset value.

All personnel within the restricted area must be monitored to assure that they do not absorb excessive amounts of radiation. Devices such as pocket dosimeters and film badges are the usual means of monitoring. Often both are worn. Pocket dosimeters may be direct reading or remote reading.

Access Control. Permanent facilities are usually separated from unrestricted areas by shielded walls. Sometimes, particularly in on-site radiographic examination, access barriers may be only ropes or sawhorses. In such instances, the entire perimeter around the work area should be under continual surveillance by radiographic personnel. Signs that carry a symbol designated by the U.S. Government must be posted around any high-radiation area. This helps to inform casual bystanders of the potential hazard, but should never be assumed to prevent unauthorized entry into the danger zone. In fact, no interlock, no radiation alarm, and no other safety device

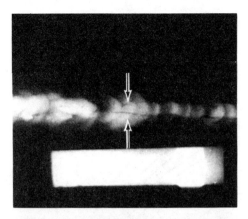

**Figure 15.21—Undercut at the Root of a Weld.
Top: Radiograph; Bottom: Metallographic Section**

should be considered a substitute for constant vigilance on the part of radiographic personnel.

EDDY CURRENT TESTING

General Description

EDDY CURRENT TESTING (ET) is an electromagnetic nondestructive testing method in which eddy current flow is induced in the test piece. Changes in the flow of the eddy currents caused by variations in the test piece or discontinuities in a weld are detected by a nearby coil or coils and measured by suitable instruments. The method can be used to detect certain discontinuities in welds. However, the results can be affected by variations in dimensions of the test piece or the testing arrangement, and by variations in the physical and metallurgical properties of the test piece.

Applications

EDDY CURRENT TESTING is primarily used for continuous inspection of seamless and welded piping and tubing during production. Testing of ferromagnetic steel, austenitic stainless steel, copper alloy, and nickel alloy tubular products are covered by ASTM specifications.

Eddy Currents

EDDY CURRENTS ARE circulating alternating electrical currents in a conducting material induced by an adjacent alternating magnetic field. Generally, the currents are inducted in the component by making it the core of an ac induction coil, as shown in Figure 15.22. A crack in a welded seam can disrupt the eddy current flow and the magnetic field produced by that current.

There are two ways of measuring changes that occur in the magnitude and distribution of these currents. Either the resistive component or the inductive component of impedance of the exciting coil, or of a secondary coil, can be measured. Electronic equipment is available for measuring either or both of these components. The current distribution within a test piece may be changed by the presence of inhomogeneities. If a component is homogeneous, free from discontinuities, and has an undistorted grain structure, the mean free path of an electron passing through will be maximum in length. A crack, inclusion, cavity, or other conditions in an otherwise homogeneous material will cause a back scattering of electron flow, and thereby shorten their mean free paths. If an electric wave is considered instead of electrons, the same factors that would impede the flow of a single electron will also impede the passage of a wave front. The wave front will be totally or partially reflected or absorbed, or both. The various relationships between conductivity and such factors as impurities, cracks, grain size, hardness, and strength, have been reported in the literature.

Electromagnetic Properties of Eddy Current Coils

WHEN AN ENERGIZED coil is brought near a metal test piece, eddy currents are induced into the piece. Those currents set up magnetic fields that act in opposition to the original magnetic field. The impedance, Z, of the exciting coil or any coil in close proximity to the specimen is affected by the presence of the induced eddy currents in the test piece. When the path of the eddy currents in the test piece is distorted by the presence of discontinuities or other inhomogeneties (Figure 15.22), the apparent impedance of the coil is altered. This change in impedance can be measured, and is useful in giving indications of discontinuities or other variations in structure.

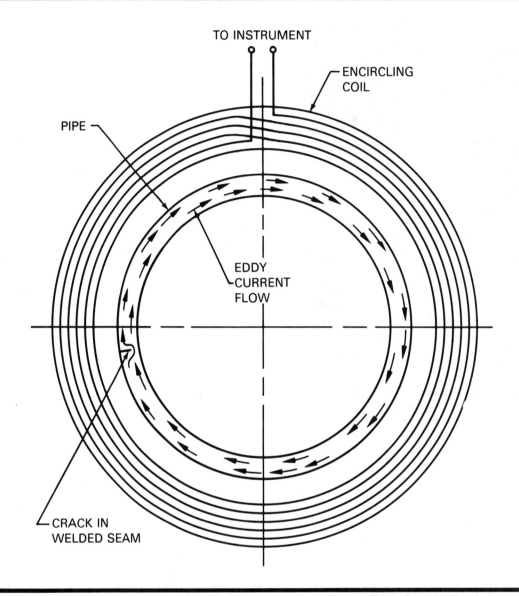

Figure 15.22—Effect of a Crack in Eddy Current Flow During Eddy Current Testing

The test coil used to induce the eddy currents determines, to a large extent, the information which can be obtained during the test. The basic electrical variables of an eddy current testing system are the *ac* voltage, E, the current flowing through the coil, I, and the coil impedance, Z, that are related by the equation:

$$I = E/Z$$

The impedance of the coil is affected by its magnetic field. Any changes in that field will affect the current flow in the coil. Also, the distance between the coil and the test piece affects the impedance of the coil and the eddy current flow. This distance must be held constant to detect discontinuities in a moving test piece.

Impedance changes also affect the phase relationships between the voltage across the coil and the current through the coil. This provides a basis for sorting out the effects of variations in spacing and other variables. Phase changes can be observed by means of a cathode ray tube or other display.

Properties of Eddy Currents

IN GENERATING EDDY currents, the test piece is brought into the field of a coil carrying alternating current. The

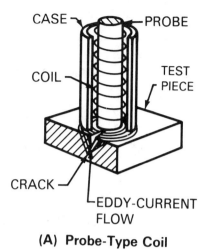

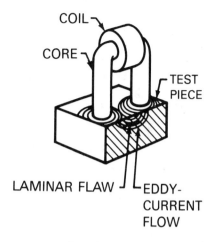

(A) Probe-Type Coil

(B) Horseshoe-Shape or U-Shape Coil

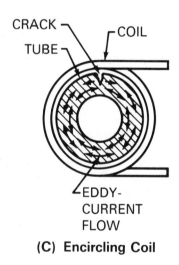

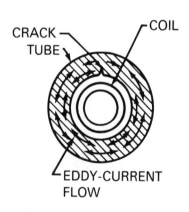

(C) Encircling Coil

(D) Internal or Bobbin-Type Coil

Figure 15.23—Types and Applications of Coils Used in Eddy-Current Inspection

coil may encircle the part, may be in the form of a probe, or in the case of tubular shapes, may be wound to fit inside a tube or pipe. Typical applications are shown in Figure 15.23.

The eddy current in the metal test piece also sets up a magnetic field, which opposes the original magnetic field. The impedance of the exciting coil or of a second coil coupled to the first and in close proximity to the test piece is affected by the presence of the induced eddy cur-rents. A second coil is often used as a convenience, and is called a *sensing* or *pickup coil*. In the case of a crack or an unwelded seam, the discontinuity must be oriented nearly normal to the eddy current flow to disturb it. The change in coil impedance caused by the presence of a dis-continuity can be measured, and is used to give an indi-cation of the extent of defects.

Subsurface discontinuities may also be detected, but eddy currents decrease with depth.

Alternating Current Saturation

A HIGH AC magnetizing force may be used to simultaneously saturate (magnetically) a test piece and create an eddy current signal for detection equipment. This increases the penetration of the eddy currents and suppresses the influence of certain disturbing magnetic variables.

All ferromagnetic materials that have been magnetically saturated will retain a certain amount of magnetization, called the residual field, when the external magnetizing force has been removed. This residual field may be large or small, depending upon the nature of the magnetizing force applied and material retentivity.

Demagnetization is necessary whenever the residual field (1) may affect the operation or accuracy of instruments when placed in service, (2) interferes with inspection of the part at low field strengths or with proper functioning of the part, or (3) might cause particles to be attracted and held to the surface of moving parts, particularly parts running in oil, resulting in undue wear. There are many ways to demagnetize an object, the most common being to pass current directly through the test piece. The selected method should give the required degree of removal of the residual field.

Skin Effects

EDDY CURRENTS ARE strongest near the surface of the test piece. This effect is shown in Figure 15.24. The term standard depth refers to that depth where the eddy current density is approximately 37 percent of that at the surface. Depth of current penetration varies inversely with the electrical conductivity and magnetic permeability of the metal and with the frequency of the alternating eddy currents. Table 15.6 shows typical standard depths of penetration for several metals and magnetizing current frequencies. Normally, a part being inspected must have a thickness of at least two or three times the standard depth before thickness ceases to have an effect on eddy current response.

Electromagnetic Testing

ELECTROMAGNETIC TESTING CONSISTS of observing the interaction between electromagnetic fields and metal test pieces. The three things required for an electromagnetic test are

(1) A coil or coils carrying an alternating current
(2) A means of measuring the electrical properties of the coil or coils
(3) A metal part to be tested

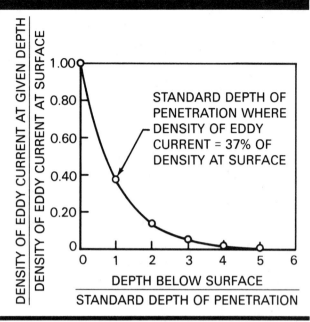

Figure 15.24—Variation in Eddy Current Density as a Function of Depth

The test coils, being specialized sensing elements, are in some way similar to lenses in an optical system. Their design is a fundamental consideration depending upon the nature of the test. Probe coils that are brought up against the surface to be tested are used in testing a variety of metallic shapes for discontinuities. Annular coils encircle the part and are used especially for inspecting tubing, rods, bars, wires, and small parts.

Electromagnetic testing involves (1) interaction between applied and induced electromagnetic fields, and (2) imparting of energy into the test piece much like the transmission of x-rays, heat, or ultrasound. Upon entering the test piece, a portion of the electromagnetic energy produced by the test coil is absorbed and converted into heat through the action of resistivity and, if the conductor is magnetic, hysteresis. The remainder of the energy is stored in the electromagnetic field. As a result, the electrical properties of the test coil are altered by the properties of the part under test. Hence, the current flowing in the coil reflects certain information about the part, namely dimensions; mechanical, metallurgical, and chemical properties; and presence of discontinuities. The character of the interaction between the applied and induced electromagnetic fields is determined by two basically distinct phenomena within the test part: (1) the induction of eddy currents in the metal by the applied field, and (2) the action of the applied field upon the magnetic domains, if any, of the part.

Obviously, only the first phenomenon can act in the case of nonferromagnetic metals. In the case of ferromagnetic metals, both phenomena are present; however,

Table 15.6
Typical Depths of Eddy Current Penetration with Magnetizing Current Frequency

Metal	Standard (37%) Depth of Penetration											
	1 kHz		4 kHz		16 kHz		64 kHz		250 kHz		1 MHz	
	in.	mm	in.	mm	in.	mm	in.	mm	in.	mm	in.	mm
Aluminum, Type 6061 T6	0.126	3.2	0.063	1.6	0.032	0.8	0.016	0.4	0.008	0.2	0.0040	0.1
Aluminum, Type 7075 T6	0.144	3.6	0.072	1.8	0.036	0.9	0.018	0.5	0.009	0.2	0.0046	0.1
Copper	0.082	2.1	0.041	1.0	0.021	0.5	0.010	0.3	0.005	0.1	0.0026	0.07
Lead	0.292	7.4	0.146	3.7	0.073	1.9	0.037	0.9	0.018	0.5	0.0092	0.2
Magnesium	0.134	3.4	0.066	1.7	0.033	0.8	0.017	0.4	0.008	0.2	0.0042	0.1
Stainless steel, Type 304	0.516	13.1	0.257	6.5	0.130	0.3	0.065	1.7	0.031	0.8	0.0165	0.4
Zirconium	0.445	11.3	0.222	5.6	0.112	2.8	0.056	1.4	0.028	0.7	0.0141	0.4
High alloy steel*	0.020	0.5	0.0095	0.2	0.0049	0.1	0.0025	0.1	0.001	0.03	0.0006	0.01

*The standard depths are for tests without magnetic saturation. When saturated, the values for ferromagnetic steels are approximately the same as those for austenitic stainless steel.

the second usually has the stronger influence. This accounts for the basic difference in principle between the testing of ferromagnetic and nonferromagnetic metals.

Among the physical and metallurgical variables that affect electromagnetic tests in metals are the following:

(1) Physical shape, external dimensions, and thickness of the part
(2) Distance between the part and the electromagnetic coil
(3) Plating or coating thickness, if present
(4) Chemical composition
(5) Distribution of alloying or impurity elements (influenced by heat treatment of the part)
(6) Lattice dislocations caused by mechanical working
(7) Temperature
(8) Inhomogeneities and most types of discontinuities
(9) Residual and applied stresses in ferromagnetic metals

In practice, many and sometimes all of the above factors may vary simultaneously. It is difficult under such conditions to obtain a meaningful response from the magnetic flux set up within the test piece because several variables may have affected the test signal. Because of the above factors, test features and standards that duplicate the test condition need to be evaluated. The resulting voltage, which is the variable usually sensed by electromagnetic testing devices, must be very carefully analyzed to isolate the pertinent effects from any extraneous effects.

Associated with any electromagnetic test signal are three important attributes: amplitude, phase, and frequency. The test signal may contain either a single frequency (that selected for the test), or a multitude of frequencies (harmonics of the test signal frequency). In the latter case, the test signal frequency is referred to as the "fundamental frequency." In addition, there are amplitude and phase factors associated with each harmonic frequency. The engineer has available a number of techniques that make use of all this information, thereby permitting discrimination between test variables. The important techniques are amplitude discrimination, phase discrimination, harmonic analysis, coil design, choice of test frequency, and magnetic saturation.

Calibration and Quality Standards

WHEN USING ELECTROMAGNETIC test methods for inspection of metals, it is essential that adequate standards are available to (1) make sure that the equipment is functioning properly and is picking up discontinuities, and (2) ascertain whether the discontinuities are cause for rejection of the part.

It is not the discontinuity itself that is detected by the test equipment but rather the effect that it has on the eddy currents in the piece being inspected. It is necessary, therefore, to correlate the change in eddy currents with the cause of the change. For this reason, standards must be available when calibrating an electromagnetic testing unit. The standards must contain either natural or artificial imperfections that can accurately reproduce the exact change in electromagnetic characteristics expected when production items are tested. Such standards are usually considered equipment calibration standards; that is, they demonstrate that the equipment is picking up any discontinuities for which the piece is being inspected. These standards are not only used to facilitate the initial adjustment or calibration of the test instrument but also to check periodically on the reproducibility of the measurements.

It is not enough just to be able to locate discontinuities in a test piece; the inspector must be able to determine if any discontinuity is severe enough to be cause for rejection. For this purpose, quality assurance standards are required against which the test instrument can be calibrated to show the limits of acceptability or rejectability for any type of discontinuity. Quality assurance standards may be either actual production items representing the limits of acceptability or prepared samples containing artifical discontinuities to serve the same purpose. The types of reference discontinuities that must be used for a particular application are usually given in the product specification.

Several discontinuities that have been used for reference standards include filed transverse notches, milled or electrical-discharge-machine longitudinal and transverse notches, and drilled holes.

Detectable Discontinuities

BASICALLY, ANY DISCONTINUITY that appreciably alters the normal flow of eddy currents can be detected by eddy-current testing. With encircling-coil inspection of either solid cylinders or tubes, surface discontinuities having a combination of predominantly longitudinal and radial dimensional components are readily detected. When discontinuities of the same size are located at progressively greater depth, beneath the surface of the part being inspected, they become increasingly difficult to detect. Discontinuities can be detected at depths greater than 1/2 in. (13 mm) only with special equipment designed for this purpose.

Laminar discontinuities, such as those sometimes found in welded tubes, may not alter the flow of the eddy currents enough to be detected unless the discontinuity extends to the outside or inside surface, or exists in a weld with outward bent fibers caused by upsetting during welding. A similar difficulty could arise for the detection of a thin planar discontinuity that is oriented substantially perpendicular to the axis of the cylinder.

Regardless of the limitations, a majority of objectionable discontinuities can be detected by eddy-current inspection at high travel speed and at low cost. Some of the discontinuities that are readily detected are seams, laps, cracks, slivers, scabs, pits, slugs, open welds, missed welds, misaligned welds, black or gray oxide weld penetrators, and pinholes.

ULTRASONIC TESTING

General Description

ULTRASONIC TESTING (UT) is a nondestructive method in which beams of high frequency sound waves are introduced into a test object to detect and locate surface and internal discontinuities. A sound beam is directed into the test object on a predictable path, and is reflected at interfaces or other interruptions in material continuity. The reflected beam is detected and analyzed to define the presence and location of discontinuities.

The detection, location, and evaluation of discontinuities is possible because (1) the velocity of sound through a given material is nearly constant, making distance measurements possible, and (2) the amplitude of a reflected sound pulse is nearly proportional to the size of the reflector.

Applications

ULTRASONIC TESTING CAN be used to detect cracks, laminations, shrinkage cavities, pores, slag inclusions, incomplete fusion or bonding, incomplete joint penetration, and other discontinuities in weldments and brazements. With proper techniques, the approximate position and depth of the discontinuity can be determined, and in some cases, the approximate size of the discontinuity.

Advantages

THE PRINCIPLE ADVANTAGES of UT compared to other NDT methods for weldments are as follows:

(1) Can detect discontinuities in thick sections
(2) Has relatively high sensitivity to small discontinuities
(3) Able to determine depth of internal discontinuities and to estimate their size and shape
(4) Can adequately inspect from one surface
(5) Equipment can be moved to the job site
(6) Is nonhazardous to personnel or other equipment

Limitations

THE PRINCIPLE LIMITATIONS of UT are as follows:

(1) Setup and operation require trained and experienced technicians, especially for manual examinations.
(2) Weldments that are rough, irregular in shape, very small, or thin are difficult or impossible to inspect; this includes fillet welds.
(3) Discontinuities at the surface are difficult to detect.
(4) A couplant is needed between the sound transducers and the weldment to transmit the ultrasonic wave energy.
(5) Reference standards are required to calibrate the equipment and to evaluate the size of discontinuities.
(6) Reference standards should reflect the item to be examined with respect to design, material specifications, and heat treatment condition.

Basic Equipment

MOST ULTRASONIC TESTING systems use the following basic components:

(1) An electronic signal generator (pulser) that produces bursts of alternating voltage.

(2) A sending transducer that emits a beam of ultrasonic waves when alternating voltage is applied.

(3) A couplant to transmit the ultrasonic energy from the transducer to the test piece and vice versa.

(4) A receiving transducer to convert the sound waves to alternating voltage. This transducer may be combined with the sending transducer.

(5) An electronic device to amplify and demodulate or otherwise change the signal from the receiving transducer.

(6) A display or indicating device to characterize or record the output from the test piece.

(7) An electronic timer to control the operation.

The basic components are shown in block form in Figure 15.25.

Equipment operating in a pulse-echo method with video presentation is most commonly used for hand scanning of welds. The pulse-echo equipment produces repeated bursts of high frequency sound with a time interval between bursts to receive signals from the test piece and from any discontinuities in the weld or base metal. The pulse rate is usually between 100 and 5000 pulses per second.

In the video presentation, the time base line is located horizontally along the bottom of a cathode ray tube (CRT) screen, with a vertical initial pulse indication at the left side of the base line. An *A scan*, indicates that the time lapse between pulses is represented by the horizontal direction, Figure 15.25, and the relative amplitude of the returning signal is represented by the degree of vertical deflection on the CRT screen. The screen is usually graduated in both horizontal and vertical directions to facilitate measurements of pulse displays.

A search unit is used to direct a sound beam into the test object. It consists of a holder and a transducer. The transducer element is usually a piezoelectric crystalline substance. When excited with high frequency electrical energy, the transducer produces mechanical vibrations at a natural frequency. A transducer also can receive physical vibrations and transform them into low energy electrical impulses. In the pulse-echo mode, the ultrasonic unit senses reflected impulses, amplifies them, and presents them as spikes called *pips* on the CRT screen, Figure 15.25. The horizontal location of a reflector pip on the screen is proportional to the distance the sound has traveled in the test piece. This makes it possible to determine location of reflectors by using horizontal screen graduations as a distance measuring ruler.

Sound Behavior

SOUND PASSES THROUGH most metals in a fairly well-defined beam. Initially, the sound beam has a cross section approximately the size of the transducer element. It propagates with slight divergence in a fairly straight line. As the sound beam travels through the metal, there is some attenuation or decrease in energy. The beam will continue to propagate until it reaches a boundary within the object being tested. Either partial or complete reflection of the sound beam takes place at a boundary.

The behavior of sound at UT frequencies resembles visible light in several ways:

(1) Divergence of the beam can be controlled by focusing.

(2) The beam will reflect predictably from surfaces of different densities.

(3) The beam will refract at an interface between materials of different density.

On the other hand, the behavior differs from light in that different vibrational modes and velocities can occur in the same medium.

Test Frequency

THE SOUND WAVE frequencies used in weld inspection are between 1 and 6 MHz, which are beyond the audible range. Most weld testing is performed at 2.25 MHz. Higher frequencies, i.e., 5 MHz, will produce small, sharp sound beams useful in locating and evaluating discontinuities in thin wall weldments.

Wave Form

THERE ARE THREE basic modes of propagating sound through metals: (1) longitudinal, (2) transverse, and (3) surface waves. In the first two modes, waves are propagated by the displacement of successive atoms or molecules in the metal.

Longitudinal waves, sometimes called straight or compressional, represent the simplest wave mode. This wave form exists when the motions of the particles are parallel to the direction of sound beam propagation, as shown in Figure 15.26(A). Longitudinal waves have relatively high velocity and relatively short wave length. As a result, the energy can be focused into a sharp beam with a minimum of divergence. They are readily propagated in water.

Another mode is the transverse wave, also called shear wave, in which the principal particle motion is perpendicular to the direction of sound beam propagation, as shown in Figure 15.26(B). The velocity of these waves is approximately half that of longitudinal waves. Advan-

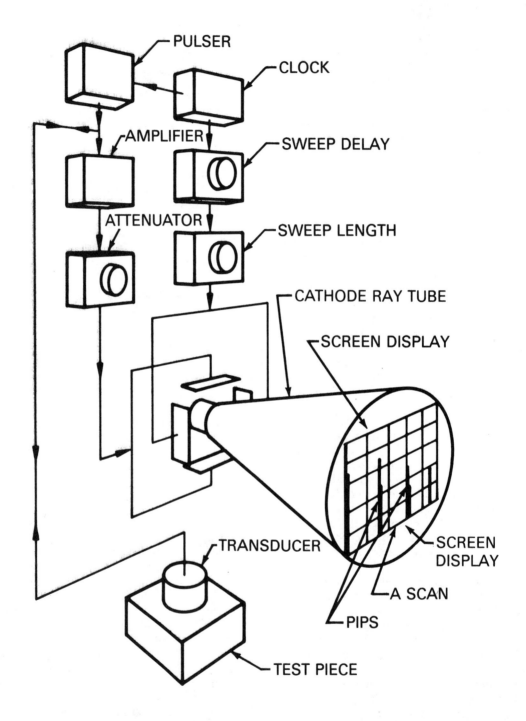

Figure 15.25—Block Diagram, Pulse-Echo Flaw Detector

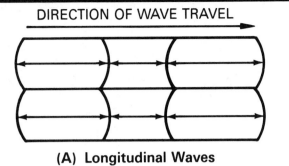

(A) Longitudinal Waves

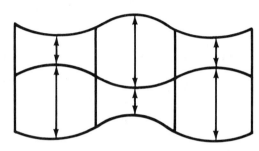

(B) Shear Waves

Note: Internal arrows represent physical movement of particles within the material.

Figure 15.26—Direction of Atomic Vibration for Longitudinal and Shear Waves

tages of this wave form are lower velocities that allow easier electronic timing, and greater sensitivity to small indications. On the other hand, these waves are more easily dispersed and cannot be propagated in a liquid medium (water).

The third mode is surface waves, sometimes referred to as Rayleigh waves. In this mode, ultrasonic waves are propagated along the metal surface, similar to waves on the surface of water. These surface waves have little movement below the surface of a metal. Therefore, they are not used for examination of welded and brazed joints.

Longitudinal wave ultrasound is generally limited in use to detecting inclusions and lamellar type discontinuities in base metal. Shear waves are most valuable in the detection of weld discontinuities because of their ability to furnish three dimensional coordinates for discontinuity locations, orientations, and characteristics. The sensitivity of shear waves is also about double that of longitudinal waves for the same frequency and search unit size.

The zones in the base metal adjacent to a weld should first be tested with longitudinal waves to ensure that the base metal does not contain discontinuities that would interfere with shear wave evaluation of the weld.

Coupling

A LIQUID OR hydraulic material is used for transmission of ultrasonic waves into the test object. Some of the more common couplants are water, light oil, glycerine, and cellulose gum powder mixed with water. Couplants can become a housekeeping problem, so care must be used to prevent accidents. Also, couplants, or the various solvents used to remove them, can be detrimental to repair welding or subsequent operations. Several proprietary couplants are available that minimize these problems.

A weldment must be smooth and flat to allow intimate coupling. Weld spatter, slag, and other irregularities should be removed.Depending on the testing technique, it may be necessary to remove the weld reinforcement.

Calibration

ULTRASONIC TESTING (UT) is basically a comparative evaluation. The horizontal (time) and the vertical (amplitude) dimensions on the CRT screen of the test unit are related to distance and size, respectively. It is necessary to establish a zero starting point for these variables, and to calibrate an ultrasonic unit to some basic standard before use.

Various test blocks are used to assist in calibration of the equipment.[10] Known reflecting areas can simulate typical discontinuities. Notches substitute for surface cracks, side-drilled holes for slag inclusions or internal cracks, and angulated flat-bottomed holes for small areas of incomplete fusion. The test block material must be similar in acoustic qualities to the metal being tested.

The International Institute of Welding (IIW) test block is widely used as a calibration block for ultrasonic testing of steel welds. This and other test blocks are used to calibrate an instrument for sensitivity, resolution, linearity, angle of sound propagation, and distance and gain calibrations.

Test Procedures

MOST ULTRASONIC TESTING of welds is done following a specific code or procedure. An example of such a procedure is that contained in *AWS D1.1, Structural Welding Code—Steel* for testing groove welds in structures.

---

10. Standard test blocks are shown in *ASTM E164, Standard practice for ultrasonic contact examination of weldments*. Philadelphia, PA: American Society for Testing and Materials, latest edition.

ASTM E164, Standard Practice for Ultrasonic Contact Examination of Weldments[11] covers examination of specific weld configurations in wrought ferrous and aluminum alloys to detect weld discontinuities. Recommended procedures for testing butt, corner, and T-joints are given for weld thicknesses from 0.5 to 8 in. (13 to 200 mm). Procedures for calibrating the equipment and appropriate calibration blocks are included. Other ASTM standards cover testing procedures with various ultrasonic inspection methods for inspection of pipe and tubing.[12]

To ultrasonically test a welded joint properly, the search unit must be manipulated in one or more specific patterns to adequately cover the through-thickness and length of the joint. In most cases, the joint must be scanned from two or more directions to ensure that the beam will intercept any discontinuities that exist. A typical search pattern for testing butt joints is shown in Figure 15.27. Similar procedures for butt, corner, and T-joints are illustrated in *ASTM E164, Standard Practice for Ultrasonic Contact Examination of Weldments*.

Evaluation of Weld Discontinuities

THE RELIABILITY OF ultrasonic examination depends greatly upon the interpretive ability of the ultrasonic testing technician. With the use of proper inspection techniques, significant information concerning a discontinuity can be learned from the signal response and display on the CRT screen.

There are six basic items of information available through ultrasonic testing which describe a weld discontinuty, depending upon the sensitivity of the test. These are as follows:

(1) The returned signal amplitude is a measure of the reflecting area. (See Figure 15.28.)

(2) Discontinuity length is determined by search unit travel in lengthwise direction. (See Figure 15.29.)

(3) The location of the discontinuity in the weld cross section can be estimated. (See Figure 15.30.)

(4) The orientation, and to some degree, the shape of the discontinuity can be determined by comparing signal sizes when tested from different directions. (See Figure 15.31.)

11. Available from the American Society for Testing and Materials 1916 Race Street, Philadelphia, PA 19102.

12. Procedures for UT of boiler and pressure vessel components are given in the *ASME Boiler and Pressure Vessel Code, Section V, Nondestructive Examination. Section XI, Inservice Inspection Requirements for Nuclear Power Plants*, gives methods for locating, sizing, and evaluating discontinuites for continuing service life and fracture mechanics analysis.

(5) The reflected pulse shape and sharpness can be used as an indicator of discontinuity type. (See Figure 15.32.)

(6) The height of the discontinuity within the weld can be estimated by the coordination of the travel distance of the search unit to and from the weld with the rise and fall of the signal. (See Figure 15.33.)

These six items of information concerning or describing a weld discontinuity can be used in the following manner. The first two items, the returned signal size and length, can be used as a basis for accepting or rejecting a single discontinuity in a weld. The third item, the location of the discontinuity within the cross section of the weld, is useful information when making a repair or evaluating a discontinuity for severity. Each of these first three items of information is essential in the proper inspection of welding for acceptance or repair.

The latter three items of information, namely orientation, pulse shape, and dynamic envelope, can be used to interpret more accurately the nature of the discontinuity. This information is of value in determining if the welding procedures are under control, and whether the component can continue in service.

Operator Qualifications

THE OPERATOR IS the key to successful ultrasonic testing. Generally speaking, UT requires more training and experience than the other nondestructive testing methods, with the possible exception of radiographic testing. Many critical variables are controlled by the operators. Therefore, the accuracy of an ultrasonic examination depends largely on knowledge and ability of the operator. For this reason, most standards require ultrasonic technicians to meet the requirements of ASNT-TC-1A.[13] Experience in welding and other nondestructive testing methods is helpful.

Reporting

CAREFUL TABULATION OF information on a report form is necessary for a meaningful test. A report form from ANSI/AWS D1.1, *Structural Welding Code—Steel* recommended for use on building and bridge weldments is shown in Figure 15.34. The welding inspector should be familiar with the kinds of data that must be recorded and evaluated so that a satisfactory determination of weld acceptability can be obtained.

13. Refer to SNT-TC-1A, *Personnel qualification and certification in nondestructive testing*. Columbus, Ohio: American Society for Nondestructive Testing.

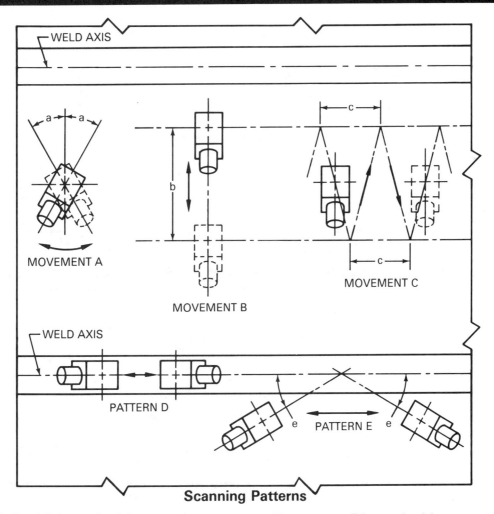

Scanning Patterns

Longitudinal Discontinuities

Movements A, B, and C are combined as one scanning pattern.

Scanning Movement A. Rotation angle a = 10 degrees.

Scanning Movement B. Scanning distance b shall be such that the section of weld being tested is covered.

Scanning Movement C. Progression distance c shall be approximately one-half the transducer width.

Transverse Discontinuities

Scanning pattern D (when welds are ground flush).

Scanning pattern E (when weld reinforcement is not ground flush).

Scanning angle e = 15 degrees max.

Notes:

1. Testing patterns are all symmetrical around the weld axis with the exception of pattern D which is conducted directly over the weld axis.

2. Testing from both sides of the weld axis is to be made wherever mechanically possible.

Figure 15.27—Typical Search Pattern for Ultrasonic Testing

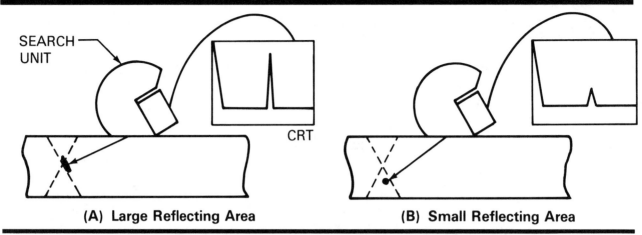

(A) Large Reflecting Area **(B) Small Reflecting Area**

Figure 15.28—The Size of the Reflecting Area Indicated by PIP Height

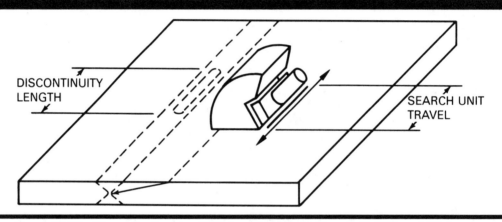

Figure 15.29—Discontinuity Length Determined from Search Unit Travel Distance

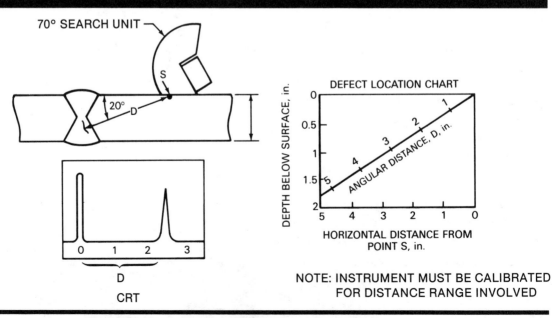

NOTE: INSTRUMENT MUST BE CALIBRATED FOR DISTANCE RANGE INVOLVED

Figure 15.30—Estimating the Location of Indication

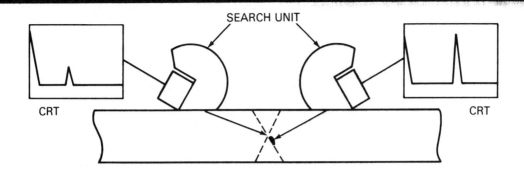

Figure 15.31—Discontinuity Orientation Indicated by PIP Height

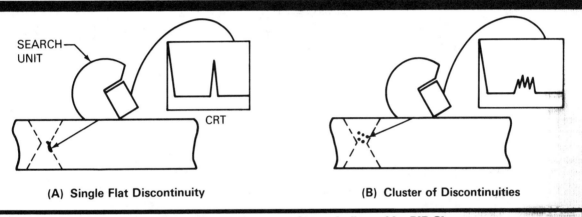

(A) Single Flat Discontinuity

(B) Cluster of Discontinuities

Figure 15.32—Discontinuity Type and Shape Indicated by PIP Shape

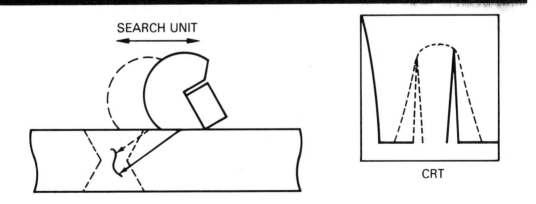

Figure 15.33—Estimating the Height of the Indication within the Weld

ACOUSTIC EMISSION TESTING

General Description

ACOUSTIC EMISSION TESTING (AET) consists of the detection of acoustic signals produced by plastic deformation or crack formation during loading or thermal stressing of metals. These signals are present in a wide frequency spectrum along with ambient noise from many other sources. Transducers, strategically placed on a structure, are activated by arriving acoustic signals. Ambient noise in the composite signal is significantly reduced by suitable filtering methods, and any source of significant signals is located by triangulation based on

REPORT OF ULTRASONIC TESTING OF WELDS

Project _____ Report no. _____

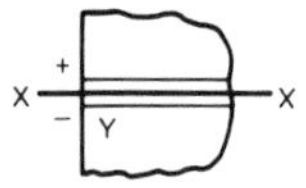

Weld identification _____
Material thickness _____
Weld joint AWS _____
Welding process _____
Quality requirements — section no. _____
Remarks _____

Line number	Indication number	Transducer angle	From Face	Leg*	Decibels				Discontinuity					Discontinuity elevation	Remarks
					Indication level (a)	Reference level (b)	Attenuation factor (c)	Indication rating (d)	Length	Angular distance (sound path)	Depth from "A" surface	Distance From X	Distance From Y		
1															
2															
3															
4															
5															
6															
7															
8															
9															
10															
11															
12															
13															
14															
15															
16															
17															
18															
19															
20															
21															
22															
23															
24															
25															
26															

We, the undersigned, certify that the statements in this record are correct and that the welds were prepared and tested in accordance with the requirements of 6C of AWS D1.1, (_____) Structural Welding Code.
year

Test date _____   Manufacturer or contractor _____

Inspected by _____   Authorized by _____

Date _____

Figure 15.34—Sample Form for Reporting Ultrasonic Test Results

the arrival times of these signals at several strategically placed transducers.

Acoustic emissions can be used to assess weld quality by monitoring during or after welding, or both. In weldments, regions having incomplete penetration, cracking, porosity, inclusions, or other discontinuities can be identified by detecting acoustic emissions originating at these regions. During welding processes, acoustic emissions are caused by many things, including plastic deformation, melting, friction, solidification, solid phase transformation, and cracking. Monitoring of acoustic emissions during welding can even include, in some instances, automatic feedback control of the welding process. In large-scale automatic welding, the readout equipment can be conveniently located near the welding controls or in a quality-monitoring area.

Location of Acoustic Emissions

POSITIONS OF ACOUSTIC sources along a welded joint can be presented in a variety of ways. One technique displays the number of events versus distance along the weld on an oscilloscope screen or an X-Y plotter. Another technique uses a digital-line printer that gives the time of the event, its location, and its intensity. This information facilitates appraisal of the severity of each source. After the acoustic emission sources are graded, other nondestructive testing methods may be used to evaluate the indications.

Applications

Monitoring During Continuous Welding. Detection and location of acoustic emission sources in weldments during fabrication may provide information related to the integrity of the weld.[14] Such information may be used to direct repair procedures on the weld or as a guide for application of other NDT methods. A major attribute of AET for in-process monitoring of welds is the ability of the method to provide immediate real-time information on weld integrity. This feature makes the method useful for lower welding costs by pinpointing defects that can be repaired at the most convenient point in the production process. Acoustic emission activity from discontinuities in the weldment is stimulated by the thermal stresses from the welding process. The resulting activity is detected by sensors in the vicinity of the weld that convert the acoustic signals into electronic signals.

Acoustic emission monitoring for the evaluation of quality and the control of a welding process requires preliminary studies for each application to establish operating conditions such as the number, location, and mounting of sensors; gain settings; filtering; data presentation; and data interpretation. These studies normally include correlation with other nondestructive and destructive methods of inspection.

The monitoring equipment normally consists of sensors, electronic instrumentation, and recording devices. Monitoring acoustic emission during welding may require specialized apparatus because of severe environmental factors and interfering noise sources.

Acoustic emission data may be accumulated during the welding process. Because of the delay between weld fusion and acoustic emission activity, monitoring must continue for some time following welding to acquire all significant data. The monitoring time after welding increases with increasing weld heat input, ranging from 10 seconds for manual gas tungsten arc welding (approximately 100 A) to more than 2 minutes for submerged arc welding (600 to 800 A). The time should be established during developmental monitoring of trial welds.

Observable conditions that occur in conjunction with unusual acoustic emission activity are recorded to aid in later interpretation of the data. This would include clean up or chipping and grinding by the welder, for example.

In general, acoustic emission weld data must be evaluated against a base line obtained from (1) known acceptable welds of a given type using the specific acoustic emission system, and (2) signals from the same weld type known to contain defects. Significant weld discontinuities may be characterized by increases in the acoustic emission event count, the rate of events, the intensity, or the peak amplitude.

Monitoring During Resistance Spot Welding. The acoustic emissions produced during the making of a spot weld can be related to weld quality characteristics such as the strength and size of the nugget, the amount of expulsion, and the amount of cracking. Therefore, in-process acoustic emission monitoring can be used both as an examination method and as a means for providing feedback control.[15] The resistance spot welding process consists of several stages. These are the set-down of the electrodes, squeeze time, weld time, forging, hold time, and lift-off. Many acoustic emission signals are produced during each of these stages. Often, these signals can be identified with respect to the nature of their source.

Most of the emission signal features can be related to factors of weld quality. The emissions occurring during set-down and squeeze time can often be related to the condition of the electrodes and the surface of the parts. The large but brief signal at current initiation can be related to the initial contact resistance and the cleanliness of the parts.

---

14. For detailed information, refer to *ASTM E749, Standard practice for acoustic emission monitoring during continuous welding.* Philadelphia, PA: American Society for Testing and Materials.

15. Additional information may be found in *ASTM E751, Standard practice for acoustic emission monitoring during resistance spot welding.* Philadelphia, PA: American Society for Testing and Materials.

During welding current flow, signals are produced by plastic deformation, nugget expansion, friction, melting, and expulsion. Those signals caused by expulsion (spitting or flashing, or both), generally have large amplitudes, and can be distinguished from the rest of the acoustic emission associated with nugget formation.

Following termination of the welding current, some metals exhibit appreciable noise during solidification, which can be related to nugget size and inclusions. As the nugget cools during the hold period, acoustic emission can result from solid-phase transformations and cracking.

During the electrode lift-off stage, separation of the electrode from the part produces signals that can be related to the condition of the electrode as well as the cosmetic condition of the weld. The acoustic emission response corresponding to each stage can be separately detected and analyzed.

A measure of the cumulative acoustic emissions during resistance spot welding cannot be expected to relate clearly to weld quality. On the other hand, by using both time discrimination and multiple detection levels, the various segments of acoustic emissions can be separately measured and related to various indicators of quality. Commercial instrumentation is available that is capable of separately monitoring several of the acoustic-emission segments. For instance, the expulsion count, phase-transformation count, and cracking count can be monitored and recorded for each resistance spot weld, giving a permanent record of quality.

Acoustic emission surveillance of spot-welding can take two forms, namely monitoring and control. The purpose of on-line monitoring is to identify and segregate unacceptable welds for quality evaluation purposes. On-line control, on the other hand, utilizes the acoustic emission instrumentation to complete a feedback loop between the process and the source of welding current by automatically adjusting one or two process variables to compensate for deteriorating welding conditions.

Monitoring During Proof Testing. AET methods have been applied to welded pressure vessels and other welded structures during proof testing. A sound vessel stops emitting signals when the load is reduced, and does not emit further bursts until the previous load has been exceeded. A growing crack continues to emit signals as it is loaded. Location of suspect areas in such structures is a well established AET technique. Locating systems have been developed, some providing sophisticated analysis of the signal data collected.

RECORDING OF RESULTS

WITH ANY TYPE of inspection, defective areas in a weldment must be identified in some manner to assure that they will be located and repaired properly. Many identification methods are available. In addition to logging the results by type, size, and location, the defective area should be marked directly on the weldment. The following rules should be followed:

(1) Marking should be positive, clear, and in a color contrasting with the metal.

(2) Appropriate personnel should be familiar with the marking system.

(3) The color used for marking defects should not be used on the weldments for any other purpose.

(4) The marking material should withstand exposure to handling or further processing until the defective area can be repaired, but should not damage the weldment during subsequent processing or repair.

(5) If service conditions dictate, the same inspection should be repeated and the results carefully recorded.

After repair is completed, the same inspection should be repeated and the results carefully recorded.

PROOF TESTING

Purpose

MANY WELDED COMPONENTS are nondestructively examined by proof testing during or subsequent to fabrication. The purpose of these tests include

(1) Assurance of safe operation
(2) Detection of any weakness of the design
(3) Exposure of quality deficiencies
(4) Prevention of in-service failures

Proof testing is achieved by applying one or more tests that exceed actual service requirements by a predetermined factor of safety. This may involve overloading the component or testing for leaks, or both.

Load Testing

WELDED COMPONENTS CAN be proof tested by applying specific loads without failure or permanent deformation. Such tests are usually designed to subject the parts to stresses exceeding those anticipated during service. However, the stresses are maintained below or at the minimum specified yield strength of the metal. Many load test requirements and application details are mandated by codes, specifications, and contractual documents that apply to individual product forms.

Structural members are often proof tested by demonstrating their ability to carry loads equal to or larger than any anticipated service conditions. This can be accomplished by statically loading with a testing machine, sand bags, or scrap iron, or by dynamically loading with special testing equipment. Acceptance is based on freedom from cracking or objectionable permanent deformation.

Hydrostatic Testing

CLOSED CONTAINERS ARE usually proof tested by filling them with water and applying a predetermined test pressure. For components built in accordance with the *ASME Boiler and Pressure Vessel Code*, this pressure is 150 percent of design pressure. For other components, the test pressure may be based upon a percentage of the minimum yield strength. After a fixed holding time, the container is inspected for soundness by visually checking for leakage, or by monitoring the hydrostatic pressure. Visual inspection can be enhanced by an indicating system applied to the outside of the vessel, such as light blue chalk that turns dark blue in the presence of a small amount of water. To increase sensitivity, an enhancing material, such as water-soluble colored or fluorescent dyes can be added to the test water for detection of small leaks by developers or ultraviolet light.

Open containers (e.g., storage tanks) may also be hydrostatically tested by filling them with water or partially submerging them into water (e.g., ship barges). The hydrostatic pressure exerted against any boundary is governed by the head of water.

Hydrostatic testing is a relatively safe operation because water is practically noncompressible and, therefore, stores little energy. A small leak results in a meaningful pressure drop that limits the driving force available to propagate a crack. However, three questions should be considered to conduct safe hydrostatic tests:

(1) Are the foundation and the support structure strong enough to hold the water-filled container? This is of special importance if the containers are designed to hold a gas or a light-weight liquid.

(2) Are there any pockets where energy can build up in the form of compressed air?

(3) Is the metal at a temperature where it has adequate notch toughness to assure that a relatively small leak or discontinuity will not propagate into a catastrophic brittle fracture?

Pneumatic Testing

PNEUMATIC TESTING IS similar to hydrostatic testing except that compressed air is used to pressurize a closed vessel. This type of test is primarily used for small units that can be submerged in water during the tests. The water provides a convenient leak indicator (air bubbles)—and is an effective energy absorber in the event that the container fails.

Other applications include units mounted on foundations that are not able to support the weight associated

with hydrostatic tests, and vessels where water or other liquid may be harmful or cannot be adequately removed. An example of the latter is a plate-fin heat exchanger designed for cryogenic service.

Acceptance of pneumatic tests is based on freedom from leakage. Small leaks are seldom detected without some indicating devices. If units cannot be submerged in water, spraying with a soap or detergent solution and checking for bubbles is an effective procedure. This procedure is called a soap bubble test. For special applications, sound detection devices are available that report and locate all but the smallest air leaks.

During any pneumatic test, large amounts of energy may be stored in compressed air or gas in a large volume or under high pressure, or both. A small leak or rupture can easily grow into a catastrophic failure, and can endanger adjacent life and property. All pneumatic testing should follow a written test procedure for the product to be tested, and the procedure should include safety precautions.

Spin Testing

WELDED COMPONENTS THAT rotate in service can be proof tested by spinning them at speeds above the design values to develop desired stresses from centrifugal forces. Visual and other nondestructive testing plus dimensional measurements are employed to determine the acceptability of the parts. Spin testing must be done in a safe enclosure in case the component should rupture.

Leak Testing

WHEN FREEDOM FROM leakage is of primary importance and a high pressure test is not desirable or possible, a number of low-pressure leak testing techniques are available. All are based upon the principal of filling a container with a product that has a low viscosity and has the ability to penetrate through very small openings. For instance, if a container is designed to hold water or oil, a low-pressure pneumatic test may be an acceptable leak test.

To increase the effectiveness and accuracy of such leak tests, tracer gases are often used. These include freon for standard tests and helium for critical applications. Leak detection instruments (calibrated sniffers) are used to detect the presence of the tracer gas escaping through any leaks in the vessel.

Vacuum Box Testing

PRESSURE LEAK TESTING requires the ability to pressurize one side and inspect from the other side of the component. This represents a limitation for components that cannot be pressurized. However, a vacuum box test can be used. This involves coating a test area with a soap or detergent film and placing a gasketed transparent box over the area to be inspected. After evacuating the box to a partial vacuum of not less than 2 psi (14 kPa), the inspector looks for any bubble formations that indicate the presence of a leak. For critical service, such as LNG tanks, vacuum levels as high as 8 psi (55 kPa) may be selected.

Mechanical Stress Relieving

PROOF-TESTING OPERATIONS CAN reduce the residual stresses associated with welding. In the as-welded condition, any weldment will have peak residual stresses equal to or slightly below the yield strength of the metal. When proof testing the weldments, the yield strength may be exceeded in highly stressed areas, and the metal will plastically deform if it has adequate ductility at ambient temperature. After the load is removed, the peak residual stresses in those areas will be lower than the original stresses. This result tends to improve product reliability.

Hardness Testing

HARDNESS TESTING CAN be used by inspectors as both a nondestructive test and as part of a series of destructive tests on a production test piece. It is essential to document the hardness of all destructively tested weld samples.

There are three static hardness tests commonly used to measure the effects of processing on metals and for quality control in production. They are the Brinell, Rockwell, and Vickers hardness tests. All of these tests are based on the size of an indentation made with a particular indenter design under a specified load. The choice of method depends on the size and finish of the area to be tested, the composition and thickness of the test object, and the end use of the object.

In the case of carbon and alloy steels, there is an approxiamte relationship between Brinell, Rockwell, and Vickers hardness numbers, and between hardness number and tensile strength. These data are presented in tabular form in ASTM A370, *Standard Methods and Definitions for Mechanical Testing of Steel Products*.[16] With Brinell hardness testing, the approximate tensile strength in psi of carbon and low alloy steels is 500 times the hardness number. The Brinell test is commonly used for quality control in industry but the impression is relatively large compared to the other tests. Therefore, it can only be used for obtaining hardness on a relatively large area, and where the impression on the surface is not objectionable. Where marking will affect the end use of

the product, Rockwell or Vickers tests can be applied with small surface indentations. Because these methods can be applied to small or narrow areas, they can be used for survey work to locate variations in hardness across a weld.

For the welding inspector, the most important difference involves the size of the indenter or penetrator, which may range from a 10 mm diameter ball to a small diamond tip. The size of the indenter has little effect on the test results for reasonably homogenous base metal or weld metal, but indenter size is of major importance when testing a narrow heat affected zone or weld bead that contains several metallurgically differing zones. A small indenter may detect narrow areas of different hardness while the 10 mm ball produces an average value for the entire zone. Any specification requirement that includes heat-affected zone hardness tests should specify the type of test and the indenter to be employed.

In hardness testing, specimen preparation is important for reliable results. The surface should be flat, reasonably free from scratches, and must be normal to the applied load for uniform indentations. With thin, soft metals, the testing method must produce a shallow indentation that is not restricted by the anvil of the testing machine. This can be accomplished, particularly with a Rockwell testing machine with a small indenter and a light load.

Brinell Hardness Test. The Brinell hardness test consists of impressing a hardened steel ball into the test surface using a specified load for a definite time. Following this, the diameter of the impression is accurately measured and converted to a hardness number from a table. Stationary machines impress a 10 mm ball into the test object. The load for steel is 3000 kg and for softer metals, 500 or 1500 kg.

Two diameters of the impression are measured at 90 degrees by using a special Brinell microscope graduated in tenths of millimeters. The mean diameter is used to determine the Brinell hardness number from the table.

To test larger components, portable Brinell equipment is available consisting of a 7 or 10 mm ball and a calibrated reference bar. A hammer blow is used to simultaneously indent both the reference bar and the material being tested. The hardness of the material being tested can be determined by inserting the hardness of the reference bar and the diameters of the two indentations into a formula or by using a special slide rule. The accuracy of this method is enhanced by selecting a reference bar of approximately the same hardness as the unknown test material. Testing should be done to the requirements of *ASTM E10, Standard Test Method for Brinell Hardness of Metallic Materials*.[17]

16. Available from the American Society for Testing and Materials, 1916 Race Street, Philadelphia, PA.

17. Available from the American Society for Testing and Materials, 1916 Race Street, Philadelphia, PA.

Rockwell Hardness Test. The Rockwell hardness test measures the depth of residual penetration made by a small hardened steel ball or a diamond cone. The test is performed by applying a minor load of 10 kg to seat the penetrator in the surface of the specimen and hold it in position. The machine dial is turned to a "set" point, and a major load is applied. After the pointer comes to rest, the major load is released, with minor load remaining.

The Rockwell hardness number is read directly on the dial. Hardened steel balls of 1/8 or 1/16 in. (3.2 mm or 1.6 mm) diameter are used for soft metals, and a cone-shaped diamond penetrator is used for hard metals. Testing is conducted in accordance with *ASTM E18, Standard Test Method for Rockwell Hardness and Rockwell Superficial Hardness of Metallic Materials.*

DESTRUCTIVE TESTS

DESTRUCTIVE TESTS OF weldments are not normally employed as an inspection method, but those destructive tests specified by the contract documents or relevant standards are required. Certain standard tests published in *AWS B4.0, Standard Methods for Mechanical Testing of Welds* are extensively used in the destructive testing of weldments. Addition information is presented in Chapter 12 of this volume.

FILLER METALS

THE MANUFACTURE AND sale of filler metals is covered by the AWS A5 series of standards. These standards specify the number and type of destructive tests of filler metals or deposited weld metals. However, the purchase documents specify the frequency of testing and define the lot size. The tests normally involve the following:

(1) Chemical composition of filler metal or deposited weld metal
(2) Tensile and impact tests of the undiluted weld metal
(3) Macro and break test of fillet welds

The particular tests to be conducted depend on the product and standard. The relevant AWS filler metal specification should be reviewed in each case.

STANDARD PRODUCT MANUFACTURING

CERTAIN WELDED METAL products produced on a dedicated mill are certified based on results of destructive tests of samples from a lot. Steel pipe and tubing are typical examples in which transverse weld tensile and bend tests must be conducted on pipe selected from each heat of steel. These tests are in addition to the visual, proof, dimensional and other nondestructive tests required by the specification. It is the inspector's job to review all relevant specification requirements and conduct or observe all inspections needed to assure acceptable product quality.

CUSTOM FABRICATED PRODUCTS

UNDER CERTAIN CONDITIONS production sampling and impact testing of welds and heat affected zones are required by the *Boiler and Pressure Vessel Code* and by the *American Bureau of Shipping Rules for Building and Classing Steel Vessels.* Other codes or contract specifications may require special destructive tests of production material to certify a portion of the fabricated products.

Frequently customer specifications for stainless steel clad products require chemical analysis to assure that the calculated ferrite number is within specification limits. Tests for ferrite number can also be conducted nondestructively using calibrated magnetic instruments.[18]

METALLOGRAPHIC EXAMINATION

METALLOGRAPHIC TESTS CAN be used to determine:

(1) The soundness of the joint
(2) The distribution of nonmetallic inclusions in the joint
(3) The number of weld passes
(4) The location and depth of weld penetration
(5) The extent of the heat affected zone
(6) The metallurgical structure in the weld metal and heat affected zone

Macroscopic examination involves visual examination of as-polished or polished-and-etched surfaces by the unaided eye or at magnifications up to 10 power. Etching of the surface can reveal gross structure and weld bead configuration. Pores, cracks, and inclusions are better observed on a polished surface. Microscopic examination is carried out by observing a highly polished and etched surface at magnifications of about 50 times or higher.

---

18. Refer to *AWS A4.2 Standard procedures for calibrating magnetic instruments to measure the delta ferrite content of austenitic stainless steel weld metal.* Miami, FL: American Welding Society, latest edition.

Samples are obtained by sectioning test welds or production control welds, including run-off tabs. They can be prepared by cutting, machining, or grinding to reveal the desired surface and can be subjected to further preparation as needed to reveal the desired structure. Prior to conducting metallographic examinations, the entire heat-affected zone created by any thermal cutting process used during sample preparation must be mechanically removed.

Macroscopic Examination

MACROETCHING IS USED to reveal the heterogeneity of metals and alloys. Typical applications of macroetching in the welding of metals are the study of weld structure; measurement of joint penetration; dilution of filler metal by base metal; and presence of slag, flux, porosity, and cracks in weld and heat-affected zones. When macroscopic examination is used as an inspection procedure, sampling should be done in an early stage of manufacturing to permit corrective action to be taken if necessary.

Sample preparation need not be elaborate. Any method of preparing a smooth surface with a minimum amount of cold work is satisfactory. Cross sections may be faced on a lathe or a shaper. The usual procedure is to take a roughing cut, followed by a finish cut with sharp tools. This should provide a smooth surface and remove cold work from prior operations. Grinding is usually conducted in the same manner, using free-cutting wheels and light finishing cuts. When fine detail is required, the specimen should be polished through a series of metallographic papers.

After surface preparation, the sample is cleaned carefully with suitable solvents. Any grease, oil, or other residue will produce uneven etching. Once cleaned, the sample surface should not be touched or contaminated in any way.

Recommended solutions and procdures for macroetching are given in *ASTM E340, Standard Method for Macroetching Metals and Alloys*. Caution must be observed in handling chemicals and mixing solutions. Many of the etchants are strong acids that require special handling and storage.[19] In all cases, the various chemicals should be added slowly to the water or other solvent while stirring.

Microscopic Examination

IN EXAMINING FOR exceedingly small discontinuities or for metallurgical structure at high magnification, specimens are polished, etched and examined by microscope to reveal the microstructure of the base metal, heat-affected zone, fusion zone, and weld metal. Procedures for selection, cutting, mounting, and polishing metallographic specimens are given in *ASTM E3, Standard Methods of Preparation of Metallographic Specimens*. Recommended chemical solutions for etching various metals and alloys and safety precautions in handling etching chemicals are given in *ASTM E407, Standard Method for Microetching Metals and Alloys*.

BRAZED JOINTS

GENERAL CONSIDERATIONS

INSPECTION OF A completed assembly or subassembly is the last step in the brazing operation, and is essential for assuring satisfactory and uniform quality of a brazed unit. The design of a brazement is extremely important to the inspection operation. Where practical, the design should permit easy and adequate examination of completed joints. An intelligent choice of brazing processes, brazing filler metal, joint design, and cleaning methods will also assist in the inspection process. The inspection method chosen to evaluate a final brazed component should depend on the service requirements. In many cases, the inspection methods are specified by the user or by regulatory codes.

Testing and inspection of brazed joints can be conducted on procedure qualification test joints or on a finished brazed assembly, and the tests may be either destructive or nondestructive. Preproduction and workmanship samples are often used for comparison purposes during production. These may be sample specimens made during the development of the brazing procedure or samples taken from actual production. In any case, they should show the minimum acceptable production quality of the brazed joint.

TESTING METHODS

BRAZED JOINTS AND completed brazements can be tested both nondestructively and destructively. Nondestructive test methods that may be used are visual examination, liquid penetrant, radiographic, and ultrasonic as well as proof and leak testing. Destructive testing methods

---

19. The users of chemical etchants must be familar with the proper handling and storage of the chemicals and the appropriate safety precautions. Refer to Sax, N.I. *Dangerous properties of industrial materials*, latest ed. New York: Van Nostrand Reinhold Co.

include metallographic examination and peel, tension, shear, and torsion tests. All of these testing methods are discussed in the *Welding Handbook*, Vol. 2, 7th Edition.

SUPPLEMENTARY READING LIST

American Society of Mechanical Engineers (ASME). *Boiler and pressure vessel code.* New York: American Society of Mechanical Engineers (latest ed.).

American Society for Metals (ASM). *Metals handbook,* 8th ed., Vol. 11. Metals Park, OH: American Society for Metals, 1976.

American Society for Nondestructive Testing. (ASNT). No.SNT-TC-1A, *Personnel qualification and certification in nondestructive testing.* Columbus, OH: The American Society for Nondestructive Testing.

American Welding Society (AWS). *B1.0, Guide for the nondestructive inspection of welds.* Miami, FL: American Welding Society (latest ed.).

————*D1.1, Structural welding code—steel.* Miami, FL: American Welding Society (latest ed.).

————*Welding inspection,* 2nd ed. Miami, FL: American Welding Society, 1980.

Nichols, R. W., ed. *Non-destructive examination in relation to structural integrity.* London: Applied Science Publishers, Ltd., 1979.

Society for Nondestructive Testing. *Nondestructive testing handbook,* Vols. 1 and 2. Columbus, Ohio: Society for Nondestructive Testing, 1982.

Stout, R. D. *Hardness as an index of the weldability and service performance of steel weldments.* New York: Welding Research Council Bulletin 189, Nov. 1973.

SAFE PRACTICES

PREPARED BY A COMMITTEE CONSISTING OF:

G. R. Spies, *Chairman*
The B.O.C. Group, Inc.

G. C. Barnes
Alloy Rods Co.

K. L. Brown
The Lincoln Electric Company

W. Beisner
Hobart Brothers Company

O. J. Fisher
Babcock and Wilcox Co.

W. S. Howes
NEMA

C. Philp
Handy and Harman Co.

WELDING HANDBOOK COMMITTEE MEMBER:
A. F. Manz
A. F. Manz Associates

CHAPTER 16

SAFE PRACTICES

GENERAL WELDING SAFETY

THIS CHAPTER COVERS the basic elements of safety general to all welding, cutting, and related processes. It includes safety procedures common to a variety of applications. However, it does not cover all safety aspects of every welding process; especially not those involving sophisticated technology. For this reason, those chapters in the handbook devoted to specific processes should also be referenced for additional important safety information.

Safety is an important consideration in all welding, cutting, and related work. No activity is satisfactorily completed if someone is injured. The hazards that may be encountered and the practices that will minimize personal injury and property damage are discussed here.

MANAGEMENT SUPPORT

THE MOST IMPORTANT component of an effective safety and health program is management support and direction. Management must clearly state objectives and demonstrate its commitment to safety and health by consistent execution of safe practices.

Management must designate approved areas where welding and cutting operations may be carried on safely. When these operations must be done in other than designated areas, management must assure that proper procedures to protect personnel and property are established and followed.

Management must be certain that only approved welding, cutting, and allied equipment is used. Such equipment includes torches, regulators, welding machines, electrode holders, and personal protective devices. Adequate supervision must be provided to assure that all equipment is properly used and maintained.

TRAINING

THOROUGH AND EFFECTIVE training is a key aspect of a safety program. Adequate training is mandated under provisions of the U.S. Occupational Safety and Health Act (OSHA), especially those of the Hazard Communication Standard (29 CFR 1910.1200). Welders and other equipment operators perform most safely when they are properly trained in the subject.[1] Proper training includes instruction in the safe use of equipment and processes, and the safety rules that must be followed. Personnel need to know and understand the rules and the consequences of disobeying them. For example, welders must be trained to position themselves while welding or cutting so that their heads are not in the gases or fume plume.[2]

Before work begins, users must always read and understand the manufacturer's instructions on safe practices for the materials and equipment, and also the material safety data sheets.

Certain AWS specifications call for precautionary labels on consumables and equipment. These labels concerning the safe use of the products should be read and followed. A typical label is illustrated in Figure 16.1.

Manufacturers of welding consumables must provide on request a *Material Safety Data Sheet* (MSDS) that identifies those materials present in their products that have hazardous physical or health properties. The MSDS provides the OSHA permissible exposure limit,

---

1. The term *welder* is intended to include all welding and cutting personnel, brazers, and solderers.

2. Fume plume is the smoke-like cloud containing minute solid particles arising directly from the area of melting metal. In distinction to a gas, fumes are metallic vapors that have condensed to solid and are often associated with a chemical reaction, such as oxidation.

WARNING: PROTECT yourself and others. Read and understand this label.

FUMES AND GASES can be dangerous to your health. ARC RAYS can injure eyes and burn. ELECTRIC SHOCK can KILL.

- **Read and understand the manufacturer's instructions and your employer's safety practices.**

- **Keep your head out of the fumes.**

- **Use enough ventilation, exhaust at the arc, or both, to keep fumes and gases from your breathing zone and the general area.**

- **Wear correct eye, ear, and body protection.**

- **Do not touch live electrical parts.**

- **See American National Standard Z49.1 "Safety in Welding and Cutting" published by the American Welding Society, 550 N.W. LeJeune Rd., Miami, Florida 33126; OSHA Safety and Health Standards, 29 CFR 1910, available from U.S. Government Printing Office, Washington, DC 20402.**

DO NOT REMOVE THIS LABEL

Figure 16.1—Minimum Warning Label for Arc Welding Processes and Equipment

called the Threshold Limit Value (TLV)[3], and any other exposure limit used or recommended by the manufacturer.

Employers that use consumables must make the applicable MSDS readily available to their employees, as well as train them to read and understand the contents. The MSDS contain important information about the ingredients contained in welding electrodes, rods, and fluxes, the composition of fumes which may be generated in their use, and means to be followed to protect the welder and others from hazards which might be involved.

Under OSHA Hazard Communication Standard 29 CFR 1910.1200, employers are responsible for the training of employees with respect to hazardous materials used in their workplace. Many welding consumables are included in the definition of hazardous materials according to this standard. Welding employers must comply with the communication and training requirements of this standard.

The proper use and maintenance of the equipment must also be taught. For example, defective or worn electrical insulation cannot be tolerated in arc welding or cutting, nor can defective or worn hoses be used in oxyfuel gas welding and cutting, brazing, or soldering. Proper training in equipment operation is fundamental to safe operation.

Persons must be trained to recognize safety hazards. If they are to work in an unfamiliar situation or environment, they must be thoroughly briefed on the potential hazards involved. For example, consider a person who must work in a confined space. If the ventilation is poor and an air-supplied helmet is required, the need and instructions for its proper use must be thoroughly explained to the employee. The consequences of improperly using the equipment must be covered. When employees believe that the safety precautions for a given task are not adequate, or not understood, they should question their supervisors before proceeding.

GENERAL HOUSEKEEPING

GOOD HOUSEKEEPING IS essential to avoid injuries. A welder's vision is often restricted by necessary eye protection. Persons passing a welding station must shield their eyes from the flame or arc radiation. The limited vision of the welder and passers-by makes them vulnerable to tripping over objects on the floor. Therefore,

---

3. TLV is a registered trademark of the American Conference of Governmental and Industrial Hygienists (ACGIH).

welders and supervisors must always make sure that the area is clear of tripping hazards. Management must lay out the production area so that gas hoses, cables, mechanical assemblies, and other equipment do not cross walkways or interfere with routine tasks.

When work is above ground or floor level, safety rails or lines must be provided to prevent falls as a result of restricted vision from eye protection devices. Safety lines and harnesses can be helpful to restrict workers to safe areas, and to catch them in case of a fall.

Unexpected events, such as fire and explosions, do occur in industrial environments. All escape routes must be identified and kept clear so that orderly, rapid, and safe evacuation of an area can take place. Storage of goods and equipment in evacuation routes should be avoided. If an evacuation route must be temporarily blocked, employees who would normally use that route must be trained to use an alternate route.

PROTECTION IN THE GENERAL AREA

EQUIPMENT, MACHINES, CABLES, hoses, and other apparatus should always be placed so that they do not present a hazard to personnel in passageways, on ladders, or on stairways. Warning signs should be posted to designate welding areas, and to specify that eye protection must be worn.

Protective Screens

PERSONS IN AREAS adjacent to welding and cutting must be protected from radiant energy and hot spatter by (1) flame-resistant screens or shields, or (2) suitable eye and face protection and protective clothing. Appropriate radiation-protective, semi-transparent materials are permissible.

Where operations permit, work stations should be separated by noncombustible screens or shields, as shown in Figure 16.2. Booths and screens should permit circulation of air at floor level as well as above the screen.

Wall Reflectivity

WHERE ARC WELDING or cutting is regularly carried on adjacent to painted walls, the walls should be painted

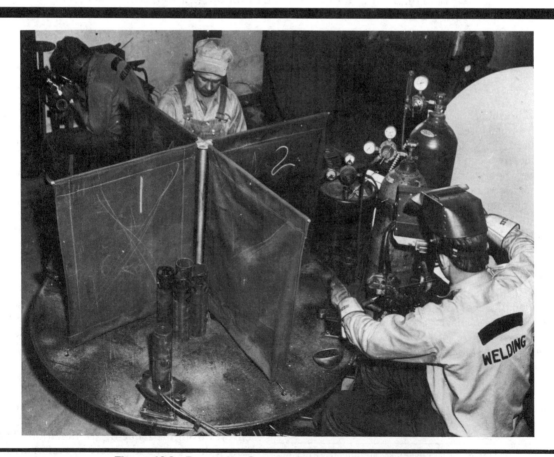

Figure 16.2—Protective Screens Between Work Stations

with a finish having low reflectivity of ultraviolet radiation.[4] Finishes formulated with certain pigments, such as titanium dioxide or zinc oxide, have low reflectivity to ultraviolet radiation. Color pigments may be added if they do not increase reflectivity. Pigments based on powdered or flaked metals are not recommended because they reflect ultraviolet radiation.

PUBLIC DEMONSTRATIONS

PERSONS PUTTING ON public demonstrations involving observation of arc or oxyfuel gas welding or cutting processes are responsible for the safety of observers and the general public. Observers are not likely to have the necessary protective equipment to let them observe demonstrations safely. For exhibits involving observation of arc or oxyfuel gas welding and cutting processes, appropriate eye protection for both observers and passers-by is mandatory.

Fume exposure must be controlled by appropriate ventilation. Electric cables and hoses must be routed to avoid audience exposure to possible electric shock or tripping hazards. Protection must be provided against fires from fuels, combustibles, and overheated apparatus and wiring.

Fire extinguishers must be on hand in case of fire. Combustible materials must be removed from the area or shielded from flames, sparks, and molten metal.

Safety precautions at public events should be passive types, that is, they should not require the audience to take action to protect itself. For example, a protective, moveable transparent screen allows an audience to observe a welding operation with the screen in place. After welding is completed, the screen can be moved to allow the audience to observe the completed weld. Additional information is given in *ANSI/ASC Z49.1, Safety in Welding and Cutting.*

FIRE

IN MOST WELDING, cutting, and allied processes, a high-temperature heat source is present. Open flames, electric arcs, hot metal, sparks, and spatter are ready sources of ignition. Many fires are started by sparks, which can travel horizontally up to 35 feet from their source and fall much greater distances. Sparks can pass through or lodge in cracks, holes, and other small openings in floors and walls.

The risk of fire is increased by combustibles in the work area, or by welding or cutting too close to combustibles that have not been shielded. Materials most commonly ignited are combustible floors, roofs, partitions, and building contents including trash, wood, paper, tex-

tiles, plastics, chemicals, and flammable liquids and gases. Outdoors, the most common combustibles are dry grass and brush.

The best protection against fire is to perform welding and cutting in specially designated areas or enclosures of noncombustible construction that are kept free of combustibles. Combustibles should always be removed from the work area or shielded from the operation.

Fuel for engine driven equipment should be stored and used with care. Equipment manufacturer's instructions should be followed because fuels and their vapors are combustible and can be explosive under some conditions.

Fuel gases, such as acetylene or propane, are other common flammables often found in cutting and welding areas. Special attention should be given to fuel gas cylinders, hoses, and apparatus to prevent gas leakage.

Combustibles that cannot be removed from the area should be covered with tight fitting, flame resistant material. These include combustible walls and ceilings. Floors should be free of combustible materials for a radius of 35 feet around the work area. All doorways, windows, cracks, and other openings should be covered with a flame-resistant material or, if possible, the work area should be enclosed with portable flame-resistant screens.

If welding or cutting is to be performed on or adjacent to a metal wall, ceiling, or partition, combustibles on the other side must be moved to a safe location. If this cannot be done, a fire watcher should be stationed where the combustibles are located. Heat from welding can be conducted through metal partitions and ignite combustibles on the opposite side. A thorough examination for evidence of fire should be made before leaving the work area. Fire inspection should be continued for at least 30 minutes after the operation is completed.

Welding or cutting should not be performed on material having a combustible coating or internal structure, as in walls or ceilings. Hot scrap or slag must not be placed in containers holding combustible materials. Suitable fire extinguishers should always be available nearby.

Welding, brazing, or cutting should not be performed on combustible floors or platforms that may readily be ignited by heat from the operation. Welders must be alert for traveling vapors of flammable liquids. Vapors are often heavier than air and can travel along floors and in depressions for distances of several hundred feet from where the flammable liquid is stored. Light vapors can travel along ceilings to adjacent rooms.

HOT WORK PERMIT SYSTEM

WHEN WELDING, CUTTING, or similar hot working operations are to be performed in areas not normally assigned for such operations, a hot work permit system

---

4. For further guidance, see *Ultraviolet Reflectance of Paint.* Miami: American Welding Society, 1976.

should be used. The purpose of the hot work permit system is to alert area supervisors to an extraordinary danger of fire that will exist at a particular time. The permit system should include a check list of safety precautions that includes an inspection for fire extinguishers, establishment of fire watches if necessary, search for flammable materials, and safety instructions for personnel in the area who are not involved in the hot work.

EXPLOSION

FLAMMABLE GASES, VAPORS, and dusts, when mixed with air or oxygen in certain proportions, present danger of explosion as well as fire. To prevent explosions, avoid all sources of ignition. Welding, brazing, soldering, cutting, or operating equipment that can produce heat or sparks must not be done in atmospheres containing flammable gases, vapors, or dusts. Such flammables must be kept in leak-tight containers or be well removed from the work area. Heat or sparks may cause otherwise low-volatile materials to produce flammable vapors.

Hollow containers must be vented before applying heat. Heat must not be applied to a container that has held an unknown material, a combustible substance, or a substance that may form flammable vapors on heating.[5] The container must first be thoroughly cleaned or filled with an inert gas. Heat should never be applied to a workpiece that is covered by an unknown substance, nor to a substance that may form flammable or toxic vapors on heating.

Adequate eye and body protection must be worn when operations involve a risk of explosion.

BURNS

BURNS OF THE eye or body are serious hazards of welding, brazing, soldering, and cutting. Eye, face, and body protection for the operator and others in the work area are required to prevent burns from ultraviolet and infrared radiation, sparks, and spatter.

Eye and Face Protection

Arc Welding and Cutting. Welding helmets or handshields containing appropriate filter lenses and cover plates must be used by welders and welding operators and nearby personnel when viewing an arc.[6] Suggested shade numbers of filter plates for various welding, braz-

---

5. Additional information is given in *AWS F4.1, Recommended safe practices for the preparation for welding and cutting of containers and piping that have held hazardous substances*. Miami: American Welding Society (latest edition).
6. Standards for welding helmets, handshields, face shields, goggles, and spectacles are given in ANSI Z87.1, *Practice for occupational and educational eye and face protection*. New York: American National Standards Institute (latest edition).

ing, soldering, and thermal cutting operations are given in Table 16.1.

Safety spectacles, goggles, or other suitable eye protection must also be worn during welding and cutting operations. Such devices must have full conforming side shields when there is danger of exposure to injurious rays or to flying particles from grinding or chipping operations. Spectacles and goggles may have clear or colored lenses, depending on the intensity of the radiation that may come from adjacent welding or cutting operations when the welding helmet is raised or removed. Number 2 filter lenses are recommended for general purpose protection.

Oxyfuel Gas Welding and Cutting, Submerged Arc Welding. Safety goggles with filter lenses (Table 16.1) and full conforming side shields must be worn while performing oxyfuel gas welding and cutting. During submerged arc welding, the arc is covered by flux and not readily visible; hence, an arc welding helmet is not needed. However, because the arc occasionally flashes through the flux burden, the operator should wear tinted safety glasses.

Torch Brazing and Soldering. Safety spectacles with or without side shields and with appropriate filter lenses are recommended for torch brazing and soldering. As with oxyfuel gas welding and cutting, a bright yellow flame may be visible during torch brazing. A filter similar to that used with those processes should be used for torch brazing.

Resistance, Induction, Salt-Bath, Dip, Infrared Welding, and Brazing. Operators and helpers engaged in these processes must wear safety spectacles, goggles, and a face shield to protect their eyes and face from spatter. Filter lenses are not necessary but may be used for comfort.

Protective Clothing

STURDY SHOES OR boots, and heavy clothing similar to that in Figure 16.3 should be worn to protect the whole body from flying sparks, spatter, and radiation burns. Woolen clothing is preferable to cotton because it is not so readily ignited. Cotton clothing, if used, should be chemically treated to reduce its combustibility. Clothing treated with nondurable flame retardants must be retreated after each washing or cleaning. Clothing or shoes of synthetic or plastic materials which can melt and cause severe burns should not be worn. Outer clothing should be kept reasonably free of oil and grease, especially in an oxygen-rich atmosphere.

Cuffless pants and covered pockets are recommended to avoid spatter or spark entrapment. Pockets should be emptied of flammable or readily ignitable materials

Table 16.1
Suggested Viewing Filter Plates

Operation	Plate Thickness		Welding Current, A	Lowest Shade Number	Comfort Shade Number[a]
	in.	mm			
Shielded metal arc welding	–	–	Under 60	7	–
			60-160	7	10
			160-250	10	12
			250-550	11	14
Gas metal arc and flux cored arc welding	–	–	Under 60	7	–
			60-160	10	11
			160-250	10	12
			250-500	10	14
Gas tungsten arc welding	–	–	Under 50	8	10
			50-150	8	12
			150-500	10	14
Plasma arc welding	–	–	Under 20	6	6-8
			20-100	8	10
			100-400	10	12
			400-800	11	14
Oxyfuel gas welding (steel)[b]	Under 1/8	3.2	–	–	4, 5
	1/8-1/2	3.2-12.7	–	–	5, 6
	Over 1/2	12.7	–	–	6, 8
Plasma arc cutting[c]	–	–	Under 300	8	9
			300-400	9	12
			400-800	10	14
Air-carbon arc cutting	–	–	Under 500	10	12
			500-1000	11	14
Oxyfuel gas cutting (steel)[b]	Under 1	25	–	–	3, 4
	1-6	25-100	–	–	4, 5
	Over 6	150	–	–	5, 6
Torch brazing	–	–	–	–	3, 4
Torch soldering	–	–	–	–	2

a. To select the best shade for the application, first select a dark shade. If it is difficult to see the operation properly, select successively lighter shades until the operation is sufficiently visible for good control. However, do not go below the lowest recommended number, where given.
b. With oxyfuel gas welding or cutting, the flame emits strong yellow light. A filter plate that absorbs yellow or sodium wave lengths of visible light should be used for good visibility.
c. The suggested filters are for applications where the arc is clearly visible. Lighter shades may be used where the arc is hiden by the work or submerged in water.

before welding because they may be ignited by sparks or weld spatter and result in severe burns. Pants should be worn outside of shoes. Protection of the hair with a cap is recommended, especially if a hairpiece is worn. Flammable hair preparations should not be used.

Durable gloves of leather or other suitable material should always be worn. Gloves not only protect the hands from burns and abrasion, but also provide insulation from electrical shock. A variety of special protective clothing is also available for welders. Aprons, leggings, suits, capes, sleeves, and caps, all of durable materials, should be worn when welding overhead or when special circumstances warrant additional protection of the body.

Sparks or hot spatter in the ears can be particularly painful and serious. Properly fitted, flame-resistant ear

Figure 16.3—Typical Protective Clothing for Arc Welding

plugs should be worn whenever operations pose such risks.

NOISE

EXCESSIVE NOISE, PARTICULARLY continuous noise at high levels, can damage hearing. It may cause either temporary or permanent hearing loss. U.S. Department of Labor Occupational Safety and Health Administration regulations describe allowable noise exposure levels. Requirements of these regulations may be found in General Industry Standards, 29 CFR 1910.95.

In welding, cutting, and allied operations, noise may be generated by the process or the equipment, or both.[7]

7. Additional information is presented in *Arc welding and cutting noise*. Miami: American Welding Society, 1979.

Processes that tend to have high noise levels are air carbon arc and plasma arc cutting. Engine-driven generators sometimes emit a high noise level, as do some high-frequency and induction welding power sources.

MACHINERY GUARDS

WELDERS AND OTHER workers must be protected from injury by machinery and equipment that they are operating or by other machinery operating in the work area. Moving components and drive belts must be covered by guards to prevent physical contact.

Because welding helmets and dark filter lenses restrict the visibility of welders, they may be even more susceptible than ordinary workers to injury from unseen, unguarded machinery. Therefore, special attention is required to this hazard.

When repairing machinery by welding or brazing, the power to the machinery must be disconnected, locked out, and tagged to prevent inadvertent operation and injury. Welders assigned to work on equipment with safety devices removed should fully understand the hazards involved, and the steps to be taken to avoid injury. When the work is completed, the safety devices must be replaced. Rotating and automatic welding machines, fixtures, and welding robots must be equipped with appropriate guards or sensing devices to prevent operation when someone is in the danger area.

Pinch points on welding and other mechanical equipment can also result in serious injury. Examples of such equipment are resistance welding machines, robots, automatic arc welding machines, jigs, and fixtures. To avoid injury with such equipment, the machine should be equipped so that both of the operator's hands must be at safe locations when the machine is actuated. Otherwise, the pinch points must be suitably guarded mechanically. Metalworking equipment should not be located where a welder could accidentally fall into or against it while welding. During maintenance of the equipment, pinch points should be blocked to prevent them from closing in case of equipment failure. In very hazardous situations, an observer should be stationed to prevent someone from turning the power on until the repair is completed.

FUMES AND GASES

WELDERS, WELDING OPERATORS, and other persons in the area must be protected from overexposure to fumes and gases produced during welding, brazing, soldering, and cutting. Overexposure is exposure that is hazardous to health, and exceeds the permissible limits specified by a government agency, such as the U.S. Department of Labor, Occupational Safety and Health Administration (OSHA), Regulations 29 CFR 1910.1000, or other recognized authority, such as the American Conference of Governmental Industrial Hygienists (ACGIH) in its publications *Threshold Limit Values for Chemical Substances* and *Physical Agents in the Workroom Environment*. Persons with special health problems may have unusual sensitivity that requires even more stringent protection.

Fumes and gases are usually a greater concern in arc welding than in oxyfuel gas welding, cutting, or brazing because a welding arc may generate a larger volume of fume and gas, and greater varieties of materials are usually involved.

Protection from excess exposure is usually accomplished by ventilation. Where exposure would exceed permissible limits with available ventilation, respiratory protection must be used. Protection must be provided not only for the welding and cutting personnel but also for other persons in the area.

ARC WELDING

Nature and Sources

FUMES AND GASES from arc welding and cutting cannot be classified simply. Their composition and quantity depend upon the base metal composition; the process and consumables used; coatings on the work, such as paint, galvanizing, or plating; contaminants in the atmosphere, such as halogenated hydrocarbon vapors from cleaning and degreasing activities; and other factors.

In welding and cutting, the composition of the fume usually differs from the composition of the electrode or consumables. Reasonably expected fume constituents from normal operations include products of volatilizaton, reaction, or oxidation of consumables, base metals, coatings, and atmospheric contaminants. Reasonably expected gaseous products include carbon monoxide, carbon dioxide, fluorides, nitrogen oxides, and ozone.

The quantity and chemical composition of air contaminants change substantially with the process and with a wide range of variables inherent in each process. During arc welding, the arc energy and temperature depend on the process as well as the welding variables.

Therefore, fumes and gases are generated in varying degrees in different welding operations.

Welding fume is a product of vaporization, oxidation, and condensation of components in the consumable and, to some degree, the base metal. The electrode, rather than the base metal, is usually the major source of fume. However, significant fume contituents can originate from the base metal if it contains alloying elements or a coating that is volatile at elevated temperatures.

Various gases are generated during welding. Some are a product of the decomposition of fluxes and electrode coatings. Others are formed by the action of arc heat or ultraviolet radiation emitted by the arc on atmospheric constituents and contaminants. Potentially hazardous gases include carbon monoxide, oxides of nitrogen, ozone, and phosgene or other decomposition products of chlorinated hydrocarbons, such as phosgene.

Helium and argon, although chemically inert and nontoxic, are simple asphyxiants, and could dilute the atmospheric oxygen concentration to potentially harmful low levels. Carbon dioxide (CO_2) and nitrogen can also cause asphyxiation.

Ozone may be generated by ultraviolet radiation from welding arcs. This is particularly true with gas shielded arcs, especially when argon is used. Photochemical reactions between ultraviolet radiation and chlorinated hydrocarbons result in the production of phosgene and other decomposition products.

The arc heat is responsible for the formation of oxides of nitrogen from atmospheric nitrogen. Hence, nitrogen oxides may be produced by a welding arc or other high temperature heat sources. Thermal decomposition of carbon dioxide and inorganic carbonate compounds by an arc results in the formation of carbon monoxide. Levels can be significant when CO_2 is used as the shielding gas.

Reliable estimates of fume and gas composition cannot be made without considering the nature of the welding process and system being examined. For example, aluminum and titanium are normally arc welded in an atmosphere of argon or helium, or mixtures of the two gases. The arc creates relatively little fume but may emit an intense radiation that can produce ozone. Inert gas shielded arc welding of steels also creates a relatively low fume level.

Arc welding of steel in oxidizing environments, however, generates considerable fume and can produce carbon monoxide and oxides of nitrogen. The fumes generally consist of discrete particles of amorphous slags containing iron, manganese, silicon, and other metallic constituents, depending on the alloy system involved. Chromium and nickel compounds are found in fumes when stainless steels are arc welded.

Some covered and flux cored electrodes are formulated with fluorides. The fumes associated with those electrodes can contain significantly more fluorides than oxides.

Factors Affecting Generation Rates

THE RATE OF generation of fumes and gases during arc welding of steel depends on numerous variables. Among these are

(1) Welding current
(2) Arc voltage (arc length)
(3) Type of metal transfer or welding process
(4) Shielding gas

These variables are interdependent and can have a substantial effect on total fume generation.

Welding current. In general, fume generation rate increases with increased welding current. The increase, however, varies with the process and electrode type. Certain covered, flux-cored, and solid wire electrodes exhibit a nonproportional increase in fume generation rate with increasing current.

Studies have shown that fume generation rates with covered electrodes are proportional to the welding current raised to a power.[8] The exponent is 2.24 for E6010 electrodes and 1.54 for E7018 electrodes. Similar trends were reported in other studies.

Flux cored and solid electrode fume generation rates are more complexly related to welding current. Welding current levels affect the type of metal droplet transfer. As a result, the fume generation rate can decrease with increasing current until some minimum is reached. Then, it will increase in a relatively proportional fashion.

An increase in current can increase ultraviolet radiation from the arc. Therefore, the generation of gases formed photochemically by this radiation, such as ozone, can be expected to increase as welding current is increased. Measurements of ozone concentration during gas metal arc and gas tungsten arc welding have shown such behavior.

Arc Voltage (Arc Length). Arc voltage and arc length are directly related. For a given arc length there is a corresponding arc voltage, mostly dependent upon the type of electrode, welding process, and power supply. In general, increasing arc voltage (arc length) increases the fume generation rate for all open arc welding processes. The levels differ somewhat for each process or electrode type.

Type of Metal Transfer. When steel is joined with gas metal arc welding using a solid wire electrode, the mode of metal transfer depends upon the current and voltage. At low welding current and voltage, short-circuiting transfer takes place, that is, droplets are deposited during short circuits between the electrode and molten weld pool. As the current or voltage is increased, metal transfer changes to globular type where large globules of metal are projected across the arc into the weld pool. At high currents, transfer changes to a spray mode where fine metal droplets are propelled rapidly across the arc. Fume generation rate appears to follow a transition also. It is relatively high during short-circuiting transfer because of arc turbulence. As the transition current is approached in an argon-rich shielding gas, the fume rate decreases and then increases again as spray transfer is achieved. In the spray region, the rate of fume generation is proportional to welding current.

For other welding processes, the type of metal transfer does not vary substantially with current or voltage. In these cases, fume generation follows the relationship for welding current changes.

Shielding Gas. When gas metal arc welding or flux cored arc welding with certain electrodes, shielding gas must be used. The type of shielding gas affects both the composition of the fume and its rate of generation. It also affects the kind of gases found in the welding environment. For example, the fume generation rate is higher with CO_2 shielding than with argon-rich shielding. The rate of fume formation with argon-oxygen or argon-CO_2 mixtures increases with the oxidizing potential of the mixture.

For welding processes where inert gas shielding is used, such as gas tungsten arc or plasma arc welding, the fume generation rate varies with the type of gas or gas mixture. For example, there can be more fume with helium than with argon shielding.

By-product gases also vary with shielding gas composition. The rate of formation of ozone depends upon the wave lengths and intensity of the ultraviolet rays generated in the arc; ozone is more commonly found with argon-rich gases than with carbon dioxide. Oxides of nitrogen are present in the vicinity of any open arc process. Carbon monoxide is commonly found around CO_2 shielded arcs.

Welding Process. Studies of the relative fume generation rates of consumable electrode processes for welding on mild steel have shown definite trends. Considering the ratio of the weight of fumes generated per weight of metal deposited, covered electrodes and self-shielded flux-cored electrodes produce the most fume. Gas-shielded flux cored electrodes produce less fume, and solid wire electrodes produce an even smaller amount. The submerged arc welding process consistently pro-

---

8. Refer to *Fumes and gases in the welding environment*. Miami: American Welding Society, 1979.

duces the least fumes because the fume is captured in the flux and slag cover.

Consumables. With a specific process, the fume rate depends upon the composition of the consumables. Some components of covered and flux-cored electrodes are designed to decompose and form protective gases during welding. Hence, they will generate relatively high fume levels.

Because the constituents of many covered and flux cored electrodes are proprietary, the fume generation rates of electrodes of the same AWS classification produced by different manufacturers can differ substantially. The only reliable method for comparing filler metals is actual product testing to determine specific fume generation characteristics.

OXYFUEL GAS WELDING AND CUTTING

THE TEMPERATURES ENCOUNTERED in oxyfuel gas welding and cutting are lower than those found in electric arc processes. Consequently, the quantity of fumes generated is normally lower. The gases formed are reaction products of fuel gas combustion and of chemical reactions between the gases and other materials present. The fumes generated are the reaction products of the base metals, coatings, filler metals, fluxes, and the gases being used. In oxyfuel gas cutting of steel, the fumes produced are largely oxides of iron.

Fume constituents of greater hazard may be expected when coatings such as galvanizing, paint primers, or cadmium plating are present. Gases of greatest concern include oxides of nitrogen, carbon monoxide, and carbon dioxide. Oxides of nitrogen may be present in especially large amounts when oxyfuel gas cutting stainless steels using either the chemical flux or the iron powder process.

EXPOSURE FACTORS

Position of the Head

THE SINGLE MOST important factor influencing exposure to fume is the position of the welder's head with respect to the fume plume. When the head is in such a position that the fume envelops the face or helmet, exposure levels can be very high. Therefore, welders must be trained to keep their heads to one side of the fume plume. In some cases, the work can be positioned so the fume plume rises to one side.

Type of Ventilation

VENTILATION HAS A significant influence on the amount of fumes in the work area, and hence the welder's expo-

sure. Ventilation may be local, where the fumes are extracted near the point of welding, or general, where the shop air is changed or filtered. The appropriate type will depend on the welding process, the material being welded, and other shop conditions. Adequate ventilation is necessary to keep the welder's exposure to fumes and gases within safe limits.

Shop Size

THE SIZE OF the welding or cutting enclosure is important. It affects the background fume level. Fume exposure inside a tank, pressure vessel, or other confined space will certainly be higher than in a high-bay fabrication area.

Background Fume Level

THE BACKGROUND FUME level depends on the number and type of welding stations and the duty cycle for each station.

Design of Welding Helmet

THE EXTENT TO which the helmet curves under the chin towards the chest affects the amount of fume exposure. Close-fitting helmets can be effective in reducing exposure.

Base Metal and Surface Condition

THE TYPE OF base metal being welded influences both the constituents and the amount of fume generated. Surface contaminants or coatings may contribute significantly to the hazard potential of the fume. Paints containing lead and platings containing cadmium generate dangerous fumes during welding and cutting. Galvanized material evolves zinc fume.

VENTILATION

THE BULK OF fume generated during welding and cutting consists of small particles that remain suspended in the atmosphere for a considerable time. As a result, fume concentration in a closed area can build up over time, as can the concentration of any gases evolved or used in the process. The particles eventually settle on the walls and floor, but the settling rate is low compared to the generation rate of the welding or cutting processes. Therefore, fume concentration must be controlled by ventilation.

Adequate ventilation is the key to control of fumes and gases in the welding environment.[9] Natural,

9. Additional information is available in *Welding fume control with mechanical ventilation*, 2nd Ed., San Francisco: Fireman's Fund Insurance Companies, 1980 (Available from the American Welding Society.)

mechanical, or respirator ventilation must be provided for all welding, cutting, brazing, and related operations. The ventilation must ensure that concentrations of hazardous air-borne contaminants are maintained below recommended levels. These levels must be no higher than the allowable levels specified by the U.S. Occupational Safety and Health Administration or other applicable authorities.

Many ventilation methods are available. They range from natural convection to localized devices, such as air-ventilated welding helmets. Examples of ventilation include:

(1) Natural ventilation
(2) General area mechanical ventilation
(3) Overhead exhaust hoods
(4) Portable local exhaust devices
(5) Downdraft tables
(6) Cross draft tables
(7) Extractors built into the welding equipment
(8) Air-ventilated helmets

General Ventilation

IN MOST CASES, general ventilation is more effective in protecting personnel in adjacent areas than in protecting the welders themselves. Such ventilation may occur naturally outdoors, or when the shop doors and windows are open. It is acceptable when precautions are taken by the welder to keep his or her breathing zone away from the fume plume, and when sampling of the atmosphere shows that concentrations of contaminants do not exceed permissible limits. Natural ventilation often meets these criteria when all of the following conditions are present:

(1) Space of more than 10,000 ft³ per welder is provided.
(2) Ceiling height is more than 16 feet.
(3) Welding is not done in a confined area.
(4) The general welding area is free of partitions, balconies, or other structural barriers that significantly obstruct cross ventilation. (General area refers to a building or a room in a building, not a welding booth or screened area that is used to provide protection from welding radiation.)
(5) Toxic materials with low permissible exposure limits are not present as deliberate constituents. (The employer should refer to the Material Safety Data Sheet provided by the material supplier.) When natural ventilation is not sufficient, fans may be used to force and direct the required amount of air through a building or work room.

The effectiveness of general ventilation, natural or forced, is dependent upon the design of the system.

Work areas where fresh air is introduced and contaminated air is exhausted must be arranged so that the welding fumes and gases are carried away, not concentrated in dead zones. In some cases, the fresh air supply may be located so that incoming air provides the required protection for the welders as well as for personnel in the general area.

Air movement should always be from either side across the welder. This makes it easier for the welder to keep out of the plume and to keep fumes and gases from entering the welding helmet. Air should not blow toward the face or back of the welder because it may force the fume behind the helmet.

General mechanical ventilation may be a necessary supplement to local ventilation to maintain the background level of air-borne contaminants within acceptable limits. Local ventilation is usually necessary to provide satisfactory health hazard control at the individual welding station.

Local Ventilation

GENERAL VENTILATION MAY control contamination levels in the broad area, but in many cases, it will not provide the local control needed to protect the welder. Local exhaust ventilation is usually the most effective way of providing protection at the local work station. It provides efficient and economical fume control, and may be applied by one of the following methods:

A Fixed Open or Enclosing Hood. This consists of a top and at least two sides placed to surround the welding or cutting operation. It must have sufficient air flow and velocity to keep contaminant levels at or below permissible limits.[10]

A Movable Hood with a Flexible Duct. The hood is positioned by the welder as close as practicable to the point of welding, as illustrated in Figure 16.4. It should have sufficient air flow to produce a velocity of 100 ft/min (30 m/min) in the zone of welding.

Air flow requirements range from 150 ft³/min (4 m³/min) when the hood is positioned at 4 to 6 in. (100 to 150 mm) from the weld, to 600 ft³/min (17 m³/min) at 10 to 12 in. (250 to 300 mm) from the weld. This is particularly applicable for bench work but may be used for any location, provided the hood is moved as required. An air velocity of 100 ft/min (31 m/min) will not disturb the torch gas shield during gas shielded arc welding, provided adequate gas flow rates are used.

---

10. See *Industrial ventilation, a manual of recommended practice.* Cincinnati: American Conference of Governmental Industrial Hygienists (latest edition).

Figure 16.4—Movable Hood Positioned Near the Welding Arc

Higher air velocities may disturb the gas shield and render it less effective.

Crossdraft or Downdraft Table.   A crossdraft table is a welding bench with the exhaust hood placed to draw air horizontally across the table. The welder should face in a direction perpendicular to the air flow so that the air flow is across his or her body. A downdraft table has a grill to support the work above an exhaust hood that draws the air downward and away from the welder's head.

Gun-Mounted Fume Removal Equipment.   This equipment extracts the fumes at the point of welding and results in an almost smokeless environment. The exhaust rate must be set so that is does not interfere with the shielding gas pattern provided by the welding pro-

cess. The flux cored arc welding process produces large quantities of fume, and virtually all of the fume can be collected using a gun-mounted fume removal device, as shown in Figure 16.5.

Where permissible, air cleaners that have high efficiencies in the collection of submicron particles may be used to recirculate a portion of ventilated air that would otherwise be exhausted. Some air cleaners do not remove gases. Therefore, the filtered air must be monitored to assure that harmful gas concentrations do not exceed safe limits.

A Water Table Used for Oxyfuel Gas and Plasma Arc Cutting Operations.   This is a cutting table filled with water near or in contact with the bottom surface of the work. Much fume emerging from the cut is captured in the water.

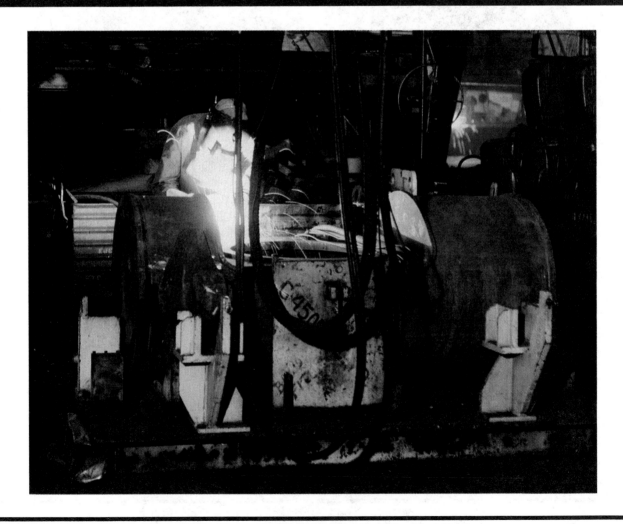

Figure 16.5—Removal of Fume with a Gun-Mounted Device

Respiratory Protective Equipment

WHERE NATURAL OR mechanical ventilation is not adequate or where very toxic materials require a supplement to ventilation, respiratory protective equipment must be used.[11] Airline respirators or face masks that give protection against all contaminants are generally preferred. Air-supplied welding helmets are also available commercially. Filter-type respirators, approved by the U.S. Bureau of Mines for metal fume, give adequate protection against particulate contaminants that are less toxic than lead, provided they are used and maintained correctly. Their general use is not recommended, however, because of the difficulty in assuring proper use and maintenance. They will not protect against mercury vapor, carbon monoxide, or nitrogen dioxide. For these hazards an airline respirator, hose mask, or gas mask is required.

SPECIAL VENTILATION SITUATIONS

Welding in Confined Spaces

SPECIAL CONSIDERATION MUST be given to the safety and health of welders and other workers in confined spaces.[12] Gas cylinders must be located outside of the confined space to avoid possible contamination of the space with leaking gases or volatiles. Welding power sources should also be located outside to reduce danger of engine exhaust and electric shock.

A means for removing persons quickly in case of emergency must be provided. Safety belts and lifelines, when used, should be attached to the worker's body in a manner that avoids the possibility of the person becoming jammed in the exit. A trained helper should be stationed outside the confined space with a preplanned rescue procedure to be put into effect in case of emergency.

In addition to keeping airborne contaminants in breathing atmospheres at or below recommended limits, ventilation in confined spaces must also (1) assure adequate oxygen for life support (at least 19.5 percent by volume), (2) prevent accumulation of an oxygen-enriched atmosphere, (i.e. not over 23.5 percent by volume), and (3) prevent accumulation of flammable mixtures. Asphyxiation can quickly result in unconsciousness and death without warning if oxygen is not present in sufficient concentration to support life. Air contains approximately 21 percent oxygen by volume. A confined space must not be entered unless it is

well ventilated, or the welder is properly trained to work in such spaces and is wearing an approved air-supplied breathing apparatus.[13] A similarly equipped second person must be present.

Confined spaces should be tested before entering for (1) toxic or flammable gases and vapors, and (2) adequate or excess oxygen. The tests should be made with instruments approved by the U.S. Bureau of Mines. Heavier-than-air gases, such as argon, methylacetylene-propadiene, propane, and carbon dioxide, may accumulate in pits, tank bottoms, low areas, and near floors. Lighter-than-air gases, such as helium and hydrogen, may accumulate in tank tops, high areas, and near ceilings. The precautions for confined spaces also apply to those areas. If practical, a continuous monitoring system with audible alarms should be used for work in a confined space.

Oxygen-enriched atmospheres pose great danger to occupants of confined areas. They are especially hazardous at oxygen concentrations above 25 percent. Materials that burn normally in air may flare up violently in an oxygen-enriched atmosphere. Clothing may burn fiercely; oil or grease-soaked clothing or rags may catch fire spontaneously; paper may flare into flame. Very severe and fatal burns can result.

Protection in confined spaces must be provided not only for welders but also for other persons in the enclosure. Only clean, respirable air must be used for ventilation. Oxygen, other gases, or mixtures of gases must never be used for ventilation.

When welding, cutting, or related processes are performed in confined areas where adequate and proper ventilation cannot be provided and there is immediate danger to life and health, positive pressure self-contained breathing apparatus must be used. It must have an emergency air supply of at least five minutes duration in the event that the main source fails.

Welding of Containers

WELDING OR CUTTING on the outside or inside of containers or vessels that have held dangerous substances presents special hazards. Flammable or toxic vapors may be present, or may be generated by the applied heat. The immediate area outside and inside the container should be cleared of all obstacles and hazardous materials.[14] When repairing a container in place, entry of hazardous substances released from the floor or the soil beneath the container must be prevented. The required

11. For additional information refer to ANSI Z88.2, *Practices for respiratory protection.* New York: American National Standards Institute (latest edition).

12. For further precautions see ANSI Z117.1, *Safety requirements for working in tanks and other confined spaces. (latest edition).*

13. Air-supplied respirators or hose masks are those accepted by the U.S. Bureau of Mines or other recognized agency.

14. For more complete procedures, refer to AWS F4.1, *Recommended safe practices for the preparation for welding and cutting containers and piping that have held hazardous substances.* Miami: American Welding Society (latest edition).

personal and fire protection equipment must be available, serviceable, and in position for immediate use.

When welding or cutting inside of vessels that have held dangerous materials, the precautions for confined spaces must also be observed.

Gases generated during welding should be discharged in a safe and environmentally acceptable manner in accordance with Government rules and regulations. Provisions must be made to prevent pressure buildup inside containers. Testing for gases, fume, and vapors should be conducted periodically to ensure that recommended limits are maintained during welding.

An alternative method of providing safe welding of containers is to fill them with an inert medium such as water, inert gas, or sand. When using water, the level should be kept to within a few inches of the point where the welding is to be done. The space above the water should be vented to allow the heated air to escape. With inert gas, the responsible individual needs to know the percentage of inert gas that must be present in the tank to prevent fire or explosion, and how to safely produce and maintain a safe atmosphere during welding.

Highly Toxic Materials

CERTAIN MATERIALS, WHICH are sometimes present in consumables, base metals, coatings, or atmospheres for welding or cutting operations, have permissible exposure limits of 1.0 mg/m³ or less. Among such materials are the following metals and their compounds:

(1) Antimony
(2) Arsenic
(3) Barium
(4) Beryllium
(5) Cadmium
(6) Chromium
(7) Cobalt
(8) Copper
(9) Lead
(10) Manganese
(11) Mercury
(12) Nickel
(13) Selenium
(14) Silver
(15) Vanadium

Base metals and filler metals that may release some of these materials as fume during welding or cutting are given in Table 16.2.

Manufacturer's Material Safety Data Sheets should be consulted to determine if any of these highly toxic materials are present in welding filler metals and fluxes being used. Material Safety Data Sheets should be requested from suppliers. However, welding filler metals and fluxes are not the only source of these materials. They

Table 16.2
Possible Toxic Materials Evolved During Welding or Thermal Cutting

Base or Filler Metal	Evolved Metals or Their Compounds
Carbon and low alloy steels	Chromium, manganese, vanadium
Stainless steels	Chromium, manganese, nickel
Manganese steels and hardfacing materials	Chromium, cobalt, manganese, nickel, vanadium
High copper alloys	Beryllium, chromium, copper, lead, nickel
Coated or plated steel or copper	Cadmium*, chromium, copper, lead, nickel, silver

*When cadmium is a constituent in a filler metal, a warning label must be affixed to the container or coil. Refer to ANSI/ASC Z49.1, *Safety in Welding and Cutting.* New York: American National Standards Institute (latest edition).

may also be present in base metals, coatings, or other sources in the work area. Radioactive materials under Nuclear Regulatory Commission jurisdiction require special considerations.

When toxic materials are encountered as designated constituents in welding, brazing, or cutting operations, special ventilation precautions must be taken to assure that the levels of these contaminants in the atmosphere are at or below the limits allowed for human exposure. All persons in the immediate vicinity of welding or cutting operations involving toxic materials must be similarly protected. Unless atmospheric tests under the most adverse conditions establish that exposure is within acceptable concentrations, the following precautions must be observed.

Confined Spaces. Whenever any toxic materials are encountered in confined space operations, local exhaust ventilation and respiratory protection must be used.

Indoors. When any toxic materials are encountered in indoor operations, local exhaust mechanical ventilation must be used. When beryllium is encountered, respiratory protection in addition to local exhaust ventilation is essential.

Outdoors. Whenever any toxic materials are encountered in outdoor operations, respiratory protection approved by the Mine Safety and Health Association (MSHA), the National Institute of Occupational Safety and Health (NIOSH), or other approving authority may be required.

Persons should not consume food in areas where fumes that contain materials with very low allowable exposure limits may be generated. They should also practice good personal hygiene, such as washing hands before touching food, to prevent ingestion of toxic contaminants.

Fluorine Compounds

FUMES AND GASES from fluorine compounds can be dangerous to health, and can burn the eyes and skin on contact. Local mechanical ventilation or respiratory protection must be provided when welding, brazing, cutting, or soldering in confined spaces involving fluxes, coatings, or other material containing fluorine compounds.

When such processes are employed in open spaces, the need for local exhaust ventilation or respiratory protection will depend upon the circumstances. Such protection is not necessary when air samples taken in breathing zones indicate that all fluorides are within allowable limits. However, local exhaust ventilation is always desirable for fixed-location production welding and for all production welding of stainless steels when filler metals or fluxes containing fluorides are used.

Zinc

FUMES CONTAINING ZINC compounds may produce symptoms of nausea, dizziness, or fever (sometimes called metal fume fever). Welding or cutting where zinc may be present in consumables, base metals, or coatings should be done as described for fluorine compounds.

Cleaning Compounds

CLEANING COMPOUNDS OFTEN require special ventilation precautions because of their possible toxic or flammable properties. Manufacturers' instructions should be followed before welding or cutting is done with cleaned consumables or on cleaned based metal.

Chlorinated Hydrocarbons

DEGREASING OR CLEANING operations involving chlorinated hydrocarbons must be so located that vapors from such operations do not enter the atmosphere surrounding molten weld metal or the welding arc. A reaction product having an objectionable, irritating odor, and containing highly toxic phosgene gas is produced when such vapors enter the atmosphere of arc welding operations. Low levels of exposure can produce feelings of nausea, dizziness, and weakness. High exposures may produce serious health impairment.

Cutting of Stainless Steel

CUTTING STAINLESS STEEL by oxyfuel gas, gas shielded arc, or plasma arc cutting should be done using local mechanical ventilation to remove the fumes generated. Fume from plasma arc cutting done under water is mostly captured in the water.

MEASUREMENT OF EXPOSURE

THE AMERICAN CONFERENCE OF GOVERNMENTAL INDUSTRIAL HYGIENISTS (ACGIH) and the U.S. Department of Labor, Occupational Health and Safety Administration (OSHA) have established allowable limits of airborne contaminants. They are called threshold limit values (TLV)[15] or permissible exposure limits (PEL).

The TLV is the concentration of an airborne substance to which most workers may be repeatedly exposed, day after day, without adverse effect. In adapting these to the working environment, a TLV-TWA (Threshold Limit Value-Time Weighted Average) quantity is defined. TLV-TWA is the time-weighted average concentration for a normal 8-hour workday or 40-hour workweek to which nearly all workers may be repeatedly exposed without adverse effect. TLV-TWA values should be used as guides in the control of health hazards, and should not be interpreted as sharp lines between safe and dangerous concentrations.

TLV's are revised annually as necessary. They may or may not correspond to OSHA permissible exposure limits (PEL) for the same materials. In many cases, current ACGIH values for welding materials are more stringent than OSHA levels.

The only way to assure that airborne contaminant levels are within the allowable limits is to take air samples at the breathing zones of the personnel involved. An operator's actual on-the-job exposure to welding fume should be measured following the guidelines provided in *ANSI/AWS F1.1, Method for Sampling Airborne Particulates Generated by Welding and Allied Processes*. This document describes how to obtain an accurate breathing zone sample of welding fume for a particular welding operation. Both the amount of fume and the composition of the fume can be determined in a single test using this method. Multiple samples are recommended for increased accuracy. When a helmet is worn, the sample should be collected inside the helmet in the welder's breathing zone.

---

15. TLV is a registered trademark of the American Conference of Governmental Industrial Hygienists.

HANDLING OF COMPRESSED GASES

GAS CYLINDERS AND CONTAINERS

GASES USED IN welding and cutting operations are packaged in containers called cylinders.[16] Only cylinders designed and maintained in accordance with U.S. Department of Transportation (DOT) specifications may be used in the United States. The use of other cylinders may be extremely dangerous and is illegal. Cylinders requiring periodic retest under DOT regulations may not be filled unless the retest is current.

Filling

CYLINDERS MAY BE filled only with the permission of the owner. They should only be filled by recognized gas suppliers or those with the proper training and facilities to do so. Filling one cylinder from another is dangerous and should not be attempted by anyone not qualified to do so. Combustible or incompatible combinations of gases must never be mixed in cylinders.

Usage and Storage

WELDING MUST NOT be performed on gas cylinders. Cylinders must not be allowed to become part of an electrical circuit because arcing may result. Cylinders containing shielding gases used in conjunction with arc welding must not be grounded. Electrode holders, welding torches, cables, hoses, and tools should not be stored on gas cylinders, which might cause arcing or interference with valve operation. Arc-damaged gas cylinders may rupture and result in injury or death.

Cylinders must not be used as work rests or rollers. They should be protected from bumps, falls, falling objects, and weather; and should not be dropped. Cylinders should not be kept in passageways where they might be struck by vehicles. They should be kept in areas where temperatures do not fall below −20°F (−30°C) nor exceed 130°F (55°C). Any of these exposures, misuses, or abuses could damage them to the extent that they might fail with serious consequences.

Cylinders must not be hoisted using ordinary slings or chains. A proper cradle or cradle sling that securely retains the cylinder should be used. Electromagnets should not be used to handle cylinders.

Cylinders must always be secured by the user against falling during either use or storage. Acetylene and liquefied gas cylinders should always be stored and used in the upright position. Other cylinders are preferably stored and used in the upright position, but this is not essential in all circumstances.

Before using gas from a cylinder, the contents should be identified by the label thereon. Contents must never be identified by any other means such as cylinder color, banding, or shape. These may vary among manufacturers, geographical areas, or product lines and could be completely misleading. The label on the cylinder is the only proper notice of the contents. If a label is not on a cylinder, the contents should not be used and the cylinder should be returned to the supplier.

A valve protection cap is provided on many cylinders to protect the safety device and the cylinder valve. This cap should always be in place except when the cylinder is in use. The cylinder should never be lifted manually or by hoist by the valve protection cap. The threads that secure these valve protection caps are intended only for that purpose, and may not be capable of supporting full cylinder weight. The caps should always be threaded completely onto the cylinders and hand tightened.

Gas cylinders and other containers must be stored in accordance with all state and local regulations and the appropriate standards of OSHA and the National Fire Protection Association (NFPA). Safe handling and storage procedures are discussed in the *Handbook of Compressed Gases.*

Withdrawal of Gas

MANY GASES IN high-pressure cylinders are filled to pressures of 2000 psig or more. Unless the equipment to be used with a gas is designed to operate at full cylinder pressure, an approved pressure-reducing regulator must be used to withdraw gas from a cylinder or manifold. Simple needle valves should never be used. A pressure-relief or safety valve, rated to function at less than the maximum allowable pressure of the welding equipment, should also be employed. The valve function is to prevent failure of the equipment at pressures in excess of working limits if the regulator should fail in service.

Valves on cylinders containing high pressure gas, particularly oxygen, should always be opened slowly to avoid the high temperature of adiabatic recompression, which can occur if the valves are opened rapidly. In the case of oxygen, the heat can ignite the valve seat which, in turn, may cause the metal to melt or burn. The cylinder valve outlet should point away from the operator and other persons when opening the valve to avoid injury should a fire occur.

Prior to connecting a gas cylinder to a pressure regulator or a manifold, the valve outlet should be cleaned of dirt, moisture, and other foreign matter by first wiping it with a clean, oil-free cloth. Then the valve should be

---

16. For additional information on compressed gases and safe practices, refer to the Compressed Gas Association's, *Handbook of compressed gases.* New York: Van Nostrand Reinhold.

opened momentarily and closed immediately. This is known as "cracking the cylinder valve". Fuel gas cylinders must never be cracked near sources of ignition (i.e., sparks and flames), while smoking, nor in confined spaces.

A regulator should be drained of gas pressure prior to connecting it to a cylinder and also after closing the cylinder valve upon shutdown of operation. The outlet threads on cylinder valves are standardized for specific gases so that only regulators or manifolds with similar threads can be attached.[17] Standard valve thread connections for gases normally used for welding, brazing, and allied processes are given in Table 16.3.

It is preferable not to open valves on low pressure fuel gas cylinders more than one turn. This usually provides adequate flow and permits rapid closure of the cylinder valve in an emergency. High pressure cylinder valves, on the other hand, usually must be opened fully to backseat the packing and prevent packing leaks during use.

The cylinder valve should be closed after each use of a cylinder and when an empty cylinder is to be returned to the supplier. This prevents loss of product through leaks that might develop and go undetected while the cylinder is unattended, and also avoids hazards that might be caused by leaks. It also prevents backflow of contaminants into the cylinder. It is advisable to return cylinders to the supplier with about 25 psi (172 kPa) of contents remaining. This prevents possible contamination by the atmosphere during shipment.

Pressure Relief Devices

ONLY TRAINED PERSONNEL should be allowed to adjust pressure relief devices on cylinders. These devices are intended to provide protection in the event the cylinder is subjected to a hostile environment, usually fire or other source of heat. Such environments may raise the pressure within cylinders. To prevent cylinder pressures from exceeding safe limits, the safety devices are designed to relieve the contents.

CRYOGENIC CYLINDERS AND TANKS

CRYOGENIC CYLINDERS AND tanks are used to store at very low temperatures those liquids that change to gases at normal conditions of temperature and pressure. So-called cryogenic liquids for commercial purposes include oxygen, nitrogen, and argon, although other gases may be handled as cryogenic liquids.

Cylinders and tanks for storing cryogenic liquids are usually double-walled vessels that are evacuated and insulated between the walls. They are designed to contain liquids at very low temperatures with a minimum of heat gain. Heat gain from the atmosphere causes evaporation of the product. Liquid containers hold a greater amount of product for a given volume than high pressure gas cylinders. For safety, containers of liquid must be handled carefully. They must always be maintained in an upright position. Whenever they are moved, a cylinder handling truck designed for this purpose must be used. They should not be rolled on a bottom edge, as is often done with high pressure cylinders.

Failure to properly handle these cylinders can result in rupture of either the inner or outer cylinder wall with consequent loss of vacuum and rapid rise of internal pressure. This will result in activation of the cylinder's protective devices and loss of the cylinder contents. The cylinder protection devices must never be tampered with nor deactivated. Over pressurization could result in explosive failure.

Damage to the internal walls or fittings of a cryogenic-liquid container is often evidenced by visible frosting on the exterior of the container. Whenever frosting appears on the exterior of a cryogenic-liquid container, personnel should be kept clear and the gas supplier notified. In general, people should stay clear of the container until the frost disappears. When the frost disappears it means that the contents have evaporated and any dangerous internal pressure has been relieved.

Although the contents of cryogenic-liquid containers are predominantly in liquid form at a very low temperature, the product withdrawn from these containers should be a gas at room temperature. The conversion takes place within a vaporizer system that evaporates the liquid and warms the gas to atmospheric temperature.

In some cases, the user may want to withdraw liquid from the container. When liquid is withdrawn, protective clothing should be worn to prevent bodily contact with the cold product. Loose-fitting insulated gloves and an adequate face shield are essential. Contact of liquid

Table 16.3
Compressed Gas Association Standard and Alternate Valve Thread Connections for Compressed Gas Cylinders

Gas	Connection Number
Acetylene	510 or 300
Argon	580
Butane	510
Carbon dioxide (CO$_2$)	320
Helium	580
Hydrogen	350
Methylacetylene-propadiene (MPS)	510
Nitrogen	580 or 555
Oxygen	540
Propane	510
Propylene	510

17. Refer to ANSI/CGA V-1, *Compressed gas cylinder valve outlet and inlet connections*. New York: Compressed Gas Association.

with the skin will cause burns similar to heat burns, and prolonged contact will cause severe frostbite. Injuries of this nature should be treated in the same manner as injuries from exposure to *low* temperatures.

Compared to their normal properties, the properties of many materials are drastically different at cryogenic liquid temperatures. Many metals, including carbon steel, and most elastomers, such as rubber, become extremely brittle. When cryogenic liquids are to be withdrawn from cylinders, all materials in the transfer line must have satisfactory properties at the low temperatures.

Liquid oxygen may react with explosive violence on contact with asphalt or similar bituminous materials. Therefore, liquid oxygen tanks must never be mounted on such surfaces, and liquid oxygen must not be allowed to contact them. Liquid oxygen tanks must always be installed on concrete pads.

REGULATORS

A PRESSURE REDUCING regulator should always be used when withdrawing gas from gas cylinders for welding or cutting operations.[18]

Pressure reducing regulators must be used only for the gas and pressure given on the label. They should not be used with other gases, or at other pressures even though the cylinder valve outlet threads may be the same. The threaded connections to the regulator must not be forced. Improper fit of threads between a gas cylinder and regulator or between the regulator and hose indicates that an improper combination of devices is being used.

Use of adaptors to change the cylinder connection thread is not recommended because of the danger of using an incorrect regulator or of contaminating the regulator. For example, gases that are oil-contaminated can deposit an oily film on the internal parts of the regulator. This film can contaminate oil-free gas or result in fire or explosion in the case of oxygen.

The threads and connection glands of regulators should be inspected before use for dirt or damage. If a hose or cylinder connection leaks, it should not be forced with excessive torque. Damaged regulators and components should be repaired by properly trained mechanics or returned to the manufacturer for repair.

A suitable valve or flowmeter should be used to control gas flow from a regulator. The internal pressure in a regulator should be drained before it is connected to or removed from a gas cylinder or manifold.

MANIFOLDS

A MANIFOLD IS used when gas is needed without interruption or at a higher delivery rate than can be supplied from a single cylinder. A manifold must be designed for the specific gas and operating pressure, and be leak tight. The manifold components should be approved for such purpose, and used only for the gas and pressure for which they are approved. Oxygen and fuel gas manifolds must meet specific design and safety requirements.[19]

Piping and fittings for acetylene and methylacetylene-propadiene (MPS) manifolds must not be unalloyed copper or alloys containing 70 percent or more copper. These fuel gases react with copper under certain conditions to form unstable copper acetylide. This compound may detonate under shock or heat.

Manifolded piping systems must contain an appropriate over-pressure relief valve. Each fuel gas cylinder lead should incorporate a backflow check valve and a flash arrester. Backflow check valves must also be installed in each line at each station outlet where both fuel gas and oxygen are provided for a welding, cutting, or preheating torch.

Piping Systems

UNLESS it is known that a piping system is specifically designed and constructed to withstand the full cylinder pressure or tank pressure of the compressed gas source supplying it, the piping system must always be protected with safety pressure relief devices sufficient to prevent development of pressure in the system beyond the capacity of the weakest element.

Such pressure relief devices may be relief valves or bursting discs. A pressure reducing regulator must never be solely relied upon to prevent over pressurization of the system. A pressure relief device must be located in every section of the system which could be exposed to the full source supply pressure while isolated from other protective relief devices (such as by a closed valve).

Some pressure regulators have integral safety relief valves. These valves are designed for the protection of the regulator only, and should not be relied upon to protect the downstream system.

In cryogenic piping systems, relief devices should be located in every section of the system where liquified gas may become trapped. Upon warming, such liquids vaporize to gas, and in a confined space, the gas pressure can increase dramatically.

Pressure relief devices protecting fuel gas piping systems or other hazardous gas systems should be vented to safe locations.

---

18. Gas regulators should meet the requirements of E-4, *Standard for gas regulators for welding and cutting.* New York: Compressed Gas Association, and other code regulations.

19. Refer to *ANSI/NFPA 51, Oxygen-fuel gas systems for welding, cutting, and allied processes.* San Francisco: National Fire Protection Association, latest edition, for information on manifold and piping systems.

OXYGEN

OXYGEN IS NONFLAMMABLE but it supports the combustion of flammable materials. It can initiate combustion and vigorously accelerate it. Therefore, oxygen cylinders and liquid oxygen containers should not be stored in the vicinity of combustibles nor with cylinders of fuel gas. Oxygen should never be used as a substitute for compressed air. Pure oxygen supports combustion more vigorously than air, which contains only 20 percent oxygen. Therefore the identification of oxygen and air should be differentiated.

Oil, grease, and combustible dusts may spontaneously ignite on contact with oxygen. Hence, all systems and apparatus for oxygen service must be kept free of any combustibles. Valves, piping, or system components that have not been expressly manufactured for oxygen service must be cleaned and approved for this service before use.[20]

Apparatus that has been manufactured expressly for oxygen service, and is usually so labeled, must be kept in the clean condition as originally received. Oxygen valves, regulators, and apparatus should never be lubricated with oil. If lubrication is required, the type of lubricant and the method of applying the lubricant should be specified in the manufacturer's literature. If it is not, then the device should be returned to the manufacturer or authorized representative for service.

Oxygen must never be used to power compressed air tools. These are almost always lubricated with oil. Similarly, oxygen must not be used to blow dirt from work and clothing because they are often contaminated with oil, or grease, or combustible dust.

Only clean clothing should be worn when working with oxygen systems. Oxygen must not be used to ventilate confined spaces. Severe burns may result from ignition of clothing or the hair in an oxygen-rich atmosphere.

FUEL GASES

FUEL GASES COMMONLY used in oxyfuel gas welding (OFW) and cutting (OFC) are acetylene, methylacetylene-propadiene (MPS), natural gas, propane, and propylene. Hydrogen is used in a few applications. Gasoline is sometimes used as fuel for oxygen cutting. It vaporizes in the torch. These gases should always be referred to by name.

Acetylene in cylinders is dissolved in a solvent so that it can be safely stored under pressure. In the free state, acetylene should never be used at pressures over 15 psig (100,000 Pa) because it can dissociate with explosive violence at higher pressures.

Acetylene and MPS should never be used in contact with silver, mercury, or alloys containing 70 percent or more copper. These gases react with these metals to form unstable compounds that may detonate under shock or heat. Valves on fuel gas cylinders should never be opened to clean the valve outlet, especially not near possible sources of flame ignition or in confined spaces.

When fuel gases are used for a brazing furnace atmosphere, they must be burned or vented to a safe location. Prior to filling a furnace or retort with fuel gas, the equipment must first be purged with a nonflammable gas, such as nitrogen or argon, to prevent formation of an explosive air-fuel mixture.

Special attention must be given when using hydrogen. Flames of hydrogen may be difficult to see and parts of the body, clothes, or combustibles may come in contact with hydrogen flames.

FUEL GAS FIRES

THE BEST PROCEDURES for avoiding fire from a fuel gas or liquid is to keep it contained within the system, that is, to prevent leaks. All fuel systems should be checked carefully for leaks upon assembly and at frequent intervals thereafter. Fuel gas cylinders should be examined for leaks, especially at fuse plugs, safety devices, and valve packing. One common source of fire in welding and cutting is ignition of leaking fuel by flying sparks or spatter.

In the event of a fuel fire, one of the most effective means for controlling the fire is to shut off the fuel valve, if accessible. A fuel gas valve should not be opened beyond the point necessary to provide adequate flow. In this way it can be shut off quickly in an emergency. In most cases, this is less than one turn of the handle. If the immediate valve controlling the burning gas is inaccessible, another upstream valve may cut off the flow of gas.

Most fuel gases in cylinders are in liquid form or dissolved in liquids. Therefore, the cylinders should always be used in the upright position to prevent liquid surges into the system.

A fuel gas cylinder can develop a leak and sometimes result in a fire. In case of fire, the fire alarm should be sounded, and trained fire personnel should be summoned immediately. A small fire in the vicinity of a cylinder valve or a safety device should be extinguished, if possible, by closing the valve or by the use of water, wet cloths, or fire extinguishers. If the leak cannot be stopped, the cylinder should be removed by trained fire personnel to a safe outdoor location, and the supplier notified. A warning sign should be posted, and no smoking or other ignition sources should be permitted in the area.

In the case of a large fire at a fuel gas cylinder, the fire alarm should be actuated, and all personnel should be evacuated from the area. The cylinder should be kept

---

20. Refer to G4.1, *Cleaning equipment for oxygen service*. New York: Compressed Gas Association.

wet by fire personnel with a heavy stream of water to keep it cool. It is usually better to allow the fire to continue to burn and consume the issuing gas rather than attempt to extinguish the flame. If the fire is extinguished, there is danger that the escaping gas may reignite with explosive violence.

SHIELDING GASES

ARGON, HELIUM, NITROGEN, and carbon dioxide (CO_2) are used for shielding with some welding processes. All, except carbon dioxide, are used as brazing atmospheres. They are odorless and colorless and can displace air needed for breathing.

Confined spaces filled with these gases must be well ventilated before personnel enter them. If there is any question, the space should be checked first for adequate oxygen concentration with an oxygen analyzer. If an analyzer is not available, an air-supplied respirator should be worn by anyone entering the space. Containers of these gases should not be placed in confined spaces, as discussed previously.

ELECTRICAL SAFETY

ELECTRIC SHOCK

ELECTRIC SHOCK CAN cause sudden death. Injuries and fatalities from electric shock in welding and cutting operations can occur if proper precautionary measures are not followed. Most welding and cutting operations employ some type of electrical equipment. For example, automatic oxyfuel gas cutting machines use electric motor drives, controls, and systems.

Some electrical accidents may not be avoidable, such as those caused by lightning. However, the majority are avoidable, including those caused by lack of proper training.

Shock Mechanism

ELECTRIC SHOCK OCCURS when an electric current of sufficient magnitude to create an adverse effect passes through the body. The severity of the shock depends mainly on the amount of current, the duration of flow, the path of flow, and the state of health of the person. The current is caused to flow by the applied voltage. The amount of current depends upon the applied voltage and the resistance of the body path. The frequency of the current may also be a factor when alternating current is involved.

Shock currents greater than about 6 milliamperes (mA) are considered primary because they are capable of causing direct physiological harm. Steady state currents between 0.5 and 6 mA are considered secondary shock currents. Secondary shock currents are defined as those capable of causing involuntary muscular reactions without normally causing direct physiological harm. The 0.5 mA level is called the perception threshold because it is the point at which most people just begin to feel the tingle from the current. The level of current sensation varies with the weight of the individual and to some extent between men and women.

Shock Sources

MOST ELECTRICAL EQUIPMENT can be shock hazards if improperly installed, used, or maintained. Shock can occur from lightning-induced voltage surges in power distribution systems. Even earth grounds can attain high potential relative to true ground during severe transient phenomena. Such circumstances, however, are rare.

In welding and cutting work, most electrical equipment is powered from ac sources of between 115 and 575 V, or by engine-driven generators. Most welding is done with less then 100 arc volts. (Fatalities have resulted with equipment operating at less than 80 volts.) Some arc cutting methods operate at over 400 V, and electron beam welding machines at up to about 150 kV. Most electric shock in the welding industry occurs as the result of accidental contact with bare or poorly insulated conductors operating at such voltages. Therefore, welders must take precautions against contacting bare elements in the welding circuit, as well as those in the primary circuits.

Electrical resistance is usually reduced in the presence of water or moisture. Electrical hazards are often more severe under such circumstances. When arc welding or cutting is to be done under damp or wet conditions including heavy perspiration, the welder must wear dry gloves and clothing in good condition to prevent electric shock. The welder should be protected from electrically conductive surfaces, including the earth, by rubber-soled shoes as a minimum, and preferably by an insulating layer such as a rubber mat or dry wooden board. Similar precautions against accidental contact with bare conducting surfaces must be taken when the welder is required to work in a cramped kneeling, sitting, or lying position. Rings and jewelry should be removed before welding to decrease the possibility of electric shock.

WEARERS OF PACEMAKERS

THE TECHNOLOGY OF heart pacemakers and the extent to which they are influenced by other electrical devices is constantly changing. It is impossible to make a general statement concerning the possible effects of welding operations on such devices. Wearers of pacemakers or other electronic equipment vital to life should check with the device manufacturer or their doctor to determine whether any hazard exists.

EQUIPMENT SELECTION

ELECTRIC SHOCK HAZARDS are minimized by proper equipment installation and maintenance, good operator practice, proper operator clothing and body protection, and the use of equipment designed for the job and situation. Equipment should meet applicable NEMA or ANSI standards, such as American National Standards Institute's ANSI/UL 551, *Safety Standard for Transformer Type Arc Welding Machines*, latest edition.

If a significant amount of welding and cutting work is to be done under electrically hazardous conditions, automatic machine controls that reduce the no-load (open circuit) voltage to a safe level are recommended. When special welding and cutting processes require open-circuit voltages higher than those specified in *ANSI/NEMA Publication EW-1, Electrical Arc Welding Apparatus*, insulation and operating procedures that are adequate to protect the welder from these higher voltages must be provided.

PERSONNEL TRAINING

A GOOD SAFETY training program is essential. Employees must be fully instructed in electrical safety by a competent person before being allowed to commence operations. As a minimum, this training should include the points covered in *ANSI/ASC Z49.1, Safety in Welding and Cutting* (published by the American Welding Society). Persons should not be allowed to operate electrical equipment until they have been properly trained.

INSTALLATION

EQUIPMENT SHOULD BE installed in a clean, dry area. When this is not possible, it should be adequately guarded from dirt and moisture. Installation must be done to the requirements of *ANSI/NFPA 70, National Electric Code*, and local codes. This includes necessary disconnects, fusing, and type of incoming power lines.

Terminals for welding leads and power cables must be shielded from accidental contact by personnel or by metal objects, such as vehicles and cranes. Connections between welding leads and power supplies may be guarded using (1) dead front construction and recepta-cles for plug connections, (2) terminals located in a recessed opening or under a non-removable hinged cover, (3) insulating sleeves, or (4) other equivalent mechanical means.

GROUNDING

THE WORKPIECE BEING welded and the frame or chassis of all electrically powered machines must be connected to a good electrical ground. Grounding can be done by locating the workpiece or machine on a grounded metal floor or platen, or by connecting it to a properly grounded building frame or other satisfactory ground. Chains, wire ropes, cranes, hoists, and elevators must not be used as grounding connectors nor to carry welding current.

The work lead is not the grounding lead. The work-lead connects the work terminal on the power source to the workpiece. A separate lead is required to ground the workpiece or power source work terminal.

Care should be taken when connecting the grounding circuit. Otherwise, the welding current may flow through a connection intended only for grounding, and may be of higher magnitude than the grounding conductor can safely carry. Special radio-frequency grounding may be necessary for arc welding machines equipped with high-frequency arc initiating devices.[21]

Connections for portable control devices, such as push buttons to be carried by the operator, must not be connected to circuits with operating voltages above about 120 V. Exposed metal parts of portable control devices operating on circuits above 50 V must be grounded by a grounding conductor in the control cable. Controls using intrinsically safe voltages below 30 V are recommended.

CABLES AND CONNECTIONS

ELECTRICAL CONNECTIONS MUST be tight, and checked periodically for tightness. Magnetic work clamps must be free of adherent metal particles and spatter on contact surfaces. Coiled welding leads should be spread out before use to avoid overheating and damage to the insulation. Jobs alternately requiring long and short leads should be equipped with insulated cable connectors so that idle lengths can be disconnected when not needed.

Equipment, cables, fuses, plugs, and receptacles must be used within their current carrying and duty cycle capacities. Operation of apparatus above the current rating or the duty cycle results in overheating and rapid deterioration of insulation and other parts. Actual welding current may be higher than that shown by indicators on the welding machine if welding is done with short

---

21. See EW-1, *Electric arc welding power sources*, Section 10.5.6, National Electrical Manufacturers Association.

leads or low voltage, or both. High currents are likely with general purpose welding machines when they are used with processes that use low arc voltage, such as gas tungsten arc welding.

Welding leads should be the flexible type of cable designed especially for the rigors of welding service. Insulation on cables used with high voltages or high-frequency oscillators must provide adequate protection. The recommendations and precautions of the cable manufacturer should always be followed. Cable insulation must be kept in good condition, and cables repaired or replaced promptly when necessary.

OPERATIONS

WELDERS SHOULD NOT allow the metal parts of electrodes, electrode holders, or torches to touch their bare skin or any wet covering of the body. Dry gloves in good condition must always be worn. The insulation on electrode holders must be kept in good repair. Electrode holders should not be cooled by immersion in water. If water-cooled welding guns or holders are used, they should be free of water leaks and condensation that would adversely affect the welder's safety. Welders should not drape nor coil the welding leads around their bodies.

A welding circuit must be de-energized to avoid electric shock while the electrode, torch, or gun is being changed or adjusted. One exception concerns covered electrodes with shielded metal arc welding. When the circuit is energized, covered electrodes must be changed with dry welding gloves, not with bare hands. In any case, de-energization of the circuit is desirable for optimum safety even with covered electrodes.

When a welder has completed the work or has occasion to leave the work station for an appreciable time, the welding machine should be turned off. Similarly, when the machine is to be moved, the input power supply should be electrically disconnected at the source. When equipment is not in use, exposed electrodes should be removed from the holder to eliminate the danger of accidental electrical contact with persons or conducting objects. Also, welding guns of semiautomatic welding equipment should be placed so that the gun switch cannot be operated accidentally.

MULTIPLE ARC WELDING OPERATIONS

THERE CAN BE increased danger of electrical shock when several welders are welding on a large metal structure, such as a building frame or ship, and the structure is part of the return welding circuits. Proper electrical contact must exist at all joints in the structure. Sparking or heating at any point in the structure makes it unsuitable as a return circuit.

Where two or more welders are working on the same structure and one is likely to touch simultaneously the exposed parts of more than one electrode holder, the welding machines must be connected to minimize shock hazard. It is preferable that all dc welding machines be connected with the same polarity. A test lamp or voltmeter can be used to determine whether the polarities are matched. It is preferable to connect all single-phase ac welding machines to the same phase of the supply circuit with the same instantaneous polarity. These precautions minimize the potential difference between electrode holders.

In some cases, the preferable connections may not be possible. Welding may require the use of both dc polarities, or supply circuit limitations may require distribution of ac welding machines among the phases of the supply circuit. In such cases, the no-load voltage between electrode holders or welding guns may be twice the normal voltage. Because of the increased voltage, the welders and other area personnel must be instructed to avoid simultaneous contact with more than one electrode holder, welding gun, or installed electrode.

MODIFICATION AND MAINTENANCE

ONLY QUALIFIED PERSONNEL should perform equipment modification and maintenance. Commutators on rotating welding machines should be kept clean to prevent excessive arcing. Rectifier type welding machines should be inspected frequently for accumulations of dust or lint that interfere with ventilation. Louvers and internal electrical coil ventilating ducts should be similarly inspected. It is good practice to occasionally blow out the welding machine with clean, dry compressed air at low pressure. Adequate safety precautions such as proper eye protection should be taken. Air filters in the ventilating systems of electrical components are not recommended unless provided by the manufacturer of the welding machine. The filters should be inspected as recommended by the manufacturer. The reduction of air flow resulting from the accumulation of dust on the air filter can subject internal components to an overheating condition and subsequent failure. Machines that have become wet should be thoroughly dried and properly retested before being used.

PREVENTION OF FIRES

FIRES RESULTING FROM electric welding equipment are generally caused by overheating of electrical components, flying sparks or spatter from the welding or cutting operation, or mishandling of fuel in engine-driven equipment. Most precautions against electrical shock are also applicable to the prevention of fires caused by overheating of equipment. Avoidance of fire from sparks and spatter was covered previously.

The fuel systems of engine-driven equipment must be in good condition. Leaks must be repaired promptly. Engine-driven machines must be turned off before refueling, and any fuel spills should be wiped up and fumes allowed to dissipate before the engine is restarted. Otherwise, the ignition system, electrical controls, spark producing components, or engine heat may start a fire.

PROCESSES

THE BROAD AREAS of welding safety that are applicable to most welding, cutting, brazing, or soldering processes have been addressed previously. The precautions and procedures that are unique to the particular processes are discussed here. All applicable information must be considered to provide complete safety precautions and procedures for each process.

OXYFUEL GAS WELDING AND CUTTING

Torches

ONLY APPROVED WELDING and cutting torches should be used.[22] They should be kept in good working order, and serviced at regular intervals by the manufacturer or qualified technicians. A torch must be used only with the fuel gas for which it is designed. The fuel gas and oxygen pressures should be those recommended by the torch manufacturer.

The procedures recommended by the manufacturer should be followed when lighting and extinguishing the torch. The torch should be lighted only with a friction lighter, pilot light, or similar ignition source. Matches or cigarette lighters should not be used.

Hoses

HOSES USED SHOULD be only those specified for oxyfuel gas welding and cutting systems. Generally, these hoses are manufactured in accordance with *Specification 1P-7 for Rubber Welding Hose*, published by the Compressed Gas Association and the Rubber Manufacturers Association. Fuel gas hose is usually red with left-hand threaded fittings. Green hose with right-hand threaded fittings is generally used for oxygen.[23] Hoses should be free of oil and grease, and in good condition. When parallel lengths are strapped together for convenience, no more than 4 in. (100 mm) of any 12 in. (300 mm) section of hose should be covered.

Only proper ferrules and clamps should be used to secure hose to fittings. Long runs of hose should be avoided. Excess hose should be coiled to prevent kinks and tangles, but it should not be wrapped around cylinders or cylinder carts while in use.

Backfire and Flashback

A BACKFIRE DURING welding and cutting is a momentary retrogression of the flame back into the tip. It usually results in a momentary flame-out followed by reignition of the normal tip flame, and is accompanied by a pop or bang, depending upon the size of tip. In severe cases, the hot combustion products within the tip may be forced back into the torch and even the hoses. Occasionally, (but especially with oxygen), such backfires ignite the inner liner of the hose and result in burn-through of the hose wall. Such backfires can cause injury. Furthermore, when the hose ruptures, the flow of gases into the area is continuous until the valve at the tank is closed.

A flashback is an occurrence initiated by a backfire where the flame continues to burn inside the equipment instead of being re-established at the tip. Flashbacks result in very rapid internal heating of the equipment, and can quickly destroy it. A flashback is usually recognized by a whistling or squealing sound. The equipment will heat up rapidly and sparks may issue from the tip. The flashback should be extinguished by turning off the torch valves as quickly as possible. Different manufacturers may recommend shutting off either the fuel or oxygen first, but the most important concern is to get both valves closed quickly.

Backfires and flashbacks are not ordinarily a concern if the apparatus is operated in accordance with the manufacturer's instructions. Generally, they occur from allowing a tip to become overheated by flame backwash, forcing the tip into the work, or providing insufficient gas flow for the size of the tip. If frequent backfiring or flashbacks are experienced, the cause should be investigated. There is probably something wrong with the equipment or operation.

To prevent backfires and flashbacks, hose lines should be purged before lighting oxyfuel gas equipment. Purging flushes out any combustible oxygen-fuel or air-fuel gas mixtures in the hoses. It is done by first opening either the fuel or oxygen valve on the torch and allowing

---

22. Oxyfuel gas torches should meet the requirements of E-5, *Torch standard for welding and cutting*. New York: Compressed Gas Association, and appropriate government regulations.

23. Hose connections must comply with E-1, *Standard connections for regulators, torches and fitted hose for welding and cutting equipment*. New York: Compressed Gas Association.

that gas to flow for several seconds to clear that hose of any possible gas mixtures. That valve is then closed, and the other valve is opened to allow the other gas to flow for a similar period of time. Purging should always be done before any welding or cutting tip is lighted. The purge stream must not be directed towards any flame or source of ignition. Torches should not be purged in confined spaces because of possible explosion of accumulated gases.

Hose Line Safety Devices

REVERSE-FLOW CHECK VALVES and flashback arresters for hose line service are available. These devices can prevent backflow of gases and flashbacks into hoses provided they are operating properly. They must be used strictly in accordance with the manufacturer's instructions, and maintained regularly in accordance with the manufacturer's recommendations.

Shutdown Procedures

WHEN OXYFUEL GAS operations are completed, the equipment should always be completely shut down and the gas pressures drained from the system. Cylinder supply valves must be closed. The equipment should not be left unattended until the shutdown has been completed.

VENTILATED STORAGE

OXYFUEL GAS CYLINDERS or equipment connected to cylinders must always be stored in well ventilated spaces, and should not be stored in confined areas or unventilated cabinets. Even small gas leaks in confined spaces can result in explosive mixtures that might be ignited with disastrous results. For the same reason, oxygen gas cylinders should never be transported in enclosed vehicles, particularly not in closed vans or the trunks of automobiles.

ARC WELDING AND CUTTING

THE POTENTIAL HAZARDS of arc welding are fumes and gases, electric shock, infrared and ultraviolet radiation, burns, fire, explosion, and sometimes noise. These hazards have been described previously. Precautions to avoid injury or death from these hazards must be followed. The safety precautions for arc welding also apply to arc cutting.

Noise during arc cutting operations can be high. Prolonged exposure to high noise levels can lead to hearing damage. Where necessary, ear protection must be provided for the operator and others in the area.

Plasma arc cutting is a particularly noisy and a high fume generating process. Two common accessories are available for mechanized plasma arc cutting of plate to aid in fume and noise control. One accessory is a water table, which is simply a cutting table filled with water to the bottom surface of the plate or above the plate. In the later case, cutting is done under water using a special torch, to minimize noise and reduce radiation. The high-speed gases emerging from the plasma jet produce turbulence in the water. This action traps almost all of the fume particles in the water.

Another accessory is a water muffler to reduce noise. The muffler is a nozzle attached to a special torch body that produces a curtain of water around the front of the torch. It is always used in conjunction with a water table. The combination of a water curtain at the top of the plate and a water table contacting the bottom of the plate encloses the arc in a noise-reducing shield. The noise output is reduced by roughly 20 dB. This equipment should not be confused with cutting variations using water injection or water shielding.

RESISTANCE WELDING

THE MAIN HAZARDS of resistance welding processes and equipment are as follows:

(1) Electric shock from contact with high voltage terminals or components
(2) Ejection of small particles of molten metal from the weld
(3) Crushing of some part of the body between the electrodes or other moving components of the machine

Mechanical Considerations

Guarding. Initiating devices on resistance welding equipment, such as push buttons and switches, must be positioned or guarded to prevent the operator from inadvertently activating them.

With some machines, the operator's hands can be expected to pass under the point of operation during loading and unloading. These machines must be effectively guarded by proximity-sensing devices, latches, blocks, barriers, dual hand controls, or similar accessories that prevent (1) the hands from passing under the point of operation, or (2) the ram from moving while the hands are under the point of operation.

Static Safety Devices. Press, flash, and upset welding machines should have static safety devices to prevent movement of the platen or head during maintenance or setup for welding. Pins, blocks, and latches are examples of such devices.

Portable Welding Machines. The support system of suspended portable welding gun equipment, with the exception of the gun assembly, must be capable of withstanding the total mechanical shock load in the event of failure of any component of the system. Devices such as cables, chains, or clamps are considered satisfactory.

Guarding should be provided around the mounting and actuating mechanism of the movable arm of a portable welding gun if it can cause injury to the operator's hands. If suitable guarding cannot be achieved, the gun should have two handles and two operating switches that must be actuated to energize the machine.

Stop Buttons. One or more emergency stop buttons should be provided on all welding machines, with a minimum of one at each operator position.

Guards. Eye protection against expelled metal particles must be provided by a guard of suitable fire-resistant material or by the use of approved personal protective eye wear. For flash welding equipment, flash guards of suitable fire-resistant material must be provided to control flying sparks and molten metal.

Electrical Considerations

ALL EXTERNAL WELD initiating control circuits should operate at a maximum of about 120 V for stationary equipment and about 36 V for portable equipment.

Resistance welding equipment containing high voltage capacitors must have adequate electrical insulation and must be completely enclosed. All enclosure doors must be provided with suitable interlocks that are wired into the control circuit. The interlocks must effectively interrupt power and discharge all high voltage capacitors when the door or panel is open. In addition, a manually operated switch or suitable positive device should be provided to assure complete discharge of all high voltage capacitors. The doors or panels must be kept locked except during maintenance.

The welding transformer secondary should be grounded by one of the following methods:

(1) Permanent grounding of the welding secondary circuit
(2) Connection of a grounding reactor across the secondary winding with a reactor tap to ground

As an alternative on stationary machines, an isolation contactor may be used to open all of the primary lines.

The grounding of one side of the secondary windings on multiple spot welding machines can cause undesirable transient currents to flow between the transformers when they are either connected to different primary phases or have different secondary voltages, or both. A similar condition can also exist with portable spot welding guns when several units are used to weld the same assembly or another one that is nearby. Such situations require use of a grounding reactor or isolation contactor.

Installation

ALL EQUIPMENT SHOULD be installed in conformance with the *ANSI/NFPA 70, National Electric Code* (latest edition). The equipment should be installed by qualified personnel under the direction of a competent technical supervisor. Prior to production use, the equipment should be inspected by competent safety personnel to ensure that it is safe to operate.

BRAZING AND SOLDERING

HAZARDS ENCOUNTERED WITH brazing and soldering operations are similar to those associated with welding and cutting processes. Brazing and soldering operations may be done at temperatures where some elements in the filler metal will vaporize. Personnel and property must be protected against hot materials, gases, fumes, electrical shock, radiation, and chemicals.

It is essential that adequate ventilation be provided so that personnel do not inhale gases and fumes generated during brazing or soldering. Some filler metals and base metals contain toxic materials such as cadmium, beryllium, zinc, mercury, or lead that vaporize during brazing. Fluxes contain chemical compounds of fluorine, chlorine, and boron that are harmful if they are inhaled or contact the eyes or skin. Suitable ventilation to avoid these hazards was described previously.

Brazing Atmospheres

FLAMMABLE GASES ARE sometimes used as atmospheres for furnace brazing operations. These include combusted fuel gas, hydrogen, and dissociated ammonia. Prior to introducing such atmospheres, the furnace or retort must be purged of air by safe procedures recommended by the furnace manufacturer.

Adequate area ventilation must be provided to exhaust and discharge to a safe place explosive or toxic gases that may emanate from furnace purging and brazing operations. Local environmental regulations should be consulted when designing the exhaust system.

Dip Brazing and Soldering

IN DIP BRAZING and soldering, the parts to be immersed in the bath must be completely dry. The presence of moisture on the parts will cause an instantaneous generation of steam that may expel the contents of the dip pot with explosive force and create a serious burn hazard. Predrying of the parts will prevent this problem. If supplementary flux is necessary, it must be adequately dried to remove not only moisture but also water of hydration to avoid explosion hazards.

Solder Flux

SOME FLUXES, SUCH as the rosin, petrolatum, and reaction types, give off considerable smoke, the amount depending on the soldering temperature and the duration of heating. The American Conference of Governmental Industrial Hygienists has established a safe threshold limit value for decomposition products of rosin core solder of 0.1 mg/m³ aliphatic aldehydes, measured as formaldehyde. Suitable ventiltion must be provided for soldering operations to meet this requirement.

Other fluxes give off fumes that are harmful if breathed in any but small quantities. Prolonged inhalation of halides and some of the newer organic fluxes should be avoided. The aniline type fluxes and some of the amines also evolve fumes that are harmful, and can cause dermatitis. Fluorine in flux can be dangerous to health, cause burns, and be fatal if ingested. Thus, proper ventilation must be provided in the work area to remove fume from soldering operations.

HIGH FREQUENCY WELDING

HIGH FREQUENCY GENERATORS are electrical devices and require all usual safety precautions in handling and repairing such equipment. Voltages are in the range from 400 to 20,000 V and are lethal. These voltages may be either low or high frequency. Proper care and safety precautions should be taken while working on high frequency generators and their control systems. Units must be equipped with safety interlocks on access doors and with automatic safety grounding devices to prevent operation of the equipment when access doors are open. The equipment should not be operated with panels or high voltage covers removed nor with interlocks and grounding devices blocked. This equipment should not be confused with high frequency arc stabilization equipment used in GTAW.

The output high-frequency primary leads should be encased in metal ducting and should not be operated in the open. Induction coils and contact systems should always be properly grounded for operator protection. High frequency currents are more difficult to ground than low frequency currents, and grounding lines must be kept short and direct to minimize inductive impedance. The magnetic field from the output system must not induce heat in adjacent metallic sections and cause fires or burns.

Injuries from high frequency power, especially at the upper range of welding frequencies, tend to produce severe local surface tissue damage. However, they are not likely to be fatal because current flow is shallow.

High frequency welding stations often emit a loud steady whine that can cause permanent hearing loss. Ear protection is essential under these circumstances.

ELECTRON BEAM WELDING

THE PRIMARY HAZARDS associated with electron beam welding equipment are electric shock, x-radiation, fumes and gases, and damaging visible radiation. Precautionary measures must be taken at all times to assure that proper protective procedures are always observed. *AWS F2.1, Recommended Safe Practices for Electron Beam Welding and Cutting* and *ANSI/ASC Z49.1, Safety in Welding and Cutting* (latest edition) give the general safety requirements that should be strictly adhered to at all times.

Electric Shock

EVERY ELECTRON BEAM welding machine operates at voltages above 20 kV. These voltages can cause fatal injury, regardless of whether the machine is referred to as a low voltage or a high voltage machine. The manufacturers of electron beam welding equipment, in meeting various regulatory requirements, produce machines that are well-insulated against high voltage. However, precautions should be exercised with all systems when high voltage is present. The manufacturer's instructions should be followed for operation and maintenance of the equipment.

X-rays

THE X-RAYS GENERATED by an electron beam welding machine are produced when electrons, traveling at high velocity, collide with matter. The majority of x-rays are produced when the electron beam impinges upon the workpiece. Substantial amounts are also produced when the beam strikes gas molecules or metal vapor in the gun column and work chamber. Underwriters Laboratories and OSHA regulations have established firm rules for permissible x-ray exposure levels, and producers and users of equipment must observe these rules.

Generally, the steel walls of the chamber are adequate protection in systems up to 60 kV, assuming proper design. High-voltage machines utilize lead lining to block x-ray emission beyond the chamber walls. Leaded glass windows are employed in both high and low voltage electron beam systems. Generally, the shielded vacuum chamber walls provide adequate protection for the operator.

In the case of nonvacuum systems, a radiation enclosure must be provided to assure the safety of the operator and other persons in the area. Thick walls of high-density concrete or other similar material may be employed in place of lead, especially for large radiation enclosures on nonvacuum installations. In addition, special safety precautions should be imposed to prevent personnel from accidentally entering or being trapped inside the enclosure when equipment is in operation.

A complete x-ray radiation survey of the electron beam equipment should always be made at the time of installation and at regular intervals thereafter. This should be done by personnel trained to make a proper radiation survey to assure initial and continued compliance with all radiation regulations and standards applicable to the site where the equipment is installed.

Fumes and Gases

IT IS UNLIKELY that the very small amount of air in a high vacuum electron beam chamber would be sufficient to produce ozone and oxides of nitrogen in harmful concentrations. However, nonvacuum and medium vacuum electron beam systems are capable of producing these by-products, as well as other types of airborne contaminants in concentrations above acceptable levels.

Adequate area ventilation must be employed to reduce concentrations of airborne contaminants around the equipment within permissible exposure limits. Proper exhausting techniques should be employed to maintain residual concentrations in the enclosure within those limits.

Visible Radiation

DIRECT VIEWING OF intense radiation emitted by molten weld metal can be harmful to eyesight. Viewing of the welding operation should be done through a filter lens commonly used for arc welding.

LASER BEAM WELDING AND CUTTING

THE BASIC HAZARDS associated with laser operation are

(1) Eye damage from the beam including burns of the cornea or retina, or both
(2) Skin burns from the beam
(3) Respiratory system damage from hazardous materials evolved during operation
(4) Electrical shock
(5) Chemical hazards
(6) Contact with cryogenic coolants

Laser manufacturers are required to qualify their equipment with the U.S. Bureau of Radiological Health (BRH). Electrical components should be in compliance with NEMA standards. User action is governed by OSHA requirements. In all cases, *American National Standard Z136.1, Safe Use of Lasers* (latest edition), should be followed.

Eye Protection

EYE INJURY IS readily caused by laser beams. With laser beams operating at visible or near infrared wavelengths, even a five milliwatt beam can inflict retinal damage. Safety glasses are available that are substantially transparent to visible light but are opaque to specific laser beam outputs. Selective filters for ruby, Nd-YAG, and other laser systems are available. Glasses appropriate to the specific laser system must be used. At longer infrared wavelengths (such as that of a CO_2 laser), ordinarily transparent materials such as glass are opaque. Clear safety glasses with side shields may be used with these systems, and the only light reaching the eye will be from incandescence of the workpiece. Nevertheless, extreme brilliance can result if a plasma is generated at high power, and filter lenses should then be used for viewing the operation.

Burns

LASER BURNS CAN be deep and very slow to heal. Exposure must be avoided by appropriate enclosure of the beam or by methods that prevent beam operation unless the beam path is unobstructed. This is particularly important for nonvisible beams that provide no external evidence of their existence unless intercepted by a solid.

Electric Shock

HIGH VOLTAGES AS well as large capacitor storage devices are associated with lasers. Therefore, the possibility for lethal electrical shock is always present. Electrical system enclosures should have appropriate interlocks on all access doors and provisions for discharging capacitor banks before entry. The equipment should be appropriately grounded.

Fumes and Gases

HAZARDOUS PRODUCTS MAY be generated from interaction of the beam and the workpiece. For example, plastic materials used for "burn patterns" to identify beam shape and distribution in high power CO_2 laser systems can generate highly toxic vapors if irradiated in an oxygen-lean atmosphere.

In deep penetration welding, fine metal fume can arise from the joint. Also, intense plasma generation can produce ozone. Consequently, adequate ventilation and exhaust provisions for laser work areas are necessary.

FRICTION WELDING

FRICTION WELDING MACHINES are similar to machine tool lathes in that one workpiece is rotated by a drive system. They are also similar to hydraulic presses in that one workpiece is forced against the other. Therefore, safe practices for lathes and power presses should be used as guides for the design and operation of friction welding machines.

Machines should be equipped with appropriate mechanical guards and shields as well as two-hand operating switches and electrical interlocks. These devices should be designed to prevent operation of the machine when the work area, rotating drive, and force system are accessible to the operator or others.

Operating personnel should wear appropriate eye protection and safety apparel commonly used with machine tool operations. In any case, applicable Occupational Safety and Health Administration (OSHA) standards should be strictly observed.

EXPLOSION WELDING

EXPLOSIVES AND EXPLOSIVE devices are a part of explosion welding. Such materials and devices are inherently dangerous, but there are safe methods for handling them. However, if the materials are misused, they can kill or injure, and destroy or damage property.

Explosive materials should be handled and used only by trained personnel who are experienced in that field. Handling and safety procedures must comply with all applicable federal, state, and local regulations. Federal jurisdiction on the sale, transport, storage, and use of explosives is through the U.S. Bureau of Alcohol, Tobacco, and Firearms; the Hazardous Materials Regulation Board of the U.S. Department of Transportation; the Occupational Safety and Health Administration; and the Environmental Protection Agency. Many states and local governments require a blasting license or permit, and some cities have special explosive requirements.

The Institute of Makers of Explosives provides educational publications to promote the safe handling, storage, and use of explosives. The National Fire Protection Association provides recommendations for safe manufacture, storage, handling, and use of explosives.[24]

ULTRASONIC WELDING

WITH HIGH-POWER ULTRASONIC equipment, high voltages are present in the frequency converter, the welding head, and the coaxial cable connecting these components. Consequently, the equipment should not be operated with the panel doors open or housing covers removed. Door interlocks are usually installed to prevent introduction of power to the equipment when the high voltage circuitry is exposed. The cables are shielded fully and present no hazard when properly connected and maintained.

Because of hazards associated with application of clamping force, the operator should not place hands or arms in the vicinity of the welding tip when the equipment is energized. For manual operation, the equipment should be activated by dual palm buttons that meet the requirements of OSHA. Both buttons must be pressed simultaneously to actuate a weld cycle, and both must be released before the next cycle is initiated. For automated systems in which the weld cycle is sequenced with other operations, guards should be installed for operator protection. Such hazards can be further minimized by setting the welding stroke to the minimum that is compatible with workpiece clearance.

THERMIT WELDING

THERMIT MIX, IN the crucible or on the workpieces, can lead to rapid formation of steam when the chemical reaction for thermit welding takes place. This may cause ejection of molten metal from the crucible. Therefore, the thermit mix should be stored in a dry place, the crucible should be dry, and moisture should not be allowed to enter the system before or during welding.

The work area should be free of combustible materials that may be ignited by sparks or small particles of molten metal. The area should be well ventilated to avoid the buildup of fumes and gases from the reaction. Starting powders and rods should be protected against accidental ignition.

Personnel should wear appropriate protection against hot particles or sparks. This includes full face shields with filter lenses for eye protection and headgear. Safety boots are recommended to protect the feet from hot sparks. Clothing should not have pockets or cuffs that might catch hot particles.

Preheating should be done using the safety precautions applicable to oxyfuel gas equipment and operations.

THERMAL SPRAYING

THE POTENTIAL HAZARDS to the health and safety of personnel involved in thermal spraying operations and to persons in the immediate vicinity are as follows:

(1) Electrical shock
(2) Fire
(3) Fumes and gases
(4) Dust
(5) Arc radiation
(6) Noise

These hazards are not unique to thermal spraying methods. For example, flame spraying has hazards similar to those associated with the oxyfuel gas welding and cutting processes. Likewise, arc spraying and plasma spraying are similar in many respects to gas metal arc and plasma arc welding, respectively. Safe practices for

---

24. See ANSI/NFPA 495, *Manufacture, transportation, storage, and use of explosive materials.* New York: American National Standards Institute (latest edition).

these processes should be followed when thermal spraying with similar equipment. However, thermal spraying does generate dust and fumes to a greater degree.[25]

Fire Prevention

AIRBORNE FINELY DIVIDED solids, especially metal dusts, must be treated as explosives. To minimize danger from dust explosions, spray booths must have adequate ventilation.

A wet collector of the water-wash type is recommended to collect the spray dust. Bag or filter type collectors are not recommended. Good housekeeping in the work area should be maintained to avoid accumulation of metal dusts, particularly on rafters, tops of booths, and in floor cracks.

Paper, wood, oily rags, and other combustibles in the spraying area can cause a fire, and should be removed before the equipment is operated.

Protection of Personnel

THE GENERAL REQUIREMENTS for the protection of thermal spray operators are the same as for welders.

Eye Protection. Helmets, hand shields, face shields, or goggles should be used to protect the eyes, face, and neck during all thermal spraying operations. Safety goggles should be worn at all times. Table 16.4 is a guide for the selection of the proper filter shade number for viewing a specific spraying operation.

Table 16.4
Recommended Eye Filter Plates for Thermal Spraying Operations

Operation	Filter Shade Numbers
Wire flame spraying (except molybdenum)	2 to 4
Wire flame spraying of molybdenum	3 to 6
Flame spraying of metal powder	3 to 6
Flame spraying of exotherimics or ceramics	4 to 8
Plasma and arc spraying	9 to 12
Fusing operations	4 to 6

Respiratory Protection. Most thermal spraying operations require that respiratory protective devices be used by the operator. The nature, type, and magnitude of the fume and gas exposure determine which respiratory protective device should be used. All devices selected for use should be of a type approved by the U.S. Bureau of

25. Additional information may be found in *Thermal spraying: practice, theory, and application.* Miami: American Welding Society, 1985.

Mines, National Institute for Occupational Safety and Health, or other approving authority for the purpose intended.

Ear Protection. Ear protectors or properly fitted soft rubber ear plugs should be worn to protect the operator from the high-intensity noise from the process. Federal, state, and local codes should be checked for noise protection requirements.

Protective Clothing. Appropriate protective clothing required for a thermal spraying operation will vary with the size, nature, and location of the work to be performed. When working in confined spaces, flame-resistant clothing and gauntlets should be worn. Clothing should be fastened tightly around the wrists and ankles to keep dusts from contacting the skin.

The intense ultraviolet radiation of plasma and electric arc spraying can cause skin burns through normal clothing. Protection against radiation during arc spraying is practically the same as that for normal arc welding at equivalent current levels.

ADHESIVE BONDING

ADEQUATE SAFETY PRECAUTIONS must be observed with adhesives. Corrosive materials, flammable liquids, and toxic substances are commonly used in adhesive bonding. Therefore, manufacturing operations should be carefully supervised to ensure that proper safety procedures, protective devices, and protective clothing are being used. All federal, state, and local regulations should be complied with, including OSHA Regulation 29CRF 1900.1000, *Air Contaminants*.

General Requirements

Flammable Materials. All flammable materials, such as solvents, should be stored in tightly sealed drums and issued in suitably labeled safety cans to prevent fires during storage and use. Solvents and flammable liquids should not be used in poorly ventilated, confined areas. When solvents are used in trays, safety lids should be provided. Flames, sparks, or spark-producing equipment must not be permitted in the area where flammable materials are being handled. Fire extinguishers should be readily available.

Toxic Materials. Severe allergic reactions can result from direct contact, inhalation, or ingestion of phenolics and epoxies as well as most catalysts and accelerators. The eyes or skin may become sensitized over a long period of time even though no signs of irritation are visible. Once workers are sensitized to a particular type of adhesive, they may no longer be able to work near it because of allergic reactions. Careless handling of adhesives by production workers may expose others to toxic materials if proper safety rules are not observed. For example, coworkers may touch tools, door knobs, light switches, or other objects contaminated by careless workers.

For the normal individual, proper handling methods that eliminate skin contact with an adhesive should be sufficient. It bis mandatory that protective equipment, barrier creams, or both be used to avoid skin contact with certain types of formulations.

Factors to be considered in determining the extent of precautionary measures to be taken include:

(1) The frequency and duration of exposure
(2) The degree of hazard associated with a specific adhesive
(3) The solvent or curing agent used
(4) The temperature at which the operations are performed
(5) The potential evaporation surface area exposed at the work station

All these elements should be evaluated in terms of the individual operation.

Precautionary Procedures

A NUMBER OF measures are recommended in the handling and use of adhesives and auxiliary materials.

Personal Hygiene. Personnel should be instructed in proper procedures to prevent skin contact with solvents, curing agents, and uncured base adhesives. Showers, wash bowls, mild soaps, clean towels, refatting creams, and protective equipment should be provided.

Curing agents should be removed from the hands with soap and water. Resins should be removed with soap and water, alcohol, or a suitable solvent. Any solvent should be used sparingly and be followed by washing with soap and water. In case of allergic reaction or burning, prompt medical aid should be obtained.

Work Area. Areas in which adhesives are handled should be separated from other operations. These areas should contain the following facilities in addition to the proper fire equipment:

(1) Sink with running water
(2) Eye shower or rinse fountain
(3) First aid kit
(4) Ventilating facilities

Ovens, presses, and other curing equipment should be individually vented to remove fume. Vent hoods should be provided at mixing and application stations.

Protective Devices. Plastic or rubber gloves should be worn at all times when working with potentially toxic adhesives. Contaminated gloves must not contact objects that others may touch with their bare hands. Those gloves should be discarded or cleaned using procedures that remove the particular adhesive. Cleaning may require solvents, soap and water, or both. Hands, arms, face, and neck should be coated with a commercial barrier ointment or cream. This type of material may provide short-term protection and facilitate removal of adhesive components by washing.

Full face shields should be worn for eye protection whenever the possibility of splashing exists, otherwise glasses or goggles should be worn. In case of irritation, the eyes should be immediately flushed with water and then promptly treated by a physician.

Protective clothing should be worn at all times by those who work with the adhesives. Shop coats, aprons, or coveralls may be suitable, and they should be cleaned before reuse.

SUPPLEMENTARY READING LIST

American National Standards Institute. ANSI/NFPA 51B-1977, *Cutting and welding processes.* Quincy, MA: National Fire Protection Association.

———. ANSI/NFPA 51-1983, *Oxygen-fuel gas systems for welding, cutting and allied processes.* Quincy, MA: National Fire Protection Association.

———. ANSI/ASC Z49.1, *Safety in welding and cutting.* Miami: American Welding Society.

American Society for Metals. *Metals handbook,* Vol. 4. Heat Treating, 9th Ed. Metals Park, OH: American Society for Metals (1981): 389-416.

American Welding Society. *Arc welding and cutting noise.* Miami, FL: American Welding Society (1979).

———. *Arc welding safely.* Miami: American Welding Society

———. *Characterization of arc welding fumes.* Miami, FL: American Welding Society (1983).

———. *Effects of welding on health I, II, III, and IV.* Miami, FL: American Welding Society (1979, 1981, 1983).

———. *Fumes and gases in the welding environment.* Miami, FL: American Welding Society (1979).

————. *Oxyfuel gas welding, cutting, and heating safely.* Miami, American Welding Society.

————. AWS F2.1-78, *Recommended safe practices for electron beam welding and cutting.* Miami, FL: American Welding Society (1978).

————. AWS F4.1-80, *Recommended safe practices for the preparation for welding and cutting of containers that have held hazardous substances.* Miami, FL: American Welding Society (1980).

————. AWS C2.1-73, *Recommended safe practices for thermal spraying.* Miami, FL: American Welding Society (1973).

————. *The welding environment.* Miami, FL: American Welding Society (1973).

————. *Ultraviolet reflectance of paint.* Miami, FL: American Welding Society (1976).

————. *Welding fume control, a demostration project.* Miami, FL: American Welding Society (1982).

Balchin, N. C. *Health and safety in welding and allied processes,* 3rd Ed. England: The Welding Institute (1983).

Barthold, L. O. et al., *Electrostatic effects of overhead transmission lines, Part I-Hazards and Effects, IEEE Transactions, Power Apparatus and Systems,* Vol. PAS-91, (1972): 422-444.

Compressed Gas Association, Incorporated. *Handbook of compressed gases,* 2nd Ed. New York, NY: Van Nostrand Reinhold Co. (1981).

————. *Handling acetylene cylinders in fire situations,* SB-4. New York, NY: Compressed Gas Association (1972).

————. *Safe Handling of Compressed Gases in Containers,* P-1, New York, NY: Compressed Gas Association (1974).

Dalziel, Charles F. Effects of electric current on man, *ASEE Journal.* June 1973: 18-23.

Fireman's Fund Insurance Companies. *Welding fume control with mechanical ventilation,* 2nd Ed. San Francisco, CA: Fireman's Fund Insurance Companies (1981).

The Welding Institute. *The facts about fume.* England: The Welding Institute (1976).

TERMS AND DEFINITIONS

PREPARED BY A COMMITTEE CONSISTING OF:

G. E. Metzger, Chairman
AFWAL/MLLS

W. L. Green, 1st Vice Chairman
Ohio State University

E. A. Harwart, 2nd V. Chairman
Consultant

E. J. Seal, Secretary
American Welding Society

J. T. Biskup
Canadian Welding Bureau

W. F. Brown
Westinghouse Hanford

C. D. Burnham
Consultant

R. J. Christoffel
General Electric Company

G. B. Coates
General Electric Company

Michael D. Cooper
Hobart School of Welding Technology

J. E. Greer
General Railroad

M. J. Grycko
Packer Engineering Associates

D. E. Hamilton
Trinity Engrg. Testing Corporation

A. Ray Hollins, Jr.
Duke Power Company

M. J. Houle
National Board of Boiler and Pressure Vessel Inspectors

Stephen R. Morse
Deere and Company

L. C. Northard
Tennessee Valley Authority

D. H. Orts
Armco, Incorporated

J. J. Stanczak
Steel Detailers & Designers

J. J. Vagi
Babcock and Wilcox

J. L. Wilk
Coast to Coast Construction Company

WELDING HANDBOOK COMMITTEE MEMBER:
J. R. Hannahs
Midmark Corporation

APPENDIX A

TERMS AND DEFINITIONS

INTRODUCTION

THE COMMITTEE ON Definitions and Symbols was formed by the American Welding Society to establish standard terms and symbols to aid in the communication of welding information. *Standard Welding Terms and Definitions*, ANSI/AWS A3.0, is the major product of work done by the Subcommittee on Definitions in support of this purpose. This appendix is a selection of terms and definitions, and other related information, from the 1985 edition of ANSI/AWS A3.0; the reader is referred to the latest edition of ANSI/AWS A3.0 for the most current and complete welding vocabulary, including illustrations and commentary.

The standard terms and definitions published here are those that should be used in both the oral and the written language of welding; but, their use is particularly important in the writing of standards (codes, specifications, recommended practices, methods, classifications, and guides) and all other documents concerning welding. Because ANSI/AWS A3.0, from which this glossary is extracted, is intended to be a comprehensive compilation of welding terminology, nonstandard terms used in the welding industry are also included. All terms are either standard or nonstandard, and nonstandard terms are labeled as such.

To make this glossary most useful, the terms are arranged in the conventional dictionary letter-by-letter alphabetical sequence. It is the policy of the American Welding Society to use only generic terms and definitions in this glossary. The numerous proprietary brand and trademark names commonly used to describe welding processes, equipment, and filler metals are not included.

The illustrations shown in this appendix are not in the order of reference from the alphabetical listing. Instead, they are grouped by general categories. The first seven illustrations cover joint type and configuration. Figures A8 through A11 involve welding position and bead sequence. Figures A12 through A14 illustrate weld nomenclature. The final figure, Figure A15, is the Master Chart of Welding and Allied Processes. The two tables in this appendix follow the figures. These tables contain the official abbreviations of the processes shown in Figure A15.

The Master Chart of Welding and Allied Processes is a visual display of a hierarchy of processes, with the highest generic level (least specific) in the center, and the most specific in boxes around the perimeter. Some of the process variations are also included in the boxes. For historical reasons, the basis of classification is not entirely consistent; the determining factors include the energy source, capillary flow, and the physical state of the base material at the time of welding (i.e. liquid versus solid).

GLOSSARY

A

activated rosin flux. A rosin base flux containing an additive that increases wetting by the solder.

actual throat. The shortest distance between the weld root and the face of a fillet weld.

adhesive bonding (ABD). A materials joining process in which an adhesive is placed between the faying surfaces. The adhesive solidifies to produce an adhesive bond.

air carbon arc cutting (AAC). An arc cutting process that melts base metals by the heat of a carbon arc and removes the molten metal by a blast of air.

arc blow. The deflection of an electric arc from its normal path because of magnetic forces.

arc brazing (AB). A brazing process that uses an electric arc to provide the heat. See **carbon arc brazing.**

arc cutting (AC). A group of cutting processes that melt the base metal with the heat of an arc between an electrode and the base metal. See Figure A15, Master Chart of Welding and Allied Processes.

arc cutting gun (gas metal arc cutting). A device used in semiautomatic, machine, and automatic arc cutting to transfer current, guide the consumable electrode, and direct the shielding gas.

arc force. The axial force developed by an arc plasma.

arc gouging. An arc cutting process variation used to form a bevel or groove.

arc oxygen cutting. A nonstandard term for **oxygen arc cutting.**

arc seam weld. A seam weld made by an arc welding process. See Figure A7(A).

arc spot weld. A spot weld made by an arc welding process. See Figure A7(D).

arc spraying (ASP). A thermal spraying process using an arc between two consumable electrodes of surfacing materials as a heat source and a compressed gas to atomize and propel the surfacing material to the substrate.

arc strike. A discontinuity consisting of any localized remelted metal, heat affected metal, or change in the surface profile of any part of a weld or base metal resulting from an arc.

arc welding (AW). A group of welding processes that produces coalescence of metals by heating them with an arc, with or without the application of pressure, and with or without the use of filler metal.

arc welding electrode. A component of the welding circuit through which current is conducted and which terminates at the arc.

arc welding gun. A device used in semiautomatic, machine, and automatic arc welding to transfer current, guide the consumable electrode, and direct the shielding gas.

as-welded. The condition of weld metal, welded joints, and weldments after welding, but prior to any subsequent thermal, mechanical, or chemical treatments.

autogenous weld. A fusion weld made without the addition of filler metal.

automatic welding. Welding with equipment that performs the welding operation without adjustment of the controls by a welding operator. The equipment may or may not load and unload the workpieces. See also **machine welding.**

B

back bead. A weld bead resulting from a back weld pass.

backfire. The momentary recession of the flame into the welding tip or cutting tip followed by immediate reappearance or complete extinction of the flame.

back gouging. The removal of weld metal and base metal from the other side of a partially welded joint to facilitate complete fusion and complete joint penetration upon subsequent welding from that side.

backhand welding. A welding technique in which the welding torch or gun is directed opposite to the progress of welding.

backing. A material or device placed against the back side of the joint, or at both sides of a weld in electroslag and electrogas welding, to support and retain molten weld metal. The material may be partially fused or remain unfused during welding and may be either metal or nonmetal. See Figure A4(D).

backing bead. A weld bead resulting from a backing pass.

backing filler metal. A nonstandard term for **consumable insert.**

backing pass. A weld pass made for a backing weld.

backing ring. Backing in the form of a ring, generally used in the welding of pipe.

backing shoe. A nonconsumable backing device used in electroslag and electrogas welding.

backing weld. Backing in the form of a weld. See Figure A12(D).

backstep sequence. A longitudinal sequence in which weld passes are made in the direction opposite to the progress of welding. See Figure A11(A).

back weld. A weld made at the back of a single groove weld. See Figure A12(C).

balling up. The formation of globules of molten brazing filler metal or flux due to lack of wetting of the base metal.

base material. The material to be welded, brazed, soldered, or cut. See also **base metal** and **substrate**.

base metal. The metal to be welded, brazed, soldered, or cut. See also **base material** and **substrate**.

bead weld. A nonstandard term for **surfacing weld**.

bevel. An angular edge preparation. See Figures A4(B) and A5(B).

bevel angle. The angle formed between the prepared edge of a member and a plane perpendicular to the surface of the member. See Figure A6.

bevel groove weld. A type of groove weld. See Figures A4(B) and A5(B).

bit. That part of the soldering iron, usually made of copper, that actually transfers heat (and sometimes solder) to the joint.

blacksmith welding. A nonstandard term for **forge welding**.

block sequence. A combined longitudinal and cross-sectional sequence for a continuous multiple pass weld in which separated increments are completely or partially welded before intervening increments are welded. See Figure A11(B).

blowhole. A nonstandard term for **porosity**.

bond. See bonding force, covalent bond, mechanical bond, and metallic bond.

bond coat (thermal spraying). A preliminary (or prime) coat of material that improves adherence of the subsequent spray deposit.

bonding force. The force that holds two atoms together; it results from a decrease in energy as two atoms are brought closer to one another.

bond line. The cross section of the interface between thermal spray deposits and substrate, or between adhesive and adherend in an adhesive bonded joint.

bottle. A nonstandard term for **gas cylinder**.

boxing. The continuation of a fillet weld around a corner of a member as an extension of the principal weld. See Figure A11(C).

braze. A weld produced by heating an assembly to the brazing temperature using a filler metal having a liquidus above 840°F (450°C) and below the solidus of the base metal. The filler metal is distributed between the closely fitted faying surfaces of the joint by capillary action.

braze interface. The interface between filler metal and base metal in a brazed joint.

brazement. An assembly whose component parts are joined by brazing.

brazer. One who performs a manual or semiautomatic brazing operation.

braze welding. A welding process variation in which a filler metal, having a liquidus above 840°F (450°C) and below the solidus of the base metal, is used. Unlike brazing, in braze welding the filler metal is not distributed in the joint by capillary action.

brazing (B). A group of welding processes that produce coalescence of materials by heating them to the brazing temperature in the presence of a filler metal having a liquidus above 840°F (450°C) and below the solidus of the base metal. The filler metal is distributed between the closely fitted faying surfaces of the joint by capillary action.

brazing alloy. A nonstandard term for **brazing filler metal**.

brazing filler metal. The metal that fills the capillary joint clearance and has a liquidus above 840°F (450°C) but below the solidus of the base metals.

brazing operator. One who operates machine or automatic brazing equipment.

brittle nugget. A nonstandard term used to describe a faying plane failure in a resistance weld peel test.

bronze welding. A nonstandard term for **braze welding**.

buildup. A surfacing variation in which surfacing metal is deposited to achieve the required dimensions. See also **buttering**.

burner. A nonstandard term for **oxygen cutter**.

burning. A nonstandard term for **oxygen cutting**.

burn through. A nonstandard term for excessive melt through or a hole.

burn through weld. A nonstandard term for a **seam weld** or **spot weld**.

buttering. A surfacing variation that deposits surfacing metal on one or more surfaces to provide metallurgically compatible weld metal for the subsequent completion of the weld. See also **buildup**.

butt joint. A joint between two members aligned approximately in the same plane. See Figure A1(A).

button. That part of a weld, including all or part of the nugget, that tears out in the destructive testing of spot, seam, or projection welded specimens.

butt weld. A nonstandard term for a weld in a butt joint.

C

carbon arc brazing (CAB). A brazing process that produces coalescence of metals by heating them with an electric arc between two carbon electrodes. The filler metal is distributed in the joint by capillary action.

carbon arc cutting (CAC). An arc cutting process that severs base metals by melting them with the heat of an arc between a carbon electrode and the base metal.

carbon arc welding (CAW). An arc welding process that produces coalescence of metals by heating them with an arc between a carbon electrode and the base metal. No shielding is used. Pressure and filler metal may or may not be used.

carbonizing flame. A nonstandard term for **reducing flame**.

caulk weld. A nonstandard term for **seal weld**.

chain intermittent weld. An intermittent weld on both sides of a joint in which the weld increments on one side

are approximately opposite those on the other side. See Figure A11(D).

chamfer. A nonstandard term for **bevel**.

chemical flux cutting (FOC). An oxygen cutting process that severs base metals using a chemical flux to facilitate cutting.

chill ring. A nonstandard term for **backing ring**.

clad brazing sheet. A metal sheet on which one or both sides are clad with brazing filler metal.

coalescence. The growing together or growth into one body of the materials being welded.

coated electrode. A nonstandard term for **covered electrode**.

coating density. A nonstandard term for **spray deposit density ratio**.

coextrusion welding (CEW). A solid-state welding process that produces coalescence of the faying surfaces by heating and forcing base metals through a extrusion die.

cold crack. A crack which develops after solidification is complete.

cold soldered joint. A joint with incomplete coalescence caused by insufficient application of heat to the base metal during soldering.

cold welding (CW). A solid-state welding process in which pressure is used at room temperature to produce coalescence of metals with substantial deformation at the weld. See also **diffusion welding**, **forge welding** and **hot pressure welding**.

complete fusion. Fusion which has occurred over the entire base metal surface intended for welding and between all adjoining weld beads.

complete joint penetration. A penetration by weld metal for the full thickness of the base metal in a joint with a groove weld. See Figures A13(C) and A13(D).

complete penetration. A nonstandard term for **complete joint penetration**.

concavity. The maximum distance from the face of a concave fillet weld perpendicular to a line joining the weld toes.

cone. The conical part of an oxyfuel gas flame next to the orifice of the tip.

constricted arc (plasma arc welding and cutting). A plasma arc column that is shaped by a constricting nozzle orifice.

consumable insert. Preplaced filler metal that is completely fused into the joint root and becomes part of the weld.

contact resistance (resistance welding). Resistance to the flow of electric current between two workpieces or an electrode and a workpiece.

contact tube. A device which transfers current to a continuous electrode.

convexity. The maximum distance from the face of a convex fillet weld perpendicular to a line joining the weld toes.

copper brazing. A nonstandard term for brazing with a copper filler metal.

cored solder. A solder wire or bar containing flux as a core.

corner-flange weld. A flange weld with only one member flanged at the joint.

corner joint. A joint between two members located approximately at right angles to each other. See Figure A1(B).

corona (resistance welding). The area sometimes surrounding the nugget of a spot weld at the faying surface which provides a degree of solid-state welding.

CO_2 welding. A nonstandard term for **gas metal arc welding.**

covalent bond. A primary bond arising from the reduction in energy associated with overlapping half-filled orbitals of two atoms.

covered electrode. A composite filler metal electrode consisting of a core of a bare electrode or metal cored electrode to which a covering sufficient to provide a slag layer on the weld metal has been applied. The covering may contain materials providing such functions as shielding from the atmosphere, deoxidation, and arc stabilization and can serve as a source of metallic additions to the weld.

crack. A fracture type discontinuity characterized by a sharp tip and high ratio of length and width to opening displacement.

crater. A depression at the termination of a weld bead.

cutting attachment. A device for converting an oxyfuel gas welding torch into an oxygen cutting torch.

cutting blowpipe. A nonstandard term for **cutting torch.**

cutting nozzle. A nonstandard term for **cutting tip.**

cutting tip. The part of an oxygen cutting torch from which the gases issue.

cutting torch (arc). A device used in air carbon arc cutting, gas tungsten arc cutting, and plasma arc cutting to control the position of the electrode, to transfer current, and to control the flow of gases.

cutting torch (oxyfuel gas). A device used for directing the preheating flame produced by the controlled combustion of fuel gases and to direct and control the cutting oxygen.

cylinder manifold. A multiple header for interconnection of gas or fluid sources with distribution points.

D

defect. A discontinuity or discontinuities that by nature or accumulated effect (for example, total crack length) render a part or product unable to meet minimum applicable acceptance standards or specifications. This term designates rejectability. See also **discontinuity** and **flaw.**

deposit (thermal spraying). A nonstandard term for spray deposit.

deposited metal. Filler metal that has been added during welding.

deposition efficiency (arc welding). The ratio of the weight of deposited metal to the net weight of filler metal consumed, exclusive of stubs.

deposition efficiency (thermal spraying). The ratio of the weight of spray deposit to the weight of the surfacing material sprayed, usually expressed in percent.

deposition sequence. A nonstandard term for **weld pass sequence.**

depth of fusion. The distance that fusion extends into the base metal or previous pass from the surface melted during welding.

diffusion bonding. A nonstandard term for **diffusion brazing** and **diffusion welding**.

diffusion brazing (DFB). A brazing process that produces coalescence of metals by heating them to brazing temperature and by using a filler metal or an in situ liquid phase. The filler metal may be distributed by capillary action or may be placed or formed at the faying surfaces. The filler metal is diffused with the base metal to the extent that the joint properties have been changed to approach those of the base metal. Pressure may or may not be applied.

diffusion welding (DFW). A solid-state welding process that produces coalescence of the faying surfaces by the application of pressure at elevated temperature. The process does not involve macroscopic deformation, melting, or relative motion of the workpieces. A solid filler metal may or may not be inserted between the faying surfaces. See also **cold welding, forge welding,** and **hot pressure welding**.

dilution. The change in chemical composition of a welding filler metal caused by the admixture of the base metal or previous weld metal in the weld bead. It is measured by the percentage of base metal or previous weld metal in the weld bead.

dip brazing (DB). A brazing process using the heat furnished by a molten chemical or metal bath. When a molten chemical bath is used, the bath may act as a flux. When a molten metal bath is used, the bath provides the filler metal.

dip soldering (DS). A soldering process using the heat furnished by a molten metal bath which provides the solder filler metal.

direct current electrode negative (DCEN). The arrangement of direct current arc welding leads in which the workpiece is the positive pole and the electrode is the negative pole of the welding arc.

direct current electrode positive (DCEP). The arrangement of direct current arc welding leads in which the workpiece is the negative pole and the electrode is the positive pole of the welding arc.

direct current reverse polarity. A nonstandard term for **direct current electrode positive**.

direct current straight polarity. A nonstandard term for **direct current electrode negative**.

discontinuity. An interruption of the typical structure of a weldment, such as a lack of homogeneity in the mechanical, metallurgical, or physical characteristics of the material or weldment. A discontinuity is not necessarily a defect. See also **defect** and **flaw**.

double-bevel-groove weld. A type of groove weld. See Figure 5B.

double-flare-bevel-groove weld. A weld in grooves formed by a member with a curved surface in contact with a planar member. See Figure A5(F).

double-flare-V-groove weld. A weld in grooves formed by two members with curved surfaces. See Figure A5(G).

double-J-groove weld. A type of groove weld. See Figure A5(D).

double-square-groove weld. A type of groove weld. See Figure A5(A).

double-U-groove weld. A type of groove weld. See Figure A5(E).

double-V-groove weld. A type of groove weld. See Figure A5(C).

double-welded joint. A fusion welded joint that is welded from both sides. See Figure A5.

downhand. A nonstandard term for **flat position**.

drag (thermal cutting). The offset distance between the actual and straight line exit points of the gas stream or cutting beam measured on the exit surface of the material.

E

edge-flange weld. A flange weld with two members flanged at the location of welding.

edge joint. A joint between the edges of two or more parallel or nearly parallel members. See Figure A1(E).

edge weld. A weld in an edge joint.

edge weld size. The weld metal thickness measured at the weld root.

effective throat. The minimum distance minus any convexity between the weld root and the face of a fillet weld.

electric arc spraying. A nonstandard term for **arc spraying.**

electric bonding. A nonstandard term for surfacing by thermal spraying.

electric brazing. A nonstandard term for **arc brazing** and **resistance brazing.**

electrode. See welding electrode.

electrode extension. For gas metal arc welding, flux cored arc welding, and submerged arc welding, the length of unmelted electrode extending beyond the end of the contact tube.

electrode force. The force between the electrodes in making spot, seam, or projection welds by resistance welding.

electrode holder. A device used for mechanically holding the electrode while conducting current to it.

electrode indentation (resistance welding). The depression formed on the surface of workpieces by electrodes.

electrode lead. The electrical conductor between the source of arc welding current and the electrode holder.

electrogas welding (EGW). An arc welding process that produces coalescence of metals by heating them with an arc between a continuous filler metal electrode and the work. Molding shoes are used to confine the molten weld metal for vertical position welding. The electrodes may be either flux cored or solid. Shielding may or may not be obtained from an externally supplied gas or mixture.

electron beam cutting (EBC). A cutting process that uses the heat obtained from a concentrated beam composed primarily of high velocity electrons which impinge upon the workpieces; it may or may not use an externally supplied gas.

electron beam gun. A device for producing and accelerating electrons. Typical components include the emitter (also called the filament or cathode) which is heated to produce electrons via thermionic emission, a cup (also called the grid or grid cup), and the anode.

electron beam welding (EBW). A welding process that produces coalescence of metals with the heat obtained from a concentrated beam composed primarily of high velocity electrons impinging on the joint.

electroslag welding (ESW). A welding process that produces coalescence of metals with molten slag that melts the filler metal and the surfaces of the workpieces. The weld pool is shielded by this slag which moves along the full cross section of the joint as welding progresses. The process is initiated by an arc that heats the slag. The arc is then extinguished by the conductive slag, which is kept molten by its resistance to electric current passing between the electrode and the workpieces.

end return. A nonstandard term for **boxing.**

erosion (brazing). A condition caused by dissolution of the base metal by molten filler metal resulting in a reduction in the thickness of the base metal.

explosion welding (EXW). A solid-state welding process that affects coalescence by high velocity movement together with the workpieces produced by a controlled detonation.

F

face reinforcement. Weld reinforcement at the side of the joint from which welding was done. See Figures A12(A) and A12(C). See also **root reinforcement.**

faying surface. That mating surface of a member that is in contact with or in close proximity to another member to which it is to be joined.

ferrite number. An arbitrary, standardized value designating the ferrite content of an austenitic stainless steel weld metal. It should be used in place of percent ferrite or volume percent ferrite on a direct replacement basis.

filler metal. The metal to be added in making a welded, brazed, or soldered joint. See also **brazing filler metal, consumable insert, solder, welding electrode, welding rod,** and **welding wire.**

filler wire. A nonstandard term for **welding wire.**

fillet weld. A weld of approximately triangular cross section joining two surfaces approximately at right angles to each other in a lap joint, T-joint, or corner joint. See Figure A12(E).

fillet weld break test. A test in which the specimen is loaded so that the weld root is in tension.

fillet weld leg. The distance from the joint root to the toe of the fillet weld.

fillet weld size. For equal leg fillet welds, the leg lengths of the largest isosceles right triangle which can be inscribed within the fillet weld cross section. For unequal leg fillet welds, the leg lengths of the largest right triangle that can be inscribed within the fillet weld cross section.

fillet weld throat. See **actual throat**, **effective throat**, and **theoretical throat**.

firecracker welding. A variation of the shielded metal arc welding process in which a length of covered electrode is placed along the joint in contact with the workpieces. During the welding operation, the stationary electrode is consumed as the arc travels the length of the electrode.

fisheye. A discontinuity found on the fracture surface of a weld in steel that consists of a small pore or inclusion surrounded by an approximately round, bright area.

flame cutting. A nonstandard term for **oxygen cutting**.

flame propagation rate. The speed at which a flame travels through a mixture of gases.

flame spraying (FLSP). A thermal spraying process in which an oxyfuel gas flame is the source of heat for melting the surfacing material. Compressed gas may or may not be used for atomizing and propelling the surfacing material to the substrate.

flange weld. A weld made on the edges of two or more members to be joined, usually light gage metal, at least one of the members being flanged.

flange weld size. The weld metal thickness measured at the weld root.

flare-bevel-groove weld. A weld in a groove formed by a member with a curved surface in contact with a planar member. See Figures A4(G) and A5(F).

flare-V-groove weld. A weld in a groove formed by two members with curved surfaces. See Figures A4(H) and A5(G).

flash. Material that is expelled from a flash weld prior to the upset portion of the welding cycle.

flash butt welding. A nonstandard term for **flash welding**.

flash coat. A thin coating usually less than 0.002 in. (0.05 mm) in thickness.

flash welding (FW). A resistance welding process that produces coalescence at the faying surfaces of a butt joint by a flashing action and by the application of pressure after heating is substantially completed. The flashing action, caused by the very high current densities at small contacts between the parts, forcibly expels the material from the joint as the parts are slowly moved together. The weld is completed by a rapid upsetting of the workpieces.

flat position. The welding position used to weld from the upper side of the joint; the face of the weld is approximately horizontal. See Figures A8(A) and A9(A).

flaw. A near synonym for discontinuity but with an undesirable connotation. See also **defect** and **discontinuity**.

flow brightening (soldering). Fusion of a metallic coating on a base metal.

flux. Material used to prevent, dissolve, or facilitate removal of oxides and other undesirable surface substances.

flux cored arc welding (FCAW). An arc welding process that produces coalescence of metals by heating them with an arc between a continuous filler metal electrode and the work. Shielding is provided by a flux contained within the tubular electrode. Additional shielding may or may not be obtained from an externally supplied gas or gas mixture. See also **flux cored electrode**.

flux cored electrode. A composite filler metal electrode consisting of a metal tube or other hollow configuration containing ingredients to provide such functions as shielding atmosphere, deoxidation, arc stabilization, and slag formation. Minor amounts of alloying materials may be included in the core. External shielding may or may not be used.

flux cover (metal bath dip brazing and dip soldering). A layer of molten flux over the molten filler metal bath.

flux oxygen cutting. A nonstandard term for **chemical flux cutting**.

forehand welding. A welding technique in which the welding torch or gun is directed toward the progress of welding.

forge welding (FOW). A solid-state welding process that produces coalescence of metals by heating them in air in a forge and by applying pressure or blows sufficient to cause permanent deformation at the interface. See also **cold welding, diffusion welding, hot pressure welding,** and **roll welding.**

friction welding (FRW). A solid-state welding process that produces coalescence of materials under compressive force contact of workpieces rotating or moving relative to one another to produce heat and plastically displace material from the faying surfaces.

furnace brazing (FB). A brazing process in which the workpieces are placed in a furnace and heated to the brazing temperature.

furnace soldering (FS). A soldering process in which the workpieces are placed in a furnace and heated to the soldering temperature.

fused spray deposit (thermal spraying). A self-fluxing spray deposit that is subsequently heated to coalescence within itself and with the substrate.

fusion. The melting together of filler metal and base metal (substrate), or of base metal only, which results in coalescence. See also **depth of fusion.**

fusion face. A surface of the base metal that will be melted during welding.

fusion welding. Any welding process that uses fusion of the base metal to make the weld.

fusion zone. The area of base metal melted as determined on the cross section of a weld.

G

gap. A nonstandard term for **joint clearance** and **root opening.**

gas brazing. A nonstandard term for **torch brazing.**

gas cutter. A nonstandard term for **oxygen cutter.**

gas cutting. A nonstandard term for **oxygen cutting.**

gas cylinder. A portable container used for transportation and storage of a compressed gas.

gas gouging. A nonstandard term for **oxygen gouging.**

gas laser. A laser in which the lasing medium is a gas.

gas metal arc cutting (GMAC). An arc cutting process in which metals are severed by melting them with the heat of an arc between a continuous filler metal electrode and the workpiece. Shielding is obtained entirely from an externally supplied gas.

gas metal arc welding (GMAW). An arc welding process that produces coalescence of metals by heating them with an arc between a continuous filler metal electrode and the workpieces. Shielding is obtained entirely from an externally supplied gas.

gas pocket. A nonstandard term for **porosity.**

gas regulator. A device for controlling the delivery of gas at some substantially constant pressure.

gas shielded arc welding. A general term used to describe **flux cored arc welding** (when gas shielding is employed), **gas metal arc welding,** and **gas tungsten arc welding.**

gas torch. A nonstandard term for **cutting torch** and **welding torch.**

gas tungsten arc cutting (GTAC). An arc cutting process in which metals are severed by melting them with an arc between a single tungsten electrode and the workpiece. Shielding is obtained from a gas.

gas tungsten arc welding (GTAW). An arc welding process that produces coalescence of metals by heating them with an arc between a tungsten electrode (nonconsumable) and the workpieces. Shielding is obtained from a gas. Pressure may or may not be used, and filler metal may or may not be used.

gas welding. A nonstandard term for **oxyfuel gas welding.**

globular transfer (arc welding). The transfer of molten metal in large drops from a consumable electrode across the arc. See also **short circuiting transfer** and **spray transfer.**

gouging. The forming of a bevel or groove by material removal. See also **arc gouging, back gouging,** and **oxygen gouging.**

groove angle. The total included angle of the groove between workpieces. See Figure A6.

groove face. That surface of a joint member included in the groove. See Figure A3.

groove radius. The radius used to form the shape of a J- or U-groove weld. See Figure A6.

groove weld. A weld made in a groove between the workpieces. See Figures A4 and A5.

groove weld size. The joint penetration of a groove weld. See Figure A13.

groove weld throat. A nonstandard term for **groove weld size.**

ground connection. An electrical connection of the welding machine frame to the earth for safety. See also **workpiece connection** and **workpiece lead.**

ground lead. A nonstandard term for **workpiece lead.**

gun. See **arc cutting gun, arc welding gun, electron beam gun, resistance welding gun, soldering gun,** and **thermal spraying gun.**

H

hammer welding. A nonstandard term for **cold welding** and **forge welding.**

hard solder. A nonstandard term for silver-base brazing filler metals.

heat affected zone. That portion of the base metal that has not been melted, but whose mechanical properties or microstructure have been altered by the heat of welding, brazing, soldering, or cutting.

high frequency resistance welding. A group of resistance welding process variations that uses high frequency welding current to concentrate the welding heat at the desired location.

horizontal fixed position (pipe welding). The position of a pipe joint in which the axis of the pipe is approximately horizontal, and the pipe is not rotated during welding. See Figure A10(C).

horizontal position (fillet weld). The position in which welding is performed on the upper side of an approximately horizontal surface and against an approximately vertical surface. See Figure A9(B).

horizontal position (groove weld). The position of welding in which the weld axis lies in an approximately horizontal plane and the weld face lies in an approximately vertical plane. See Figure A8(B).

horizontal rolled position (pipe welding). The position of a pipe joint in which the axis of the pipe is approximately horizontal, and welding is performed in the flat position by rotating the pipe. See Figure A10(A).

hot crack. A crack that develops during solidification.

hot pressure welding (HPW). A solid-state welding process that produces coalescence of metals with heat and application of pressure sufficient to produce macrodeformation of the base metal. Vacuum or other shielding media may be used. See also **diffusion welding** and **forge welding.**

hydrogen brazing. A nonstandard term for any brazing process that takes place in a hydrogen atmosphere.

I

impulse (resistance welding). A group of pulses occurring on a regular frequency separated only by an interpulse time.

inclined position. The position of a pipe joint in which the axis of the pipe is at an angle of approximately 45 degrees to the horizontal, and the pipe is not rotated during welding. See Figure A10(D).

inclined position (with restriction ring). The position of a pipe joint in which the axis of the pipe is at an angle of approximately 45 degrees to the horizontal, and a restriction ring is located near the joint. The pipe is not rotated during welding. See Figure A10(E).

included angle. A nonstandard term for **groove angle.**

induction brazing (IB). A brazing process in which the heat required is obtained from the resistance of the workpieces to induced electric current.

induction soldering (IS). A soldering process in which the heat required is obtained from the resistance of the workpieces to induced electric current.

induction welding (IW). A welding process that produces coalescence of metals by the heat obtained from the resistance of the workpieces to the flow of induced high frequency welding current with or without the application of pressure. The effect of the high frequency welding current is to concentrate the welding heat at the desired location.

inert gas. A gas that normally does not combine chemically with the base metal or filler metal. See also **protective atmosphere.**

inert gas metal arc welding. A nonstandard term for gas metal arc welding.

inert gas tungsten arc welding. A nonstandard term for gas tungsten arc welding.

infrared brazing (IRB). A brazing process in which the heat required is furnished by infrared radiation.

infrared soldering (IRS). A soldering process in which the heat required is furnished by infrared radiation.

intergranular penetration. The penetration of a filler metal along the grain boundaries of a base metal.

interpass temperature. In a multipass weld, the temperature of the weld metal before the next pass is started.

iron soldering (INS). A soldering process in which the heat required is obtained from a soldering iron.

J

J-groove weld. A type of groove weld. See Figures A4(E) and A5(D).

joint. The junction of members or the edges of members which are to be joined or have been joined. See Figure A1.

joint clearance. The distance between the faying surfaces of a joint. In brazing, this distance is referred to as that which is present before brazing, at the brazing temperature, or after brazing is completed.

joint efficiency. The ratio of the strength of a joint to the strength of the base metal, expressed in percent.

joint penetration. The depth a weld extends from its face into a joint, exclusive of reinforcement.

joint root. That portion of a joint to be welded where the members approach closest to each other. In cross section, the joint root may be either a point, a line, or an area. See Figure A2.

joint type. A weld joint classification based on the five basic arrangements of the component parts such as butt joint, corner joint, edge joint, lap joint, and T-joint. See Figure A1.

K

kerf. The width of the cut produced during a cutting process.

L

lamellar tear. A terrace-like fracture in the base metal with a basic orientation parallel to the wrought surface. It is caused by the high stress in the thickness direction that results from welding.

land. A nonstandard term for **root face.**

lap joint. A joint between two overlapping members in parallel planes. See Figure A1(D).

laser. A device that produces a concentrated coherent light beam by stimulating electronic or molecular transitions to lower energy levels. Laser is an acronym for **light amplification by stimulated emission of radiation.**

laser beam cutting (LBC). A thermal cutting process that severs materials by melting or vaporizing them with the heat obtained from a laser beam, with or without the application of gas jets to augment the removal of material.

laser beam welding (LBW). A welding process that produces coalescence of materials with the heat obtained from the application of a concentrated coherent light beam impinging upon the joint.

lead burning. A nonstandard term for the welding of lead.

liquation. The separation of a low melting constituent of an alloy from the remaining constituents, usually apparent in alloys having a wide melting range.

locked-up stress. A nonstandard term for **residual stress.**

longitudinal crack. A crack with its major axis orientation approximately parallel to the weld axis.

M

machine welding. Welding with equipment that performs the welding operation under the constant observation and control of a welding operator. The equipment may or may not load and unload the workpieces. See also **automatic welding.**

macroetch test. A test in which the specimen is prepared with a fine finish and etched to give a clear definition of the weld.

manual welding. A welding operation performed and controlled completely by hand. See also **automatic welding, machine welding,** and **semiautomatic arc welding.**

mask (thermal spraying). A device for protecting a substrate surface from the effects of blasting or adherence of a spray deposit.

mechanical bond (thermal spraying). The adherence of a spray deposit to a roughened surface by the mechanism of particle interlocking.

melt-through. Visible root reinforcement produced in a joint welded from one side.

metal arc cutting (MAC). Any of a group of arc cutting processes that serves metals by melting them with the heat of an arc between a metal electrode and the base metal. See also **gas metal arc cutting** and **shielding metal arc cutting.**

metal cored electrode. A composite filler metal electrode consisting of a metal tube or other hollow configuration containing alloying materials. Minor amounts of ingredients providing such functions as arc stabilization and fluxing of oxides may be included. External shielding gas may or may not be used.

metal electrode. A filler or nonfiller metal electrode used in arc welding or cutting that consists of a metal wire or rod that has been manufactured by any method and that is either bare or covered.

metallic bond. The principal bond that holds metals together and is formed between base metals and filler metals in all welding processes. This is a primary bond arising from the increased spatial extension of the valence electron wave functions when an aggregate of metal atoms is brought close together. See also **bonding force** and **covalent bond.**

metallizing. A nonstandard term for **thermal spraying.**

metallurgical bond. A nonstandard term for **metallic bond.**

metal powder cutting (POC). An oxygen cutting process that severs metals through the use of powder, such as iron, to facilitate cutting.

MIG welding. A nonstandard term for **flux cored arc welding** and **gas metal arc welding.**

mixing chamber. That part of a welding or cutting torch in which a fuel gas and oxygen are mixed.

molten weld pool. A nonstandard term for **weld pool.**

multiport nozzle (plasma arc welding and cutting). A constricting nozzle containing two or more orifices located in a configuration to achieve a degree of control over the arc shape.

N

neutral flame. An oxyfuel gas flame in which the portion used is neither oxidizing nor reducing. See also **oxidizing flame** and **reducing flame.**

nontransferred arc (plasma arc welding and cutting, and plasma spraying). An arc established between the electrode and the constricting nozzle. The workpiece is not in the electrical circuit. See also **transferred arc.**

nozzle. A device that directs shielding media.

nugget. The weld metal joining the workpieces in spot, roll spot, seam, or projection welds.

nugget size (resistance welding). The diameter of a spot or projection weld or width of a seam weld measured in the plane of the faying surfaces.

O

orifice gas (plasma arc welding and cutting). The gas that is directed into the torch to surround the electrode. It becomes ionized in the arc to form the plasma and issues from the orifice in the torch nozzle as the plasma jet.

oven soldering. A nonstandard term for **furnace soldering.**

overhead position. The position in which welding is performed from the underside of the joint. See Figures A8(D) and A9(D).

overlap. The protrusion of weld metal beyond the weld toes or weld root. See Figure A14(B).

overlap (resistance seam welding). The portion of the preceding weld nugget remelted by the succeeding weld.

overlaying. A nonstandard term for **surfacing.**

oxidizing flame. An oxyfuel gas flame having an oxidizing effect due to excess oxygen. See also **neutral flame** and **reducing flame.**

oxyacetylene welding (OAW). An oxyfuel gas welding process that produces coalescence of metals by heating them with a gas flame or flames obtained from the combustion of acetylene with oxygen. The process may be used with or without the application of pressure and with or without the use of filler metal.

oxyfuel gas cutting (OFC). A group of cutting processes used to sever metals by means of the chemical reaction of oxygen with the base metal at elevated temperatures. The necessary temperature is maintained by means of gas flames obtained from the combustion of a specified fuel gas and oxygen. See also **oxygen cutting**.

oxyfuel gas spraying. A nonstandard term for **flame spraying**.

oxyfuel gas welding (OFW). A group of welding processes that produces coalescence by heating materials with an oxyfuel gas flame or flames, with or without the application of pressure, and with or without the use of filler metal.

oxygas cutting. A nonstandard term for **oxygen cutting**.

oxygen arc cutting (AOC). An oxygen cutting process used to sever metals by means of the chemical reaction of oxygen with the base metal at elevated temperatures. The necessary temperature is maintained by an arc between a consumable tubular electrode and the base metal.

oxygen cutter. One who performs a manual oxygen cutting operation.

oxygen cutting (OC). A group of cutting processes used to sever or remove metals by means of the chemical reaction between oxygen and the base metal at elevated temperatures. In the case of oxidation-resistant metals, the reaction is facilitated by the use of a chemical flux or metal powder. See also **chemical flux cutting, metal powder cutting, oxyfuel gas cutting, oxygen arc cutting, and oxygen lance cutting**.

oxygen cutting operator. One who operates machine or automatic oxygen cutting equipment.

oxygen gouging. An application of oxygen cutting in which a bevel or groove is formed.

oxygen grooving. A nonstandard term for **oxygen gouging**.

oxygen lance. A length of pipe used to convey oxygen to the point of cutting in oxygen lance cutting.

oxygen lance cutting (LOC). An oxygen cutting process used to sever metals with oxygen supplied through a consumable lance. The preheat to start the cutting is obtained by other means.

oxygen lancing. A nonstandard term for **oxygen lance cutting**.

oxyhydrogen welding (OHW). An oxyfuel gas welding process that produces coalescence of materials by heating them with a gas flame or flames obtained from the combustion of hydrogen with oxygen, without the application of pressure and with or without the use of filler metal.

P

parallel welding. A resistance welding secondary circuit variation in which the secondary current is divided and conducted through the workpieces and electrodes in parallel electrical paths to simultaneously form multiple resistance spot, seam, or projection welds.

parent metal. A nonstandard term for **base metal**.

partial joint penetration. Joint penetration that is intentionally less than complete. See Figure A13.

penetration. A nonstandard term for **joint penetration** and **root penetration**.

percussion welding (PEW). A welding process that produces coalescence at the faying surface using the heat from an arc produced by a rapid discharge of electrical energy. Pressure is applied percussively during or immediately following the electrical discharge.

pilot arc (plasma arc welding). A low current continuous arc between the electrode and the constricting nozzle to ionize the gas and facilitate the start of the welding arc.

plasma arc cutting (PAC). An arc cutting process that severs metal by melting a localized area with a constricted arc and removing the molten material with a high velocity jet of hot, ionized gas issuing from the constricting orifice.

plasma arc welding (PAW). An arc welding process that produces coalescence of metals by heating them with a constricted arc between an electrode and the workpiece (transferred arc) or the electrode and the constricting nozzle (nontransferred arc). Shielding is obtained from the hot, ionized gas issuing from the torch which may be supplemented by an auxiliary source of shielding gas. Shielding gas may be an inert gas or a mixture of gases. Pressure may or may not be used, and filler metal may or may not be supplied.

plasma metallizing. A nonstandard term for **plasma spraying**.

plasma spraying (PSP). A thermal spraying process in which a nontransferred arc of a plasma torch is utilized to create a gas plasma that acts as the source of heat for melting and propelling the surfacing material to the substrate.

plenum chamber. (plasma arc welding and cutting, and plasma spraying). The space between the inside wall of the constricting nozzle and the electrode.

plug weld. A weld made in a circular hole in one member of a joint, fusing that member to another member. A fillet-welded hole is not to be construed as conforming to this definition.

polarity. See **direct current electrode negative** and **direct current electrode positive.**

porosity. Cavity type discontinuities formed by gas entrapment during solidification.

postheating. The application of heat to an assembly after welding, brazing, soldering, thermal spraying, or thermal cutting. See also **postweld heat treatment.**

postweld heat treatment. Any heat treatment after welding.

powder cutting. A nonstandard term for **chemical flux cutting** and **metal powder cutting.**

precoating. Coating the base metal in the joint by dipping, electroplating, or other applicable means prior to soldering or brazing.

preform. Brazing or soldering filler metal fabricated in a shape or form for a specific application.

preheat. A nonstandard term for **preheat temperature.**

preheat current (resistance welding). An impulse or series of impulses that occur prior to and are separated from the welding current.

preheat temperature. A specified temperature that the base metal must attain in the welding, brazing, soldering, thermal spraying, or cutting area immediately before these operations are performed.

pressure-controlled welding. A resistance welding process variation in which a number of spot or projection welds are made with several electrodes functioning progressively under the control of a pressure-sequencing device.

pressure gas welding (PGW). An oxyfuel gas welding process that produces coalescence simultaneously over the entire area of faying surfaces by heating them with gas flames obtained from the combustion of a fuel gas and oxygen and by the application of pressure, without the use of filler metal.

pretinning. A nonstandard term for **precoating.**

procedure qualification. The demonstration that welds made by a specific procedure can meet prescribed standards.

procedure qualification record (PQR). A document providing the actual welding variables used to produce an acceptable test weld and the results of tests conducted on the weld to qualify a welding procedure specification.

process. A grouping of basic operational elements used in welding, cutting, adhesive bonding, or thermal spraying. See also Figure 15, Master Chart of Welding and Allied Processes.

projection welding (PW). A resistance welding process that produces coalescence by the heat obtained from the resistance to the flow of the welding current. The resulting welds are localized at predetermined points by projections, embossments, or intersections.

protective atmosphere. A gas or vacuum envelope surrounding the workpieces used to prevent or facilitate removal of oxides and other detrimental surface substances.

puddle. A nonstandard term for **weld pool.**

pull gun technique. A nonstandard term for **backhand welding.**

pulse (resistance welding). A current of controlled duration of either polarity through the welding circuit.

R

random intermittent welds. Intermittent welds on one or both sides of a joint in which the weld increments are made without regard to spacing.

reaction flux (soldering). A flux composition in which one or more of the ingredients reacts with a base metal upon heating to deposit one or more metals.

reaction stress. A stress that cannot exist in a member if the member is isolated as a free body without connection to other parts of the structure.

reducing atmosphere. A chemically active protective atmosphere which at elevated temperature will reduce metal oxides to their metallic state.

reducing flame. A gas flame having a reducing effect due to excess fuel gas. See also **neutral flame** and **oxidizing flame**.

reflowing. A nonstandard term for **flow brightening**.

reflow soldering. A nonstandard term for a soldering process variation in which preplaced solder is melted to produce a soldered joint or coated surface.

residual stress. Stress present in a member that is free of external forces or thermal gradients.

resistance brazing (RB). A brazing process in which the heat required is obtained from the resistance to electric current flow in a circuit of which the workpiece is a part.

resistance butt welding. A nonstandard term for **flash welding** and **upset welding**.

resistance seam welding (RSEW). A resistance welding process that produces coalescence at the faying surfaces of overlapped parts progressively along a length of a joint. The weld may be made with overlapping weld nuggets, a continuous weld nugget, or by forging the joint as it is heated to the welding temperature by resistance to the flow of the welding current. See Figure A7(B).

resistance soldering (RS). A soldering process in which the heat required is obtained from the resistance to electric current flow in a circuit of which the workpiece is a part.

resistance spot welding (RSW). A resistance welding process that produces coalescence at the faying surfaces of a joint by the heat obtained from resistance to the flow of welding current through the workpieces from electrodes that serve to concentrate the welding current and pressure at the weld area. See Figure A7(C).

resistance welding (RW). A group of welding processes that produces coalescence of the faying surfaces with the heat obtained from resistance of the work to the flow of the welding current in a circuit of which the work is a part, and by the application of pressure.

resistance welding electrode. The part(s) of a resistance welding machine through which the welding current and, in most cases, force are applied directly to the work. The electrode may be in the form of a rotating wheel, rotating roll, bar, cylinder, plate, clamp, chuck, or modification thereof.

resistance welding gun. A manipulatable device to transfer current and provide electrode force to the weld area (usually in reference to a portable gun).

reverse polarity. A nonstandard term for **direct current electrode positive**.

roll welding (ROW). A solid-state welding process that produces coalescence of metals by heating and by applying sufficient pressure with rolls to cause deformation at the faying surfaces. See also **forge welding**.

root. A nonstandard term for **joint root** and **weld root**.

root bead. A weld that extends into or includes part or all of the joint root.

root edge. A root face of zero width. See also **root face**.

root face. That portion of the groove face adjacent to the joint root. See Figure A3.

root gap. A nonstandard term for **root opening**.

root opening. The separation at the joint root between the workpieces.

root penetration. The depth that a weld extends into the joint root. See Figure A13.

root radius. A nonstandard term for **groove radius**.

root reinforcement. Weld reinforcement opposite the side from which welding was done. See Figure A12(A).

root surface. The exposed surface of a weld opposite the side from which welding was done. See Figure A12(B).

S

scarf joint. A form of butt joint.

seal coat. Material applied to infiltrate the pores of a thermal spray deposit.

seal weld. Any weld designed primarily to provide a specific degree of tightness against leakage.

seam weld. A continuous weld made between or upon overlapping members, in which coalescence may start and occur on the faying surfaces, or may have proceeded from the outer surface of one member. The continuous weld may consist of a single weld bead or a series of overlapping spot welds. See Figure A7. See also **arc seam weld** and **resistance seam welding**.

secondary circuit. That portion of a welding machine that conducts the secondary current between the secondary terminals of the welding transformer and the electrodes, or electrode and workpiece.

self-fluxing alloys (thermal spraying). Surfacing materials that wet the substrate and coalesce when heated to their melting point, without the addition of a flux.

semiautomatic arc welding. Arc welding with equipment that controls only the filler metal feed. The advance of the welding is manually controlled.

series welding. A resistance welding secondary circuit variation in which the secondary current is conducted through the workpieces and electrodes or wheels in a series electrical path to simultaneously form multiple resistance spot, seam, or projection welds.

set down. A nonstandard term for **upset**.

shadow mask. A thermal spraying process variation in which an area is partially shielded during thermal spraying, thus permitting some overspray to produce a feathering at the coating edge.

sheet separation (resistance welding). The gap surrounding the weld between faying surfaces, after the joint has been welded in spot, seam, or projection welding.

shielded metal arc cutting (SMAC). A metal arc cutting process in which metals are severed by melting them with the heat of an arc between a covered metal electrode and the base metal.

shielded metal arc welding (SMAW). An arc welding process that produces coalescence of metals by heating them with an arc between a covered metal electrode and the workpieces. Shielding is obtained from decomposition of the electrode covering. Pressure is not used, and filler metal is obtained from the electrode.

shielding gas. Protective gas used to prevent atmospheric contamination.

short circuiting transfer (arc welding). Metal transfer in which molten metal from a consumable electrode is deposited during repeated short circuits. See also **globular transfer** and **spray transfer**.

shoulder. A nonstandard term for **root face**.

shrinkage stress. A nonstandard term for **residual stress**.

shrinkage void. A cavity type discontinuity normally formed by shrinkage during solidification.

silver alloy brazing. A nonstandard term for brazing with a silver-base filler metal.

silver soldering. A nonstandard term for brazing with a silver-base filler metal.

single-bevel-groove weld. A type of groove weld. See Figure A4(B).

single-flare-bevel-groove weld. A weld in a groove formed by a member with a curved surface in contact with a planar member. See Figure A4(G).

single-flare-V-groove weld. A weld in a groove formed by two members with curved surfaces. See Figure A4(H).

single impulse welding. A resistance welding process variation in which spot, projection, or upset welds are made with a single impulse.

single-J-groove weld. A type of groove weld. See Figure A4(E).

single-port nozzle. A constricting nozzle containing one orifice, located below and concentric with the electrode.

single-square-groove weld. A type of groove weld. See Figure A4(A).

single-U-groove weld. A type of groove weld. See Figure A4(F).

single-V-groove weld. A type of groove weld. See Figure A4(C).

single-welded joint. A fusion welded joint that is welded from one side only. See Figure A4.

skull. The unmelted residue from a liquated filler metal.

slag inclusion. Nonmetallic solid material entrapped in weld metal or between weld metal and base metal.

slot weld. A weld made in an elongated hole in one member of a joint fusing that member to another member. The hole may be open at one end. A fillet welded slot is not to be construed as conforming to this definition.

slugging. The act of adding a separate piece or pieces of material in a joint before or during welding that results in a welded joint not complying with design, drawing, or specification requirements.

soft solder. A nonstandard term for **solder**.

solder. A filler metal used in soldering that has a liquidus not exceeding 840°F (450°C).

soldering (S). A group of welding processes that produces coalescence of materials by heating them to the soldering temperature and by using a filler metal having a liquidus not exceeding 840°F (450°C) and below the solidus of the base metals. The filler metal is distributed between the closely fitted faying surfaces of the joint by capillary action.

soldering gun. An electrical soldering iron with a pistol grip and a quick heating, relatively small bit.

soldering iron. A soldering tool having an internally or externally heated metal bit usually made of copper.

solder interface. The interface between filler metal and base metal in a soldered joint.

solid-state welding (SSW). A group of welding processes that produces coalescence at temperatures essentially below the melting point of the base metal without the addition of a brazing filler metal. Pressure may or may not be used.

spacer strip. A metal strip or bar prepared for a groove weld and inserted in the joint root to serve as a backing and to maintain the root opening during welding. It can also bridge an exceptionally wide root opening due to poor fit.

spit. A nonstandard term for **flash**.

split pipe backing. Backing in the form of a pipe segment used for welding round bars.

spool. A filler metal package consisting of a continuous length of welding wire in coil form wound on a cylinder (called a barrel) which is flanged at both ends. The flange contains a spindle hole of smaller diameter than the inside diameter of the barrel.

spot weld. A weld made between or upon overlapping members in which coalescence may start and occur on the faying surfaces or may proceed from the outer surface of one member. The weld cross section (plan view) is approximately circular. See Figure A7. See also **arc spot weld** and **resistance spot welding**.

spray deposit. The coating or layer of surfacing material applied by a thermal spraying process.

spray deposit density ratio (thermal spraying). The ratio of the density of the spray deposit to the theoretical density of a surfacing material, usually expressed as percent of theoretical density.

spray transfer (arc welding). Metal transfer in which molten metal from a consumable electrode is propelled axially across the arc in small droplets. See also **globular transfer** and **short circuiting transfer**.

square-groove weld. A type of groove weld. See Figures A4(A) and A5(A).

stack cutting. Thermal cutting of stacked metal plates arranged so that all the plates are severed by a single cut.

staggered intermittent weld. An intermittent weld on both sides of a joint in which the weld increments on one side are alternated with respect to those on the other side.

standoff distance. The distance between a nozzle and the workpiece.

stick electrode. A nonstandard term for **covered electrode**.

stick electrode welding. A nonstandard term for **shielded metal arc welding**.

stickout. A nonstandard term for **electrode extension**.

stopoff. A material used on the surfaces adjacent to the joint to limit the spread of soldering or brazing filler metal.

straight polarity. A nonstandard term for **direct current electrode negative**.

stranded electrode. A composite filler metal electrode consisting of stranded wires that may mechanically enclose materials to improve properties, stabilize the arc, or provide shielding.

stress relief cracking. Intergranular cracking in the heat affected zone or weld metal that occurs during the exposure of weldments to elevated temperatures during postweld heat treatment or high temperature service.

stress relief heat treatment. Uniform heating of a structure or a portion thereof to a sufficient temperature to relieve the major portion of the residual stresses, followed by uniform cooling.

stringer bead. A type of weld bead made without appreciable weaving motion. See also **weave bead**.

stub. The short length of welding rod or consumable electrode that remains after its use for welding.

stud arc welding (SW). An arc welding process that produces coalescence of metals by heating them with an arc between a metal stud, or similar part, and the other workpiece. When the surfaces to be joined are properly heated, they are brought together under pressure. Partial shielding may be obtained by the use of a ceramic ferrule surrounding the stud. Shielding gas or flux may or may not be used.

stud welding. A general term for joining a metal stud or similar part to a workpiece. Welding may be accomplished by arc, resistance, friction, or other process with or without external gas shielding.

submerged arc welding (SAW). An arc welding process that produces coalescence of metals by heating them with an arc or arcs between a bare metal electrode or electrodes and the workpieces. The arc and molten metal are shielded by a blanket of granular, fusible material on the workpieces. Pressure is not used, and filler metal is obtained from the electrode and sometimes from a supplemental source (welding rod, flux, or metal granules).

substrate. Any material to which a thermal spray deposit is applied.

suck-back. A nonstandard term for a **concave root surface**.

surface expulsion. Expulsion occurring at an electrode-to-workpiece contact rather than at the faying surface.

surfacing. The application by welding, brazing, or thermal spraying of a layer or layers of material to a surface to obtain desired properties or dimensions, as opposed to making a joint.

surfacing material. The material that is applied to a base metal or substrate during surfacing.

surfacing metal. The metal that is applied to a base metal or substrate during surfacing. See also **surfacing material**.

surfacing weld. A weld applied to a surface, as opposed to making a joint, to obtain desired properties or dimensions.

sweat soldering. A soldering process variation in which two or more parts that have been precoated with solder are reheated and assembled into a joint without the use of additional solder.

synchronous timing (resistance welding). The initiation of each half cycle of welding transformer primary current on an accurately timed delay with respect to the polarity reversal of the power supply.

T

tacker. A nonstandard term for a tack welder.

tack weld. A weld made to hold parts of a weldment in proper alignment until the final welds are made.

theoretical throat. The distance from the beginning of the joint root perpendicular to the hypotenuse of the largest right triangle that can be inscribed within the cross section of a fillet weld. This dimension is based on the assumption that the root opening is equal to zero.

thermal cutting (TC). A group of cutting processes that melts the base metal. See also **arc cutting, electron beam cutting, laser beam cutting** and **oxygen cutting**.

thermal spraying (THSP). A group of processes in which finely divided metallic or nonmetallic surfacing materials are deposited in a molten or semimolten condition on a substrate to form a spray deposit. The surfacing material may be in the form of powder, rod, or wire. See also **arc spraying, flame spraying,** and **plasma spraying**.

thermal spraying gun. A device for heating, feeding, and directing the flow of a surfacing material.

thermal stress. Stress resulting from nonuniform temperature distribution.

thermit crucible. The vessel in which the thermit reaction takes place.

thermit mixture. A mixture of metal oxide and finely divided aluminum with the addition of alloying metals as required.

thermit mold. A mold formed around the workpieces to receive the molten metal.

thermit reaction. The chemical reaction between metal oxide and aluminum that produces superheated molten metal and a slag containing aluminum oxide.

thermit welding (TW). A welding process that produces coalescence of metals by heating them with superheated liquid metal from a chemical reaction between a metal oxide and aluminum, with or without the application of pressure. Filler metal is obtained from the liquid metal.

thermocompression bonding. A nonstandard term for hot pressure welding.

throat of a fillet weld. See actual throat, effective throat, and theoretical throat.

throat of a groove weld. A nonstandard term for groove weld size.

TIG welding. A nonstandard term for **gas tungsten arc welding.**

tinning. A nonstandard term for **precoating.**

T-joint. A joint between two members located approximately at right angles to each other in the form of a T. See Figure A1(C).

toe crack. See Figure A14(A).

torch brazing (TB). A brazing process in which the heat required is furnished by a fuel gas flame.

torch soldering (TS). A soldering process in which the heat required is furnished by a fuel gas flame.

torch tip. See **cutting tip** and **welding tip.**

transferred arc (plasma arc welding). A plasma arc established between the electrode and the workpiece.

transverse crack. A crack with its major axis oriented approximately perpendicular to the weld axis.

twin carbon arc brazing. A nonstandard term for carbon arc brazing.

U

U-groove weld. A type of groove weld. See Figures A4(F) and A5(E).

ultrasonic coupler (ultrasonic soldering and ultrasonic welding). Elements through which ultrasonic vibration is transmitted from the transducer to the tip.

ultrasonic soldering. A soldering process variation in which high frequency vibratory energy is transmitted through molten solder to remove undesirable surface films and thereby promote wetting of the base metal. This operation is usually accomplished without a flux.

ultrasonic welding (USW). A solid-state welding process that produces coalescence of materials by the local application of high frequency vibratory energy as the workpieces are held together under pressure.

underbead crack. A crack in the heat affected zone generally not extending to the surface of the base metal. See Figure A14(A).

undercut. A groove melted into the base metal adjacent to the weld toe or weld root and left unfilled by weld metal. See Figure A14(B)C.

underfill. A depression on the weld face or root surface extending below the adjacent surface of the base metal. See Figure A14(C).

upset. Bulk deformation resulting from the application of pressure in welding. The upset may be measured as a percent increase in interface area, a reduction in length, a percent reduction in lap joint thickness, or a reduction in cross wire weld stack height.

upset butt welding. A nonstandard term for **upset welding.**

upset distance. The total loss of axial length of the workpieces from the initial contact to the completion of the weld. In flash welding, the upset distance is equal to the platen movement from the end of flash time to the end of upset.

upset welding (UW). A resistance welding process that produces coalescence over the entire area of faying surfaces or progressively along a butt joint by the heat obtained from the resistance to the flow of welding current through the area where those surfaces are in contact. Pressure is used to complete the weld.

V

vacuum brazing. A nonstandard term for various brazing processes that take place in a chamber or retort below atmospheric pressure.

vertical position. The position of welding in which the weld axis is approximately vertical. See Figures A8(C) and A9(C).

vertical position (pipe welding). The position of a pipe joint in which welding is performed in the horizontal position and the pipe may or may not be rotated. See Figure A10(B).

V-groove weld. A type of groove weld. See Figures A4(C) and A5(C).

W

wave soldering (WS). An automatic soldering process where workpieces are passed through a wave of molten solder. See also **dip soldering.**

wax pattern (thermit welding). Wax molded around the workpieces to the form desired for the completed weld.

weave bead. A type of weld bead made with transverse oscillation.

weld. A localized coalescence of metals or nonmetals produced either by heating the materials to the welding temperature, with or without the application of pressure, or by the application of pressure alone, with or without the use of filler metal.

weldability. The capacity of a material to be welded under the imposed fabrication conditions into a specific, suitably designed structure and to perform satisfactorily in the intended service.

weld axis. A line through the length of a weld, perpendicular to and at the geometric center of its cross section. See Figures A8 and A9.

weld bead. A weld resulting from a pass. See also **stringer bead** and **weave bead.**

weld bonding. A resistance spot welding process variation in which the spot weld strength is augmented by adhesive at the faying surfaces.

weld brazing. A joining method that combines resistance welding with brazing.

weld crack. A crack located in the weld metal or heat affected zone.

welder. One who performs a manual or semiautomatic welding operation.

welder performance qualification. The demonstration of a welder's ability to produce welds meeting prescribed standards.

weld face. The exposed surface of a weld on the side from which welding was done. See Figures A12(A) and A12(E).

welding. A materials joining process used in making welds. See also Figure A15, Master Chart of Welding and Allied Processes.

welding blowpipe. A nonstandard term for **welding torch.**

welding current. The current in the welding circuit during the making of a weld.

welding cycle. The complete series of events involved in the making of a weld.

welding electrode. A component of the welding circuit through which current is conducted and that terminates at the arc, molten conductive slag, or base metal. See also **arc welding electrode, flux cored electrode, metal cored electrode, metal electrode, resistance welding electrode,** and **stranded electrode.**

welding ground. A nonstandard term for **workpiece connection.**

welding leads. The workpiece lead and electrode lead of an arc welding circuit.

welding machine. Equipment used to perform the welding operation: For example, spot welding machine, arc welding machine, and seam welding machine.

welding operator. One who operates machine or automatic welding equipment.

welding position. See **flat position, horizontal fixed position, horizontal position, horizontal rolled position, inclined position, overhead position,** and **vertical position.** See Figures A8 and A9.

welding procedure. The detailed methods and practices involved in the production of a weldment. See also **welding procedure specification.**

welding procedure specification (WPS). A document providing in detail the required variables for a specific application to assure repeatability by properly trained welders and welding operators.

welding rod. A form of welding filler metal, normally packaged in straight lengths, that does not conduct electrical current.

welding sequence. The order of making the welds in a weldment.

welding tip. That part of an oxyfuel gas welding torch from which the gases issue.

welding torch (arc). A device used in the gas tungsten and plasma arc welding processes to control the position of the electrode, to transfer current to the arc, and to direct the flow of shielding and plasma gas.

welding torch (oxyfuel gas). A device used in oxyfuel gas welding, torch brazing, and torch soldering for directing the heating flame produced by the controlled combustion of fuel gases.

welding wheel. A nonstandard term for **resistance welding electrode.**

welding wire. A form of welding filler metal, normally packaged as coils or spools, that may or may not conduct electrical current depending upon the welding process with which it is used. See also **welding electrode** and **welding rod.**

weld interface. The interface between weld metal and base metal in a fusion weld, between base metals in a solid-state weld without filler metal or between filler metal and base metal in a solid-state weld with filler metal, and in a braze.

weld line. A nonstandard term for **weld interface.**

weldment. An assembly whose component parts are joined by welding.

weld metal. That portion of a weld that has been melted during welding.

weldor. A nonstandard term for **welder.**

weld pass. A single progression of welding or surfacing along a joint or substrate. The result of a pass is a weld bead, layer, or spray deposit.

weld pass sequence. The order in which the weld passes are made.

weld penetration. A nonstandard term for **joint penetration** and **root penetration.**

weld pool. The localized volume of molten metal in a weld prior to its solidification as weld metal.

weld puddle. A nonstandard term for **weld pool.**

weld reinforcement. Weld metal in excess of the quantity required to fill a joint. See also **face reinforcement** and **root reinforcement.**

weld root. The points, as shown in cross section, at which the back of the weld intersects the base metal surfaces. See Figure A12(B).

weld size. See **edge weld size, fillet weld size, flange weld size,** and **groove weld size.**

weld tab. Additional material on which the weld may be initiated or terminated.

weld throat. See **actual throat, effective throat,** and **theoretical throat.**

weld toe. The junction of the weld face and the base metal. See Figure A12(E).

wetting. The phenomenon whereby a liquid filler metal or flux spreads and adheres in a thin continuous layer on a solid base metal.

wiped joint. A joint made with solder having a wide melting range and with the heat supplied by the molten solder poured onto the joint. The solder is manipulated with a hand-held cloth or paddle so as to obtain the required size and contour.

work connection. A nonstandard term for **workpiece connection.**

work lead. A nonstandard term for **workpiece lead.**

workpiece. The part being welded, brazed, soldered, or cut.

workpiece connection. The connection of the workpiece lead to the workpiece.

workpiece lead. The electrical conductor between the arc welding current source and the workpiece connection.

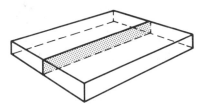

Applicable Welds

Bevel-groove	U-groove
Flare-bevel-groove	V-groove
Flare-V-groove	Edge-flange
J-groove	Braze
Square-groove	

(A) Butt Joint

Applicable Welds

Fillet	Corner-flange
Bevel-groove	Edge-flange
Flare-bevel-groove	Plug
Flare-V-groove	Slot
J-groove	Spot
Square-groove	Seam
U-groove	Projection
V-groove	Braze

(B) Corner Joint

Applicable Welds

Fillet	Slot
Bevel-groove	Spot
Flare-bevel-groove	Seam
J-groove	Projection
Square-groove	Braze
Plug	

(C) T-Joint

Applicable Welds

Fillet	Slot
Bevel-groove	Spot
Flare-bevel-groove	Seam
J-groove	Projection
Plug	Braze

(D) Lap Joint

Applicable Welds

Bevel-groove	V-groove
Flare-bevel-groove	Edge
Flare-V-groove	Corner-flange
J-groove	Edge-flange
Square-groove	Seam
U-groove	

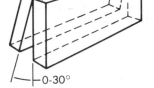

0-30°

(E) Edge Joint

Figure A1—Joint Types

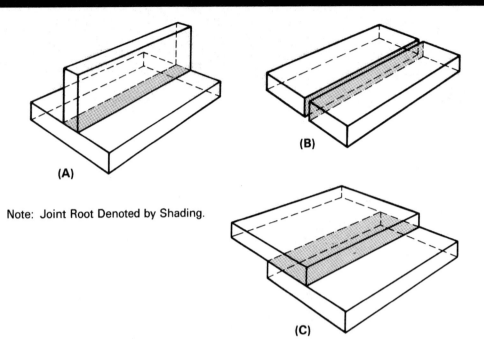

(A)

(B)

(C)

Note: Joint Root Denoted by Shading.

Figure A2—Joint Root

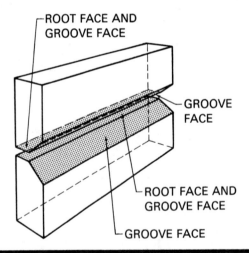

ROOT FACE AND
GROOVE FACE

GROOVE
FACE

ROOT FACE AND
GROOVE FACE

GROOVE FACE

Figure A3—Groove Face and Root Face

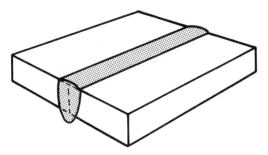

(A) Single-Square-Groove Weld

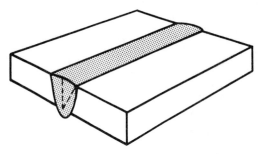

(B) Single-Bevel-Groove Weld

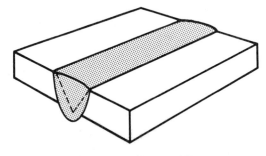

(C) Single-V-Groove Weld

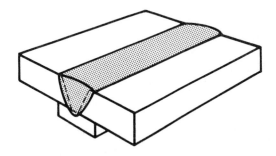

(D) Single-V-Groove Weld with Backing

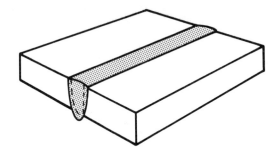

(E) Single-J-Groove Weld

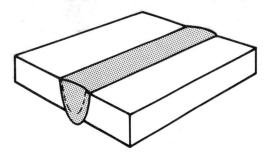

(F) Single-U-Groove Weld

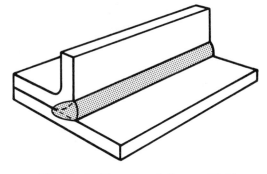

(G) Single-Flare-Bevel Groove Weld

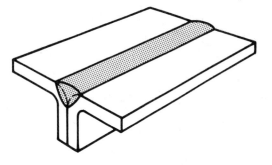

(H) Single-Flare-V-Groove Weld

Figure A4—Single-Groove Weld Joints

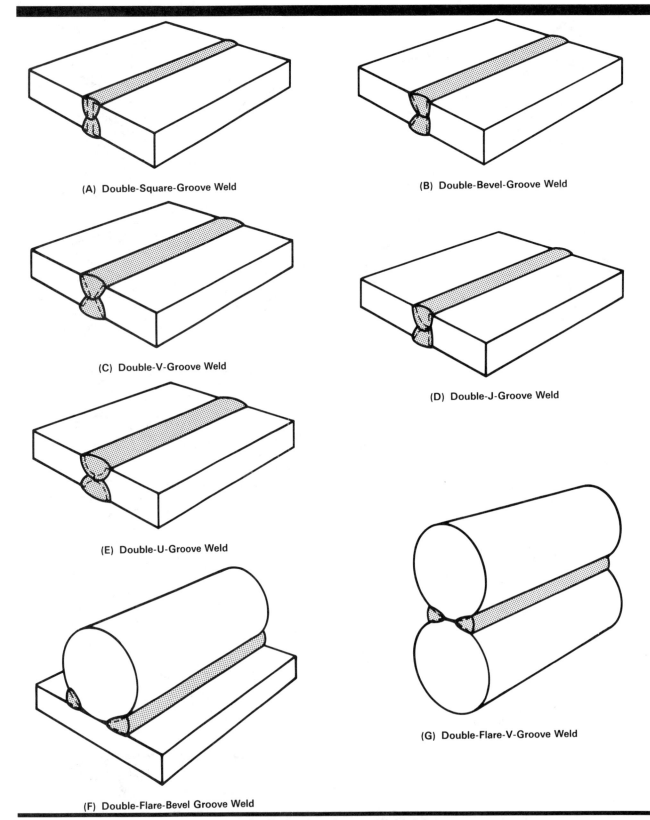

(A) Double-Square-Groove Weld

(B) Double-Bevel-Groove Weld

(C) Double-V-Groove Weld

(D) Double-J-Groove Weld

(E) Double-U-Groove Weld

(F) Double-Flare-Bevel Groove Weld

(G) Double-Flare-V-Groove Weld

Figure A5—Double-Groove Weld Joints

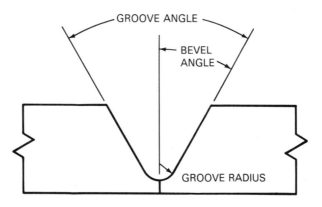

Figure A6—Bevel Angle, Groove Angle, Groove Radius

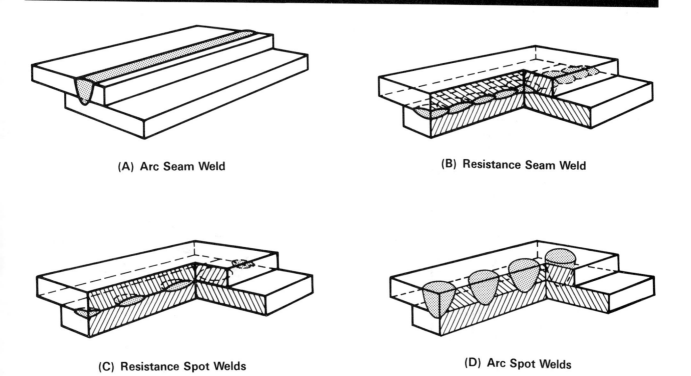

(A) Arc Seam Weld

(B) Resistance Seam Weld

(C) Resistance Spot Welds

(D) Arc Spot Welds

Figure A7—Seam Welds and Spot Welds

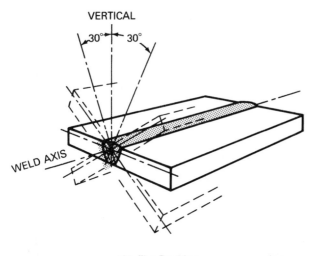

(A) Flat Position
1G Position

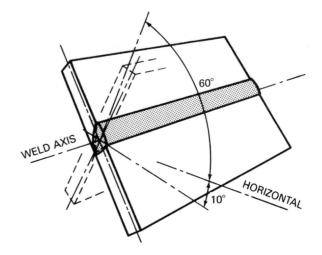

(B) Horizontal Position
2G Position

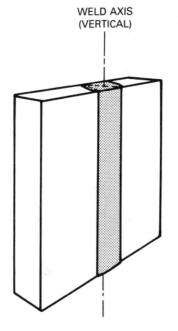

(C) Vertical Position
3G Position

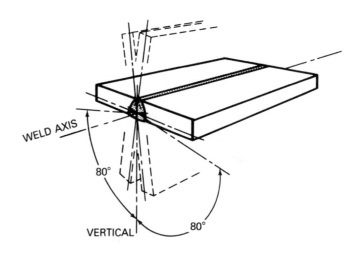

(D) Overhead Position
4G Position

Figure A8—Welding Position—Groove Welds

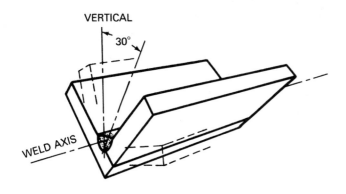

(A) Flat Position
1F Position

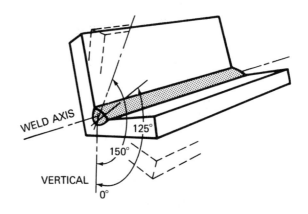

(B) Horizontal Position
2F Position

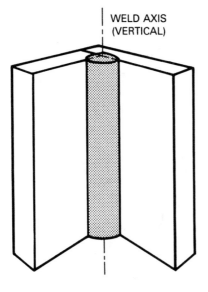

(C) Vertical Position
3F Position

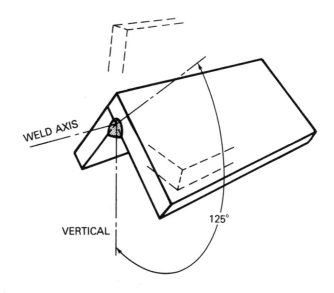

(D) Overhead Position
4F Position

Figure A9—Welding Position—Fillet Welds

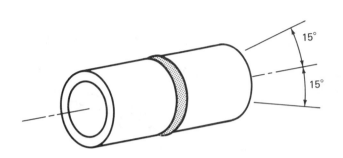

(A) Hoizontal Rolled Position
1G Position

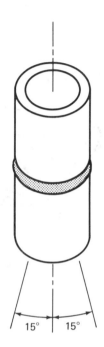

(B) Vertical Position
2G Position

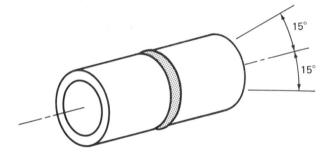

(C) Horizontal Position
5G Position

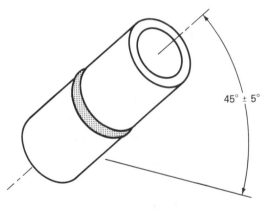

(D) Inclined Position
6G Position

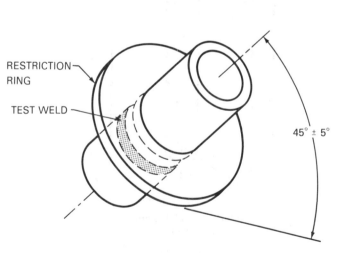

RESTRICTION RING

TEST WELD

(E) Inclined Position with Restriction Ring
6GR Position

Figure A10—Welding Position—Pipe Welding

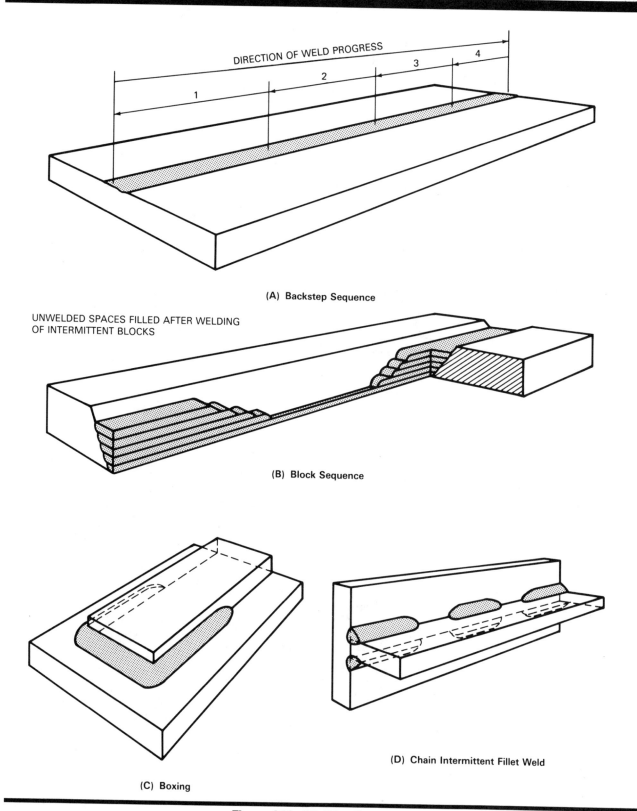

(A) Backstep Sequence

UNWELDED SPACES FILLED AFTER WELDING
OF INTERMITTENT BLOCKS

(B) Block Sequence

(C) Boxing

(D) Chain Intermittent Fillet Weld

Figure A11—Welding Sequence

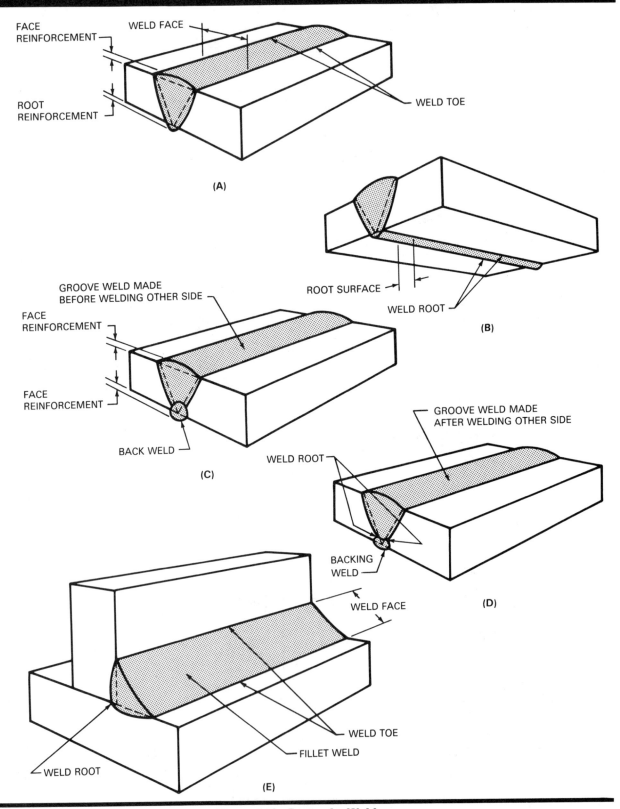

Figure A12—Parts of a Weld

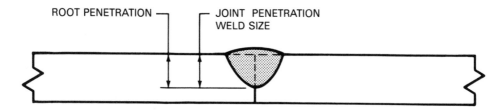

(A) Incomplete Joint Penetration or Partial Joint Penetration

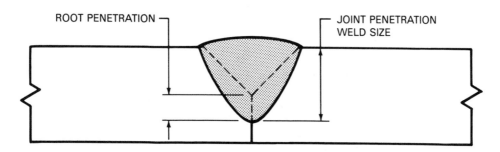

(B) Incomplete Joint Penetration or Partial Joint Penetration

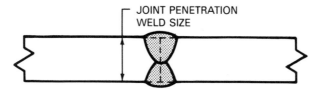

(C) Complete Joint Penetration

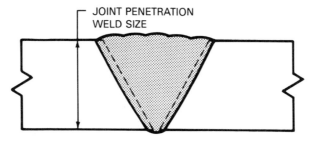

(D) Complete Joint Penetration

Figure A13—Joint Penetration

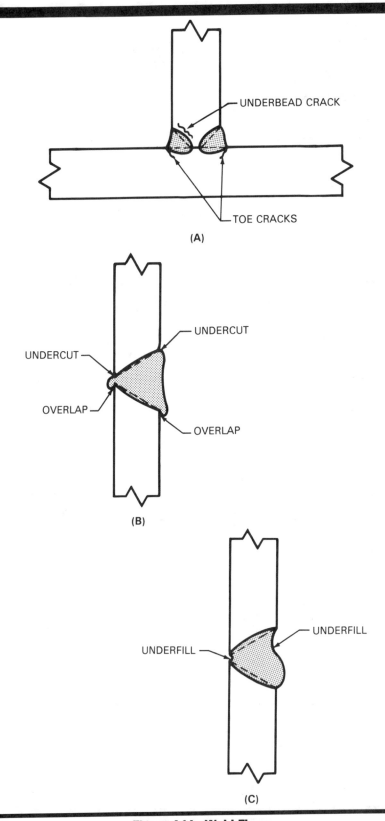

Figure A14—Weld Flaws

MASTER CHART OF WELDING AND ALLIED PROCESSES

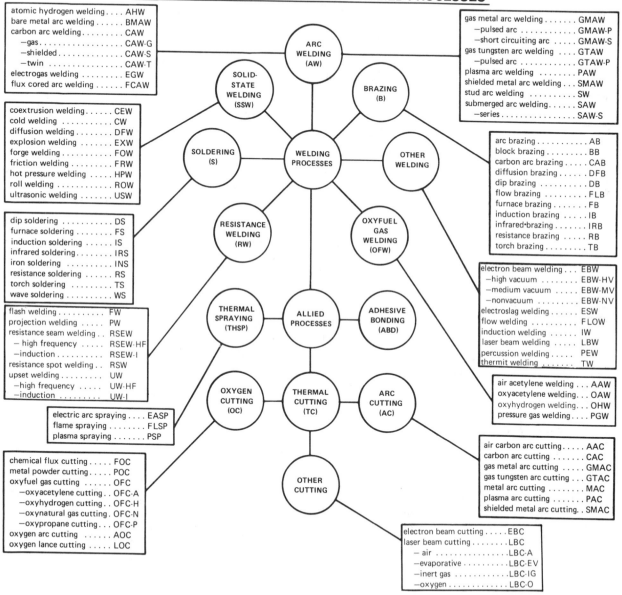

atomic hydrogen welding	AHW
bare metal arc welding	BMAW
carbon arc welding	CAW
—gas	CAW-G
—shielded	CAW-S
—twin	CAW-T
electrogas welding	EGW
flux cored arc welding	FCAW

coextrusion welding	CEW
cold welding	CW
diffusion welding	DFW
explosion welding	EXW
forge welding	FOW
friction welding	FRW
hot pressure welding	HPW
roll welding	ROW
ultrasonic welding	USW

dip soldering	DS
furnace soldering	FS
induction soldering	IS
infrared soldering	IRS
iron soldering	INS
resistance soldering	RS
torch soldering	TS
wave soldering	WS

flash welding	FW
projection welding	PW
resistance seam welding	RSEW
— high frequency	RSEW-HF
—induction	RSEW-I
resistance spot welding	RSW
upset welding	UW
—high frequency	UW-HF
—induction	UW-I

electric arc spraying	EASP
flame spraying	FLSP
plasma spraying	PSP

chemical flux cutting	FOC
metal powder cutting	POC
oxyfuel gas cutting	OFC
—oxyacetylene cutting	OFC-A
—oxyhydrogen cutting	OFC-H
—oxynatural gas cutting	OFC-N
—oxypropane cutting	OFC-P
oxygen arc cutting	AOC
oxygen lance cutting	LOC

gas metal arc welding	GMAW
—pulsed arc	GMAW-P
—short circuiting arc	GMAW-S
gas tungsten arc welding	GTAW
—pulsed arc	GTAW-P
plasma arc welding	PAW
shielded metal arc welding	SMAW
stud arc welding	SW
submerged arc welding	SAW
—series	SAW-S

arc brazing	AB
block brazing	BB
carbon arc brazing	CAB
diffusion brazing	DFB
dip brazing	DB
flow brazing	FLB
furnace brazing	FB
induction brazing	IB
infrared brazing	IRB
resistance brazing	RB
torch brazing	TB

electron beam welding	EBW
—high vacuum	EBW-HV
—medium vacuum	EBW-MV
—nonvacuum	EBW-NV
electroslag welding	ESW
flow welding	FLOW
induction welding	IW
laser beam welding	LBW
percussion welding	PEW
thermit welding	TW

air acetylene welding	AAW
oxyacetylene welding	OAW
oxyhydrogen welding	OHW
pressure gas welding	PGW

air carbon arc cutting	AAC
carbon arc cutting	CAC
gas metal arc cutting	GMAC
gas tungsten arc cutting	GTAC
metal arc cutting	MAC
plasma arc cutting	PAC
shielded metal arc cutting	SMAC

electron beam cutting	EBC
laser beam cutting	LBC
— air	LBC-A
—evaporative	LBC-EV
—inert gas	LBC-IG
—oxygen	LBC-O

Figure A15—Master Chart of Welding and Allied Processes

Table A1
Designation of Welding and Allied Processes by Letters

Welding and Allied Processes	Letter Designation	Welding and Allied Processes	Letter Designation
adhesive bonding	ABD	induction	RSEW-I
arc welding	AW	resistance spot welding	RSW
atomic hydrogen welding	AHW	upset welding	UW
bare metal arc welding	BMAW	high frequency	UW-HF
carbon arc welding	CAW	induction	UW-I
gas	CAW-G	soldering	S
shielded	CAW-S	dip soldering	DS
twin	CAW-T	furnace soldering	FS
electrogas welding	EGW	induction soldering	IS
flux cored arc welding	FCAW	infrared soldering	IRS
gas metal arc welding	GMAW	iron soldering	INS
pulsed arc	GMAW-P	resistance soldering	RS
short circuiting arc	GMAW-S	torch soldering	TS
gas tungsten arc welding	GTAW	wave soldering	WS
pulsed arc	GTAW-P	solid-state welding	SSW
plasma arc welding	PAW	coextrusion welding	CEW
shielded metal arc welding	SMAW	cold welding	CW
stud arc welding	SW	diffusion welding	DFW
submerged arc welding	SAW	explosion welding	EXW
series	SAW-S	forge welding	FOW
brazing	B	friction welding	FRW
arc brazing	AB	hot pressure welding	HPW
block brazing	BB	roll welding	ROW
carbon arc brazing	CAB	ultrasonic welding	USW
diffusion brazing	DFB	thermal cutting	TC
dip brazing	DB	arc cutting	AC
flow brazing	FLB	air carbon arc cutting	AAC
furnace brazing	FB	carbon arc cutting	CAC
induction brazing	IB	gas metal arc cutting	GMAC
infrared brazing	IRB	gas tungsten arc cutting	GTAC
resistance brazing	RB	metal arc cutting	MAC
torch brazing	TB	plasma arc cutting	PAC
other welding processes		shielded metal arc cutting	SMAC
electron beam welding	EBW	electron beam cutting	EBC
high vacuum	EBW-HV	laser beam cutting	LBC
medium vacuum	EBW-MV	air	LBC-A
nonvacuum	EBW-NV	evaporative	LBC-EV
electroslag welding	ESW	inert gas	LBC-IG
flow welding	FLOW	oxygen	LBC-O
induction welding	IW	oxygen cutting	OC
laser beam welding	LBW	chemical flux cutting	FOC
percussion welding	PEW	metal powder cutting	POC
thermit welding	TW	oxyfuel gas cutting	OFC
oxyfuel gas welding	OFW	oxyacetylene cutting	OFC-A
air acetylene welding	AAW	oxyhydrogen cutting	OFC-H
oxyacetylene welding	OAW	oxynatural gas cutting	OFC-N
oxyhydrogen welding	OHW	oxypropane cutting	OFC-P
pressure gas welding	PGW	oxygen arc cutting	AOC
resistance welding	RW	oxygen lance cutting	LOC
flash welding	FW	thermal spraying	THSP
projection welding	PW	arc spraying	ASP
resistance seam welding	RSEW	flame spraying	FLSP
high frequency	RSEW-HF	plasma spraying	PSP

Table A2
Alphabetical Cross Reference to Table A1

Letter Designation	Welding and Allied Processes	Letter Designation	Welding and Allied Processes
AAC	air carbon arc cutting	IRB	infrared brazing
AAW	air acetylene welding	IRS	infrared soldering
ABD	adhesive bonding	IS	induction soldering
AB	arc brazing	IW	induction welding
AC	arc cutting	LBC	laser beam cutting
AHW	atomic hydrogen welding	LBC-A	laser beam cutting-air
AOC	oxygen arc cutting	LBC-EV	laser beam cutting-evaporative
ASP	arc spraying	LBC-IG	laser beam cutting-inert gas
AW	arc welding	LBC-O	laser beam cutting-oxygen
B	brazing	LBW	laser beam welding
BB	block brazing	LOC	oxygen lance cutting
BMAW	bare metal arc welding	MAC	metal arc cutting
CAB	carbon arc brazing	OAW	oxyacetylene welding
CAC	carbon arc cutting	OC	oxygen cutting
CAW	carbon arc welding	OFC	oxyfuel gas cutting
CAW-G	gas carbon arc welding	OFC-A	oxyacetylene cutting
CAW-S	shielded carbon arc welding	OFC-H	oxyhydrogen cutting
CAW-T	twin carbon arc welding	OFC-N	oxynatural gas cutting
CEW	coextrusion welding	OFC-P	oxypropane cutting
CW	cold welding	OFW	oxyfuel gas welding
DB	dip brazing	OHW	oxyhydrogen welding
DFB	diffusion brazing	PAC	plasma arc cutting
DFW	diffusion welding	PAW	plasma arc welding
DS	dip soldering	PEW	percussion welding
EBC	electron beam cutting	PGW	pressure gas welding
EBW	electron beam welding	POC	metal powder cutting
EBW-HV	electron beam welding-high vacuum	PSP	plasma spraying
EBW-MV	electron beam welding-medium vacuum	PW	projection welding
EBW-NV	electron beam welding-nonvacuum	RB	resistance brazing
EGW	electrogas welding	RSEW	resistance seam welding
ESW	electroslag welding	RSEW-HF	resistance seam welding-high frequency
EXW	explosion welding	RSEW-I	resistance seam welding-induction
FB	furnace brazing	RSW	resistance spot welding
FCAW	flux cored arc welding	ROW	roll welding
FLB	flow brazing	RW	resistance welding
FLOW	flow welding	S	soldering
FLSP	flame spraying	SAW	submerged arc welding
FOC	chemical flux cutting	SAW-S	series submerged arc welding
FOW	forge welding	SMAC	shielded metal arc cutting
FRW	friction welding	SMAW	shielded metal arc welding
FS	furnace soldering	SSW	solid-state welding
FW	flash welding	SW	stud arc welding
GMAC	gas metal arc cutting	TB	torch brazing
GMAW	gas metal arc welding	TC	thermal cutting
GMAW-P	gas metal arc welding—pulsed arc	THSP	thermal spraying
GMAW-S	gas metal arc welding—short circuiting arc	TS	torch soldering
GTAC	gas tungsten arc cutting	TW	thermit welding
GTAW	gas tungsten arc welding	USW	ultrasonic welding
GTAW-P	gas tungsten arc welding-pulsed arc	UW	upset welding
HPW	hot pressure welding	UW-HF	upset welding-high frequency
IB	induction brazing	UW-I	upset welding-induction
INS	iron soldering	WS	wave soldering

METRIC PRACTICE GUIDE

PREPARED BY A COMMITTEE CONSISTING OF:

E. D. Brandon, Chairman
Sandia National Laboratories

WELDING HANDBOOK COMMITTEE MEMBER:
J. R. Hannahs
Midmark

APPENDIX B

METRIC PRACTICE GUIDE

INTRODUCTION

THIS APPENDIX WAS prepared to guide Handbook readers in the use of metric units for welding. Acquiring familiarity with these units is essential in light of their expanding use by American industry as a replacement for U.S. customary units. In fact, measurement practices all over the world are currently in various stages of transition—a condition that is expected to continue during the next several years. This transition includes a change from both the U.S. customary units and the noncoherent metric units to the International System of Units, officially abbreviated SI, from the French "Système Internationale d'Unités."

This appendix is an extraction from the AWS document, A1.1, *Metric Practice Guide for the Welding Industry*, prepared by the AWS Committee on Metric Practice. For more complete coverage of the SI system and methods of handling the transition period, Handbook readers should refer to A1.1, latest edition.

1. SCOPE

This metric practice guide defines the International System of Units (SI). It contains the base units, supplementary units, derived SI units, and rules for their use in AWS documents and by the welding industry. It also contains factors and rules for converting from U.S. customary units to SI units and recommendations to industry for managing the transition.

2. FEATURES OF SI

SI is a modernized metric system of measurement. It was formally established in 1960 as the International System of Units and is officially recognized by all industrial nations. It has features that make it superior to the U.S. customary system and to other metric systems. These features are the following:

2.1 Absolute Base. An absolute base is one in which force is not defined by action of gravity. An absolute sys-

tem has several advantages, the greatest being simplicity of calculation. In SI units, a force of one newton gives a mass of one kilogram an acceleration of one meter per second squared. In contrast, in European metric practice, a force of one kilogram force (gravitational system) accelerates the same mass 9.806 650 meters per second squared, whereas in U.S. customary units, one pound force accelerates a mass of one slug 32.174 feet per second squared.

2.2 Coherence. Coherence is the characteristic which relates any derived unit to any other, or to the base units from which it is formed, without the use of conversion factors.

In SI units, a force of one newton applied through a distance of one meter does work equivalent to one joule, which equals the work produced by one watt of power in one second.

In U.S. customary units, a force of one pound applied through a distance of one inch produces energy equivalent to 0.000 107 Btu, which is the same as produced by one horsepower in 0.000 505 hours.

2.3 Unique Units. Another desirable characteristic of SI is its use of only one unit for each physical quantity. The SI units for force, energy, and power are the same regardless of whether the process is mechanical, electrical, or thermal. Power, whether in engines or air conditioners, is measured in watts. By contrast, the customary system has 9 commonly used units for area, 25 units for energy, 26 units for length, and so on.

2.4 Decimal System. SI is a decimal system, and thus easier to use because it is easier to work in multiples of ten and in decimal notation than in the fractions and decimalized fraction equivalents common to the customary system.

The above combination of features makes SI an excellent system suitable for all kinds of measurements. Though there remain areas that can and no doubt will be improved, the SI System is practical for universal application and is rapidly becoming the commonly used world measurement system.

3. SI UNITS AND SYMBOLS

SI consists of seven base units, two supplementary units, a series of derived units consistent with the base and supplementary units, and a series of prefixes for the formation of multiples of the various units.

3.1 Base Units. SI base units and their symbols are shown in Table B.1. The National Bureau of Standards defines the SI base units as follows:

Table B.1
SI Base Units and Symbols

Quantity	Unit	Symbol
length	meter	m
mass	kilogram	kg
time	second	s
electric current	ampere	A
thermodynamic temperature	kelvin	K
luminous intensity	candela	cd
amount of substance	mole	mol

meter.[1] The length of the path traveled by light in vacuum during a time interval of 1/299 792 458 of a second.

---

1. The traditional American spelling of the words "meter" and "liter" are used in this volume. In Europe these words are spelled "metre" and "litre", and therefore the reader is likely to encounter both spellings in the literature.

kilogram. The mass equal to the mass of the international prototype of the kilogram.

second. The duration equal to 9192 631 770 periods of the radiation corresponding to the transition between the two hyperfine levels of the ground state of the cesium-133 atom.

ampere. That constant current which, if maintained in two straight, parallel conductors of infinite length, of negligible cross section, and placed one meter apart in a vacuum, would produce between these conductors a force equal to 2×10^{-7} newtons per meter of length.

kelvin. The thermodynamic temperature that is the fraction 1/273.16 of the thermodynamic temperature of the triple point of water.

candela. The luminous intensity, in a given direction, of a source that emits monochromatic radiation of frequency 540×10^{12} hertz and that has a radiant intensity in that direction of (1/683) watt per steradian.

mole. The amount of substance of a system which contains as many elementary entities as there are atoms in 0.012 kilogram of carbon-12 (when the mole is used, the elementary entities must be specified and may be atoms, molecules, ions, electrons, other particles, or specified groups of such particles).

3.2 Supplementary Units. These units consist of two purely geometric units, the radian and the steradian, as follows:

radian. The plane angle between two radii of a circle that cut off on the circumference an arc equal in length to the radius. Symbol: rad

steradian. The solid angle which, having its vertex in the center of a sphere, cuts off an area of the surface of a sphere equal to that of a square with sides of length equal to the radius of the sphere. Symbol: sr

3.3 Derived Units. Any additional quantity can be measured using combinations of the base and supplementary units just given. Some examples of derived units are shown in Table B.2.

3.4 Prefixes. SI prefixes should be used to indicate orders of magnitude, thus simplifying numeric terms and providing a convenient substitute for writing powers of ten as generally preferred in computation. For example, 16 800 meters or 16.8×10^3 meters becomes 16.8 kilometers. Correct usage of prefixes is discussed in paragraph 7.1. A summary of prefixes, factors, and symbols is shown in Table B.3.

Table B.2
Formulas for SI Derived Units

Quantity	Unit	Symbol	Formula
acceleration-linear	meter per second squared		m/s^2
-angular	radian per second squared		rad/s^2
area	square meter		m^2
capacitance	farad	F	$A \cdot s/V$
conductivity (thermal)	watt per meter kelvin		$W/(m \cdot K)$
density	kilogram per meter cubed		kg/m^3
electromotive force	volt	V	W/A
electricity (quantity)	coulomb	C	$A \cdot s$
energy, work, heat, and impact strength	joule	J	$N \cdot m$
field strength, electric	volt per meter		V/m
force	newton	N	$kg \cdot m/s^2$
frequency	hertz	Hz	s^{-1}
inductance	henry	H	$V \cdot s/A$
luminous flux	lumen	lm	$cd \cdot sr$
illumination	lux	lx	lm/m^2
magnetic flux	weber	Wb	$V \cdot s$
magnetic flux density, magnetic induction	tesla	T	Wb/m^2
power	watt	W	J/s
pressure, stress	pascal	Pa	N/m^2
resistance	ohm	Ω	V/A
velocity-linear	meter per second		m/s
-angular	radian per second		rad/s
volume	cubic meter		m^3

Table B.3
Summary of Prefixes for SI Units

Exponential Expression	Multiplication Factor	Prefix	Symbol
10^{18}	1 000 000 000 000 000 000	exa	E
10^{15}	1 000 000 000 000 000	peta	P
10^{12}	1 000 000 000 000	tera	T
10^9	1 000 000 000	giga	G
10^6	1 000 000	mega	M
10^3	1 000	kilo	K
10^2	100	hecto*	h
10	10	deka*	da
10^{-1}	0.1	deci*	d
10^{-2}	0.01	centi*	c
10^{-3}	0.001	milli	m
10^{-6}	0.000 001	micro	μ
10^{-9}	0.000 000 001	nano	n
10^{-12}	0.000 000 000 001	pico	p
10^{-15}	0.000 000 000 000 001	femto	f
10^{-18}	0.000 000 000 000 000 001	atto	a

*Nonpreferred. Prefixes should be selected in steps of 10^3 so that the resultant number before the prefix is between 0.1 and 1000 (see 7.1). These prefixes should not be used for units of linear measurement, but may be used for higher order units. For example, the linear measurement, decimeter, is nonpreferred, but square decimeter is acceptable.

4. OTHER UNITS USED WITH SI

There are certain units which, although not part of SI, are in widespread use and are acceptable for use with SI. These units are shown in Table B.4.

Table B.4
Summary of NONSTANDARD SI Units
in Common Usage

Unit	Symbol	Value in SI units
minute	min	1 min = 60 s
hour	h	1 h = 60 min = 3600 s
day	d	1 d = 24 h = 1440 min = 86 400 s
degree (angular)	°	1° = (π/180) rad = 0.0175 rad
bar	bar	1 bar = 0.1 MPa = 10^5 Pa
liter	L	1 L = 0.001 m^3 = 1 dm^3 = 1000 cm^3
degree Celsius	°C	1 °C = 1K (incremental) = K − 273.15
angstrom	Å	1 Å = 0.1 nm = 10^{-10}m

5. UNITS PERTAINING TO WELDING

The recommended SI units to be used in welding nomenclature are shown in Table B.5. The selection of these terms was based on the use of (1) SI base units where practicable, (2) numbers of reasonable size, and (3) accepted units currently in use or anticipated to be used.

Factors for converting from the commonly used U.S. customary units to these recommended units are given in 6.2.

6. CONVERSIONS

6.1 General. Those units used in all branches of engineering with the appropriate conversion factors are shown in Table B.6.

6.2 Special Conversions for Welding. Terms that are in common usage in the welding industry with their conversions are shown in Table B.7.

6.3 SI Equivalents for Electrode and Fillet Sizes. The standard diameters of welding electrodes and their approximate SI equivalents are shown in Table B.8. Also shown in Table B.8 are approximate equivalents of fillet weld sizes for drawings and specifications. *These values are for conversion only and are not intended for new designs where a more rational series for sizing may be used.*

6.4 Rules for Conversion and Rounding. Conversion and rounding are necessary only during a transition period. In manufacturing practice, this occurs most frequently when designing is done in SI units and fabrication is done in conventional units. In this case, the conversion is from SI to U.S. customary units. The necessity for conversion and rounding disappears when all steps can be done in one system.

Exact conversion from one system to another usually results in numbers which are inconvenient to use. Also, the intended precision is exaggerated when the conversion results in more decimal places than are necessary.

The degree of accuracy of the converted number is based on the intended or necessary precision of the product. The precision should be determined by the designer or user. The guidelines given herein may then be applied to arrive at appropriate numerical equivalents.

6.4.1 Inch-Millimeter Conversion. Exact conversion from inches to millimeters often results in unnecessarily long decimal numbers. Showing more decimal places than necessary leads to misinterpretation, uses valuable space, and increases the possibility of error. The numbers should be rounded to eliminate insignificant decimal places, consistent with the accuracy required. The rounding of equivalent millimeter dimensions should be handled as described here.

6.4.1.1 Nominal Dimensions. The closest practical indication of equivalent inch and millimeter values occurs when the millimeter value is shown to one less decimal place than its inch equivalent. For example, 0.365 in. equals 9.27 mm. However, fractional inch conversions may exaggerate the intended precision. For example, the rule may not be applicable when changing 1 7/8 in. to 47.63 mm unless the precision of 1 7/8 in. was 1.875 inches. Some dimensions must be converted more accurately to ensure interchangeability of parts. The methods described in 6.4.3 will accomplish this requirement.

6.4.1.2 Tolerances. The following round off criteria should be used when it is necessary to ensure the physical and functional interchangeability of parts made and inspected using either system of measurements, and when inch dimensions are converted to millimeter equivalents and shown on dual dimensioned drawings:

(1) **Basic and Maximum-Minimum Dimensions.** Basic dimensions are inherently precise and should be converted exactly. When the function of a feature requires that the maximum and minimum limits in millimeters be within the inch limits, maximum limits are rounded down and minimum limits are rounded up.

(2) **Dimensions Without Tolerance.** Untoleranced dimensions are converted to exact millimeter

Table B.5
Units Pertaining to Welding

Property	Unit	Symbol
area dimensions	square millimeter	mm^2
current density	ampere per square millimeter	A/mm^2
deposition rate	kilogram per hour	kg/h
electrical resistivity	ohm meter	$\Omega \cdot m$
electrode force (upset, squeeze, hold)	newton	N
flow rate (gas and liquid)	liter per minute	L/min
fracture toughness	meganewton meter $^{-3/2}$	$MN \cdot m^{-3/2}$
impact strength	joule	$J = N \cdot m$
linear dimensions	millimeter	mm
power density	watt per square meter	W/m^2
pressure (gas and liquid)	kilopascal	$kPa = 1000\ N/m^2$
tensile strength	megapascal	$MPa = 1\,000\,000\ N/m^2$
thermal conductivity	watt per meter kelvin	$W/(m \cdot K)$
travel speed	millimeter per second	mm/s
volume dimensions	cubic millimeter	mm^3
wire feed rate	millimeter per second	mm/s

Table B.6
Conversion Factors for Common Engineering Terms

Property	To Convert From	To	Multiply By
acceleration (angular)	revolution per minute squared	rad/s^2	$1.745\ 329 \times 10^{-3}$
acceleration (linear)	in./min^2	m/s^2	$7.055\ 556 \times 10^{-6}$
	ft/min^2	m/s^2	$8.466\ 667 \times 10^{-5}$
	in./min^2	mm/s^2	$7.055\ 556 \times 10^{-3}$
	ft/min^2	mm/s^2	$8.466\ 667 \times 10^{-2}$
	ft/s^2	m/s^2	$3.048\ 000 \times 10^{-1}$
angle, plane	degree	rad	$1.745\ 329 \times 10^{-2}$
	minute	rad	$2.908\ 882 \times 10^{-4}$
	second	rad	$4.848\ 137 \times 10^{-6}$
area	in.^{2}	m^2	$6.451\ 600 \times 10^{-4}$
	ft^2	m^2	$9.290\ 304 \times 10^{-2}$
	yd^2	m^2	$8.361\ 274 \times 10^{-1}$
	in.^{2}	mm^2	$6.451\ 600 \times 10^{2}$
	ft^2	mm^2	$9.290\ 304 \times 10^{4}$
	acre (U.S. Survey)	m^2	$4.046\ 873 \times 10^{3}$
density	pound mass per cubic inch	kg/m^3	$2.767\ 990 \times 10^{4}$
	pound mass per cubic foot	kg/m^3	$1.601\ 846 \times 10$
energy, work, heat, and impact energy	foot pound force	J	$1.355\ 818$
	found poundal	J	$4.214\ 011 \times 10^{-2}$
	Btu*	J	$1.054\ 350 \times 10^{3}$
	calorie*	J	$4.184\ 000$
	watt hour	J	$3.600\ 000 \times 10^{3}$

(continued)

Table B.6 (Continued)

Property	To Convert From	To	Multiply By
force	kilogram-force	N	9.806 650
	pound-force	N	4.448 222
impact strength	(see energy)		
length	in.	m	$2.540\ 000 \times 10^{-2}$
	ft	m	$3.048\ 000 \times 10^{-1}$
	yd	m	$9.144\ 000 \times 10^{-1}$
	rod (U.S. Survey)	m	5.029 210
	mile (U.S. Survey)	km	1.609 347
mass	pound mass (avdp)		
	metric ton	kg	$4.535\ 924 \times 10^{-1}$
	ton (short, 2000 lbm)	kg	$1.000\ 000 \times 10^{3}$
		kg	$9.071\ 847 \times 10^{2}$
	slug	kg	$1.459\ 390 \times 10$
power	horsepower (550 ft lbf/s)	W	$7.456\ 999 \times 10^{2}$
	horsepower (electric)	W	$7.460\ 000 \times 10^{2}$
	Btu/min*	W	$1.757\ 250 \times 10$
	calorie per minute*	W	$6.973\ 333 \times 10^{-2}$
	foot pound-force per minute	W	$2.259\ 697 \times 10^{-2}$
pressure	pound force per square inch	kPa	6.894 757
	bar	kPa	$1.000\ 000 \times 10^{2}$
	atmosphere	kPa	$1.013\ 250 \times 10^{2}$
	kip/in.^{2}	kPa	$6.894\ 757 \times 10^{3}$
temperature	degree Celsius, $t_{°C}$	K	$t_K = t_{°C} + 273.15$
	degree Fahrenheit, $t_{°F}$	K	$t_K = (t_{°F} + 459.67)/1.8$
	degree Rankine, $t_{°R}$	K	$t_K = t_{°R}/1.8$
	degree Fahrenheit, t_F	°C	$t_{°C} = (t_F - 32)/1.8$
	kelvin, t_K	°C	$t_{°C} = t_K - 273.15$
tensile strength (stress)	ksi	MPa	6.894 757
torque	inch pound force	N·m	$1.129\ 848 \times 10^{-1}$
	foot pound force	N·m	1.355 818
velocity (angular)	revolution per minute	rad/s	$1.047\ 198 \times 10^{-1}$
	degree per minute	rad/s	$2.908\ 882 \times 10^{-4}$
	revolution per minute	deg/min	$3.600\ 000 \times 10^{2}$
velocity (linear)	in./min	m/s	$4.233\ 333 \times 10^{-4}$
	ft/min	m/s	$5.080\ 000 \times 10^{-3}$
	in./min	mm/s	$4.233\ 333 \times 10^{-1}$
	ft/min	mm/s	5.080 000
	mile/hour	km/h	1.609 344
volume	in.^{3}	m^3	$1.638\ 706 \times 10^{-5}$
	ft^3	m^3	$2.831\ 685 \times 10^{-2}$
	yd^3	m^3	$7.645\ 549 \times 10^{-1}$
	in.^{3}	mm^3	$1.638\ 706 \times 10^{4}$
	ft^3	mm^3	$2.831\ 685 \times 10^{7}$
	in.^{3}	L	$1.638\ 706 \times 10^{-2}$
	ft^3	L	$2.831\ 685 \times 10$
	gallon	L	3.785 412

*thermochemical

Table B.7
Conversions for Common Welding Terms*

Property	To Convert From	To	Multiply By
area dimensions	in.^{2}	mm^2	6.451 600 $\times$ 10^2
(mm^2)	mm^2	in.^{2}	1.550 003 $\times$ 10^{-3}
current density	A/in.^{2}	A/mm^2	1.550 003 $\times$ 10^{-3}
(A/mm^2)	A/mm^2	A/in.^{2}	6.451 600 $\times$ 10^2
deposition rate**	lb/h	kg/h	0.45**
(kg/h)	kg/h	lb/h	2.2**
electrical resistivity	$\Omega \cdot$ cm	$\Omega \cdot$ m	1.000 000 $\times$ 10^{-2}
($\Omega \cdot$ m)	$\Omega \cdot$ m	$\Omega \cdot$ cm	1.000 000 $\times$ 10^2
electrode force	pound-force	N	4.448 222
(N)	kilogram-force	N	9.806 650
	N	lbf	2.248 089 $\times$ 10^{-1}
flow rate	ft^3/h	L/min	4.719 475 $\times$ 10^{-1}
(L/min)	gallon per hour	L/min	6.309 020 $\times$ 10^{-2}
	gallon per minute	L/min	3.785 412
	cm^3/min	L/min	1.000 000 $\times$ 10^{-3}
	L/min	ft^3/h	2.118 880
	cm^3/min	ft^3/h	2.118 880 $\times$ 10^{-3}
frature toughness	ksi $\cdot$ in.$^{1/2}$	MN $\cdot$ m$^{-3/2}$	1.098 855
(MN $\cdot$ m$^{-3/2}$)	MN $\cdot$ m$^{-3/2}$	ksi $\cdot$ in.$^{1/2}$	0.910 038
heat input	J/in.	J/m	3.937 008 $\times$ 10
(J/m)	J/m	J/in.	2.540 000 $\times$ 10^{-2}
impact energy	foot pound force	J	1.355 818
linear measurements	in.	mm	2.540 000 $\times$ 10
(mm)	ft	mm	3.048 000 $\times$ 10^2
	mm	in.	3.937 008 $\times$ 10^{-2}
	mm	ft	3.280 840 $\times$ 10^{-3}
power density	W/in.^{2}	W/m^2	1.550 003 $\times$ 10^3
(W/m^2)	W/m^2	W/in.^{2}	6.451 600 $\times$ 10^{-4}
	psi	Pa	6.894 757 $\times$ 10^3
	lb/ft^2	Pa	4.788 026 $\times$ 10
	N/mm^2	Pa	1.000 000 $\times$ 10^6
pressure (gas and liquid)	kPa	psi	1.450 377 $\times$ 10^{-1}
(kPa)	kPa	lb/ft^2	2.088 543 $\times$ 10
	kPa	N/mm^2	1.000 000 $\times$ 10^{-3}
	torr	kPa	1.333 22 $\times$ 10^{-1}
	(mm Hg at 0 °C)		
	micron	kPa	1.333 22 $\times$ 10^{-4}
	(μm Hg at 0 °C)		
	kPa	torr	7.500 64 $\times$ 10
	kPa	micron	7.500 64 $\times$ 10^3
	psi	kPa	6.894 757
tensile strength	lb/ft^2	kPa	4.788 026 $\times$ 10^{-2}
(MPa)	N/mm^2	MPa	1.000 000
	MPa	psi	1.450 377 $\times$ 10^2
	MPa	lb/ft^2	2.088 543 $\times$ 10^4
	MPa	N/mm^2	1.000 000
thermal conductivity	cal/(cm $\cdot$ s $\cdot$ °C)	W/(m $\cdot$ K)	4.184 000 $\times$ 10^2
(W/(m $\cdot$ K))			
travel speed, wire feed	in./min	mm/s	4.233 333 $\times$ 10^{-1}
speed (mm/s)	mm/s	in./min	2.362 205

*Preferred units are given in parentheses
**Approximate conversion

Table B.8
Approximate Equivalents for Electrode Diameters and Fillet Weld Legs

Electrode size in.	mm	(approximate equivalents)	Fillet size in.	mm
0.030	0.8		1/8	3
0.035	0.9		5/32	4
0.040	1.0		3/16	5
0.045	1.1		1/4	6
1/16	1.6		5/16	8
5/64	2.0		3/8	10
3/32	2.4		7/16	11
1/8	3.2		1/2	13
5/32	4.0		5/8	16
3/16	4.8		3/4	19
1/4	6.4		1	25

equivalents and rounded to equivalent or better precision, depending on the purpose of the dimension.

(3) **Toleranced Dimensions.** The normal practice for toleranced dimensions is to use Method A as described in 6.4.1.2.2. However, when the function of a feature requires that the millimeter equivalents be within the inch dimension tolerance limits in all cases, Method B is used as described 6.4.1.2.3.

6.4.1.2.1 Number of Decimal Places in Tolerances.
Table B.9 lists the criteria for retaining decimal places in millimeter equivalents of inch tolerances. The number of decimal places is determined by the inch tolerance span.

6.4.1.2.2 Round-Off Method A.
This method produces rounded millimeter limits which will not vary from the inch limits by more than five percent. Thus, for a dimension with a tolerance of 0.005 in., the maximum amount that the rounded millimeter can be greater or less than the inch limit is 0.000 050 in. To determine the millimeter equivalents of inch dimensions by Method A, certain steps should be followed.

The first step in this method is to determine the maximum and minimum limits in inches. Next, determine the tolerance span in inches. Proceed by converting the inch limit dimensions to millimeter values using exact millimeter equivalents. Finally, based on the tolerance span in inches, establish the number of decimal places to be retained using Table B.9 and the millimeter values rounded according to the rounding rules given in 6.4.3. See the example in Table B.10.

6.4.1.2.3 Round-Off Method B.
This method is used when the resulting millimeter values must be within the inch value limits. In extreme cases, this method may result in the lower limit millimeter value being greater than the lower inch value by a maximum of five percent.

Table B.9
Millimeter Value Round-Off Using Inch Tolerance Span

Inch Tolerance Span		Round off Millimeter Value to	
At Least	Less Than	These Decimal Places	
0.000 04	0.0004	4 places	0.00XX
0.0004	0.004	3 places	0.0XX
0.004	0.04	2 places	0.XX
0.04	0.4	1 place	X.X
0.4 and over		Whole number	XX

Example: The span of a + 0.005 to − 0.003 in. tolerance is 0.008. Since 0.008 is between 0.004 and 0.04, two decimal places are retained in individually converting 0.005 and 0.003.

Table B.10
Comparison of Round-off Methods A and B

Inch dimension:	1.934-1.966 in.
Tolerance span:	0.032 in.
Conversion:	1.934 in. = 49.1236 mm (exactly) 1.966 in. = 49.9364 mm (exactly)
Table B.9:	0.032 lies between 0.004 and 0.04; therefore, the millimeter values are to be rounded to two decimal places.
Method A:	Rounding off 49.1236 and 49.9364 to two decimal places via the method shown in 6.4.3 gives 49.12 mm and 49.94 mm, respectively.
	Method A gives a tolerance span of 0.82 mm.
Method B:	Rounding to within the inch tolerance limits requires the 49.1236 mm limit to be rounded up, giving 49.13 mm as the lower limit, and the 49.9364 mm limit to be rounded down, giving 49.93 mm as the upper limit.
	Method B gives a tolerance span of 0.80 mm.

Note: The tolerance span of 0.032 in. equals 0.8128 mm. In this example, Method A would increase the tolerance span by 0.0072 mm (0.88 percent); whereas, Method B would decrease the tolerance span by 0.0128 mm (1.6 percent).

Similarly, the upper limit millimeter value may be smaller than the upper inch by a maximum of five percent. Thus, the tolerance span may be reduced by ten percent of the original design inch tolerance; however, it is very unlikely that the five percent maximum will occur at both limits simultaneously. To determine the millimeter equivalents of inch dimensions by Method B, certain steps should be followed.

The first step in this method is to determine the maximum and minimum limits in inches. Next, determine the tolerance span in inches. Proceed by converting the inch tolerances to exact millimeter equivalents. Finally, based on the tolerance span in inches, establish the number of decimal places to be retained in the millimeter values using Table B.9. If rounding is required, the millimeter values should be rounded to fall within the inch tolerance limits; that is, to the lower value for the upper limit and to the next higher value for the lower limit. See the example in Table B.10.

6.4.2 Other Conversions. To establish meaningful and equivalent converted values, a careful determination should be made of the number of significant digits to be retained so as not to sacrifice or exaggerate the precision of the value. To convert a pressure of 1000 psi to 6.894 757 MPa is not practical because the value does not warrant expressing the conversion using six decimal places. Applying the policy of paragraph 6.4.2.2 a practical conversion would be 7 MPa. The intended precision of a value can be established from the specified tolerance or by an understanding of the equipment, process, or accuracy of the measuring device.

6.4.2.1 Temperature. All temperatures expressed in whole numbers of degrees Fahrenheit are converted to the nearest 0.1 kelvin or degree Celsius. Fahrenheit temperatures indicated to be approximate, maximum, or minimum, or to have a tolerance of $\pm 5°F$ or more, are converted to the nearest whole number in kelvin or degrees Celsius. Fahrenheit temperatures having a tolerance of plus or minus 100°F or more are converted to the nearest 10 kelvin or degrees Celsius.

EXAMPLE: $100 \pm 5°F = 38 \pm 3°C = 311 \pm 3$ K

$1000 \pm 100°F = 540 \pm 30°C = 810 \pm 30$ K

6.4.2.2 Pressure or Stress Conversion. In most cases, stress values are converted from ksi to the nearest one megapascal. Pressure or stress values having an uncertainty of more than two percent may be converted without rounding by the approximate factors.

1 psi = 7 kPa   1 ksi = 7 MPa

6.4.3 Round-Off Rules. When the next digit beyond the last digit to be retained is less than five, the last digit retained is not changed.

EXAMPLE: 4.463 25 rounded to three decimal places is 4.463

6.4.3.1 When the digits beyond the last digit to be retained amount to more than five followed by zeros, the last digit retained is increased by one.

EXAMPLE: 8.376 52 rounded to three decimal places is 8.377

6.4.3.2 Where the digit beyond the last digit to be retained is exactly five followed by zeros (expressed or implied), the last digit to be retained, if even, is unchanged, but if odd, is increased by one.

EXAMPLE: 4.365 00 becomes 4.36 when rounded to two decimal places

4.355 00 also becomes 4.36 when rounded to two decimal places

6.4.3.3 The final rounded value is obtained from the precise value to be rounded, not from a series of successive roundings. To maintain precision during conversion, the millimeter equivalent value is carried out to at least one extra decimal place. Generally, it is best to use exact values and round off only the final result.

7. STYLE AND USAGE

7.1 Application and Usage of Prefixes.

7.1.1 Prefixes should be used with SI units to indicate orders of magnitude. Prefixes provide convenient substitutes for using powers of 10, and they eliminate insignificant digits.

Correct	*Incorrect*
12.3 km	12,300 m, 12.3×10^3 m

7.1.2 Prefixes in steps of 1000 are recommended. The use of the prefixes hecto, deka, deci, and centi should be avoided.

Correct	*Incorrect*
mm, m, km	hm, dam, dm, cm

7.1.3 Prefixes should be chosen so that the numerical value lies between 0.1 and 1000. For special situations such as tabular presentations, the same unit, multiple, or submultiple may be used even though the numerical value exceeds the range of 0.1 to 1000.

7.1.4 Multiple and hyphenated prefixes should not be used.

Correct	Incorrect
pF, GF, GW	$\mu\mu$F, Mkg, kMW, G-W

7.1.5 It is generally desirable to use only base and derived units in the denominator. Prefixes are used with the numerator unit to give numbers of appropriate size (see 7.1.3).

Preferred	Nonpreferred
200 J/kg, 5 kg/m^3	0.2 J/g, 1 kg/mm

7.1.6 Prefixes are to be attached to the base SI units with the exception of the base unit for mass, the kilogram, which contains a prefix. In this case, the required prefix is attached to gram.

7.1.7 Prefixes should not be mixed unless magnitudes warrant a difference.

Correct	Incorrect
5 mm long $\times$ 100 mm high	5 mm long $\times$ 0.1 m high

Exception

4 mm diam x 50 m long

7.2 Use of Nonpreferred Units.

7.2.1 The units from different systems should not be mixed.

Correct	Incorrect
kilogram per meter cubed (kg/m^3)	kilogram per gallon (kg/gal)

7.2.2 The use of non-SI units should be limited to those which have been approved by the National Bureau of Standards such as for temperature, time, and angle (see Table B.4). Other non-SI units should not be used.

7.3 Mass, Force, and Weight.
The unit of mass is the kilogram. The kilogram-force is non-SI and is not to be used.

The unit of force is the newton (N).

Correct	Incorrect
N/m^2 or Pa, N m/s or W	kgf/m^2

The common term *weight* should not be used in technical literature.

7.4 Temperature.
The SI unit for temperature is the kelvin. The degree Celsius is, however, widely used in engineering work and may be used where considered necessary or desirable for clearer understanding. The degree Celsius was formerly called the "degree centigrade." The degree Fahrenheit should not be used.

7.5 Time.
The SI unit for time is the second. The use of minute, hour, and day is permissible but not preferred.

7.6 Angles.
The SI unit for plane angles is the radian. The SI unit for solid angles is the steradian. The degree (deg) may be used where appropriate or convenient and should be decimalized.

EXAMPLE: 5.8 deg

Angular minutes and seconds should not be used.

7.7 Stress and Pressure.
The SI unit for pressure and stress is the pascal, which is newton per meter squared (N/m^2). The bar is an acceptable but nonpreferred unit of pressure.

Other pressure and stress terms such as the following should not be used.

Incorrect

kilogram-force per square centimeter (kgf/cm2)
torr
pound-force per square inch (psi)

7.8 Capitalization.
Celsius is the only SI unit name which is always capitalized. Other SI unit names are capitalized only at the beginning of a sentence. Examples: newton, pascal, meter, kelvin, hertz, degree Celsius.

SI unit symbols are not capitalized except for the several units derived from a proper name.

EXAMPLES:
(1) A (ampere), K (kelvin), W (watt), N (newton), J (joule), etc.
(2) m (meter), kg (kilogram), etc.
Only three prefix symbols are capitalized, namely, T (tera), G (giga), and M (mega).

7.9 Plurals.
Unit symbols are the same for singular and plural. Unit names form their plurals in the usual manner.

EXAMPLE: 50 newtons (50 N) 25 grams (25 g)

7.10 Punctuation.
Periods are not to be used after SI unit symbols except at the end of a sentence.

Periods (not commas) are used as decimal markers. Periods are not used in unit symbols or in conjunction with prefixes.

Correct	Incorrect
5.7 mm	5.7 m.m.

A raised dot is used to indicate the product of two unit symbols. A space is used between unit names to indicate the product.

EXAMPLE: meter second (m·s)   kilogram meter (kg·m)

A slash or solidus (/) or a negative exponent is used as follows to indicate the quotient of two unit symbols.

(1) Only one solidus should be used in a combination.
(2) A solidus and a negative exponent can be used together, even together with parentheses, for complicated cases.
(3) Numerical values are to be used so that the denominator is unity.

Correct	Incorrect
m/s, m·s⁻¹, 4 m/s, m/s²,	m/0.1 s, m/s/s,
m·kg/(s³·A)	m·kg/s³/A, kg/s/m²

Correct: m/s, m·s^{-1}, 4 m/s, m/s^2, m·kg/(s^3·A)

Incorrect: m/0.1 s, m/s/s, m·kg/s^3/A, kg/s/m^2

The word "per" is used to indicate the quotient of two unit names.

EXAMPLE: meter per second squared (m/s^2)

7.11 Number Grouping.

7.11.1 Numbers made up of five or more digits should be written with a space separating each group of three digits counting both to the left and right of the decimal point. With four digit numbers, the spacing is optional.

7.11.2 Commas are not used between the groups of three digits.

Correct	Incorrect
1420 462.1; 0.045 62	
1452 or 1 452	1,420,462.1; 0.04562

7.12 Miscellaneous Styling.

7.12.1 A space is to be used between the numerical value and the unit symbol.

Correct	Incorrect
4 mm	4mm

7.12.2 Unit symbols and names are never used together in a single expression:

Correct	Incorrect
meter per second (m/s)	meter/s

7.12.3 Numbers are expressed as decimals, not as fractions. The decimal should be preceded by a zero when the number is less than unity.

Correct	Incorrect
0.5 kg, 1.74 m	1/2 kg, .5 kg, 1 3/4 m

7.12.4 SI unit symbols should be printed in upright type rather than slanted, script, etc.

7.12.5 Typed rather than hand-drawn prefixes should be used when possible. The spelled word may be used in preference to the use of hand-drawn symbols.

7.12.6 When it is necessary or desirable to use U.S. customary units in an equation or table, SI units or quantities should be restated in a separate equation, table, or column in a table. As an alternate, a note may be added to the equation or table giving the factors to be used in converting the calculated result in U.S. customary to preferred SI units. The SI equivalents may follow and be inserted in parentheses.

8. COMMONLY USED METRIC CONVERSIONS

Commonly used tables for converting length, pressure and stress, and temperatures between SI and U.S. Customary Units are provided in Tables B.11, B.12, and B.13, respectively.

Table B.11
Inch and Millimeter Decimal Equivalents of Fractions of an Inch

1 in. = 25.4 mm exactly

To convert inches to millimeters, multiply the inch value by 25.4.

To convert millimeters to inches, divide the millimeter value by 25.4.

Inch		Millimeter	Inch		Millimeter
Fraction	Decimal		Fraction	Decimal	
1/64	0.015 625	0.396 875	33/64	0.515 625	13.096 875
1/32	0.031 250	0.793 750	17/32	0.531 250	13.493 750
3/64	0.046 875	1.190 625	35/64	0.546 875	13.890 625
1/16	0.062 500	1.587 500	9/16	0.562 500	14.287 500
5/64	0.078 125	1.984 375	37/64	0.578 125	14.684 375
3/32	0.093 750	2.381 250	19/32	0.593 750	15.081 250
7/64	0.109 375	2.778 125	39/64	0.609 375	15.478 125
1/8	0.125 000	3.175 000	5/8	0.625 000	15.875 000
9/64	0.140 625	3.571 875	41/64	0.640 625	16.271 875
5/32	0.156 250	3.968 750	21/32	0.656 250	16.668 750
11/64	0.171 875	4.365 625	43/64	0.671 875	17.065 625
3/16	0.187 500	4.762 500	11/16	0.687 500	17.462 500
13/64	0.203 125	5.159 375	45/64	0.703 125	17.859 375
7/32	0.218 750	5.556 250	23/32	0.718 750	18.256 250
15/64	0.234 375	5.953 125	47/64	0.734 375	18.653 125
1/4	0.250 000	6.350 000	3/4	0.750 000	19.050 000
17/64	0.265 625	6.746 875	49/64	0.765 625	19.446 875
9/32	0.281 250	7.143 750	25/32	0.781 250	19.843 750
19/64	0.296 875	7.540 625	51/64	0.796 875	20.240 625
5/16	0.312 500	7.937 500	13/16	0.812 500	20.637 500
21/64	0.328 125	8.334 375	53/64	0.828 125	21.034 375
11/32	0.343 750	8.731 250	27/32	0.843 750	21.431 250
23/64	0.359 375	9.128 125	55/64	0.859 375	21.828 125
3/8	0.375 000	9.525 000	7/8	0/875 000	22.225 000
25/64	0.390 625	9.921 875	57/64	0.890 625	22.621 875
13/32	0.406 250	10.318 750	29/32	0.906 250	23.018 750
27/64	0.421 875	10.715 625	59/64	0.921 875	23.415 625
7/16	0.437 500	11.112 500	15/16	0.937 500	23.812 500
29/64	0.453 125	11.509 375	61/64	0.953 125	24.209 375
15/32	0.468 750	11.906 250	31/32	0.968 750	24.606 250
31/64	0.484 375	12.303 125	63/64	0.984 375	25.003 125
1/2	0.500 000	12.700 000	1	1.000 000	25.400 000

Table B.12
Pressure and Stress Equivalents—Pounds-Force per Square Inch to Kilopascals and Thousand Pounds-Force per Square Inch to Megapascals

1 psi = 6894.757 Pa

To convert psi to pascals, multiply the psi value by 6.894×10^3.
To convert pascals to psi, divide the pascal value by $6.894\ 757 \times 10^3$.

psi ksi	0	1	2	3	4	5	6	7	8	9
					kPa MPa					
0	0.0000	6.8948	13.7895	20.6843	27.5790	34.4738	41.3685	48.2633	55.1581	62.0528
10	68.9476	75.8423	82.7371	89.6318	96.5266	103.4214	110.3161	117.2109	124.1056	131.0004
20	137.8951	144.7899	151.6847	158.5794	165.4742	172.3689	179.2637	186.1584	193.0532	199.9480
30	206.8427	213.7375	220.6322	227.5270	234.4217	241.3165	248.2113	255.1060	262.0008	268.8955
40	275.7903	282.6850	289.5798	296.4746	303.3693	310.2641	317.1588	324.0536	330.9483	337.8431
50	344.7379	351.6326	358.5274	365.4221	372.3169	379.2116	386.1064	393.0012	399.8959	406.7907
60	413.6854	420.5802	427.4749	434.3697	441.2645	448.1592	455.0540	461.9487	468.4835	475.7382
70	482.6330	489.5278	496.4225	503.3173	510.2120	517.1068	524.0015	530.8963	537.7911	544.6858
80	551.5806	558.4753	565.3701	572.2648	579.1596	586.0544	592.9491	599.8439	606.7386	613.6334
90	620.5281	627.4229	634.3177	641.2124	648.1072	655.0019	661.8967	668.7914	675.6862	682.5810
100	689.4757									

Notes: This table may be used to obtain SI equivalents of values expressed in psi or ksi. SI values are usually expressed in kPa when original value is in psi and in MPa when original value is in ksi.

This table may be extended to values below 1 or above 100 psi (or ksi) by manipulation of the decimal point and addition.

Example: To convert 135 ksi to MPa, add the separate conversions of 100, 30, and 5 ksi (689.4757 + 206.8427 + 34.4738) to get 930.7922 MPa which rounds to 931 MPa.

Copyright ASTM. Reprinted with permission.

Table B.13
Conversions for Farenheit–Celsius Temperature Scales

Find the number to be converted in the center (boldface) column. If converting Fahrenheit degrees, read the Celsius equivalent in the column headed "C°." If converting Celsius degrees, read the Fahrenheit equivalent in the column headed "°F."

°C		°F	°C		°F	°C		°F	°C		°F
−273	−459		−40	−40	−40	24.4	76	168.8	199	390	734
−268	−450		−34	−30	−22	25.6	78	172.4	204	400	752
−262	−440		−29	−20	−4	26.7	80	176.0	210	410	770
−257	−430		−23	−10	14	27.8	82	179.6	216	420	788
−251	−420		−17.8	0	32	28.9	84	183.2	221	430	806
−246	−410		−16.7	2	35.6	30.0	86	186.8	227	440	824
−240	−400		−15.6	4	39.2	31.1	88	190.4	232	450	842
−234	−390		−14.4	6	42.8	32.2	90	194.0	238	460	860
−229	−380		−13.3	8	46.4	33.3	92	197.6	243	470	878
−223	−370		−12.2	10	50.0	34.4	94	201.2	249	480	896
−218	−360		−11.1	12	53.6	35.6	96	204.8	254	490	914
−212	−350		−10.0	14	57.2	36.7	98	208.4	260	500	932
−207	−340		−8.9	16	60.8	37.8	100	212.0	266	510	950
−201	−330		−7.8	18	64.4	43	110	230	271	520	968
−196	−320		−6.7	20	68.0	49	120	248	277	530	986
−190	−310		−5.6	22	71.6	54	130	266	282	540	1004
−184	−300		−4.4	24	75.2	60	140	284	288	550	1022
−179	−290		−3.3	26	78.8	66	150	302	293	560	1040
−173	−280		−2.2	28	82.4	71	160	320	299	570	1058
−168	−270	−454	−1.1	30	86.0	77	170	338	304	580	1076
−162	−260	−436	0.0	32	89.6	82	180	356	310	590	1094
−157	−250	−418	1.1	34	93.2	88	190	374	316	600	1112
−151	−240	−400	2.2	36	96.8	93	200	392	321	610	1130
−146	−230	−382	3.3	38	100.4	99	210	410	327	620	1148
−140	−220	−364	4.4	40	104.0	100	212	414	332	630	1166
−134	−210	−346	5.6	42	107.6	104	220	428	338	640	1184
−129	−200	−328	6.7	44	111.2	110	230	446	343	650	1202
−123	−190	−310	7.8	46	114.8	116	240	464	349	660	1220
−118	−180	−292	8.9	48	118.4	121	250	482	354	670	1238
−112	−170	−274	10.0	50	122.0	127	260	500	360	680	1256
−107	−160	−256	11.1	52	125.6	132	270	518	366	690	1274
−101	−150	−238	12.2	54	129.2	138	280	536	371	700	1292
−96	−140	−220	13.3	56	132.8	143	290	554	377	710	1310
−90	−130	−202	14.4	58	136.4	149	300	572	382	720	1328
−84	−120	−184	15.6	60	140.0	154	310	590	388	730	1346
−79	−110	−166	16.7	62	143.6	160	320	608	393	740	1364
−73	−100	−148	17.8	64	147.2	166	330	626	399	750	1382
−68	−90	−130	18.9	66	150.8	171	340	644	404	760	1400
−62	−80	−112	20.0	68	154.4	177	350	662	410	770	1418
−57	−70	−94	21.1	70	158.0	182	360	680	416	780	1436
−51	−60	−76	22.2	72	161.6	188	370	698	421	790	1454
−46	−50	−58	23.3	74	165.2	193	380	716	427	800	1472
432	810	1490	738	1360	2480	1043	1910	3470	1349	2460	4460
438	820	1508	743	1370	2498	1049	1920	3488	1354	2470	4478
443	830	1526	749	1380	2516	1054	1930	3506	1360	2480	4496
449	840	1544	754	1390	2534	1060	1940	3524	1366	2490	4514
454	850	1562	760	1400	2552	1066	1950	3542	1371	2500	4532

Table B.13 (Continued)

Find the number to be converted in the center (boldface) column. If converting Fahrenheit degrees, read the Celsius equivalent in the column headed "F°." If converting Celsius degrees, read the Fahrenheit equivalent in the column headed "°F."

°C		°F	°C		°F	°C		°F	°C		°F
460	**860**	1580	766	**1410**	2570	1071	**1960**	3560	1377	**2510**	4550
466	**870**	1598	771	**1420**	2588	1077	**1970**	3578	1382	**2520**	4568
471	**880**	1616	777	**1430**	2606	1082	**1980**	3596	1388	**2530**	4586
477	**890**	1634	782	**1440**	2624	1088	**1990**	3614	1393	**2540**	4604
482	**900**	1652	788	**1450**	2642	1093	**2000**	3632	1399	**2550**	4622
488	**910**	1670	793	**1460**	2660	1099	**2010**	3650	1404	**2560**	4640
493	**920**	1688	799	**1470**	2678	1104	**2020**	3668	1410	**2570**	4568
499	**930**	1706	804	**1480**	2696	1110	**2030**	3686	1416	**2580**	4676
504	**940**	1724	810	**1490**	2714	1116	**2040**	3704	1421	**2590**	4694
510	**950**	1742	816	**1500**	2732	1121	**2050**	3722	1427	**2600**	4712
516	**960**	1760	821	**1510**	2750	1127	**2060**	3740	1432	**2610**	4730
521	**970**	1778	827	**1520**	2768	1132	**2070**	3758	1438	**2620**	4748
527	**980**	1796	832	**1530**	2786	1138	**2080**	3776	1443	**2630**	4766
532	**990**	1814	838	**1540**	2804	1143	**2090**	3794	1449	**2640**	4784
538	**1000**	1832	843	**1550**	2822	1149	**2100**	3812	1454	**2650**	4802
543	**1010**	1850	849	**1560**	2840	1154	**2110**	3830	1460	**2660**	4820
549	**1020**	1868	854	**1570**	2858	1160	**2120**	3848	1466	**2670**	4838
554	**1030**	1886	860	**1580**	2876	1166	**2130**	3866	1471	**2680**	4856
560	**1040**	1904	866	**1590**	2894	1171	**2140**	3884	1477	**2690**	4874
566	**1050**	1922	871	**1600**	2912	1177	**2150**	3902	1482	**2700**	4892
571	**1060**	1940	877	**1610**	2930	1182	**2160**	3920	1488	**2710**	4910
577	**1070**	1958	882	**1620**	2948	1188	**2170**	3938	1493	**2720**	4928
582	**1080**	1976	888	**1630**	2966	1193	**2180**	3956	1499	**2730**	4946
588	**1090**	1994	893	**1640**	2984	1199	**2190**	3974	1504	**2740**	4964
593	**1100**	2012	899	**1650**	3002	1204	**2200**	3992	1510	**2750**	4982
599	**1110**	2030	904	**1660**	3020	1210	**2210**	4010	1516	**2760**	5000
604	**1120**	2048	910	**1670**	3038	1216	**2220**	4028	1521	**2770**	5018
610	**1130**	2066	916	**1680**	3056	1221	**2230**	4046	1527	**2780**	5036
616	**1140**	2084	921	**1690**	3074	1227	**2240**	4064	1532	**2790**	5054
621	**1150**	2102	927	**1700**	3092	1232	**2250**	4082	1538	**2800**	5072
627	**1160**	2120	932	**1710**	3110	1238	**2260**	4100	1543	**2810**	5090
632	**1170**	2138	938	**1720**	3128	1243	**2270**	4118	1549	**2820**	5108
638	**1180**	2156	943	**1730**	3146	1249	**2280**	4136	1554	**2830**	5126
643	**1190**	2174	949	**1740**	3164	1254	**2290**	4154	1560	**2840**	5144
649	**1200**	2192	954	**1750**	3182	1260	**2300**	4172	1566	**2850**	5162
654	**1210**	2210	960	**1760**	3200	1266	**2310**	4190	1571	**2860**	5180
660	**1220**	2228	966	**1770**	3218	1271	**2320**	4208	1577	**2870**	5198
666	**1230**	2246	971	**1780**	3236	1277	**2330**	4226	1582	**2880**	5216
671	**1240**	2264	977	**1790**	3254	1282	**2340**	4244	1588	**2890**	5234
677	**1250**	2282	982	**1800**	3272	1288	**2350**	4262	1593	**2900**	5252
682	**1260**	2300	988	**1810**	3290	1293	**2360**	4280	1599	**2910**	5270
688	**1270**	2318	993	**1820**	3308	1299	**2370**	4298	1604	**2920**	5288
693	**1280**	2336	999	**1830**	3326	1304	**2380**	4316	1610	**2930**	5306
699	**1290**	2354	1004	**1840**	3344	1310	**2390**	4334	1616	**2940**	5324
704	**1300**	2372	1010	**1850**	3362	1316	**2400**	4352	1621	**2950**	5342
710	**1310**	2390	1016	**1860**	3380	1321	**2410**	4370	1627	**2960**	5360
716	**1320**	2408	1021	**1870**	3398	1327	**2420**	4388	1632	**2970**	5278
721	**1330**	2426	1027	**1880**	3416	1332	**2430**	4406	1638	**2980**	5398
727	**1340**	2444	1032	**1890**	3434	1338	**2440**	4424	1643	**2990**	5414
732	**1350**	2462	1038	**1900**	3452	1343	**2450**	4442	1649	**3000**	5432

$$C = 5/9(F - 32) \qquad F = 9/5\,C + 32$$

WELDING HANDBOOK
INDEX OF MAJOR SUBJECTS

	Eighth Edition Volume	Seventh Edition Volume	Sixth Edition Section	Chapter
L				
Laser beam cutting		3		6
Laser beam welding		3		6
Launch vehicles			5	92
Lead, welding of		4		11
Liners, applied			5	93
Liners, clad steel			5	93
Liquid penetrant testing	1			12
Low alloy steels, welding of		4		1
M				
Magnesium and magnesium alloys, welding of		4		9
Magnetic fields, influence on arcs	1			2
Magnetic particle testing	1			12
Maraging steels		4		4
Martensitic stainless steels		4		2
Mechanical properties		1		5
Mechanical testing	1			15
Mechanical treatments of weldments		1		6
Melting rates (electrode)	1			2
Metal powder cutting		2		13
Metal transfer	1			2
Metallurgy, general	1			4
Metallurgy of brazing and soldering	1			4
Metallurgy of surfacing alloys		2		14
Metallurgy of welding	1			4
Metals, physical properties	1			2
Metric practice guide	1			App. B
Molybdenum, welding of		4		11
N				
Narrow gap welding		2		4
NEMA power source requirements		2		1
Nickel and nickel alloys, welding of		4		6
Nondestructive examination (testing)	1			15
Nondestructive examination symbols	1			6
Nuclear power plants			5	87
O				
Oxyfuel gas cutting		2		13
Oxyfuel gas welding		2		10
Oxygen cutting		2		13

Z

INDEX